Differential Equations

Differential Equations

Amrinder Pal Singh
Assistant Professor
Department of Basic and Applied Sciences
Punjabi University, Patiala (Punjab)

I.K. International Pvt. Ltd.
NEW DELHI

Published by

I.K. International Pvt. Ltd.
4435-36/7, Ansari Road, Daryaganj
New Delhi–110 002 (India)
E-mail: info@ikinternational.com
Website: www.ikbooks.com

ISBN: 978-93-90620-54-8

Published by Krishan Makhijani for I.K. International Pvt. Ltd., 4435-36/7, Ansari Road, Daryaganj, New Delhi–110 002 and Printed by Rekha Printers Pvt. Ltd., Okhla Industrial Area, Phase II, New Delhi–110 020.

Dedicated

to

My Dad

whose memories had always been an inspiration

Preface

Differential equations occur in connection with numerous natural phenomena and much of our understanding of nature comes from our ability to solve such equations. Taking note of importance of differential equations, this text is designed to give a simple, lucid, rigorous and comprehensive account of fundamental notions of ordinary differential equations, partial differential equations, Laplace and Fourier transforms and their applications.

This book has three sections. Section A deals with ordinary differential equations, section B is devoted to study of partial differential equations and section C gives introduction to Laplace and Fourier transforms, Fourier series and their applications in finding solutions of differential equations.

Section A has fifteen chapters, giving systematic introduction to basic terms of theory of differential equations including their formulation, classification and different types of solutions (chapter 1), some basic methods of finding solution of ordinary differential equations of the first order with first degree (chapter 2), notion of exact differential equations (chapter 3), first order linear differential equations and Bernoulli's equation (chapter 4), first order higher degree differential equations (chapter 5), geometrical meaning of differential equations, singular solution and Clairaut's differential equation (chapter 6), orthogonal trajectories (chapter 7), homogeneous linear differential equations with constant coefficients (chapter 8), non-homogeneous linear differential equations with constant coefficients and their solution using operator methods (chapter 9), method of variation of parameters and undetermined coefficients (chapter 10), equations reducible to linear differential equations with constant coefficients (chapter 11), simultaneous differential equations and their solution in symmetrical form, total differential equations with condition of integrability and their solution methods (chapter 12), power series and Frobenius methods for solution of linear differential equations of the second order (chapter 13) and two important second order linear differential equations – Legendre differential equations and Bessel's differential equations and their solution in terms of two special functions – Legendre polynomials and Bessel's function of first and second kind, respectively (chapter 14). Finally, chapter 15 is devoted to applications of the theory of ordinary differential equations in formulation and solution of various practical problems, viz., motion under gravity in a resisting medium, escape velocity of a particle, atmospheric pressure above the sea level, Newton's law of cooling, decay of radioactive materials, simple harmonic motion, chemical reactions and many more.

Section B has of ten chapters and discusses important terms used in theory of formulation and solution of partial differential equations (chapter 1), classification of solutions of partial differential equations, Lagrange's method for solution of linear partial differential equations of the order one and its applications in finding integral surfaces passing through a given curve (chapter 2), some standard methods for the solution of nonlinear partial differential equations (chapter 3), Charpit's method and Jacobi's method (chapter 4), solution of homogeneous linear partial differential equations with constant coefficients (chapter 5) and non-homogeneous linear partial differential equations with constant coefficients (chapter 6), partial differential equations reducible to equations with constant coefficients (chapter 7), classification of linear partial differential equations of second order (chapter 8) and Monge's method for solution of nonlinear partial differential equations of the second order (chapter 9). Finally, chapter 10 discusses applications of partial differential equations to many real physical problems, viz., one- and two-dimensional wave equations, one- and two-dimensional heat equations, diffusion equations, temperature distribution in a given medium, Laplace's equation and many more.

Section C consists of nine chapters and is devoted to the study of Laplace and inverse Laplace transforms, definition and properties of beta and gamma functions and formulae for Laplace and inverse Laplace transforms of some basic mathematical functions (chapter 1), Laplace transforms of derivatives and integrals (chapter 2), first and second shifting theorems, Laplace transforms of unit step function and Dirac-delta function (chapter 3), differentiation and integration of Laplace transforms (chapter 4), convolution theorem and its applications in finding inverse Laplace transform of product of two functions and Laplace transform of periodic functions (chapter 5) and applications of Laplace transforms in finding solution of ordinary and partial differential equations subject to boundary conditions (chapter 6). Since Fourier series are useful in the study of problems of heat conduction, mechanics, concentration of chemicals and pollutants, electrostatics and acoustics, etc. Chapter 7 is devoted to introduction to Fourier series, Fourier series of even and odd functions, half range Fourier cosine and sine series and applications of Fourier series in finding solution of heat equation, wave equation and Laplace equation, etc. Chapter 8 discusses Fourier transforms (integral transform similar to Laplace transform) and inverse Fourier transforms, finite Fourier cosine and sine transforms, properties of Fourier transforms, relation between Fourier and Laplace transforms, convolution theorem for Fourier transforms and Parseval's identity for Fourier transforms. Finally, Chapter 9 discusses applications of Fourier transforms in finding solution of problems like temperature distribution in a given medium, heat equation, wave equation, Laplace equation and many more.

The text contains simplified examples with suitable exercises. Hints for selected problems are also provided.

Designed for undergraduate and postgraduate students of mathematics, this book can be prescribed as an introductory book on differential equations for those working in the area of science and engineering and preparing for various competitive examinations. This book includes almost all the methods for finding solution of ordinary differential equations and partial differential equations with applications. The text also contains the topics of Laplace transforms and Fourier transforms and their applications in finding solutions of differential equations.

Amrinder Pal Singh

Acknowledgements

I offer my sincere thanks to my senior professor Dr. Rakesh Kumar for his continuous support and encouragement. I am indebted to all authors listed in the bibliography for providing a good and exhaustive literature on the subject due to which the writing of this text has been possible. I am also grateful to my family and friends for their continuous encouragement.

Any comments and suggestions for improving the text will be appreciated.

Amrinder Pal Singh

Contents

SECTION-A

Ordinary Differential Equations

1

CHAPTER

Some Basic Concepts

Definition: An equation involving derivatives of one or more dependent variables with respect to one or more independent variables is called a *differential equation*.

Differential equations occur in connection with numerous problems. Many of the laws of nature find their applications in the language of differential equations. In other words, differential equations are the language of nature, for example, the problem of determining the motion of a projectile, the problem of determining the vibrations of a wire, Newton's law of universal gravitation, Maxwell's field equations, the motion of the planets and the refraction of light. The mathematical formulation of such problems give rise to differential equations. So, much of our understanding of nature comes from our ability to solve such equations. Now we begin by classifying differential equations according to whether there is one or more than one independent variables are involved.

Definition: A differential equation involving ordinary derivatives of one or more dependent variables with respect to a single independent variable is called an *ordinary differential equation*.

Definition: A differential equation involving partial derivatives of one or more dependent variables with respect to more than one independent variable is called a *partial differential equation*.

$$\frac{d^2y}{dx^2} = 6x^2 \tag{1.1}$$

$$\frac{d^4y}{dx^4} + 5\frac{d^2z}{dx^2} + 3z = \sin x \tag{1.2}$$

$$x\frac{\partial u}{\partial x} + y\frac{\partial u}{\partial y} = nu \tag{1.3}$$

In these given differential equations, Eq. (1.1), is an ordinary differential equation involving one dependent variable and Eq. (1.2) is an ordinary differential equation involving two dependent variables y and z. Equation (1.3) is a partial differential equation involving two independent variables, x and y.

We can further classify differential equations according to their order and degree also.

Definition: The order of a differential equation is the order of the highest order derivative occurring in the differential equation. In the above given differential equations, Eqs. (1.1), (1.2) and (1.3) are of order 2, 4 and 1 respectively.

Definition: The degree of a differential equation is the degree (or power) of the highest order derivative occurring in the differential equation after the equation has been made free of radicals and fractions in its derivatives.

In above given differential equations, degree of each equation is 1. Based on the degree of a differential equation we can further classify them as linear and non-linear differential equations.

Definition: A differential equation is linear, when the dependent variable and its derivative occur only in the first degree and no product of the dependent variable and/or its derivatives of various orders occur. Further, no transcendental function of dependent variable and/or its derivatives is present. The general form of linear ordinary differential equation of order n, in the dependent variable y and the independent variable x is

$$a_0(x)\frac{d^n y}{dx^n} + a_1(x)\frac{d^{n-r} y}{dx^{n-1}} + \dots + a_{n-1}(x)\frac{dy}{dx} + a_n(x)y = r(x)$$

where, $a_0(x)$ is not identically zero.

Definition: A differential equation which is not linear, is called a nonlinear differential equation.

For example, the following ordinary differential equations are both linear.

$$\frac{d^2y}{dx^2} + 5\frac{dy}{dx} + 4y = 0 \tag{1.4}$$

$$\frac{d^4y}{dx^4} + x^2\frac{d^2y}{dx^2} + x^3\frac{dy}{dx} = xe^x \tag{1.5}$$

The following ordinary differential equations are all nonlinear.

$$\frac{d^2y}{dx^2} + 5\frac{dy}{dx} + 6y^2 = 0 \tag{1.6}$$

$$\frac{d^2y}{dx^2} + 5\left(\frac{dy}{dx}\right)^3 + 6y = 0 \tag{1.7}$$

$$\frac{d^2y}{dx^2} + 5y\frac{dy}{dx} + 6y = 0 \tag{1.8}$$

Linear ordinary differential equations are further classified according to the nature of the coefficients of the dependent variables and their derivatives. For example, Eq. (1.4) is said to be linear with constant coefficients while Eq. (1.5) is linear with variable coefficients. We can also denote the derivatives as

$$\frac{dy}{dx} = y', \ \frac{d^2y}{dx^2} = y'', \ \frac{d^3y}{dx^3} = y''', \text{ etc.}$$ ▲▲

EXAMPLE 1.1

Differential Equation	Order	Degree	Linear/Nonlinear
$y'' + 16y = 2x$	2	1	Linear
$y'y'' + y^2 = x^2$	2	1	Nonlinear
$x^2y^2 - xy' + 6y = \log x$	2	1	Linear
$y' = \sqrt{x} + \sqrt{y}$	1	1	Nonlinear
$(y'')^2 = 1 + (y')^3$	2	2	Nonlinear
$y' = \sin y$	1	1	Nonlinear
$(y')^2 + 3xy' + y = 0$	1	2	Nonlinear
$[1 + (y')]^{1/2} = y''$	2	2	Nonlinear

Formation of a Differential Equation

Let y and x be the dependent and the independent variables respectively. The equation $f(x, y, c) = 0$ containing one arbitrary constant c, represents a family of curves. An ordinary differential equation of the family is formed in an attempt to eliminate the arbitrary constant c. ▲▲

EXAMPLE 1.2

Form the differential equation of simple harmonic motion given by $y = A\cos(nx + \alpha)$.

Solution: To eliminate the constants A and α, differentiating it twice, we have

$$\frac{dy}{dx} = -nA\sin(nx + \alpha) \quad \text{and} \quad \frac{d^2y}{dx^2} = -n^2A\cos(nx + 2) = -n^2y$$

Thus,
$$\frac{d^2y}{dx^2} + n^2y = 0$$

is the desired differential equation.

EXAMPLE 1.3

Obtain the differential equation of all circles of radius r and centre (a, b).

Solution: The equation of circle is

$$(x - a)^2 + (y - b)^2 = r^2 \qquad \text{(i)}$$

where, a, b and r are constants.

Differentiating twice, we get

$$x - a + (y - b)\frac{dy}{dx} = 0$$

and
$$1 + (y - b)^2 \frac{d^2y}{dx^2} + \left(\frac{dy}{dx}\right)^2 = 0$$

∴
$$y - b = -\frac{1+\left(\frac{dy}{dx}\right)^2}{\frac{d^2y}{dx^2}}$$

and
$$x - a = -(y - b)\frac{dy}{dx} = \frac{\frac{dy}{dx}\left[1+\left(\frac{dy}{dx}\right)^2\right]}{\frac{d^2y}{dx^2}}$$

Substituting these in Eq. (i), we get

$$[1 + (dy/dx)^2]^3 = a^2 [d^2y/dx^2]^2$$

EXAMPLE 1.4

Find the differential equation of the system of curves $y = ax^2 + b\cos nx + c$. where, a, b and c are arbitrary constants.

Solution: We have

$$y = ax^2 + b\cos nx + c \tag{i}$$

Differentiating thrice w.r.t. x, we get

$$\frac{dy}{dx} = 2ax - bn\sin nx \tag{ii}$$

$$\frac{d^2y}{dx^2} = 2a - bn^2\cos nx \tag{iii}$$

and
$$\frac{d^3y}{dx^3} = bn^3\sin nx \tag{iv}$$

Eliminating a between Eqs. (ii) and (iii), we get

$$x\frac{d^2y}{dx^2} - \frac{dy}{dx} = -bx(nx\cos nx + \sin nx) \tag{v}$$

Eliminating b between Eqs. (iv) and (v) by dividing, we get

$$\frac{\dfrac{d^3y}{dx^3}}{x\dfrac{d^2y}{dx^2}-\dfrac{dy}{dx}} = \frac{-n^2 \sin nx}{nx \cos nx + \sin nx}$$

This is the required differential equation.

Note: We have eliminated three constants to get a differential equation of the third order.

▲▲

Solution of a Differential Equation

A solution of a differential equation is a relation between the dependent and independent variables, not involving the derivatives such that the relation and the derivatives obtained from it satisfy the given differential equation.

For example, the function $y = \sin 2x$ is a solution of the differential equation

$$y' = 2\cos 2x \text{ in the interval } -\infty < x < \infty.$$

Solution of a differential equation is also called an integral of the differential equation.

A solution of a differential equation which contains as many arbitrary constants as order of the differential equation is called *general solution*. Other solutions, obtained by giving particular values to the arbitrary constants in the general solution, are called *particular solutions*. In some cases, there exists a solution, which cannot be obtained from the general solution, such a solution is called a *singular solution*. For example, it can be verified that $y = -x^2/4$ is also a solution of the differential equation

$$(y')^2 + xy' = y$$

with general solution $y = cx + c^2$. Therefore, the parabola $4y + x^2 = 0$ is the singular solution of this differential equation. ▲▲

EXAMPLE 1.5

Show that $y = A\cos x + B\sin x$ is a solution of the differential equation

$$d^2y/dx^2 + y = 0.$$

Solution: We have

$$dy/dx = -A\sin x + B\cos x$$
$$d^2y/dx^2 = -A\cos x - B\sin x$$

Clearly $d^2y/dx^2 + y = 0$ which is the given differential equation.

EXAMPLE 1.6

Show that

$$Ax^2 + By^2 = 1 \quad \text{(i)}$$

is the solution of the differential equation.

$$x\left[y\frac{d^2y}{dx^2}+\left(\frac{dy}{dx}\right)^2\right]-y\frac{dy}{dx}=0.$$

Solution: We note that the differential equation does not contain the arbitrary constants A, B and is of the second order.

Differentiating Eq. (i) twice, we get

$$Ax+By\frac{dy}{dx}=0 \qquad \text{(ii)}$$

and

$$A+B\left[\left(\frac{dy}{dx}\right)^2+y\frac{d^2y}{dx^2}\right]=0 \qquad \text{(iii)}$$

On eliminating A and B from Eqs. (ii) and (iii), we get

$$x\left[y\frac{d^2y}{dx^2}+\left(\frac{dy}{dx}\right)^2\right]-y\frac{dy}{dx}=0$$

which is the given differential equation.

EXERCISE 1.1

1. Find the order and degree of the following differential equations. State also whether they are linear or nonlinear.
   (a) $y''+3y'+4y=0$ (b) $(y'')^2+3xy'+y=0$
   (c) $y'=\log y$ (d) $[1+y']^{1/2}=y'''$
   (e) $yy''+t^2y'+4y=\cos t$ (f) $x^2y''+xy'+3y=5x$
2. Eliminate the arbitrary constants and obtain the differential equation satisfied by the following functions.
   (a) $y=cx+c^2$
   (b) $y=Ae^x+Be^{-x}+c$
   (c) $y=c\sin x$
   (d) $y=a\cos\theta x+b\sin\theta x$; θ fixed constant
   (e) $y=c\cosh(x/c)$; $c\neq 0$
   (f) $y=(a/x^2)+bx$
3. Find the differential equation of a family of circles passing through the origin and having centres on the x-axis.

4. Find the differential equation of all straight lines in a plane.
5. By eliminating the constant c, obtain the differential equation of which $y^2 = 4a(x + a)$ is the solution.
6. Find the differential equation of the family of curves

$$y = e^x(A \cos x + B \sin x)$$

where, A and B are arbitrary constants.

7. Verify that $y = x \log x - x$ is a solution of the differential equation $y' = \log x$.
8. Find all values of m for which $y = e^{mx}$ is a solution of the differential equation $y'' - 4y' + y = 0$.

ANSWERS

Exercise 1.1

1. (a) 2, 1, linear
   (b) 2, 2, nonlinear
   (c) 1, 1, nonlinear
   (d) 3, 2, nonlinear
   (e) 2, 1, nonlinear
   (f) 2, 1, linear

2. (a) $x\dfrac{dy}{dx} + \left(\dfrac{dy}{dx}\right)^2 = y$
   (b) $\dfrac{d^3y}{dx^3} - \dfrac{dy}{dx} = 0$
   (c) $(\tan x)y' - y = 0$
   (d) $y'' + \theta^2 y = 0$
   (e) $y \log\{[1 + (y')^2]^{1/2} + y'\} = x[1 + (y')^2]^{1/2}$
   (f) $x^2y'' + 2xy' - 2y = 0$

3. $2xy\dfrac{dy}{dx} + x^2 - y^2 = 0$

4. $\dfrac{d^2y}{dx^2} = 0$

5. $y = \left[1 - \left(\dfrac{dy}{dx}\right)^2\right] = 2x\dfrac{dy}{dx}$

6. $\dfrac{d^2y}{dx^2} - \dfrac{2dy}{dx} + 2y = 0$

8. $m = 2 \pm \sqrt{3}$

2
CHAPTER

Solution of Differential Equations with First Order and First Degree

In chapter 1, we discussed some basic terms involving a differential equation and formation of a differential equation if its solution is given. In this and the subsequent chapters, we shall consider the methods of finding solution of a differential equation. In this direction, we will first start with some basic methods for finding solution of an ordinary differential equation with first order and first degree of the form $\frac{dy}{dx} = f(x,y)$.

Equations in which Variables are Separable

A first order ordinary differential equation is said to be in variable separable form if it is possible, by using elementary algebraic manipulations, to arrange so that all the dependent variables, say y are on one side and all the independent variables, say x are on the other side. The property of being separable is rather special. But it is pleasing that many of the differential equations of mathematical physics turns out to be separable. To find solution of a differential equation in variable separable form after separating the variables, we can simply integrate both sides.

➲ Working Rule

Let the given differential equation be in the form

$$\frac{dy}{dx} = f(x, y) \tag{2.1}$$

Step-1: Write Eq. (2.1) as $g(x)dx = h(y)dy$ after separating terms of x and y on each side.

Step-2: Integrate both sides to get the general solution, that is,

$$\int g(x)dx = \int h(y)dy + c$$

$$\Rightarrow \quad G(x) = H(x) + c$$

where, c is constant of integration. ▲▲

EXAMPLE 2.1

Solve $y(1 + x)dx + x(1 + y)dy = 0$.

Solution: We can write the given differential equation as

$$\frac{1+x}{x}dx = -\frac{1+y}{y}dy$$

The solution is

$$\int \frac{1+x}{x}dx = \int -\frac{1+y}{y}dy + c$$

$$\Rightarrow \quad \log x + x = -\log y - y + c$$

$$\Rightarrow \quad x + y + \log xy = c$$

which is the required general solution containing the arbitrary constant c.

EXAMPLE 2.2

Solve $y - x\dfrac{dy}{dx} = a\left(y^2 + \dfrac{dy}{dx}\right)$.

Solution: The differential equation may be written as

$$\frac{dy}{y(1-ay)} = \frac{dx}{x+a}$$

$$\Rightarrow \quad \left(\frac{1}{y} + \frac{a}{1-ay}\right)dy = \frac{dx}{x+a}$$

Integrating we get

$$\log y - \log(1 - ay) = \log(x + a) + \log c$$

$$\Rightarrow \quad \log\left(\frac{y}{1-ay}\right) = \log c(x + a)$$

$$\therefore \quad y = c(1 - ay)(x + a).$$

EXAMPLE 2.3

Solve $\sqrt{1+x^2}\sqrt{1+y^2}\, dx + xy\,dy = 0$.

Solution: We can write the given differential equation as

$$\frac{\sqrt{1+x^2}}{x}dx + \frac{1}{2}\frac{2y}{\sqrt{1+y^2}}dy = 0$$

On integration, we get

$$\int \frac{\sqrt{1+x^2}}{x} dx + \sqrt{1+y^2} = c$$

$$\Rightarrow \quad \sqrt{1+x^2} + \frac{1}{2}\log\left|\frac{\sqrt{1+x^2}-1}{\sqrt{1+x^2}+1}\right| + \sqrt{1+y^2} = c$$

as

$$\int \frac{\sqrt{1+x^2}}{x} dx = \int \frac{t^2}{t^2-1} dt \qquad \left\{ \begin{array}{ll} \text{Put } 1 + x^2 = t^2 \\ \Rightarrow \quad x dx = t dt \end{array} \right.$$

$$= \int \left(1 + \frac{1}{t^2-1}\right) dt$$

$$= t + \frac{1}{2}\log\left|\frac{t-1}{t+1}\right|$$

$$= \sqrt{1+x^2} + \frac{1}{2}\log\left|\frac{\sqrt{1+x^2}-1}{\sqrt{1+x^2}+1}\right|.$$

EXAMPLE 2.4

Find a curve for which the tangent at each point makes a minimum angle α with the radius vector.

Solution: Let ϕ be the angle between the radius vector and the tangent at a point of the curve then according to the statement, we have

$$\tan \phi = r\frac{d\theta}{dr}$$

Given $\phi = a$

$$\tan a = r\frac{d\theta}{dr} \Rightarrow \frac{dr}{r} = \cot \alpha d\theta$$

On integrating, we have

$$\log r = \theta \cot a + c \Rightarrow r = e^{\theta \cot a + c}$$

$$\therefore \qquad r = ae^{\theta \cot a} \qquad (\because e^c = a)$$

which is the required curve.

EXAMPLE 2.5

Reduce the differential equation $y'' + e^{2y}(y')^3 = 0$ to a lower order equation and find the solution of the differential equation.

Solution: Let $\qquad y' = u$

$$\therefore \qquad y'' = \frac{du}{dx} = \frac{du}{dy}\frac{dy}{dx} = \frac{du}{dy}y' = \frac{du}{dy}u$$

The given differential equation becomes

$$u\frac{du}{dy} + e^{2y}u^3 = 0$$

$$\Rightarrow \qquad \frac{du}{u^2} + e^{2y}dy = 0$$

On integrating, we get

$$\frac{-1}{u} + \frac{e^{2y}}{2} = c$$

$$\Rightarrow \qquad -\frac{dx}{dy} + \frac{e^{2y}}{2} = c$$

$$\Rightarrow \qquad -dx + \frac{e^{2y}}{2}dy = c\,dy$$

Again integrating, we obtain

$$-x + \frac{e^{2y}}{y} = cy + d$$

where c and d are arbitrary constants. ▲▲

Equations Reducible to Separable Form

Let the given differential equation be of the form

$$\frac{dy}{dx} = f(ax + by + c)$$

Substituting $ax + by + c = z$, we get

$$a + b\frac{dy}{dx} = \frac{dz}{dx}$$

$$\Rightarrow \qquad \frac{dy}{dx} = \frac{1}{b}\left(\frac{dz}{dx} - a\right)$$

The given differential equation simplifies to

$$\frac{1}{b}\left(\frac{dz}{dx} - a\right) = f(z)$$

$$\Rightarrow \qquad \frac{dz}{a + bf(z)} = dx$$

which is in variable separable form and can be solved by simply integrating both sides. ▲▲

EXAMPLE 2.6

Solve $\dfrac{dy}{dx} = (4x + y + 1)^2$.

Solution: Put $4x + y + 1 = z \quad \Rightarrow \quad \dfrac{dy}{dx} = \dfrac{dz}{dx} - 4$

Hence, the given equation becomes

$$\frac{dz}{dx} - 4 = z^2 \quad \Rightarrow \quad \frac{dz}{4+z^2} = dx$$

Integrating it, we get

$$\frac{1}{2}\tan^{-1}\left(\frac{z}{2}\right) = x + c$$

$$\Rightarrow \qquad 4x + y + 1 = 2\tan(2x + c').$$

where $c' = 2c$, is an arbitrary constant.

EXAMPLE 2.7

Solve $\dfrac{dy}{dx} = \sin(x + y) + \cos(x + y)$.

Solution: Put $x + y = z \quad \Rightarrow \quad \dfrac{dy}{dx} = \dfrac{dz}{dx} - 1$

The given equation becomes

$$\frac{dz}{dx} - 1 = \sin z + \cos z$$

$$\Rightarrow \qquad dx = \frac{dz}{1 + \sin z + \cos z}$$

Integrating it, we get

$$x = \int \frac{dz}{1 + \cos z + \sin z} + c$$

$$= \frac{1}{2}\int \frac{dz}{2\cos^2 z/2 + 2\sin z/2 \cos z/2} + c$$

$$= \frac{1}{2}\int \frac{\sec^2 z/2}{1 + \tan z/2}\,dz + c$$

$$= \log[1 + \tan z/2] + c$$

$$\therefore \qquad x = \log\left[1 + \tan\frac{(x + y)}{2}\right] + c.$$

EXAMPLE 2.8

Solve $\frac{x+y-a}{x+y-b}\frac{dy}{dx} = \frac{x+y+a}{x+y+b}$.

Solution: Put $x + y = z \Rightarrow \frac{dy}{dx}=\frac{dz}{dx}-1$

The given equation becomes

$$\frac{z-a}{z-b}\left(\frac{dz}{dx}-1\right) = \frac{z+a}{z+b}$$

$$\Rightarrow \quad \frac{dz}{dx}-1 = \frac{z+a}{z+b}\times\frac{z-b}{z-a} = \frac{z^2+(a-b)z-ab}{z^2+(b-a)z-ab}$$

$$\Rightarrow \quad \frac{dz}{dx} = \frac{z^2+(a-b)z-ab}{z^2+(b-a)z-ab}+1$$

$$\Rightarrow \quad \frac{dz}{dx} = \frac{2(z^2-ab)}{z^2+(b-a)z-ab}$$

$$\Rightarrow \quad \frac{z^2-ab+(b-a)z}{z^2+ab}dz = 2dx$$

$$\Rightarrow \quad \left[1+\frac{(b-a)z}{z^2-ab}\right]dz = 2dx$$

Integrating it, we get

$$z + \frac{1}{2}(b-a)\log|z^2-ab| = 2x + c$$

$$\Rightarrow \quad (b-a)\log|(x+y)^2-ab| = 2(x-y+c)$$

which is the required solution.

EXAMPLE 2.9

Solve $\frac{dy}{dx}-x\tan(y-x) = 1$.

Solution: Put $y - x = z \Rightarrow \frac{dy}{dx} = \frac{dz}{dx}+1$

The given equation becomes

$$\left(\frac{dz}{dx}+1\right)-x\tan z = 1$$

$$\Rightarrow \quad \frac{dz}{\tan z} = x\,dx$$

Integrating it, we get $\log \sin z = \frac{x^2}{2} + \log c$

$$\Rightarrow \quad \sin(y - x) = ce^{x^2/2}$$

which is the required solution.

EXAMPLE 2.10

Solve $\frac{xdx + ydy}{xdy - ydx} = \sqrt{\frac{a^2 - x^2 - y^2}{x^2 + y^2}}$.

Solution: Put $x = r\cos\theta$, $y = r\sin\theta$

$$\Rightarrow \quad \frac{dx}{d\theta} = \frac{dr}{d\theta}\cos\theta - r\sin\theta, \quad \frac{dy}{d\theta} = \frac{dr}{d\theta}\sin\theta + r\cos\theta$$

$$\therefore \quad \frac{dy}{dx} = \frac{dy/d\theta}{dx/d\theta} = \frac{\frac{dr}{d\theta}\sin\theta + r\cos\theta}{\frac{dr}{d\theta}\cos\theta - r\sin\theta}$$

The given equation becomes

$$\frac{r\cos\theta\left(\frac{dr}{d\theta}\sin\theta + r\cos\theta\right) + r\sin\theta\left(\frac{dr}{d\theta}\cos\theta + r\sin\theta\right)}{r\cos\theta\left(\frac{dr}{d\theta}\sin\theta + r\cos\theta\right) - r\sin\theta\left(\frac{dr}{d\theta}\cos\theta - r\sin\theta\right)}$$

$$= \sqrt{\frac{a^2 - r^2}{r^2}} \quad \text{as } x^2 + y^2 = r^2$$

$$\Rightarrow \quad \frac{r\frac{dr}{d\theta}(\cos^2\theta + \sin^2\theta)}{r^2(\cos^2\theta + \sin^2\theta)} = \frac{\sqrt{a^2 - r^2}}{r}$$

$$\Rightarrow \quad \frac{dr}{d\theta} = \sqrt{a^2 - r^2}$$

$$\Rightarrow \quad \frac{dr}{\sqrt{a^2 - r^2}} = d\theta$$

Integrating it, we get

$$\sin^{-1}\left(\frac{r}{a}\right) = \theta + c$$

$$\Rightarrow \quad r = a\sin(\theta + c)$$

$$\therefore \qquad \sqrt{x^2+y^2} = a\sin\left[\tan^{-1}\left(\frac{y}{x}\right)+c\right]$$

which is the required solution. ▲▲

EXERCISE 2.1

1. Solve $(1 - x)dy + (1 - y)dx = 0$.
2. Solve $\dfrac{dy}{dx} = e^{3x-2y} + x^2e^{-2y}$.
3. Solve $\dfrac{dy}{dx} = \dfrac{x(2\log x+1)}{\sin y + y\cos y}$.
4. Solve $\dfrac{dy}{dx}\tan y = \sin(x + y) + \sin(x - y)$.
5. Solve $\operatorname{cosec} x \log y\, dy + x^2y^2 dx = 0$.
6. Solve $\cos y \log(\sec x + \tan x)dx = \cos x \log(\sec y + \tan y)dy$.
7. Solve by substituting $y' = u$

$$y'' + (1 + y^{-1})(y')^2 = 0.$$

8. Solve the initial value problem

$$y' = \frac{x}{y} - \frac{x}{1+y};\ y(0) = 2.$$

9. Solve the initial value problem $L\dfrac{dI}{dt} + RI = 0$; $I(0) = I_0$ where, I, R, L are, respectively, the current, resistance and inductance in an electrical circuit. (R and L being constants).
10. Find the curve in which the angle between the radius vector and the tangent is m-times the vectorial angle and which passes through the point $(a, \pi/2m)$.
11. Find the curve $y = f(x)$ through the origin for which $y'' = y'$ and the tangent at the origin is $y = x$.
12. The drop of liquid evaporates at a rate proportional to its area of surface. If the radius initially is 4 mm and 5 minutes later, the radius is reduced to 2 mm, find the radius of the drop as a function of time.
13. A particle moves on a straight line so that its acceleration is equal to four times its velocity. At time $t = 0$, its displacement from origin is 1 m and its velocity is 1.5 m/sec. Find an approximation to the time when the displacement is 10 m.
14. Solve $(x + y)^2 \dfrac{dy}{dx} = a^2$.
15. Solve $x dy - y dx = (x^2 - y^2)^{1/2} dx$.

16. Solve $\frac{y}{x}\frac{dy}{dx}+\frac{x^2+y^2-1}{2(x^2+y^2)+1}=0$ (*Hint:* Put $x^2 + y^2 = z$).

17. Solve $\frac{dy}{dx} = \cos(x + y + 1)$.

18. Solve $x^4\frac{dy}{dx}+x^2y+\text{cosec}(xy)=0$.

19. Solve $\sqrt{(1+x^2+y^2+x^2y^2)}+xy\frac{dy}{dx}=0$.

20. Solve $(2ax + x^2)\frac{dy}{dx} = a^2 + 2ax$.

ANSWERS

Exercise 2.1

1. $(1 - x)(1 - y) = c$.
2. $3e^{2y} = 2(e^{3x} + x^3) + 6c$.
3. $y \sin y = x^2 \log x + c$.
4. $\sin y = -2 \cos x + c$.
5. $(1 + \log y)y + x^2 \cos x - 2x \sin x - 2 \cos x = c$.
6. $[\log(\sec x + \tan x)]^2 - [\log(\sec y + \tan y)]^2 = c$.
7. $(y - 1)e^y = cx + d$.
8. $3y^2 + 2y^3 = 3x^2 + 28$.
9. $I = I_0 e^{-(R/L)t}$.
10. $r^m = a^m \sin m\theta$.
11. $y = e^x - 1$.
12. $r(t) = 4 - 0.4t$.
13. $y = (3e^{4t} + 5)/8$, 0.8 sec.
14. $a \tan\left(\frac{y+c}{a}\right) = x + y$.
15. $y = x \sin(\log cx)$.
16. $x^2 + 2y^2 - 3\log(x^2 + y^2 + 2) + c = 0$.
17. $x = \text{cosec}(x + y + 1) - \cot(x + y + 1) + c$.
18. $\cos(xy) + \frac{1}{2x^2} = c$.

19. $\log x - \log[1 + \sqrt{1+x^2}] + \sqrt{1+x^2} + \sqrt{1+y^2} = c.$

20. $x(x + 2a)^3 = e^{2(x+c)a}.$ ▲▲

HOMOGENEOUS EQUATIONS

Before discussing homogeneous equations, we will define a homogeneous function.

Definition: A function $f(x, y)$ of two variables is said to be homogeneous of degree n (a real number) in x and y if

$$f(\lambda x, \lambda y) = \lambda^n f(x, y) \text{ for all } \lambda > 0.$$

If $f(x, y)$ is a homogeneous function then we can also write it in the form $x^n g(y/x)$ or $y^n g(x/y)$.

Some examples of homogeneous functions are the following

$f(x, y)$	Degree of homogeneity
$x^2 + xy$	2
$\tan^{-1}(y/x)$	0
$1/(x + y)$	–1
$\sqrt{x}/\sqrt{x^2+y^2}$	–1/2
$(x + y)/(x - y)$	0
$(x + y)/(\sqrt{x}+\sqrt{y})$	1/2

The function $f(x, y) = (x^2 + y)/(x^2 + y^2)$ is not homogeneous although its denominator is homogeneous with degree 2.

Note: From the above examples, we can note that in a homogeneous function combined degree of each term in numerator/denominator is same and degree of homogeneity = degree of numerator – degree of denominator.

Now we will define a homogeneous equation.

Definition: A differential equation of the form

$$\frac{dy}{dx} = \frac{f_1(x, y)}{f_2(x, y)}$$

where, $f_1(x, y)$ and $f_2(x, y)$

are homogeneous functions of x, y and of the same degree, is said to homogeneous. Such equations can be solved by putting

$$y = vx \text{ (or } x = vy)$$

where, y (or x) is changed to another variable v.

➲ Working Rule

Step-1: Put $y = vx$ (or $x = vy$).

Step-2: Transform the given equation in terms of v and x (or v and y).

Step-3: The resulting equation will be in variable separable form. Separate the variables and integrate to get the required general solution. ▲▲

EXAMPLE 2.11

Solve $x^2dy + y(x + y)dx = 0$.

Solution: The given differential equation can be written as

$$\frac{dy}{dx} = -\frac{y(x+y)}{x^2}$$

which is homogeneous. Putting $y = vx$, we get

$$v + x\frac{dv}{dx} = -v(1 + v)$$

$$\Rightarrow \quad \frac{dv}{v^2+2v} = \frac{-dx}{x} \Rightarrow \frac{1}{2}\left(\frac{1}{v}-\frac{1}{v+2}\right)dv = -\frac{dx}{x}$$

On integrating, we have

$$\frac{1}{2}\log\frac{v}{v+2} = -\log x + \log c$$

$$\Rightarrow \quad \log\sqrt{\frac{v}{v+2}}+\log x = \log c$$

$$\Rightarrow \quad \log\left(x\sqrt{\frac{v}{v+2}}\right) = \log c$$

$$\Rightarrow \quad \log\left(x\sqrt{\frac{y}{y+2x}}\right) = \log c$$

$$\Rightarrow \quad x\sqrt{\frac{y}{y+2x}} = c \Rightarrow x^2y = c^2(y + 2x)$$

which is the required general solution.

EXAMPLE 2.12

Solve $y - x\dfrac{dy}{dx} = x + y\dfrac{dy}{dx}$.

Solution: The given equation can be written as

$$\frac{dy}{dx} = \frac{y-x}{y+x}$$

Put $y = vx$, $\frac{dy}{dx} = v + x\frac{dv}{dx}$, we get

$$v + x\frac{dv}{dx} = \frac{v-1}{v+1}$$

$$\Rightarrow \quad x\frac{dv}{dx} = -\frac{1+v^2}{v+1} \quad \Rightarrow \quad \frac{v+1}{1+v^2}dv = -\frac{dx}{x}$$

Integrating it, we get

$$\frac{1}{2}\log(1+v^2) + \tan^{-1}v = -\log x + \log c$$

$$\Rightarrow \quad \frac{1}{2}\log\frac{(1+v^2)x^2}{c^2} = -\tan^{-1}v$$

$$\Rightarrow \quad \log\frac{x^2+y^2}{c^2} = -2\tan^{-1}(y/x)$$

$$\Rightarrow \quad x^2 + y^2 = c^2e^{-2\tan^{-1}(y/x)}$$

which is the required solution.

EXAMPLE 2.13

Solve $x\frac{dy}{dx} = y(\log y - \log x + 1)$.

Solution: The given equation can be written as

$$\frac{dy}{dx} = \frac{y[\log y/x+1]}{x}$$

Put $y = vx$, $\frac{dy}{dx} = v + x\frac{dv}{dx}$ to get

$$v + x\frac{dv}{dx} = v[\log v + 1]$$

$$\Rightarrow \quad \frac{dv}{v\log v} = \frac{dx}{x}$$

On integrating, we get

$$\log(\log v) = \log x + \log c$$

$$\Rightarrow \quad \log v = cx$$

$$\Rightarrow \quad v = e^{cx}$$

$$\Rightarrow \quad y = xe^{cx}$$

which is the required solution.

EXAMPLE 2.14

Solve $(1 + e^{x/y})\,dx + e^{x/y}(1 - x/y)\,dy = 0.$

Solution: The given equation can be written as

$$\frac{dx}{dy} = \frac{e^{x/y}(1-x/y)}{1+e^{x/y}}$$

which is homogeneous equation. Putting $x = vy$, we get

$$v + y\frac{dv}{dy} = \frac{e^{v}(1-v)}{1+e^{v}}$$

$$\Rightarrow \qquad y\frac{dv}{dy} = -\frac{e^{v}(1-v)}{1+e^{v}} - v = -\frac{v+e^{v}}{1+e^{v}}$$

$$\Rightarrow \qquad -\frac{dy}{y} = \frac{1+e^{v}}{v+e^{v}}dv = \frac{d(v+e^{v})}{v+e^{v}}$$

On integrating, we get $\log y = \log(v + e^{v}) + \log c$

$$\Rightarrow \qquad y(v + e^{v}) = -c \quad \Rightarrow \quad x + ye^{x/y} = c' \text{ (say)}$$

which is the required solution.

EXAMPLE 2.15

Solve $\left(x\tan\frac{y}{x} - y\sec^2\frac{y}{x}\right)dx + x\sec^2\frac{y}{x}\,dy = 0.$

Solution: The given equation can be written as

$$\frac{dy}{dx} = \frac{y\sec^2\frac{y}{x} - x\tan\frac{y}{x}}{x\sec^2\frac{y}{x}} = \frac{y}{x} - \frac{\tan\frac{y}{x}}{\sec^2\frac{y}{x}}$$

Putting $y = vx$, we get

$$v + x\frac{dv}{dx} = v - \frac{\tan v}{\sec^2 v}$$

$$\Rightarrow \qquad \frac{\sec^2 v}{\tan v}dv = \frac{-dx}{x}$$

On integrating, we get

$$\log\tan v = -\log x + \log c$$

$$\Rightarrow \qquad x\tan\frac{y}{x} = c$$

which is the required solution.

EXAMPLE 2.16

Solve $x\cos\frac{y}{x}(ydx + xdy) = y\sin\frac{y}{x}(xdy - ydx)$.

Solution: The given equation can be written as

$$\frac{dy}{dx} = \frac{-y\left[y\sin\frac{y}{x} + x\cos\frac{y}{x}\right]}{x\left[x\cos\frac{y}{x} - y\sin\frac{y}{x}\right]}$$

$$\Rightarrow \quad \frac{dy}{dx} = \frac{y}{x}\,\frac{y\sin(y/x) + x\cos(y/x)}{y\sin(y/x) - x\cos(y/x)}$$

Putting $y = vx$, we get

$$v + x\frac{dv}{dx} = \frac{v(v\sin v + \cos v)}{(v\sin v - \cos v)}$$

$$\Rightarrow \quad v\frac{dv}{dx} = \frac{v(v\sin v + \cos v)}{v\sin v + \cos v} - v = \frac{2v\cos v}{v\sin v - \cos v}$$

$$\Rightarrow \quad \left(\frac{v\sin v - \cos v}{v\cos v}\right)dv = 2\frac{dx}{x}$$

$$\Rightarrow \quad \left(\tan v - \frac{1}{v}\right)dv = 2\frac{dx}{x}$$

Integrating it, we get

$$\log\sec v - \log v = 2\log x + \log c$$

$$\Rightarrow \quad \sec y(x) = cxy.$$

which is the required solution. ▲▲

Equations Reducible to the Homogeneous Form

The equation of the form $\frac{dy}{dx} = \frac{ax + by + c}{a'x + b'y + c'}$

is not homogeneous but it can be reduced to a homogeneous form as follows.

Case I: When $\frac{a}{a'} \neq \frac{b}{b'}$

Change the variables x, y to X, Y related by the equations

$$x = X + h,\ y = Y + K$$

where, h, k are the constants to be chosen so as to make the given equation homogeneous.

We have $dx = dX$, $dy = dY$

The equation becomes

$$\frac{dY}{dX} = \frac{aX + bY + (ah + bk + c)}{a'X + b'Y + (a'h + b'k + c')}$$

This equation will be homogeneous if

$$ah + bk + c = 0, a'h + b'k + c' = 0$$

$$\therefore \quad h = \frac{bc' - b'c}{ab' - b'a}, K = \frac{ca' - c'a}{ab' - b'a}$$

which are meaningful except when

$$ab' - a'b = 0, \text{ i.e., when } a/a' = b/b'$$

The homogeneous equation $\frac{dy}{dx} = \frac{aX + bY}{a'X + b'Y}$ can now be solved by means of the substitution $y = vX$.

Case II: When $\frac{a}{a'} = \frac{b}{b'}$

Let, $$\frac{a}{a'} = \frac{b}{b'} = r, \text{ say}$$

$$\therefore \quad a = a'r, \ b = b'r$$

The equation now becomes

$$\frac{dy}{dx} = \frac{r(a'x + b'y) + c}{a'x + b'y + c'}$$

We put $a'x + b'y = z$ so that $a' + b'\frac{dy}{dx} = \frac{dz}{dx}$ and obtain $\frac{dz}{dx} = b'\frac{rz + e}{z + c'} + a'$ which is in variable sparable form.

➲ Working Rule

When $a/a' \neq b/b'$

Step 1: Put $x = X + h$, $y = Y + k$ in the given differential equation.

Step 2: Equate the constant terms in the numerator and denominator to zero.

Step 3: Solve these resulting equations for h and k.

Step 4: Solve the resulting homogeneous equation in X and Y by putting $Y = vX$ (or $X = vY$).

Step 5: Replace X by $x - h$ and Y by $y - K$.

Step 6: Substitute the values of h and K in the general solution obtained in Step 5. ▲▲

EXAMPLE 2.17

Solve $\dfrac{dy}{dx}+\dfrac{x-y-2}{x-2y-3}=0$.

Solution: Putting $x = X + h$, $y = Y + k$, the above equation becomes

$$\frac{dY}{dX}+\frac{X-Y+(h-K-2)}{X-2Y+(h-2k-3)}=0$$

To eliminate h and K, we have

$$h - K - 2 = 0,\ h - 2k - 3 = 0$$

$$\Rightarrow \qquad h = 1,\ K = -1$$

$$\therefore \qquad \frac{dY}{dX}+\frac{X-Y}{X-2Y} = 0$$

Put $Y = vX$ so that $\dfrac{dY}{dX} = v + X\dfrac{dv}{dX}$, we get

$$v+X\frac{dv}{dX}+\frac{1-v}{1-2v} = 0$$

$$\Rightarrow \qquad X\frac{dv}{dX}+\frac{1-2v^2}{1-2v} = 0$$

$$\Rightarrow \qquad \frac{1-2v}{1-2v^2}dv+\frac{dX}{X} = 0$$

$$\Rightarrow \qquad \left[\frac{1}{1-2v^2}-\frac{2v}{1-2v^2}\right]dv+\frac{dX}{X} = 0$$

$$\Rightarrow \qquad \left[\frac{1}{2}\frac{1}{\left(\frac{1}{2}-v^2\right)}+\frac{1}{2}\frac{-4v}{1-2v^2}\right]dv+\frac{dX}{X} = 0$$

On integrating, we get

$$\frac{1}{2}\frac{1}{1/\sqrt{2}}\log\left[\frac{\frac{1}{\sqrt{2}}+v}{\frac{1}{\sqrt{2}}-v}\right]+\frac{1}{2}\log(1-2v^2)+\log X = \log c$$

$$\Rightarrow \qquad \log\left(\frac{X+\sqrt{2Y}}{X-\sqrt{2Y}}\right)^{1/\sqrt{2}} + \log(X^2 - 2Y^2) = \log c$$

which is the required solution with $X = x - 1$ and $Y = y + 1$.

EXAMPLE 2.18

Solve $\dfrac{dy}{dx} = \dfrac{x - y + B}{2x + 2y + 5}$.

Solution: Put $x - y = z$

$$\therefore \qquad 1 - \frac{dy}{dx} = \frac{dz}{dx}$$

The given equation becomes

$$1 - \frac{dz}{dx} = \frac{z+3}{2z+5}$$

$$\Rightarrow \qquad \frac{dz}{dx} = \frac{z+2}{2z+5}$$

$$\Rightarrow \qquad \left(2 + \frac{1}{z+2}\right)dz = dx$$

On integrating it, we get

$$2z + \log(z + 2) = x + c$$

$$\Rightarrow \qquad x - 2y + \log(x - y + 2) = c$$

which is the required solution. ▲▲

EXERCISE 2.2

1. Solve the differential equation

$$(x^2 + 4y^2 + xy)\,dx - x^2\,dy = 0.$$

2. Solve the initial value problem

$$(3xy + y^2)\,dx + (x^2 + xy)\,dy = 0;\ y(1) = 1.$$

3. Solve $(x^2 + y^2)\dfrac{dy}{dx} = xy$.

4. Solve $x\dfrac{dy}{dx} + \dfrac{y^2}{x} = y$.

5. Solve $x\frac{dy}{dx} - y = x\sqrt{x^2 + y^2}$.

6. Solve $\frac{dy}{dx} = \frac{y}{x} - \tan\frac{y}{x}$.

7. Solve $[x\tan(y/x) - y\sec^2(y/x)]dx - x\sec^2(y/x)\,dy = 0$.

8. Solve $ye^{x/y}dx = (xe^{x/y} + y^2)\,dy$.

9. Solve $xy(\log x/y)\,dx + (y^2 - x^2)(\log x/y)\,dy = 0$.

10. Solve $\frac{dy}{dx} + \frac{x^2 + 3y^2}{3x^2 + y^2} = 0$.

11. Solve $(6x^2 + 2y^2)\,dx - (x^2 + 4xy)\,dy = 0$.

12. Solve $\frac{dy}{dx} + \frac{x + 2y + 3}{2x + 3y + 4}$.

13. Solve $(2x + y + 3)\,dx = (2y + x + 1)\,dy$.

14. Find the solution of the differential equation

$$(y - x + 1)\,dy - (y + x + 2)\,dx = 0.$$

15. Solve $(3y + 2x + 4)\,dx - (4x + 6y + 5)\,dy = 0$.

16. Solve $(3y - 7x + 7)\,dx + (7y - 3x + 3)\,dy = 0$.

ANSWERS

Exercise 2.2

1. $2y = x\tan[2\log|cx|]$.
2. $x^2y(2x + y) = 3$.
3. $cy = e^{x^2/2y^2}$.
4. $cx = e^{x/y}$.
5. $y = x\sin h(x + c)$.
6. $\sin(y/x) = cx$.
7. $x\tan(y/x) = c$.
8. $e^{x/y} = y + c$.
9. $\log y - \frac{x^2}{4y^2}\left(x\log\frac{y}{x} + 1\right) = c$.
10. $\log(x + y) + 2xy(x + y)^{-2} = c$.
11. $(2x + y)(2y - 3x) = cx$.

12. $\frac{2+\sqrt{3}}{2}\log\left(\sqrt{3}(y+2)-x+1\right) - \frac{2-\sqrt{3}}{2}\log\left(\sqrt{3}(y+2)+x-1\right) = c.$

13. $\left(x+y+\frac{4}{3}\right)(x - y + 2)^3 = c.$

14. $2(y - x + 1)^2 - (2x + 1)^2 = c.$

15. $21x - 42y + 9\log(14x + 21y + 22) = c.$

16. $(x - y - 1)^2(x + y - 1)^3 = c.$

3

CHAPTER

Exact Differential Equations

In chapter 2, we have learnt about some basic methods for the solution of differential equations of the first order with first degree. In this chapter, we will discuss a special class of equations known as exact differential equations. After defining exact and non-exact differential equations, we will study the methods for their solutions.

Definition: Let $f(x, y)$ be a function of two real variables x and y such that f has continuous first order partial derivatives then the total differential df of the function $f(x, y)$ is defined as

$$\boxed{df = \frac{\partial f}{\partial x}dx + \frac{\partial f}{\partial y}dy}$$

For example, consider $f(x, y) = xy^2 + 2x^3y$

Then $$\frac{\partial f}{\partial x} = y^2 + 6x^2y, \quad \frac{\partial f}{\partial y} = 2xy + 2x^3$$

and the total differential df is given by

$$df = (y^2 + 6x^2y)\,dx + (2xy + 2x^3)\,dy$$

Definition: A differential equation of the form $M(x, y)\,dx + N(x, y)\,dy = 0$ is said to be exact if its left hand member is the exact differential of some function $f(x, y)$, that is

$$\boxed{df = Mdx + Ndy = 0}$$

Its solution, therefore, is

$$f(x, y) = c$$

where, c is constant of integration.

For example, the differential equation $y^2dx + 2xy\,dy = 0$ is an exact differential equation since $y^2dx + 2xy\,dy = d(xy^2)$. On the other hand, the more simple appearing equation $ydx + 2xdy = 0$ obtained from the above equation by dividing by y, is not exact.

It is always not possible to say that the given differential equation is exact by writing it as exact differential of some function. So we need a simple test to determine whether or not a given differential equation is exact. This is given by the following theorem.

Theorem 3.1 The necessary and sufficient condition for the differential equation $Mdx + Ndy = 0$ to be exact is

$$\boxed{\frac{\partial M}{\partial y} = \frac{\partial N}{\partial x}}.$$

***Proof:* Condition is necessary**

Let the equation $Mdx + Ndy = 0$ be exact

Therefore, we can write

$$Mdx + Ndy = df \tag{3.1}$$

where, $f(x, y)$ is some function of x and y

But we know that

$$df = \frac{\partial f}{\partial x}dx + \frac{\partial f}{\partial y}dy \tag{3.2}$$

Equating coefficients of dx and dy in Eqs. (3.1) and (3.2), we get

$$M = \frac{\partial f}{\partial x} \quad \text{and} \quad N = \frac{\partial f}{\partial y}$$

$$\Rightarrow \qquad \frac{\partial M}{\partial y} = \frac{\partial^2 f}{\partial y \partial x} \quad \text{and} \quad \frac{\partial N}{\partial x} = \frac{\partial^2 f}{\partial x \partial y}$$

Since
$$\frac{\partial^2 f}{\partial y \partial x} = \frac{\partial^2 f}{\partial x \partial y}$$

Hence,
$$\frac{\partial M}{\partial y} = \frac{\partial N}{\partial x}$$

which is the necessary condition for exactness.

Condition is sufficient: Let us assume $\frac{\partial M}{\partial y} = \frac{\partial N}{\partial x}$ holds.

In this case, we have to show that

$$Mdx + Ndy = 0 \text{ is exact.}$$

Let
$$\int Mdx = f$$

where y is supposed constant while performing integration with respect to x.

Then, $$\frac{\partial}{\partial x}\left(\int M dx\right) = \frac{\partial f}{\partial x}$$

i.e., $$M = \frac{\partial f}{\partial x}$$

$$\Rightarrow \qquad \frac{\partial M}{\partial y} = \frac{\partial^2 f}{\partial y \partial x}$$

by using assumption, we obtain

$$\frac{\partial N}{\partial x} = \frac{\partial^2 f}{\partial y \partial x} = \frac{\partial^2 f}{\partial x \partial y}$$

Integrating both sides w.r.t. x (taking y as constant)

$$N = \frac{\partial f}{\partial y} + \phi(y)$$

where $\phi(y)$ is a function of y alone.

$$\therefore \qquad Mdx + Ndy = \frac{\partial f}{\partial x}dx + \left(\frac{\partial f}{\partial y} + \phi(y)\right)dy$$

$$= \left(\frac{\partial f}{\partial x}dx + \frac{\partial f}{\partial y}dy\right) + \phi(y)dy$$

$$= df + d\left(\int \phi(y)dy\right)$$

$$= d\left[f + \int \phi(y)dy\right]$$

which shows that $Mdx + Ndy = 0$ is exact.

Note: From the above, if $Mdx + Ndy = 0$ is exact then $Mdx + Ndy = 0$ becomes

$$d\left[f + \int \phi(y)dy\right] = 0$$

On integration, we have

$$f + \int \phi(y)dy = c$$

Since $f = \int_{y\text{-constant}} Mdx$ and $\phi(y) =$ terms of N not containing x.

$\therefore$ The general solution is

$$\int_{y\text{-constant}} M dx + \int (\text{terms of } N \text{ not containing } x)\, dy = c$$

provided $\dfrac{\partial M}{\partial y} = \dfrac{\partial N}{\partial x}$.

Remark: We can also write formula for solution of an exact differential equation

$$M dx + N dy = 0$$

as $\displaystyle\int_{x\text{-constant}} N dy + \int (\text{terms of } M \text{ not containing } y) dx = c$ which is same as the earlier written formula. ▲▲

EXAMPLE 3.1

Show that the equation

$$(x^4 - 2xy^2 + y^4)\, dx - (2x^2y - 4xy^3 + \sin y)\, dy = 0$$

is exact and also solve it.

Solution: Here, $M = x^4 - 2xy^2 + y^4$, $N = -(2x^2y - 4xy^3 + \sin y)$

$$\frac{\partial M}{\partial y} = -4xy + 4y^3, \quad \frac{\partial N}{\partial x} = -4xy + 4y^3$$

$$\Rightarrow \quad \frac{\partial M}{\partial y} = \frac{\partial N}{\partial x}.$$

Thus, the given equation is exact.

Therefore, the solution will be

$$\int_{y\text{-constant}} M dx + \int (\text{terms of } N \text{ not containing } x)\, dy = c$$

$$\Rightarrow \quad \int_{y\text{-constant}} (x^4 - 2xy^2 + y^4)\, dx + \int (-\sin y)\, dy = c$$

$$\Rightarrow \quad \frac{x^5}{5} - x^2y^2 + xy^4 + \cos y = c.$$

EXAMPLE 3.2

Solve $(1 + e^{x/y})dx + e^{x/y}\left(1 - \dfrac{x}{y}\right)dy = 0$.

Solution: Here, $M = 1 + e^{x/y}$, $N = e^{x/y}\left(1 - \dfrac{x}{y}\right)$

$$\frac{\partial M}{\partial y} = e^{x/y}\left(-\frac{x}{y^2}\right), \quad \frac{\partial N}{\partial x} = e^{x/y}\frac{1}{y}\left(1-\frac{x}{y}\right)+e^{x/y}\left(\frac{-1}{y}\right) = e^{x/y}\left(\frac{-x}{y^2}\right)$$

$$\therefore \quad \frac{\partial M}{\partial y} = \frac{\partial N}{\partial x}$$

Thus, the given equation is exact. Its solution is

$$\int_{y\text{-constant}} (1+e^{x/y})dx + \int (0)\,dy = c$$ as there is no term in N not containing x.

$$\Rightarrow \quad x + ye^{x/y} = c.$$

Note: We have solved the above example in Chapter 2 by using $x = vy$ and got the same solution as found here.

EXAMPLE 3.3

Solve $xdx + ydx + \dfrac{xdy - ydx}{x^2+y^2} - 0.$

Solution: The given equation is

$$\left(x-\frac{y}{x^2+y^2}\right)dx+\left(y+\frac{x}{x^2+y^2}\right)dy = 0$$

Here, $M = x-\dfrac{y}{x^2+y^2}$, $N = y+\dfrac{x}{x^2+y^2}$

$$\frac{\partial M}{\partial y} = \frac{y^2-x^2}{(x^2+y^2)^2} \qquad \frac{\partial N}{\partial x} = \frac{y^2-x^2}{(x^2+y^2)^2}$$

$$\Rightarrow \quad \frac{\partial M}{\partial y} = \frac{\partial N}{\partial x}$$

Therefore, the given equation is exact. Its solution will be

$$\int_{y\text{-constant}} \left(x-\frac{y}{x^2+y^2}\right)dx+\int (y)\,dy = c$$

$$\Rightarrow \quad \frac{x^2}{2}-y\cdot\frac{1}{y}\tan^{-1}\frac{x}{y}+\frac{y^2}{2} = c$$

$$\Rightarrow \quad x^2+y^2-2\tan^{-1}\frac{x}{y} = c'$$

where $c' = 2c$ is constant of integration. ▲▲

EXERCISE 3.1

1. Check the equation

$$(3x^2 + 2e^y)\,dx + (2xe^y + 3y^2)\,dy = 0$$

for exactness. If it is exact, find the solution.

2. Solve the initial value problem

$$e^x(\cos y\,dx - \sin y\,dy) = 0;\ y(0) = 0.$$

3. If exact, find the solution of the equation

$$\cos x(\cos x - \sin a \sin y)\,dx + \cos y(\cos y - \sin a \sin x)\,dy = 0$$

where, a is an arbitrary constant.

4. Solve $(y^2e^{xy^2} + 4x^3)\,dx + (2xye^{xy^2} - 3y^2)\,dy = 0$.
5. Find the solution of the equation

$$\left[y\left(1+\frac{1}{x}\right)+\cos y\right]dx + [x + \log x - x\sin y]\,dy = 0.$$

6. Solve $\dfrac{dy}{dx} + \dfrac{y\cos x + \sin y + y}{\sin x + x\cos y + x} = 0.$
7. Solve $[\cos x \tan y + \cos(x + y)]dx + [\sin x \sec^2 y + \cos(x + y)]\,dy = 0.$
8. Determine for what values of a and b, the following differential equation is exact and obtain the general solution of the exact equation.

$$(y + x^3)\,dx + (ax + by^3)\,dy = 0$$

9. Under what condition, $(ax + y)\,dx + (Kx + by)\,dy = 0$ is exact?
10. Find the relation between a and b, c and d such that

$(a\sinh x \cos y + b\cosh x \sin y)\,dx + (c\sinh x \cos y + d\cosh x \sin y)\,dy = 0$ is exact.

11. Solve $(x^2 - 2xy - y^2)\,dx - (x + y^2)\,dy = 0.$

ANSWERS

Exercise 3.1

1. Exact, $x^3 + 2xe^y + y^3 = c$.
2. $e^x \cos y = 1$.
3. $2(x + y) + \sin 2x + \sin 2y - 4\sin a \sin y \sin x = c$.
4. $e^{xy^2} + x^4 - y^3 = c$.
5. $(x + \log x)y + x\cos y = c$.
6. $y\sin x + (\sin y + y)x = c$.

7. $\sin x \tan y + \sin(x + y) = c.$

8. $a = 1$, b is arbitrary, $xy + \frac{x^4}{4} + \frac{6y^4}{4} = c.$

9. $k = 1$, a, b are arbitrary.

10. $a = -d$, $b = c$

11. $x^3 - y^3 = 3xy(x + y) + c.$

▲▲

INTEGRATING FACTORS

We have seen that Eq. (3.2) is not exact also for this equation

$$M = y \quad N = 2r$$

$$\frac{\partial M}{\partial y} = 1 \quad \frac{\partial N}{\partial x} = 2$$

$$\therefore \qquad \frac{\partial M}{\partial y} \neq \frac{\partial N}{\partial x}$$

But on multiplying Eq. (3.2) by y, we get Eq. (3.1) which is exact. So, any factor which on multiplication makes a non-exact equation an exact equation is called integrating factor, denoted by IF. So, for Eq. (3.2) IF $= y$. Now we give the formal definition of integrating factor.

Definition: If the first order differential equation

$$Mdx + Ndy = 0 \text{ is not exact,}$$

then we multiply it by an appropriate factor $\phi(x, y)$ such that the new equation

$$M(x, y)\,\phi(x, y)\,dx + N(x, y)\,\phi(x, y)\,dy = 0$$

is exact. The factor $\phi(x, y)$ is called an integrating factor.

Note: Integrating factor of the equation $Mdx + Ndy = 0$ is not unique.

Determination of an integrating factor for an arbitrary first order differential equation is difficult. However, we can find integrating factor in some special cases, which we will be studying in the succeeding discussion.

Case I: Integrating factors found by inspection

In a number of cases, the integrating factor can be found after regrouping the terms of the equation and recognizing each group as being a part of an exact differential. In this connection, the following integrable combinations prove quite useful.

$$x\,dy + y\,dx = d(x, y)$$

$$\frac{xdy - ydx}{x^2} = d\left(\frac{y}{x}\right)$$

$$\frac{ydx - xdy}{y^2} = d\left(\frac{x}{y}\right)$$

$$\frac{xdy - ydx}{x^2 + y^2} = d\left(\tan^{-1}\frac{y}{x}\right) = -d\left(\tan^{-1}\frac{x}{y}\right)$$

$$\frac{ydx - xdy}{xy} = d\left(\log\frac{x}{y}\right) = -d\left(\log\frac{y}{x}\right)$$

$$\frac{xdy + ydx}{x^2y^2} = -d\left(\frac{1}{xy}\right)$$

$$\frac{xdx + ydy}{x^2 + y^2} = d\left(\frac{1}{2}\log(x^2 + y^2)\right)$$

$$\frac{xdy - ydx}{x^2 - y^2} = d\left(\frac{1}{2}\log\frac{x + y}{x - y}\right)$$

EXAMPLE 3.4

Solve $(y^2e^x + 2xy)\,dx - x^2\,dy = 0$.

Solution: Here, $M = y^2e^x + 2xy \qquad N = -x^2$

$$\frac{\partial M}{\partial y} = 2ye^x + 2x \qquad \frac{\partial N}{\partial x} = -2x$$

Clearly, $\frac{\partial M}{\partial y} \neq \frac{\partial N}{\partial x}$. Hence, the equation is not exact.

On multiplying by $1/y^2$, the given equation becomes

$$\left(e^x + \frac{2x}{y}\right)dx - \frac{x^2}{y^2}dy = 0$$

$$\Rightarrow \qquad e^x dx + \frac{2xydx - x^2dy}{y^2} = 0$$

$$\Rightarrow \qquad d(e^x) + d\left(\frac{x^2}{y}\right) = 0$$

$$\Rightarrow \qquad d\left(e^x + \frac{x^2}{y}\right) = 0$$

Now, the equation is exact and its solution is $e^x + \frac{x^2}{y} = c$.

EXAMPLE 3.5

Solve $ydx - xdy + (1 + x^2)dx + x^2 \sin y\, dy = 0$

Solution: The given equation is not exact.

Multiply the given equation by $1/x^2$

$$\frac{ydx - xdy}{x^2} + \left(\frac{1}{x^2} + 1\right)dx + \sin y\, dy = 0$$

$$\Rightarrow \quad d\left(\frac{-y}{x}\right) + d\left(\frac{-1}{x} + x\right) + d(-\cos y) = 0$$

$$\Rightarrow \quad d\left(\frac{-y}{x} - \frac{1}{x} + x - \cos y\right) = 0$$

Now, the given equation is exact and its solution is

$$\frac{-y}{x} - \frac{1}{x} + x - \cos y = c$$

$$\rightarrow \quad x^2 - y - 1 - x\cos y = cx.$$

EXAMPLE 3.6

Solve $y(2x^2y + e^x)dx - (e^x + y^3)dy = 0.$

Solution: The given equation is not exact and it can be written as

$$(ye^x dx - e^x dy) + 2x^2y^2 dx - y^3 dy = 0$$

Multiply this equation by $1/y^2$

$$\frac{ye^x dx - e^x dy}{y^2} + 2x^2 dx - ydy = 0$$

$$\Rightarrow \quad d\left(\frac{e^x}{y}\right) + d\left(\frac{2x^3}{3}\right) - d\left(\frac{y^2}{2}\right) = 0$$

$$\Rightarrow \quad d\left(\frac{e^x}{y} + \frac{2x^3}{3} - \frac{y^2}{2}\right) = 0$$

Now, the equation is exact and its solution is

$$\frac{e^x}{y} + \frac{2x^3}{3} - \frac{y^2}{2} = c$$

$$\Rightarrow \quad 6e^x + 4x^3y - 3y^3 = 6cy.$$

Case II: If the equation $Mdx + Ndy = 0$ is homogeneous in x and y, i.e., if M and N are homogeneous functions of the same degree in x and y then $\dfrac{1}{Mx + Ny}$ is an integrating factor (IF) provided $Mx + Ny \neq 0$.

Proof: The given differential equation may be written as

$$Mdx + Ndy = \frac{1}{2}\left[(Mx+Ny)\left(\frac{dx}{x}+\frac{dy}{y}\right)+(Mx-Ny)\left(\frac{dx}{x}-\frac{dy}{y}\right)\right]$$

$$\therefore \quad \frac{1}{Mx+Ny}(Mdx + Ndy) = \frac{1}{2}\left[\frac{dx}{x}+\frac{dy}{y}+\frac{Mx-Ny}{Mx+Ny}\left(\frac{dx}{x}-\frac{dy}{y}\right)\right] \qquad (3.3)$$

Since M, N are homogeneous functions of the same degree, (say n) in x and y therefore, we can write

$$M = x^n \phi_1\left(\frac{y}{x}\right), N = x^n \phi_2\left(\frac{y}{x}\right)$$

$$\therefore \quad \frac{Mx-Ny}{Mx+Ny} = \frac{x^{n+1}\phi_1\left(\frac{y}{x}\right)-x^n y\phi_2\left(\frac{y}{x}\right)}{x^{n+1}\phi_1\left(\frac{y}{x}\right)+x^n y\phi_2\left(\frac{y}{x}\right)}$$

$$= \frac{\phi_1\left(\frac{y}{x}\right)-\frac{y}{x}\phi_2\left(\frac{y}{x}\right)}{\phi_1\left(\frac{y}{x}\right)+\frac{y}{x}\phi_2\left(\frac{y}{x}\right)} = f\left(\frac{y}{x}\right), \text{ say}$$

So, Eq. (3.3) reduces to

$$\frac{1}{Mx+Ny}(Mdx + Ndy) = \frac{1}{2}\left[d(\log x)+d(\log y)+f\left(\frac{y}{x}\right)(d(\log x)-d(\log y))\right]$$

$$= \frac{1}{2}\left[d(\log xy)-f\left(\frac{y}{x}\right)d\left(\log\frac{y}{x}\right)\right]$$

$$= \frac{1}{2}\left[d(\log xy)-\phi\left(\log\frac{y}{x}\right)d\left(\log\frac{y}{x}\right)\right]$$

$$\text{where, } \phi\left(\log\frac{y}{x}\right) = f\left(\frac{y}{x}\right)=f\left(e^{\log\frac{y}{x}}\right)$$

$$= \frac{1}{2}\left[d\left(\log xy-\psi\left(\log\frac{y}{x}\right)\right)\right]$$

$$\text{where } \psi\left(\log\frac{y}{x}\right) = \phi\left(\log\frac{y}{x}\right)d\left(\log\frac{y}{x}\right)$$

which is an exact differential.

Therefore, $Mdx + Ndy = 0$ becomes an exact differential equation on multiplying with integrating factor $\dfrac{1}{Mx+Ny}$.

Note: If $Mx + Ny = 0 \Rightarrow \dfrac{M}{N} = \dfrac{-y}{x}$

Then $Mdx + Ndy = 0$ can be written as

$$\frac{M}{N}dx + dy - 0$$

$$\Rightarrow \qquad \frac{-y}{x}dx + dy = 0 \Rightarrow \frac{dy}{y} - \frac{dx}{x} = 0$$

$$\Rightarrow \qquad d(\log|y|) - d(\log|x|) = 0$$

$$\Rightarrow \qquad d\left(\log\left|\frac{y}{x}\right|\right) = 0$$

On integration, we have

$$\log\left|\frac{y}{x}\right| = \log c$$

$$\therefore \qquad |y| = c|x|$$

which is the required solution.

EXAMPLE 3.7

Solve $(x^2y - 2xy^2)\,dx - (x^3 - 3x^2y)\,dy = 0$.

Solution: Here, $M = x^2y - 2xy^2$, $N = 3x^2y - x^3$

$$\frac{\partial M}{\partial y} = x^2 - 4xy, \quad \frac{\partial N}{\partial x} = 6xy - 3x^2$$

Clearly, $\dfrac{\partial M}{\partial y} \neq \dfrac{\partial N}{\partial x}$, hence the given equation is not exact.

We can see easily that M and N are homogeneous functions of x and y with same degree.

$$\therefore \qquad \text{IF} = \frac{1}{Mx+Ny} = \frac{1}{x^2y^2}$$

Multiplying the given equation by IF, we get

$$\left(\frac{1}{y} - \frac{2}{x}\right)dx + \left(\frac{3}{y} - \frac{x}{y^2}\right)dy = 0$$

Here, we have

$$M = \frac{1}{y} - \frac{2}{x}, N = \frac{3}{y} - \frac{x}{y^2}$$

$$\frac{\partial M}{\partial y} = \frac{-1}{y^2}, \quad \frac{\partial N}{\partial x} = \frac{-1}{y^2}$$

$$\therefore \quad \frac{\partial M}{\partial y} = \frac{\partial N}{\partial x}$$

Hence, the equation is exact and its solution is

$$\int_{y\text{-constant}} \left(\frac{1}{y} - \frac{2}{x}\right) dx + \int \left(\frac{3}{y}\right) dy = c$$

$$\Rightarrow \quad \frac{x}{y} + \log \frac{y^3}{x^2} = c$$

EXAMPLE 3.8

Solve $x^2y\,dx - (x^3 + y^3)dy = 0$.

Solution: Here, $M = x^2y \qquad N = -x^3 - y^3$

Since M and N are homogeneous functions with same degree,

$$\therefore \quad \text{IF} = \frac{1}{Mx + Ny} = \frac{-1}{y^4}$$

Multiplying the given equation by IF, we get

$$\frac{-x^2}{y^3} dx + \left(\frac{x^3}{y^4} + \frac{1}{y}\right) dy = 0$$

We can check easily that this equation is exact and its solution is

$$\int_{y\text{-constant}} \left(\frac{-x^2}{y^3}\right) dx + \int \left(\frac{1}{y}\right) dy = c$$

$$\Rightarrow \quad \frac{-x^3}{3y^3} + \log y = c$$

$$\Rightarrow \quad x^3 = 3y^3 (\log y - c).$$

EXAMPLE 3.9

Solve the following differential equation

$$(y^3 - 2yx^2)\,dx + (2xy^2 - x^3)\,dy = 0.$$

Solution: Here, $M = y^3 - 2yx^2$, $N = 2xy^2 - x^3$

Since M and N are homogeneous functions with some degree therefore,

$$\frac{1}{Mx+Ny} = \frac{1}{3xy(y^2-x^2)}$$

Multiplying the given equation by IF, we get

$$\frac{y^2-2x^2}{3x(y^2-x^2)}dx + \frac{(2y^2-x^2)}{3y(y^2-x^2)}dy = 0$$

We can check easily that this equation is exact and its solution is

$$\int\limits_{y\text{-constant}} \frac{y^2-2x^2}{3x(y^2-x^2)}dx + \int\left(\frac{1}{3y}\right)dy = c$$

$$\Rightarrow \quad \frac{1}{3}\int\limits_{y\text{-constant}} \left(\frac{1}{x} - \frac{x}{y^2-x^2}\right)dx + \frac{1}{3}\log y = c$$

$$\Rightarrow \quad \frac{1}{3}\left[\log x + \frac{1}{2}\log(y^2-x^2)\right] + \frac{1}{3}\log y = c$$

$$\Rightarrow \quad 2\log xy + \log(y^2-x^2) = 6c$$

$$\Rightarrow \quad x^2y^2(y^2-x^2) = e^{6c} = c', \text{ (say)}.$$

Case III: If the equation $Mdx + Ndy = 0$ is of the form $f_1(xy)\,y\,dx + f_2(xy)\,x\,dy = 0$ then $\frac{1}{Mx-Ny}$ is an integrating factor provided $Mx - Ny \neq 0$.

Proof: The given equation may be written as

$$Mdx + Ndy = \frac{1}{2}\left[(Mx+Ny)\left(\frac{dx}{x}+\frac{dy}{y}\right) + (Mx-Ny)\left(\frac{dx}{x}-\frac{dy}{y}\right)\right]$$

$$\therefore \quad \frac{1}{Mx-Ny}(Mdx + Ndy) = \frac{1}{2}\left[\frac{Mx+Ny}{Mx-Ny}\left(\frac{dx}{x}+\frac{dy}{y}\right) + \frac{dx}{x} - \frac{dy}{y}\right] \tag{3.4}$$

Since $Mdx + Ndy = 0$ is of the form

$$f_1(xy)\,y\,dx + f_2(xy)\,x\,dy = 0 \text{ therefore}$$

$$M = f_1(xy)\,y,\ N = f_2(xy)\,x$$

$$\therefore \quad \frac{Mx+Ny}{Mx-Ny} = \frac{f_1(xy)+f_2(xy)}{f_1(xy)-f_2(xy)} = \phi(xy), \text{ say}$$

So, Eq. (3.4) reduces to

$$\frac{1}{Mx-Ny}(Mdx + Ndy) = \frac{1}{2}[\phi(xy)(d(\log x) + d(\log y)) + d(\log x) - d(\log y)]$$

$$= \frac{1}{2}\left[\phi(xy)d(\log xy) + d\left(\log\frac{x}{y}\right)\right]$$

$$= \frac{1}{2}\left[\psi(\log xy)d(\log xy) + d\left(\log\frac{x}{y}\right)\right]$$

where, $\psi(\log xy) = \phi(xy) = \phi(e^{\log xy})$

$$= \frac{1}{2}\left[d\left(\psi(\log xy) + \log\frac{x}{y}\right)\right]$$

which is an exact differential.

Therefore, $Mdx + Ndy = 0$ becomes an exact differential equation on multiplying with the integrating factor $\dfrac{1}{Mx-Ny}$.

Note: If $Mx - Ny = 0 \Rightarrow \dfrac{M}{N} = \dfrac{y}{x}$

Then $Mdx + Ndy = 0$ can be written as

$$\frac{M}{N}dx + dy = 0$$

$$\Rightarrow \quad \frac{y}{x}dx + dy = 0 \Rightarrow \frac{dx}{x} + \frac{dy}{y} = 0$$

$$\Rightarrow \quad d(\log|x|) + d(\log|y|) = 0$$

$$\Rightarrow \quad d(\log|xy|) = 0$$

On integration, we have

$$\log|xy| = \log c$$

$$\therefore \quad |xy| = c$$

which is the required solution.

EXAMPLE 3.10

Solve $(x^2y^2 + xy + 1)ydx + (x^2y^2 - xy + 1)xdy = 0$.

Solution: Here, $M = x^2y^3 + xy^2 + y$, $N = x^3y^2 - x^2y + x$

$$\frac{\partial M}{\partial y} = 3x^2y^2 + 2xy + 1, \quad \frac{\partial N}{\partial x} = 3x^2y^2 - 2xy + 1$$

Clearly, $\frac{\partial M}{\partial y} \neq \frac{\partial N}{\partial x}$, hence the given equation is not exact.

We can see easily that M and N are of the form $yf_1(xy)$ and $xf_2(xy)$, respectively.

$$\therefore \qquad \text{IF} = \frac{1}{Mx - Ny} = \frac{1}{2x^2y^2}$$

Multiplying the given equation by IF, we get

$$\frac{1}{2}\left(y + \frac{1}{x} + \frac{1}{x^2y}\right)dx + \frac{1}{2}\left(x - \frac{1}{y} + \frac{1}{xy^2}\right)dy = 0$$

which can be shown to be exact. So solution of this exact equation is

$$\int_{y\text{-constant}} \frac{1}{2}\left(y + \frac{1}{x} + \frac{1}{x^2y}\right)dx + \int\left(\frac{-1}{2y}\right)dy = c$$

$$\Rightarrow \qquad \frac{1}{2}\left[xy + \log x - \frac{1}{xy}\right] - \frac{1}{2}\log y = c$$

$$\Rightarrow \qquad xy - \frac{1}{xy} + \log\frac{x}{y} = 2c = c' \text{ (say)}$$

EXAMPLE 3.11

Solve $y(xy + 2x^2y^2)dx + x(xy - x^2y^2)dy = 0$.

Solution: The given equation is of the form

$$f_1(xy)y\,dx + f_2(xy)x\,dy = 0$$

$$\therefore \qquad \text{IF} = \frac{1}{Mx - Ny} = \frac{1}{3x^3y^3}$$

Multiplying the given equation by IF, we get

$$\left(\frac{1}{3x^2y} + \frac{2}{3x}\right)dx + \left(\frac{1}{3xy^2} - \frac{1}{3y}\right)dy = 0$$

which is exact and its solution is

$$\int_{y\text{-constant}} \left(\frac{1}{3x^2y} + \frac{2}{3x}\right)dx + \int\left(-\frac{1}{3y}\right)dy = c$$

$$\Rightarrow \qquad \frac{-1}{3xy} + \frac{2}{3}\log x - \frac{1}{3}\log y = c$$

$$\Rightarrow \qquad \frac{-1}{xy} + \log \frac{x^2}{y} = 3c = c' \text{ (say)}$$

EXAMPLE 3.12

Solve $(1 + xy)y\,dx + (1 - xy)x\,dy = 0$.

Solution: The given equation is of the form

$$f_1(xy)\,y\,dx + f_2(xy)\,x\,dy = 0$$

$$\therefore \qquad \text{IF} = \frac{1}{Mx - Ny} = \frac{1}{2x^2y^2}$$

Multiplying the given equation by IF, we get

$$\left(\frac{1}{2x^2y} + \frac{1}{2x}\right)dx + \left(\frac{1}{2xy^2} - \frac{1}{2y}\right)dy = 0$$

which is an exact equation and its solution is

$$\int_{y\text{-constant}} \left(\frac{1}{2x^2y} + \frac{1}{2x}\right)dx + \int\left(\frac{-1}{2y}\right)dy = c$$

$$\Rightarrow \qquad \frac{-1}{2xy} + \frac{1}{2}\log x - \frac{1}{2}\log y = c$$

$$\Rightarrow \qquad \log \frac{x}{y} - \frac{1}{xy} = 2c = c' \text{ (say)}$$

Case IV: For the equation $Mdx + Ndy = 0$

(a) If $\dfrac{\dfrac{\partial M}{\partial y} - \dfrac{\partial N}{\partial x}}{N}$ is a function of x only, $f(x)$ (say) then $e^{\int f(x)dx}$ is an integrating factor.

(b) If $\dfrac{\dfrac{\partial M}{\partial y} - \dfrac{\partial N}{\partial x}}{M}$ is a function of y only, $g(y)$ (say) then $e^{-\int g(y)dy}$ is an integrating factor.

Proof:

(a) Multiplying the given equation by $e^{\int f(x)dx}$, we get

$$Me^{\int f(x)dx}dx + Ne^{\int f(x)dx}dy = 0 \qquad (3.5)$$

Therefore $e^{\int f(x)dx}$ is an integrating factor of the given equation, if Eq. (3.5) is exact,

i.e., if $\dfrac{\partial}{\partial y}\left(Me^{\int f(x)dx}\right)=\dfrac{\partial}{\partial x}\left(Ne^{\int f(x)dx}\right)$

i.e., if $e^{\int f(x)dx}\dfrac{\partial M}{\partial y} = \dfrac{\partial N}{\partial x}e^{\int f(x)dx}+Ne^{\int f(x)dx}f(x)$

i.e., if $\dfrac{\partial M}{\partial y}-\dfrac{\partial N}{\partial x} = Nf(x)$

i.e., if $\dfrac{\dfrac{\partial M}{\partial y}-\dfrac{\partial N}{\partial x}}{N} = f(x)$ which is a function of x only.

(b) Multiplying the given equation by $e^{-\int g(y)dy}$, we get

$$Me^{\int -g(y)dy}dx+Ne^{\int -g(y)dy}dy = 0 \tag{3.6}$$

Therefore, $e^{\int -g(y)dy}$ is an integration factor of the given equation if Eq. (3.6) is exact.

i.e., if $\dfrac{\partial}{\partial y}\left(Me^{\int -g(y)dy}\right)=\dfrac{\partial}{\partial x}\left(Ne^{\int -g(y)dy}\right)$

i.e., if $\dfrac{\partial M}{\partial y}e^{\int -g(y)dy}-Me^{\int -g(y)dy}g(y)=\dfrac{\partial N}{\partial x}e^{\int -g(y)dy}$

i.e., if $\dfrac{\partial M}{\partial y}-\dfrac{\partial N}{\partial x} = Mg(y)$

i.e., if $\dfrac{\dfrac{\partial M}{\partial y}-\dfrac{\partial N}{\partial x}}{M} = g(y)$, which is a function of y only.

Note: Generally, while finding integrating factor, we start with this rule.

EXAMPLE 3.13

Solve the differential equation $(5x^3 + 12x^2 + 6y^2)dx + 6xy\,dy = 0$.

Solution: Here, $M = 5x^3 + 12x^2 + 6y^2$, $N = 6xy$

$$\frac{\partial M}{\partial y}-\frac{\partial N}{\partial x} = 12y - 6xy = 6y \neq 0$$

But $\dfrac{\dfrac{\partial M}{\partial y}-\dfrac{\partial N}{\partial x}}{N} = \dfrac{1}{x} = f(x)$, a function of x alone. Therefore

$$\text{IF} = e^{\int f(x)\,dx} = e^{\int \frac{1}{x}dx} = e^{\log x} = x$$

Multiplying the given equation by IF, we get

$$(5x^4 + 12x^3 + 6xy^2)\,dx + 6x^2y\,dy = 0$$

which is an exact equation and its solution is

$$\int_{y\text{-constant}} (5x^4 + 12x^3 + 6xy^2)dx + \int (0)dy = c$$

$$\Rightarrow \qquad x^5 + 3x^4 + 3x^2y^2 = c$$

EXAMPLE 3.14

Solve $(xy^2 - e^{1/x^3})\,dx - x^2y\,dy = 0$.

Solution: Here, $M = xy^2 - e^{1/x^3}$, $N = -x^2y$

$$\frac{\partial M}{\partial y} - \frac{\partial N}{\partial x} = 2xy + 2xy = 4xy \neq 0$$

But $\dfrac{\dfrac{\partial M}{\partial y}-\dfrac{\partial N}{\partial x}}{N} = \dfrac{-4}{x} = f(x)$, a function of x only. Therefore

$$\text{IF} = e^{\int -\frac{4}{x}dx} = e^{-4\log x} = x^{-4}$$

Multiplying the given equation by IF, we get

$$\left(\frac{y^2}{x^3} - \frac{1}{x^4}e^{1/x^3}\right)dx - \frac{y}{x^2}dy = 0$$

which is an exact equation and its solution is

$$\int_{y\text{-constant}} \left(\frac{y^2}{x^3} - \frac{1}{x^4}e^{1/x^3}\right)dx + \int (0)dy = c$$

$$\Rightarrow \qquad -\frac{y^2}{2x^2} + \frac{e^{1/x^3}}{3} = c.$$

EXAMPLE 3.15

Solve $\left(y+\frac{1}{3}y^3+\frac{1}{2}x^2\right)dx+\frac{1}{4}(x+xy^2)dy=0$.

Solution: Here, $M = y+\frac{1}{3}y^3+\frac{1}{2}x^2$, $N=\frac{1}{4}(x+xy^2)$

$$\frac{\partial M}{\partial y}-\frac{\partial N}{\partial x} = 1+y^2-\frac{1}{4}(1+y^2)=\frac{3}{4}(1+y^2)\neq 0$$

But $\dfrac{\dfrac{\partial M}{\partial y}-\dfrac{\partial N}{\partial x}}{N} = \dfrac{3}{x} = f(x)$, a function of x only.

Therefore, IF $= e^{\int \frac{3}{x}dx} = e^{3\log x} = x^3$

Multiplying the given equation by IF, we get

$$\left(x^3y+\frac{1}{3}x^3y^3+\frac{1}{2}x^5\right)dx+\frac{1}{4}x^4(1+y^2)dy = 0$$

which can be shown to be exact and its solution is

$$\int\limits_{y\text{ constant}} \left(x^3y+\frac{1}{3}x^3y^3+\frac{1}{2}x^5\right)dx+\int(0)dy=c$$

$\Rightarrow$ $$x^4y(3+y^2)+x^6=c$$

EXAMPLE 3.16

Solve $(3x^2y^4 + 2xy)\,dx + (2x^3y^3 - x^2)\,dy = 0$.

Solution: Here, $M = 3x^2y^4 + 2xy$, $N = 2x^3y^3 - x^2$

$\therefore$ $$\frac{\partial M}{\partial y}-\frac{\partial N}{\partial x} = 12x^2y^3 + 2x - 6x^2y^3 + 2x$$
$$= 2x(3x^2y^3+2)\neq 0$$

But $\dfrac{\dfrac{\partial M}{\partial y}-\dfrac{\partial N}{\partial x}}{M} = \dfrac{2}{y} = g(y)$, a function of y only

Therefore,

$$\text{IF} = e^{\int -g(y)dy} = e^{\int -\frac{2}{y}dy} = e^{-2\log y} = y^{-2}$$

Multiplying the given equation by IF, we get

$$\left(3x^2y^2 + \frac{2x}{y}\right)dx + \left(2x^3y - \frac{x^2}{y^2}\right)dy = 0$$

which can be shown to be exact and its solution is

$$\int_{y\text{-constant}} \left(3x^2y^2 + \frac{2x}{y}\right)dx + \int (0)\,dy = 0$$

$$\Rightarrow \qquad x^3y^2 + \frac{x^2}{y} = c$$

$$\Rightarrow \qquad x^3y^3 + x^2 = cy.$$

EXAMPLE 3.17

Solve $(y\log y)\,dx + (x - \log y)\,dy = 0.$

Solution: Here, $M = y\log y,\ N = x - \log y$

$$\therefore \qquad \frac{\partial M}{\partial y} - \frac{\partial N}{\partial x} = 1 + \log y - 1 = \log y \neq 0$$

But $\dfrac{\dfrac{\partial M}{\partial y} - \dfrac{\partial N}{\partial x}}{M} = \dfrac{1}{y} = g(y)$ = a function of y only.

Therefore, IF $= e^{\int -\frac{1}{y}dy} = e^{-\log y} = y^{-1}$

Multiplying the given equation by IF, we get

$$\log y\,dx + \frac{1}{y}(x - \log y)\,dy = 0$$

which is an exact equation and its solution is

$$\int_{y\text{-constant}} \log y\,dx + \int \left(\frac{-\log y}{y}\right)dy = c$$

$$\Rightarrow \qquad x\log y - \frac{1}{2}(\log y)^2 = c$$

EXAMPLE 3.18

Solve the differential equation $(3x^2y^3e^y + y^3 + y^2)\,dx + (x^3y^3e^y - xy)\,dy = 0.$

Solution: Here, $M = 3x^2y^3e^y + y^3 + y^2,\ N = x^3y^3e^y - xy$

$$\therefore \quad \frac{\partial M}{\partial y} - \frac{\partial N}{\partial x} = 9x^2y^2e^y + 3x^2y^3e^y + 3y^2 + 2y - 3x^2y^3e^y + y$$

$$= 9x^2y^2e^y + 3y^2 + 3y \neq 0$$

But $\dfrac{\dfrac{\partial M}{\partial y} - \dfrac{\partial N}{\partial x}}{M} = \dfrac{3}{y} = g(y)$, a function of y only. Therefore

$$\text{IF} = e^{\int \frac{-3}{y} dy} = e^{-3 \log y} = y^{-3}.$$

Multiplying the given equation by IF, we get

$$\left(3x^2e^y + 1 + \frac{1}{y}\right)dx + \left(x^3e^y - \frac{x}{y^2}\right)dy = 0$$

which is an exact equation and its solution is

$$\int_{y\text{-constant}} \left(3x^2e^y + 1 + \frac{1}{y}\right)dx + \int (0)dy = c$$

$$\Rightarrow \quad x^3e^y + x + \frac{x}{y} = c$$

$$\Rightarrow \quad yx^3e^y + xy + x = cy.$$

EXAMPLE 3.19

Solve $\cos x \cos y dx - 2 \sin x \sin y dy = 0$.

Solution: Here, $M = \cos x \cos y$, $N = -2 \sin x \sin y$

$$\therefore \quad \frac{\partial M}{\partial y} - \frac{\partial N}{\partial x} = -\cos x \sin y + 2 \cos x \sin y$$

$$= \cos x \sin y \neq 0$$

But $\dfrac{\dfrac{\partial M}{\partial y} - \dfrac{\partial N}{\partial x}}{N} = -\dfrac{1}{2}\dfrac{\cos x}{\sin x} = f(x)$, a function of x only.

Therefore, $\text{IF} = e^{\int \frac{-1}{2} \frac{\cos x}{\sin x} dx} = e^{\frac{-1}{2} \log(\sin x)} = \dfrac{1}{\sqrt{\sin x}}$

Multiplying the given equation by IF, we get

$$\frac{\cos x \cos y}{\sqrt{\sin x}} dx - 2\sqrt{\sin x} \sin y \, dy = 0$$

which is an exact equation and its solution is

$$\int_{y\text{-constant}} \cos y \frac{\cos x}{\sqrt{\sin x}} dx + \int (0)\, dy = c$$

$$\Rightarrow \qquad 2\cos y\sqrt{\sin x} = c$$

EXAMPLE 3.20

Solve the differential equation

$$(x+2y^3)\frac{dy}{dx} = y.$$

Solution: The given equation can be written as

$$y\,dx - (x + 2y^3)\,dy = 0$$

Here, $M = y$, $N = -(x + 2y^3)$

$$\therefore \qquad \frac{\partial M}{\partial y} - \frac{\partial N}{\partial x} = 1 + 1 = 2 \neq 0$$

But $\dfrac{\dfrac{\partial M}{\partial y} - \dfrac{\partial N}{\partial x}}{M} = \dfrac{2}{y} = g(y)$, a function of y only.

Therefore, $\qquad \text{IF} = e^{\int \frac{-2}{y} dy} = e^{-2\log y} = y^{-2}$

Multiplying the given equation by IF, we get

$$\frac{1}{y}dx - \left(\frac{x}{y^2} + 2y\right)dy = 0$$

which is an exact equation and its solution is

$$\int_{y\text{-constant}} \frac{1}{y}dx + \int (-2y)dy = c$$

$$\Rightarrow \qquad \frac{x}{y} - y^2 = c$$

$$\Rightarrow \qquad x - y^3 = cy$$

Case V: If the given equation is of the form $x^a y^b(my\,dx + nx\,dy) + x^{a'}y^{b'}(m'y\,dx + n'x\,dy) = 0$; then $x^h y^k$ is an integrating factor, where,

$$\frac{a+h+l}{m} = \frac{b+k+l}{n}, \frac{a'+h+l}{m'} = \frac{b'+k+l}{n'}$$

Proof: Multiplying the given equation by $x^h y^k$; we get

$$x^{a+b} y^{b+k} (my\,dx + nx\,dy) + x^{a'+h} y^{b'+k} (m'y\,dx + n'x\,dy) = 0$$

$$\Rightarrow (mx^{a+h} y^{b+k+1} + m'x^{a'+h} y^{b'+k+1})dx + (nx^{a+h+1} y^{b+k} + n'x^{a'+h+l} y^{b'+k})dy = 0 \quad (3.7)$$

Therefore, $x^h y^k$ is an integrating factor of the given equation if Eq. (3.7) is exact

i.e., if $\frac{\partial}{\partial y}(mx^{a+h} y^{b+x+1} + m'x^{a'+h} y^{b'+k+l}) = \frac{\partial}{\partial x}(nx^{a+h+l} y^{b+k} + n'x^{a'+h+1} y^{b'+k})$

i.e., if $m(b + k + l)x^{a+h} y^{b+k} + m'(b' + k + l)x^{a'+h} y^{b'+k} = n(a + h + l)x^{a+h} y^{b+k} + n'(a' + h + l)x^{a'+h} y^{b'+k}$

i.e., if $m(b + k + l) = n(a + h + l)$ and $m'(b' + k + l) = n'(a' + h + l)$

i.e., if $\frac{a+h+l}{m} = \frac{b+k+l}{n}$ and $\frac{a'+h+l}{m'} = \frac{b'+k+l}{n'}$

EXAMPLE 3.21

Solve $(2x^2y - 3y^4)dx + (3x^3 + 2xy^3)dy = 0$.

Solution: Rewriting the given equation as

$$x^2(2y\,dx + 3x\,dy) + y^3(-3y\,dx + 2x\,dy) = 0$$

and comparing with

$$x^a y^b (my\,dx + nx\,dy) + x^{a'} y^{b'} (m'y\,dx + n'x\,dy) = 0$$

We have

$$a = 2,\ b = 0,\ m = 2,\ n = 3,\ a' = 0,\ b' = 3,\ m' = -3,\ n' = 2$$

Hence the integrating factor must be $x^h y^k$.
Multiplying the original equation by the IF, we get

$$(2x^{h+2} y^{k+1} - 3x^h y^{k+4})dx + (3x^{h+3} y^k + 2x^{h+1} y^{k+3})dy = 0$$

Here, $M = 2x^{h+2} y^{k+1} - 3x^h y^{k+4}$

$$N = 3x^{h+3} y^k + 2x^{h+1} y^{k+3}$$

Then,

$$\frac{\partial M}{\partial y} = 2(k + 1)x^{h+2} y^k - 3(k + 4)x^h y^{k+3}$$

and

$$\frac{\partial N}{\partial x} = 3(h + 3)x^{h+2} y^k + 2(h + 1)x^h y^{k+3}$$

for exactness $\frac{\partial M}{\partial y} = \frac{\partial N}{\partial x}$.

$\therefore$ By comparing, we get

$$2(k + 1) = 3(h + 3) \text{ and } -3(k + 4) = 2(h + 1)$$

Solving it, we get

$$h = \frac{-49}{13}, k = \frac{-23}{13}$$

$$\therefore \quad \text{IF} = x^{-49/13}y^{-23/13}$$

Multiplying by the IF, the given equation must become exact.

$$(2x^{-23/13}y^{16/13} - 3x^{-49/13}y^{29/13})\,dx + (3x^{10/13}y^{-23/13} + 2x^{-36/13}y^{16/13})\,dy = 0$$

Its solution will be

$$\int_{y\text{-constant}} (2x^{-23/13}y^{16/13} - 3x^{-49/13}y^{29/13})\,dx + \int (0)\,dy = 0$$

$$\Rightarrow \quad \frac{2x^{-10/13}}{-10/13}y^{16/13} - \frac{3x^{-36/13}}{-36/13}y^{29/13} = c$$

$$\Rightarrow \quad 5x^{-36/13}y^{29/13} - 12x^{-10/13}y^{16/13} = c' \text{ (say).}$$

EXAMPLE 3.22

Solve $y(xy + 2x^2y^2)\,dx + x(xy - x^2y^2)\,dy = 0$.

Solution: Rewriting the given equation as

$$xy\,(y\,dx + x\,dy) + x^2y^2\,(2y\,dx - x\,dy) = 0$$

and comparing with

$$x^a y^b\,(my\,dx + nx\,dy) + x^{a'}y^{b'}\,(m'y\,dx + n'x\,dy) = 0$$

We have

$$a = b = 1,\ m = n = 1,\ a' = b' = 2,\ m' = 2,\ n' = -1$$

Hence, the integrating factor must be $x^h y^k$.
Multiplying the original equation by the IF, we get

$$(x^{h+1}y^{k+2} + 2x^{h+2}y^{k+3})\,dx + (x^{h+2}y^{k+1} - x^{h+3}y^{k+2})\,dy = 0$$

Here, $M = x^{h+1}y^{k+2} + 2x^{h+2}y^{k+3}$

$$N = x^{h+2}y^{k+1} - x^{h+3}y^{k+2}$$

Then, $$\frac{\partial M}{\partial y} = (k + 2)\,x^{h+1}y^{k+1} + 2(k + 3)\,x^{h+2}y^{k+2}$$

and $$\frac{\partial N}{\partial x} = (h + 2)x^{h+1}y^{k+1} - (h + 3)x^{h+2}y^{k+2}$$

for exactness $\frac{\partial M}{\partial y} = \frac{\partial N}{\partial x}$.

$\therefore$ By comparing, we get

$$k + 2 = h + 2 \text{ and } 2(k + 3) = -(h + 3)$$

Solving it, we get $h = k = -3$

$$\therefore \qquad \text{IF} = x^{-3}y^{-3}$$

Multiplying by the IF, the given equation must become exact.

$$\left(\frac{1}{x^2y} + \frac{2}{x}\right)dx + \left(\frac{1}{xy^2} - \frac{1}{y}\right)dy = 0$$

Its solution will be

$$\int_{y\text{-constant}} \left(\frac{1}{x^2y} + \frac{2}{x}\right)dx + \int\left(\frac{-1}{y}\right)dy = c$$

$$\Rightarrow \qquad \frac{-1}{xy} + 2\log x - \log y = c$$

$$\Rightarrow \qquad \log\frac{x^2}{y} - \frac{1}{xy} = c$$

EXAMPLE 3.23

Solve $(y^2 + 2x^2y)\,dx + (2x^3 - xy)\,dy = 0$.

Solution: The given equation is not exact. It can be put in the form

$$y(ydx - xdy) + x^2(2ydx + 2xdy) = 0$$

Let x^hy^k be an integrating factor.
Multiplying the original equation by it, we get

$$(x^hy^{k+2} + 2x^{h+2}y^{k+1})\,dx + (2x^{h+3}y^k - x^{h+1}y^{k+1})\,dy = 0$$

Here, $M = x^hy^{k+2} + 2x^{h+2}y^{k+1}$ and $N = 2x^{h+3}y^k - x^{h+1}y^{k+1}$

$$\therefore \qquad \frac{\partial M}{\partial y} = (k + 2)x^hy^{k+1} + 2(k + 1)x^{h+2}y^k$$

and $$\frac{\partial N}{\partial x} = 2(h+3)x^{h+2}y^k - (h+1)x^h y^{k+1}$$

for exactness $\frac{\partial M}{\partial y} = \frac{\partial N}{\partial x}$.

On comparing, we get

$$k + 2 = -(h+1) \text{ and } 2(k+1) = 2(h+3)$$

Solving it, we get $h = \frac{-5}{2}$, and $k = \frac{-1}{2}$

$$\therefore \qquad \text{IF} = x^{-3/2}y^{-1/2}$$

Multiplying by the IF, the given equation must become exact.

$$(x^{-3/2}y^{3/2} + 2x^{-1/2}y^{1/2})dx + (2x^{1/2}y^{-1/2} - x^{-3/2}y^{1/2})dy = 0$$

Its solution will be

$$\int_{y\text{-constant}} (x^{-3/2}y^{3/2} + 2x^{-1/2}y^{1/2})dx + \int (0)dy = c$$

$$\Rightarrow \qquad \frac{x^{-3/2}y^{3/2}}{-3/2} + \frac{2x^{1/2}y^{1/2}}{1/2} = c$$

$$\Rightarrow \qquad 4x^{1/2}y^{1/2} - \frac{2}{3}x^{-3/2}y^{3/2} = c.$$

▲▲

EXERCISE 3.2

1. Show that $\frac{1}{y^2}, \frac{1}{xy}$ are integrating factors of $y\,dx - x\,dy = 0$. Find solution in each case also.
2. By inspection, obtain an integrating factor and solve the differential equation
$$x\,dx + y\,dy + 2(x^2 + y^2)dx = 0.$$
3. Solve the differential equation
$$y\,dx - x\,dy + e^{1/x}dx = 0$$
by finding an integrating factor by inspection.
4. Solve the differential equation
$$x(1 + y^2)dy + y(1 + x^2)dx = 0$$
by finding an integrating factor by inspection.

5. Solve the differential equation

$$y\,dy\,(x^2 + y^2 + a^2) + x\,dx(x^2 + y^2 - a^2) = 0.$$

6. Solve the differential equation

$$(2xy + x^2)y' = 3y^2 + 2xy.$$

7. Solve $(xy \sin xy + \cos xy)\,y\,dx + (xy \sin xy - \cos xy)\,x\,dy = 0$.
8. Solve $(x^4y^4 + x^2y^2 + xy)\,y\,dx + (x^4y^4 - x^2y^2 + xy)\,x\,dy = 0$.
9. Solve $(x^2 + y^2 + 2x)\,dx + 2y\,dy = 0$.
10. Solve the differential equation

$$(y^4 + 2y)\,dx + (xy^3 + 2y^4 - 4x)\,dy = 0.$$

11. Solve $(xy^2 - x^2)\,dx + (3x^2y^2 + x^2y + y^2)\,dy = 0$.
12. Solve the following equations, by using a suitable integrating factor

$$ydx - xdy + \log x dx = 0.$$

13. Solve $2ydx - x(2\log x - y)\,dy = 0$. Using suitable integrating factor.
14. Find the integrating factor and hence, solve the differential equation

$$dx + e^{(y-x)}\,dy = 0.$$

15. Solve $(2x + y)\,dy - (x + 2y)\,dx = 0$.
16. Solve $y^2 dx + x(x - y)\,dy = 0$.
17. Solve the initial value problem

$$2xy\,dx + (x^2 + \pi \cos \pi y)\,dy = 0; \; y(1) = 1.$$

18. Solve $(\cos x + y \sin x)\,dx = (\cos x)\,dy$; $y(\pi) = 0$.
19. Solve the following differential equation

$$(2ydx + 3xdy) + 2xy(3ydx + 4xdy) = 0.$$

20. Solve the following differential equation

$$x(3ydx + 2xdy) + 8y^4(ydx + 3xdy) = 0.$$

ANSWERS

Exercise 3.2

1. $x = cy$ in both the cases.
2. IF $= 1/(x^2 + y^2)$, $x^2 + y^2 = ce^{-4x}$.
3. IF $= 1/x^2$, $y + xe^{1/x} = cx$.
4. IF $= 1/(xy)$, $\log(x^2y^2) + x^2 + y^2 = c$.

5. $\frac{1}{2}(x^2 + y^2)^2 + a^2(y^2 - x^2) = c.$

6. $x^3 = cy(x + y).$

7. $x = cy \cos xy.$

8. $\frac{1}{2}x^2y^2 - \frac{1}{xy} + \log \frac{x}{y} = c.$

9. $e^x(x^2 + y^2) = c.$

10. $xy + y^2 + \frac{2x}{y^2} = c.$

11. $4x^5y + 2x^4y^2 + \frac{4}{3}x^3y^3 + y^4x^3 = c.$

12. $y + cx + \log x + 1 = 0.$

13. $4y \log y = y^2 + c.$

14. $\text{IF} = e^x, e^x + e^y = c.$

15. $\text{IF} = \frac{1}{y^2 - x^2}, (y - x)^3 = c(y + x).$

16. $\text{IF} = \frac{1}{x^2 y}, y = ce^{y/x}.$

17. Exact, $x^2y + \sin \pi y = 1.$

18. Exact, $y \cos x + \sin x.$

19. $x^2y^3 + 2x^2y^4 = c.$

20. $x^2y^2(x + 4y^4) = c.$

4

CHAPTER

Bernoulli's Linear and Differential Equations

In chapter 1, we gave the definition of the linear ordinary differential equation of the order n. In this chapter, we will discuss the linear ordinary differential equations of the first order and equations of the first order reducible to the linear form.

Definition: A *first order ordinary differential equation* is linear in the dependent variable y and the independent variable x if it can be written in the form

$$\frac{dy}{dx} + P(x)y = Q(x) \tag{4.1}$$

For example, the equation

$$x\frac{dy}{dx} + (x + 1)y = x^3$$

is a first order linear differential equation for it can be written as

$$\frac{dy}{dx} + \left(1 + \frac{1}{x}\right)y = x^2$$

with $P(x) = 1 + \dfrac{1}{x}$ and $Q(x) = x^2$.

For finding solution of Eq. (4.1), write it as

$$[P(x)y - Q(x)]\,dx + dy = 0 \tag{4.2}$$

On comparing with $Mdx + Ndy = 0$, we have

$$M = P(x)y - Q(x),\ N = 1$$

$$\frac{\partial M}{\partial y} - \frac{\partial N}{\partial x} = P(x)$$

which implies Eq. (4.2) is not exact. So, we will multiply Eq. (4.2) by integrating factor obtained as under.

$$\frac{\dfrac{\partial M}{\partial y}-\dfrac{\partial N}{\partial x}}{N} = P(x), \text{ therefore IF } = e^{\int P(x)dx}$$

$$e^{\int P(x)dx}\frac{dy}{dx}+e^{\int P(x)dx}P(x)y = e^{\int P(x)dx}Q(x)$$

$$\Rightarrow \qquad \frac{d}{dx}\left[e^{\int P(x)dx}y\right] = Q(x)e^{\int P(x)dx}$$

Integrating this we obtain the solution of Eq. (4.1) in the form

$$e^{\int P(x)dx}y = \int e^{\int P(x)dx}Q(x)dx + c$$

where c is an arbitrary constant.

▲▲

EXAMPLE 4.1

Solve $\left(\dfrac{e^{-2\sqrt{x}}}{\sqrt{x}}-\dfrac{y}{\sqrt{x}}\right)\dfrac{dx}{dy} = 1.$

Solution: The given equation can be written as

$$\frac{dy}{dx}+\frac{y}{\sqrt{x}} = \frac{e^{-2\sqrt{x}}}{\sqrt{x}}$$

$$\therefore \qquad \text{IF} = e^{\int \frac{1}{\sqrt{x}}dx} = e^{2\sqrt{x}}$$

Hence, the solution of the given equation is

$$ye^{2\sqrt{x}} = \int e^{2\sqrt{x}}\frac{e^{-2\sqrt{x}}}{\sqrt{x}}dx + c$$

$$\Rightarrow \qquad ye^{2\sqrt{x}} = \int \frac{1}{\sqrt{x}}dx + c$$

$$\Rightarrow \qquad ye^{2\sqrt{x}} = 2\sqrt{x} + c$$

EXAMPLE 4.2

Solve $(y\log y)\,dx + (x - \log y)\,dy = 0.$

Solution: The given equation can be written as

$$\frac{dy}{dx} + \frac{x}{y\log y} = \frac{1}{y}$$

$$\therefore \qquad \text{IF} = e^{\int \frac{1}{y\log y}dy} = e^{\log(\log y)} = \log y$$

Hence, the solution of the given equation is

$$\log y \cdot x = \int \log y \frac{1}{y} dy + c$$

$$\Rightarrow \qquad x\log y = \frac{(\log y)^2}{2} + c$$

$$\Rightarrow \qquad x = \frac{1}{2}\log y + c(\log y)^{-1}.$$

EXAMPLE 4.3

Solve $(1 + y^2) + (x - e^{\tan^{-1}y})\dfrac{dy}{dx} = 0.$

Solution: The given equation can be written as

$$\frac{dx}{dy} + \frac{x}{1+y^2} = \frac{e^{\tan^{-1}y}}{1+y^2}$$

$$\therefore \qquad \text{IF} = e^{\int \frac{1}{1+y^2}dy} = e^{\tan^{-1}y}$$

Hence, the solution of the given equation is

$$x \cdot e^{\tan^{-1}y} = \int e^{\tan^{-1}y} \cdot \frac{e^{\tan^{-1}y}}{1+y^2} dy + c$$

$$\Rightarrow \qquad xe^{\tan^{-1}y} = \int \frac{e^{2\tan^{-1}y}}{1+y^2} dy + c.$$

Let $\tan^{-1}y = t$, $\dfrac{1}{1+y^2}dy = dt$

$$\therefore \qquad xe^{\tan^{-1}y} = \int e^{2t} dt + c$$

$$\Rightarrow \qquad xe^{t} = \frac{e^{2t}}{2} + c$$

$$\Rightarrow \qquad xe^{\tan^{-1}y} = \frac{1}{2}e^{2\tan^{-1}y} + c$$

EXAMPLE 4.4

Solve $\sqrt{a^2+x^2}\dfrac{dy}{dx}+y=\sqrt{a^2+x^2}-x.$

Solution: The given equation can be written as

$$\frac{dy}{dx}+\frac{y}{\sqrt{a^2+x^2}} = \frac{\sqrt{a^2+x^2}-x}{\sqrt{a^2+x^2}}$$

$$\therefore \qquad \text{IF} = e^{\int \frac{1}{\sqrt{a^2+x^2}}dx} = e^{\log \frac{x+\sqrt{a^2+x^2}}{a}}$$

$$= \frac{x+\sqrt{a^2+x^2}}{a}$$

Hence, the solution of the given equation is

$$\frac{x+\sqrt{a^2+x^2}}{a}\cdot y = \int \frac{x+\sqrt{a^2+x^2}}{a}\cdot\frac{\sqrt{a^2+x^2}-x}{\sqrt{a^2+x^2}}dx+c$$

$$= \int \frac{(a^2+x^2)-x^2}{a\sqrt{a^2+x^2}}dx+c$$

$$= \int \frac{a}{\sqrt{a^2+x^2}}dx+c$$

$$= \sinh^{-1}\frac{x}{a}+c.$$

EXAMPLE 4.5

Solve $r\sin\theta\, d\theta + (r^3 - 2r^2\cos\theta + \cos\theta)\, dr = 0.$

Solution: The given equation can be written as

$$\sin\theta\frac{d\theta}{dr}+\frac{1}{r}(1-2r^2)\cos\theta = -r^2$$

Put $\cos\theta = y$ so that $-\sin\theta\dfrac{d\theta}{dr}=\dfrac{dy}{dr}$

Thus, the given equation becomes

$$\frac{-dy}{dr}+\frac{1}{r}(1-2r^2)y = -r^2$$

$$\Rightarrow \qquad \frac{dy}{dr}+\left(2r-\frac{1}{r}\right)y = r^2$$

$\therefore$ $$\text{IF} = e^{\int\left(2r-\frac{1}{r}\right)dr} = e^{r^2-\log r} = \frac{1}{r}e^{r^2}$$

Hence, the solution of the given equation is

$$\frac{1}{r}e^{r^2}y = \int \frac{1}{r}e^{r^2}r^2 dr + c$$

$\Rightarrow$ $$\frac{ye^{r^2}}{r} = \int e^{r^2} r\,dr + c$$

$\Rightarrow$ $$\frac{ye^{r^2}}{r} = \frac{e^{r^2}}{2} + c$$

$\Rightarrow$ $$2e^{r^2}\cos\theta = re^{r^2} + 2cr$$

$\Rightarrow$ $$r(1 + 2c\,e^{-r^2}) = 2\cos\theta.$$

EXAMPLE 4.6

The initial value problem governing the current I flowing in a series RL circuit when a voltage $V(t) = t$ is applied, is given by

$$IR + L\frac{dI}{dt} = t; \quad t \geq 0, I(0) = 0$$

where, R and L are constants. Find the current $I(t)$ at time t.

Solution: The given equation can be written as

$$\frac{dI}{dt} + \frac{R}{L}I = \frac{1}{L}t$$

$\therefore$ $$\text{IF} = e^{\int \frac{R}{L}dt} = e^{Rt/L}$$

Hence, the solution of the given equation is

$$Ie^{Rt/L} = \int e^{Rt/L}\frac{1}{L}t\,dt + c$$

$$= \frac{1}{L}\left[\frac{e^{Rt/L}}{R/L}t - \frac{e^{Rt/L}}{(R/L)^2}\right] + c$$

$\therefore$ $$I = \frac{1}{R}\left[t - \frac{L}{R}\right] + ce^{-Rt/L}$$

Using $I(0) = 0$, we have $c = L/R^2$

$\therefore$ $$I = \frac{t}{R} + \frac{L}{R^2}[e^{-Rt/L} - 1]$$

▲▲

EXERCISE 4.1

Find the general solution of the following first order linear differential equations.

1. $xy^1 + y = \sin x$.

2. $(x - a)\dfrac{dy}{dx} + 3y = 12(x - a)^3$, $(x > a > 0)$.

3. $\cot 3x \dfrac{dy}{dx} - 3y = \cos 3x + \sin 3x$, $0 < x < \pi/2$.

4. $3x(1 - x^2)y^2 \dfrac{dy}{dx} + (2x^2 - 1)y^3 = ax^3$.

5. $x \log x \dfrac{dy}{dx} + y = (\log x)^2$

6. $e^{-y} \sec^2 y\, dy = dx + x dy$.

7. $\dfrac{dy}{dx} = \dfrac{x + y\cos x}{1 + \sin x}$.

8. $(1 + y^2)dx = (\tan^{-1} y - x)dy$.

9. $\dfrac{dy}{dx} = y\cos x + \cos x \sin^2 x$.

10. $\dfrac{dy}{dx} - \dfrac{1+3x^2}{x(1+x^2)} y = x\dfrac{1-x^2}{1+x^2}$, given $y = 0$, when $x = 1$.

11. $\dfrac{dy}{dx} = \dfrac{y}{2y\log y + y - x}$.

ANSWERS

Exercise 4.1

1. $xy = c - \cos x$.

2. $y = 2(x - a)^3 + \dfrac{c}{(x-a)^3}$.

3. $y\cos 3x = \dfrac{1}{12}[6x - \cos 6x - \sin 6x] + c$.

4. $y^2 = cx + cx\sqrt{1-x^2}$.

5. $y = \log x + c/\log x$.

6. $xe^y = c + \tan y$.

7. $y[1 + \sin x] = c - x^2/2$.

8. $x = \tan^{-1} y - 1 + ce^{\tan^{-1}} y.$
9. $y + \sin^2 x + 2\sin x + 2 = c \cdot e^{\sin x}$
10. $y = \frac{-1x}{2}(1 + x^2) + x^2.$
11. $y = x + x^{-1} + cx^{-2}.$

▲▲

BERNOULLI EQUATIONS

We now consider a rather special type of equations that can be reduced to a linear differential equation by an appropriate substitution.

Definition: An equation of the form

$$\frac{dy}{dx} + P(x)y = Q(x)y^n \tag{4.3}$$

is called a Bernoulli differential equation.

If $n - 0$ or 1, then the equation is linear. For all other values of n, the equation is non-linear and which is reducible to the linear form.

On dividing by y^n, we get

$$y^{-n}\frac{dy}{dx} + P(x)y^{1-n} = Q(x) \tag{4.4}$$

If we let $z = y^{1-n}$, then $\frac{dz}{dx} = (1 - n)y^{-n}\frac{dy}{dx}$ and Eq. (4.4) transforms into

$$\frac{1}{1-n}\frac{dz}{dx} + P(x)z = Q(x)$$

$$\Rightarrow \qquad \frac{dz}{dx} + (1 - n)P(x)z = (1 - n)Q(x)$$

Letting $P_1(x) = (1 - n)P(x)$ and $Q_1(x) = (1 - n)Q(x)$, this may be written as

$$\frac{dz}{dx} + P_1(x)z = Q_1(x)$$

which is linear in z and can be solved easily.

▲▲

EXAMPLE 4.7

Solve $\frac{dy}{dx} + y\cos x = y^n \sin 2x.$

Solution: The given equation may be written as

$$\frac{1}{y^n}\frac{dy}{dx} + \frac{1}{y^{n-1}}\cos x = 2\sin x\cos x$$

Let $\frac{1}{y^{n-1}} = z \quad \therefore \ (1-n)\frac{1}{y^n}\frac{dy}{dx} = \frac{dz}{dx}$

$$\therefore \qquad \frac{1}{1-n}\frac{dz}{dx} + z\cos x = 2\sin x\cos x$$

$$\Rightarrow \qquad \frac{dz}{dx} + z(1-n)\cos x = 2(1-n)\sin x\cos x\,dx.$$

This is a linear equation, whose

$$\text{IF} = e^{\int(1-n)\cos x\,dx} = e^{(1-n)\sin x}$$

Hence, solution of equation is

$$e^{(1-n)\sin x}z = \int e^{(1-n)\sin x}\,2(1-n)\sin x\cos x\,dx$$

Put $(1-n)\sin x = t \ \therefore \ (1-n)\cos x\,dx = dt$

$$\therefore \qquad ze^t = \int e^t\frac{2t}{1-n}dt + c$$

$$\Rightarrow \qquad ze^t = \frac{2e^t}{1-n}(t-1) + c$$

$$\Rightarrow \qquad z = \frac{2}{1-n}(t-1) + ce^{-t}$$

$$\Rightarrow \qquad \frac{1}{y^{n-1}} = \frac{2}{1-n}[(1-n)\sin x - 1] + ce^{-(1-n)\sin x}$$

EXAMPLE 4.8

Solve the differential equation

$$\frac{dy}{dx} - y = y^2(\sin x + \cos x).$$

Solution: The given equation may be written as

$$y^{-2}\frac{dy}{dx} - y^{-1} = \sin x + \cos x$$

Let $y^{-1} = z \quad \therefore \ -y^{-2}\frac{dy}{dx} = \frac{dz}{dx}$

$$\therefore \qquad -\frac{dz}{dx} - z = \sin x + \cos x$$

$$\Rightarrow \qquad \frac{dz}{dx} + z = -(\sin x + \cos x)$$

This is a linear equation with

$$\text{IF} = e^{\int 1 \cdot dx} = e^x$$

Hence, solution of equation is

$$e^x \cdot z = \int -(\sin x + \cos x)\, e^x\, dx + c$$

$$\Rightarrow \qquad e^x z = -e^x \sin x + c$$

$$\therefore \qquad y = \frac{1}{(e^{-x} - \sin x)}$$

EXAMPLE 4.9

Solve $xy(1 + xy^2)\dfrac{dy}{dx} = 1$.

Solution: The given equation may be written as

$$\frac{dx}{dy} - yx = y^3x^2$$

$$\Rightarrow \qquad x^{-2}\frac{dx}{dy} - yx^{-1} = y^3$$

Let $x^{-1} = z \quad \therefore -x^{-2}\dfrac{dx}{dy} = \dfrac{dz}{dy}$

$$\therefore \qquad \frac{dz}{dy} + yz = -y^3$$

This is a linear equation with

$$\text{IF} = e^{\int y\,dy} = e^{y^2/2}$$

Hence, solution of equation is

$$e^{y^2/2} \cdot z = \int e^{y^2/2}(-y^3)\,dy + c$$

Let $y^2/2 = t \quad \Rightarrow \quad y\,dy = dt$

$$\therefore \qquad \int e^t z = \int e^t(-2t)\,dt + c$$

$$e^t z = -2e^t(t - 1) + c$$

$$z = 2(1-t) + ce^{-t}$$

$$\therefore \quad \frac{1}{x} = 2(1 - y^{2/2}) + ce^{-y^2/2}$$

EXAMPLE 4.10

Solve $\frac{dz}{dx} + \left(\frac{z}{x}\right)\log z = \frac{z}{x}(\log z)^2$.

Solution: The given equation may be written as

$$\frac{1}{z}\frac{dz}{dx} + \frac{1}{x}\log z = \frac{1}{x}(\log z)^2$$

Put $\log z = t$ so that $\frac{1}{z}\frac{dz}{dx} = \frac{dt}{dx}$

$$\therefore \quad \frac{dt}{dx} + \frac{t}{x} = \frac{t^2}{x}$$

$$\Rightarrow \quad t^{-2}\frac{dt}{dx} + \frac{t^{-1}}{x} = \frac{1}{x}$$

Let $t^{-1} = u \therefore -t^{-2}\frac{dt}{dx} = \frac{du}{dx}$

$$\therefore \quad -\frac{du}{dx} + \frac{u}{x} = \frac{1}{x}$$

$$\Rightarrow \quad \frac{du}{dx} - \frac{u}{x} = \frac{-1}{x}$$

This is a linear equation with IF $= e^{\int \frac{-1}{x}dx}$

$$= e^{-\log x} = x^{-1}$$

Hence, solution of equation is

$$x^{-1} \cdot u = \int x^{-1} \cdot \left(\frac{-1}{x}\right) dx + c$$

$$\Rightarrow \quad x^{-1}u = x^{-1} + c$$

$$\therefore \quad (\log z)^{-1} = 1 + cx.$$

EXAMPLE 4.11

Solve $\frac{dy}{dx} = e^{x-y}(e^x - e^y)$

Solution: The given equation may be written as

$$\frac{dy}{dx} = \frac{e^x}{e^y}(e^x - e^y)$$

$$\Rightarrow \qquad e^y\frac{dy}{dx} + e^y e^x = e^x \cdot e^x$$

Put $e^y = z \therefore e^y\frac{dy}{dx} = \frac{dz}{dx}$

Hence, equation becomes

$$\frac{dz}{dx} + ze^x = e^x e^x$$

This is a linear equation with IF $= e^{\int e^x dx} = e^{e^x}$

Hence, solution of equation is

$$e^{e^x} \cdot z = \int e^{e^x} e^x e^x dx + c$$

Let $e^x = t \therefore e^x dx = dt$

$$\therefore \qquad e^t z = \int e^t t\,dt + c$$

$$\Rightarrow \qquad z = (t - 1) + ce^{-t}$$

$$\therefore \qquad e^y = (e^x - 1) + ce^{-e^x}$$

▲▲

EXERCISE 4.2

1. Solve $x\frac{dy}{dx} + y = y^2 \log x$.
2. Solve $\frac{dy}{dx} - \frac{\tan y}{1+x} = (1 + x)e^x \sec y$.
3. Solve $y - x\frac{dy}{dx} = a\left(y^2 + \frac{dy}{dx}\right)$.
4. Solve the following differential equation $\frac{dy}{dx} + x\sin 2y = x^3 \cos^2 y$.
5. Solve the following differential equation $xy - \frac{dy}{dx} = y^3 e^{-x^2}$.
6. Solve the following differential equation

$$2\frac{dy}{dx} - y\sec x = y^3 \tan x.$$

7. Solve $\frac{dy}{dx} = \frac{x^2 + y^2 + 1}{2xy}$.

8. Solve $\frac{dy}{dx} = \frac{y}{x - \sqrt{xy}}$.

9. Solve $\frac{dy}{dx} - \frac{\tan y}{1 + x} = (1 + x) e^x \sec y$.

10. Solve $y(2xy + e^x)\, dx = e^x\, dy$.

ANSWERS

Exercise 4.2

1. $y(cx + \log ex) = 1$.
2. $\sec x = y(\tan x + c)$.
3. $(x + a) = y(ax + c)$.
4. $\tan y = \frac{1}{2}(x^2 - 1) + ce^{-x^2}$.
5. $e^{x^2} = y^2(2x + c)$.
6. $\frac{-1}{y^2}(\sec x + \tan x) = \sec x + \tan x - x + c$.
7. $y^2 = x^2 + cx - 1$.
8. $\sqrt{x} = \sqrt{y}\,(\log \sqrt{y} + c)$.
9. $\sin y = (1 + x)(e^x + c)$.
10. $e^x = y(c - x^2)$.

5

CHAPTER

First Order Higher Degree Differential Equations

So far, we have discussed differential equations of the first order and first degree. Now we shall study differential equations of the first order and higher degree. Such differential equations will contain only the first differential coefficient dy/dx but it will occur in a degree higher than one. The most general form of a differential equation of the first order but not of the first degree is

$$\left(\frac{dy}{dx}\right)^n + A_1\left(\frac{dy}{dx}\right)^{n-1} + A_2\left(\frac{dy}{dx}\right)^{n-2} + \ldots + A_{n-1}\left(\frac{dy}{dx}\right)^n + A_n = 0$$

or

$$p^n + A_1 p^{n-1} + A_2 p^{n-2} + \ldots + A_{n-1} p + A_n = 0$$

where, $p = dy/dx$ and $A_1, A_2, \ldots A_n$ are functions of x and y.

Since it is a differential equation of the first order, its general solution will contain only one arbitrary constant.

Now we shall consider the various methods of solving the differential equations of the above type.

Equations Solvable for p

Suppose a differential equation of the first order and of degree n can be reduced into n linear factors in p of the type

$$(p - f_1(x, y))(p - f_2(x, y)) \ldots (p - f_n(x, y)) = 0.$$

Equating each of the factors to zero, we have

$$p = f_1(x, y), p = f_2(x, y), \ldots p = f_n(x, y)$$

Now, these n equations of the first order and first degree can be solved easily. Let their solutions be

$$\phi_1(x, y, c_1) = 0, \phi_2(x, y, c_2) = 0, \ldots \phi_n(x, y, c_n) = 0$$

where, $c_1, c_2, \ldots c_n$ are arbitrary constants.

Because in any of the above solutions the arbitrary constants are free to take any real value, so without any loss of generality, we can replace the arbitrary constant c_1, c_2, ... c_n by a single arbitrary constant c. Thus, the n solutions of the given differential equation are

$$\phi_1(x, y, c) = 0,\ \phi_2(x, y, c) = 0,\ \dots\ \phi_n(x, y, c) = 0$$

Combining the above equations, we get a single solution as

$$\phi_1(x, y, c)\,\phi_2(x, y, c)\ \dots\ \phi_n(x, y, c) = 0$$

▲▲

EXAMPLE 5.1

Solve $p^2 + 2py \cot x = y^2$.

Solution: Solving for p, we obtain

$$p = \frac{-2y \cot x \pm \sqrt{4y^2 \cot^2 x + 4y^2}}{2}$$

$$= -y \cot x \pm y \operatorname{cosec} x$$

$$= y(-\cot x \pm \operatorname{cosec} x)$$

Thus, we have

$$\frac{dy}{dx} = y(-\cot x + \operatorname{cosec} x) \tag{1}$$

and

$$\frac{dy}{dx} = y(-\cot x - \operatorname{cosec} x) \tag{2}$$

In both the equations, the variables are separable. Equation (1) gives

$$\frac{dy}{y} = (-\cot x + \operatorname{cosec} x)\,dx$$

Integrating, we get

$$\log y = -\log \sin x + \log \tan \frac{x}{2} + \log c$$

$$\Rightarrow \quad \log y = \log \frac{c \tan \frac{x}{2}}{\sin x}$$

$$\therefore \quad y = \frac{c}{2\cos^2 \frac{x}{2}} = \frac{c}{1 + \cos x}$$

Again Eq. (2) gives

$$\frac{dy}{y} = (-\cot x - \operatorname{cosec} x)\,dx$$

Integrating, we get

$$\log y = -\log \sin x - \log \tan \frac{x}{2} + \log c$$

$$\Rightarrow \qquad \log y = \log \frac{c}{\sin x \tan \dfrac{x}{2}}$$

$$\therefore \qquad y = \frac{c}{2\sin^2 \dfrac{x}{2}} = \frac{c}{1-\cos x}$$

Thus, the solutions of Eqs. (1) and (2) are

$$y - \frac{c}{1+\cos x} = 0, \quad y - \frac{c}{1-\cos x} = 0$$

The required solution of the given differential equation, therefore, is

$$\left(y - \frac{c}{1+\cos x}\right)\left(y - \frac{c}{1-\cos x}\right) = 0.$$

EXAMPLE 5.2

Solve $\dfrac{dy}{dx} - \dfrac{dx}{dy} = \dfrac{x}{y} - \dfrac{y}{x}$.

Solution: Given equation can be written as

$$p - \frac{1}{p} = \frac{x}{y} - \frac{y}{x}$$

or

$$p^2 + p\left(\frac{y}{x} - \frac{x}{y}\right) - 1 = 0$$

Factorising $\left(p + \dfrac{y}{x}\right)\left(p - \dfrac{x}{y}\right) = 0$

Thus, we have

$$p + \frac{y}{x} = 0 \qquad (1)$$

and

$$p - \frac{x}{y} = 0 \qquad (2)$$

Equation (1) gives $\dfrac{dy}{dx} + \dfrac{y}{x} = 0 \Rightarrow xdy + ydx = 0$

$$\Rightarrow \qquad d(xy) = 0$$

Integrating, we get $xy = c$

Again, Eq. (2) gives $\dfrac{dy}{dx} - \dfrac{y}{x} = 0 \Rightarrow xdx - ydy = 0$

Integrating, we get $x^2 - y^2 = c$

Thus, the solutions of Eqs. (1) and (2) are

$$xy - c = 0,\ x^2 - y^2 - c = 0$$

Therefore, the required solution of the given differential equation is

$$(xy - c)(x^2 - y^2 - c) = 0.$$

EXAMPLE 5.3

Solve $xyp^2 - (x^2 + y^2)p + xy = 0$.

Solution: Solving the given differential equation for p, we get

$$p = \frac{dy}{dx} = \frac{(x^2+y^2) \pm \sqrt{(x^2+y^2)^2 - 4x^2y^2}}{2xy}$$

$$= \frac{(x^2+y^2) \pm (x^2-y^2)}{2xy}$$

Therefore, the component equations are

$$\frac{dy}{dx} = \frac{x}{y} \tag{1}$$

and

$$\frac{dy}{dx} = \frac{y}{x} \tag{2}$$

In each of the above differential equations, the variables are separable.

Equation (1) gives $xdx - ydy = 0$

Integrating, we get $x^2 - y^2 = c$

Again, Eq. (2) gives $\frac{dx}{x} - \frac{dy}{y} = 0$

Integrating, we get

$$\log x - \log y = \log c$$

$\Rightarrow$

$$\frac{x}{y} = c$$

or

$$x = cy$$

Thus, the solutions of Eqs (1) and (2) are

$$x^2 - y^2 - c = 0,\ x - cy = 0$$

Therefore, the required solution of the given differential equation is

$$(x^2 - y^2 - c)(x - cy) = 0.$$

EXAMPLE 5.4

Solve $3p^2y^2 - 2xyp + 4y^2 - x^2 = 0$.

Solution: Multiplying throughout by 3, we get

$$9p^2y^2 - 6xyp + 12y^2 - 3x^2 = 0$$

or

$$(3yp - x)^2 - 4(x^2 - 3y^2) = 0 \tag{1}$$

Put $x^2 - 3y^2 = u^2$

$$\therefore \quad 2x - 6y\frac{dy}{dx} = 2u\frac{du}{dx}$$

$$\Rightarrow \quad x - 3yp = u\frac{du}{dx}$$

Putting in Eq. (1), we get

$$u^2\left(\frac{du}{dx}\right)^2 - 4u^2 = 0$$

$$\therefore \quad \frac{du}{dx} = \pm 2$$

Therefore, we have

$$\frac{du}{dx} = 2 \quad \text{and} \quad \frac{du}{dx} = -2$$

Integrating, we get

$$u = 2x + c \quad \text{and} \quad u = -2x + c$$

$$\Rightarrow \quad x^2 - 3y^2 = (2x + c)^2 \quad \text{and} \quad x^2 - 3y^2 = (-2x + c)^2$$

Therefore, the required solution of the given differential equation is

$$x^2 - 3y^2 = (c \pm 2x)^2.$$

EXAMPLE 5.5

Solve $x^2p^3 + y(1 + x^2y)p^2 + y^3p = 0$.

Solution: The given differential equation is

$$p(x^2p^2 + (y + x^2y^2)p + y^3) = 0$$

or

$$p((x^2p^2 + yp) + (x^2y^2p + y^3)) = 0$$

or

$$p(p(x^2p + y) + y^2(x^2p + y) = 0$$

or

$$p(x^2p + y)(p + y^2) = 0$$

Equating each factor to zero, the component equations are

$$p = 0 \tag{1}$$

$$x^2p + y = 0 \tag{2}$$

and

$$p + y^2 = 0 \tag{3}$$

Equation (1) gives $\frac{dy}{dx} = 0$ or $y = c$

Equation (2) gives $\frac{dy}{dx} = \frac{-y}{x^2}$ or $\frac{dx}{x^2} + \frac{dy}{y} = 0$

Integrating, we have $\frac{-1}{x} + \log y = \log c$

or $$\log \frac{y}{c} = \frac{1}{x} \quad \text{or} \quad y = ce^{1/x}.$$

Again, Eq. (3) gives $\frac{dy}{dx} = -y^2$ or $dx + \frac{dy}{y^2} = 0$

Integrating, we have $x - \frac{1}{y} = c$ or $y(x - c) = 1$

Thus, the solutions of Eqs. (1), (2) and (3) are

$$y - c = 0,\ y - ce^{1/x} \text{ and } y(x - c) - 1 = 0$$

Therefore, the required solution of the given differential equation is

$$(y - c)(y - ce^{1/x})(y(x - c) - 1) = 0.$$

EXAMPLE 5.6

Solve $\left(1 - y^2 + \frac{y^4}{x^2}\right)p^2 - 2\frac{y}{x}p + \frac{y^2}{x^2} = 0.$

Solution: Multiplying throughout by x^2, the given differential equation becomes

$$(x^2 - x^2y^2 + y^4)p^2 - 2xyp + y^2 = 0$$

or $$x^2p^2 - 2xyp + y^2 = p^2(x^2y^2 - y^4)$$

or $$(xp - y)^2 = p^2y^2(x^2 - y^2)$$

$\therefore$ $$xp - y = \pm py\sqrt{x^2 - y^2}$$

or $$p[x \mp y\sqrt{x^2 - y^2}] = y$$

or $$\frac{dx}{dy} = \frac{x \mp y\sqrt{x^2 - y^2}}{y}$$

Put $x = vy$, $\frac{dx}{dy} = v + y\frac{dv}{dy}$

$$\therefore \qquad v + y\frac{dv}{dy} = \frac{vy \mp y^2\sqrt{v^2-1}}{y} = v \mp y\sqrt{v^2-1}$$

or
$$\frac{dv}{dy} = \mp\sqrt{v^2-1} \quad \text{or} \quad \frac{dv}{\sqrt{v^2-1}} = \mp dy$$

Integrating, we get

$$\cosh^{-1} v = c \mp y$$

or
$$\log[v+\sqrt{v^2-1}] = c \mp y$$

$$\therefore \qquad \log\frac{x+\sqrt{x^2-y^2}}{y} = c \mp y.$$

EXAMPLE 5.7

Solve $x^2\left(\frac{dy}{dx}\right)^2 - 2xy\left(\frac{dy}{dx}\right) + 2y^2 - x^2 = 0.$

Solution: Solving the given differential equation for dy/dx we have

$$\frac{dy}{dx} = \frac{2xy \pm \sqrt{4x^2y^2 - 4x^2(2y^2-x^2)}}{2x^2}$$

$$= \frac{2xy \pm 2x\sqrt{y^2-2y^2+x^2}}{2x^2} = \frac{y \pm \sqrt{x^2-y^2}}{x}$$

Each of the above differential equations is homogenous.

So, putting $y = vx$; $\frac{dy}{dx} = v + x\frac{dv}{dx}$, we have

$$v + x\frac{dv}{dx} = \frac{vx \pm \sqrt{x^2-v^2x^2}}{x} = v \pm \sqrt{1-v^2}$$

$$\therefore \qquad x\frac{dv}{dx} = \pm\sqrt{1-v^2} \quad \text{or} \quad \frac{dv}{\sqrt{1-v^2}} = \pm\frac{dx}{x}$$

Integrating,we get

$$\sin^{-1} v = \pm\log x \pm \log c$$

or
$$\sin^{-1}(y/x) = \pm\log(cx)$$

which are the required solutions.

EXAMPLE 5.8

Solve $x^2p^2 - 2xyp + y^2 = x^2y^2 + x^4$.

Solution: The given differential equation is

$$x^2p^2 - 2xyp + y^2 - x^2y^2 - x^4 = 0$$

Solving for p, we get

$$p = \frac{dy}{dx} = \frac{2xy \pm \sqrt{4x^2y^2 - 4x^2(y^2 - x^2y^2 - x^4)}}{2x^2}$$

$$= \frac{y \pm \sqrt{y^2 - y^2 + x^2y^2 + x^4}}{x} = \frac{y \pm x\sqrt{x^2 + y^2}}{x}$$

Putting $y = vx$ so that $dy/dx = v + x(dy/dx)$, we have

$$v + x\frac{dy}{dx} = \frac{vx \pm x^2\sqrt{1+v^2}}{x} = v \pm x\sqrt{1+v^2}$$

or $$\frac{dv}{dx} = \pm\sqrt{1+v^2} \quad \text{or} \quad dx = \pm\frac{dv}{\sqrt{1+v^2}}$$

Integrating, we have $x + c = \pm\sinh^{-1} v$

or $$x + c = \pm\sinh^{-1}(y/x)$$

or $$y/x = \sinh(\pm(x + c)) = \pm\sinh(x + c)$$

or $$y^2 - x^2\sinh^2(x + c) = 0$$

which is the required solution. ▲▲

EXERCISE 5.1

Solve the following differential equations:

1. $x^2\left(\frac{dy}{dx}\right)^2 + xy\frac{dy}{dx} - 6y^2 = 0$.
2. $xyp^2 + p(3x^2 - 2y^2) - 6xy = 0$.
3. $p^2 - 2p\cosh x + 1 = 0$.
4. $y^2 + xyp - x^2y^2 = 0$.
5. $(x - y)^2p^2 - 3y(x - y)p + 2y^2 + xy - x^2 = 0$.
6. $(p - xy)(p - x^2)(p - y^2) = 0$.
7. $p(p - y) = x(x + y)$.
8. $x(dy/dx)^2 + (y - x)(dy/dx) - y = 0$.
9. $y = x(p + \sqrt{1+p^2})$.

10. $p^3 + 2xp^2 - y^2p^2 - 2xy^2p = 0.$
11. $p^2 - 2p\sinh x - 1 = 0.$

ANSWERS

Exercise 5.1

1. $(x^2y - c)(y - cx^2) = 0.$
2. $(y - cx^2)(y^2 + 3x^2 - c) = 0.$
3. $(y - e^x + c)(y + e^{-x} + c) = 0.$
4. $(y^2 - ce^{(1+\sqrt{5})x})(y^2 - ce^{(1-\sqrt{5})x}) = 0.$
5. $\left[\frac{-1}{2}\log(v^2 + v - 1) + \frac{3}{2\sqrt{5}}\log\frac{2v+1-\sqrt{5}}{2v+1+\sqrt{5}} - \log(cx)\right]$

$$\left[\tan^{-1} v - \frac{1}{2}\log(1+v^2) - \log(cx)\right] = 0$$

where $v = y/x$.

6. $(y - ce^{\frac{1}{2}x^2})(3y - x^3 - c)(xy + cy + 1] = 0.$
7. $(y + x + 1 - ce^x)(2y + x^2 - c) = 0.$
8. $(y - x - c)(xy - c) = 0.$
9. $x^2 + y^2 = cx.$
10. $(y - c)(y + x^2 - c)(xy + cy + 1) = 0.$
11. $(y - e^x - c)(y - e^{-x} - c) = 0.$ ▲▲

Equations Solvable for *y*

Suppose that the given differential equation on solving for y gives

$$y = f(x, p) \tag{1}$$

Differentiating with respect to x, we obtain

$$p = \frac{dy}{dx} = \phi\left(x, p, \frac{dp}{dx}\right)$$

which is a differential equation in two variables x and p.

Suppose that it is possible to solve this equation, let the solution be

$$F(x, p, c) = 0 \tag{2}$$

where c is an arbitrary constant.

The solution of Eq. (1) may be exhibited in either of the two forms. We may eliminate p between Eqs. (1) and (2) and obtain $g(x, y, c) = 0$ as the required solution or we may solve Eqs. (1) and (2) for x, y and obtain

$$x = f_1(p, c),\ y = f_2(p, c)$$

as the required solution, where p is the parameter.

Note: This method is especially useful for equations which do not contain x. ▲▲

EXAMPLE 5.9

Solve $y + px = p^2x^4$.

Solution: Differentiating with respect to x, we have

$$\frac{dy}{dx} + p + x\frac{dp}{dx} = 4x^3p^2 + 2x^4p\frac{dp}{dx}$$

$$\Rightarrow \quad 2p - 4x^3p^2 - 2x^4p\frac{dp}{dx} + x\frac{dp}{dx} = 0$$

$$\Rightarrow \quad 2p(1 - 2x^3p) + x(1 - 2x^3p)\frac{dp}{dx} = 0$$

$$\Rightarrow \quad \left(2p + x\frac{dp}{dx}\right)(1 - 2x^3p) = 0$$

Rejecting $1 - 2x^3p = 0$, we have

$$2p + x\frac{dp}{dx} = 0$$

$$\Rightarrow \quad \frac{dp}{p} + \frac{2dx}{x} = 0$$

Integrating, we have

$$\log p = 2\log x = \log c$$

$$\Rightarrow \quad px^2 = c \Rightarrow p = c/x^2$$

Substituting the value of p in the given differential equation so as to eliminate p, we obtain

$$y + c/x = c^2$$

$$\Rightarrow \quad xy = c^2x - c.$$

EXAMPLE 5.10

Solve $y - 2px = f(xp^2)$.

Solution: Differentiating with respect to x, we get

$$\frac{dy}{dx} - 2p - 2x\frac{dp}{dx} = f'(xp^2)\left[p^2 + 2xp\frac{dp}{dx}\right]$$

$$\Rightarrow \quad -p - 2x\frac{dp}{dx} = f'(xp^2)p\left(p + 2x\frac{dp}{dx}\right)$$

$$\Rightarrow \quad (1 + pf'(xp^2))\left(p + 2x\frac{dp}{dx}\right) = 0$$

Rejecting $1 + pf'(xp^2) = 0$, we have

$$p + 2x\frac{dp}{dx} = 0 \quad \Rightarrow \quad \frac{dx}{x} + 2\frac{dp}{p} = 0$$

Integrating, we have

$$\log x + 2\log p = \log c^2$$

$$\Rightarrow \quad xp^2 = c^2 \therefore p = c/\sqrt{x}$$

Putting the value of p in the given differential equation, so as to eliminate p, we obtain

$$y - \frac{2c}{\sqrt{x}}x = f\left(x\frac{c^2}{x}\right)$$

$$\therefore \quad y = 2c\sqrt{x} + f(c^2).$$

EXAMPLE 5.11

Solve $y - x = x\dfrac{dy}{dx} + \left(\dfrac{dy}{dx}\right)^2$.

Solution: The given differential equation can be written as

$$y = x(1 + p) + p^2$$

Differentiating with respect to x, we have

$$p = \frac{dy}{dx} = 1 + p + x\frac{dp}{dx} + 2p\frac{dp}{dx}$$

$$\Rightarrow \quad 1 + x\frac{dp}{dx} + 2p\frac{dp}{dx} = 0$$

$$\Rightarrow \quad \frac{dp}{dx}(x + 2p) + 1 = 0$$

$$\Rightarrow \quad \frac{dx}{dp} + x = -2p$$

which is a linear differential equation with

$$\text{IF} = e^{\int 1dp} = e^p$$

∴ Its solution is

$$xe^p = \int -2p e^p dx + c$$

$$\Rightarrow \quad xe^p = -2(pe^p - e^p) + c$$

$$\Rightarrow \quad x = -2(p-1) + ce^{-p} \qquad (1)$$

Substituting the value of x in the given differential equation, we get

$$y = (1+p)[-2(p-1) + ce^{-p}] + p^2$$

$$\Rightarrow \quad y = c(1+p)e^{-p} + 2 - p^2 \qquad (2)$$

Equations (1) and (2) gives required solution where p is a parameter.

EXAMPLE 5.12

Solve $xp^2 - 2yp + x = 0$.

Solution: The given equation can be written as

$$2yp = x(p^2+1)$$

$$\Rightarrow \quad 2y = x\left(p + \frac{1}{p}\right)$$

Differentiating with respect to x, we get

$$\frac{2dy}{dx} = 2p = \left(p + \frac{1}{p}\right) + x\left(1 - \frac{1}{p^2}\right)\frac{dp}{dx}$$

$$\Rightarrow \quad \left(p - \frac{1}{p}\right) = x\left(1 - \frac{1}{p^2}\right)\frac{dp}{dx}$$

$$\Rightarrow \quad \left(p - \frac{1}{p}\right)\left(p - x\frac{dp}{dx}\right) = 0.$$

Rejecting $1 - \dfrac{1}{p^2} = 0$ we have,

$$p - x\frac{dp}{dx} = 0$$

$$\Rightarrow \quad \frac{dp}{p} - \frac{dx}{x} = 0$$

Integrating, we have

$$\log p - \log x = \log c$$

$$\therefore \quad p = cx$$

Putting for p in the given differential equation, we get

$$2y = x\left(cx + \frac{1}{cx}\right)$$

$$\Rightarrow \qquad 2cy = c^2x^2 + 1.$$

EXAMPLE 5.13

Solve $y = \sin p - p\cos p$.

Solution: Differentiating with respect to x, we have

$$\frac{dy}{dx} = p = (\cos p - \cos p + p\sin p)\frac{dp}{dx}$$

$$\Rightarrow \qquad 1 = \sin p\,\frac{dp}{dx}$$

which is in variable separable form, hence integrating it, we get

$$x = -\cos p + c$$

$$\therefore \qquad \cos p = c - x$$

Substituting the value of cos p and sin p in the given differential equation so as to eliminate p, we obtain

$$y = \sqrt{1-(c-x)^2} - (c-x)\cos^{-1}(c-x)$$

$$\Rightarrow \qquad \cos\left(\left(\sqrt{1-(c^2+2cx-x^2} - y\right)/(c-x)\right) = c - x$$

which is the required solution of the given differential equation.

EXAMPLE 5.14

Solve $p^3 + xp^2 = y$.

Solution: The given differential equation is

$$y = xp^2 + p^3 \tag{1}$$

Differentiating with respect to x, we get

$$p = \frac{dy}{dx} = p^2 + 2xp\frac{dp}{dx} + 3p^2\frac{dp}{dx}$$

$$\Rightarrow \qquad p\left[(1-p) - (2x+3p)\frac{dp}{dx}\right] = 0$$

Rejecting $p = 0$, we have

$$(1-p)\frac{dx}{dp} - 2x = 3p$$

$$\Rightarrow \qquad \frac{dx}{dp} - \frac{2}{1-p}x = \frac{3p}{1-p}$$

$$\Rightarrow \qquad \frac{dx}{dp} + \frac{2}{p-1}x = \frac{-3p}{p-1}$$

which is a linear differential equation with

$$\text{IF} = e^{\int \frac{2}{p-1}dx} = e^{2\log(p-1)} = (p-1)^2$$

Hence, its solution is given by

$$x(p-1)^2 = \int \frac{-3p}{p-1}(p-1)^2 dp + c$$

$$\Rightarrow \qquad x(p-1)^2 = -3\int p(p-1)\,dp + c$$

$$\Rightarrow \qquad x(p-1)^2 = -3\int (p^2 - p)\,dp + c$$

$$\Rightarrow \qquad x(p-1)^2 = -p^3 + \frac{3}{2}p^2 + c$$

$$\therefore \qquad x = \frac{1}{2}(2c + 3p^2 - 2p^3)/(p-1)^2 \qquad (2)$$

Substituting this value of x in Eq. (1), we get

$$y = p^2\,\frac{1}{2}\,\frac{2c + 3p^2 - 2p^3}{(p-1)^2} + p^3$$

$$\Rightarrow \qquad y = \frac{2cp^2 + 3p^4 - 2p^5 + 2p^5 - 4p^4 + 2p^3}{2(p-1)^2}$$

$$\Rightarrow \qquad y = \frac{2cp^2 + 2p^3 - p^4}{2(p-1)^2} \qquad (3)$$

Equations (2) and (3) give the required solution where p is a parameter.

EXAMPLE 5.15

Solve $x^2 + p^2x - yp = 0$.

Solution: Solving the given differential equation for y, we get

$$y = px + \frac{x^2}{p} \qquad (1)$$

Differentiating with respect to x, we get

$$p = \frac{dy}{dx} = p + \frac{xdp}{dx} + \frac{2x}{p} - \frac{x^2}{p^2}\frac{dp}{dx}$$

$$\Rightarrow \qquad (p^2 - x)\frac{dp}{dx} + 2p = 0$$

$$\Rightarrow \qquad \frac{dx}{dp} - \frac{x}{2p} = -\frac{1}{2}p$$

which is a linear differentiating equation of the first order and of first degree in x and p with

$$\text{IF} = e^{\int \frac{-1}{2p}dp}$$

$$= e^{-\frac{1}{2}\log p} = \frac{1}{\sqrt{p}}$$

Therefore, its solution is

$$x\frac{1}{\sqrt{p}} = \int -\frac{1}{2}p\frac{1}{\sqrt{p}}dp + c$$

$$\Rightarrow \qquad x\frac{1}{\sqrt{p}} = -\frac{p^{3/2}}{3} + c$$

$$\therefore \qquad x = c\sqrt{p} - \frac{1}{3}p^2 \qquad (2)$$

Putting the value of x from Eq. (2) in Eq. (1), we get

$$y = p\left(c\sqrt{p} - \frac{1}{3}p^2\right) + \frac{1}{p}\left(c\sqrt{p} - \frac{1}{3}p^2\right)^2 \qquad (3)$$

Thus, Eqs. (2) and (3) give the general solution of the differential equation in parametric form having p as parameter. ▲▲

EXERCISE 5.2

Solve the following differential equations:

1. $y + px = p^2x^4$.
2. $y = 2px + p^n$.
3. $4p^3 + 3xp = y$.
4. $x^2 + p^2x = yp$.
5. $y = x(p + \sqrt{1+p^2})$.
6. $y = a\sqrt{1+p^2}$.

7. $y = xp^2 + p$.
8. $y = 3x + a \log p$.
9. $y = p \tan p + \log \cos p$.
10. $p^3 - 4xyp + 8p^2 = 0$.
11. $y - 2px = \tan^{-1}(xp^2)$.

ANSWERS

Exercise 5.2

1. $xy = c^2x - c$.
2. $x = c/p^2 - np^{n-1}/(n+1),\ y = 2c/p + (n-1)p^n/(n+1)$.
3. $x = -(12/7)p^2 + cp^{-3/2},\ y = -(8/7)p^3 + 3cp^{-1/2}$.
4. $x = \dfrac{-1}{3}p^2 + cp^{1/2},\ y = \dfrac{\left(\dfrac{-1}{3}p^2 + cp^{1/2}\right)^2}{p} + p\left(\dfrac{-1}{3}p^2 + cp^{1/2}\right)$.
5. $x^2 + y^2 - 2xc = 0$.
6. $x = a \log (y + \sqrt{y^2 - a^2}) + c$.
7. $x(p-1)^2 = c - p + \log p$ and given equation.
8. $y = 3x + a \log \dfrac{3}{1 - ce^{3x/a}}$.
9. $x = \tan p + c,\ y = p \tan p + \log \cos p$.
10. $y = c(x - c)^2$.
11. $y = 2\sqrt{cx} + \tan^{-1} c$.

▲▲

Equations Solvable for *x*

Suppose that the given differential equation on solving for x, gives

$$x = f(y, p) \tag{1}$$

Differentiating with respect to y, we obtain

$$\frac{1}{p} = \frac{dx}{dy} = \phi(y, p, dp/dy)$$

which is a differential equation in two variables y and p. Suppose that it is possible to solve this equation. Let the solution be

$$F(y, p, c) = 0 \tag{2}$$

where c is an arbitrary constant.

The solution of Eq. (1) may be exhibited in either of the two forms. We may either eliminate p between Eqs. (1) and (2) and obtain $g(x, y, c) = 0$ as the required solution or we may solve Eqs. (1) and (2) for x, y and obtain

$$x = f_1(p, c),\ y = f_2(p, c)$$

as the required solution, where p is the parameter.

Note: This method is especially useful for equations which do not contain y. ▲▲

EXAMPLE 5.16

Solve $y = 2px + y^2p^3$.

Solution: Solving it for x, we obtain

$$x = \frac{y - y^2p^3}{2p} = \frac{y}{2p} - \frac{y^2p^2}{2}$$

Differentiating with respect to y, we obtain

$$\frac{dx}{dy} = \frac{1}{p} = \frac{1}{2p} \frac{-y}{2p^2}\frac{dp}{dy} - yp^2 - py^2\frac{dp}{dy}$$

$$\Rightarrow \quad \frac{1}{2p} + p^2y + \left(\frac{y}{2p^2} + py^2\right)\frac{dp}{dy} = 0$$

$$\Rightarrow \quad \frac{1+2p^3y}{2p} + \frac{y(1+2p^3y)}{2p^2}\frac{dp}{dy} = 0$$

Rejecting $\dfrac{1+2p^3y}{2p} = 0$, we have

$$1 + \frac{y}{p}\frac{dp}{dy} = 0$$

$$\Rightarrow \quad \frac{dy}{y} + \frac{dp}{p} = 0$$

Integrating, we have

$$\log y + \log p = \log c$$

$$\Rightarrow \quad yp = c$$

$$\therefore \quad p = c/y$$

Substituting the value of p in the given equation, we obtain

$$y = \frac{2cx}{y} + y^2\frac{c^3}{y^3}$$

$$\Rightarrow \qquad y^2 = 2cx + c^3$$

which is the required solution.

EXAMPLE 5.17

Solve $x = y + p^2$.

Solution: Differentiating the given equation with respect to y, we get

$$\frac{dx}{dy} = \frac{1}{p} = 1 + 2p\frac{dp}{dy}$$

$$\Rightarrow \qquad dy = \frac{2p^2}{1-p}dp = -2\left(p+1+\frac{1}{p-1}\right)dp$$

Integrating, we get

$$y = -2\left[\frac{1}{2}p^2 + p + \log(p-1)\right] + c$$

$$\Rightarrow \qquad y = c - [p^2 + 2p + 2\log(p-1)] \tag{1}$$

Substituting this in the given differential equation, we obtain

$$x = c - [2p + 2\log(p-1)] \tag{2}$$

Equations (1) and (2) together give the required solution in the parametric form having p as parameter.

EXAMPLE 5.18

Solve $y^2 \log y = xyp + p^2$.

Solution: Solving for x, the given differential equation can be written as

$$xyp = y^2 \log y - p^2$$

$$\Rightarrow \qquad x = \frac{y\log y}{p} - \frac{p}{y}$$

Differentiating with respect to y, we get

$$\frac{dx}{dy} = \frac{1}{p} = \frac{p(1+\log y) - y\log y\dfrac{dp}{dy}}{p^2} - \frac{y\dfrac{dp}{dy} - p}{y^2}$$

$$\Rightarrow \qquad \frac{1}{p} = \frac{1}{p} + \frac{1}{p}\log y - \frac{y}{p^2}\log y\frac{dp}{dy} - \frac{1}{y}\frac{dp}{dy} + \frac{p}{y^2}$$

$$\Rightarrow \qquad \frac{1}{y}\frac{dp}{dy}\left(1 + \frac{y^2}{p^2}\log y\right) = \frac{p}{y^2}\left(1 + \frac{y^2}{p^2}\log y\right)$$

$$\Rightarrow \quad \left(\frac{1}{y}\frac{dp}{dy} - \frac{p}{y^2}\right)\left(1 + \frac{y^2}{p^2}\log y\right) = 0$$

Rejecting $1 + \frac{y^2}{p^2}\log y = 0$, we have

$$\frac{1}{y}\frac{dp}{dy} - \frac{p}{y^2} = 0$$

$$\Rightarrow \quad \frac{dp}{p} - \frac{dy}{y} = 0$$

Integrating, we have

$$\log p - \log y = \log c$$

$$\therefore \quad \frac{p}{y} = c \Rightarrow p = cy$$

Substituting the value of p in the given equation, we get

$$y^2 \log y = xycy + c^2y^2$$

$$\Rightarrow \quad \log y = cx + c^2.$$

EXAMPLE 5.19

Solve $p^2y + 2px - y = 0$.

Solution: Solving the given differential equation for x, we get

$$2px = y - p^2y$$

$$\Rightarrow \quad x = \frac{y}{2p} - \frac{yp}{2}$$

Differentiating with respect to y, we get

$$\frac{dx}{dy} = \frac{1}{p} = \frac{1}{2p} - \frac{y}{2p^2}\frac{dp}{dy} - \frac{p}{2} - \frac{y}{2}\frac{dp}{dy}$$

$$\Rightarrow \quad \frac{1}{2p} + \frac{p}{2} + \frac{y}{2}\frac{dp}{dy}\left(\frac{1}{p^2} + 1\right) = 0$$

$$\Rightarrow \quad \frac{p}{2}\left(\frac{1}{p^2} + 1\right) + \frac{y}{2}\frac{dp}{dy}\left(\frac{1}{p^2} + 1\right) = 0$$

$$\Rightarrow \quad \left(\frac{1}{p^2} + 1\right)\left(\frac{p}{2} + \frac{y}{2}\frac{dp}{dy}\right) = 0$$

Rejecting $\frac{1}{p^2}+1 = 0$, we have

$$\frac{p}{2}+\frac{y}{2}\frac{dp}{dy} = 0 \quad \Rightarrow \quad \frac{dy}{y}+\frac{dp}{p} = 0$$

Integrating, we get

$$\log y + \log p = \log c$$

$$\Rightarrow \qquad yp = c \quad \therefore p = c/y$$

Substituting the value of p in the given differential equation, we get

$$\frac{c^2}{y^2}\cdot y+\frac{2c}{y}x - y = 0$$

$$\Rightarrow \qquad \frac{c^2}{y}+\frac{2cx}{y} - y = 0$$

$$\Rightarrow \qquad y^2 - 2cx - c^2 = 0$$

which is the required solution.

EXAMPLE 5.20

Solve $axyp^2 + (2x - b)p - y = 0$.

Solution: The given equation can be written as

$$2x = \frac{y}{p}+b-ayp$$

Differentiating with respect to y, we get

$$2\frac{dx}{dy} = \frac{2}{p} = \frac{1}{p} - \frac{y}{p^2}\frac{dp}{dy} - a\left(p+y\frac{dp}{dy}\right)$$

$$\Rightarrow \qquad \frac{1}{p^2}\left(p+\frac{ydp}{dy}\right)+a\left(p+\frac{ydp}{dy}\right) = 0$$

$$\Rightarrow \qquad \left(\frac{1}{p^2}+a\right)\left(p+\frac{ydp}{dy}\right) = 0$$

Rejecting $\frac{1}{p^2} + a = 0$, we have

$$p+\frac{ydp}{dy} = 0 \quad \Rightarrow \quad \frac{dp}{p}+\frac{dy}{y} = 0$$

Integrating, we have

$$\log p + \log y = \log c$$

$$\Rightarrow \qquad py = c \Rightarrow p = \frac{c}{y}$$

On putting the value of p in the given differential equation, we get

$$2x = \frac{y^2}{c} + b - ac$$

$$\Rightarrow \qquad 2xc = y^2 + bc - ac^2$$

$$\therefore \qquad ac^2 + (2x - b)c - y^2 = 0.$$

EXAMPLE 5.21

Solve $p = \tan\left(x - \dfrac{p}{1+p^2}\right)$.

Solution: Solving for x, we have

$$x = \tan^{-1}p + \frac{p}{1+p^2} \tag{1}$$

Differentiating with respect to y, we get

$$\frac{dx}{dy} = \frac{1}{p} = \frac{1}{1+p^2}\frac{dp}{dy} + \frac{(1+p^2)\cdot 1 - 2p^2}{(1+p^2)^2}\frac{dp}{dy}$$

$$\Rightarrow \qquad \frac{1}{p} = \frac{2(1+p^2) - 2p^2}{(1+p^2)^2}\frac{dp}{dy}$$

$$\Rightarrow \qquad \frac{1}{p} = \frac{2}{(1+p^2)^2}\frac{dp}{dy}$$

$$\Rightarrow \qquad dy = \frac{2p}{(1+p^2)^2}dp$$

Integrating, we have

$$y = \frac{-1}{1+p^2} + c \tag{2}$$

Equations (1) and (2) together constitute the general solution in the parametric form having p as a parameter.

EXAMPLE 5.22

Solve $x + \dfrac{p}{\sqrt{1+p^2}} = a$

Solution: Solving the given differential equation for x, we get

$$x = a - \frac{p}{\sqrt{1+p^2}}$$

Differentiating with respect to y, we get

$$\frac{dx}{dy} = \frac{1}{p} = -\frac{\sqrt{1+p^2}\frac{dp}{dy} - p\frac{2p}{2\sqrt{1+p^2}}\frac{dp}{dy}}{(1+p^2)}$$

$$\Rightarrow \quad \frac{1}{p} = \frac{1+p^2-p^2}{(1+p^2)^{3/2}}\frac{dp}{dy}$$

$$\Rightarrow \quad \frac{1}{p} = -\frac{1}{(1+p^2)^{3/2}}\frac{dp}{dy}$$

$$\Rightarrow \quad dy = \frac{-p}{(1+p^2)^{3/2}}dp$$

Integrating, we get

$$y = -\frac{(1+p^2)^{-1/2}}{-1/2} + c$$

$$\Rightarrow \quad y = \frac{1}{\sqrt{1+p^2}} + c$$

$$\therefore \quad (y-c)^2 = \frac{1}{1+p^2}$$

From the given differential equation, we have

$$(x-a)^2 = \frac{p^2}{1+p^2} = \frac{(p^2+1)-1}{1+p^2} = 1 - \frac{1}{1+p^2}$$

$$\Rightarrow \quad (x-a)^2 = 1-(y-c)^2$$

$$\therefore \quad (x-a)^2 + (y-c)^2 = 1$$

which is the required solution of the given differential equation. ▲▲

EXERCISE 5.3

Solve the following differential equations:

1. $ap^2 + py - x = 0.$

2. $\left(\frac{dy}{dx}\right)^3 - 4xy\frac{dy}{dx} + 8y^2 = 0.$
3. $4(xp^2 + yp) = p^4.$
4. $4p^3 + 3xp = y.$
5. $ap^3 = a + bp.$
6. $x = y + a\log p.$
7. $p^3 - 4xyp + 8y^2 = 0.$
8. $\left(\frac{dy}{dx}\right)^3 \cdot y^2 - 2x\frac{dy}{dx} + y = 0.$
9. $yp^2 - 2xp + y = 0.$
10. $y + px = a^4p^2.$

ANSWERS

Exercise 5.3

1. $x = \frac{(c - a\cosh^{-1} p)}{\sqrt{p^2 - 1}} p, y = \frac{(c - a\cosh^{-1} p)}{\sqrt{p^2 - 1}} - ap.$
2. $64y = c(4x - c)^2.$
3. $y = 4c(xyc + 1).$
4. $y = \frac{-8}{7}p^3 + cp^{-1/2}, x = \frac{-12}{7}p^2 + \frac{c}{3}p^{-3/2}.$
5. $x = \frac{a}{p^3} + \frac{b}{p^2}, y = \frac{3a}{2p^2} + \frac{2b}{p} + c.$
6. $y = c - a\log(p - 1), x = c + a\log\frac{p}{p-1}.$
7. $y = c(c - x)^2.$
8. $2cx = x^3 + y^2.$
9. $y^2 = 2cx - c^2.$
10. $y = \frac{-c}{x} + c^2.$

6

CHAPTER

Singular Solution and Clairaut's Differential Equations

In chapter 1, we defined solution of a differential equation, known as general solution. A solution which is different from general solution in the sense that it cannot be deduced from the general solution and contains no arbitrary constant is known as *singular solution*. We will start this chapter firstly giving geometrical meaning of a differential equation. Then, we discuss method of finding singular solution and solving Clairaut's differential equation.

Geometrical meaning of a differential equation of the first order.

Consider a differential equation

$$f(x, y, dy/dx) = 0 \tag{1}$$

and let

$$\phi(x, y, c) = 0 \tag{2}$$

be its general solution where c is the arbitrary constant.

For each value of c, Eq. (2) represents a curve so that Eq. (2) represents the equation of a family of curves, c being the parameter for the family. Hence, we say that every differential equation of the first order represents a family of curves.

Now, let us take a point (x, y) in the plane and substitute the coordinates of this point in Eqs. (1) and (2) and solve the resulting equations in dy/dx and c. Then, the values of c, so obtained are the values of the parameter for the particular curves of the family (2) through (x, y) and the values of dy/dx are the slopes of the tangents to these curves at the point (x, y). Naturally, the degree of c in Eq. (2) must be equal to that of dy/dx in Eq. (1), which is known as the *degree of the differential equation*.

Singular Solutions

Sometimes, a differential equation of degree higher than one possesses a solution which does not contain any arbitrary constant and which cannot be obtained from the general solution of that differential equation by giving particular values to the arbitrary constant. The solutions of differential equations other than the general and particular solutions, are known as singular solutions.

The singular solution of a differential equation is given by the envelope of the family of curves represented by that differential equation. In this connection, we have the following result.

Whenever the family of curves

$$\phi(x, y, c) = 0 \tag{1}$$

represented by the differential equation

$$f(x, y, dx/dy) = 0 \tag{2}$$

possesses an envelope, the equation of the envelope is the singular solution of the differential Eq. (2).

Suppose that the family of curves (1) possesses an envelope. For any point $P(x, y)$ on the envelope, there exists a curve of the family (1). Say, $\phi(x, y, c_1) = 0$, which touches the envelope at (x, y). The values of x, y, dy/dx for the curve at P, satisfy the given differential Eq. (2). Also, the values of x, y, dy/dx at P for the envelope are the same as for the curve. Thus, we see that the values of x, y, dy/dx at every point of the envelope satisfy the given differential Eq. (2). This solution does not contain any arbitrary constant and in general, cannot be obtained from Eq. (1) by giving any particular value to the arbitrary constant c. Hence, the equation of the envelope is a solution of the differential Eq. (2); known as singular solution.

Method of Finding Singular Solution

Let $F(c) = 0$ be an algebraic equation. The discriminant of this equation is the simplest function of the coefficients; the vanishing of which represents the condition that the equation should have two equal roots. Thus, the discriminant of the quadratic $Ax^2 + Bx + c = 0$ is $B^2 - 4AC$. The equation $B^2 - 4AC = 0$ is called the discriminant relation.

Let

$$f(x, y, dy/dx) = 0 \tag{1}$$

be a given differential equation and let

$$\phi(x, y, c) = 0 \tag{2}$$

be its general solution.

Now, it is known that the envelope of any family of curves $\phi(x, y, c) = 0$ is contained in the locus obtained on eliminating c between (2) and

$$\frac{\partial \phi(x,y,c)}{\partial c} = 0 \tag{3}$$

Let the c elimination (known as c-discriminant) between Eqs. (2) and (3) be

$$\psi(x, y) = 0 \tag{4}$$

Since Eq. (4) may sometimes represent loci other than the envelope, therefore it is necessary to verify, if any part of the locus represented by Eq. (4) is or is not a solution of the given differential Eq. (1). Only those parts will constitute the singular solution, which satisfy Eq. (1).

From the theory of equations, we know that Eq. (4) represents the locus of the point (x, y) such that at least two of the corresponding values of c are equal, that is, such that at least two of the curves of the family through (x, y) coincide. Now if two curves of the family coincide, their tangents must also coincide and so the corresponding values of p, i.e., dy/dx given by Eq. (1) must also coincide. Hence, we see that the envelope and hence the singular solution is also contained in the locus obtained an eliminating p between

$$f(x, y, p) = 0 \tag{5}$$

and

$$\frac{\partial f(x, y, p)}{\partial p} = 0 \tag{6}$$

The p-eliminant between (5) and (6) is called the *p-discriminant relation* and the envelope and hence, the singular solution is also contained in it.

Note: For the first degree differential equations, there will be no singular solution.

While solving a differential equation of the first order but of higher degree, we must find the general solution and singular solution if asked for completely solve the given differential equation.

Working Rule for Finding the Singular Solution

Let the given differential equation be

$$f(x, y, dy/dx) = f(x, y, p) = 0 \tag{1}$$

Find the general solution of Eq. (1) and let it be

$$\phi(x, y, c) = 0 \tag{2}$$

Find the c-discriminant relation by eliminating c between

$$\phi(x, y, c) = 0 \quad \text{and} \quad \frac{\partial \phi(x, y, c)}{\partial c} = 0$$

Also, find the p-discriminant relation by eliminating p between

$$f(x, y, p) = 0 \quad \text{and} \quad \frac{\partial f(x, y, p)}{\partial p} = 0.$$

The locus common to both the c-discriminant and the p-discriminant relations give us the singular solution of Eq. (1) provided it satisfies equation (1). ▲▲

EXAMPLE 6.1

Find the general and singular solutions of

$$9p^2(2-y)^2 = 4(3-y).$$

Solution: The given equation is

$$3\frac{dy}{dx} = 3p = \pm\frac{2\sqrt{3-y}}{2-y}$$

Separating the variables, we get

$$dx = \pm\frac{3}{2}\frac{2-y}{\sqrt{3-y}}dy$$

Substituting $3-y=t^2$ and integrating, we get

$$x + c = \pm\frac{3}{2}\int\frac{(t^2-1)(-2t\,dt)}{t} = \pm(t^3-3t)$$

$\Rightarrow$ $$x + c = \pm t(t^2-3)$$

$\Rightarrow$ $$(x+c)^2 = y^2(3-y)$$

This is the general solution. The singular solution can be found as under.

The general solution can be written as

$$c^2 + 2xc + x^2 - y^2(3-y) = 0$$

This is a quadratic equation in the parameter c and so its envelope (c-discriminant relation) is

$$4x^2 - 4\{x^2 - y^2(3-y)\} = 0$$

$\Rightarrow$ $$y^2(3-y) = 0$$

As $y = 0$ does not satisfy the original differential equation, therefore, the required singular solution is $3-y=0$.

Alternately, we can also find the singular solution by using p-discrminant relation. The original equation is a quadratic equation in the parameter p so its envelope (p-discriminant relation) is

$$0 + 4.9(2-y)^2\,4(3-y) = 0$$

$\Rightarrow$ $$(2-y)^2(3-y) = 0$$

As $2-y=0$ do not satisfy the given differential equation, therefore, the singular solution is $3-y=0$.

Clearly, the singular solution is the part which is common to the c-discriminant relation and p-discriminant relation.

EXAMPLE 6.2

Find the general and singular solutions of the differential equation $x^3p^2 + x^2yp + a^3 = 0$.

Solution: Solving the given differential equation for y, we get

$$x^2yp = -x^3p^2 - a^3 \tag{1}$$

$$\Rightarrow \quad y = -xp - a^3x^{-2}p{-}1$$

Differentiating with respect to x, and writing p for dy/dx, we get

$$\frac{dy}{dx} = p = -p - x\frac{dp}{dx} + 2a^3x^{-3}p^{-1} + a^3x^{-2}p^{-2}\frac{dp}{dx}$$

$$\Rightarrow \quad 2p - 2a^3x^{-3}p^{-1} + x\frac{dp}{dx} - a^3x^{-2}p^{-2}\frac{dp}{dx} = 0$$

$$\Rightarrow \quad \left(2p + x\frac{dp}{dx}\right) - a^3x^{-3}p^{-2}\left(2p + x\frac{dp}{dx}\right) = 0$$

$$\Rightarrow \quad \left(2p + x\frac{dp}{dx}\right)\left(1 - \frac{a^3}{p^2x^3}\right) = 0$$

Rejecting $1 - \dfrac{a^3}{p^2x^3} = 0$, we have

$$2p + x\frac{dp}{dx} = 0$$

$$\Rightarrow \quad \frac{dp}{p} + 2\frac{dx}{x} = 0$$

Integrating, we get

$$\log p + 2\log x = \log c$$

$$\Rightarrow \quad px^2 = c \quad \therefore p = c/x^2$$

Substituting the value of p in the given differential equation, so as to eliminate p, we get

$$x^3\left(\frac{c^2}{x^4}\right) + x^2y\left(\frac{c}{x^2}\right) + a^3 = 0$$

$$\Rightarrow \quad \frac{c^2}{x} + cy + a^3 = 0$$

$$\Rightarrow \quad c^2 + cxy + a^3x = 0 \tag{2}$$

which is the general solution of the given differential Eq. (1).

From Eq. (2), the c-discriminant relation is

$$x^2y^2 - 4a^3x = 0$$

$$\Rightarrow \qquad x(xy^2 - 4a^3) = 0$$

From Eq. (1), the p-discriminant relation is

$$x^4y^2 - 4a^3x^3 = 0$$

$$\Rightarrow \qquad x^3(xy^2 - 4a^3) = 0$$

Now $x(xy^2 - 4a^3) = 0$ appears in both the c-discriminant and p-discriminant relations and it can be easily seen that

$$x = 0,\ xy^2 - 4a^3 = 0$$

satisfy the given differential equation.

Hence, the singular solutions are

$$x = 0,\ xy^2 - 4a^3 = 0.$$

EXAMPLE 6.3

Find the general and singular solutions of

$$27y - 8p^3 = 0.$$

Solution: Solving the given differential equation for p, we have

$$p = \frac{dy}{dx} = \frac{3}{2}y^{1/3}$$

$$\Rightarrow \qquad \frac{2}{3}y^{-1/3}dy = dx$$

Integrating, we have

$$y^{2/3} = x + c \quad \Rightarrow \quad y^2 = (x + c)^3 \tag{1}$$

which is the required general solution of the given differential equation.

For finding the singular solution, differentiate Eq. (1) partially with respect to c

$$0 = 3(x + c)^2 \quad \Rightarrow \quad x + c = 0 \tag{2}$$

By eliminating c between Eqs. (1) and (2), we get the c-discriminant relation as

$$y^2 = 0 \quad \Rightarrow \quad y = 0$$

which satisfies the given differential equation (1).

Hence, the singular solution is $y = 0$.

EXAMPLE 6.4

Solve and examine for singular solution

$$(8p^3 - 27)x = 12p^2y \tag{1}$$

Solution: Solving the given differential equation for y, we get

$$y = \frac{(8p^3 - 27)x}{12p^2}$$

$$\Rightarrow \qquad y = \frac{2}{3}px - \frac{9}{4}\frac{x}{p^2} \qquad (2)$$

Differentiating with respect to x, we get

$$p = \frac{dy}{dx} = \frac{2}{3}p + \frac{2}{3}x\frac{dp}{dx} - \frac{9}{4}\left\{\frac{p^2 - x\cdot 2p\frac{dp}{dx}}{p^4}\right\}$$

$$\Rightarrow \qquad \frac{1}{3}p - \frac{2}{3}x\frac{dp}{dx} + \frac{9}{4}\frac{1}{p^2} - \frac{9}{2}\frac{x}{p^3}\frac{dp}{dx} = 0$$

$$\Rightarrow \qquad \left(\frac{4p^3 + 27}{12p^2}\right) - \left(\frac{4p^3 + 27}{6p^3}\right)x\frac{dp}{dx} - 0$$

$$\Rightarrow \qquad \left(\frac{4p^3 + 27}{6p^2}\right)\left(\frac{1}{2} - \frac{x}{p}\frac{dp}{dx}\right) = 0$$

which gives $4p^3 + 27 = 0$

and

$$\frac{1}{2} - \frac{x}{p}\frac{dp}{dx} = 0$$

$$\Rightarrow \qquad \frac{dx}{x} - 2\frac{dp}{p} = 0$$

Integrating, we have

$$\log x - 2\log p = \log c$$

$$\Rightarrow \qquad p^2 = cx$$

Substituting value of p^2 in Eq. (1), we get

$$[8(cx)^{3/2} - 27]x = 12cxy$$

$$\Rightarrow \qquad [8(cx)^{3/2} - 27 - 12cy]x = 0$$

which gives $x = 0$

and

$$8(cx)^{3/2} - 27 - 12cy = 0$$

$$\Rightarrow \qquad 8(cx)^{3/2} = 27 + 12cy$$

$$\Rightarrow \qquad 64c^3x^3 = (12cy + 27)^2$$

$$\Rightarrow \qquad x^3 = \frac{144c^2}{64c^3}\left(y + \frac{27}{12c}\right)^2$$

$$\Rightarrow \qquad x^3 = \frac{9}{4c}\left(y + \frac{9}{4c}\right)^2$$

Replacing $\frac{9}{4c}$ by c', we get

$$x^3 = c'(y + c')^2.$$

Hence, the general solution of Eq. (1) is given by

$$x[x^3 - c'(y + c')^2] = 0$$

For finding singular solution, partially differentiate Eq. (2) with respect to p

$$0 = \frac{2}{3}x - \frac{9}{4}x \times \frac{-2}{p^3}$$

$$\Rightarrow \qquad 0 = \frac{2}{3}x + \frac{9}{2}\frac{x}{p^3}$$

$$\Rightarrow \qquad 0 = 4p^3x + 27x$$

$$\Rightarrow \qquad 0 = (4p^3 + 27)x$$

which gives

$$x = 0 \text{ and } 4p^3 + 27 = 0$$

Clearly, $x = 0$ does not satisfy Eq. (1), so

$$4p^3 + 27 = 0 \text{ gives } p = \frac{-3}{4^{1/3}}$$

Substituting the value of p in Eq. (2), we have

$$y = \frac{2}{3} \times \frac{-3}{4^{1/3}}x - \frac{9}{4}\frac{x}{\frac{9}{4^{2/3}}}$$

$$\Rightarrow \qquad y = \frac{-2x}{4^{1/3}} - \frac{x}{4^{1/3}}$$

$$\Rightarrow \qquad 4^{1/3}y = -3x$$

$$\Rightarrow \qquad 4y^3 = -27x^3 \quad \Rightarrow \quad 4y^3 + 27x^3 = 0$$

which is the required singular solution.

Note: While finding general solution, we have rejected $4p^3 + 27 = 0$

In fact $4p^3 + 27 = 0$ on integration gives us the required singular solution as

$$4p^3 + 27 = 0 \Rightarrow p = \frac{-3}{4^{1/3}}$$

$$\Rightarrow \quad dy = \frac{-3}{4^{1/3}}dx$$

Integrating, we get

$$y = \frac{3}{4^{1/3}}x \Rightarrow 4y^3 + 27x^3 = 0$$

We have not used any arbitrary constant of integration as singular solution does not contain any arbitrary constant.

EXAMPLE 6.5

Find the general and singular solutions of the differential equation $xp^2 - 2yp + 4x = 0$.

Solution: Solving the given differential equation for p, we get

$$p = \frac{2y \pm \sqrt{4y^2 - 16x^2}}{2x} = \frac{y \pm \sqrt{y^2 - 4x^2}}{x}$$

$$\Rightarrow \quad \frac{dy}{dx} = \frac{y \pm \sqrt{y^2 - 4x^2}}{x}$$

By putting $y = vx$, $\frac{dy}{dx} = v + x\frac{dv}{dx}$, we get

$$v + x\frac{dv}{dx} = \frac{vx \pm \sqrt{v^2x^2 - 4x^2}}{x}$$

$$\Rightarrow \quad x\frac{dv}{dx} = \pm\sqrt{v^2 - 4} \Rightarrow \frac{dx}{x} = \pm\frac{dv}{\sqrt{v^2 - 4}}$$

On integration, we get

$$\log x = \log c \pm \log (v + \sqrt{v^2 - 4})$$

$$\Rightarrow \quad \log x = \log [c(v + \sqrt{v^2 - 4})], \log x = \log\left[\frac{c}{v + \sqrt{v^2 - 4}}\right]$$

$$\Rightarrow \quad x = c(v + \sqrt{v^2 - 4}), x = \frac{c}{v + \sqrt{v^2 - 4}}$$

$$\Rightarrow \quad x^2 = c(y + \sqrt{y^2 - 4x^2}),\ y + \sqrt{y^2 - 4x^2} = c$$

$$\Rightarrow \quad (x^2 - cy)^2 = (y^2 - 4x^2)c^2,\ (y - c)^2 = y^2 - 4x^2$$

$$\Rightarrow \quad x^4 - 2c^2x^2y + 4c^2x^2 = 0,\ 4x^2 - 2cy + c^2 = c$$

$$\Rightarrow \quad x^2(x^2 - 2cy + 4c^2) = 0,\ 4x^2 - 2cy + c^2 = 0$$

Clearly, $x = 0$ does not satisfy the given differential equation, so,

$$c_1^2x^2 - 2c_1y + 4 = 0 \text{ where } 1/c = c_1$$

and

$$c_1^2x^2 - 2c_1y + 4 = 0 \text{ where } c/4 = c_1$$

Hence, the required general solution is

$$c_1^2x^2 - 2c_1y + 4 = 0.$$

For finding singular solution we have c-discriminant relation as

$$4y^2 - 16x^2 = 0 \quad \Rightarrow \quad y = \pm 2x$$

and by using the given differential equation p-discriminant relation as

$$4y^2 - 16x^2 = 0 \Rightarrow y = \pm 2x$$

which clearly satisfies the given differential equation.

Hence, the required singular solution is $y = \pm 2x$.

EXAMPLE 6.6

Solve and examine for singular solution, the equation

$$xp^2 - (x - a)^2 = 0 \tag{1}$$

Solution: Solving the given differential equation for p, we have

$$p = \frac{dy}{dx} = \pm \frac{x - a}{\sqrt{x}} = \pm (x^{1/2} - ax^{-1/2})$$

$$\Rightarrow \quad dy = \pm(x^{1/2} - ax^{-1/2})\,dx$$

Integrating, we get

$$y + c = \pm\left(\frac{2}{3}x^{3/2} - 2ax^{1/2}\right)$$

$$\Rightarrow \quad y + c = \pm\frac{2}{3}x^{1/2}(x - 3a)$$

$$\Rightarrow \quad 9(y + c)^2 = 4x(x - 3a)^2 \tag{2}$$

which is the required general solution of Eq. (1).

For finding singular solution, differentiate Eq. (2) partially with respect to c

$$18(y + c) = 0 \Rightarrow y + c = 0 \tag{3}$$

By eliminating c between Eqs. (2) and (3), we get the c-discriminant relation as

$$4x(x-3a)^2 = 0 \Rightarrow x(x-3a)^2 = 0 \quad (4)$$

Now, the given differential Eq. (1) can be written as

$$xp^2 + 0 - p - (x-a)^2 = 0$$

which is a quadratic in p. Therefore, the p-discriminant relation is

$$0 + 4x(x-a)^2 = 0 \quad \Rightarrow \quad x(x-a)^2 = 0 \quad (5)$$

Now, the common factor in Eqs. (4) and (5) is $x = 0$

Clearly, $x = 0$ satisfies the given differential Eq. (1).

Hence, the required singular solution of Eq. (1) is $x = 0$.

EXAMPLE 6.7

Find the general and singular solutions of

$$y^2(1+p^2) = r^2 \quad (1)$$

Solution: Solving the given differential equation for p, we have

$$p = \frac{dy}{dx} = \pm\frac{\sqrt{r^2-y^2}}{y}$$

$$\Rightarrow \quad dx = \pm\frac{y}{\sqrt{r^2-y^2}}dy$$

Integrating, we get

$$x + c = \pm(r^2-y^2)^{1/2}$$

$$\Rightarrow \quad (x+c)^2 = r^2 - y^2$$

$$\Rightarrow \quad y^2 + (x+c)^2 - r^2 = 0 \quad (2)$$

which is the general solution of the given differential equation.

For finding singular solution of Eq. (1), differentiate Eq. (2) partially with respect to c, we get

$$0 + 2(x+c) = 0 \quad \Rightarrow \quad x + c = 0 \quad (3)$$

By eliminating c between Eqs. (2) and (3), we get the c-discriminant relation as

$$y^2 = r^2$$

which clearly satisfies the given differential Eq. (1).

Hence, $y = \pm r$ is the required singular solution of Eq. (1).

EXAMPLE 6.8

Find the general and singular solutions of $3xy = 2px^2 - 2p^2$.

Solution: Solving the given differential equation for y, we have

$$3y = 2px - 2p^2x^{-1} \quad (1)$$

Differentiating Eq. (1) with respect to x, we get

$$3p = 3\frac{dy}{dx} = 2p + 2x\frac{dp}{dx} + 2x^{-2}p^2 - 4px^{-1}\frac{dp}{dx}$$

$$\Rightarrow \qquad p - 2x\frac{dp}{dx} - \frac{2p^2}{x^2} + \frac{4p}{x}\frac{dp}{dx} = 0$$

$$\Rightarrow \qquad p\left(1 - \frac{2p}{x^2}\right) - 2x\frac{dp}{dx}\left(1 - \frac{2p}{x^2}\right) = 0$$

$$\Rightarrow \qquad \left(1 - \frac{2p}{x^2}\right)\left(p - 2x\frac{dp}{dx}\right) = 0$$

which gives $1 - \dfrac{2p}{x^2} = 0$

and $\qquad p - 2x\dfrac{dp}{dx} = 0 \Rightarrow \dfrac{dx}{x} - \dfrac{2dp}{p} = 0$

Integrating, we obtain

$$\log x - 2\log p = -\log c$$

$$\Rightarrow \qquad p^2 = cx \qquad (2)$$

By eliminating p from Eqs. (1) and (2), we have

$$3xy = 2px^2 - 2cx$$

$$\Rightarrow \qquad x(3y + 2c) = 2px^2$$

$$\Rightarrow \qquad 3y + 2c = 2px \Rightarrow (3y + 2c)^2 = 4p^2x^2$$

$$\Rightarrow \qquad (3y + 2c)^2 = 4cx^3 \qquad (3)$$

which is the general solution of the given differential Eq. (1).

For finding singular solution, write Eq. (3) as

$$4c^2 + 4c(3y - x^3) + 9y^2 = 0$$

which is a quadratic in c. So, the c-discriminant relation is

$$16(3y - x^3)^2 - 144y^2 = 0$$

$$\Rightarrow \qquad (3y - x^3)^2 - 9y^2 = 0$$

$$\Rightarrow \qquad x^6 - 6x^3y = 0 \Rightarrow x^3(x^3 - 6y) = 0$$

Clearly, $x = 0$ does not satisfy Eq. (1), but

$$y = x^3/6 \text{ satisfies Eq. (1)}$$

Hence, $x^3 - 6y = 0$ is the required singular solution of Eq. (1).

Note: The factor $1-\frac{2p}{x^2}=0$, which we have rejected while finding general solution, on integration without any arbitrary constant of integration gives us the same singular solution as

$$2dy = x^2dx$$

$$\Rightarrow \qquad 2y = \frac{x^3}{3} \Rightarrow x^3 - 6y = 0.$$ ▲▲

EXERCISE 6.1

1. Find the general and singular solutions of
$$y^2 - 2pxy + p^2(x^2 - 1) = m^2.$$
2. Investigate for singular solutions
$$4x(x-1)(x-2)p^2 - (3x^2 - 6x + 2)^2 = 0.$$
3. Find the general and singular solutions of
$$p^2 + y^2 = 1.$$
4. Find the general and singular solutions of
$$9p^2(1-y)^2 = 4(2-y).$$
5. Show that the equation $x^2(y - px) = p^2y$ has no singular solution.
6. Solve and examine for singular solutions the differential equation
$$(y - px)^2 + a^2y = 0.$$
7. Find the general and singular solutions of
$$(xp - y)^2 = p^2 - 1.$$
8. Solve and examine for singular solutions of the differential equation
$$(px - y)(x - py) = 2p.$$
9. Find the general and singular solutions of
$$xp^2 - 2yp + x = 0.$$
10. Solve and examine for singular solutions the differential equation
$$y = 2px + y^2p^3.$$
11. Find the general and singular solutions of
$$ayp^2 + (2x - b)p - y = 0.$$

ANSWERS

Exercise 6.1

1. $(y - cx)^2 = m^2 + c^2$; $y^2 + m^2x^2 = m^2$.
2. $(y + c)^2 = x^3 - 3x^2 + 2x$; $x = 0$, $x = 1$, $x = 2$.
3. $y = \cos(x + c)$; $y = \pm 1$.
4. $(x + c)^2 = (y + 1)^2(2 - y)$; $y = 2$.
6. $(y - cx)^2 + a^2x = 0$; $4xy = a^2$.
7. $(cx - y)^2 = c^2 - 1$; $x^2 - y^2 = 1$.
8. $x^2 + y^2 - c(x^2 - y^2) - 1 + c^2 = 0$; $x^4 - 2x^2y^2 + y^4 - x(x^2 + y^2) + 4 = 0$.
9. $2cy = c^2x^2 + 1$; $y = \pm x$.
10. $y^2 = 2cx + c^3$; $4x^3 + 27y^2 = 0$.
11. $ac^2 + (2x - b)c - y^2 = 0$; $a^2y^2 + x^2 = 0$.

▲▲

Clairaut's Equation

The differential equation

$$y = px + f(p) \tag{1}$$

is known as Clairaut's equation. Here $f(p)$ is a function of p only. To solve Eq. (1), we differentiate Eq. (1) with respect to x and obtain

$$\frac{dy}{dx} = p = p + x\frac{dp}{dx} + f'(p)\frac{dp}{dx}$$

$$\Rightarrow \qquad (x + f'(p))\frac{dp}{dx} = 0$$

$$\Rightarrow \qquad \frac{dp}{dx} = 0 \quad \text{and} \quad x + f'(p) = 0 \tag{2}$$

Now $\frac{dp}{dx} = 0$ on integrating gives

$$p = \text{constant} = c, \text{ say} \tag{3}$$

By eliminating p between Eqs. (1) and (3), we obtain

$$y = cx + f(c) \tag{4}$$

which is the required general solution of Eq. (1).

For finding singular solution, differentiate Eq. (4) partially with respect to c

$$0 = x + f'(c) \tag{5}$$

On eliminating c between Eqs. (4) and (5), we get envelope of Eq. (4) which is the required singular solution.

Now, if we differentiate Eq. (1) partially with respect to p, we obtain

$$0 = x + f'(p) \tag{6}$$

which is same as Eq. (2).

Equations (4) and (5) differ from the Eq. (1) and (6) only in having p instead of c. Therefore, the c-discriminant relation which is the c-eliminate between Eq. (4) and (5) and the p-discriminant relation which is the p-eliminate between Eq. (1) and (6) are identical and either of them gives us the required singular solution. Hence, the singular solution of the Clairaut's Eq. (1) is obtained by eliminating p between Eq. (1) and (2).

Note: To obtain the general solution of a differential equation in Clairaut's form simply replace p by c.

▲▲

EXAMPLE 6.9

Find general solution of $y = px + \log p$.

Solution: The given equation is in Clairaut's form so the required general solution will be

$$y = cx + \log c$$

obtained on changing p to the arbitrary constant c.

EXAMPLE 6.10

Obtain the general and singular solutions of the nonlinear equation $y = xy' + (y')^2$.

Solution: The given equation is in Clairaut's form

$$y = px + p^2$$

so its general solution is

$$y = cx + c^2 \tag{1}$$

For finding singular solution, differentiate Eq. (1) partially with respect to c

$$0 = x + 2c \quad \Rightarrow \quad c = -x/2 \tag{2}$$

Eliminate c between Eqs. (1) and (2)

$$y = \frac{-x^2}{2} + \frac{x^2}{4} = \frac{-x^2}{4}$$

$\therefore$ $x^2 + 4y = 0$ is the required singular solution of the given differential equation.

EXAMPLE 6.11

Solve

$$x^2(y - px) = p^2y.$$

Solution: Let $x^2 = u$, $y^2 = v$

$\Rightarrow$ $$2xdx = du,\ 2ydy = dv$$

$$\Rightarrow \qquad \frac{y}{x}\frac{dy}{dx} = \frac{dv}{du} \Rightarrow \frac{y}{x}p = P \quad \Rightarrow \quad p = \frac{x}{y}P$$

Substituting this value of p in the given differential equation, we obtain

$$x^2\left(y - \frac{x^2}{y}P\right) = \frac{x^2}{y}P^2$$

$$\Rightarrow \qquad y^2 - x^2P = P^2$$

$$\Rightarrow \qquad v - uP = P^2$$

$$\therefore \qquad v = uP + P^2$$

which is in Clairaut's form. Therefore, the general solution is

$$v = uc + c^2 \tag{1}$$

$$\Rightarrow \qquad y^2 = cx^2 + c^2$$

For finding singular solution, differentiate Eq. (1) partially with respect to c

$$0 = u + 2c \quad \Rightarrow \quad c = \frac{-u}{2} \tag{2}$$

Eliminating c between Eqs. (1) and (2), we obtain

$$v = \frac{-u^2}{2} + \frac{u^2}{4} = \frac{-u^2}{4}$$

$$\Rightarrow \qquad u^2 + 4v = 0 \quad \Rightarrow \quad x^4 + 4y^2 = 0$$

which is the required singular solution.

EXAMPLE 6.12

Find the general and singular solutions of

$$y^2 - 2pxy + p^2(x^2 - 1) = m^2.$$

Solution: The given differential equation can be written as

$$(y - px)^2 = p^2 + m^2$$

$$\Rightarrow \qquad y - px = \pm\sqrt{p^2 + m^2}$$

$$\Rightarrow \qquad y = px \pm \sqrt{p^2 + m^2}$$

which is in Clairaut's form. Hence, the general solution is

$$y = cx \pm \sqrt{c^2 + m^2}$$

$$\Rightarrow \quad (y - cx)^2 = c^2 + m^2$$

$$\Rightarrow \quad c^2(x^2 - 1) - 2cxy + y^2 - m^2 = 0 \tag{1}$$

For finding singular solution, calculate c-discriminant relation using Eq. (1)

$$4x^2y^2 - 4(x^2 - 1)(y^2 - m^2) = 0$$

$$\Rightarrow \quad x^2m^2 + y^2 - m^2 = 0$$

$$\Rightarrow \quad y^2 + m^2x^2 = m^2$$

which is the required singular solution.

EXAMPLE 6.13

Find the general and singular solutions of

$$y = px + a/p$$

Solution: The given differential equation is in Clairaut's form and so the general solution is

$$y = cx + a/c \tag{1}$$

where, c is an arbitrary constant.

For finding singular solution, differentiate Eq. (1) partially with respect to c

$$0 = x - \frac{a}{c^2} \quad \Rightarrow \quad c^2x = a \tag{2}$$

By eliminating c between Eq. (1) and (2), we obtain

$$cy = c^2x + a = 2a$$

$$\Rightarrow \quad y^2c^2 = 4a^2$$

$$\Rightarrow \quad y^2\left(\frac{a}{x}\right) = 4a^2$$

$$\Rightarrow \quad y^2 = 4ax$$

which is the required singular solution.

EXAMPLE 6.14

Solve $(x^2 + y^2)(1 + p)^2 - 2(x + y)(1 + p)(x + yp) + (x + yp)^2 = 0$.

Solution: The given equation can be written as

$$x^2 + y^2 - 2(x+y)\frac{(x+yp)}{(1+p)} + \left(\frac{x+yp}{1+p}\right)^2 = 0$$

Put $x^2 + y^2 = v \quad \therefore 2x + 2yp = \dfrac{dv}{dx}$

and
$$x + y = u \quad \therefore 1 + p = \frac{du}{dx}$$

$$\therefore \qquad \frac{x+yp}{1+p} = \frac{1}{2}\frac{dv}{du} = \frac{1}{2}P$$

Making the above substitutions, we get

$$v - 2u\frac{P}{2} + \frac{P^2}{4} = 0$$

$$\Rightarrow \qquad v = up - P^{2/4}$$

which is in Clairaut's form and hence its general solution is

$$v = uc - \frac{1}{4}c^2 \tag{1}$$

$$\Rightarrow \qquad x^2 + y^2 = c(x+y) - \frac{1}{4}c^2.$$

For finding, singular solution differentiate Eq. (1) partially with respect to c

$$0 = u - \frac{c}{2} \Rightarrow c = 2u \tag{2}$$

Eliminate c between Eqs. (1) and (2)

$$v = u(2u) - \frac{1}{4}(4u^2) = u^2$$

$$\therefore \qquad x^2 + y^2 = (x+y)^2$$

$$\Rightarrow \qquad xy = 0$$

which is the required singular solution.

EXAMPLE 6.15

Solve the equation $x^2p^2 + yp(2x+y) + y^2 = 0$ by using the substitution $y = u$, $xy = v$ and find its singular solution.

Solution: We have

$$y = u \qquad x = v/u$$

$$\Rightarrow \qquad \frac{dy}{du} = 1 \qquad \frac{dx}{du} = \frac{u\dfrac{dv}{du} - v}{u^2}$$

These give

$$p = \frac{dy}{dx} = \frac{u^2}{u\dfrac{dv}{du} - v} = \frac{u^2}{uP - v}$$

Making substitutions in the given differential equation, we shall obtain

$$\frac{v^2}{u^2}\frac{u^4}{(uP-v)^2}+u\frac{u^2}{uP-v}\left(2\frac{v}{u}+u\right)+u^2=0$$

$$\Rightarrow \qquad v^2+u\left(\frac{2v}{u}+u\right)(uP-v)+(uP-v)^2=0$$

$$\Rightarrow \qquad v^2+(u^2+2v)(uP-v)+(uP-v)^2=0$$

$$\Rightarrow \qquad v^2+u^3P-u^2v+2uvP-2v^2+u^2P^2+v^2-2uvP=0$$

$$\Rightarrow \qquad u^3P-u^2v+u^2P^2=0$$

$$\Rightarrow \qquad uP-v+P^2=0$$

$$\therefore \qquad v=uP+P^2$$

This is Clairaut's equation. Its general solution is $v = uc + c^2$

$$\Rightarrow \qquad xy = cy + c^2 \tag{1}$$

To obtain the singular solution, we differentiate Eq. (1) partially with respect to c and obtain

$$0 = y + 2c \tag{2}$$

Eliminating c between Eqs. (1) and (2), we obtain

$$xy = \left(\frac{-y}{2}\right)y+\frac{y^2}{4}$$

$$\Rightarrow \qquad xy = \frac{-y}{2}$$

$$\Rightarrow \qquad y(y+4x)=0$$

Thus, $y = 0$, $y + 4x = 0$ are both singular solutions.

EXAMPLE 6.16

Find the general and singular solutions of

$$y = x\left(\frac{dy}{dx}\right)+a\sqrt{b^2+\left(\frac{dy}{dx}\right)^2}.$$

Solution: The given differential equation is

$$y = xp + a\sqrt{b^2+p^2}$$

which is in Clairaut's form. Therefore, its general solution is

$$y = cx + a\sqrt{b^2+c^2} \tag{1}$$

$$\Rightarrow \quad (y - cx)^2 = a^2(b^2 + c^2)$$

$$\Rightarrow \quad c^2(x^2 - a^2) - 2xyc + y^2 - b^2 = 0$$

Hence, the c-discriminant relation is

$$4x^2y^2 - 4(x^2 - a^2)(y^2 - b^2) = 0$$

$$\Rightarrow \quad b^2x^2 + a^2y^2 - a^2b^2 = 0$$

$$\Rightarrow \quad \frac{x^2}{a^2} + \frac{y^2}{b^2} = 1$$

which is the required singular solution.

EXAMPLE 6.17

Solve the differential equation

$$(px^2 + y^2)(px + y) = (p + 1)^2$$

by reducing it to Clairaut's form and find its singular solution.

Solution: Let us put $u = x + y$ and $v = xy$

so that $$\frac{du}{dx} = 1 + \frac{dy}{dx} = 1 + p$$

and $$\frac{dv}{dx} = y + x\frac{dy}{dx} = y + xp$$

$$\therefore \quad P = \frac{dv}{du} = \frac{y + xp}{1 + p}$$

$$\Rightarrow \quad P(1 + p) = y + xp \quad \Rightarrow \quad p(P - x) = y - P$$

$$\Rightarrow \quad p = \frac{y - P}{P - x}$$

Substituting this value of p in the given differential equation, we get

$$\left(x^2\left(\frac{y-P}{P-x}\right) + y^2\right)\left(x\left(\frac{y-P}{P-x}\right) + y\right) = \left(\frac{y-P}{P-x} + 1\right)^2$$

$$\Rightarrow \quad (x^2y - x^2P + y^2P - y^2x)(xy - xP + yP - xy) = (y - P + P - x)^2$$

$$\Rightarrow \quad [P(y^2 - x^2) - yx(y - x)]\,P(y - x) = (y - x)^2$$

$$\Rightarrow \quad [P(y + x) - yx]P = 1$$

$$\Rightarrow \quad [Pu - v]P = 1$$

$$\Rightarrow \quad Pu - v = 1/P$$

$$\Rightarrow \quad v = Pu - 1/P$$

which is of Clairaut's form. Hence, its general solution is $v = cu - 1/c$

$$\Rightarrow \quad c^2u - cv - 1 = c$$

$$\Rightarrow \quad c^2(x + y) - cxy - 1 = 0 \qquad (1)$$

The c-discriminant relation by using Eq. (1) is

$$x^2y^2 + 4(x + y) = 0$$

which is the required singular solution.

EXAMPLE 6.18

Reduce the differential equation $(px - y)(x - py) = 2p$ to Clairaut's form by substituting $x^2 = u$ and $y^2 = v$ and find its complete primitive and the singular solution.

Solution: We put $\quad u = x^2 \quad \text{and} \quad v = y^2$

so that $\quad du = 2xdx \quad \text{and} \quad dv = 2ydy$

$$\therefore \qquad P = \frac{dv}{du} = \frac{y}{x}\frac{dy}{dx} = \frac{y}{x}p$$

$$\Rightarrow \qquad p = xP/y$$

Substituting the value of p in the given differential equation, we get

$$\left(\frac{x^2P}{y} - y\right)(x - xP) = 2\frac{xP}{y}$$

$$\Rightarrow \qquad x(x^2P - y^2)(1 - P) = 2xP$$

$$\Rightarrow \qquad (x^2P - y^2)(1 - P) = 2P$$

$$\Rightarrow \qquad (uP - v)(1 - P) = 2P$$

$$\Rightarrow \qquad uP \quad v = \frac{2P}{1-p}$$

$$\Rightarrow \qquad v = uP - \frac{2P}{1-p}$$

which is in Clairaut's form. Hence, its general solution is

$$v = uc - \frac{2c}{1-c}$$

$$\Rightarrow \qquad y^2 = c^2x^2 - \frac{2c}{1-c}$$

$$\Rightarrow \qquad y^2(1 - c) = c^2x^2(1 - c) - 2c$$

$$\Rightarrow \qquad c^2x^2 - c(x^2 + y^2 - 2) + y^2 = 0 \qquad (1)$$

Now by using Eq. (1), the c-discriminant relation is

$$(x^2 + y^2 - 2)^2 - 4x^2y^2 = 0$$

$$\Rightarrow \qquad (x^2 + y^2 - 2 + 2xy)(x^2 + y^2 - 2 - 2xy) = 0$$

$$\Rightarrow \qquad [(x + y)^2 - 2][(x - y)^2 - 2] = 0$$

Hence, the required singular solutions are

$$x + y + \sqrt{2} = 0, x + y - \sqrt{2} = 0, x - y + \sqrt{2} = 0$$

and $$x - y - \sqrt{2} = 0.$$

EXAMPLE 6.19

Reduce the differential equation $y = 2px + y^2p^3$ to Clairaut's form by putting $y^2 = v$ and hence find its general and singular solutions.

Solution: We put $$y^2 = v$$
so that

$$2y\frac{dy}{dx} = \frac{dv}{dx} \Rightarrow 2yp = P$$

$\therefore$ $$p = P/2y$$

Substituting this value of p in the given differential equation, we get

$$y = 2x\left(\frac{P}{2y}\right) + y^2\left(\frac{P^3}{8y^3}\right)$$

$\Rightarrow$ $$y^2 = xP + \frac{P^3}{8}$$

$\Rightarrow$ $$v = xP + \frac{P^3}{8}$$

which is in Clairaut's form. Hence, its general solution is

$$v = xc + \frac{c^3}{8} \Rightarrow y^2 = cx + \frac{c^3}{8} \qquad (1)$$

For finding singular solution, differentiate Eq. (1) partially with respect to c, we obtain

$$0 = x + \frac{3c^2}{8} \Rightarrow c^2 = \frac{-8x}{3} \qquad (2)$$

Eliminate c between Eqs. (1) and (2), we get

$$y^2 = c\left(x + \frac{1}{8}c^2\right)$$

$\Rightarrow$ $$y^4 = c^2\left(x + \frac{-1}{3}x\right)^2 = c^2\frac{4}{9}x^2$$

$\Rightarrow$ $$y^4 = \frac{4}{9} \times \frac{-8}{3}x^3 = \frac{-32}{27}x^3$$

$\Rightarrow$ $$27y^4 + 32x^3 = 0$$

which is the required singular solution.

EXAMPLE 6.20

Solve and test for singular solutions

$$p^3 - 4pxy + 8y^2 = 0.$$

Solution: We put $y = v^2$; so that

$$p = \frac{dy}{dx} = 2v\frac{dv}{dx} = 2vP$$

Substituting the value of p in the given differential equation, we get

$$8v^3P^3 - 8vPxy + 8y^2 = 0$$

$$\Rightarrow \quad 8v^3P^3 - 8xv^3P + 8v^4 = 0$$

$$\Rightarrow \quad P^3 - xP + v = 0$$

$$\Rightarrow \quad v = xP - P^3$$

which is in Clairaut's form. Hence, its general solution is

$$v = xc - c^3 \quad \Rightarrow \quad \sqrt{y} - xc - c^3$$

$$\Rightarrow \quad y = c^2(x - c^2)^2$$

$$\Rightarrow \quad y = c_1(x - c_1)^2 \tag{1}$$

where, $c_1 = c^2$

which is the required general solution.

For finding singular solution, differentiate Eq. (1) partially with respect to c_1, we obtain

$$0 = (x - c_1)^2 - 2c_1(x - c_1)$$

$$\Rightarrow \quad (x - c_1)(x - 3c_1) = 0 \tag{2}$$

Now, we shall eliminate c_1 between Eqs. (1) and (2) to get the c-discriminant relation.

when $\quad x - c_1 = 0 \Rightarrow y = 0$

and when $x - 3c_1 = 0 \Rightarrow c_1 = x/3$

Equation (1) gives $\quad y = \frac{1}{3}x\left(x - \frac{1}{3}x\right)^2 = \frac{4}{27}x^3$

$$\Rightarrow \quad 27y - 4x^3 = 0$$

Hence, $y = 0$ and $y = \frac{4}{27}x^3$ are the required singular solutions.

EXAMPLE 6.21

Reduce the equation

$$xyp^2 - (x^2 + y^2 - 1)p + xy = 0$$

to Clairaut's form by substituting $x^2 = u$ and $y^2 = v$. Hence, show that the equation represents a family of conics touching the four sides of a square.

Solution: Now $\qquad x^2 = u$ and $y^2 = v$

$\therefore \qquad 2xdx = du$ and $2ydy = dv$

$$\therefore \qquad P = \frac{dv}{du} = \frac{y}{x}\frac{dy}{dx} = \frac{y}{x}p$$

$$\Rightarrow \qquad p = \frac{xP}{y}$$

Substituting the value of p in the given differentiable equation, we get

$$xy\frac{x^2P^2}{y^2} - (x^2 + y^2 - 1)\frac{xP}{y} + xy = 0$$

$$\Rightarrow \qquad \frac{x}{y}x^2P^2 - \frac{x}{y}(x^2 + y^2 - 1)P + \frac{x}{y}y^2 = 0$$

$$\Rightarrow \qquad x^2P^2 - (x^2 + y^2 - 1)P + y^2 = 0$$

$$\Rightarrow \qquad uP^2 - (u + v - 1)P + v = 0$$

$$\Rightarrow \qquad u(P^2 - P) - v(P - 1) + P = 0$$

$$\Rightarrow \qquad uP(P - 1) - v(P - 1) + P = 0$$

$$\Rightarrow \qquad uP - v + \frac{P}{P-1} = 0$$

$$\Rightarrow \qquad v = uP + \frac{P}{P-1}$$

which is in Clairaut's form. Hence, its general solution is

$$v = uc + \frac{c}{c-1}$$

$$\Rightarrow \qquad y^2 = cx^2 + \frac{c}{c-1}$$

$$\Rightarrow \qquad (c - 1)y^2 = c(c - 1)x^2 + c$$

$$\Rightarrow \qquad c^2x^2 - (x^2 + y^2 - 1)c + y^2 = 0 \tag{1}$$

For finding singular solution, c-discriminant relation by using Eq. (1) is

$$(x^2 + y^2 - 1)^2 - 4x^2y^2 = 0$$

$$\Rightarrow \qquad (x^2 + y^2 - 1)^2 - (2xy)^2 = 0$$

$$\Rightarrow \qquad (x^2 + y^2 - 1 + 2xy)(x^2 + y^2 - 1 - 2xy) = 0$$

$$\Rightarrow \qquad [(x + y)^2 - 1][(x - y)^2 - 1] = 0$$

$$\Rightarrow \qquad (x + y + 1)(x + y - 1)(x - y + 1)(x - y - 1) = 0$$

Hence, the required singular solutions are

$$x + y + 1 = 0,\ x + y - 1 = 0,\ x - y + 1 = 0 \text{ and } x - y - 1 = 0$$

These equations are the equations of four sides of a square.

Hence, the given differential equation represents conic given by $c^2x^2 - (x^2 + y^2 - 1)c + y^2 = 0$ touching the four sides of a square.

EXAMPLE 6.22

Transform the equation

$$(2x^2 + 1)\left(\frac{dy}{dx}\right)^2 + (x^2 + 2xy + y^2 + 2)\frac{dy}{dx} + 2y^2 + 1 = 0$$

to Claircut's form by the substitutions $x + y = u$ and $xy = v$ and interpret it. Find its singular solution also.

Solution: The given differential equation can be written as

$$(2x^2 + 1)p^2 + (x^2 + 2xy + y^2 + 2)p + 2y^2 + 1 = 0$$

We put $x + y = u$ and $xy - 1 = v$

So that $1 + \dfrac{dy}{dx} = \dfrac{du}{dx}$ and $y + x\dfrac{dy}{dx} = \dfrac{dv}{dx}$

$$\therefore \qquad P = \frac{dv}{du} = \frac{y + xp}{1 + p}$$

The given differential equation gives

$$p^2x^2 + (p^2x^2 + 2pxy + y^2) + (x^2 + y^2)p + y^2 + (p^2 + 2p + 1) = 0$$

$$\Rightarrow \quad p^2x^2 + (px + y)^2 + (x^2 + y^2)p + y^2 + (p + 1)^2 = 0$$

$$\Rightarrow \quad x^2p(p + 1) + y^2(p + 1) + (px + y)^2 + (p + 1)^2 = 0$$

$$\Rightarrow \quad (p + 1)(x^2p + y^2 + p + 1) + (px + y)^2 = 0$$

$$\Rightarrow \quad (p + 1)(x(px + y) + y(px + y) - xy(p + 1) + p + 1) + (px + y)^2 = 0$$

$$\Rightarrow \quad (p + 1)((x + y)(px + y) - (xy - 1)(p + 1)) + (px + y)^2 = 0$$

$$\Rightarrow \quad (p + 1)(x + y)(px + y) - (xy - 1)(p + 1)^2 + (px + y)^2 = 0$$

$$\Rightarrow \quad (x + y)\left(\frac{px + y}{p + 1}\right) - (xy - 1) + \left(\frac{px + y}{p + 1}\right)^2 = 0$$

$$\Rightarrow \quad P(x + y) - (xy - 1) + P^2 = 0$$

$$\Rightarrow \quad Pu - v + P^2 = 0$$

$$\Rightarrow \quad v = Pu - P^2$$

which is in Clairaut's form.

Hence, its solution is

$$v = cu + c^2 \tag{1}$$

$$\Rightarrow \qquad xy - 1 = c(x + y) + c^2$$

c-discriminant relation using Eq. (1) is

$$u^2 + 4v = 0$$

$$\Rightarrow \qquad (x + y)^2 + 4(xy - 1) = 0$$

which is the required singular solution.

EXAMPLE 6.23

Solve $p = \sin(y - xp)$. Also, find its singular solution.

Solution: The given differential equation can be written as

$$y = px + \sin^{-1}p$$

which is in Clairaut's form. Hence, its general solution is

$$y = cx + \sin^{-1}c \tag{1}$$

For finding singular solution, differentiate Eq. (1) partially with respect to c to obtain

$$0 = x + \frac{1}{\sqrt{1-c^2}} \quad \Rightarrow \quad x^2 = \frac{1}{1-c^2}$$

$$\Rightarrow \qquad x^2(1 - c^2) = 1 \quad \Rightarrow \quad c^2 = \frac{x^2-1}{x^2} \tag{2}$$

Eliminate c between Eqs. (1) and (2) to get c-discriminant as

$$y = \sqrt{x^2-1} + \sin^{-1}\left(\frac{\sqrt{x^2-1}}{x}\right)$$

which is the required singular solution. ▲▲

EXERCISE 6.2

Solve the following differential equations for general solution:

1. $xp^2 - yp + a = 0$.
2. $p = \log(px - y)$.
3. $\sin px \cos y - \cos px \sin y = p$.
4. $\cos y \cos px + \sin y \sin px = p$.
5. $(x - a)p^2 + (x - y)p - y = 0$.
6. $p^2(x^2 - a^2) - 2pxy + y^2 + a^2 = 0$.

7. $y = x\dfrac{dy}{dx} + e^{dy/dx}$.
8. $y + xp\log p = (2 + 3\log p)p^3$.
9. $y + 2\left(\dfrac{dy}{dx}\right)^2 = (x+1)\dfrac{dy}{dx}$.
10. $(px + y)^2 = py^2$.

Solve the following differential equations for general and singular solutions

11. $y = xy' + (y')^3$.
12. $y = xy' - (1/y')$.
13. $y = xy' - (y')^2/2$.
14. $y = xy' - e^{2/y'}$.
15. $y = xy' + e^{-y'}$.
16. Find the general and singular solutions of $y = x(dy/dx) + a\sqrt{1+(dy/dx)^2}$.
17. Prove that, for the equation

$$y = px - (1 + p^2)^{1/2}$$

the envelope (singular solution) has the equation

$$9y^2(1 - y^2) = 4(2 - y).$$

18. Solve $y = 2x\dfrac{dy}{dx} + y^2\left(\dfrac{dy}{dx}\right)^3$ and find its singular solution also.
19. Show that the singular solution of

$$px + y = p^2x^4 \text{ is } 4x^2y + 1 = 0.$$

20. Find the general and singular solutions of

$$3p^2e^4 - px + 1 = 0.$$

ANSWERS

Exercise 6.2

1. $xc^2 - yc + a = 0$.
2. $c = \log(cx - y)$.
3. $y = cx - \sin^{-1}c$.
4. $y = cx - \cos^{-1}c$.
5. $(x - a)c^2 + (x + y)c - y = 0$.
6. $c^2(x^2 - a^2) - 2cxy + y^2 + a^2 = 0$.

7. $y = cx + e^c$.
8. $x = 3p^2 + cp^{-1}$; $y = 2p^3 - c \log p$.
9. $y = cx + (c - 2c^2)$.
10. $xy = cy - c^2$ {***Hint:*** Put $u = y$, $v = xy$}
11. $y = cx - c^3$; $y^2 = 4x^3/27$.
12. $y = cx - (1/c)$; $y^2 + 4x = 0$.
13. $y = cx - (c^2/2)$; $2y = x^2$.
14. $y = cx - e^{2/c}$; $y = \{\log(x/2) - 1\}(x/2)$; $(x > 0)$.
15. $y = cx + e^{-c}$; $y = (1 - \log x)x$ $(x > 0)$.
16. $y = cx + a\sqrt{1+c^2}$; $x^2 + y^2 = a^2$.
18. $y^2 = 2cx + \underline{c}^3$; $27y^4 + 32x^3 = 0$.
19. $3c^2e^{-y} - xce^{-y} + 1 = 0$; $x^2 - 12e^y = 0$.

7

CHAPTER

Orthogonal Trajectories

In this chapter, we consider problems that give rise to some of the types of first order ordinary differential equations. First, we will formulate the problem mathematically then we solve the resulting equation. We know that the general solution of a first order differential equation contains one arbitrary constant. For each value of the arbitrary constant, we obtain an integral curve. The totality of all these curves forms a one parameter family of curves. Thus, the general solution of a first order ordinary differential equation represents one parameter family of curves. We now, formulate the problem of finding another one parameter family of curves which is orthogonal to the given one parameter family of curves. The problem of finding the orthogonal family of curves arises in many physical situations such as in a two-dimensional electric field the lines of force (flux lines) and the equipotential curves are orthogonal to each other in fluid flow, the stream lines and the equipotential lines (lines of constant velocity potential) are orthogonal trajectories. Likewise, the lines of heat flow for a body are perpendicular to the isothermal curves.

Definition: A curve which cuts every member of a given family of curves according to a given law is called a *trajectory* of the given family. We shall consider only the case when each trajectory cuts every member of a given family at a constant angle.

Definition: A curve that intersects every member of a given family of curves at right angles is called an orthogonal trajectory of the given family.

Definition: A curve that intersects every member of a given family of curves at a constant angle other than right angle is called an *oblique trajectory* of the given family.

For example, the concentric circles $x^2 + y^2 = r^2$; with centre at the origin and radius r are orthogonal to the radial lines $y = mx$. In polar form, these lines are $\theta = \alpha$; where α is a constant. Therefore, the one parameter family of integral curves $x^2 + y^2 = r^2$ are the orthogonal trajectories of the one parameter family of straight lines $y = mx$. We have defined that two curves intersect orthogonally if the angle of intersection between the curves is a right angle; that is, the tangents to the curves at the point of intersection are perpendicular. Based on this fact we have the following method for finding orthogonal trajectories for a given family of curves. In terms of Cartesian coordinates; let the given family of curves be

$$f(x, y, c) = 0 \tag{1}$$

where c is a parameter.

Differentiating Eq. (1) with respect to x, we get

$$\frac{\partial f}{\partial x} + \frac{\partial f}{\partial y}\frac{dy}{dx} = 0 \tag{2}$$

Eliminating c between Eqs. (1) and (2), we will obtain the differential equation of the given family of curves (1). Let it be

$$\phi\left(x, y, \frac{dy}{dx}\right) = 0 \tag{3}$$

Let (X, Y) be the coordinates of any point on a trajectory so that dY/dX is the slope of the tangent at the point (X, Y). Let dy/dx is the slope of the tangent to a curve which is a member of the given family (1) for a given value of c at any point (x, y).

For orthogonal trajectory, at the point of intersection of the above two tangents, we must have

$$\frac{dy}{dx}\frac{dY}{dX} = -1 \quad \Leftrightarrow \quad \frac{dy}{dx} = \frac{-dX}{dY}$$

Thus, the differential equation of the family of orthogonal trajectories is

$$\phi\left(X, Y, \frac{-dX}{dY}\right) = 0$$

In the usual rotations, we write it as

$$\phi\left(x, y, -\frac{dx}{dy}\right) = 0$$

Since, in polar coordinates (r, θ) slope is given by $r\dfrac{d\theta}{dr}$.

So if $\phi(r, \theta, dr/d\theta) = 0$ is the differential equation for a given family of curves $f(r, \theta, c) = 0$ then the differential equation of the required orthogonal trajectories will be given by

$$\phi\left(r, \theta, -r^2\frac{d\theta}{dr}\right) = 0.$$

➲ Working Rule

1. Form the differential equation of the family of curves.
2. Write $\dfrac{-1}{dy/dx}$ for $\dfrac{dy}{dx}$ or $-r^2\dfrac{d\theta}{dr}$ for $\dfrac{dr}{d\theta}$.
3. Solve the new differential equation to get the equation of orthogonal trajectory.

Self Orthogonal Families

If every member of a family of curves cuts all other members of the family at right angle, then the family is the system of orthogonal trajectories of itself and hence, it is called *self-orthogonal family*.

Thus, if the differential equation of the family is identical with the differential equation for its orthogonal trajectory, then the family is self-orthogonal.

EXAMPLE 7.1

Show that the orthogonal trajectory of a system of concurrent straight lines is a system of concentric circles and conversely.

Solution: Let the equation of a system of concurrent straight lines be

$$y = mx \tag{1}$$

where, m is the parameter.

Clearly, all these straight lines pass through the origin and are thus concurrent.

First, we have to show that the orthogonal trajectory of the family of straight lines (1) is a system of concentric circles.

Differentiating Eq. (1) with respect to x, we get

$$\frac{dy}{dx} = m \tag{2}$$

Eliminating m between Eqs. (1) and (2), we get

$$y = x\frac{dy}{dx} \tag{3}$$

which is the differential equation of the given family of straight lines (1).

Writing $-dx/dy$ for dy/dx in Eq. (3), the differential equation of the orthogonal trajectories of the family of straight lines (1) is

$$y = x\left(\frac{-dx}{dy}\right)$$

$$\Rightarrow \qquad x\,dx + y\,dy = 0$$

Integrating, we get

$$x^2 + y^2 = r^2$$

where, r is a parameter.

This is the equation of a family of concentric circles having origin as their common centre. Conversely, we have to show that the orthogonal trajectory of the family of concentric circles with common centre origin is a system of concurrent straight lines.

Let the equation of a system of concentric circles be

$$x^2 + y^2 = r^2 \tag{4}$$

where, r is the parameter. Clearly, all these circles have origin as their common centre and thus these are concentric.

Differentiating Eq. (4) with respect to x, we get

$$x + y\left(\frac{dy}{dx}\right) = 0 \tag{5}$$

Since Eq. (5) is independent of the parameter c, so it is the differential equation of the family of concentric circles (4).

Replacing dy/dx by $-dx/dy$ in Eq. (5), the differential equation of the orthogonal trajectories of the family (4) will be

$$x + y\left(\frac{-dx}{dy}\right) = 0$$

$$\Rightarrow \qquad \frac{dx}{x} = \frac{dy}{y}$$

Integrating, we get

$$\log x + \log m = \log y$$

$$\Rightarrow \qquad y = mx$$

where, m is the parameter.

This is the equation of a family of concurrent straight lines passing through the origin.

EXAMPLE 7.2

Find the equation of the system of orthogonal trajectories of a series of confocal and co-axial parabolas

$$r = 2a\,(1 + \cos\theta).$$

Solution: Differentiating

$$r = 2a\,(1 + \cos\theta) \tag{1}$$

with respect to θ, we obtain

$$\frac{dr}{d\theta} = -2a \sin \theta \qquad (2)$$

Eliminating a between Eqs. (1) and (2), we obtain

$$\frac{r}{dr/d\theta} = \frac{1 + \cos \theta}{-\sin \theta}$$

which is the differential equation of the given family. Changing

$$dr/d\theta \text{ to } -r^2 d\theta/dr$$

we obtain

$$\frac{r}{-r^2 d\theta/dr} = \frac{1 + \cos \theta}{-\sin \theta}$$

$$\Rightarrow \qquad \frac{1}{r}\frac{dr}{d\theta} = \frac{1 + \cos \theta}{\sin \theta}$$

as the differential equation of the family of orthogonal trajectories. We can write it as

$$\frac{dr}{dr} = (\text{cosec } \theta + \cot \theta) d\theta$$

Integrating, we get

$$\log r = \log \tan \frac{0}{2} + \log \sin \theta + \log c$$

$$\Rightarrow \qquad r = c \sin \theta \tan \frac{\theta}{2}$$

$$= 2c \sin^2 \frac{\theta}{2} = c(1 - \cos \theta)$$

which is the required family of orthogonal trajectories of the family (1).

EXAMPLE 7.3

Find the orthogonal trajectories of the curves

$$\frac{x^2}{a^2 + \lambda} + \frac{y^2}{b^2 + \lambda} = 1$$

where, λ being the parameter of the family.

Solution: Differentiating the given equation with respect to x, we obtain

$$\frac{x}{a^2 + \lambda} + \frac{y}{b^2 + \lambda}\frac{dy}{dx} = 0$$

$$\Rightarrow \qquad \lambda = -\frac{b^2x + a^2y\dfrac{dy}{dx}}{x + y\dfrac{dy}{dx}}$$

$$\therefore \qquad a^2 + \lambda = \frac{(a^2 - b^2)x}{x + y\dfrac{dy}{dx}}, \; b^2 + \lambda = \frac{(a^2 - b^2)y\dfrac{dy}{dx}}{x + y\dfrac{dy}{dx}}$$

On substituting these values in the given equation, we get

$$\left(x - y\frac{dx}{dy}\right)\left(x + y\frac{dy}{dx}\right) = a^2 - b^2 \tag{1}$$

which is the differential equation of the given family.

Changing dy/dx to $-dx/dy$ in Eq. (1); we obtain

$$\left(x + y\frac{dy}{dx}\right)\left(x - y\frac{dx}{dy}\right) = a^2 - b^2$$

which is the same as Eq. (1). Thus, we see that the given family is self-orthogonal, that is, every member of the given family cuts every other member of the same family orthogonally.

EXAMPLE 7.4

Find the orthogonal trajectories of the system of circles touching a given straight line at a given point.

Solution: Let the circle touch the x-axis at the point (0, 0) so that its centre is (0, a) on y-axis, where a is the radius. Hence, its equation is

$$x^2 + (y - a)^2 = a^2$$

$$\Rightarrow \qquad x^2 + y^2 - 2ay = 0$$

where, a is the parameter.

Differentiating with respect to x, we get

$$2x + 2(y - a)\frac{dy}{dx} = 0$$

$$\Rightarrow \qquad x + \left(y - \frac{x^2 + y^2}{2y}\right)\frac{dy}{dx} = 0$$

$$\Rightarrow \qquad 2xy + (y^2 - x^2)\frac{dy}{dx} = 0 \tag{2}$$

This is the differential equation of the family of circles (1). Orthogonal trajectory of (1) is obtained by replacing $\frac{dy}{dx}$ by $\frac{-1}{dy/dx}$; whose equation becomes

$$2xy + (y^2 - x^2)\left(-\frac{dx}{dy}\right) = 0$$

$$\Rightarrow \quad 2xy\frac{dy}{dx} - (y^2 - x^2) = 0$$

$$\Rightarrow \quad 2y\frac{dy}{dx} - \frac{y^2}{x} = -x$$

Put $y^2 = v \therefore 2y\frac{dy}{dx} = \frac{dv}{dx}$

Hence, the equation becomes

$$\frac{dv}{dx} - \frac{1}{x}v = -x$$

This is a linear equation with

$$\text{IF} = e^{\int \frac{-1}{x}dx}$$

$$= e^{-\log x} = \frac{1}{x}$$

Thus, its solution will be

$$v \cdot \frac{1}{x} = \int \frac{1}{x}(-x)\,dx$$

$$\Rightarrow \quad y^2\frac{1}{x} = -x + c$$

$$\Rightarrow \quad y^2 + x^2 = cx$$

$$\Rightarrow \quad x^2 + y^2 - 2bx = 0 \quad \text{where} \quad c = 2b$$

$$\therefore \quad (x - b)^2 + y^2 = b^2.$$

This represents a family of circles touching y-axis at (0, 0) and whose radius is b. So that its centre is $(b, 0)$.

EXAMPLE 7.5

Find the differential equation satisfied by the system of parabolas $y^2 = 4a(x + a)$; and show that orthogonal trajectories of the system belong to the system itself.

Solution: The given equation of the family of parabolas is

$$y^2 = 4a(x + a) \tag{1}$$

where, a is the parameter.

Differentiating Eq. (1) with respect to x; we get

$$2y\frac{dy}{dx} = 4a \Rightarrow y\frac{dy}{dx} = 2a \tag{2}$$

Eliminating a from Eqs. (1) and (2); we get the differential equation of the system given by Eq. (1) as

$$y^2 = 2y\frac{dy}{dx}\left(x+\frac{1}{2}y\frac{dy}{dx}\right)$$

$$\Rightarrow \quad y^2 = 2xy\frac{dy}{dx} + y^2\left(\frac{dy}{dx}\right)^2 \tag{3}$$

Replacing $\frac{dy}{dx}$ by $\frac{-dx}{dy}$ from Eq. (3), we get the differential equation of the system of orthogonal trajectories of (1) as

$$y^2 = 2xy\left(\frac{-dx}{dy}\right) + y^2\left(\frac{-dx}{dy}\right)^2$$

Multiplying above by $\left(\frac{dx}{dy}\right)^2$, we get

$$y^2\left(\frac{dy}{dx}\right)^2 = -2xy\frac{dy}{dx} + y^2$$

$$\Rightarrow \quad y^2 = 2xy\frac{dy}{dx} + y^2\left(\frac{dy}{dx}\right)^2 \tag{4}$$

Comparing Eqs. (3) and (4); we conclude that both are identical, hence the family (1) is self-orthogonal.

EXAMPLE 7.6

Find the orthogonal trajectories of the family of co-axial circles

$$x^2 + y^2 + 2gx + c = 0$$

where, g is the parameter.

Solution: Differentiating the given equation with respect to x, we have

$$x + y\frac{dy}{dx} + g = 0 \tag{1}$$

Eliminating g from the given equation and (1), we have

$$x^2 + y^2 - 2x\left(x + y\frac{dy}{dx}\right) + c = 0$$

$$\Rightarrow \qquad y^2 - x^2 - 2xy\frac{dy}{dx} + c = 0$$

Changing dy/dx to $-dx/dy$; we see that the differential equation of the required family of orthogonal trajectories is

$$y^2 - x^2 + 2xy\frac{dx}{dy} + c = 0$$

$$\Rightarrow \qquad 2xy\frac{dx}{dy} - x^2 = -c - y^2$$

Putting $x^2 = t \therefore 2x\frac{dx}{dy} = \frac{dt}{dy}$

We will obtain a linear differential equation with t as dependent variable.

$$y\frac{dt}{dy} - t = -c - y^2$$

$$\Rightarrow \qquad \frac{dt}{dy} - \frac{t}{y} = \frac{-c}{y} - y$$

with $\qquad \text{IF} = e^{\int \frac{-1}{y}dy} = e^{-\log y} = \frac{1}{y}$

Thus, its solution is given by

$$t\frac{1}{y} = \int \frac{1}{y}\left(\frac{-c}{y} - y\right)dy + d$$

$$\Rightarrow \qquad x^2 \cdot \frac{1}{y} = \frac{c}{y} - y + d$$

$$\Rightarrow \qquad x^2 = c - y^2 + dy$$

$$\Rightarrow \qquad x^2 + y^2 + 2fy - c = 0$$

where, $d = -2f$

which represents a co-axial system of circles orthogonal to the given system with f as parameter.

EXAMPLE 7.7

Find the orthogonal trajectory of the family of curves

$$r^n = a\sin n\theta.$$

Solution: We have

$$n \log r = \log a + \log \sin n\theta$$

Differentiating with respect to θ, we have

$$\frac{n}{r}\frac{dr}{d\theta} = \frac{n \cos n\theta}{\sin n\theta} \quad \Rightarrow \quad \frac{1}{r}\frac{dr}{d\theta} = \cot n\theta$$

which is the differential equation of the given family of curves.

Replacing $dr/d\theta$ by $-r^2 d\theta/dr$, we obtain

$$\frac{1}{r}\left(-r^2 \frac{d\theta}{dr}\right) = \cot n\theta$$

$$\Rightarrow \qquad \tan n\theta d\theta + \frac{dr}{r} = 0$$

Integrating, we get

$$-\frac{1}{n} \log \cos n\theta + \log r = \log c$$

$$\Rightarrow \qquad n \log r = n \log c + \log \cos n\theta$$

$$\Rightarrow \qquad r^n = b \cos n\theta, \text{ where } n \log c = b.$$

which is the required orthogonal trajectory.

EXAMPLE 7.8

Find the orthogonal trajectories of the family of parabolas $y^2 = 4ax$; where a is the variable parameter.

Solution: Differentiating the given equation

$$y^2 = 4ax \tag{1}$$

with respect to x, we get

$$2y\frac{dy}{dx} = 4a \tag{2}$$

Eliminating a between Eqs. (1) and (2), we get

$$y = 2x\frac{dy}{dx} \tag{3}$$

which is the differential equation of the given family of parabolas (1).

So to obtain the differential equation of the orthogonal trajectories, write $-dx/dy$ for dy/dx in Eq. (3). Thus, the differential equation of the required family of orthogonal trajectories is

$$y = 2x\left(\frac{-dx}{dy}\right) \quad \Rightarrow \quad 2x dx + y dy = 0$$

Integrating, we get

$$x^2 + \frac{y^2}{2} = c^2 \quad \Rightarrow \quad 2x^2 + y^2 = b^2$$

where $b^2 = 2c^2$

which is the required family of orthogonal trajectories.

EXAMPLE 7.9

Show that the orthogonal trajectories of the family of conics $y^2 - x^2 + 4xy - 2cx = 0$ consists of a family of cubics, with the common asymptote $x + y = 0$.

Solution: The given family of conics can be written as

$$y^2 - x^2 + 4xy = 2cx \tag{1}$$

where c is the parameter.

Differentiating Eq. (1) with respect to x, we have

$$y\frac{dy}{dx} - x + 2x\frac{dy}{dx} + 2y = c \tag{2}$$

Eliminating c between Eqs. (1) and (2), we get

$$\frac{y\dfrac{dy}{dx} - x + 2x\dfrac{dy}{dx} + 2y}{y^2 - x^2 + 4xy} = \frac{1}{2x}$$

$$\Rightarrow \quad 2\left[xy\frac{dy}{dx} - x^2 + 2x^2\frac{dy}{dx} + 2xy\right] = y^2 - x^2 + 4xy$$

$$\Rightarrow \quad 2x(y + 2x)\frac{dy}{dx} = x^2 + y^2 \tag{3}$$

which is the differential equation of the given family of curves (1).

Now, to obtain the differential equation of the family of orthogonal trajectories, write $-dx/dy$ for dy/dx in Eq. (3). Hence, we obtain

$$2x(y + 2x)\left(\frac{-dx}{dy}\right) = x^2 + y^2$$

$$\Rightarrow \quad \frac{dy}{dx} = \frac{-2x(y + 2x)}{x^2 + y^2}$$

Let $y = vx \therefore \dfrac{dy}{dx} = v + x\dfrac{dv}{dx}$

$$\Rightarrow \quad v + x\frac{dv}{dx} = \frac{-2x^2(v+2)}{x^2(1+v^2)}$$

$$\Rightarrow \quad x\frac{dv}{dx} = \frac{-2(v+2)}{1+v^2} - v = \frac{-3v - v^3 - 4}{1+v^2}$$

$$\Rightarrow \quad -\frac{dx}{x} = \frac{1+v^2}{v^2+3v+4}dv$$

Integrating, we have

$$-\log x = \frac{1}{3}\log(v^3 + 3v + 4) - \log c$$

$$\Rightarrow \quad c = x(v^3 + 3v + 4)^{1/3}$$

$$\Rightarrow \quad c^3 = x^3(v^3 + 3v + 4)$$

$$\Rightarrow \quad c^3 = y^3 + 3x^2y + 4x^3$$

which is the family of orthogonal trajectories of the given family of conic curves (1).

This equation may be written as

$$(y + x)(y^2 - yx + 4x^2) = c^3 \tag{4}$$

Clearly, $x + y = 0$ is the only real asymptote of Eq. (4). Since it does not have the parameter c, therefore it is the common asymptote of all the members which belong to Eq. (4).

EXAMPLE 7.10

Find the orthogonal trajectory of $r = e^{a\theta}$.

Solution: The given equation can be written as

$$\log r = a\theta \tag{1}$$

Differentiating Eq. (1) with respect to θ, we get

$$\frac{1}{r}\frac{dr}{d\theta} = a \tag{2}$$

Eliminating a between Eqs. (1) and (2), we get

$$\frac{\log r}{\frac{1}{r}\frac{dr}{d\theta}} = \theta \Rightarrow \frac{1}{r}\frac{dr}{d\theta} = \frac{\log r}{\theta}$$

which is the differential equation of the family of curves (1). Putting $-r^2\frac{d\theta}{dr}$ for $\frac{dr}{d\theta}$; the differential equation of the orthogonal trajectories is

$$-r\frac{d\theta}{dr} = \frac{\log r}{\theta}$$

$$\Rightarrow \qquad \frac{\log r}{r} dr = -\theta d\theta$$

Integrating, we have

$$\frac{(\log r)^2}{2} = \frac{-\theta^2}{2} + \frac{c^2}{2}$$

$$\Rightarrow \qquad \theta^2 + (\log r)^2 = c^2$$

which is the required family of orthogonal trajectories.

EXAMPLE 7.11

Find the orthogonal trajectories of the curve

$$y = \frac{x^3 - a^3}{3x}$$

where, a is the parameter.

Solution: The given equation can be written as

$$x^3 - 3xy = a^3 \qquad (1)$$

Differentiating (1) with respect to x, we obtain

$$3x^2 - 3y - 3x\frac{dy}{dx} = 0 \qquad (2)$$

which is the differential equation of the family of curves (1).

Writing $\frac{-dx}{dy}$ for $\frac{dy}{dx}$ in Eq. (2), we get

$$3x^2 - 3y - 3x\left(\frac{-dx}{dy}\right) = 0 \qquad (3)$$

which is the differential equation of the family of orthogonal trajectories Eq. (3) gives

$$x\frac{dx}{dy} + x^2 = y$$

Let $x^2 = u \quad \therefore \quad x\frac{dx}{dy} = \frac{1}{2}\frac{du}{dy}$

$$\Rightarrow \qquad \frac{1}{2}\frac{du}{dy} + u = y$$

$$\Rightarrow \qquad \frac{du}{dy} + 2u = 2y$$

which is a linear differential equation with

$$\text{IF} = e^{\int 2dy} = e^{2y}$$

Therefore, its solution is given by

$$ue^{2y} = \int 2ye^{2y}dy + b$$

$$\Rightarrow \qquad ue^{2y} = 2\left[y\frac{e^{2y}}{2} - \frac{e^{2y}}{4}\right] + b$$

$$\Rightarrow \qquad x^2 = y - \frac{1}{2} + be^{-2y}$$

which is the required family of orthogonal trajectories with b as parameter. ▲▲

EXERCISE 7.1

1. Find the orthogonal trajectories of the family of semi-cubical parabolas $ay^2 = x^3$; where, a is a variable parameter.
2. Find the orthogonal trajectories of

$$\frac{x^2}{a^2} + \frac{y^2}{a^2 + \lambda} = 1$$

   where, λ is an arbitrary parameter.
3. Find the orthogonal trajectories of the family of parabolas $y = ax^2$.
4. Find the orthogonal trajectory of the family of cardioides

$$r = a(1 + \cos\theta)$$

   where, a is the parameter.
5. Prove that the orthogonal trajectories of the curves. $A = r^2 \cos\theta$ are the curves

$$B = r\sin^2\theta$$

6. Find the orthogonal trajectories of the family curves

$$x^{2/3} + y^{2/3} = \lambda^{2/3}.$$

7. Find the orthogonal trajectories of

$$r = c\sin^2\theta$$

   where, c is the parameter.

8. Prove that the orthogonal trajectories of the rectangular hyperbola $xy = a^2$ are

$$x^2 - y^2 = c^2$$

9. Find the equation of the family of curves that are orthogonal to the family of curves

$$ax^2 + y^2 = 1$$

10. Find the orthogonal trajectories of the family of curves given by

$$r(1 + \cos\theta) = 2a$$

where, a is the parameter.

11. Find the orthogonal trajectories of

$$r^n \sin(n\theta) = a^n.$$

12. Find the orthogonal trajectories of the family of circles passing through the points (0, 2) and (0, –2).

13. Determine the orthogonal trajectories of the family of curves $e^{2x}\sin 2y = c$ where, c is the parameter.

14. Find the orthogonal trajectories of the family of curves

$$r = c(\sec\theta + \tan\theta).$$

15. Find the orthogonal trajectories of the family of curves

$$r = c(\operatorname{cosec}\theta + \cot\theta).$$

16. Show that the family of curves

$$\frac{x^2}{c} + \frac{y^2}{c+2} + 1 = 0$$

is self-orthogonal.

17. Find the orthogonal trajectories of the family of curves

$$r^2 = c\sin(2\theta).$$

18. Find the orthogonal trajectories of the family of curves

$$r = c\theta^2.$$

19. Show that the family of curves

$$r^n = a\sec n\theta \quad \text{and} \quad r^n = b\operatorname{cosec} n\theta$$

are orthogonal.

20. Find the orthogonal trajectories of the family of curves

$$x^2 + y^2 - ay = 1$$

where, a is the parameter.

ANSWERS

Exercise 7.1

1. $3y^2 + 2x^2 = c^2$.
2. $x^2 + y^2 + c = 2a^2 \log x$.
3. $x^2 + 2y^2 = c^2$.
4. $r = c(1 - \cos\theta)$.
6. $y^{4/3} - x^{4/3} = c^{4/3}$.
7. $r^2 = a\cos\theta$.
9. $y^2 = ce^{x^2+y^2}$.
10. $\dfrac{2c}{r} = 1 - \cos\theta$.
11. $r^n \cos n\theta = c^n$.
12. $x^2 + (y - 2c)^2 = 4(c^2 - 1);\ |c| > 1$.
13. $e^{2x}\cos 2y = a$.
14. $r = ae^{-\sin\theta}$.
15. $r = ae^{-\cos\theta}$.
17. $r^2 = a\cos(2\theta)$.
18. $r = ae^{-\theta^2/4}$.
20. $x^2 + y^2 + cx + 1 = 0$.

8

CHAPTER

Homogeneous Linear Differential Equations

Linear differential equations originate in a variety of applications in science and engineering. Equations with constant coefficients arise in the theory of electric circuits, vibrations, simple harmonic motion, etc. Variable coefficients equations arise in many areas of physics, electric circuits, mathematical modeling of physical problems, etc. We have studied in chapter 1 that linear differential equations are the equations in which the dependent variable and its derivative appear only in the first degree and are not multiplied together. A linear ordinary differential equation of order n is written as

$$a_0(x)\frac{d^n y}{dx^n} + a_1(x)\frac{d^{n-1} y}{dx^{n-1}} + \ldots + a_{n-1}(x)\frac{dy}{dx} + a_n(x)y = r(x)$$

or

$$a_0(x)y^{(n)} + a_1(x)y^{(n-1)} + \ldots + a_{n-1}(x)y' + a_n(x)y = r(x) \qquad (1)$$

where, y is the dependent variable and x is the independent variable and $a_0(x) \neq 0$.

If $r(x) = 0$, then it is called homogeneous differential equation, otherwise it is called a non-homogeneous differential equation. If $a_i(x)$, $\{i = 1, 2, \ldots n)$ are constants then the Eq. (1) is linear differential equation with constant coefficients otherwise it is known as linear differential equation with variable coefficients.

For example,

$$y'' + 4y' + 3y = 0 \qquad (2)$$

$$y'' + 2y' + y = x^2 e^x \qquad (3)$$

$$x^2y'' + xy' + (x^2 - 4)y = 0 \qquad (4)$$

$$(1 - x^2)y'' - 2xy' + 20y = \sin x \qquad (5)$$

Equations (2) and (3) are linear differential equations with constant coefficients and Eqs. (4) and (5) are variable coefficients second order equations.

Equations (2) and (4) are homogeneous differential equations while Eqs. (3) and (5) are non-homogeneous differential equations. In practice, equations with the coefficients as the functions of x with no restriction placed on their nature do not usually have solutions expressible in terms of elementary functions and even when they do, it is very difficult to find them. However, if the coefficients are constant then the solution in terms of elementary functions can be easily obtained. Moreover, the linear equations with constant coefficients are of great practical importance as well as of theoretical interest. Before discussing methods of solution for linear differential equations with constant coefficients, we will define some terms.

Definition: Let $f_1(x), f_2(x), \ldots f_n(x)$ be n functions

Then

$$c_1f_1(x) + c_2f_2(x) + \ldots + c_nf_n(x)$$

where, $c_1, c_2, \ldots c_n$ are constants, a linear combination of the given functions.

Definition: Let $f_1(x), f_2(x), \ldots f_n(x)$ be n functions. Then, these functions are said to be linearly independent on some interval I where they are defined, if the equation

$$c_1f_1(x) + c_2f_2(x) + \ldots + c_nf_n(x) = 0$$

implies $c_1 = c_2 = \ldots = c_n = 0$.

These functions are said to be linearly dependent on I if at least one c_i, $(i = 1, 2, -n)$ is non-zero.

Definition: Let $f_1(x), f_2(x), \ldots f_n(x)$ be n functions. The Wronskian of these functions is denoted by $W(f_1, f_2, \ldots f_n)$ and is defined by

$$W(f_1, f_2, \ldots f_n) = \begin{vmatrix} f_1 & f_2 & - & f_n \\ f_1' & f_2' & - & f_n' \\ - & - & - & - \\ f_1^{(n-1)} & f_2^{(n-1)} & - & f_n^{(n-1)} \end{vmatrix}$$

The Wronskian of the n functions exist on I if all the functions $f_1, f_2, \ldots f_n$ are differentiable $n - 1$ times on the interval I. If any one or more functions are not differentiable then the Wronskian does not exist on I.

Note: From the theory of linear algebra, if $W = 0$, then $f_1, f_2, \ldots f_n$ are linearly dependent on I otherwise linearly independent on I.

Consider the homogeneous linear differential equation with constant coefficients of order n

$$a_0 \frac{d^n y}{dx^n} + a_1 \frac{d^{n-1}y}{dx^{n-1}} + \ldots + a_{n-1}\frac{dy}{dx} + a_n y = 0 \tag{6}$$

where, $a_0, a_1, a_2, \ldots a_n$ are constants.

Let y_1 and y_2 be any two solutions of Eq. (6)

$$\therefore \quad a_0 \frac{d^n y_1}{dx^n} + a_1 \frac{d^{n-1} y}{dx^{n-1}} + \ldots + a_{n-1} \frac{dy_1}{dx} + a_n y_1 = 0 \tag{7}$$

and
$$a_0 \frac{d^n y_2}{dx^n} + a_1 \frac{d^{n-1} y_2}{dx^{n-1}} + \ldots + a_{n-1} \frac{dy_2}{dx} + a_n y_2 = 0 \tag{8}$$

If c_1, and c_2 are any two arbitrary constants, we have

$$a_0 \frac{d^n (c_1 y_1 + c_2 y_2)}{dx^n} + a_1 \frac{d^{n-1} (c_1 y_1 + c_2 y_2)}{dx^{n-1}} + \ldots + a_{n-1} \frac{d(c_1 y_1 + c_2 y_2)}{dx} + a_n (c_1 y_1 + c_2 y_2)$$

$$= c_1 \left\{ a_0 \frac{d^n y_1}{dx^n} + a_1 \frac{d^{n-1} y_1}{dx^{n-1}} + \ldots + a_{n-1} \frac{dy_1}{dx} + a_n y_1 \right\}$$

$$+ \; c_2 \left\{ a_0 \frac{d^n y_2}{dx^n} + a_1 \frac{d^{n-1} y_2}{dx^{n-1}} + \ldots + a_{n-1} \frac{dy_2}{dx} + a_n y_2 \right\}$$

$= c_1(0) + c_2(0) = 0$ (By using Eqs. (7) and (8))

This shows that if y_1, y_2 are any two solutions of Eq. (6) then their linear combination, that is, $c_1 y_1 + c_2 y_2$ is also a solution of Eq. (6) where, c_1, c_2 are arbitrary constants.

We can also generalize the above result as since the general solution of a differential equation of the nth order contains n arbitrary constants therefore we can deduce that if $y_1, y_2, \ldots y_n$ be any n solutions of Eq. (6) then

$$c_1 y_1 + c_2 y_2 + \ldots + c_n y_n = u, \text{ say}$$

is the general solution of Eq. (6) where $c_1, c_2 \ldots c_n$ are n arbitrary constants. Moreover, if the n solutions $y_1, y_2, \ldots y_n$ are also linearly independent then $y_1, y_2, \ldots y_n$ are called the fundamental solutions of Eq. (6).

Let v be any particular solution of non-homogeneous linear differential equation with constant coefficients, that is,

$$a_0 \frac{d^n y}{dx^n} + a_1 \frac{d^{n-1} y}{dx^{n-1}} + \ldots + a_{n-1} \frac{dy}{dx} + a_n y = r(x) \tag{9}$$

$$\therefore \quad a_0 \frac{d^n v}{dx^n} + a_1 \frac{d^{n-1} v}{dx^{n-1}} + \ldots + a_{n-1} \frac{dv}{dx} + a_n v = r(x) \tag{10}$$

From the above, since u is a solution of Eq. (6); we must have

$$a_0 \frac{d^n u}{dx^n} + a_1 \frac{d^{n-1} u}{dx^{n-1}} + \ldots + a_{n-1} \frac{du}{dx} + a_n u = r(x) \tag{11}$$

On adding Eqs. (10) and (11), we obtain

$$a_0 \frac{d^n(u+v)}{dx^n} + a_1 \frac{d^{n-1}(u+v)}{dx^{n-1}} + \ldots + a_{n-1} \frac{d(u+v)}{dx} + a_n(u+v) = r(x)$$

This shows that $y = u + v$ is the general solution of Eq. (9). It follows that in order to be able to solve the differential Eq. (9), we have to obtain first the fundamental solution u of its corresponding homogeneous part (6) and any particular solution v of (9) and then

$$y = u + v = c_1y_1 + c_2y_2 + \ldots + c_ny_n + v$$

is the general solution of Eq. (9).

The part $c_1y_1 + c_2y_2 + \ldots + c_ny_n = u$ is called the complementary, functions (CF) and v the particular integral (PI).

Therefore, the complete solution (CS) is

$$y = \text{CF} + \text{PI}$$

Thus, in order to solve Eq. (9), we have first to learn to solve Eq. (6).

Operators: The part d/dx of the symbol dy/dx can be thought of as an operator such that when it operates on y, the result is the derivative of y. From this point of view, the symbol d/dx is called an operator and the function y is called the operand. We can similarly think of d^2/dx^2, d^3/dx^3 etc., as operators. For the sake of convenience, the operators d/dx, d^2/dx^2, d^3/dx^3, etc., are denoted by D, D^2, D^3, etc. The index of D indicates the number of times the operation of differentiation is to be carried out. So Eq. (9) can be written in operator form as

$$a_0D^ny + a_1D^{n-1}y + \ldots + a_{n-1}Dy + a_ny = r(x)$$

$$\Rightarrow \quad (a_0D^n + a_1D^{n-1} + \ldots + a_{n-1}D + a_n)y = r(x) \tag{12}$$

$$\Rightarrow \quad f(D)y = r(x)$$

where,

$$f(D) = a_0D^n + a_1D^{n-1} + \ldots + a_{n-1}D + a_n$$

is a polynomial in D of degree n.

Regarding D as a number, we can factorize $f(D)$ so that, from this point of view, we may have

$$f(D) \equiv (D - m_1)(D - m_2) \ldots (D - m_n)$$

So, Eq. (12) becomes

$$(D - m_1)(D - m_2) \ldots (D - m_n)y = r(x) \tag{13}$$

Since we know that in order to solve Eq. (13), we have to first solve its corresponding homogeneous linear differential equation which is

$$(D - m_1)(D - m_2) \ldots (D - m_n)y = 0$$

We will denote

$$f(D) = a_0D^n + a_1D^{n-1} + \ldots + a_{n-1}D + a_n = 0$$

by AE which is called the auxiliary equation. Rules for finding general solution of the homogeneous linear differential equation of order n with constant coefficients are illustrated in the following examples by reducing the second order linear equations to the first order linear equations. ▲▲

EXAMPLE 8.1

Factorising the differential operator and reducing it to the first order equations, solve the differential equation

$$\frac{d^2y}{dx^2} - 4\frac{dy}{dx} - 5y = 0.$$

Solution: In the operator notation, the given differential equation can be written as

$$(D^2 - 4D - 5)y = (D - 5)(D + 1)y = 0$$

Let $(D + 1)y = u$, so we obtain

$$(D - 5)u = 0$$

$$\Rightarrow \quad \frac{du}{dx} - 5u = 0$$

which is a first order equation with solution $\log u = 5x + \log A$

$$\therefore \quad u = Ae^{5x}$$

Hence, $\quad (D + 1)y = Ae^{5x}$

$$\Rightarrow \quad \frac{dy}{dx} + y = Ae^{5x}$$

which is a first order linear equation with

$$\text{IF} = e^{\int dx} = e^x$$

Hence, we have the solution as

$$ye^x = \int Ae^{5x}e^x\,dx + B$$

$$\Rightarrow \quad ye^x = \frac{Ae^{6x}}{6} + B$$

$$\therefore \quad y = A'e^{5x} + Be^{-x}$$

which is the required general solution.

Note: We have the auxiliary equation (AE) for the given differential equation as

$$D^2 - 4D - 5 = 0$$

$$\Rightarrow \quad D = 5, -1$$

So we have two distinct roots of the AE. On seeing the general solution we can conclude that the general solution is obtained as the linear combination of e^{5x} and e^{-x} that is exponential function raised to power first root multiplied by the independent variable x and exponential function raised to power the second root multiplied by x.

$$\therefore \qquad y = A'e^{5x} + Be^{-x}$$

where, A' and B are arbitrary constants.

EXAMPLE 8.2

Factorizing the differential operator and reducing it to first order equations, solve the differential equation.

$$4\frac{d^2y}{dx^2} + 12\frac{dy}{dx} + dy = 0$$

Solution: In the operator notation, the given differential equation can be written as

$$(4D^2 + 12D + 9)y = (2D + 3)(2D + 3)y = 0$$

Let $(2D + 3)y = u$, so we obtain

$$(2D + 3)u = 0$$

$$\Rightarrow \qquad 2\frac{du}{dx} + 3u = 0$$

The solution of this equation is

$$u = Ae^{-3x/2}$$

Hence,

$$(2D + 3)y = Ae^{-3x/2}$$

$$\Rightarrow \qquad \frac{dy}{dx} + \frac{3}{2}y = \frac{A}{2}e^{-3x/2}$$

which is a linear differential equation of the first order with

$$\text{IF} = e^{\int \frac{3}{2}dx} = e^{3x/2}.$$

The solution is given by

$$ye^{3x/2} = \int \frac{A}{2}e^{-3x/2}e^{3x/2}dx + B$$

$$\Rightarrow \qquad y = (A'x + B)e^{-3x/2}$$

which is the required general solution.

Note: We have the auxiliary equation (AE) for the given differential equation as

$$4D^2 + 12D + 9 = 0$$

$$\Rightarrow \qquad D = -3/2, -3/2$$

So we have two repeated roots of the AE. Therefore by using the previous example, we have

$$y = Ae^{-3x/2} + Bxe^{-3x/2}$$

Here we have multiplied the second term by x for making $e^{-3x/2}$, $xe^{-3/2x}$ linearly independent. We can conclude that if the roots of AE are repeated then in general solution increase the power of x in the succeeding terms by 1 as compared to the previous term.

Now we will discuss the case, when the roots of the AE are imaginary.

EXAMPLE 8.3

Find the solution of the differential equation

$$\frac{d^2y}{dx^2} + 2\frac{dy}{dx} + 2y = 0.$$

Solution: In the operator notation, the given differential equation can be written as

$$(D^2 + 2D + 2)y = 0$$

Therefore, its auxiliary equation is

$$D^2 + 2D + 2 = 0$$

$$D = \frac{-2 \pm \sqrt{4-8}}{2} = -1 \pm i$$

By using the previous examples, we can have the general solution as

$$y = Ae^{(-1+i)x} + Be^{(-1-i)x}$$

as the roots are distinct.

By making use of Euler's theorem, which states that $e^{\pm i\theta} = \cos\theta \pm i\sin\theta$,

We have

$$\begin{aligned} y &= e^{-x}(Ae^{ix} + Be^{-ix}) \\ &= e^{-x}[A(\cos x + i\sin x) + B(\cos x - i\sin x)] \\ &= e^{-x}[(A+B)\cos x + (A-B)i\sin x] \\ &= e^{-x}[A'\cos x + B'\sin x] \end{aligned}$$

where we have written, $A + B = A'$, $(A - B)i = B'$

Note: We can summarize the results of above examples for a second order equation

$$a_0\frac{d^2y}{dx^2} + a_1\frac{dy}{dx} + a_2y = 0$$

written in operator form

$$(a_0D^2 + a_1D + a_2)y = 0$$

with AE:

$$a_0D^2 + a_1D + a_2 = 0 \text{ as under.}$$

(i) If AE has distinct real roots α, β then general solution is

$$y = Ae^{\alpha x} + Be^{\beta x}$$

(ii) If AE has repeated real roots α, α then general solution is

$$y = Ae^{\alpha x} + Bxe^{\alpha x}$$

(iii) If AE has complex roots $\alpha \pm i\beta$ then general solution is

$$y = e^{\alpha x}(A\cos \beta x + B\sin \beta x)$$

In general, we can say for a differential equation of order n

(a) corresponding to a root α (repeated r times), the corresponding solution is

$$(c_1 + c_2x + \ldots + c_rx^{r-1})e^{\alpha x}$$

containing r arbitrary constants.

(b) if $\alpha + i\beta$, $\alpha - i\beta$ are conjugate imaginary roots, each repeated r times, then the corresponding solution is

$$e^{\alpha x}[(p_1 + p_2x + \ldots + p_rx^{r-1})\cos \beta x + (q_1 + q_2x + \ldots + q_rx^{r-1})\sin \beta x]$$

with r arbitrary constants p_i and q_j each.

EXAMPLE 8.4

Solve $\dfrac{d^3y}{dx^3} - 9\dfrac{dy}{dx} = 0.$

Solution: In operator notation, the given differential equation can be written as

$$(D^3 - 9D)y = 0$$

Its AE is

$$D^3 - 9D = 0$$

$\Rightarrow$ $$D(D^2 - 9) = 0$$

$\therefore$ $$D = 0, 3, -3$$

Hence, the general solution is

$$y = c_1e^{0x} + c_2e^{3x} + c_3e^{-3x}$$

i.e., $$y = c_1 + c_2e^{3x} + c_3e^{-3x}$$

EXAMPLE 8.5

Solve $\frac{d^3y}{dx^3} + 6\frac{d^2y}{dx^2} + 11\frac{dy}{dx} + 6y = 0.$

Solution: In operator notation, the given differential equation can be written as

$$(D^3 + 6D^2 + 11D + 6)y = 0$$

Its AE is

$$D^3 + 6D^2 + 11D + 6 = 0$$

Clearly, $D = -1$ is satisfying this equation as

$$-1 + 6 - 11 + 6 = 0$$

$\therefore$ $D = -1$ is its first root. The remaining two roots can be found as under

−1	1	6	11	6
		−1	−5	−6
	1	5	6	0

$\therefore$ $$D^2 + 5D + 6 = 0 \Rightarrow D = -2, -3$$

So we have three distinct roots −1, −2, −3
Hence, the general solution is

$$y = c_1e^{-x} + c_2e^{-2x} + c_3e^{-3x}.$$

EXAMPLE 8.6

Solve $\frac{d^2y}{dx^2} + 6\frac{dy}{dx} + 9y = 0.$

Solution: In operator notation, the given differential equation can be written as

$$(D^2 + 6D + 9)y = 0$$

Its AE is

$$D^2 + 6D + 9 = (D + 3)^2 = 0$$

$\Rightarrow$ $$D = -3, -3$$

In this case, we have two repeated roots.
Hence, its general solution is

$$y = c_1e^{-3x} + c_2xe^{-3x}.$$

EXAMPLE 8.7

Find the solution of the initial value problem

$$\frac{d^2y}{dx^2} + 4\frac{dy}{dx} + 13y = 0; \quad y(0) = 0, y'(0) = 1.$$

Solution: In operator notation, the given differential equation can be written as

$$(D^2 + 4D + 13)y = 0$$

Its AE is

$$D^2 + 4D + 13 = 0$$

$$\therefore \quad D = \frac{-4 \pm \sqrt{16 - 52}}{2} = -2 \pm 3i$$

Hence, the general solution is

$$y = e^{-2x}(c_1 \cos 3x + c_2 \sin 3x)$$

Substituting in the initial conditions, we obtain

$$y(0) = 0 = A \quad \therefore\ y = Be^{-2x} \sin 3x$$

$$y' = Be^{-2x}(3 \cos 3x - 2 \sin 3x)$$

$$\therefore \quad y'(0) = 1 = 3B \quad \Rightarrow \quad B = 1/3$$

Hence, the solution of the initial value problem is

$$y = (e^{-2x} \sin 3x)/3.$$

EXAMPLE 8.8

Solve the differential equation

$$\frac{d^4y}{dx^4} - 2\frac{d^3y}{dx^3} + 5\frac{d^2y}{dx^2} - 8\frac{dy}{dx} + 4y = 0.$$

Solution: In operator notation, the given differential equation can be written as

$$(D^4 - 2D^3 + 5D^2 - 8D + 4)y = 0$$

Its AE is

$$D^4 - 2D^3 + 5D^2 - 8D + 4 = 0$$

$$\Rightarrow \quad (D - 1)^2(D^2 + 4) = 0$$

$$\therefore \quad D = 1, 1, 2i, -2i$$

The root 1, being repeated twice. Therefore, its general solution is

$$y = (c_1 + c_2x)e^x + c_3 \cos 2x + c_4 \sin 2x.$$

EXAMPLE 8.9

Solve $(D^3 + 1)y = 0$.

Solution: AE is

$$D^3 + 1 = (D + 1)(D^2 - D + 1) = 0$$

$$\therefore \qquad D = -1, \frac{1}{2} \pm \frac{\sqrt{3}i}{2}$$

Hence, its general solution is

$$y = c_1 e^{-x} + e^{x/2}\left[c_2 \cos\frac{\sqrt{3}}{2}x + c_3 \sin\frac{\sqrt{3}}{2}x\right]$$

EXAMPLE 8.10

Solve $\dfrac{d^4y}{dx^4} + m^4 y = 0.$

Solution: In operator notation, the given equation can be written as

$$(D^4 + m^4)y = 0$$

Its AE is

$$D^4 + m^4 = 0$$

$$\Rightarrow \qquad (D^4 + 2D^2m^2 + m^4) - 2D^2m^2 = 0$$

$$\Rightarrow \qquad (D^2 + m^2)^2 - (\sqrt{2}Dm)^2 = 0$$

$$\Rightarrow \qquad (D^2 + m^2 + \sqrt{2}Dm)(D^2 + m^2 - \sqrt{2}Dm) = 0$$

$$\therefore \qquad D = -\frac{m}{\sqrt{2}} \pm i\frac{m}{\sqrt{2}}, \frac{m}{\sqrt{2}} \pm i\frac{m}{\sqrt{2}}$$

Hence, the general solution of the equation will be

$$y = e^{(-m/\sqrt{2})x}\left[c_1 \cos\frac{m}{\sqrt{2}}x + c_2 \sin\frac{m}{\sqrt{2}}x\right] + e^{(m/\sqrt{2})x}\left[c_3 \cos\frac{m}{\sqrt{2}}x + c_4 \sin\frac{m}{\sqrt{2}}x\right]$$

EXAMPLE 8.11

Solve $\dfrac{d^4y}{dx^4} - 4\dfrac{d^3y}{dx^3} + 8\dfrac{d^2y}{dx^2} - 8\dfrac{dy}{dx} + 4y = 0.$

Solution: In operator notation, the given differential equation can be written as

$$(D_4 - 4D^3 + 8D^2 - 8D + 4)y = 0$$

Its AE is

$$D^4 - 4D^3 + 8D^2 - 8D + 4 = 0$$

$$\Rightarrow \qquad (D^2 - 2D + 2)^2 = 0$$

$$\Rightarrow \qquad D = 1 \pm i, 1 \pm i$$

Here, imaginary root $1 \pm i$ is being repeated twice.
Hence, the solution of the equation is

$$y = e^x[(c_1 + c_2x)\cos x + (c_3 + c_4x)\sin x].$$

EXAMPLE 8.12

Solve $(D^4 - w^4)y = 0$

Solution: Its AE is

$$D^4 - w^4 = 0 \Rightarrow D^2 = w^2, -w^2$$

$$\therefore \quad D = \pm w, \pm wi$$

Hence, the general solution is given by

$$y = c_1e^{wx} + c_2e^{-wx} + c_3\cos wx + c_4\sin wx.$$

EXAMPLE 8.13

Solve the differential equation

$$y^{iv} + 32y'' + 256y = 0.$$

Solution: In operator notation, the given differential equation can be written as

$$(D^4 + 32D^2 + 256)y = 0$$

Its AE is

$$D^4 + 32D^2 + 256 = 0$$

$$\Rightarrow \quad (D^2 + 16)^2 = 0$$

$$\therefore \quad D = \pm 4i, \pm 4i$$

The roots of this equation $\pm 4i$ are being repeated twice. Therefore, the general solution is

$$y = (c_1 + c_2x)\cos 4x + (c_3 + c_4x)\sin 4x.$$

EXAMPLE 8.14

Solve $(D^3 - 4D^2 + 4D)y = 0$.

Solution: Its AE is

$$D^3 - 4D^2 + 4D = 0$$

$$\Rightarrow \quad D(D^2 - 4D + 4) = 0$$

$$\therefore \quad D = 0, 2, 2$$

Hence, its general solution is

$$y = c_1e^{0x} + (c_2 + c_3x)e^{2x}$$

i.e.,

$$y = c_1 + (c_2 + c_3x)e^{2x}.$$

EXAMPLE 8.15

Solve the differential equation

$$(D^3 + 3D^2 + 3D + 1)y = 0$$

Solution: Its AE is

$$D^3 + 3D^2 + 3D + 1 = 0$$

Clearly, $D = -1$ is satisfying this equation as

$$-1 + 3 - 3 + 1 = 0$$

Therefore, one root is –1. The remaining two roots can be found as

$$\begin{array}{r|rrrr} -1 & 1 & 3 & 3 & 1 \\ & & -1 & -2 & -1 \\ \hline & 1 & 2 & 1 & \underline{|0} \end{array}$$

$$D^2 + 2D + 1 = 0$$

$$\Rightarrow \qquad D = -1, -1$$

Thus, the root –1 is repeated thrice.
Hence, its general solution is

$$y = (c_1 + c_2x + c_3x^2)e^{-x}.$$

EXAMPLE 8.16

Show that e^x, e^{2x}, e^{3x} are the fundamental solutions of $y''' - 6y'' + 11y' - 6y = 0$ on any interval I of the set of real numbers.

Solution: Substituting $y = e^x, e^{2x}, e^{3x}$ we find that they satisfy the differential equation $y''' - 6y'' + 11y' - 6y = 0$.

The Wronskian of these functions is

$$W(x) = \begin{vmatrix} e^x & e^{2x} & e^{3x} \\ e^x & 2e^{2x} & 3e^{3x} \\ e^x & 4e^{2x} & 9e^{3x} \end{vmatrix} = e^{6x}\begin{vmatrix} 1 & 1 & 1 \\ 1 & 2 & 3 \\ 1 & 4 & 9 \end{vmatrix} = 2e^{6x} \neq 0$$

Therefore, the solutions are linearly independent and they form a set of fundamental solutions on any interval, I of the set of real numbers.

EXAMPLE 8.17

Find a differential equation of second order for which 1, e^{-2x} are its solutions.

Solution: For a second order equation, linear combination of 1, e^{-2x} will also be its solution. Thus,

$$y = c_1 + c_2e^{-2x} = c_1e^{0x} + c_2e^{-2x}.$$

Therefore, 0 and –2 are roots of the AE.

$\therefore$ AE is $$(D-0)(D-2) = D^2 - 2D = 0$$

Hence, the required differential equation is

$$(D^2 - 2D)y = y'' - 2y' = 0.$$

EXAMPLE 8.18

Find a differential equation of the second order for which $e^{(5+3i)x}; e^{(5-3i)x}$ are its solutions.

Solution: Clearly, linear combination of these two given solutions will also be a solution.

$\therefore$ $$y = c_1 e^{(5+3i)x} + c_2 e^{(5-3i)x}$$

Therefore $5 + 3i$ are roots of the AE.

$\therefore$ AE is $$[D-(5+3i)][D-(5-3i)] = 0$$

$\Rightarrow$ $$(D-5)^2 - (3i)^2 = 0$$

$\Rightarrow$ $$D^2 - 10D + 25 + 9 = 0$$

$\Rightarrow$ $$D^2 - 10D + 34 = 0$$

Hence, the required differential equation is

$$(D^2 - 10D + 34)y = 0$$

i.e. $$y'' - 10y' + 34y = 0.$$

EXAMPLE 8.19

Find a homogeneous linear differential equation of the lowest order with real constant coefficients with particular solution

$$e^{-x} + \cos 5x + 3\sin 5x.$$

Solution: Clearly, roots of AE are –1, $\pm 5i$

Therefore, AE will be

$$(D-1)(D^2 + 5^2) = 0$$

$\Rightarrow$ $$D^3 - D^2 + 25D - 25 = 0$$

Hence, the required differential equation will be

$$y''' - y'' - 25y' + 25 = 0.$$

EXAMPLE 8.20

Find a homogeneous linear differential equation of the lowest order with real constant coefficients with particular solution $3\cos 2x + 5\sinh 3x$.

Solution: The given particular solution can be written as

$$3\cos 2x + 5\left(\frac{e^{3x} - e^{-3x}}{2}\right)$$

Clearly, roots of AE are $\pm 2i, \pm 3$

Therefore, AE will be

$$(D^2 + 4)(D^2 - 9) = 0$$

$$\Rightarrow \quad D^4 - 5D^2 - 36 = 0$$

Hence, the required differential equation will be

$$y^{iv} - 5y'' - 36y = 0.$$

▲▲

EXERCISE 8.1

Solve the following differential equations:

1. $\dfrac{d^2y}{dx^2} - 5\dfrac{dy}{dx} - 6y = 0.$
2. $4\dfrac{d^2y}{dx^2} + 4\dfrac{dy}{dx} + y = 0.$
3. $2\dfrac{d^2y}{dx^2} + 3\dfrac{dy}{dx} + 4y = 0.$
4. $\dfrac{d^2y}{dx^2} + a^2y = 0.$
5. $(8D^2 - 6D - 5)y = 0.$
6. $\dfrac{d^4y}{dx^4} + 4y = 0.$
7. $\dfrac{d^4y}{dx^4} - y = 0.$
8. $(D^2 + D + 1)^2(D - 2)y = 0.$
9. $[D^2 - 2aD + (a^2 + b^2)]y = 0.$
10. $(D^2 + 9D)y = 0.$
11. $(D^2 - 6D + 18)y = 0.$
12. $y''' - 2y'' + 4y' - 8y = 0.$
13. $y^{iv} + 2y'' + y = 0.$
14. $y''' + y' = 0.$
15. $y''' + 4y'' + 5y' + 2y = 0.$

Solve the initial value problems:

16. $y'' - y' - 12y = 0;\ y(0) = y,\ y'(0) = -5.$
17. $4y''' - 4y'' - 9y' + 9y = 0;\ y(0) = 1,\ y'(0) = 0,\ y''(0) = 0.$
18. $y^{iv} - 3y''' = 0;\ y(0) = 2,\ y'(0) = 5,\ y''(0) = 15,\ y'''(0) = 27.$

Solve the following boundary value problems:

19. $y'' + 25y = 0;\ y(0) = 1,\ y(\pi) = -1.$
20. $y'' - 36y = 0;\ y(0) = 2,\ y(16) = 1/e.$
21. $y''' + \pi^2 y' = 0;\ y(0) = 0,\ y(1) = 0,\ y'(0) + y'(1) = 0.$
22. Find the intervals on which the three functions 1, cos x, sec $x (x > 0)$ are linearly independent.
23. Are the functions e^x, sinh x, cosh x, linearly independent for $x \in (0, \infty)$.
24. Are the functions 1, cos x, sin x linearly independent for $x \in (0, \infty)$.
25. Find a differential equation of the second order for which the following functions are solutions:

    (i) e^{3x}, e^{-2x}

    (ii) e^{-x}, xe^{-x}

    (iii) $e^{-(a + ib)x}, e^{-(a - ib)x}$.

26. Find a homogeneous linear differential equation of lowest order with real constant coefficients which has the following particulars solution

    (i) $5 + e^x + 2e^{3x}$

    (ii) $xe^{-x} + e^{2x}$

    (iii) $x^2 e^{2x} + 2e^{-2x}$.

ANSWERS

Exercise 8.1

1. $y = c_1 e^{-x} + c_2 e^{6x}$.
2. $y = (c_1 + c_2 x) e^{-x/2}$.
3. $y = e^{-3x/4}\left[c_1 \cos\frac{\sqrt{23}}{4}x + c_3 \sin\frac{\sqrt{23}}{4}x\right].$
4. $y = c_1 \cos ax + c_2 \sin ax$.
5. $y = c_1 e^{-x/2} + c_2 e^{3x/4}$.
6. $y = e^{-x}(c_1 \cos x + c_2 \sin x) + e^x (c_1 \cos x + c_2 \sin x)$.
7. $y = c_1 e^x + c_2 e^{-x} + c_3 \cos x + c_4 \sin x$.
8. $y = e^{-x/2}\left[(c_1 + c_2 x)\cos\frac{\sqrt{3}}{2}x + (c_3 + c_4 x)\sin\frac{\sqrt{3}}{2}x\right] + c_5 e^{2x}.$
9. $y = e^{ax} [c_1 \cos bx + c_2 \sin bx]$.
10. $y = c_1 + c_2 e^{-9x}$.

11. $y = e^{3x}[c_1 \cos 3x + c_2 \sin 3x]$.
12. $y = c_1 e^{2x} + c_2 \cos 2x + c_3 \sin 2x$.
13. $y = (c_1 + c_2 x) \cos x + (c_3 + c_4 x) \sin x$.
14. $y = c_1 + c_2 \cos x + c_3 \sin x$.
15. $y = c_1 e^{-2x} + (c_2 + c_3 x) e^{-x}$.
16. $5y = e^{4x} + 3e^{-3x}$.
17. $y = (9e^x - 5e^{3x/2} + e^{-3x/2})/5$.
18. $y = 1 + 2x + 3x^2 + e^{3x}$.
19. $y = \cos x + c_2 \sin 5x$, c_2 arbitrary.
20. $y = [(2e^2 - 1) e^{-6x} - e^{6x}]/(e^2 - 1)$.
21. $y = c_1 \sin \pi x$; c_i arbitrary.
22. $W = -2\tan^3 x$, linearly independent on $(0, \pi/2)$, $\left((2n-1)\dfrac{\pi}{2}, (2n+1)\dfrac{\pi}{2}\right)$; $n = 1, 2, \ldots$.
23. Dependent, $W = 0$.
24. $W = 1$, linearly independent.
25. (i) $y'' - y' - 6y = 0$
    (ii) $y'' + 2y' + y = 0$
    (iii) $y'' + 2ay' + (a^2 + b^2)y = 0$.
26. (i) $y''' - 4y'' + 3y' = 0$
    (ii) $y''' - 3y' - 2y = 0$
    (iii) $y^{iv} - 4y''' + 16y' - 16y - 0$.

9

CHAPTER

Non-homogeneous Linear Differential Equations

We have learnt in chapter 8 that the general solution of a non-homogeneous linear differential equation

$$a_0 \frac{d^n y}{dx^n} + a_1 \frac{d^{n-1}y}{dx^{n-1}} + \ldots a_{n-1}\frac{dy}{dx} + a_n y = r(x)$$

where, $a_0 \neq 0$, $r(x) \neq 0$ and $a_0, a_1, \ldots a_n$ are constants, is $y = y_c + y_p$.

Here, y_c and y_p are respectively, known as the complementary function (CF) and particular integral (PI) or particular solution. It was seen in the earlier chapter that when $r(x) = 0$, the general solution is $y = y_c$ and we have also discussed the method of finding the complementary function. It only remains to find the particular function. In this chapter, we will discuss methods of finding y_p depending on the nature of $r(x)$.

Definition: $\frac{1}{f(D)}f(x)$ is that function of x, not containing arbitrary constants which when operated upon by $f(D)$ gives $r(x)$ that is

$$f(D)\left\{\frac{1}{f(D)}r(x)\right\} = r(x)$$

Therefore, $\frac{1}{f(D)}r(x)$ can be regarded as the inverse operator of $f(D)$. Clearly, $\frac{1}{f(D)}r(x)$ satisfies the equation $f(D)y = r(x)$, hence the particular integral of this equation will be $\frac{1}{f(D)}r(x)$.

Now, we will prove two important relations

(I) To show that $\frac{1}{D}r(x) = \int r(x)dx$

Let $$\frac{1}{D}r(x) = u \qquad (1)$$

Operating by D, we obtain

$$D\left\{\frac{1}{D}r(x)\right\} = Du \Rightarrow r(x) = Du$$

$$\therefore \qquad r(x) = \frac{du}{dx}$$

Integrating, we get

$$u = \int r(x)dx \qquad (2)$$

No arbitrary constant being added since $u = \frac{1}{D}r(x)$ contains no arbitrary constant.

Thus, Eqs. (1) and (2) give $\frac{1}{D}r(x) = \int r(x)dx$.

(II) To show that $\frac{1}{D-a}r(x) = e^{ax}\int r(x)e^{-ax}dx$

Let $$\frac{1}{D-a}r(x) = u \qquad (1)$$

Operating by $D - a$, we obtain

$$(D-a)\frac{1}{D-a}r(x) = (D-a)u$$

$$\Rightarrow \qquad r(x) = \frac{du}{dx} - au$$

which is a linear differential equation with

$$\text{IF} = e^{\int -adx} = e^{-ax} = e^{-ax}$$

and its solution is given by

$$ue^{-ax} = \int r(x)e^{-ax}dx \qquad (2)$$

No constant being added as $u = \frac{1}{D-a}r(x)$ does not contain any arbitrary constant.

Thus, Eqs. (1) and (2) give $\frac{1}{D-a}r(x) = e^{ax}\int r(x)e^{-ax}dx$. ▲▲

EXAMPLE 9.1

Solve the differential equation $\frac{d^2y}{dx^2} + 9y = \sec 3x$.

Solution: We have equation $(D^2 + 9)y = \sec 3x$

For finding y_c, we consider $(D^2 + 9)y = 0$

Its AE is

$$D^2 + 9 = 0 \Rightarrow D = \pm 3i$$

$$\therefore \quad y_c = c_1 \cos 3x + c_2 \sin 3x$$

Now, particular integral is given by

$$y_p = \frac{1}{D^2+9}\sec 3x = \frac{1}{(D+3i)(D-3i)}\sec 3x$$

$$= \frac{1}{6i}\left[\frac{1}{D-3i} - \frac{1}{D+3i}\right]\sec 3x$$

$$= \frac{1}{6i}\left[\frac{1}{D-3i}\sec 3x - \frac{1}{D+3i}\sec 3x\right]$$

$$= \frac{1}{6i}\left[e^{3ix}\int \sec 3x\, e^{-3ix}dx - e^{-3ix}\int \sec 3x\, e^{3ix}dx\right] \quad \text{\{by using above result (II))}$$

Now, $\int e^{-3ix}\sec 3x\,dx = \int (\cos 3x - i\sin 3x)\sec 3x\,dx$

$$= \int (1 - i\tan 3x)\,dx = x + \frac{i}{3}\log \cos 3x$$

Also, $\int e^{3ix}\sec 3x\,dx = \int (\cos 3x + i\sin 3x)\sec 3x\,dx$

$$= \int (1 + i\tan 3x)\,dx = x - \frac{i}{3}\log \cos 3x$$

$$\therefore \quad y_p = \frac{1}{6i}\left[e^{3ix}\left(x + \frac{i}{3}\log\cos 3x\right) - e^{-3ix}\left(x - \frac{i}{3}\log\cos 3x\right)\right]$$

$$= \frac{1}{6i}\left[(\cos 3x + i\sin 3x\left(x + \frac{i}{3}\log\cos 3x\right) - (\cos 3x - i\sin 3x)\left(x - \frac{i}{3}\log\cos 3x\right)\right]$$

$$= \frac{1}{6i}\left[2ix\sin 3x + \frac{2i}{3}\cos 3x \log\cos 3x\right]$$

$$= \frac{x}{3}\sin 3x + \frac{\cos 3x}{9}\log\cos 3x$$

∴ The complete solution of the given equation is

$$y = y_c + y_p = c_1 \cos x + c_2 \sin x + \frac{x}{3}\sin 3x + \frac{\cos 3x}{9}\log \cos 3x.$$

EXAMPLE 9.2

Solve $\dfrac{d^2y}{dx^2} - y = e^{-x}$.

Solution: We have equation $(D^2 - 1)y = e^{-x}$

For finding y_c, we consider $(D^2 - 1)y = 0$

Its AE is

$$D^2 - 1 = 0 \Rightarrow D = \pm 1$$

∴ $$y_c = c_1 e^x + c_2 e^{-x}$$

Now, particular integral is given by

$$y_p = \frac{1}{D^2-1}e^{-x} = \frac{1}{(D+1)(D-1)}e^{-x}$$

$$= \frac{1}{2}\left[\frac{1}{D-1} - \frac{1}{D+1}\right]e^{-x}$$

$$= \frac{1}{2}\left[\frac{1}{D-1}e^{-x} - \frac{1}{D+1}e^{-x}\right]$$

$$= \frac{1}{2}\left[e^x\int e^{-x}\cdot e^{-x}dx - e^{-x}\int e^{-x}\cdot e^{+x}dx\right] \quad \text{\{by using result (II)\}}$$

$$= \frac{1}{2}\left[e^x\int e^{-2x}dx - e^{-x}x\right] = \frac{1}{2}\left[e^x\left(\frac{e^{-2x}}{-2}\right) - xe^{-x}\right]$$

$$= \frac{e^{-x}}{-4} - \frac{xe^{-x}}{2}$$

∴ The complete solution of the given equation is

$$y = c_1 e^x + c_2 e^{-x} - \frac{(x+2)}{4}e^{-x}.$$

EXAMPLE 9.3

Solve $\dfrac{d^2y}{dx^2} + 3\dfrac{dy}{dx} + 2y = e^{2x}$.

Solution: The given equation can be written as

$$(D^2 + 3D + 2) = e^{2x}$$

For finding y_c, we consider

$$(D^2 + 3D + 2)y = 0$$

Its AE is

$$D^2 + 3D + 2 = (D + 1)(D + 2) = 0$$

$\therefore$ $$D = -1, -2$$

Thus, $$y_c = c_1e^{-x} + c_2e^{-2x}$$

Now, particular integral is given by

$$y_p = \frac{1}{(D+1)(D+2)}e^{-2x}$$

Now, $\frac{1}{D+2}e^{-2x} = e^{-2x}\int e^{-2x}e^{2x}dx = xe^{-2x}$

Again $$\frac{1}{(D+1)(D+2)}e^{-2x} = \frac{1}{D+1}xe^{-2x}$$

$$= e^{-x}\int xe^{-2x}e^{x}dx$$

$$= e^{-x}\int xe^{-x}dx = e^{-x}\left[\frac{xe^{-x}}{-1} - \frac{e^{-x}}{1}\right]$$

$$= -e^{-2x}(x + 1)$$

Hence, the complete solution of the given equation is

$$y = c_1e^{-x} + c_2'e^{-2x} - xe^{-2x} \text{ where } c_2' = c_2 - 1.$$

Rules for Finding the Particular Integral

By using results I and II, we have one method for finding the particular integral but as this method involves integration terms depending upon the number of factors of $f(D)$, this method becomes lengthy. In this section, we will give methods depending upon the form of $r(x)$.

III Rule for finding the particular integral when $r(x)$ is of the form e^{ax}–

i.e., $$\frac{1}{f(D)}e^{ax}, f(a) \neq 0.$$

Prove that $\frac{1}{f(D)}e^{ax} = \frac{1}{f(a)}e^{ax}$

By simple differentiation, we know that

$$D(e^{ax}) = ae^{ax},\ D^2(e^{ax}) = a^2 e^{ax},\ \ldots\ D^n(e^{ax}) = a^n e^{ax}$$

Since $f(D)$ is a polynomial in D

$$\therefore \qquad f(D)e^{ax} = f(a)e^{ax}$$

Operating by $\dfrac{1}{f(D)}$, we obtain

$$\frac{1}{f(D)}\{f(D)e^{ax}\} = \frac{1}{f(D)}\{f(a)e^{ax}\}$$

$$\Rightarrow \qquad e^{ax} = f(a)\frac{1}{f(D)}e^{ax}\ \{\because f(a) \text{ is constant}\}$$

$$\Rightarrow \qquad \frac{1}{f(a)}e^{ax} = \frac{1}{f(D)}e^{ax} \text{ provided } f(a) \neq 0.$$

Note: If $f(a) = 0$, the above rule fails and we proceed further.
Since $f(a) = 0$ therefore, $D - a$ is a factor of $f(D)$. Let $f(D) = (D - a)\,\phi(D)$, where $\phi(a) \neq 0$.

Then,
$$\frac{1}{f(D)}e^{ax} = \frac{1}{D-a}\frac{1}{\phi(D)}e^{ax} = \frac{1}{D-a}\frac{1}{\phi(a)}e^{ax} \quad \{\because \phi(a) \neq 0\}$$

$$= \frac{1}{\phi(a)}\frac{1}{D-a}e^{ax} = \frac{1}{\phi(a)}e^{ax}\int e^{ax}\cdot e^{-ax}dx \quad \text{\{by using Rule II\}}$$

$$= \frac{1}{\phi(a)}xe^{ax}$$

$$\therefore \qquad \frac{1}{f(D)}e^{ax} = x\frac{1}{\phi(a)}e^{ax} \qquad (1)$$

Now,
$$x\frac{1}{\frac{d}{dD}[f(D)]}e^{ax} = x\frac{1}{\frac{d}{dD}[(D-a)\phi(D)]}e^{ax}$$

$$= x\frac{1}{1\cdot\phi(D) + (D-a)\phi'(D)}e^{ax}$$

$$= x\frac{1}{\phi(a) + (a-a)\phi'(a)}e^{ax} = x\frac{e^{ax}}{\phi(a)} \qquad (2)$$

Thus, Eqs. (1) and (2) give

$$\frac{1}{f(D)}e^{ax} = x\frac{1}{\frac{d}{dD}[f(D)]}e^{ax}$$

➲ Working Rule

If $r(x) = e^{ax}$ then for finding y_p, put $D = a$ in $\dfrac{r(x)}{f(D)}$, provided $f(a) \neq 0$.

EXAMPLE 9.4

Solve $\dfrac{d^2y}{dx^2} - 2K\dfrac{dy}{dx} + K^2y = e^x$.

Solution: We have $(D^2 - 2KD + K^2)y = e^x$

For finding y_c, AE is

$$D^2 - 2KD + K^2 = 0$$

$$\Rightarrow \qquad (D - K)^2 = 0$$

Hence, $D = K, K$.

Therefore, $\qquad y_c = (c_1 + c_2x)e^{kx}$

Now, $\qquad y_p = \dfrac{1}{(D-K)^2}e^x = \dfrac{1}{(1-K)^2}e^x$ (by using Rule III)

Hence, the general solution of the given differential equation is

$$y = (c_1 + c_2x)e^{Kx} + \frac{1}{(1-K)^2}e^x.$$

EXAMPLE 9.5

Solve $(D + 2)(D - 1)^2y = e^{-2x} + 2\sinh x$.

Solution: We have equation

$$(D + 2)(D - 1)^2y = e^{-2x} + e^x + e^{-x}$$

For finding y_c, AE is

$$(D + 2)(D - 1)^2 = 0 \Rightarrow D = -2, 1, 1$$

Therefore, $y_c = (c_1 + c_2x)e^x + c_3e^{-2x}$.

Now, $\qquad y_p = \dfrac{e^{-2x} + e^x + e^{-x}}{(D+2)(D-1)^2}$

Let us evaluate each of these terms separately.

$$\frac{1}{(D+2)(D-1)^2}e^{-2x} = \frac{1}{D+2}\left[\frac{1}{(D-1)^2}e^{-2x}\right]$$

$$= \frac{1}{D+2}\left[\frac{1}{(-2-1)^2}e^{-2x}\right] = \frac{1}{9}\frac{1}{D+2}e^{-2x}$$

$$= \frac{1}{9}x \cdot \frac{1}{\frac{d}{dD}[D+2]} e^{-2x}$$

$\{\because$ On putting $D = -2, f(D) = D + 2 = 0\} \Rightarrow$ Rule III fails

$$= \frac{1}{9}\frac{xe^{-2x}}{1}$$

$$\frac{1}{(D+2)(D-1)^2}e^x = \frac{1}{1+2}\left[\frac{1}{(D-1)^2}e^x\right] = \frac{1}{3}x\frac{1}{2(D-1)}e^x$$

$$= \frac{1}{3}x\frac{x}{2.1}e^x \quad \{\because \text{again Rule III fails}\}$$

$$= \frac{x^2}{6}e^x$$

and $$\frac{1}{(D+2)(D-1)^2}e^{-x} = \frac{1}{(-1+2)(-1-1)^2}e^{-x} = \frac{e^{-x}}{4}$$

Hence, $$y_p = \frac{x}{9}e^{-2x} + \frac{x^2}{6}e^x + \frac{e^{-x}}{4}$$

$\therefore$ The general solution is

$$y = (c_1 + c_2e^x) + c_3e^{-2x} + \frac{x}{9}e^{-2x} + \frac{x^2}{6}e^x + \frac{e^{-x}}{4}.$$

EXAMPLE 9.6

Solve $\frac{d^4y}{dx^4} - \frac{d^2y}{dx^2} = 2.$

Solution: We have equation $(D^4 - D^2)y = 2e^{0x}$

For finding y_c, AE is

$$D^4 - D^2 = 0 \Rightarrow D^2(D^2 - 1) = 0$$

$$\Rightarrow \quad D = 0, 0, 1, -1$$

Hence, $y_p = c_1 + c_2x + c_3e^x + c_4e^{-x}$.

Now, $$y_p = \frac{2}{D^4 - D^2}e^{0x} = 2x\frac{1}{4D^3 - 2D}e^{ax} \quad \{\because \text{Rule III fails}\}$$

$$= 2x \cdot x\frac{1}{12D^2 - 2}e^{0x} \quad \{\because \text{again Rule III fails}\}$$

$$= 2x^2\frac{1}{0-2}e^{0x} = -x^2$$

Hence, the general solution is

$$y = c_1 + c_2 x + c_3 e^x + c_4 e^{-x} - x^2.$$

IV Rule for determining the particular integral when $r(x)$ is of the form

$$\sin(ax + b) \text{ or } \cos(ax + b)$$

Prove that $\dfrac{1}{f(D^2)}\cos(ax + b) = \dfrac{1}{f(-a^2)}; \cos(ax + b); f(-a^2) \neq 0.$

By simple differentiation, we know that

$$D(\cos(ax) - b) = -a\sin(ax + b),\ D^2(\cos(ax) - b) = -a^2\cos(ax + b)$$

$$D^3(\cos(ax) - b) = a^3\sin(ax + b),\ D^4(\cos(ax) - b) = (-a^2)^2\cos(ax + b)$$

$$(D^2)^n\cos(ax + b) = (-a^2)^n\cos(ax + b)$$

It suggests that

$$f(D^2)\cos(ax + b) = f(-a^2)\cos(ax + b)\ \{\because f(D)^2) \text{ is a polynomial in } D^2\}$$

Operating by $\dfrac{1}{f(D^2)}$, we obtain

$$\frac{1}{f(D^2)}\{f(D^2)\cos(ax+b)\} = \frac{1}{f(D^2)}\{f(-a^2)\cos(ax + b)\}$$

$$\Rightarrow \quad \cos(ax + b) = f(-a^2)\frac{1}{f(D^2)}\cos(ax + b) \quad \{\because f(-a^2) \text{ is constant}\}$$

$$\Rightarrow \quad \frac{1}{f(-a^2)}\cos(ax + b) = \frac{1}{f(D^2)}\cos(ax + b), \text{ provided } f(-a^2) \neq 0$$

Similarly, we can obtain

$$\frac{1}{f(D^2)}\sin(ax + b) = \frac{1}{f(-a^2)}\sin(ax + b), \text{ provided } f(-a^2) \neq 0.$$

Note: If $f(-a^2) = 0$, the above rule fails, and we proceed further as

$$\frac{1}{f(D^2)}\cos(ax + b) = \text{Real part of } \frac{1}{f(D^2)}e^{i(ax + b)}$$

$$= \text{Real part of } x\frac{1}{\frac{d}{dD}\{f(D^2)\}}e^{i(ax+b)} \quad \{\because f(-a^2) = 0 \Rightarrow \text{Rule III fails}\}$$

$$= \text{Real part of } x \frac{1}{\frac{d}{dD}\{f(D^2)\}} \{(\cos(ax+b) + i\sin(ax+b)\}$$

$$= x \frac{1}{\frac{d}{dD}[f(D^2)]} \cos(ax+b)$$

Similarly, we can have

$$\frac{1}{f(D^2)} \sin(ax+b) = x \frac{1}{\frac{d}{dD}[f(D^2)]} \sin(ax+b), \text{ when } f(-a^2) = 0.$$

Note: We have the above rule when $f(D)$ contains only even powers of D. Suppose now that $f(D)$ contains odd powers of D also. In this case, breaking up $f(D)$ into its even and odd parts, we may write

$$f(D) = \phi_1(D^2) + D\phi_2(D^2)$$

We have

$$\frac{1}{f(D)} \cos(ax+b) = \frac{1}{\phi_1(D^2) + D\phi_2(D^2)} \cos(ax+b)$$

$$= [\phi_1(D^2) - D\phi_2(D^2)] \frac{1}{[\phi_1(D^2) + D\phi_2(D^2)][\phi_1(D^2) - D\phi_2(D^2)} \cos(ax+b)$$

$$= [\phi_1(D^2) - D\phi_2(D^2)] \frac{1}{[\phi_1(D^2)]^2 - D^2[\phi_2(D^2)]^2} \cos(ax+b)$$

$$= [\phi_1(D^2) - D\phi_2(D^2)] \frac{1}{[\phi_1(-a^2)]^2 - (-a^2)[\phi_2(-a^2)]^2} \cos(ax+b)$$

We write $\qquad \phi_1(-a^2) = p, \ \phi_2(-a)^2 = q$

$$\therefore \quad \frac{1}{f(D)} \cos ax = \frac{1}{p^2 + a^2q^2} [\phi_1(D^2) - D\phi_2(D^2)] \cos(ax+b)$$

$$= \frac{1}{p^2 + a^2q^2} [\phi_1(-a^2) - D\phi_2(-a^2)] \cos(ax+b)$$

$$= \frac{1}{p^2 + a^2q^2} [p\cos(ax+b) - q(-a\sin(ax+b))]$$

$$= \frac{p\cos(ax+b) + qa\sin(ax+b)}{p^2 + a^2q^2}$$

The case of $\sin ax$ can be similarly treated.

➲ Working Rule

If $r(x) = \cos(ax + b)$ or $\sin(ax + b)$ then for finding y_p, put $D^2 = -a^2$ in $\dfrac{r(x)}{f(D)}$ provided the denominator $\neq 0$.

EXAMPLE 9.7

Solve $\dfrac{d^2y}{dx^2} - 5\dfrac{dy}{dx} + 6y = \sin 3x$

Solution: We have $[D^2 - 5D + 6]y = \sin 3x$

For finding y_c, AE is

$$D^2 - 5D + 6 = 0 \quad \Rightarrow \quad D = 2, 3$$

Hence, $y_c = c_1e^{2x} + c_3e^{3x}$.

Also,
$$\begin{aligned} y_p &= \frac{1}{D^2 - 5D + 6}\sin 3x \\ &= \frac{1}{-9 - 5D + 6}\sin 3x \quad \{\text{by using rule IV put } D^2 = -3^2 = -9\} \\ &= \frac{1}{-3 - 5D}\sin 3x \\ &= -(3 - 5D)\frac{1}{(3 - 5D)(3 + 5D)}\sin 3x \\ &= -(3 - 5D)\frac{1}{9 - 25D^2}\sin 3x \\ &= -(3 - 5D)\frac{1}{9 - 25(-9)}\sin 3x \\ &= \frac{-1}{234}(3 - 5D)\sin 3x \\ &= \frac{-1}{234}[3\sin 3x - 15\cos 3x] \\ &= \frac{-1}{78}(\sin 3x - 5\cos 3x) \end{aligned}$$

Thus, the required general solution is

$$y = c_1e^{2x} + c_2e^{3x} - \frac{1}{78}(\sin 3x - 5\cos 3x).$$

EXAMPLE 9.8

Solve $(D^2 + 4)y = \cos 2x$.

Solution: For finding y_c, AE is

$$D^2 + 4 = 0$$

$$\Rightarrow \qquad D = \pm 2i$$

Hence, $y_c = c_1\cos 2x + c_2\sin 2x$.

Also, $y_p = \dfrac{1}{D^2+4}\cos 2x = \dfrac{1}{-4+4}\cos 2x$ $\{\because$ by using Rule IV, put $D^2 = -2^2 = -4\}$

$= x\dfrac{1}{2D}\cos 2x$ $\{\because$ Rule IV fails$\}$

$= \dfrac{x}{2}\int \cos 2x dx$ $\{\because$ of Rule I$\}$

$= \dfrac{x}{2}\dfrac{\sin 2x}{2}$

Thus, the required general solution is

$$y = c_1\cos 2x + c_2\sin 2x + \frac{x}{4}x\sin 2x.$$

EXAMPLE 9.9

Solve $\dfrac{d^3y}{dx^3} + 4\dfrac{dy}{dx} = \sin 2x.$

Solution: We have the given equation

$$(D^3 + 4D)y = \sin 2x$$

For finding y_c, AE is

$$D^3 + 4D = D(D^2 + 4) = 0$$

$$\Rightarrow \qquad D = 0, \pm 2i$$

Hence, $y_c = c_1 + c_2\cos 2x + c_3\sin 2x$.

Now, $y_p = \dfrac{1}{D^3+4D}\sin 2x = \dfrac{1}{D^2+4}\dfrac{1}{D}\sin 2x$

$= \dfrac{1}{D^2+4}\int \sin 2x dx$ {by using Rule I}

$= \dfrac{1}{D^2+4}\dfrac{-\cos 2x}{2}$

$= \dfrac{-1}{2}\dfrac{1}{-4+4}\cos 2x$ {by using Rule IV put $D^2 = -2^2 = -4$}

$$= \frac{-1}{2} x \frac{1}{2D} \cos 2x$$

$$= -\frac{x}{4} \int \cos 2x \, dx \quad \{\text{by using Rule I}\}$$

$$= \frac{-x}{4} \frac{\sin 2x}{2}$$

Thus, the required general solution is

$$y = c_1 + c_2 \cos 2x + c_3 \sin 2x - \frac{x}{8} \sin 2x.$$

EXAMPLE 9.10

Solve the differential equation

$$(D^3 - 3D^2 + 9D - 27)y = \cos 3x.$$

Solution: For finding y_c, AE is

$$D^3 - 3D^2 + 9D - 27 = 0$$

Clearly, $D = 3$ is satisfying this as

$$(3)^3 - 3(3)^2 + 9(3) - 27 = 0$$

So one root is 3. The remaining two roots can be found as

3	1	–3	9	–27
		3	0	27
	1	0	9	0

$$D^2 + 9 = 0 \Rightarrow D = \pm 3i.$$

∴ The roots of AE are 3, $\pm 3i$.

Hence, $y_c = c_1 e^{3x} + c_2 \cos 3x + c_3 \sin 3x$.

Now, $$y_p = \frac{1}{D^3 - 3D^2 + 9D - 27} \cos 3x$$

$$= \frac{1}{D^2 \cdot D - 3D^2 + 9D - 27} \cos 3x$$

$$= \frac{1}{-9D - 3(-9) + 9D - 27} \cos 3x \quad \{\text{by using Rule IV, put } D^2 = -3^2 = -9\}$$

$$= \frac{1}{0} \cos 3x \quad \{\text{Rule fails}\}$$

$$= x \frac{1}{3D^2 - 6D + 9} \cos 3x$$

$$= x\frac{1}{3(-9)-6D+9}\cos 3x \quad \{\text{by using Rule IV, put } D^2 = -3^2 = -9\}$$

$$= x\frac{1}{-6D-18}\cos 3x$$

$$= \frac{-x}{6}(D-3)\frac{1}{D^2-9}\cos 3x$$

$$= \frac{-x}{6}\frac{D-3}{-9-9}\cos 3x = \frac{x}{108}(-3\sin 3x - 3\cos 3x)$$

Thus, the required general solution is

$$y = c_1e^{3x} + c_2\cos 3x + c_3\sin 3x - \frac{x}{36}(\sin 3x + \cos 3x).$$

EXAMPLE 9.11

Solve the differential equation

$$(D^3 - 5D^2 + 9D - 5)y = e^x + \cos(x + 2).$$

Solution: For finding y_c, AE is

$$D^3 - 5D^2 + 9D - 5 = 0$$

Clearly, $D = 1$ is satisfying this as

$$1 - 5 + 9 - 5 = 0$$

Hence, one root is 1. The remaining two roots can be found as

1	1	–5	9	–5
		1	–4	5
	1	–4	5	0

$$D^2 - 4D + 5 = 0$$

$$\Rightarrow \quad D = \frac{4 \pm \sqrt{16-20}}{2} = 2 \pm i$$

Therefore, the roots of AE are 1, $2 \pm i$.

Thus, $$y_c = c_1e^x + e^{2x}(c_2\cos x + c_3\sin x)$$

Now, $$y_p = \frac{1}{D^3 - 5D^2 + 9D - 5}(e^x + \cos(x + 2))$$

Let us evaluate, each of these terms separately.

$$\frac{1}{D^3-5D^2+9D-5}e^x = \frac{1}{1-5+9-5}e^x \quad \text{\{by using Rule III\}}$$

$$= x\frac{1}{3D^2-10D+9}e^x \quad \{\because \text{ Rule III fails}\}.$$

$$= x\frac{1}{3-10+9}e^x = \frac{x}{2}e^x$$

$$\frac{1}{D^3-5D^2+9D-5}\cos(x+2) = \frac{1}{-D-5(-1)+9D-5}\cos(x+2)$$

{by using Rule IV, put $D^2 = -1$}

$$= \frac{1}{8D}\cos(x+2)$$

$$= \frac{1}{8}\int \cos(x+2)\,dx \quad \{\because \text{ by using Rule I}\}$$

$$= \frac{1}{8}\sin(x+2).$$

Thus, the required general solution is

$$y = c_1e^x + e^{2x}(c_2\cos 2x + c_3\sin 2x) + \frac{1}{2}xe^x + \frac{1}{8}\sin x.$$

V Rule for determining the particular integral when $r(x)$ is a polynomial in x.
Let us assume $r(x) = x^n$

Here, $y_p = \dfrac{1}{f(D)}x^n = [f(D)]^{-1}x^n$.

Expand $[f(D)]^{-1}$ in ascending powers of D as far as the terms containing D^n, since $D^{n+1}(x^n) = 0$, $D^{n+2}(x^n) = 0$ and so on and operate on x^n term by term.

Note: The following binomial expansions should be remembered well.

(i) $(1+x)^{-1} = 1 - x + x^2 - x^3 + x^4 - \dots$

(ii) $(1-x)^{-1} = 1 + x + x^2 + x^3 + x^4 + \dots$

(iii) $(1+x)^{-2} = 1 - 2x + 3x^2 - 4x^3 + \dots$

(iv) $(1-x)^{-2} = 1 + 2x + 3x^2 + 4x^3 + \dots$

EXAMPLE 9.12

Solve the equation

$$\frac{d^2y}{dx^2} = a + bx + cx^2,$$

given that $\frac{dy}{dx} = 0$, when $x = 0$ and $y = d$, when $x = 0$.

Solution: We have the given equation

$$D^2y = a + bx + cx^2$$

For finding y_c, AE is

$$D^2 = 0 \Rightarrow D = 0, 0$$

Hence, $y_c = c_1 + c_2x$.

Now,

$$y_p = \frac{1}{D^2}(a + bx + cx^2)$$

$$= \frac{1}{D}\left[ax + b\frac{x^2}{2} + \frac{cx^3}{3}\right]$$

$$= \frac{ax^2}{2} + \frac{bx^3}{6} + \frac{cx^4}{12}$$

Thus, the general solution is

$$y = c_1 + c_2x + \frac{1}{2}ax^2 + \frac{1}{6}bx^3 + \frac{1}{12}cx^4 \qquad (1)$$

$$\frac{dy}{dx} = c_2 + ax + \frac{1}{2}bx^2 + \frac{1}{3}cx^3 \qquad (2)$$

Putting $x = 0$, $y = d$ and $\frac{dy}{dx} = 0$ in Eqs. (1) and (2), we get

$c_2 = 0$ and $c_1 = d$.

$$\therefore \quad y = d + \frac{1}{2}ax^2 + \frac{1}{6}bx^3 + \frac{1}{12}cx^4.$$

EXAMPLE 9.13

Find the general solution of the equation

$$y^{\text{iv}} + 3y'' = 108x^2.$$

Solution: We have the given equation

$$(D^4 + 3D^2)y = 108x^2.$$

For finding y_c, AE is

$$D^4 + 3D^2 = 0$$

$$\Rightarrow \quad D^2(D^2 + 3) = 0$$

$$\Rightarrow \quad D = 0, 0, \pm\sqrt{3}i$$

Hence, $y_c = c_1 + c_2x + c_3\cos\sqrt{3}x + c_4\sin\sqrt{3}x$.

Now,
$$y_p = \frac{108}{D^2(D^2+3)}x^2 = \frac{108}{D^2}\frac{1}{3}\left[1+\frac{D^2}{3}\right]^{-1}x^2$$
$$= \frac{36}{D^2}\left[1-\frac{D^2}{3}+\frac{D^4}{9}\right]x^2$$
$$= \frac{36}{D^2}\left[x^2-\frac{2}{3}\right]$$
$$= 36\left[\frac{x^4}{12}-\frac{x^2}{3}\right] = 3x^4 - 12x^2$$

Thus, the required general solution is
$$y = c_1 + c_2x + c_3\cos\sqrt{3}x + c_4\sin\sqrt{3}x + 3x^4 - 12x^2.$$

EXAMPLE 9.14

Solve $(D^4 - a^4)y = x^4 + \sin bx$.

Solution: For finding y_c, AE is
$$D^4 - a^4 = 0$$
$$\Rightarrow \qquad D = \pm a, \pm ai$$

Hence, $y_c = c_1e^{ax} + c_2e^{-ax} + c_3\cos ax + c_4\sin ax$.

Now,
$$y_p = \frac{1}{D^4-a^4}(x^4 + \sin bx)$$

Let us evaluate each of these terms separately.
$$\frac{1}{D^4-a^4}x^4 = \frac{-1}{a^4}\left[1-\frac{D^4}{a^4}\right]^{-1}x^4$$
$$= \frac{-1}{a^4}\left[1+\frac{D^4}{a^4}+\ldots\right]x^4$$
$$= \frac{-1}{a^4}\left[x^4+\frac{24}{a^4}\right]$$
$$\frac{1}{D^4-a^4}\sin bx = \frac{\sin bx}{b^4-a^4} \quad \{\text{by using Rule IV, put } D^2 = -b^2\}$$

Thus, the required general solution is

$$y = c_1e^{ax} + c_2e^{-ax} + c_3\cos ax + c_4\sin ax - \frac{1}{a^4}\left(x^4 + \frac{24}{a^4}\right) + \frac{1}{(b^4 - a^4)}\sin bx .$$

EXAMPLE 9.15

Solve $(D^2 + 2D + 1)y = 1 + x^2$.

Solution: We have the given equation

$$(D + 1)^2y = 1 + x^2$$

For finding y_c, AE is

$$(D + 1)^2 = 0 \Rightarrow D = -1, -1$$

Hence, $y_c = c_1e^{-x} + c_2xe^{-x}$.

Now,
$$\begin{aligned} y_p &= \frac{1}{(D+1)^2}(1+x^2) \\ &= (1 + D)^{-2}(1 + x^2) \\ &= (1 - 2D + 3D^2 - 4D^3 + \ldots)(1 + x^2) \\ &= 1 + x^2 - 2(2x) + 3(2) \\ &= x^2 - 4x + 7 \end{aligned}$$

Thus, the required general solution is

$$y = (c_1 + c_2x)e^{-x} + x^2 - 4x + 7.$$

EXERCISE 9.1

Solve the following differential equations.

1. $(D^2 + 4)y = \tan 2x$.
2. $\dfrac{d^2y}{dx^2} + y = \operatorname{cosec} x$.
3. $(D^2 - 2D + 1)y = x^2 - 1$.
4. $9\dfrac{d^2y}{dx^2} + 12\dfrac{dy}{dx} + 4y = e^{-2x/3}$.
5. $\dfrac{d^3y}{dx^3} - \dfrac{d^2y}{dx^2} - 4\dfrac{dy}{dx} + 4y = e^{2x}$.
6. $(D^2 - 2D + 1)y = (1 + e^{-x})^2$.
7. $(3D^2 - 4D + 5)y = e^x - 2e^{2x} + 3e^{3x}$.

8. $\dfrac{d^2y}{dx^2} - 2\dfrac{dy}{dx} + y = 2\cosh(x/2)$.

9. $\dfrac{d^3y}{dx^3} - 5\dfrac{d^2y}{dx^2} + 7\dfrac{dy}{dx} - 3y = e^{2x}\cosh x$.

10. $(D^2 - a^2)y = e^{ax} + e^{bx}$.

11. $(D^2 + 1)y = \cos(2x - 1)$.

12. $\dfrac{d^2y}{dx^2} - 8\dfrac{dy}{dx} + 9y = 40\sin 5x$.

13. $\dfrac{d^2y}{dx^2} - 4y = \cos^2 x$.

14. $(D^4 + 2D^3 - 3D^2)y = 3e^{2x} + 4\sin x$.

15. $(D^2 + 1)y = \sin x \sin 2x$.

16. $(D^3 + 2D^2 + 4D + 8)y = x^2$.

17. $(D^3 + 2D^2 + D)y = e^{2x} + x^2 + x$.

18. $\dfrac{d^2y}{dx^2} - 4y = e^x + \sin 2x + \cos^2 x$.

19. $(D - 2)^2 = 8(e^{2x} + \sin 2x + x^2)$.

20. $(D^4 + D^2 + 16)y = 16x^2 + 256$.

21. If $\dfrac{d^2x}{dt^2} + \dfrac{g}{b}(x - a) = 0$ where, a, b and g are positive numbers and $x = a'$ and $\dfrac{dx}{dt} = 0$, when $t = 0$, then show that $x = a + (a' - a)\cos(\sqrt{g/b}\,t)$.

22. Solve $(D^2 - 7D + 6)y = 2e^{3x}$, given that $y = 1$, $dy/dx = 0$, when $x = 0$.

23. Solve $\dfrac{d^2y}{dx^2} + 2\dfrac{dy}{dx} + 10y + 37\sin 3x = 0$ and find the value of y, when $x = \pi/2$.

 If it is given that $y = 3$ and $dy/dx = 0$, when $x = 0$.

24. Transform the differential equation $\dfrac{d^2y}{dx^2}\cos x + \dfrac{dy}{dx}\sin x - 2y\cos^2 x = 2\cos^5 x$ into one having z as independent variable where, $z = \sin x$ and solve it.

25. Solve the differential equation.

$$\frac{d^4y}{dx^4} + 2\frac{d^3y}{dx^3} - 3\frac{d^2y}{dx^2} = x^2 + 3e^{2x} + 4\sin x.$$

26. Find the solution of the equation

$$\frac{d^2y}{dx^2} - y = 1$$

which vanishes when $x = 0$ and tends to a finite limit as $x \to -\infty$.

ANSWERS

Exercise 9.1

1. $y = c_1 \cos 2x + c_2 \sin 2x - \frac{1}{4} \cos 2x \log \tan\left(x + \frac{\pi}{4}\right).$
2. $y = c_1 \cos x + c_2 \sin x - x \cos x + \log(\sin x) \sin x.$
3. $y = (c_1 + c_2 x) e^x + (x^2 + 4x + 5).$
4. $y = (c_1 + c_2 x) e^{-2x/3} + \frac{1}{8} x^2 e^{-2x/3}.$
5. $y = c_1 e^x + c_2 e^{2x} + c_3 e^{-2x} + \frac{1}{4} x e^{3x}.$
6. $y = (c_1 + c_2 x) e^x + 1 + \frac{1}{9} e^{-2x} + \frac{1}{2} e^{-x}.$
7. $y = e^{2x/3}\left[c_1 \cos \frac{\sqrt{11}}{3} x + c_2 \sin \frac{\sqrt{11}}{3} x\right] + \frac{1}{4} e^x - \frac{2}{9} e^{2x} - \frac{3}{20} e^{3x}.$
8. $y = (c_1 + c_2 x) e^{-x} + e^{x/2} + 4e^{-x/2}.$
9. $y = (c_1 + c_2 x) e^x + c_3 e^{3x} + \frac{1}{8} x e^{3x} - \frac{1}{8} x^2 e^x.$
10. $y = c_1 e^{ax} + c_2 e^{-ax} + \frac{x}{2a} e^{ax} + \frac{1}{b^2 - a^2} e^{bx}.$
11. $y = c_1 \cos x + c_2 \sin x + \frac{1}{65} [(\cos (2x - 1) - 8 \sin (2x - 1)].$
12. $y = e^{4x} [c_1 \cosh \sqrt{7x} + c_2 \sinh \sqrt{7x}] + \frac{25}{29} \cos 5x - \frac{10}{29} \sin 5x.$
13. $y = c_1 e^{2x} + c_2 e^{-2x} - \frac{1}{8}\left(1 + \frac{1}{2} \cos 2x\right).$
14. $y = c_1 + c_2 x + c_3 e^x + c_4 e^{-3x} + \frac{2}{5} (2 \sin x + \cos x).$

15. $y = c_1 \cos x + c_2 \sin x + \frac{x}{4} \sin x + \frac{1}{16} \cos 3x.$

16. $y = c_1 e^{-2x} + c_2 \cos 2x + c_3 \sin 2x + \frac{1}{8}(x^2 - x).$

17. $y = c_1 + (c_2 + c_3 x)e^{-x} + \frac{1}{6} x\left(2x^2 - 9x + 24 + \frac{1}{18} e^{2x}\right).$

18. $y = c_1 e^{2x} + c_2 e^{-2x} - \frac{1}{3} e^{x} - \frac{1}{8} \sin 2x - \frac{1}{16} \cos 2x - \frac{1}{8}.$

19. $y = (c_1 + c_2 x) e^{2x} + 4x^2 e^{2x} + \cos 2x + 2x^2 + 4x + 3.$

20. $y = e^{-(\sqrt{7/2})x}\left[c_1 \cos \frac{3}{2} x + c_2 \sin \frac{3}{2} x\right] + e^{(\sqrt{7/2})x}\left[c_3 \cos \frac{3}{2} x + c_4 \sin \frac{3}{2} x\right] + x^2 + \frac{127}{8}.$

22. $y = \frac{1}{5} e^{x} - \frac{1}{15} e^{6x} - \frac{1}{3} e^{3x}.$

23. $y = e^{-x}(c_1 \cos 3x + c_2 \sin 3x) + 6 \cos 3x - \sin 3x;\ y = 1.$

24. $\frac{d^2 y}{dz^2} - 2y = 2(1 - z^2);\ y = c_1 e^{\sqrt{2} \sin x} + c_2 e^{-\sqrt{2} \sin x} + \sin^2 x.$

25. $y = c_1 - c_2 x + c_3 e^{x} + c_4 e^{-3x} - \frac{1}{36} x^4 \quad \frac{2}{27} x^3 \quad \frac{7}{27} x^2 + \frac{3}{20} e^{2x} + \frac{2}{5} \cos x + \frac{4}{5} \sin x.$

26. $y = c_1 e^{x} + c_2 e^{-x} - 1;\ y = e^{x} - 1.$ ▲▲

VI Rule for determining y_p when $r(x) = e^{ax} V$ where V is a function of x.

To show that

$$\frac{1}{f(D)} e^{ax} V = e^{ax} \frac{1}{f(D+a)} V$$

By successive differentiation, we have

$$\begin{aligned} D(e^{ax} V) &= e^{ax} DV + Vae^{ax} = e^{ax}(D + a)V \\ D^2(e^{ax} V) &= D(e^{ax}(D + a)V) \\ &= e^{ax} D(D + a)V + ae^{ax}(D + a)V \\ &= e^{ax}(D + a)^2 V \end{aligned}$$

Similarly,

$$\begin{aligned} D^3(e^{ax} V) &= e^{ax}(D + a)^3 V \\ D^n(e^{ax} V) &= e^{ax}(D + a)^n V \end{aligned}$$

$\therefore$

$$f(D)(e^{ax} V) = e^{ax} f(D + a) V$$

Taking $\frac{1}{f(D+a)}V$ in place of V, we have

$$f(D)\left[e^{ax}\frac{1}{f(D+a)}V\right] = e^{ax}f(D+a)\left[\frac{1}{f(D+a)}V\right]$$

$$\Rightarrow \qquad f(D)\left[e^{ax}\frac{1}{f(D+a)}V\right] = e^{ax}V$$

Operating by $\frac{1}{f(D)}$, we obtain

$$e^{ax}\frac{1}{f(D+a)}V = \frac{1}{f(D)}(e^{ax}V)$$

Thus, we have the given result.

➲ Working Rule

Replace D by $D + a$ and take out e^{ax} before the operator $1/f(D)$. Then determine $\frac{1}{f(D+a)}V$ by the methods discussed earlier.

EXAMPLE 9.16

Solve $\frac{d^2y}{dx^2} - 2\frac{dy}{dx} + 4y = e^x \cos x$.

Solution: We have the given equation

$$(D^2 - 2D + 4)y = e^x \cos x$$

For finding y_c, AE is

$$D^2 - 2D + 4 = 0$$

$$\Rightarrow \qquad D = 1 \pm \sqrt{3}i$$

Hence, $\quad y_c = e^x[c_1 \cos\sqrt{3}x + c_2 \sin\sqrt{3}x]$

Now, $\quad y_p = \frac{1}{D^2 - 2D + 4}e^x \cos x$

$$= e^x \frac{1}{(D+1)^2 - 2(D+1) + 4}\cos x \quad \text{\{by using Rule VI\}}$$

$$= e^x \frac{1}{D^2+3} \cos x = e^x \frac{1}{-1^2+3} \cos x$$

$$= \frac{1}{2} e^x \cos x$$

Thus, the required general solution is

$$y = e^x [c_1 \cos \sqrt{3}x + c_2 \sin \sqrt{3}x] + \frac{1}{2} e^x \cos x.$$

EXAMPLE 9.17

Solve $\frac{d^2y}{dx^2} - 2\frac{dy}{dx} + y = x^2 e^{3x}$.

Solution: The given equation is

$$(D^2 - 2D + 1)y = x^2 e^{3x}$$

For finding y_c, AE is

$$D^2 - 2D + 1 = 0$$

$\Rightarrow$ $$(D - 1)^2 = 0$$

$\Rightarrow$ $$D = 1, 1$$

Hence, $y_c = (c_1 + c_2 x)e^x$.

Now,
$$y_p = \frac{1}{(D-1)^2} x^2 e^{3x}$$

$$= e^{3x} \frac{1}{(\overline{D+3}-1)^2} x^2 \quad \text{\{by using Rule VI\}}$$

$$= e^{3x} \frac{1}{(D+2)^2} x^2$$

$$= \frac{e^{3x}}{4} \left(1 + \frac{D}{2}\right)^{-2} x^2$$

$$= \frac{e^{3x}}{4} \left(1 - 2\left(\frac{D}{2}\right) + 3\left(\frac{D}{2}\right)^2 - \ldots\right) x^2$$

$$= \frac{e^{3x}}{4} \left(1 - D + \frac{3}{4}D^2 - \ldots\right) x^2$$

$$= \frac{e^{3x}}{4} \left(x^2 - 2x + \frac{3}{4} \times 2\right)$$

Thus, the required general solution is

$$y = (c_1 + c_2x)e^x + \frac{e^{3x}}{8}(2x^2 - 4x + 3).$$

EXAMPLE 9.18

Solve $(D^2 - 1)y = \cosh x \cos x$.

Solution: For finding y_c, AE is

$$D^2 - 1 = 0 \Rightarrow D = \pm 1$$

Hence, $y_c = c_1e^x + c_2e^{-x}$.

Now,
$$y_p = \frac{1}{D^2-1}(\cosh x \cos x)$$

$$= \frac{1}{D^2-1}\left(\frac{e^x+e^{-x}}{2}\cos x\right)$$

$$= \frac{1}{2}\frac{1}{D^2-1}e^x \cos x + \frac{1}{2}\frac{1}{D^2-1}e^{-x}\cos x$$

Let us evaluate each term separately.

$$\frac{1}{D^2-1}e^x \cos x = e^x \frac{1}{(D+1)^2-1}\cos x \quad \text{\{by using Rule VI\}}$$

$$= e^x \frac{1}{D^2+2D}\cos x$$

$$= e^x \frac{1}{-1^2+2D}\cos x$$

$$= e^x \frac{2D+1}{4D^2-1}\cos x$$

$$= e^x \frac{2(-\sin x)+\cos x}{4(-1^2)-1}$$

$$= \frac{e^x}{5}(2\sin x - \cos x)$$

Similarly, we can obtain

$$\frac{1}{D^2-1}e^{-x}\cos x = -\frac{e^{-x}}{5}[2\sin x + \cos x]$$

Hence,
$$y_p = \frac{1}{2}\frac{e^x}{5}(2\sin x - \cos x) - \frac{1}{2}\frac{e^{-x}}{5}(2\sin x + \cos x)$$
$$= \frac{2}{5}\left(\frac{e^x - e^{-x}}{2}\right)\sin x - \frac{1}{5}\left(\frac{e^x + e^{-x}}{2}\right)\cos x$$
$$= \frac{2}{5}\sinh x \sin x - \frac{1}{5}\cosh x \cos x.$$

Thus, the required general solution is
$$y = c_1 e^x + c_2 e^{-x} + \frac{2}{5}\sinh x \sin x - \frac{1}{5}\cosh x \cos x.$$

EXAMPLE 9.19

Solve $(D^4 + D^2 + 1)y = ax^2 + be^{-x}\sin 2x$.

Solution: For finding y_c, AE is
$$D^4 + D^2 + 1 = 0$$
$$\Rightarrow \quad (D^2 + 1)^2 - D^2 = 0$$
$$\Rightarrow \quad (D^2 + 1 + D)(D^2 + 1 - D) = 0$$
$$\rightarrow \quad D = \frac{-1 \pm \sqrt{3}i}{2}, \frac{1 \pm \sqrt{3}i}{2}$$

Hence, $y_c = e^{-x/2}\left[c_1 \cos\frac{\sqrt{3}}{2}x + c_2 \sin\frac{\sqrt{3}}{2}x\right] + e^{x/2}\left[c_3 \cos\frac{\sqrt{3}}{2}x + c_4 \sin\frac{\sqrt{3}}{2}x\right].$

Now,
$$y_p = \frac{1}{D^4 + D^2 + 1}(ax^2 + be^{-x}\sin 2x)$$
$$= a\frac{1}{D^4 + D^2 + 1}x^2 + b\frac{1}{D^4 + D^2 + 1}e^{-x}\sin 2x$$

Let us evaluate each term separately.
$$a\frac{1}{D^4 + D^2 + 1}x^2 = a(1 + \overline{D^2 + D^4})^{-1}x^2$$
$$= a(1 - (D^2 + D^4) + \ldots)x^2$$
$$= a(x^2 - 2)$$

and $\quad b\dfrac{1}{D^4 + D^2 + 1}e^{-x}\sin 2x = be^{-x}\dfrac{1}{(D-1)^4 + (D-1)^2 + 1}\sin 2x \quad$ {by using Rule VI}

$$= be^{-x}\frac{1}{D^4-4D^3+7D^2-6D+3}\sin 2x$$

$$= be^{-x}\frac{1}{(-4)^2-4(-4)D+7(-4)-6D+3}\sin 2x$$

$$= be^{-x}\frac{1}{10D-9}\sin 2x$$

$$= be^{-x}\frac{10D+9}{10D^2-81}\sin 2x$$

$$= be^{-x}\frac{10(2\cos 2x)+9\sin 2x}{100(-4)-81}$$

$$= \frac{-be^{-x}}{481}(20\cos 2x + 9\sin 2x)$$

Thus, the required general solution is

$$y = e^{-x/2}\left[c_1\cos\frac{\sqrt{3}}{2}x + c_2\sin\frac{\sqrt{3}}{2}x\right] + e^{x/2}\left[c_3\cos\frac{\sqrt{3}}{2}x + c_4\sin\frac{\sqrt{3}}{2}x\right]$$

$$+ a(x^2-2)\frac{-be^{-x}}{481}(20\cos 2x + 9\sin 2x)$$

EXAMPLE 9.20

Solve $(D^3 - 1)y = xe^x + \cos^2 x$.

Solution: For finding y_c, AE is

$$D^3 - 1 = 0$$

$$\Rightarrow \quad (D-1)(D^2+D+1) = 0$$

$$\Rightarrow \quad D = 1, \frac{-1\pm\sqrt{3}i}{2}$$

Hence, $y_c = c_1e^x + e^{-x/2}\left[c_2\cos\frac{\sqrt{3}}{2}x + c_3\sin\frac{\sqrt{3}}{2}x\right]$.

Also,

$$y_p = \frac{1}{D^3-1}(xe^x + \cos^2 x)$$

$$= \frac{1}{D^3-1}\left(xe^x + \frac{1+\cos 2x}{2}\right)$$

$$= \frac{1}{D^3-1}xe^x + \frac{1}{2}\frac{1}{D^3-1} + \frac{1}{2}\frac{1}{D^3-1}\cos 2x$$

Let us evaluate each term separately.

$$\frac{1}{D^3-1}xe^x = e^x\frac{1}{(D+1)^3-1}x \quad \text{\{by using Rule VI\}}$$

$$= e^x\frac{1}{D^3+3D^2+3D}x.$$

$$= e^x\frac{1}{3D}\left[1+\overline{D+\frac{1}{3}D^2}\right]^{-1}x$$

$$= \frac{e^x}{3}\frac{1}{D}\left[1-\left(D+\frac{1}{3}D^2\right)+\ldots\right]x$$

$$= \frac{e^x}{3}\frac{1}{D}[x-1] = \frac{e^x}{3}\left(\frac{x^2}{2}-x\right)$$

$$\frac{1}{2}\frac{1}{D^3-1} = \frac{1}{2}\frac{1}{D^3-1}e^{0x} = \frac{1}{2}\frac{1}{0^3-1}e^{0x} = \frac{-1}{2}$$

$$\frac{1}{2}\frac{1}{D^3-1}\cos 2x = \frac{1}{2}\frac{1}{-4D-1}\cos 2x$$

$$= \frac{-1}{2}\frac{4D-1}{16D^2-1}\cos 2x$$

$$= \frac{-1}{2}\frac{4(-2\sin 2x)-\cos 2x}{16(-4)-1}$$

$$= \frac{1}{130}(-8\sin 2x - \cos 2x)$$

Thus, the required general solution is

$$y = c_1e^x + e^{-x/2}\left(c_2\cos\frac{\sqrt{3}}{2}x + c_3\sin\frac{\sqrt{3}}{2}x\right) + \frac{e^x}{6}(x^2-2x) - \frac{1}{2} - \frac{1}{130}(8\sin 2x + \cos 2x).$$

EXAMPLE 9.21

Solve the differential equation

$$(D^2 - 4D + 4)y = e^{2x}\cos^2 x$$

Solution: For finding y_c, AE is

$$D^2 - 4D + 4 = 0$$

$$\Rightarrow \quad (D-2)^2 = 0$$

$$\Rightarrow \quad D = 2, 2$$

Hence, $y_c = (c_1 + c_2 x) e^{2x}$.

Now,
$$y_p = \frac{1}{(D-2)^2} e^{2x} \cos^2 x$$
$$= e^{2x} \frac{1}{(D+2-2)^2} \cos^2 x \quad \text{\{by using Rule VI\}}$$
$$= e^{2x} \frac{1}{D^2}\left(\frac{1+\cos 2x}{2}\right)$$
$$= \frac{e^{2x}}{2} \frac{1}{D}\left(x + \frac{\sin 2x}{2}\right)$$
$$= \frac{e^{2x}}{2}\left(\frac{x^2}{2} - \frac{\cos 2x}{4}\right)$$

Thus, the required general solution is
$$y = (c_1 + c_2 x) e^{2x} + \frac{1}{8} e^{2x} (2x^2 - \cos 2x).$$

EXAMPLE 9.22

Solve $(D^3 - D^2 - 9D + 9)y = e^x (x^2 + \cos 2x)$.

Solution: For finding y_c, AE is
$$D^3 - D^2 - 9D + 9 = 0$$

Clearly, $D = 1$ is satisfying this equation as
$$1 - 1 - 9 + 9 = 0$$

Hence, one root is 1. Now, the remaining two roots can be obtained as

1	1	–1	–9	9
		1	0	–9
	1	0	–9	0

$$D^2 - 9 = 0 \Rightarrow D = 3, -3$$

Thus, roots of AE are 1, 3, –3.
Hence, $y_c = c_1 e^x + c_2 e^{3x} + c_3 e^{-3x}$.

Now,
$$y_p = \frac{1}{D^3 - D^2 - 9D + 9} e^x (x^2 + \cos 2x)$$
$$= e^x \frac{1}{(D+1)^3 - (D+1)^2 - 9(D+1) + 9} (x^2 + \cos 2x) \quad \text{\{by using Rule VI\}}$$

$$= e^x \frac{1}{D^3 + 3D^2 + 3D - D^2 - 2D - 9D}(x^2 + \cos 2x)$$

$$= e^x \frac{1}{D^3 + 2D^2 - 8D}(x^2 + \cos 2x)$$

$$= e^x \left[\frac{1}{D^3 + 2D^2 - 8D}x^2 + \frac{1}{D^3 + 2D^2 - 8D}\cos 2x\right]$$

Let us evaluate each term separately.

$$e^x \frac{1}{D(D^2 + 2D - 8)}x^2 = e^x \frac{1}{D}\frac{-1}{8}\left(1 - \frac{2D + D^2}{8}\right)^{-1} x^2$$

$$= \frac{-e^x}{8}\frac{1}{D}\left(1 + \left(\frac{2D + D^2}{8}\right) + \left(\frac{2D + D^2}{8}\right)^2 + \ldots\right)$$

$$= \frac{-e^x}{8}\frac{1}{D}\left(1 + \frac{2D + D^2}{8} + \frac{4D^2}{64}\right)x^2$$

$$= \frac{-e^x}{8}\frac{1}{D}\left(x^2 + \frac{2(2x) + 2}{8} + \frac{4(2)}{64}\right)$$

$$= \frac{-e^x}{8}\frac{1}{D}\left(x^2 + \frac{x}{2} + \frac{1}{4} + \frac{1}{8}\right)$$

$$= \frac{-e^x}{8}\frac{1}{D}\left(x^2 + \frac{x}{2} + \frac{3}{8}\right)$$

$$= \frac{-e^x}{8}\left(\frac{x^3}{3} + \frac{x^2}{4} + \frac{3x}{8}\right)$$

and

$$e^x \frac{1}{D^3 + 2D^2 - 8D}\cos 2x = e^x \frac{1}{-4D + 2(-4) - 8D}\cos 2x$$

$$= e^x \frac{1}{-12D - 8}\cos 2x$$

$$= \frac{e^x}{-4}\frac{3D - 2}{9D^2 - 4}\cos 2x$$

$$= \frac{-e^x}{4}\frac{3(-2\sin 2x) - 2\cos 2x}{9(-4) - 4}$$

$$= \frac{-e^x}{80}(3\sin 2x + \cos 2x)$$

Thus, the required general solution is

$$y = c_1e^x + c_2e^{+3x} + c_3e^{-3x} - \frac{1}{8}e^x\left(\frac{x^3}{3}+\frac{x^2}{4}+\frac{3x}{8}\right) - \frac{-e^x}{80}(3\sin 2x + \cos 2x).$$

VII Rule to determine y_p when $r(x) = xV$ where, V is a function of x.

To show that

$$\frac{1}{f(D)}xV = x\frac{1}{f(D)}V + \frac{d}{dD}\left[\frac{1}{f(D)}\right]V.$$

By using Leibnitz theorem, that is

$$D^n(uv) = {}^nC_0D^n(u)v + {}^nC_1D^{n-1}(u)D(v) + {}^nC_2D^{n-2}(u)D^2(v) + \ldots + {}^nC_nuD^n(v)$$

We have

$$\begin{aligned} D^n(xV) &= {}^1C_0D(V)x + {}^1C_1VD(x) \\ &= xDV + V\cdot 1 = xDV + (D)'V \\ D^2(xV) &= {}^2C_0D^2(V)x + {}^2C_1D(V)D(x) \\ &= xD^2V + 2DV = xD^2V + (D^2)'V \end{aligned}$$

In general,

$$\begin{aligned} D^n(xV) &= {}^nC_0D^n(V)x + {}^nC_1D^{n-1}(V)D(x) \\ &= xD^nV + nD^{n-1}V \\ &= xD^nV + (D^n)'V \end{aligned}$$

where, $(D^n)'$ denotes derivative of D^n with respect to D.

$\therefore$ $$f(D)(xV) = xf(D)V + [f(D)]'V \quad \{\because f(D) \text{ is polynomial in } D\}$$

Operating with $\frac{1}{f(D)}$, we obtain

$$\frac{1}{f(D)}\{f(D)(xV)\} = \frac{1}{f(D)}\{xf(D)V\} + \frac{1}{f(D)}\{[f(D)]'V\}$$

$$\Rightarrow \quad xV = \frac{1}{f(D)}[xf(D)V\} + \frac{1}{f(D)}\{f(D)\}'V$$

$$\Rightarrow \quad \frac{1}{f(D)}[xf(D)V] = xV - \frac{1}{f(D)}[f(D)]'V$$

Taking V in place of $f(D)V$, we have

$$\frac{1}{f(D)}[xV] = x\frac{1}{f(D)}V - \frac{1}{f(D)}[f(D)]'\frac{1}{f(D)}V$$

$$\Rightarrow \qquad \frac{1}{f(D)}[xV] = x\frac{1}{f(D)}V - \frac{1}{[f(D)]^2}[f(D)]'V$$

$$\Rightarrow \qquad \frac{1}{f(D)}[xV] = x\frac{1}{f(D)}V + \frac{d}{dD}\left[\frac{1}{f(D)}\right]V.$$

EXAMPLE 9.23

Solve $\dfrac{d^2y}{dx^2} + 4y = x\sin x.$

Solution: We have the given equation $(D^2 + 4)y = x\sin x$

For finding y_c, AE is

$$D^2 + 4 = 0 \quad \Rightarrow \quad D = \pm 2i$$

Hence, $y_c = c_1\cos 2x + c_2\sin 2x.$

Now,
$$y_p = \frac{1}{D^2+4}(x\sin x)$$

$$= x\frac{1}{D^2+4}\sin x + \frac{d}{dD}\left(\frac{1}{D^2+4}\right)\sin x \quad \{\text{by using Rule VII}\}$$

$$= x\frac{1}{-1^2+4}\sin x + \frac{-2D}{(D^2+4)^2}\sin x$$

$$= \frac{x}{3}\sin x + \frac{-2\cos x}{(-1^2+4)^2} = \frac{x}{3}\sin x - \frac{2}{9}\cos x$$

Thus, the required general solution is

$$y = c_1\cos 2x + c_2\sin 2x + \frac{x}{3}\sin x - \frac{2}{9}\cos x.$$

EXAMPLE 9.24

Solve $(D^2 + 2D^2 + 1)y = xe^{-x}\sin x.$

Solution: For finding y_c, AE is

$$D^2 + 2D + 1 = 0$$

$$\Rightarrow \qquad (D + 1)^2 = 0 \quad \Rightarrow \quad D = -1, -1$$

Hence, $y_c = (c_1 + c_2x)e^{-x}.$

Now,
$$y_p = \frac{1}{(D+1)^2}e^{-x}(x\sin x)$$

$$= e^{-x}\frac{1}{(D-1+1)^2}(x\sin x) \quad \{\text{by using Rule VI}\}$$

$$= e^{-x}\frac{1}{D^2}x\sin x$$

$$= e^{-x}\left[x\frac{1}{D^2}\sin x + \frac{d}{dD}\left(\frac{1}{D^2}\right)\sin x\right] \quad \{\text{by using Rule VII}\}$$

$$= e^{-x}\left[x\frac{1}{-1^2}\sin x - \frac{2}{D^3}\sin x\right]$$

$$= e^{-x}\left(-x\sin x - \frac{2}{(-1^2)D}\sin x\right)$$

$$= e^{-x}(-x\sin x + 2(-\cos x))$$

$$= -e^{-x}(x\sin x + 2\cos x)$$

Thus, the required general solution is

$$y = (c_1 + c_2x)e^{-x} - e^{-x}(x\sin x + 2\cos x).$$

EXAMPLE 9.25

Solve $\frac{d^2y}{dx^2} - 4y = x\sinh x$

Solution: We have the given equation $(D^2 - 4)y = x\sinh x$. For finding y_c, AE is

$$D^2 - 4 = 0 \Rightarrow D = 2, -2$$

Hence, $y_c = c_1e^{2x} + c_2e^{-2x}$.

Now,
$$y_p = \frac{1}{D^2-4}(x\sinh x) = \frac{1}{D^2-4}x\left(\frac{e^x - e^{-x}}{2}\right)$$

$$= \frac{1}{2}\left[\frac{1}{D^2-4}e^x x - \frac{1}{D^2-4}e^{-x}x\right]$$

$$= \frac{1}{2}\left[e^x\frac{1}{(D+1)^2-4}x - e^{-x}\frac{1}{(D-1)^2-4}x\right]$$

$$= \frac{1}{2}\left[e^x\frac{1}{D^2+2D-3}x - e^{-x}\frac{1}{D^2-2D-3}x\right]$$

$$= \frac{1}{2}\left[\frac{e^x}{-3}\left(1-\frac{2D+D^2}{3}\right)^{-1}x - \frac{e^{-x}}{-3}\left(1+\frac{2D-D^2}{3}\right)^{-1}x\right]$$

$$= \frac{1}{6}\left[-e^{x}\left(1+\frac{2D+D^2}{3}+\ldots\right)x+e^{-x}\left(1-\frac{2D-D^2}{3}+\ldots\right)x\right]$$

$$= \frac{1}{6}\left[-e^{x}\left(x+\frac{2}{3}\right)+e^{-x}\left(x-\frac{2}{3}\right)\right]$$

$$= \frac{-x}{3}\left(\frac{e^x-e^{-x}}{2}\right)-\frac{2}{9}\left(\frac{e^x+e^{-x}}{2}\right)$$

$$= \frac{-x}{3}\sinh x-\frac{2}{9}\cosh x$$

Thus, the required general solution is

$$y = c_1e^{2x} + c_2e^{-2x} - \frac{x}{3}\sinh x - \frac{2}{9}\cosh x$$

EXAMPLE 9.26

Solve $\dfrac{d^4y}{dx^4} - y = x^2\sin x$.

Solution: We have the given equation

$$(D^4 - 1)y = x^2\sin x.$$

For finding y_c, AE is

$$D^4 - 1 = 0$$

$$\Rightarrow \quad D = \pm 1, \pm i$$

Hence, $y_c = c_1e^x + c_2e^{-x} + c_3\cos x + c_4\sin x$.

Now,
$$y_p = \frac{1}{D^4-1}(x^2\sin x)$$

$$= \text{Imaginary part of } \frac{1}{D^4-1}x^2e^{ix}$$

Again
$$\frac{1}{D^4-1}x^2e^{ix} = e^{ix}\frac{1}{(D+i)^4-1}x^2 \quad \text{\{by using Rule VI\}}$$

$$= e^{ix}\frac{1}{D^4+4D^3i-6D^2-4Di}x^2$$

$$= \frac{e^{ix}}{-4i}\frac{1}{D}\left[1-\frac{D^3+4D^2i-6D}{4i}\right]^{-1}x^2$$

$$= \frac{e^{ix}}{-4i}\frac{1}{D}\left[1+\frac{D^3+4D^2i-6D}{4i}+\left(\frac{D^3+4D^2i-6D}{4i}\right)^2+\ldots\right]x^2$$

$$= i\frac{e^{ix}}{4}\frac{1}{D}\left[1+\frac{4D^2i-6D}{4i}+\frac{36D^2}{-16}\right]x^2$$

$$= i\frac{e^{ix}}{4}\frac{1}{D}\left[x^2+\frac{4i(2)-6(2x)}{4i}-\frac{9}{4}(2)\right]$$

$$= i\frac{e^{ix}}{4}\frac{1}{D}\left[x^2+2-\frac{3x}{i}-\frac{9}{2}\right]$$

$$= i\frac{e^{ix}}{4}\frac{1}{D}\left[x^2+3ix-\frac{5}{2}\right]$$

$$= i\frac{e^{ix}}{4}\left[\frac{x^3}{3}+\frac{3ix^2}{2}-\frac{5}{2}x\right]$$

$$= \frac{i\cos x-\sin x}{4}\left(\frac{x^3}{3}+\frac{3i}{2}x^2-\frac{5}{2}x\right)$$

whose imaginary part is

$$\frac{\cos x}{4}\left(\frac{x^3}{3}-\frac{5}{2}x\right)-\frac{3}{8}x^2\sin x$$

Thus, the required general solution is

$$y = c_1e^x + c_2e^{-x} + c_3\cos x + c_4\sin x + \frac{\cos x}{4}\left(\frac{x^3}{3}-\frac{5}{2}x\right)-\frac{3}{8}x^2\sin x$$

EXAMPLE 9.27

Solve $(D^4 + 2D^2 + 1)y = x^2\cos x$.

Solution: For finding y_c, AE is

$$D^4 + 2D^2 + 1 = 0$$

$$\Rightarrow \quad (D^2 + 1)^2 = 0 \Rightarrow D = \pm 1, \pm 1$$

Hence, $y_c = (c_1 + c_2x)\cos x + (c_3 + c_4x)\sin x$.

Now, $$y_p = \frac{1}{(D^2+1)^2}x^2\cos x$$

$$= \text{Real part of } \frac{1}{(D^2+1)^2}x^2e^{ix}$$

Now, $\frac{1}{(D^2+1)^2}x^2e^{ix} = e^{ix}\frac{1}{((D+i)^2+1)^2}x^2$ {by using Rule VI}

$$= e^{ix}\frac{1}{(D^2+2Di)^2}x^2$$

$$= \frac{e^{ix}}{4D^2i^2}\left(1+\frac{D}{2i}\right)^{-2}x^2$$

$$= \frac{-e^{ix}}{4D^2}\left(1-\frac{D}{2}i\right)^{-2}x^2$$

$$= \frac{-e^{ix}}{4D^2}\left(1+2\frac{D}{2}i+3\left(\frac{D}{2}i\right)^2+\dots\right)x^2$$

$$= \frac{-e^{ix}}{4D^2}\left[1+Di-\frac{3}{4}D^2\right]x^2$$

$$= \frac{-e^{ix}}{4D^2}\left[x^2+2xi-\frac{3}{4}(2)\right]$$

$$= \frac{-e^{ix}}{4D^2}\left[\left(x^2-\frac{3}{2}\right)+2xi\right]$$

$$= \frac{-e^{ix}}{4D}\left[\left(\frac{x^3}{3}-\frac{3}{2}x\right)+x^{2i}\right]$$

$$= -\frac{\cos x+i\sin x}{4}\left[\left(\frac{x^4}{12}-\frac{3}{4}x^2\right)+\frac{x^3}{3}i\right]$$

whose real part is

$$= -\frac{1}{4}\left[\left(\frac{x^4-9x^2}{12}\right)\cos x-\frac{x^2}{3}\sin x\right]$$

Thus, the required general solution is

$$y = (c_1+c_2x)\cos x+(c_3+c_4x)\sin x-\frac{1}{4}\left[\left(\frac{1}{12}(x^4-9x^2)\cos x-\frac{1}{3}x^3\sin x\right)\right]$$

EXAMPLE 9.28

Solve $(D^2 - 4D + 4)y = 8x^2 e^{2x} \sin 2x$.

Solution: For finding y_c, AE is

$$D^2 - 4D + 4 = 0$$

$$\Rightarrow \quad (D-2)^2 = 0 \Rightarrow D = 2, 2$$

Hence, $y_c = (c_1 + c_2 x)e^{2x}$.

Now,
$$y_p = 8\frac{1}{(D-2)^2} e^{2x}(x^2 \sin 2x)$$

$$= 8e^{2x} \frac{1}{[(D+2)-2]^2} x^2 \sin 2x \quad \text{\{by using Rule VI\}}$$

$$= 8e^{2x} \frac{1}{D^2}(x^2 \sin 2x)$$

$$= 8e^{2x} \left[\text{Imaginary part of } \frac{1}{D^2}(x^2 e^{2ix})\right]$$

Now,
$$\frac{1}{D^2}(x^2 e^{2ix}) = e^{2ix} \frac{1}{(D+2i)^2} x^2$$

$$= \frac{e^{2ix}}{4i^2}\left(1 + \frac{D}{2i}\right)^{-2} x^2$$

$$= \frac{e^{2ix}}{-4}\left(1 - 2\left(\frac{D}{2i}\right) + 3\left(\frac{D}{2i}\right)^2 - \ldots\right)x^2$$

$$= \frac{e^{2ix}}{-4}\left(1 - Di - \frac{3}{4}D^2\right)x^2$$

$$= \frac{-e^{2ix}}{4}\left(x^2 + 2xi - \frac{3}{4}(2)\right)$$

$$= \frac{(\cos 2x + i \sin 2x)}{4}\left(\left(x^2 - \frac{3}{2}\right) + 2xi\right)$$

whose imaginary part is

$$\frac{-1}{4}\left[\left(x^2 - \frac{3}{2}\right)\sin 2x + 2x \cos 2x\right] = \frac{-1}{8}[(2x^2 - 3)\sin 2x + 4x \cos 2x]$$

$$\therefore \quad y_p = -e^{2x}[(2x^2 - 3)\sin 2x + 4x \cos 2x]$$

Thus, the required general solution is

$$y = e^{2x}[c_1 + c_2x + 3\sin 2x - 2x^2\sin 2x - 4x\cos 2x]$$

EXAMPLE 9.29

Solve $\frac{d^2y}{dx^2} + 4\frac{dy}{dx} + 4y = 2\sinh 2x.$

Solution: We have the given equation

$$(D^2 + 4D + 4)y = 2\sinh 2x$$

For finding y_c, AE is

$$D^2 + 4D + 4 = 0$$

$\Rightarrow$ $$(D + 2)^2 = 0 \Rightarrow D = -2, -2$$

Hence, $y_c = (c_1 + c_2x)e^{-2x}$.

Now, $$y_p = \frac{1}{(D+2)^2}(2\sinh 2x) = \frac{1}{(D+2)^2}(e^{2x} - e^{-2x})$$

We will find y_p by using Rule VI

Thus, $$y_p = \frac{1}{(D+2)^2}e^{2x} - \frac{1}{(D+2)^2}e^{-2x}$$

$$= \frac{1}{(2+2)^2}e^{2x} - e^{-2x}\frac{1}{(D-2+2)^2}1$$

$$= \frac{1}{16}e^{2x} - e^{-2x}\frac{1}{D^2}1$$

$$= \frac{1}{16}e^{2x} - e^{-2x}\frac{x^2}{2}$$

Thus, the required general solution is

$$y = (c_1 + c_2x)e^{-2x} + \frac{1}{16}e^{2x} - \frac{1}{2}x^2e^{-2x}.$$

EXAMPLE 9.30

Solve the differential equation

$$\frac{d^2y}{dx^2} - 3\frac{dy}{dx} + 2y = \sin e^{-x}.$$

Solution: We have the given equation

$$(D^2 - 3D + 2)y = \sin e^{-x}.$$

For finding y_c, AE is

$$D^2 - 3D + 2 = 0$$

$$\Rightarrow \quad (D-2)(D-1) = 0 \Rightarrow D = 2, 1$$

Hence, $y_c = c_1e^{2x} + c_2e^x$.

Now,
$$y_p = \frac{1}{(D-2)(D-1)}\sin e^{-x}$$

$$= \frac{1}{D-2}\frac{1}{D-1}\sin e^{-x} \quad \text{\{by using Rule II\}}$$

$$= \frac{1}{D-2}e^x\int(\sin e^{-x})e^{-x}\,dx$$

$$= \frac{1}{D-2}e^x\frac{-\cos e^{-x}}{-1}$$

$$= \frac{1}{D-2}(\cos e^{-x})e^x$$

$$= e^{2x}\int(\cos e^{-x})e^{-x}e^{-2x}dx$$

$$= e^{2x}\int(\cos e^{-x})e^{-x}\,dx$$

$$= e^{2x}\frac{\sin e^{-x}}{-1} = -e^{2x}\sin e^{-x}$$

Thus, the required general solution is

$$y = c_1e^{2x} + c_2e^x - e^{2x}\sin e^{-x}.$$

EXAMPLE 9.31

Solve $(D^2 + 1)^2y = 24x\cos x$, given the initial conditions

$$x = 0,\ y = 0,\ Dy = 0,\ D^2y = 0,\ D^3y = 12.$$

Solution: For finding y_c, AE is

$$(D^2 + 1)^2 = 0$$

$$\Rightarrow \quad D = \pm i, \pm i$$

Hence, $y_c = (c_1 + c_2x)\cos x + (c_3 + c_4x)\sin x$.

Now,
$$y_p = \frac{1}{(D^2+1)^2}(24x\cos x)$$

We will find y_p by using rule VI instead of rule VII.

Thus, y_p = Real part of $24\frac{1}{(D^2+1)^2}xe^{ix}$

Now,
$$24\frac{1}{(D^2+1)^2}xe^{ix} = 24e^{ix}\frac{1}{[(D+i)^2+1]^2}x$$
$$= 24e^{ix}\frac{1}{(D^2+2iD)}x$$
$$= 24\frac{e^{ix}}{4i^2D^2}\left(1+\frac{D}{2i}\right)^{-2}x$$
$$= -6e^{ix}\frac{1}{D^2}\left(1+\frac{D}{2}i\right)^{-2}x$$
$$= -6e^{ix}\frac{1}{D^2}\left(1+2\frac{D}{2}i+3\left(\frac{D}{2}i\right)^2+\ldots\right)x$$
$$= -6e^{ix}\frac{1}{D^2}\left(x+\frac{i}{2}\right)$$
$$= -6e^{ix}\left(\frac{x^3}{6}+\frac{i}{2}x^2\right) = -6(\cos x + i\sin x)\left(\frac{x^3}{6}+\frac{i}{2}x^2\right)$$

whose real part is $-x^3\cos x + 3x^2\sin x$.

Thus, the required general solution is

$$y = (c_1 + c_2x)\cos x + (c_3 + c_4x)\sin x - x^3\cos x + 3x^2\sin x$$

By using the initial conditions, we obtain

$$c_1 = 0 = c_2 = c_3 = c_4$$

Hence, under the given conditions, the solution will be

$$y = -x^3\cos x + 3x^2\sin x.$$

EXAMPLE 9.32

Solve $\frac{d^2y}{dx^2}+3\frac{dy}{dx}+2y = e^{e^x}$.

Solution: We have the given equation

$$(D^2 + 3D + 2)y = e^{e^x}$$

For finding y_c, AE is

$$D^2 + 3D + 2 = 0$$

$$\Rightarrow \quad (D + 1)(D + 2) = 0 \quad \Rightarrow \quad D = -1, -2$$

Hence, $y_c = c_1e^{-x} + c_2e^{-2x}$

Now,
$$y_p = \frac{1}{(D+2)(D+1)}e^{e^x}$$

$$= \frac{1}{(D+2)}\left(\frac{1}{D+1}e^{e^x}\right) \quad \text{\{by using Result II\}}$$

$$= \frac{1}{D+2}e^{-x}\int e^{e^x}e^x dx$$

$$= \frac{1}{D+2}e^{-x}(e^{e^x}e^x - e^{e^x})$$

$$= e^{-2x}\int e^{e^x}e^{-x}e^{2x}dx$$

$$= e^{-2x}\int e^{e^x}e^x dx = e^{-2x}e^{e^x}$$

Thus, the required general solution is

$$y = c_1e^{-x} + c_2e^{-2x} + e^{-2x}e^{e^x}.$$

EXAMPLE 9.33

Solve $\frac{d^4y}{dx^4} + 6\frac{d^3y}{dx^3} + 11\frac{d^2y}{dx^2} + 6\frac{dy}{dx} = 20e^{-2x}\sin x.$

Solution: We have the given equation

$$(D^4 + 6D^3 + 11D^2 + 6D)y = 20e^{-2x}\sin x$$

For finding y_c, AE is

$$D^4 + 6D^3 + 11D^2 + 6D = 0$$

$$\Rightarrow \quad D(D + 1)(D + 2)(D + 3) = 0$$

$$\Rightarrow \quad D = 0, -1, -2, -3$$

Hence, $y_c = c_1 + c_2e^{-x} + c_3e^{-2x} + c_4e^{-3x}$.

Now,
$$y_p = \frac{1}{D^4 + 6D^3 + 11D^2 + 6D}(20e^{-2x}\sin x)$$

$$= 20\frac{1}{D(D+1)(D+2)(D+3)}(e^{-2x}\sin x)$$

$$= 20e^{-2x} \frac{1}{(D-2)(D-1)D(D+1)} \sin x \quad \text{\{by using Rule VI\}}$$

$$= 20e^{-2x} \frac{1}{(D-2)(D^2-1)} (-\cos x)$$

$$= 20e^{-2x} \frac{-1}{(-1^2-1)(D-2)} \cos x$$

$$= 10e^{-2x} \frac{D+2}{D^2-4} \cos x$$

$$= 10e^{-2x} \frac{-\sin x + 2\cos x}{-1^2-4}$$

$$= 2e^{-2x} \sin x - 4e^{-2x} \cos x.$$

Thus, the required general solution is

$$y = c_1 + c_2e^{-x} + c_3e^{-2x} + c_4e^{-3x} + 2e^{-2x} \sin x - 4e^{-2x} \cos x.$$

EXAMPLE 9.34

Solve $\frac{d^2y}{dx^2} + a^2y = \cos ax$.

Solution: We have the given equation

$$(D^2 + a^2)y = \cos ax$$

For finding y_c, AE is

$$D^2 + a^2 = 0$$

$$\Rightarrow \qquad D = \pm ai$$

Hence, $y_c = c_1 \cos ax + c_2 \sin ax$.

Now, $y_p = \frac{1}{D^2+a^2} \cos ax$.

We will find y_p by using Rule VI

Now,
$$y_p = \text{Real part of } \frac{1}{D^2+a^2} e^{iax}$$

Thus,
$$\frac{1}{D^2+a^2} e^{iax} = \frac{1}{(D+ia)(D-ia)} e^{iax}$$

$$= \frac{1}{D-ia} \frac{1}{ia+ia} e^{iax}$$

$$= \frac{1}{2ia}\frac{1}{D-ia}e^{iax}$$

$$= \frac{1}{2ia}e^{iax}\frac{1}{(D+ia)-ia}1$$

$$= \frac{1}{2ia}e^{iax}\frac{1}{D}1$$

$$= \frac{-i}{2a}e^{iax}\,x$$

$$= \frac{-i}{2a}(\cos ax + i\sin ax)x$$

$$= -i\frac{x}{2a}\cos ax + \frac{x}{2a}\sin ax$$

whose real part is $\frac{x}{2a}\sin ax$

Thus, the required general solution of the given equation is

$$y = c_1\cos ax + c_2\sin ax + \frac{x}{2a}\sin ax.$$

EXERCISE 9.2

Solve the following differential equations:

1. $(D^4 + D^2 + 1)y = e^{-x/2}\cos\frac{\sqrt{3}}{2}x$.
2. $(D^2 - 1)y = \cosh x \cos x$.
3. $(D + 1)^3 y = x^2 e^{-x}$.
4. $(D^2 + 1)y = e^{-x} + \cos x + x^3 + e^x \sin x$.
5. $(D^3 - D^2 + 3D + 5)y = x^2 + e^x \cos 2x$.
6. $(D^2 - 4D + 3)y = e^{2x}\sin 3x$.
7. $(D^3 - 3D^2 + 4D - 2)y = e^x + \cos x$.
8. $(D^3 + a^2 D)y = \sin ax$.
9. $(D^2 + 4)y = \sin^2 x$.
10. $\frac{d^2y}{dx^2} + a^2 y = \sin ax$.
11. $\frac{d^2y}{dx^2} + 2\frac{dy}{dx} + y = x\cos x$.
12. $(D^2 - 1)y = x\sin x + (1 + x^2)e^x$.

13. $\dfrac{d^2y}{dx^2} - y = xe^x \sin x.$

14. $(D^2 - 3D + 2)y = \sin(e^x).$

15. $(D^2 - 1)y = \cosh x \cos x + a^x.$

16. $(D^2 + 1)y = \sin x \sin 2x.$

17. $(D^2 + 9)(D^2 + 1)y = \cos 3x.$

18. $(D^4 - 1)y = x \sin x.$

19. $\dfrac{d^2y}{dx^2} - 2\dfrac{dy}{dx} + 4y = e^x \cos x + x.$

20. $y'' + y = xe^x \sin 2x.$

21. $y'' + y = xe^x \cos 2x.$

22. $\dfrac{d^2y}{dx^2} - 2\dfrac{dy}{dx} + 3y = x^2 + \cos x.$

23. $\dfrac{d^3y}{dx^3} + y = \sin 3x - \cos^2 \dfrac{x}{2}.$

24. $\dfrac{d^2y}{dx^2} + 4y = 4 \tan 2x.$

25. $4\dfrac{d^2y}{dx^2} - 4\dfrac{dy}{dx} + 2y = \cos^2\left(x + \dfrac{1}{6}\pi\right).$

26. $\dfrac{d^2y}{dx^2} - 6\dfrac{dy}{dx} + 8y = (e^{2x} - 1)^2.$

ANSWERS

Exercise 9.2

1. $y = e^{x/2}\left[c_1 \cos\dfrac{\sqrt{3}}{2}x + c_2 \sin\dfrac{\sqrt{3}}{2}x\right] + e^{-x/2}\left[c_3 \cos\dfrac{\sqrt{3}}{2}x + c_4 \sin\dfrac{\sqrt{3}}{2}x\right]$

$$+ \frac{1}{4\sqrt{3}} xe^{-x/2}\left[\sin\frac{\sqrt{3}}{2}x + \sqrt{3}\cos\frac{\sqrt{3}}{2}x\right].$$

2. $y = c_1 e^x + c_2 e^{-x} + \dfrac{2}{5}\sinh x \sin x - \dfrac{1}{5}\cosh x \cos x.$

3. $y = (c_1 + c_2 x + c_3 x^2)e^{-x} + \dfrac{1}{60}x^5 e^{-x}.$

4. $y = c_1 \cos x + c_2 \sin x + \frac{1}{8} e^{-x} + \frac{1}{2} x \sin x + x^3 - 6x - \frac{1}{5} e^x (2 \cos x - \sin x)$.

5. $y = c_1 e^{-x} + e^x (c_2 \cos 2x + c_3 \sin 2x) + \frac{1}{16} x e^x (\sin 2x - \cos 2x)$.

6. $y = c_1 e^x + c_2 e^{3x} - \frac{1}{10} e^{2x} \sin 3x$.

7. $y = c_1 e^x + e^x (c_2 \cos x + c_3 \sin 3x) + x e^x + \frac{1}{10} (3 \sin x + \cos x)$.

8. $y = c_1 + c_2 \cos ax + c_3 \sin ax - \frac{x}{2a^2} \sin ax$.

9. $y = c_1 \cos 2x + c_2 \sin 2x + \frac{1}{8} - \frac{1}{8} x \sin 2x$.

10. $y = c_1 \cos ax + c_2 \sin ax - \frac{x}{2a} \cos ax$.

11. $y = (c_1 + c_2 x) e^{-x} + \frac{1}{2} x \sin x + \frac{1}{2} \cos x - \frac{1}{2} \sin x$.

12. $y = c_1 e^x + c_2 e^{-x} - \frac{1}{2} (x \sin x + \cos x) + \frac{e^x}{12} (2x^3 - 3x^2 + 9x)$.

13. $y = c_1 e^x + c_2 e^{-x} + \frac{1}{25} e^x [(4 - 5x) \sin x - 2(5x + 1) \cos x]$.

14. $y = c_1 e^x + c_2 e^{2x} - e^{2x} \sin (e^x)$.

15. $y = c_1 e^x + c_2 e^{-x} + \frac{1}{10} e^x (2 \sin x - \cos x) - \frac{1}{10} e^{-x} (2 \sin x + \cos x) + a^x/(\log a)^2 - 1$.

16. $y = c_1 \cos x + c_2 \sin x + \frac{1}{4} x \sin x + \frac{1}{16} \cos 3x$.

17. $y = c_1 \cos 3x + c_2 \sin 3x + c_3 \cos x + c_4 \sin x - \frac{1}{48} x \sin 3x$.

18. $y = c_1 e^x + c_2 e^{-x} + c_3 \cos x + c_4 \sin x + \frac{1}{8} x^2 \cos x + \frac{3}{8} x \sin x$.

19. $y = e^x (c_1 \cos \sqrt{3x} + c_2 \sin \sqrt{3x}) + \frac{1}{2} e^x \cos x + \frac{1}{4} \left(x + \frac{1}{2} \right)$.

20. $y = c_1 \cos x + c_2 \sin x + e^x \left(\frac{11 - 5x}{50} \sin 2x + \frac{1 - 5x}{25} \cos 2x \right)$.

21. $y = c_1 \cos x + c_2 \sin x + e^x \left(\frac{5x-1}{25} \sin 2x + \frac{11-5x}{50} \cos 2x \right).$

22. $y = e^x [c_1 \cos \sqrt{2x} + c_2 \sin \sqrt{2x}\,] + \frac{1}{4}(\cos x - \sin x) + \frac{1}{27}(9x^2 + 12x + 2).$

23. $y = c_1 e^{-x} + e^{x/2} \left[c_2 \cos \frac{\sqrt{3}}{2} x + c_3 \sin \frac{\sqrt{3}}{2} x \right] + \frac{1}{730} (\sin 3x + 27 \cos 3x)$

$- \frac{1}{2} - \frac{1}{4} (\cos x - \sin x).$

24. $y = c_1 \cos 2x + c_2 \sin 2x - \cos 2x \log (\sec 2x + \tan 2x).$

25. $y = e^{x/2} \left[c_1 \cos \frac{x}{2} + c_2 \sin \frac{x}{2} \right] + \frac{1}{4} - \frac{1}{520} [(7 + 4\sqrt{3}) \cos 2x + (4 - 7\sqrt{3}) \sin 2x].$

26. $y = c_1 e^{4x} + c_2 e^{2x} + \frac{1}{8} + \frac{1}{2} x e^{4x} + x e^{2x}.$

10

CHAPTER

Method of Variation of Parameters and Undetermined Coefficients

In chapter 9, we studied operator methods for finding y_p. We will discuss in this chapter two independent methods for finding the complete solution of a non-homogeneous linear differential equation with constant coefficients.

Method of Variation of Parameters

We shall discuss a general method of solution, called the *method of variation of parameters,* which can always be used to find a particular integral whenever the complementary function of the equation is known. Consider the second order non-homogeneous linear equation

$$a_0 \frac{d^2y}{dx^2} + a_1 \frac{dy}{dx} + a_2 y = r(x), a_0 \neq 0$$

or

$$a_0 y'' + a_1 y' + a_2 y = r(x), a_0 \neq 0 \quad (1)$$

Using the methods given earlier, the complementary function is given by

$$y_c = A_1 y_1 + B y_2 \quad (2)$$

where, A and B are arbitrary constants. As we know y_1, y_2 are two linearly independent solutions of the corresponding homogeneous equation

$$a_0 \frac{d^2y}{dx^2} + a_1 \frac{dy}{dx} + a_2 y = 0, a_0 \neq 0. \quad (3)$$

or

$$a_0 y' + a, y' + a_2\, y = 0, a_0 \neq 0.$$

The idea behind the method of variation of parameters is to vary the parameters A and B, that is, we assume A and B to be the functions of x such that

$$y = Ay_1 + By_2 \quad (4)$$

is the complete solution of Eq. (1).

Now, y contains two functions A and B, which are to be determined. Therefore, we need two equations to determine them. Differentiating Eq. (4) with respect to x, we obtain

$$y' = A'y_1 + Ay_1' + B'y_2 + By_2' \tag{5}$$

If we differentiate this equation again, then the equation would contain the second derivatives A'' and B'' of the unknown functions A and B, respectively. In order that these derivatives do not appear, we set in Eq. (5)

$$A'y_1 + B'y_2 = 0 \tag{6}$$

Thus, Eq. (5) reduces to

$$y' = Ay_1' + By_2' \tag{7}$$

Again, differentiating (7) with respect to x, we obtain

$$y'' = Ay_1'' + A'y_1' + By_2'' + B'y_2' \tag{8}$$

Substituting the expressions for y, y' and y'' in Eq. (1), we obtain

$$a_0(Ay_1'' + A'y_1' + By_2'' + B'y_2') + a_1(Ay_1' + By_2') + a_2(Ay_1 + By_2) = r(x)$$

$$\Rightarrow \quad a_0(A'y_1' + B'y_2') + A(a_0y_1'' + a_1y_1' + a_2y_1) + B(a_0y_2'' + a_1y_2' + a_2y_2) = r(x) \tag{9}$$

Since y_1 and y_2 are the solutions of the homogeneous Eq. (2), we obtain

$$a_0(A'y_1' + B'y_2') = r(x)$$

$$\Rightarrow \quad A'y_1' + B'y_2' = \frac{r(x)}{a_0} \tag{10}$$

Therefore, for finding A and B, we have two equations namely (6) and (10). So by using Cramer's rule, we obtain

$$A' = \frac{\dfrac{-r(x)}{a_0}y_2}{y_1y_2' - y_2y_1'}, \quad B' = \frac{\dfrac{r(x)}{a_0}y_1}{y_1y_2' - y_2y_1'} \tag{11}$$

We note that the Wronskian $W(y_1, y_2)$ is

$$W = \begin{vmatrix} y_1 & y_2 \\ y_1' & y_2' \end{vmatrix} = y_1y_2' - y_2y_1' \neq 0$$

Since y_1, y_2 are linearly independent solutions of the homogeneous equation. Hence we can write Eq. (11) as

$$A' = \frac{-r(x)}{a_0}\frac{y_2}{W}, \quad B' = \frac{r(x)}{a_0}\frac{y_1}{W}$$

Integrating, we obtain

$$A = -\int \frac{r(x)}{a_0}\frac{y_2}{W}dx + c_1, \quad B = \int \frac{r(x)}{a_0}\frac{y_1}{W}dx + c_2$$

Substituting in Eq. (4), we obtain the general solution which contains two arbitrary constants, c_1 and c_2.

Note: This method is also applicable for variable coefficients problems. The method can also be easily extended to equations of any order, such as for the third order equation

$$a_0y''' + a_1y'' + a_2y' + a_3y = r(x),\ a_0 \neq 0 \tag{1}$$

The complementary function is

$$y_c = Ay_1 + By_2 + Cy_3$$

where, y_1, y_2, y_3 are the linearly independent solutions of the corresponding homogeneous equation and A, B, C are arbitrary constants. Consider A, B, C as parameters we assume the complete solution as

$$y = Ay_1 + By_2 + Cy_3 \tag{2}$$

Following the procedure discussed earlier, we obtain the required three equations for determining A, B, C as

$$A'y_1 + B'y_2 + C'y_3 = 0$$

$$A'y_1' + B'y_2' + C'y_3' = 0$$

and

$$A'y_1'' + B'y_2'' + C'y_3'' = \frac{r(x)}{a_0}$$

The determinant of the coefficient matrix is the Wronskian $W(y_1, y_2, y_3) \neq 0$. We determine A, B, C and substitute in Eq. (2) to obtain the general solution of Eq. (1).

▲▲

EXAMPLE 10.1

Using the method of variation parameters, solve

$$\frac{d^2y}{dx^2} + y = \tan x.$$

Solution: We have the given equation $(D^2 + 1)y = \tan x$

For finding y_c, AE is

$$D^2 + 1 = 0 \Rightarrow D = \pm i$$

Hence, $y_c = A\cos x + B\sin x$.

Consider

$$y = A\cos x + B\sin x \tag{1}$$

as the complete solution of the given equation. For determining A and B, we obtain two equations.

$$A'\cos x + B'\sin x = 0$$

$$A'(-\sin x) + B'\cos x = \tan x$$

By using Cramer's Rule, we get

$$A' = \frac{-\tan x \sin x}{W},\ B' = \frac{\tan x \cos x}{W}$$

where the coefficient matrix, $W = \begin{vmatrix} \cos x & \sin x \\ -\sin x & \cos x \end{vmatrix} = 1$

$$\therefore \qquad A' = \frac{-\sin^2 x}{\cos x} = \frac{-(1-\cos^2 x)}{\cos x} = -(\sec x - \cos x)$$

and $\qquad B' = \sin x$

Integrating, we get

$$A = -\int (\sec x - \cos x)\, dx = -\log\{\sec x + \tan x\} + \sin x + c_1$$

and $\qquad B = -\cos x + c_2$

Substituting in Eq. (1), we obtain

$$y = c_1 \cos x + c_2 \sin x - \cos x \log\{\sec x + \tan x\}$$

which is the required general solution.

EXAMPLE 10.2

Apply the method of variation of parameters to solve $\dfrac{d^2y}{dx^2} + 4y = 4\sec^2 2x.$

Solution: We have the given equation

$$(D^2 + 4)y = 4\sec^2 2x.$$

For finding y_c, AE is

$$D^2 + 4 = 0 \Rightarrow D = \pm 2i$$

Hence, $y_c = A\cos 2x + B\sin 2x.$

Consider $\qquad y = A\cos 2x + B\sin 2x \qquad (1)$

as the complete solution of the given equation.

For determining A and B, we obtain two equations

$$A'\cos 2x + B'\sin 2x = 0$$

$$A'(-2\sin 2x) + B'(2\cos 2x) = 4\sec^2 2x$$

The coefficient matrix $W = \begin{vmatrix} \cos 2x & \sin 2x \\ -2\sin 2x & 2\cos 2x \end{vmatrix} = 2$

On solving, we get

$$A' = -\frac{4\sec^2 2x \sin 2x}{2}, \; B' = \frac{4\sec^2 2x \cos 2x}{2}$$

$$= -2\sec 2x \tan 2x, \; B' = 2\sec 2x$$

Integrating, we get

$$A = -\sec 2x + c_1,\ B = \log\{\sec 2x + \tan 2x\} + c_2$$

Substituting in Eq. (1), we get

$$y = c_1 \cos 2x + c_2 \sin 2x - 1 + \sin 2x \log\{\sec 2x + \tan 2x\}$$

which is the required general solution.

EXAMPLE 10.3

Solve by the method of variation of parameters,

$$\frac{d^2y}{dx^2} - y = \frac{2}{1+e^x}.$$

Solution: We have the given equation

$$(D^2 - 1)y = \frac{2}{1+e^x}$$

For finding y_c, AE is

$$D^2 - 1 = 0 \Rightarrow D = \pm 1$$

Hence, $y_c = Ae^x + Be^{-x}$.

Assume $$y = Ae^x + Be^{-x} \tag{1}$$

to be the complete solution of the given equation. For finding A and B we need two equations.

$$A'e^x + B'e^{-x} = 0$$

$$A'e^x + B'(-e^{-x}) = \frac{2}{1+e^x}$$

The coefficient matrix $W = \begin{vmatrix} e^x & e^{-x} \\ e^x & -e^{-x} \end{vmatrix} = -2$

On solving, we obtain

$$A' = \frac{\frac{-2}{1+e^x}e^{-x}}{-2} \qquad B' = \frac{\frac{2}{1+e^x}e^x}{-2}$$

$$= \frac{1}{e^x(1+e^x)} \qquad = \frac{e^x}{1+e^x}$$

$$= \frac{1}{e^x} - \frac{1}{1+e^x}$$

$$= e^{-x} - \frac{e^{-x}}{e^{-x}+1}$$

Integrating, we get

$$A = -e^{-x} + \log(e^{-x} + 1) + c_1,\ B = -\log(e^{x} + 1) + c_2$$

Substituting in Eq. (1), we obtain

$$y = c_1e^{x} + c_2e^{-x} - 1 + e^{x}\log(e^{-x} + 1) - e^{-x}\log(e^{x} + 1)$$

which is the required general solution.

EXAMPLE 10.4

Find the general solution of the equation $y'' + 3y' + 2y = 2e^{x}$, using the method of variation of parameters.

Solution: The given equation is

$$(D^2 + 3D + 2)y = 2e^{x}$$

For finding y_c, AE is

$$D^2 + 3D + 2 = 0$$

$$\Rightarrow \quad D = -1, -2$$

Hence, $y_c = Ae^{-x} + Be^{-2x}$.

Assume the general solution as

$$y = Ae^{-x} + Be^{-2x} \tag{1}$$

For determining A and B, we need two equations

$$A'e^{-x} + B'e^{-2x} = 0$$

$$A'(-e^{-x}) + B'(-2e^{-2x}) = 2e^{x}$$

The coefficient matrix $W = \begin{vmatrix} e^{-x} & e^{-2x} \\ -e^{-x} & -2e^{-2x} \end{vmatrix} = -e^{-3x}$

On solving, we obtain

$$A' = \frac{-2e^{x}e^{-2x}}{-e^{-3x}} \qquad B' = \frac{2e^{x}e^{-x}}{-e^{-3x}}$$

$$= 2e^{2x} \qquad\qquad = -2e^{3x}$$

Integrating, we have

$$A = e^{2x} + c_1 \quad B' = \frac{-2}{3}e^{3x} + c_2$$

Substituting in Eq. (1), we obtain

$$y = c_1 e^{-x} + c_2 e^{-2x} + e^x - \frac{2}{3} e^x$$

$$= c_1 e^{-x} + c_2 e^{-2x} + \frac{1}{3} e^x$$

which is the required general solution.

EXAMPLE 10.5

Solve by the method of variation of parameters

$$y'' - 6y'' + 9y = e^{3x}/x^2.$$

Solution: The given equation is

$$(D^2 - 6D + 9)y = e^{3x}/x^2.$$

For finding y_c, AE is

$$D^2 - 6D + 9 = 0 \Rightarrow D = 3, 3$$

Hence, $y_c = Ae^{3x} + Bxe^{3x}$

Assume the general solution as

$$y = Ae^{3x} + Bxe^{3x} \tag{1}$$

For determining A and B, we need two equations.

$$A'e^{3x} + B'xe^{3x} = 0$$

$$A'(3e^{3x}) + B'(3xe^{3x} + e^{3x}) = \frac{e^{3x}}{x^2}$$

The coefficient matrix $W = \begin{vmatrix} e^{3x} & xe^{3x} \\ 3e^{3x} & e^{3x}(3x+1) \end{vmatrix} = e^{6x}$

On solving, we obtain

$$A' = \frac{-e^{3x}}{x^2}\frac{xe^{3x}}{e^{6x}} \qquad B' = \frac{e^{3x}}{x^2}\frac{e^{3x}}{e^{6x}}$$

$$= \frac{-1}{x} \qquad\qquad = \frac{1}{x^2}$$

Integrating, we have

$$A = -\log x + c_1 \quad B = \frac{-1}{x} + c_2$$

Substituting in Eq. (1), we obtain

$$y = (-\log x + c_1)e^{3x} + \left(\frac{-1}{x} + c_2\right)xe^{3x}$$

$$= (c_1 + c_2 x)e^{3x} - e^{3x}(\log x + 1).$$

which is the required general solution.

EXAMPLE 10.6

Solve by the method of variation of parameters

$$y'' - 2y' + y = e^x \log x.$$

Solution: The given equation is

$$(D^2 - 2D + 1)y = e^x \log x$$

For finding y_c, AE is

$$D^2 - 2D + 1 = 0$$

$\Rightarrow$ $$D = 1, 1$$

Hence, $y_c = (A + B^x)e^x$.
Assume, the general solution as

$$y = Ae^x + Bxe^x \tag{1}$$

For determining A and B, consider the two equations:

$$A'e^x + B'xe^x = 0$$

$$A'e^x + B'e^x(x + 1) = e^x \log x$$

The coefficient matrix $W = \begin{vmatrix} e^x & xe^x \\ e^x & e^x(x+1) \end{vmatrix} = e^{2x}.$

On solving, we obtain

$$A' = -\frac{e^x \log x \cdot xe^x}{e^{2x}} \qquad B' = \frac{e^x \log xe^x}{e^{2x}}$$

$$= -x \log x \qquad\qquad = \log x$$

Integrating, we have

$$A = -\left(\frac{x^2}{2}\log x - \frac{x^2}{4}\right) + c_1 \quad B = (x \log x - x) + c_2$$

Substituting in Eq. (1), we obtain

$$y = \left[c_1 - \left(\frac{x^2}{2}\log x - \frac{x^2}{4}\right)\right]e^x + [c_2 + x(\log x - 1)]xe^x$$

$$= (c_1 + c_2x)e^x + \frac{1}{4}x^2e^x(2\log x - 3)$$

which is the required general solution.

EXAMPLE 10.7

Solve by using the method of variation of parameters $\frac{d^2y}{dx^2} + 3\frac{dy}{dx} + 2y = e^{e^x}$.

Solution: The given equation is

$$(D^2 + 3D + 2)y = e^{e^x}$$

For finding y_c, AE is

$$D^2 + 3D + 2 = 0$$

$$\Rightarrow \quad D = -1, -2$$

Hence, $y_c = Ae^{-x} + Be^{-2x}$.

Assume
$$y = Ae^{-x} + Be^{-2x} \tag{1}$$
be the general solution.

For determining A and B, consider the two equations

$$A'e^{-x} + B'e^{-2x} = 0$$
$$A'(-e^{-x}) + B'(-2e^{-2x}) = e^{e^x}.$$

The coefficient matrix $W = \begin{vmatrix} e^{-x} & e^{-2x} \\ -e^{-x} & -2e^{-2x} \end{vmatrix} = -e^{-3x}$

On solving, we obtain

$$A' = \frac{-e^{e^x}e^{-2x}}{-e^{-3x}} \qquad B' = \frac{e^{e^x}e^{-x}}{-e^{-3x}}$$
$$= e^{e^x}e^x \qquad\qquad = -e^{e^x}e^{2x}$$
$$= -e^{e^x}e^x e^x$$

Integrating, we have

$$A = e^{e^x} + c_1 \quad B = -e^{e^x}(e^x - 1) + c_2$$

Substituting in Eq. (1), we obtain

$$y = (e^{e^x} + c_1)e^{-x} + [e^{e^x}(1 - e^x) + c_2)e^{-2x}$$
$$= c_1e^{-x} + c_2e^{-2x} + e^{-2x}e^{e^x}.$$

which is the required general solution.

EXAMPLE 10.8

Find the general solution of the equation

$$y''' - 6y'' + 11y' - 6y = e^{-x}.$$

Solution: The given equation is

$$(D^3 - 6D^2 + 11D - 6)y = e^{-x}$$

For finding y_c, AE is

$$D^3 - 6D^2 + 11D - 6 = 0$$

$$\Rightarrow \qquad D = 1, 2, 3$$

Hence, $y_c = Ae^x + Be^{2x} + Ce^{3x}$.
Assume the general solution as

$$y = Ae^x + Be^{2x} + Ce^{3x} \tag{1}$$

For determining A, B and C, consider the three equations

$$A'e^x + B'e^{2x} + C'e^{3x} = 0$$

$$A'e^x + B'2e^{2x} + C'3e^{3x} = 0$$

and

$$A'e^x + B'4e^{2x} + C'9e^{3x} = e^{-x}$$

The coefficient matrix $W = \begin{vmatrix} e^x & e^{2x} & e^{3x} \\ e^x & 2e^{2x} & 3e^{3x} \\ e^x & 4e^{2x} & 9e^{3x} \end{vmatrix}$

$$= 2e^{6x}$$

By the Cramer's rule, we obtain

$$A' = \frac{1}{2e^{6x}}\begin{vmatrix} 0 & e^{2x} & e^{3x} \\ 0 & 2e^{2x} & 3e^{3x} \\ e^{-x} & 4e^{2x} & 9e^{3x} \end{vmatrix} = \frac{e^{4x}}{2e^{6x}} = \frac{1}{2}e^{-2x}$$

$$B' = \frac{1}{2e^{6x}}\begin{vmatrix} e^x & 0 & e^{3x} \\ e^x & 0 & 3e^{3x} \\ e^x & e^{-x} & 9e^{3x} \end{vmatrix} = -\frac{2e^{3x}}{2e^{6x}} = -e^{-3x}$$

and

$$C' = \frac{1}{2e^{6x}}\begin{vmatrix} e^x & e^{2x} & 0 \\ e^x & 2e^{2x} & 0 \\ e^x & 4e^{2x} & e^{-x} \end{vmatrix} = \frac{e^{2x}}{2e^{6x}} = \frac{1}{2}e^{-4x}$$

Integrating, we obtain

$$A = \frac{-1}{4}e^{-2x} + c_1, \; B = \frac{1}{3}e^{-3x} + c_2$$

and $$C = \frac{-1}{8}e^{-4x} + c_3$$

Substituting in Eq. (1), we obtain

$$y = \left(\frac{-1}{4}e^{-2x} + c_1\right)e^{x} + \left(\frac{1}{3}e^{-3x} + c_2\right)e^{2x} + \left(\frac{-1}{8}e^{-4x} + c_3\right)e^{3x}$$

$$= c_1e^{x} + c_2e^{2x} + c_3e^{3x} + \left(\frac{-1}{4} + \frac{1}{3} - \frac{1}{8}\right)e^{-x}$$

$$= c_1e^{x} + c_2e^{2x} + c_3e^{3x} - \frac{1}{24}e^{-x}$$

which is the required general solution. ▲▲

EXERCISE 10.1

Find the general solution of the following differential equations, using the method of variation of parameters:

1. $y'' + y = \operatorname{cosec} x.$
2. $y'' + y = \sec x.$
3. $y'' + 4y' + 4y = e^{-2x}\sin x.$
4. $y'' + 16y = 32\sec 2x.$
5. $y'' + 6y' + 9y = e^{-3x}/x.$
6. $y'' + 4y' + 8y = 16e^{-2x}\operatorname{cosec}^2 2x.$
7. $y'' + 2y' + 2y = e^{x}\tan x.$
8. $\frac{d^2y}{dx^2} + y = \frac{1}{1+\sin x}.$
9. $\frac{d^2y}{dx^2} - 3\frac{dy}{dx} + 2y = \frac{1}{1+e^{-x}}.$
10. $y''' + 4y' = \sec^2 x.$
11. $y''' - 6y'' + 12y' - 8y = e^{2x}/x.$

ANSWERS

Exercise 10.1

1. $y = c_1\cos x + c_2\sin x + \sin x \log|\sin x| - x\cos x.$
2. $y = c_1\cos x + c_2\sin x + \cos x \log|\cos x| + x\sin x.$
3. $y = (c_1 + c_2x)e^{-2x} - e^{-2x}\sin x.$
4. $y = c_1\cos 4x + c_2\sin 4x + 8\cos 2x - 4\sin 4x \log|\sec 2x + \tan 2x|.$

5. $y = (c_1 + c_2x)e^{-3x} + x(\log|x| - 1)e^{-3x}$.
6. $y = e^{-2x}(c_1\cos 2x + c_2\sin 2x) + 4e^{-2x}\cos 2x \log|\operatorname{cosec} 2x + \cot 2x| - 4e^{-2x}$.
7. $y = e^x(c_1\cos x + c_2\sin x) - e^x\cos x \log|\sec x + \tan x|$.
8. $y = c_1\cos x + c_2\sin x + \sin x\log(1 + \sin x) - x\cos x - 1$.
9. $y = (e^x + e^{2x})\log(1 + e^x) + (c_1 - 1 - x)e^x + (c_2 - x)e^{2x}$.
10. $y = c_1 + c_2\cos 2x + c_3\sin 2x - (x\cos 2x)/4 + (\sin 2x \log|\cos 2x|)/8 + \dfrac{(\log|\sec 2x + \tan 2x|)}{8}$.
11. $y = (c_1 + c_2x + c_3x^2)e^{2x} + \dfrac{(x^2\log|x|e^{2x})}{8}$.

▲▲

Method of Undetermined Coefficients

The method of variation of parameters is a direct method and always leads to a solution. However, the integration that we may need to perform to carry the technique to completion may, for some problems is rather difficult. Undetermined coefficient is a method of organized guessing and the process of carrying out this method is quite straightforward involving only techniques of college algebra and differentiation but the method applies, in general, to a rather small class of problems. In the case, when the right hand side $r(x)$ is of a special form containing exponential, polynomials, cosine and sine functions, sums or product of these functions, then the particular integral can be easily obtained by the method of undetermined coefficients as explained in the following.

If $r(x)$ is of exponential form e^{ax}, then its derivatives also contain exponential e^{ax} only, that is, if $r(x) = pe^{ax}$ where, p is a constant then we can choose the particular integral as $y_p = ce^{ax}$ where, c is a constant determined by substituting y_p in the given equation and comparing both sides of the equation. That is, the equation is identically satisfied.

If $r(x)$ is a cosine or sine function that is $\cos nx$ or $\sin nx$ then their derivatives contain both the terms $\cos nx$ and $\sin nx$. In other words, if $r(x) = p\cos nx$ or $p\sin nx$ where, p is a constant or sum of these two, then, we can choose the particular integral as $y_p = c_1\cos nx + c_2\sin nx$. The constants c_1, c_2 are determined by substituting y_p in the given equation and comparing both sides of the equation.

If $r(x)$ is of the form x_m, then its derivatives contain the terms $x^n, x^{n-1}, \ldots x, 1$. Hence, when $r(x) = px^n$, where, p is a constant, then we can choose the particular integral as

$$y_p = c_1x^n + c_2x^{n-1} + \ldots + c_{n-1}x + c_n$$

where, $c_1, c_2, \ldots c_n$ are constants determined by substituting y_p in the given equation and comparing both sides of the equation. Similarly, if $r(x)$ is of the form $e^{ax}\cos bx$ or $e^{ax}\sin bx$, then their derivatives contain the terms $e^{ax}\cos bx$ and $e^{ax}\sin bx$. Hence, when $r(x) = e^{ax}\cos bx$ or $e^{ax}\sin bx$, then we can choose the particular integral as

$$y_p = e^{ax}(c_1\cos bx + c_2\sin bx)$$

However, if any term in the choice of the particular integral is also a solution of the corresponding homogeneous equation, that is, a term in the complementary function, then we multiply this term by x or x^n (if the term in the complementary function corresponding to a multiple root of multiplying m). If $r(x)$ is the sum of a number of functions, then the combination with respect to each of the terms is included in the choice of the particular integral. The above discussion is summed up in the form of the following Table 10.1.

Table 10.1

r(x)	y_p
1. x^n	$c_1x^n + c_2x^{n-1} + \ldots + c_{n-1}x + c_n$
2. e^{ax}	ce^{ax}
3. $\sin(bx + c)$ or $\cos(bx + c)$	$c_1\cos(bx + c) + c_2\sin(bx + c)$
4. $x^n\sin(bx + c)$ or $x^n\cos(bx + c)$	$\cos(bx + c)[c_1x^n + c_2x^{n-1} + \ldots + c_{n-1}x + c_n]$ $+ \sin(bx + c)[c_1'x^n + c_2'x^{n-1} + \ldots + c_{n-1}'x + c_n']$
5. $e^{ax}\sin(bx + c)$ or $e^{ax}\cos(bx + c)$	$e^{ax}[c_1\cos(bx + c) + c_2\sin(bx + c)]$
6. $x^ne^{ax}\sin(bx + c)$ or $x^ne^{ax}\cos(bx + c)$	$e^{ax}[\cos(bx + c)(c_1x^n + c_2x^{n-1} + \ldots + c_{n-1}x$ $+ c_n) + \sin(bx + c)(c_1'x^n + c_2'x^{n-1} + \ldots +$ $c_{n-1}'x + c_n')]$

The first step for performing the method of undetermined coefficient method is to see $r(x)$ then write y_p by using above rules. Secondly, if there is any term common in y_p and the complementary function y_c then multiply the common term in y_p by x raised to power one more than that in y_c.

Note: In case of method of variation of parameter, the final solution contains both complementary function (y_c) and particular integral (y_p) but we find y_c and y_p separately in the method of undetermined coefficients and in the end, sum of these two will give us the general solution of the given non-homogeneous differential equation. The method of variation of parameters has the advantage over the method of undetermined coefficients of not involving any guessing. When $r(x)$ is $\tan x$, $\sec x$, $\cot x$, $\operatorname{cosec} x$, $\log x$, etc., method of undetermined coefficients fails, since the number of terms obtained by differentiating such functions is infinite.

EXAMPLE 10.9

Find the general solution of the differential equation $y'' - 2y' - 3y = 6e^{-x} + 8e^{x}$. We have the given equation

$$(D^2 - 2D - 3)y = 6e^{-x} + 8e^{x}.$$

Solution: For finding y_c, AE is

$$D^2 - 2D - 3 = 0$$

$$\Rightarrow \quad D = 3, -1$$

Hence, $y_c = c_1e^{3x} + c_2e^{-x}$.

We note that e^{-x} appears both as a term in y_p and the right hand side $r(x)$. The term e^x appears only in $r(x)$. Hence, we choose the particular integral as

$$y_p = Axe^{-x} + Be^x$$

We have
$$y_p' = Ae^{-x}(-x + 1) + Be^x$$
and
$$y_p'' = Ae^{-x}(-1 + x - 1) + Be^x = Ae^{-x}(x - 2) + Be^x.$$

Substituting in the given equation, we get

$$A[x - 2 - 2(-x + 1) - 3x]e^{-x} + B[1 - 2 - 3]e^x = 6e^{-x} - 8e^x$$

$$\Rightarrow \quad -4Ae^{-x} - 4Be^x = 6e^{-x} - 8e^x$$

Comparing the coefficients of e^{-x} and e^x, we get

$$A = -3/2,\ B = 2$$

The general solution is

$$y = c_1e^{3x} + c_2e^{-x} - \frac{3}{2}xe^{-x} + 2e^x.$$

EXAMPLE 10.10

Find the general solution of the equation

$$y'' + y = \sin x.$$

Solution: We have the given equation $(D^2 + 1)y = \sin x$.

For finding y_c, AE is

$$D^2 + 1 = 0 \Rightarrow D = \pm i$$

Hence, $y_c = c_1 \cos x + c_2 \sin x$.

Since, $r(x)$ contains $\sin x$, we assume

$$y_p = A\cos x + B\sin x.$$

However, since these terms appear in y_c, we multiply by x. Hence,

$$y_p = x(A\cos x + B\sin x)$$

We have
$$y_p' = (A + Bx)\cos x + (B - Ax)\sin x$$
and
$$y_p'' = (2B - Ax)\cos x - (2A + Bx)\sin x$$

Substituting in the given equation, we get

$$2A\cos x - 2B\sin x = \sin x$$

Comparing both sides, we get $A = 0$ and $B = \dfrac{-1}{2}$.

The particular integral is $y_p = \frac{-1}{2} x \sin x$.

The general solution is

$$y = c_1 \cos x + c_2 \sin x - \frac{1}{2} x \sin x.$$

EXAMPLE 10.11

Using the method of undetermined coefficients, find the general solution of the differential equation

$$y'' + y = 32x^3.$$

Solution: We have the given equation

$$(D^2 + 1)y = 32x^3.$$

For finding y_c, AE is

$$D^2 + 1 = 0$$

$\Rightarrow$ $$D = \pm i$$

Hence, $y_c = c_1 \cos x + c_2 \sin x$.
Since $r(x) = 32x^3$, we choose the particular integral as

$$y_p = ax^3 + bx^2 + cx + d$$

Substituting y_p, y'_p, y''_p in the given equation we get

$$(6ax + 2b) + (ax^3 + bx^2 + cx + d) = 32x^3.$$

Comparing the coefficients of various powers of x, we get

$$a = 32,\ b = 0,\ 6a + c = 0,\ 2b + d = 0$$

The solution of the system is

$$a = 32,\ b = 0,\ c = -192,\ d = 0$$

Therefore, $y_p = 32x^3 - 192x$.
The general solution is

$$y = c_1 \cos x + c_2 \sin x + 32x(x^2 - 6).$$

EXAMPLE 10.12

Find the general solution of the equation

$$y'' + 4y' + 4y = 12e^{-2x}.$$

Solution: We have the given equation $(D^2 + 4D + 4)y = 12e^{-2x}$.
For finding y_c, AE is

$$D^2 + 4D + 4 = 0$$

$\Rightarrow$ $$D = -2, -2$$

Hence, $y_c = (c_1 + c_2x)e^{-2x}$.

We note that e^{-2x} and xe^{-2x} are present in the complementary function y_c (due to the double root – 2) and e^{-2x} is also a term in the right hand side $r(x)$. Therefore, we choose the particular integral as

$$y_p = Ax^2e^{-2x}$$

We have

$$y_p' = A(2x - 2x^2)e^{-2x}$$

and

$$y_p'' = A(2 - 8x + 4x^2)e^{-2x}.$$

Substituting in the given equation, we get

$$[A(2 - 8x + 4x^2) + 4(2x - 2x^2) + 4x^2]e^{-2x} = 12e^{-2x}$$

$$\Rightarrow \quad 2Ae^{-2x} = 12e^{-2x}$$

Comparing both sides, we get $A = 6$.

Therefore, the particular integral is

$$y_p = 6x^2e^{-2x}.$$

The general solution is

$$y = (c_1 + c_2x + 6x^2)e^{-2x}.$$

EXAMPLE 10.13

Solve by the method of undetermined coefficients

$$y'' - y = e^{3x}\cos 2x - e^{2x}\sin 3x.$$

Solution: We have the given equation

$$(D^2 - 1)y = e^{3x}\cos 2x - e^{2x}\sin 3x.$$

For finding y_c, AE is

$$D^2 - 1 = 0$$

$$\Rightarrow \quad D = \pm 1$$

Hence, $y_c = c_1e^x + c_2e^{-x}$.

Now, $\quad y_p' = e^{3x}(a\cos 2x + b\sin 2x) + e^{2x}(c\cos 3x + d\sin 3x)$

We have

$$y_p' = e^{3x}[(3a + 2b)\cos 2x + (3b - 2a)\sin 2x] + e^{2x}[(2c + 3d)\cos 3x + (2d - 3c)\sin 3x)]$$

and $y_p'' = e^{3x}[(5a + 12b)\cos 2x + (5b - 12a)\sin 2x] + e^{2x}[(12d - 5c)\cos 3x - (12c + 5d)\cos 3x]$

Substituting these in the given equation, we get

$$e^{3x}[(4a + 12b)\cos 2x + (4b - 12a)\sin 2x] + e^{2x}(12d - 6c)\cos 3x - (12c + 6d)\sin 3x]$$
$$= e^{3x}\cos 2x - e^{2x}\sin 3x$$

Comparing both sides, we get

$$4a + 12b = 1,\ 4b - 12a = 0,\ 12d - 6c = 0 \text{ and } 12c + 6d = -1$$

$$\Rightarrow \quad a = 1/40,\ b = 3/40,\ c = -1/15 \text{ and } d = -1/30$$

The particular integral is

$$y_p = \frac{1}{40}e^{3x}[\cos 2x + 3\sin 2x] - \frac{1}{30}e^{2x}[2\cos 3x + \sin 3x]$$

The general solution is

$$y = c_1e^x + c_2e^{-x} + \frac{1}{40}e^{3x}(\cos 2x + 3\sin 2x) - \frac{1}{30}e^{2x}(2\cos 3x + \sin 3x).$$

EXAMPLE 10.14

Find the general solution of the equation

$$y'' - 4y' + 13y = 12e^{2x}\sin 3x.$$

Solution: We have the given equation

$$(D^2 - 4D + 13)y = 12e^{2x}\sin 3x$$

For finding y_c, AE is

$$D^2 - 4D + 13 = 0$$

$$\Rightarrow \quad D = 2 \pm 3i$$

Hence, $y_c = e^{2x}[c_1\cos 3x + c_2\sin 3x]$.

We note that $e^{2x}\sin 3x$ appears both in the complementary function (y_c) and the right hand side $r(x)$. Therefore we choose $y_p = xe^{2x}[A\cos 3x + B\sin 3x]$

We have,

$$y_p' = (1 + 2x)e^{2x}(A\cos 3x + B\sin 3x) + 3xe^{2x}(-A\sin 3x + B\cos 3x).$$

and

$$y_p'' = (4 + 4x)e^{2x}(A\cos 3x + B\sin 3x) + 6(1 + 2x)e^{2x}(-A\sin 3x + B\cos 3x) + 9xe^{2x}(-A\cos 3x - B\sin 3x).$$

Substituting in the given equation, we get

$$e^{2x}\cos 3x[A(4 + 4x) + 6B(1 + 2x) - 9Ax - 4A(1 + 2x) - 12Bx + 13Ax) + e^{2x}\sin 3x[B(4 + 4x) - 6A(1 + 2x) - 9Bx - 4B(1 + 2x) + 12Ax + 13Bx) = 12e^{2x}\sin 3x$$

$$\Rightarrow \quad 6Be^{2x}\cos 3x - 6Ae^{2x}\sin 3x = 12e^{2x}\sin 3x$$

Comparing both sides, we get

$$A = -2 \text{ and } B = 0.$$

Therefore, the particular integral is $y_p = -2xe^{2x}\cos 3x$.

The general solution is

$$y = e^{2x}(c_1 \cos 3x + c_2 \sin 3x - 2x \cos 3x).$$

EXAMPLE 10.15

Find the general solution of the differential equation $y''' - 2y'' - 5y' + 6y = 18e^x$.

Solution: We have the given equation

$$(D^3 - 2D^2 - 5D + 6)y = 18e^x.$$

For finding y_c, AE is

$$D^3 - 2D^2 - 5D + 6 = 0$$

$$\Rightarrow \quad D = 1, -2, 3$$

Hence, $\quad y_c = c_1e^x + c_2e^{-2x} + c_3e^{3x}$

Choose the particular integral as

$$y_p = Axe^x.$$

We have $\quad y_p' = A(1 + x)e^x,\ y_p'' = A(2 + x)e^x$, and $y_p''' = A(3 + x)e^x$.

Substituting in the given equation, we get

$$Ae^x[(3 + x) - 2(2 + x) - 5(1 + x) + 6x] = 18e^x$$

$$\Rightarrow \quad -6Ae^x = 18e^x$$

Comparing both sides, we get $A = -3$.
Hence, the particular integral is $y_p = -3xe^x$.
The general solution is

$$y = c_1e^x + c_2e^{-2x} + c_3e^{3x} - 3xe^x.$$

EXAMPLE 10.16

Find the general solution of the differential equation $y'' + 2y' + 4y = 2x^2 + 3e^{-x}$.

Solution: We have the given equation $(D^2 + 2D + 4)y = 2x^2 + 3e^{-x}$.

For finding y_c, AE is

$$D^2 + 2D + 4 = 0$$

$$\Rightarrow \quad D = -1 \pm \sqrt{3}i$$

Hence, $y_c = e^{-x}[c_1 \cos \sqrt{3}x + c_2 \sin \sqrt{3}x]$.

Choose the particular integral as

$$y_p = ax^2 + bx + c + de^{-x}.$$

We have $\quad y_p' = 2ax + b - de^{-x}$

and $\quad y_p'' = 2a + de^{-x}$

Substituting in the given equation, we get

$$4ax^2 + (4a + 4b)x + (2a + 2b + 4c) + 3de^{-x} = 2x^2 + 3e^{-x}$$

Comparing both sides, we get

$$a = 1/2,\ b = -1/2,\ c = 0 \text{ and } d = 1$$

Hence, the particular integral is

$$y_p = \frac{1}{2}x^2 - \frac{1}{2}x + e^{-x}$$

The general solution is

$$y = e^{-x}[c_1 \cos\sqrt{3}x + c_2 \sin\sqrt{3}x] + \frac{1}{2}x^2 - \frac{1}{2}x + e^{-x}.$$

EXAMPLE 10.17

Find the general solution of the differential equation $y^{iiv} + 6y''' + 12y'' + 8y' = 60e^{-2x}$.

Solution: We have the given equation

$$(D^4 + 6D^3 + 12D^2 + 8D)y = 60e^{-2x}.$$

For finding y_c, AE is

$$D(D^3 + 6D^2 + 12D + 8) = 0$$

$$\Rightarrow \qquad D = 0, -2, -2, -2.$$

Hence, $y_c = c_1 + (c_2 + c_3x + c_4x^2)e^{-2x}$.

We note that e^{-2x}, xe^{-2x} and x^2e^{-2x} are present in the complementary function y_c (due to the triple root -2) and e^{-2x} is also a term on the right hand side $r(x)$. Therefore, we choose the particular integral as

$$y_p = Ax^3e^{-2x}$$

We have

$$y_p' = A(3x^2 - 2)e^{-2x}$$
$$y_p'' = A(6x - 6x^2 + 4)e^{-2x}$$
$$y_p''' = A(6 - 12x - 12x + 12x^2 - 8)e^{-2x}$$
$$= A(12x^2 - 24x - 2)e^{-2x}$$

and

$$y_p^{iv} = A(24x - 24 - 24x^2 + 48x + 4)e^{-2x}$$
$$= A(24x^2 + 72x - 20)e^{-2x}.$$

Substituting in the given equation, we get

$$12Ae^{-2x} = 60e^{-2x}$$

and comparing both sides, we get $A = -5$

Therefore, the particular integral is

$$y_p = -5x^3e^{-2x}.$$

The general solution is

$$y = c_1 + (c_2 + c_3x + c_4x^2)e^{-2x} - 5x^3e^{-2x}.$$

EXAMPLE 10.18

Find the general solution of the differential equation.

$$y^{\text{iv}} - 16y'' = 8x + 16$$

Solution: We have the given equation

$$(D^4 - 16D^2)y = 8x + 16$$

For finding y_c, AE is

$$D^2(D^2 - 16) = 0$$

$$\Rightarrow \quad D = 0, 0, 4, -4$$

Hence, $y_c = c_1 + c_2x + c_3e^{4x} + c_4e^{-4x}$.

Since $r(x)$ contains $8x + 16$, we assume

$$y_p = ax + b$$

However, since these terms appear in y_c, we multiply by x^2. Hence,

$$y_p = x^2(ax + b) = ax^3 + bx^2.$$

We have $\quad y_p' = 3ax^2 + 2bx,\ y_p'' = 6ax + 2b,\ y_p''' = 6a$ and $y_p^{\text{iv}} = 0$.

Substituting in the given equation, we get

$$0 + 16(6ax + 2b) = 8x + 16$$

Comparing both sides, we get

$$a = -1/12 \text{ and } b = -1/2$$

Hence, the particular integral is

$$y_p = -\frac{1}{12}x^3 - \frac{1}{2}x^2.$$

The general solution is

$$y = c_1 + c_2x + c_3e^{4x} + c_4e^{-4x} - (x^3 + 6x^2)/12.$$

EXAMPLE 10.19

Find the general solution of the differential equation

$$y'' + 4y' + 4y = 6e^{-2x}\cos^2 x.$$

Solution: We have the given equation

$$(D^2 + 4D + 4)y = 3e^{-2x}(1 + \cos 2x)$$

For finding y_c, AE is

$$D^2 + 4D + 4 = 0$$

$$\Rightarrow \qquad D = -2, -2.$$

Hence, $y_c = (c_1 + c_2 x)e^{-2x}$.

We note that e^{-2x} and xe^{-2x} are present in the complementary function y_c (due to the double root -2) and e^{-2x} is also a term on the right hand side $r(x)$. Therefore, we choose the particular integral as

$$y_p = Ax^2 e^{-2x} + e^{-2x}(B\cos 2x + C\sin 2x)$$

We have $y_p' = A(2x - 2x^2)e^{-2x} + e^{-2x}[(-2B + 2C)\cos 2x + (-2B - 2C)\sin 2x]$

and

$$\begin{aligned} y_p'' &= A(2 - 8x + 4x^2)e^{-2x} + e^{-2x}[(4B - 4C - 4B - 4C)\cos 2x \\ &\qquad + (4B - 4C + 4B + 4C)\sin 2x)] \\ &= A(2 - 8x + 4x^2)e^{-2x} + e^{-2x}[-8C\cos 2x + 8B\sin 2x] \end{aligned}$$

Substituting in the given equation, we get

$$2Ae^{-2x} + e^{-2x}[(-8C - 8B + 8C + 4B)\cos 2x + (8B - 8B - 8C + 4C)\sin 2x] = 3e^{-2x} + 3\cos 2x$$

Comparing both sides, we get

$$A = 3/2,\ B = -3/4, \text{ and } C = 0$$

Hence, the particular integral is

$$\begin{aligned} y_p &= \frac{3}{2}x^2 e^{-2x} - \frac{3}{4}e^{-2x}\cos 2x \\ &= 3e^{-2x}(2x^2 - \cos 2x)/4. \end{aligned}$$

The general solution is

$$y = (c_1 + c_2 x)e^{-2x} + 3e^{-2x}(2x^2 - \cos 2x)/4.$$

▲▲

EXERCISE 10.2

Find the general solution of the following differential equations by the method of undetermined coefficients.

1. $y'' + 9y = \cos 3x$.
2. $y'' - 3y' - 10y = 1 + x^2$.
3. $y'' + 6y' + 8y = e^{-3x} + e^x$.
4. $2y'' + 3y' - 2y = 5e^{-2x} + e^x$.
5. $y'' + 25y = 50\cos 5x + 30\sin 5x$.
6. $y'' - 6y' + 13y = 6e^{3x}\sin x\cos x$.
7. $y'' + 3y' + 2y = 12e^{-x}\sin^3 x$.
8. $(D^2 - 3D + 2)y = x^2 + e^x$.

9. $(D^2 - 2D + 3)y = x^3 + \cos x$.
10. $(D^3 - 2D)\, y = e^x \sin x$.
11. $y''' - 6y'' + 12y' - 8y = 12e^{2x} + 27e^{-x}$.
12. $y''' + 3y'' - 4y = 12e^{-2x} + 9e^x$.
13. $y''' - 2y'' + 4y' - 8y = 8(x^2 + \cos 2x)$.
14. $y^{iv} - 256y = 128 \cos 4x$.
15. $y^{iv} + 3y''' + 3y'' + y' = 2x + 4$.

ANSWERS

Exercise 10.2

1. $y = c_1 \cos 3x + c_2 \sin 3x + \frac{1}{6} x \sin 3x$.
2. $y = c_1 e^{-2x} + c_2 e^{5x} - (50x^2 - 30x + 69)/500$.
3. $y = c_1 e^{-2x} + c_2 e^{-4x} - e^{-3x} + e^x/15$.
4. $y = c_1 e^{-2x} + c_2 e^{x/2} - xe^{-2x} + e^x/3$.
5. $y = c_1 \cos 5x + c_2 \sin 5x + x(-3 \cos 5x + 5 \sin 5x)$.
6. $y = e^{3x}(c_1 \cos 2x + c_2 \sin 2x) - \dfrac{-3xe^{3x} \cos 2x}{4}$.
7. $y = c_1 e^{-x} + c_2 e^{-2x} + e^{-x}[-45(\cos x + \sin x) + (\cos 3x + 3 \sin 3x)]/10$.
8. $y = c_1 e^x + c_2 e^{2x} + \frac{1}{2}\left(x^2 + 3x + \frac{7}{2} - 2xe^x\right)$.
9. $y = e^x[c_1 \cos \sqrt{2}x + c_2 \sin \sqrt{2}x] + \frac{1}{27}[9x^3 + 18x^2 + 6x - 8] + \frac{1}{4}(\cos x - \sin x)$.
10. $y = c_1 + c_2 e^{2x} - \frac{1}{2} e^x \sin x$.
11. $y = (c_1 + c_2 x + c_3 x^2) e^{2x} + 2x^3 e^{2x} - e^{-x}$.
12. $y = (c_1 + c_2 x) e^{-2x} + c_3 e^x + xe^x - 2x^2 e^{-2x}$.
13. $y = c_1 e^{2x} + c_2 \cos 2x + c_3 \sin 2x - [2(x^2 + x) + x(\cos 2x + \sin 2x)]/2$.
14. $y = c_1 e^{4x} + c_2 e^{-4x} + c_3 \cos 4x + c_4 \sin 4x - \dfrac{-x \sin 4x}{2}$.
15. $y = c_1 + (c_2 + c_3 x + c_4 x^3) e^{-x} + x^2 - 2x$.

11

CHAPTER

Equations Reducible to Linear Differential Equations with Constant Coefficients

In chapters 8, 9 and 10, we have seen how to obtain the general solution of the nth order linear differential equation with constant coefficient and observed that the general solution of such equations may be easily determined. But the general nth order linear differential equations with variable coefficient is quite a different matter, however, now we shall study two such forms of linear differential equations with variable coefficients which can be reduced to linear differential equations with constant coefficient by suitable substitutions.

The Cauchy–Euler Equation

An equation of the from

$$a_0 x^n \frac{d^n y}{dx^n} + a_1 x^{n-1} \frac{d^{n-1} y}{dx^{n-1}} + \ldots + a_{n-1} x \frac{dy}{dx} + a_n y = r(x)$$

where, $a_0, a_1, \ldots a_{n-1}, a_n$ are constants is called a Cauchy–Euler equation of order n.

To know the solution of such an equation, we made some substitution so that it may reduce to an equation for which the methods of solution are known. Such a substitution is $x = e^t$. This substitution reduces the above equation to a linear equation with constant coefficient. To illustrate this, we consider the second order Cauchy–Euler equation as

$$a_0 x^2 \frac{d^2 y}{dx^2} + a_1 x \frac{dy}{dx} + a_2 y = r(x) \tag{1}$$

Take $x = e^t$, then $\log x = t \Rightarrow \frac{dt}{dx} = \frac{1}{x}$

$$\frac{dy}{dx} = \frac{dy}{dt}\frac{dt}{dx} = \frac{1}{x}\frac{dy}{dt}$$

$$\frac{d^2y}{dx^2} = \frac{d}{dx}\left(\frac{1}{x}\frac{dy}{dt}\right) = \frac{1}{x}\frac{d}{dt}\left(\frac{dy}{dt}\right)\frac{dt}{dx} - \frac{1}{x^2}\frac{dy}{dt}$$

$$= \frac{1}{x^2}\left(\frac{d^2y}{dt^2} - \frac{dy}{dt}\right)$$

Thus,

$$x\frac{dy}{dx} = \frac{dy}{dt} \quad \text{and} \quad x^2\frac{d^2y}{dx^2} = \frac{d^2y}{dt^2} - \frac{dy}{dt}$$

and hence, Eq. (1) reduces to

$$a_0\left(\frac{d^2y}{dt^2} - \frac{dy}{dt}\right) + a_1\frac{dy}{dt} + a_2y = r(e^t)$$

$$\Rightarrow \qquad b_0\frac{d^2y}{dt^2} + b_1\frac{dy}{dt} + b_2y = R(t) \tag{2}$$

where, $b_0 = a_0$, $b_1 = a_1 - a_0$, $b_2 = a_2$ and $R(t) = r(e^t)$. Equation (2) is a linear differential equation with constant coefficients and can be solved by already known methods. In a similar way, we can solve the nth-order equation.

Notation: We have denoted $\frac{d}{dx}$ by D. Similarly, we will denote $\frac{d}{dt}$ by θ. Therefore, we have $xD = \theta$ and $x^2D^2 = \theta(\theta - 1)$ and $x^3D^3 = \theta(\theta - 1)(\theta - 2)$ etc.

Note: We can notice that in Cauchy–Euler equation the power of x and the order of derivative are same in each term. ▲▲

EXAMPLE 11.1

Solve $x^3\frac{d^3y}{dx^3} + 2x^2\frac{d^2y}{dx^2} + 2y = 10\left(x + \frac{1}{x}\right)$.

Solution: We have the given equation

$$(x^3D^3 + 2x^2D^2 + 2)y = 10\left(x + \frac{1}{x}\right)$$

By putting $x = e^t$, $t = \log x$, $\frac{dt}{dx} = \frac{1}{x}$

$$xD = \theta,\ x^2D^2 = \theta(\theta - 1)$$

and

$$x^3D^3 = \theta(\theta - 1)(\theta - 2) \text{ where } \theta = \frac{d}{dt}$$

the equation becomes

$$[\theta(\theta - 1)(\theta - 2) + 2\theta(\theta - 1) + 2]y = 10(e^t + e^{-t})$$

$$\Rightarrow \qquad (\theta^3 - \theta^2 + 2)y = 10(e^t + e^{-t})$$

For finding y_c, AE is

$$\theta^3 - \theta^2 + 2 = 0$$

$$\Rightarrow \qquad \theta = -1, 1 \pm i$$

Hence, $y_c = c_1e^{-t} + e^t[c_2\cos t + c_3\sin t]$.

Also,
$$y_p = \frac{1}{\theta^3 - \theta^2 + 2}10(e^t + e^{-t})$$

$$= 10\frac{1}{\theta^3 - \theta^2 + 2}e^t + 10\frac{1}{\theta^3 - \theta^2 + 2}e^{-t}$$

$$= 10\frac{1}{1^3 - 1^2 + 2}e^t + 10\frac{1}{(-1)^3 - (-1)^2 + 2}e^{-t}$$

$$= 5e^t + 10t\frac{1}{3\theta^2 - 2\theta}e^{-t} \quad \{\because \text{Rule fails}\}$$

$$= 5e^t + 10t\frac{1}{3(-1)^2 - 2(-1)}e^{-t}$$

$$- 5e^t + 2te^{-t}$$

Thus, the required general solution is

$$y = c_1x^{-1} + x[c_2\cos(\log x) + c_3\sin(\log x)] + 5x + 2x^{-1}\log x.$$

EXAMPLE 11.2

Solve $x^2\frac{d^2y}{dx^2} - x\frac{dy}{dx} + y = \log x.$

Solution: We have the given equation

$$(x^2D^2 - xD + 1)y = \log x.$$

By putting $x = e^t$ $t = \log x$ and $\frac{d}{dt} = \theta,\ \frac{dt}{dx} = \frac{1}{x}$

We obtain

$$[\theta(\theta - 1) - \theta + 1]y = t$$

$$\Rightarrow \qquad (\theta^2 - 2\theta + 1)y = t$$

For finding y_c, AE is

$$\theta^2 - 2\theta + 1 = 0$$

$$\Rightarrow \quad \theta = 1, 1.$$

Hence, $y_c = (c_1 + c_2 t)e^t$

Also
$$y_p = \frac{1}{(\theta-1)^2}t = (1-\theta)^{-2}t$$
$$= (1 + 2\theta + 3\theta^2 + \dots)t$$
$$= t + 2$$

Thus, the required general solution is

$$y = (c_1 + c_2 \log x)x + \log x + 2.$$

EXAMPLE 11.3

Solve $x^2\frac{d^2y}{dx^2} - x\frac{dy}{dx} - 3y = x^2 \log x.$

Solution: We have the given equation

$$(x^2D^2 - xD - 3)y = x^2 \log x$$

By putting $x = e^t$, $t = \log x$, $\frac{dt}{dx} = \frac{1}{x}$ and $\frac{d}{dt} = \theta$, we obtain

$$(\theta(\theta - 1) - \theta - 3)y = te^{2t}$$

$$\Rightarrow \quad (\theta^2 - 2\theta - 3)y = te^{2t}$$

For finding y_c, AE is

$$\theta^2 - 2\theta - 3 = 0$$

$$\Rightarrow \quad (\theta - 3)(\theta + 1) = 0$$

$$\Rightarrow \quad \theta = 3, -1.$$

Hence, $y_c = c_1e^{3t} + c_2e^{-t}$

Also,
$$y_p = \frac{1}{(\theta-3)(\theta+1)}te^{2t}$$
$$= e^{2t}\frac{1}{(\theta+2-3)(\theta+2+1)}t$$
$$= e^{2t}\frac{1}{(\theta-1)(\theta+3)}t$$

$$= \frac{-1}{3} e^{2t}(1-\theta)^{-1}\left(1+\frac{\theta}{3}\right)^{-1} t$$

$$= \frac{-1}{3} e^{2t}(1+\theta+\theta^2+\ldots)\left(1-\frac{\theta}{3}+\frac{\theta^2}{9}\right)t$$

$$= \frac{-1}{3} e^{2t}\left(1+\frac{2\theta}{3}+\ldots\right)t$$

$$= \frac{-1}{3} e^{2t}\left(t+\frac{2}{3}\right).$$

Thus, the required general solution is

$$y = c_1x^3 + \frac{c_2}{x} - \frac{1}{9}x^2(3\log x + 2).$$

EXAMPLE 11.4

Solve $(x^2D^2 + 3xD + 1)y = \dfrac{1}{(1-x)^2}$.

Solution: By putting $x = e^t$, $t = \log x$, $\dfrac{dt}{dx} = \dfrac{1}{x}$ and $\dfrac{d}{dt} = \theta$, we obtain

$$(\theta(\theta - 1) + 3\theta + 1)y = \frac{1}{(1-e^t)^2}$$

$$\Rightarrow \quad (\theta^2 + 2\theta + 1)y = \frac{1}{(1-e^t)^2}$$

For finding y_c, AE is

$$\theta^2 + 2\theta + 1 = 0$$

$$\Rightarrow \quad (\theta + 1)^2 = 0 \Rightarrow \theta = -1, -1$$

Hence, $y_c = (c_1 + c_2t)e^{-t}$.

Also,

$$y_p = \frac{1}{(\theta+1)^2}\frac{1}{(1-e^t)^2}$$

Let us put $\dfrac{1}{(\theta+1)}\dfrac{1}{(1-e^t)^2} = u$

$$\therefore \quad \frac{du}{dt} + u = \frac{1}{(1-e^t)^2}$$

which is a linear differential equation of the first order with

$$\text{IF} = e^{\int dt} = e^t.$$

Hence, solution of this equation will be

$$ue^t = \int \frac{e^t}{(1-e^t)^2}\,dt = \frac{+1}{1-e^t}$$

$$\therefore \qquad u = \frac{1}{e^t(1-e^t)}$$

$$\therefore \qquad y_p = \frac{1}{(\theta+1)}\,\frac{1}{e^t(1-e^t)} = v \text{ (say)}$$

$$\therefore \qquad \frac{dv}{dt} + v = \frac{1}{e^t(1-e^t)}$$

Proceeding as above

$$ve^t = \int e^t \frac{1}{e^t(1-e^t)}\,dt$$

$$= \int \frac{1}{1-e^t}\,dt = \int \frac{e^{-t}}{e^{-t}-1}\,dt$$

$$= -\log(e^{-t} - 1)$$

Thus,

$$y_p = \frac{-1}{e^t}\log\left(\frac{1-e^t}{e^t}\right) = \frac{1}{e^t}\log\left(\frac{e^t}{1-e^t}\right)$$

Hence, the general solution is

$$y = (c_1 + c_2 \log x)\frac{1}{x} + \frac{1}{x}\log\left(\frac{x}{1-x}\right).$$

EXAMPLE 11.5

Solve $x^2\dfrac{d^2y}{dx^2} + 4x\dfrac{dy}{dx} + 2y = e^{e^x}$.

Solution: We have the given equation

$$(x^2D^2 + 4xD + 2)y = e^x$$

By putting $x = e^t$, $t = \log x$, $\dfrac{dt}{dx} = \dfrac{1}{x}$ and $\dfrac{d}{dt} = \theta$, we obtain

$$(\theta(\theta - 1) + 4\theta + 2)y = e^{e^t}$$

$$\Rightarrow \qquad (\theta^2 + 3\theta + 2)y = e^{e^t}$$

For finding y_c, AE is

$$\theta^2 + 3\theta + 2 = 0$$

$$\Rightarrow \quad \theta = -1, -2$$

Hence, $y_c = c_1 e^{-t} + c_2 e^{-2t}$

Also, $$y_p = \frac{1}{(\theta+1)(\theta+2)} e^{e^t} = \left(\frac{1}{\theta+1} - \frac{1}{\theta+2}\right) e^{e^t}$$

Now, $$\frac{1}{\theta+1} e^{e^t} = e^{-t}\int e^{e^t} e^t \, dt = e^{-t} e^{e^t}$$

and $$\frac{1}{\theta+2} e^{e^t} = e^{-2t}\int e^{e^t} e^{2t} \, dt = e^{-2t}\int e^{e^t} e^t e^t \, dt$$

$$= e^{-2t} e^{e^t}(e^t - 1)$$

Hence, the required general solution is

$$y = c_1 e^{-t} + c_2 e^{-2t} + (e^{-t} e^{e^t} - e^{-2t} e^{e^t}(e^t - 1)]$$

$$\Rightarrow \quad y = c_1 x^{-1} + c_2 x^{-2} + x^{-2} e^x.$$

EXAMPLE 11.6

Solve $x^2 \dfrac{d^2y}{dx^2} + x\dfrac{dy}{dx} + y = \log x \sin(\log x)$.

Solution: We have the given equation

$$(x^2 D^2 + xD + 1)y = \log x \sin(\log x)$$

By putting $x = e^t$, $t = \log x$, $\dfrac{dt}{dx} = \dfrac{1}{x}$ and $\dfrac{d}{dt} = \theta$,

We obtain

$$(\theta(\theta - 1) + \theta + 1)y = t \sin t$$

$$\Rightarrow \quad (\theta^2 + 1)y = t \sin t$$

For finding y_c, AE is

$$\theta^2 + 1 = 0$$

$$\Rightarrow \quad \theta = \pm i$$

Hence, $y_c = c_1 \cos t + c_2 \sin t$.

Also, $y_p = \dfrac{1}{\theta^2+1} t \sin t$ = Imaginary part of $\left(\dfrac{1}{\theta^2+1} t e^{it}\right)$

Now,
$$\frac{1}{\theta^2+1}te^{it} = e^{it}\frac{1}{(\theta+i)^2+1}t$$
$$= e^{it}\frac{1}{\theta^2+2i\theta}t$$
$$= e^{it}\frac{1}{2i\theta}\left(1+\frac{\theta}{2i}\right)^{-1}t$$
$$= e^{it}\frac{-i}{2\theta}\left(1-\frac{\theta}{2i}+\left(\frac{\theta}{2i}\right)^2-\ldots\right)t$$
$$= e^{it}\frac{-i}{2\theta}\left(t+\frac{i}{2}\right)$$
$$= e^{it}\frac{-i}{2}\left(\frac{t^2}{2}+\frac{i}{2}t\right)$$
$$= \frac{-i\cos t+\sin t}{2}\frac{(t^2+it)}{2}$$

Whose imaginary part is
$$\frac{-t^2}{4}\cos t+\frac{t}{4}\sin t.$$

Thus, the required general solution is
$$y = c_1\cos(\log x) + c_2\sin(\log x) - \frac{1}{4}(\log x)^2\cos(\log x) + \frac{1}{4}(\log x)\sin(\log x)$$

EXAMPLE 11.7

Solve $x^2\frac{d^2y}{dx^2} - 3x\frac{dy}{dx} + y = \log x\left(\frac{\sin(\log x)+1}{x}\right)$.

Solution: We have the given equation
$$(x^2D^2 - 3xD + 1)y = \log x\left(\frac{\sin(\log x)+1}{x}\right)$$

By putting $x = e^t$, $t = \log x$, $\frac{dt}{dx}=\frac{1}{x}$ and $\frac{d}{dt}=\theta$,

We obtain

$$(\theta(\theta - 1) - 3\theta + 1)y = t\left(\frac{\sin t + 1}{e^t}\right)$$

$$\Rightarrow \qquad (\theta^2 - 4\theta + 1)y = te^{-t}(t\sin t + 1).$$

For finding y_c, AE is

$$\theta^2 - 4\theta + 1 = 0$$

$$\Rightarrow \qquad \theta = 2 \pm \sqrt{3}$$

Hence,

$$y_c = c_1 e^{(2+\sqrt{3})t} + c_2 e^{(2-\sqrt{3})t}$$

$$= e^{2t}[c_1 e^{\sqrt{3}t} + c_2 e^{-\sqrt{3}t}]$$

Also,

$$y_p = \frac{1}{\theta^2 - 4\theta + 1} e^{-t} t(\sin t + 1)$$

$$= e^{-t} \frac{1}{(\theta - 1)^2 - 4(\theta - 1) + 1} t(\sin t + 1)$$

$$= e^{-t} \frac{1}{\theta^2 - 60 + 6} t(\sin t + 1)$$

$$= e^{-t}\left[\frac{1}{\theta^2 - 6\theta + 6} t\sin t + \frac{1}{\theta^2 - 6\theta + 6} t\right]$$

Now, $\frac{1}{\theta^2 - 6\theta + 6} t\sin t$ = Imaginary part of $\frac{1}{\theta^2 - 6\theta + 6} te^{it}$.

Thus,

$$\frac{1}{\theta^2 - 6\theta + 6} te^{it} = e^{it} \frac{1}{(\theta + i)^2 - 6(\theta + i) + 6} t$$

$$= e^{it} \frac{1}{\theta^2 + (2i - 6)\theta + (5 - 6i)} t$$

$$= \frac{e^{it}}{5 - 6i}\left[1 + \frac{(2i - 6)\theta + \theta^2}{5 - 6i}\right]^{-1} t$$

$$= \frac{e^{it}}{5 - 6i}\left[1 + \frac{(2i - 6)\theta}{5 - 6i}\right]^{-1} t$$

$$= \frac{e^{it}}{5 - 6i}\left(1 - \frac{(2i - 6)\theta}{5 - 6i}\right) t$$

$$= \frac{5 + 6i}{61} e^{it}\left(t + \frac{42 + 26i}{61}\right)$$

$$= \frac{5 + 6i}{61}(\cos t + i\sin t)\left(t + \frac{42 + 26i}{61}\right)$$

whose imaginary part is

$$\frac{t}{61}(5\sin t + \cos t) + \frac{2}{3721}(27\sin t + 191\cos t)$$

and
$$\frac{1}{\theta^2 - 6\theta + 6}t = \frac{1}{6}\left(1 - \frac{6\theta - \theta^2}{6}\right)^{-1} t$$

$$= \frac{1}{6}(1+\theta)t = \frac{1}{6}(t+1)$$

Therefore,

$$y_p = e^{-t}\left[\frac{t}{61}(5\sin t + 6\cos t) + \frac{2}{3721}(27\sin t + 191\cos t) + \frac{1}{6}(t+1)\right]$$

Thus, the required general solution is

$$y = x^2(c_1 x^{\sqrt{3}} + c_2 x^{-\sqrt{3}})$$

$$+\frac{1}{x}\left[\frac{\log x}{61}(5\sin(\log x) + 6\cos(\log x)) + \frac{2}{3721}(27\sin(\log x) + 191\cos(\log x)) + \frac{1}{6}(\log x + 1)\right]$$

EXAMPLE 11.8

Find the general solution of the equation

$$x^3y''' - 3xy' + 3y = 16x + 9x^2 \log x;\ x > 0.$$

Solution: We have the given equation

$$(x^3D^3 - 3xD + 3)y = 16x + 9x^2 \log x$$

By putting $x = e^t$, $t = \log x$, $\frac{dt}{dx} = \frac{1}{x}$ and $\frac{d}{dt} = \theta$, we obtain

$$[\theta(\theta-1)(\theta-2) - 3\theta + 3]y = 16e^t + 9te^{2t}$$

$$\Rightarrow \quad (\theta^3 - 3\theta^2 - \theta + 3)y = 16e^t + 9te^{2t}$$

For finding y_c, AE is

$$\theta^3 - 3\theta^2 - \theta + 3 = 0$$

$$\Rightarrow \quad \theta = \pm 1, 3$$

Hence, $y_c = c_1e^t + c_2e^{-t} + c_3e^{3t}$.

Also,
$$y_p = \frac{1}{\theta^3 - 3\theta^2 - \theta + 3}(16e^t + 9te^{2t})$$

$$= 16\frac{1}{\theta^3 - 3\theta^2 - \theta + 3}e^t + 9\frac{1}{\theta^3 - 3\theta^2 - \theta + 3}te^{2t}$$

Now, $16\frac{1}{\theta^3-3\theta^2-\theta+3}e^t = 16\frac{1}{(1)^3-3(1)^2-(1)+3}e^t$

$$= 16t\frac{1}{3\theta^2-6\theta-1}e^t \quad \{\because \text{Rule fails}\}$$

$$= 16t\frac{1}{3(1)^2-6(1)-1}e^t = 16t\frac{e^t}{-4} = -4te^t$$

and $9\frac{1}{\theta^3-3\theta^2-\theta+3}te^{2t} = 9e^{2t}\frac{1}{(\theta+2)^3-3(\theta+2)^2-(\theta+2)+3}t$

$$= 9e^{2t}\frac{1}{\theta^3+6\theta^2+12\theta+8-3\theta^2-12\theta-12-\theta-2+3}t$$

$$= 9e^{2t}\frac{1}{\theta^3+3\theta^2-\theta-3}t$$

$$= \frac{9e^{2t}}{-3}\left(1+\frac{\theta-3\theta^2-\theta^3}{3}\right)^{-1}t$$

$$= e^{2t}\left(1-\frac{1\theta}{3}\right)t = -3e^{2t}\left(t-\frac{1}{3}\right)$$

$$= (1-3t)e^{2t}$$

Therefore, $y_p = -4te^t + (1-3t)e^{2t}$.

Hence, the required general solution is

$$y = c_1x + c_2\frac{1}{x} + c_3x^3 - 4(\log x)x + (1-3\log x)x^2.$$ ▲▲

Legendre's Linear Equation

An equation of the form

$$a_0(ax+b)^n\frac{d^ny}{dx^n} + a_1(ax+b)^{n-1}\frac{d^{n-1}y}{dx^{n-1}} + \ldots + a_{n-1}(ax+b)\frac{dy}{dx} + a_ny = r(x)$$

where $a_0, a_1, \dots a_{n-1}, a_n$ are constant and $r(x)$ is a function of x is called Legendre's linear equation.

Such equation can be reduced to linear equations with constant coefficients by the substitution

$$ax + b = e^t \qquad t = \log(ax + b) \quad \Rightarrow \quad \frac{dt}{dx} = \frac{a}{ax+b}$$

Thus, if $D = \dfrac{d}{dx}$ and $\theta = \dfrac{d}{dt}$

$$\frac{dy}{dx} = \frac{dy}{dt}\frac{dt}{dx} = \frac{a}{ax+b}\frac{dy}{dt}$$

$$\Rightarrow \qquad (ax + b)\frac{dy}{dx} = a\frac{dy}{dt}$$

That is, $(ax + b)Dy = a\theta y$

$$\frac{d^2y}{dx^2} = \frac{d}{dx}\left(\frac{a}{ax+b}\frac{dy}{dt}\right) = \frac{a}{ax+b}\frac{d}{dt}\left(\frac{dy}{dt}\right)\frac{dt}{dx} + \frac{-a}{(ax+b)^2}\frac{dy}{dt}$$

$$= \frac{a^2}{(ax+b)^2}\left(\frac{d^2y}{dt^2} - \frac{dy}{dt}\right)$$

That is, $(ax + b)^2 \dfrac{d^2y}{dx^2} = a^2\left(\dfrac{d^2y}{dt^2} - \dfrac{dy}{dt}\right)$

or $\qquad (ax + b)^2 D^2 y = a^2\theta(\theta - 1)y$

Similarly, $(ax + b)^3 Dy = a^3\theta(\theta - 1)(\theta - 2)$ and so on.

By making these substitutions in the above equation, we get a linear equation with constant coefficients.

Note: We can notice that in Legendre's equation the power of $ax + b$ and the order of derivative are same in each term. ▲▲

EXAMPLE 11.9

Solve $(x+a)^2\dfrac{d^2y}{dx^2} - 4(x+a)\dfrac{dy}{dx} + 6y = x.$

Solution: We have the given equation

$$[(x + a)^2D^2 - 4(x + a)D + 6]y = x$$

By putting

$$x + a = e^t \text{ so that } t = \log(x + a)$$

$$\frac{dt}{dx} = \frac{1}{x+a} \quad \text{and} \quad \frac{d}{dt} = \theta, \text{ we obtain}$$

$$[\theta(\theta - 1) - 4\theta + 6]y = e^t - a$$

$$\Rightarrow \quad (\theta^2 - 5\theta + 6)y = e^t - a$$

For finding y_c, AE is

$$\theta^2 - 5\theta + 6 = 0$$

$$\Rightarrow \quad \theta = 2, 3$$

Hence, $y_c = c_1e^{2t} + c_3e^{3t}$.

Also,

$$y_p = \frac{1}{\theta^2 - 5\theta + 6}(e^t - a)$$

$$= \frac{1}{(1)^2 - 5(1) + 6}e^t - \frac{1}{(0)^2 - 5(0) + 6}a$$

$$= \frac{e^t}{2} - \frac{a}{6}$$

Hence, the required general solution is

$$y = c_1(x + a)^2 + c_2(x + a)^3 + \frac{1}{2}(x+a) - \frac{1}{6}a.$$

EXAMPLE 11.10

Solve $(2x - 1)^2 \dfrac{d^2y}{dx^2} + (2x-1)\dfrac{dy}{dx} - 2y = 8x^2 - 2x + 3$.

Solution: We have the given equation

$$[(2x - 1)^2D^2 + (2x - 1)D - 2]y = 8x^2 - 2x + 3$$

By putting $2x - 1 = e^t$ so that $t = \log(2x - 1)$, $\dfrac{dt}{dx} = \dfrac{2}{2x-1}$ and $\dfrac{d}{dt} = \theta$, we obtain

$$[2^2\theta(\theta - 1) + 2\theta - 2]y = 1\left(\frac{e^t+1}{2}\right)^2 - 2\left(\frac{e^t+1}{2}\right) + 3$$

$$\Rightarrow \quad (2\theta^2 - \theta - 1)y = e^{2t} + \frac{3}{2}e^t + 2$$

For finding y_c, AE is

$$2\theta^2 - \theta - 1 = 0$$

$$\Rightarrow \quad \theta = 1, -1/2$$

Hence, $y_c = c_1e^t + c_2e^{-t/2}$.

Also, $$y_p = \frac{1}{2\theta^2 - \theta - 1}\left(e^{2t} + \frac{3}{2}e^t + 2\right)$$

$$= \frac{1}{2\theta^2 - \theta - 1}e^{2t} + \frac{3}{2}\frac{1}{2\theta^2 - \theta - 1}e^t + 2\frac{1}{2\theta^2 - \theta - 1}1$$

$$= \frac{1}{2(2)^2 - (2) - 1}e^{2t} + \frac{3}{2}\frac{1}{2(1)^2 - (1) - 1}e^t + 2\frac{1}{2(0)^2 - (0) - 1}1$$

$$= \frac{1}{5}e^{2t} + \frac{3}{2}t\frac{1}{4\theta - 1}e^t - 2 \quad \{\because \text{Rule fails}\}$$

$$= \frac{1}{5}e^{2t} + \frac{3}{2}t\frac{1}{4(1) - 1}e^t - 2 = \frac{1}{5}e^{2t} + \frac{t}{2}e^{-t} - 2$$

Hence, the required general solution is

$$y = c_1(2x - 1) + c_2(2x - 5)^{-1/2} + \frac{1}{5}(2x - 1)^2 + \frac{1}{2}(2x - 1)\log(2x - 1) - 2.$$

EXAMPLE 11.11

Solve the equation $$(3x + 1)^2y'' + (3x + 1)y' + y = 6x.$$

Solution: We have the given equation

$$[(3x + 1)^2D^2 + (3x + 1)D + 1]y = 6x$$

By putting $3x + 1 = e^t$ so that $t = \log(3x + 1)$,

$$\frac{dt}{dx} = \frac{3}{3x+1} \quad \text{and} \quad \frac{d}{dt} = \theta, \text{ we obtain}$$

$$(3^2\theta(\theta - 1) + 3\theta + 1]y = 6\left(\frac{e^{-t} - 1}{3}\right)$$

$$\Rightarrow \qquad (9\theta^2 - 6\theta + 1)y = 2(e^t - 1)$$

For finding y_c, AE is

$$9\theta^2 - 6\theta + 1 = 0 \Rightarrow \theta = 1/3, 1/3$$

Hence, $y_c = (c_1 + c_2t)e^{t/3}$

Also, $$y_p = \frac{1}{9\theta^2 - 6\theta + 1}2(e^t - 1)$$

$$= 2\frac{1}{9 - 6 + 1}e^t - 2\frac{1}{1}$$

$$= \frac{e^t}{2} - 2$$

Thus, the required general solution is

$$y = [c_1 + c_2 \log(3x+1)](3x+1)^{1/3} + \frac{3}{2}(x-1).$$ ▲▲

EXERCISE 11.1

Find the solution of the following differential equations:

1. $x^2 \frac{dy}{dx} + 2x \frac{dy}{dx} - 20y = (x+1)^2.$

2. $x^2 \frac{d^2y}{dx^2} + 2x \frac{dy}{dx} = \log x.$

3. $x^3 \frac{d^3y}{dx^3} + 2x^2 \frac{d^2y}{dx^2} + x \frac{dy}{dx} - y = \cos(\log x).$

4. $x^2 \frac{d^2y}{dx^2} + 6x \frac{dy}{dx} + 6y = (\log x)^2.$

5. $(x^4D^4 + 6x^3D^3 + 9x^2D^2 + 3xD + 1)y = (1 + \log x)^2.$

6. $x^2 \frac{d^2y}{dx^2} + 4x \frac{dy}{dx} + 2y = x + \log x.$

7. $x^3y''' + 5x^2y'' + 5xy' + y = x^2 + \log x;\ x > 0.$

8. $x^2 \frac{d^2y}{dx^2} - x \frac{dy}{dx} + 4y = \cos(\log x) + x \sin(\log x)$

9. $4x^2y'' + y = \log x;\ x > 0,\ y(1) = 0,\ y(e) = 5.$

10. $x^2y'' + 3xy' + 10y = 9x^2;\ y(1) = 5/2,\ y'(1) = 8.$

11. $(3x+2)^2 \frac{d^2y}{dx^2} + 3(3x+2) \frac{dy}{dx} - 36y = 3x^2 + 4x + 1.$

12. $(1+x)^2 \frac{d^2y}{dx^2} + (1+x) \frac{dy}{dx} + y = 4\cos[\log(1+x).$

13. $(3x+2)^2 \frac{d^2y}{dx^2} + 3(3x+2) \frac{dy}{dx} - 3y = x^2 + x + 1.$

14. $(1+2x)^2 \frac{d^2y}{dx^2} - 6(1+2x) \frac{dy}{dx} + 16y = 8(1+2x)^2$ given that $y(0) = 0,\ y'(0) = 2.$

15. $(x+2)^3y''' + (x+2)^2y'' + (x+2)y' - y = 24x^2.$

ANSWERS

Exercise 11.1

1. $y = c_1x^4 + c_2x^{-5} - \dfrac{1}{14}x^2 - \dfrac{1}{9}x - \dfrac{1}{20}$.

2. $y = c_1 + \dfrac{c_2}{x} + \dfrac{(\log x)^2}{2} - \log x$.

3. $y = c_1x + c_2\cos(\log x) + c_3\sin(\log x) + \dfrac{1}{4}\log x[\cos(\log x) - \sin(\log x)]$.

4. $y = c_1x^{-2} + c_2x^{-3x} + \dfrac{1}{108}[18(\log x)^2 - 30\log x + 19]$.

5. $y = (c_1 + c_2\log x)\cos(\log x) + (c_3 + c_4\log x)\sin(\log x) + (\log x)^3 + 2(\log x) - 3$.

6. $y = c_1x^{-1} + c_2x^{-2} + \dfrac{1}{6}x + \dfrac{1}{2}\log x - \dfrac{3}{2}$.

7. $y = \dfrac{c_1}{x} + \dfrac{1}{\sqrt{x}}\left[c_2\cos\left(\dfrac{\sqrt{3}\log x}{2}\right) + c_3\sin\left(\dfrac{\sqrt{3}\log x}{2}\right)\right] + \left(\dfrac{x^2}{21} + \log x - 2\right)$.

8. $y = x[c_1\cos(\sqrt{3}\log x) + c_2\sin(\sqrt{3}\log x)] + \dfrac{1}{13}[3\cos(\log x) - 2\sin(\log x)] + \dfrac{1}{2}x\sin(\log x)$.

9. $y = 4(\log x - 1)\sqrt{x} + \log x + 4$.

10. $y = \dfrac{1}{x}[2\cos(3\log x) + 3\sin(3\log x)] + \dfrac{x^2}{2}$.

11. $y = c_1(3x + 2)^2 + c_2(3x + 2)^{-1} + \dfrac{1}{108}[(3x + 2)^2\log(3x + 2) + 1]$.

12. $y = c_1\cos(\log(x + x) + c_2\sin(\log(x + x) + 2\log(1 + x)\sin(\log(1 + x))$.

13. $y = c_1(3x + 2)^{1/3} + c_2(3x + 2)^{-1} + \dfrac{(3x+2)^2}{405} - \dfrac{(3x+2)}{108} - \dfrac{7}{27}$.

14. $y = (1 + 2x)^2\log(1 + 2x)[1 + \log(1 + 2x)]$.

15. $y = c_1(x + 2) + (x + 2)^{1/2}\left[c_2\cos\left(\dfrac{\sqrt{3}\log(x+2)}{2}\right) + c_3\sin\left(\dfrac{\sqrt{3}\log(x+2)}{2}\right)\right] + 8(x + 2)^2 - 96(x + 2)\log(x + 2) - 96$.

12

CHAPTER

Simultaneous Differential Equations

Before this chapter, we have discussed the solution of a single linear differential equation in which y is the dependent variable and x is the independent variable. In this chapter, we consider a system of two linear first order equations in two dependent variables y_1 and y_2 and one independent variable x.

For example,

$$\frac{dx}{dt} + 5x - 2y = t, \quad \frac{dy}{dt} + 2x + y = 0$$

is a system of linear constant coefficient first order equations with x, y dependent variables and t independent variable.

The solution of such systems can be obtained by eliminating one of the variables and solving the resulting linear, second order equation for the second variable. Sometimes, elimination of one of the variables may also produce a first order equation for the second variable. We illustrate the method of obtaining the solution through the following example. ▲▲

EXAMPLE 12.1

Find the solution of the system of equations

$$\frac{dy_1}{dt} + 2\frac{dy_2}{dt} - 2y_1 - y_2 = e^{2t} \tag{1}$$

and

$$\frac{dy_2}{dt} + y_1 - 2y_2 = 0 \tag{2}$$

Solution: Method of elimination

We will eliminate one of the dependent variables directly.

Differentiating Eq. (2) with respect to t, we get

$$y_2'' + y_1' - 2y_2' = 0$$

Substituting for y_1' from Eq. (1), we obtain

$$y_2'' - 2y_2' + 2y_1 + y_2 + e^{2t} - 2y_2' = 0$$

$$\Rightarrow \qquad y_2'' - 4y_2' + 2y_1 + y_2 = -e^{2t}$$

Substituting y_1 from Eq. (2), we get

$$y_2'' - 4y_2' + 2(-y_2' + 2y_2) + y_2 = -e^{2t}$$

$$\Rightarrow \qquad y_2'' - 6y_2' + 5y_2 = -e^{2t}$$

which is a second order equation in the variable y_2 and can be written as

$$(D^2 - 6D + 5)y_2 = -e^{2t}$$

For finding y_{2c}, AE is

$$D^2 - 6D + 5 = 0$$

$$\Rightarrow \qquad D = 5, 1$$

Hence, $y_{2c} = c_1e^{5t} + c_2e^t$.

Also,
$$y_{2p} = \frac{1}{D^2 - 6D + 5}(-e^{2t})$$

$$= \frac{1}{(2)^2 - 6(2) + 5}e^{2t} = \frac{1}{3}e^{2t}.$$

Therefore, the general solution is

$$y_2 = c_1e^{5t} + c_2e^t + \frac{1}{3}e^{2t}.$$

From Eq. (2), we obtain

$$y_1 = 2y_2 - y_2' = c_2e^t - 3c_1e^{5t}$$

which is the required general solution for y_1.

EXAMPLE 12.2

Solve the system of equations

$$(2D - 4)y_1 + (3D + 5)y_2 = 3t + 2$$

and
$$(D - 2)y_1 + (D + 1)y_2 = t.$$

Solution: To solve the given system of equations, we will use the elimination method. Multiply the second equation by 2 and subtract from the first equation, we obtain

$$(D + 3)y_2 = t + 2$$

which is a linear first order equation in y_2 with

$$\text{IF} = e^{\int 3dt} = e^{3t}.$$

Therefore, the solution is

$$y_2 e^{3t} = \int (t+2)e^{3t}\,dt + c_1$$

$$\Rightarrow \qquad y_2 e^{3t} = e^{3t}\left[\frac{t+2}{3} - \frac{1}{9}\right] + c_1$$

$$\Rightarrow \qquad y_2 = c_1 e^{-3t} + \frac{1}{9}(3t+5)$$

Substituting in the second equation, we get

$$(D-2)y_1 + \left[-3c_1e^{-3t} + \frac{1}{3} + c_1e^{-3t} + \frac{1}{9}(3t+5)\right] = t$$

$$\Rightarrow \qquad (D-2)y_1 = 2c_1e^{-3t} + \frac{1}{9}(6t-8)$$

which is a linear first order equation in y_2, with

$$\text{IF} = e^{\int -2dt} = e^{-2t}.$$

Therefore, the solution is

$$y_1e^{-2t} = \int\left(2c_1e^{-3t} + \frac{1}{9}(6t-8)\right)e^{-2t} + c_2$$

$$\Rightarrow \qquad y_1e^{-2t} = -\frac{2}{5}c_1e^{-5t} + \frac{1}{9}\left[(6t-8)\frac{e^{-2t}}{-2} - \frac{6e^{-2t}}{4}\right] + c_2$$

$$\Rightarrow \qquad y_1 = c_2e^{2t} - \frac{2}{5}c_1e^{-3t} - \frac{1}{18}(6t-5).$$

EXAMPLE 12.3

Find the solution of the system of equations

$$(3D+1)y_1 + 3Dy_2 = 3t + 1$$

and

$$(D-3)y_1 + Dy_2 = 2t.$$

Solution: Method of determinants

By using Cramer's rule, we obtain

$$\begin{vmatrix} 3D+1 & 3D \\ D-3 & D \end{vmatrix} y_1 = \begin{vmatrix} 3t+1 & 3D \\ 2t & D \end{vmatrix}$$

$$(3D^2 + D - 3D^2 + 9D)y_1 = 3 - 6$$

$$10Dy_1 = -3$$

Integrating, we get

$$y_1 = \frac{-3}{10}t + c_1$$

Similarly, we have

$$\begin{vmatrix} 3D+1 & 3D \\ D-3 & D \end{vmatrix} y_2 = \begin{vmatrix} 3D+1 & 3t+1 \\ D-3 & 2t \end{vmatrix}$$

$$10Dy_2 = 6 + 2t - 3 + 9t + 3$$

$\Rightarrow$ $$10Dy_2 = 11t + 6$$

Integrating, we get

$$y_2 = \frac{11}{20}t^2 + \frac{6}{10}t + c_2$$

EXAMPLE 12.4

Solve the simultaneous equations

$$\frac{dx}{dt} + 5x - 2y = t \quad \text{and} \quad \frac{dy}{dt} + 2x + y = 0.$$

being given $x = y = 0$ when $t = 0$.

Solution: We can write the given equation as

$$(D + 5)x - 2y = t \quad \text{and} \quad 2x + (D + 1)y = 0.$$

Multiplying the first equation by 2 and operating the second equation by $D + 5$ and then subtracting, we get

$$(D^2 + 6D + 9)y = -2t$$

For finding y_c, AE is

$$D^2 + 6D + 9 = 0$$

$\Rightarrow$ $$D = -3, -3$$

Hence, $y_c = (c_1 + c_2t)e^{-3t}$.

Also, $$y_p = \frac{1}{D^2 + 6D + 9}(-2t) = \frac{-2}{9}\left(1 + \frac{6D + D^2}{9}\right)^{-1} t$$

$$= \frac{-2}{9}\left(1 - \frac{6}{9}D\right)t = \frac{-2}{9}\left(t - \frac{2}{3}\right)$$

Therefore,

$$y = (c_1 + c_2t)e^{-3t} - \frac{2t}{9} + \frac{4}{27}$$

Now, to find x, substituting y in the second equation, we get

$$2x + \left[c_2 e^{-3t} - 3(c_1 + c_2 t)e^{-3t} - \frac{2}{9} + (c_1 + c_2 t)e^{-3t} - \frac{2t}{9} + \frac{4}{27}\right] = 0$$

$$\Rightarrow \qquad x = \left(c_1 - \frac{1}{2}c_2 + c_2 t\right)e^{-3t} + \frac{t}{9} + \frac{1}{27}$$

Since $x = y = 0$ when $t = 0$, we obtain

$$0 = c_1 - \frac{1}{2}c_2 + \frac{1}{27} \text{ and } 0 = c_1 + \frac{4}{27}$$

$$\therefore \qquad c_1 = -\frac{4}{27} \quad \text{and} \quad c_2 = -\frac{2}{9}$$

Hence, the desired solutions are

$$x = \frac{-1}{27}(1 + 6t)e^{-3t} + \frac{1}{27}(1 + 3t)$$

and

$$y = \frac{-2}{27}(2 + 3t)e^{-3t} + \frac{2}{27}(2 - 3t).$$

EXAMPLE 12.5

Solve the simultaneous equations

$$\frac{dx}{dt} + \frac{dy}{dt} - 2y = 2\cos t - 7\sin t$$

and

$$\frac{dx}{dt} - \frac{dy}{dt} + 2x = 4\cos t - 3\sin t.$$

Solution: We can write the given equation as

$$Dx + (D - 2)y = 2\cos t - 7\sin t \tag{1}$$

and

$$(D + 2)x - Dy = 4\cos t - 3\sin t \tag{2}$$

Operating Eq. (1) by D and Eq. (2) by $(D - 2)$ and adding, we get

$$D^2 x + (D^2 - 4)x = -2\sin t - 7\cos t - 4\sin t - 3\cos t - 8\cos t + 6\sin t$$

$$\Rightarrow \qquad 2(D^2 - 2)x = -18\cos t$$

$$\Rightarrow \qquad (D^2 - 2)x = -9\cos t.$$

For finding y_c, AE is

$$D^2 - 2 = 0$$

$$\Rightarrow \qquad D = \sqrt{2}, -\sqrt{2}$$

Hence, $x_c = c_1e^{\sqrt{2}t} + c_2e^{-\sqrt{2}t}$.

Also, $x_p = \dfrac{1}{D^2-2}(-9\cos t) = \dfrac{-9\cos t}{-1-2} = 3\cos t$

Hence, $x = c_1e^{\sqrt{2}t} + c_2e^{-\sqrt{2}t} + 3\cos t$.

Now, substituting this value of x in Eq. (2), we get

$$Dy = (D+2)x - 4\cos t + 3\sin t$$

$$= \sqrt{2}c_1e^{\sqrt{2}t} - \sqrt{2}c_2e^{-\sqrt{2}t} - 3\sin t + 2c_1e^{\sqrt{2}t} + 2c_2e^{-\sqrt{2}t} + 6\cos t - 4\cos t + 3\sin t$$

$$\Rightarrow \quad Dy = (2+\sqrt{2})c_1e^{\sqrt{2}t} + (2-\sqrt{2})c_2e^{-\sqrt{2}t} + 2\cos t.$$

Integrating, we get

$$y = (\sqrt{2}+1)c_1e^{\sqrt{2}t} + (\sqrt{2}-1)c_2e^{-\sqrt{2}t} + 2\sin t + c_3.$$

EXAMPLE 12.6

Solve the simultaneous equations

$$\frac{dx}{dt} + 4x + 3y = t \quad \text{and} \quad \frac{dy}{dt} + 2x + 5y = e^t.$$

Solution: We can write the given equations as

$$(D+4)x + 3y = t \quad \text{and} \quad 2x + (D+5)y = e^t.$$

Eliminating y, we obtain

$$[(D+4)(D+5) - 6]x = (D+5)t - 3e^t$$

$$\Rightarrow \quad [D^2 + 9D + 14)x = 1 + 5t - 3e^t$$

For finding x_c, AE is

$$D^2 + 9D + 14 = 0$$

$$\Rightarrow \quad D = -2, -7$$

Hence, $x_c = c_1e^{-2t} + c_2e^{-7t}$.

Also, $$y_p = \frac{1}{D^2+9D+14}(1 + 5t - 3e^t)$$

$$= \frac{1}{14}\left(1+\frac{9D+D^2}{14}\right)^{-1}(1+5t) - 3\frac{1}{D^2+9D+14}e^t$$

$$= \frac{1}{14}\left(1-\frac{9D}{14}\right)(1+5t) - 3\frac{1}{1+9+14}e^t$$

$$= \frac{1}{14}\left(1+5t-\frac{45}{14}\right)-\frac{1}{8}e^t = \frac{5t}{14}-\frac{31}{196}-\frac{1}{8}e^t.$$

Therefore, $$x = c_1e^{-2t} + c_2e^{-7t} + \frac{5}{14}t - \frac{1}{8}e^t - \frac{31}{196} \quad (1)$$

Substituting x in the first equation, we get

$$3y = t - (D + 4)x$$

$$= t - \left[-2c_1e^{-2t} - 7c_2e^{-7t} + \frac{5}{14} - \frac{1}{8}e^t + 4c_1e^{-2t} + 4c_2e^{-7t} + \frac{20}{14}t - \frac{4}{8}e^t - \frac{124}{196}\right]$$

$$\Rightarrow \quad y = \frac{1}{3}\left[-2c_1e^{-2t} + 3c_2e^{-7t} - \frac{3}{7}t + \frac{5}{8}e^t + \frac{27}{98}\right] \quad (2)$$

The required solution consists of the Eqs. (1) and (2).

EXAMPLE 12.7

Solve $$\frac{dx}{dt} + \frac{2}{t}(x-y) = 1$$

and $$\frac{dy}{dt} + \frac{1}{t}(x+5y) = t$$

Solution: We can write the given equations as

$$t\frac{dx}{dt} + 2(x - y) = t \quad (1)$$

and $$t\frac{dy}{dt} + (x + 5y) = t^2 \quad (2)$$

Differentiating Eq. (1) with respect to t, we get

$$t\frac{d^2x}{dt^2} + 3\frac{dx}{dt} - 2\frac{dy}{dt} = 1.$$

Substituting the value of y, we obtain

$$t^2\frac{d^2x}{dt^2} + 8t\frac{dx}{dt} - 12x = 2t^2 + 6t.$$

which is a linear homogeneous equation and can be written as

$$[t^2D^2 + 8tD + 12]x = 2t^2 + 6t$$

By putting $t = e^u$ so that $u = \log t$, $\frac{du}{dt} = \frac{1}{t}$ and $\frac{d}{du} = \theta$, we obtain

$$[\theta(\theta - 1) + 8\theta + 12]x = 2e^{2u} + 6e^u$$

$$\Rightarrow \quad (\theta^2 + 7\theta + 12)x = 2e^{2u} + 6e^u$$

For finding x_c, AE is

$$\theta^2 + 7\theta + 12 = 0$$

$$\Rightarrow \quad \theta = -3, -4$$

Hence, $x_c = c_1 e^{-3u} + c_2 e^{-4u}$.

Also,
$$x_p = \frac{1}{\theta^2 + 7\theta + 12}(2e^{2u} + 6e^u)$$

$$= 2\frac{e^{2u}}{4 + 14 + 12} + 6\frac{e^u}{1 + 7 + 12}$$

$$= \frac{e^{2u}}{15} + \frac{3}{10}e^u.$$

Therefore,
$$x = c_1 t^{-3} + c_2 t^{-4} + \frac{t^2}{15} + \frac{3}{10}t \tag{3}$$

Substituting x in Eq. (1), we get

$$2y = t\left(-3c_1 t^{-4} - 4c_2 t^{-5} + \frac{2t}{15} + \frac{3}{10}t\right)$$

$$\Rightarrow \quad +2c_1 t^{-3} + 2c_2 t^{-4} + \frac{2t^2}{15} + \frac{6}{10}t - t$$

$$\Rightarrow \quad y = \frac{-c_1}{2t^3} - \frac{c_2}{t^4} + \frac{2t^2}{15} - \frac{1}{20}t \tag{4}$$

The required solution constants of Eqs. (3) and (4).

EXAMPLE 12.8

Solve the simultaneous equations

$$\frac{dx}{dt} = 2y, \frac{dy}{dt} = 2z \text{ and } \frac{dz}{dt} = 2x.$$

Solution: Differentiating the first equation with respect to t, we get

$$\frac{d^2x}{dt^2} = 2\frac{dy}{dt} = 4z.$$

Again, differentiating with respect to t, we get

$$\frac{d^3x}{dt^3} = 4\frac{dz}{dt} = 8x$$

$$\Rightarrow \qquad (D^3 - 8)x = 0$$

Its AE is

$$D^3 - 8 = 0 \Rightarrow (D-2)(D^2 + 2D + 4) = 0$$

$$\Rightarrow \qquad D = 2, -1 \pm i\sqrt{3}$$

∴ The solution is

$$x = c_1e^{2t} + e^{-t}(c_2\cos\sqrt{3}t + c_3\sin\sqrt{3}t) \tag{1}$$

From the first equation, we have

$$y = \frac{1}{2}\frac{dx}{dt} = c_1e^{2t} + \frac{1}{2}e^{-t}[(\sqrt{3}c_3 - c_2)\cos\sqrt{3}t - (c_3 + \sqrt{3}c_2)\sin\sqrt{3}t] \tag{2}$$

From the second equation, we have

$$z = \frac{1}{2}\frac{dy}{dt} = c_1e^{2t} - \frac{1}{2}e^{-t}[(\sqrt{3}c_2 - c_3)\sin\sqrt{3}t + (c_2 + \sqrt{3}c_3)\cos\sqrt{3}t] \tag{3}$$

Hence, the Eqs. (1), (2) and (3) taken together give the required solution. ▲▲

EXERCISE 12.1

Solve the following simultaneous equations:

1. $\frac{dx}{dt} + 2y + \sin t = 0$, $\frac{dy}{dt} - 2x - \cos t = 0$; given that $x = 0$ and $y = 1$ when $t = 0$.
2. $\frac{dx}{dt} + 2y = e^t$, $\frac{dy}{dt} - 2x = e^{-t}$.
3. $\frac{dx}{dt} + 2x - 3y = t$, $\frac{dy}{dt} - 3x + 2y = e^{2t}$.
4. $(D - 1)x + Dy = 2t + 1$, $(2D + 1)x + 2Dy = t$.
5. $(D + 3)x + Dy = \sin t$, $(D - 1)x + y = \cos t$.
6. $\frac{dx}{dt} + x = y + e^t$, $\frac{dy}{dt} + y = x + e^t$.
7. $y_1' + y_1 - 3y_2 = 6e^{-t}$; $y_2' + 2y_1 - 4y_2 = 12e^t$.
8. $(D - 2)y_1 + (D + 3)y_2 = t + 4$, $(3D + 5)y_1 + (2D + 6)y_2 = 5t + 3$.
9. $(D + 3)y_1 + (3D + 23)y_2 = e^{-2t}$, $(D + 2)y_1 + (4D + 14)y_2 = e^{2t}$.

10. $(D-1)y_1 - (D+1)y_2 = t,\ (D+1)y_1 + (2D+1)y_2 = e^t.$

11. $\dfrac{d^2x}{dt^2} - \dfrac{dy}{dt} = 2x + 2t,\ \dfrac{dx}{dt} + 4\dfrac{dy}{dt} = 3y.$

ANSWERS

Exercise 12.1

1. $x = \cos 2t - \sin 2t - \cos t,\ y = \cos 2t + \sin 2t - \sin t.$

2. $x = \frac{1}{5}e^t + \frac{2}{5}e^{-t} - c_1 \sin 2t + c_2 \cos 2t,\ y = \frac{1}{5}e^t + \frac{1}{5}e^{-t} + c_1 \cos 2t + c_2 \sin 2t.$

3. $x = c_1e^t + c_2e^{-3t} + \frac{3}{7}e^{2t} - \frac{2}{5}t + \frac{13}{25},\ y = c_1e^t - c_2e^{-3t} - \frac{4}{7}e^{2t} - \frac{3t}{5} - \frac{12}{25}.$

4. $x = -t - \frac{2}{3},\ y = \frac{1}{2}t^2 + \frac{4}{3}t + c.$

5. $x = c_1e^{-t} + c_2e^{3t} - \frac{1}{5}(\cos t - 2\sin t),\ y = 2c_1e^{-t} - 2c_2e^{3t} + \frac{1}{5}(\sin t + 2\cos t).$

6. $x = c_1 + c_2e^{-2t} + e^t,\ y = c_1 - c_2e^{-2t} + e^t.$

7. $y_1 = c_1e^t + c_2e^{2t} - 5e^{-t} - 36te^t,\ y_2 = \frac{2}{3}c_1e^t + c_2e^{2t} - 2e^{-t} - 12(1+2t)e^t.$

8. $y_1 = c_1e^{-9t} + (9t-16)/27,\ y_2 = \frac{-11}{6}c_1e^{-9t} + \frac{1}{81}(52+45t) + c_2e^{-3t}.$

9. $y_1 = \frac{-9}{4}c_1e^{-t/2} + 3c_2e^{-2t} + (1413 - 738t + 168t^2)/84,\ y_2 = c_1e^{-t/2} + c_2e^{-2t}.$
$- (207 - 114t + 28t^2)/28.$

10. $y_1 = (e^t - 2c_1 - c_2e^{-t/3} - 6 - 2t + t^2)/2,\ y_2 = c_1 + c_2e^{-t/3} + 2t - (t^2/2).$

11. $x = (c_1 + c_2t)e^t + c_3e^{-3t/2} - t,\ y = (3c_2 - c_1 - c_2t)e^t - \frac{1}{6}c_3e^{-3t/2} - \frac{1}{3}.$ ▲▲

Solution of Simultaneous Equations in Symmetrical Form

In this section, we will consider simultaneous equation of the first order and of the first degree in three variables.

Consider a set of simultaneous equations in general form

$$P_1dx + Q_1dy + R_1dz = 0 \quad (1)$$

and

$$P_2dx + Q_2dy + R_2dz = 0 \quad (2)$$

where, the coefficient are functions of x, y, z. Solving these by the method of cross multiplication, we have

$$\frac{dx}{Q_1R_2 - Q_2R_1} = \frac{dy}{R_1P_2 - R_2P_1} = \frac{dz}{P_1Q_2 - P_2Q_1}$$

which is of the from

$$\frac{dx}{P}=\frac{dy}{Q}=\frac{dz}{R} \tag{3}$$

where, P, Q, R are functions of x, y, z.

Thus, we see that the simultaneous Eqs. (1) and (2) can always be put in the symmetrical form (3). Geometrically, the equation

$$\frac{dx}{P}=\frac{dy}{Q}=\frac{dz}{R}$$

represents a system of curves in space such that the direction cosines of the tangent to it at any point (x, y, z) are proportional to P, Q and R.

Solution of Simultaneous Equations in Symmetrical Form

The following methods are used for the solution of the simultaneous equations in the form

$$\frac{dx}{P}=\frac{dy}{Q}=\frac{dz}{R}. \tag{4}$$

(a) By equating two out of three members in Eq. (4), sometimes it is possible to get an equation in only two variables. The solution of this equation gives one of the relations in the complete solution of Eq. (4). This method can be repeated to get another relation with other two members. The two relations so obtained together give the complete solution of Eq. (4).

(b) If it is possible to obtain only one relation by the above method then this relation can be used to express one variable in terms of the other two with its help, it may be possible to get an equation in two variables from which another relation of the complete solution can be obtained.

(c) Sometimes, it is possible to find multipliers l, m, n and L, M, N such that at least one of the equations

$$\frac{dx}{P}=\frac{dy}{Q}=\frac{dz}{R}=\frac{ldx+mdy+ndz}{lP+mQ+nR}=\frac{Ldx+Mdy+Ndz}{LP+MQ+NR}$$

can be easily integrated.

If l, m, n are such that $lP + mQ + nR = 0$ then we get

$$l\,dx + m\,dy + n\,dz = 0.$$

which gives another equation on integration. These two equations so obtained form the complete solution. Sometimes only one set of multipliers may serve the purpose.

Note: Any combination of the methods given above will help in getting two independent relations between the variables. ▲▲

EXAMPLE 12.9

Solve $\frac{xdx}{y^3z} = \frac{dy}{x^2z} = \frac{dz}{y^3}$.

Solution: Taking the first two members

$$\frac{xdx}{y^3z} = \frac{dy}{x^2z} \quad \Rightarrow \quad x^3dx - y^3dy = 0.$$

Integrating, we obtain

$$x^4 - y^4 = 4c_1$$

Again, taking the first and last members

$$\frac{xdx}{y^3z} = \frac{dz}{y^3} \quad \Rightarrow \quad xdx - zdz = 0$$

Integrating, we obtain

$$x^2 - z^2 = 2c_2$$

Hence, the complete solution is

$$x^4 - y^4 = 4c_1 \quad \text{and} \quad x^2 - z^2 = 2c_2.$$

EXAMPLE 12.10

Solve $\frac{dx}{x} = \frac{dy}{x+z} = \frac{dz}{-z}$.

Solution: Consider the first and last members

$$\frac{dx}{x} = \frac{dz}{-z}$$

Integrating, we obtain

$$\log x = -\log z + \log c_1$$
$$xz = c_1$$

Again, consider the first two members

$$\frac{dx}{x} = \frac{dz}{x+z}$$

Using $z = c_1/x$, we get

$$\frac{dx}{x} = \frac{dy}{x+(c_1/x)} \quad \Rightarrow \quad \left(1+\frac{c_1}{x^2}\right)dx = dy$$

Integrating, we obtain

$$x - \frac{c_1}{x} + c_2 = y$$

$$\Rightarrow \qquad y = x - z + c_2$$

Hence, the complete solution is

$$xz = c_1 \quad \text{and} \quad y - x + z = c_2.$$

EXAMPLE 12.11

Solve $\dfrac{dx}{xy} = \dfrac{dy}{y^2} = \dfrac{dz}{zxy - 2x^2}$.

Solution: Consider the first two members

$$\frac{dx}{xy} = \frac{dy}{y^2} \quad \Rightarrow \quad \frac{dx}{x} = \frac{dy}{y}.$$

Integrating, we obtain

$$\log x = \log y + \log c_1$$

$$\Rightarrow \qquad x = yc_1 \qquad (1)$$

Again, consider the last two members

$$\frac{dy}{y^2} = \frac{dz}{zxy - 2x^2}.$$

Using $x = yc_1$, we get

$$\frac{dy}{y^2} = \frac{dz}{zc_1y^2 = 2c_1^2y^2}.$$

$$\Rightarrow \qquad c_1 dy = \frac{dz}{z - 2c_1}$$

Integrating, we obtain

$$c_1 y = \log(z - 2c_1) + c_2$$

$$\Rightarrow \qquad x = \log\left(z - \frac{2x}{y}\right) + c_2$$

$$\Rightarrow \qquad x = \log(zy - 2x) - \log y + c_2 \qquad (2)$$

Hence, complete solution is given by Eqs. (1) and (2).

EXAMPLE 12.12

Solve $\dfrac{xdx}{z^2 - 2yz - y^2} = \dfrac{dy}{y+z} = \dfrac{dz}{y-z}$.

Solution: Choosing x, y, z as multipliers, we have each fraction $= \dfrac{xdx + ydy + zdz}{0}$.

$$\therefore \qquad xdx + ydy + zdz = 0$$

Integrating, we obtain

$$x^2 + y^2 + z^2 = c_1 \tag{1}$$

Again considering the last two members, we have

$$\frac{dy}{y+z} = \frac{dz}{y-z}$$

$$\Rightarrow \qquad ydy - zdy = ydz + zdz$$

$$\Rightarrow \qquad ydy - (ydz + zdy) - zdz = 0$$

Integrating, we obtain

$$y^2 - 2yz - z^2 = c_2 \tag{2}$$

Hence, the complete solution is given by Eqs. (1) and (2).

EXAMPLE 12.13

Solve $\dfrac{adx}{(b-c)yz} = \dfrac{bdy}{(c-a)zx} = \dfrac{cdz}{(a-b)xy}$.

Solution: Choosing x, y, z as multipliers, we have each fraction $= \dfrac{axdx + bydy + czdz}{0}$

$$\therefore \qquad axdx + bydy + czdz = 0$$

Integrating, we obtain

$$ax^2 + by^2 + cz^2 = c_1 \tag{1}$$

Again, choosing ax, by, cz as multipliers, we have

$$\text{each fraction} = \frac{a^2xdx + b^2ydy + c^2zdz}{0}$$

$$\therefore \qquad a^2xdx + b^2ydy + c^2zdz = 0$$

Integrating, we obtain

$$a^2x^2 + b^2y^2 + c^2z^2 = c^2 \tag{2}$$

Hence the complete solution is given by Eqs. (1) and (2).

EXAMPLE 12.14

Solve $\dfrac{dx}{y+z} = \dfrac{dy}{z+x} = \dfrac{dz}{x+y}$.

Solution: The given equation can be rewritten as

$$\frac{dx-dy}{y-x}=\frac{dy-dz}{z-y}=\frac{dx+dy+dz}{2(x+y+z)}$$

From the first two members, we have

$$\frac{dx-dy}{x-y}=\frac{dy-dz}{y-z}.$$

Integrating, we obtain

$$\log(x-y) = \log(y-z) + \log c_1$$

$$\therefore \qquad \frac{x-y}{y-z} = c_1 \qquad (1)$$

Again, from the first and the last members, we have

$$\frac{dx-dy}{y-x}=\frac{dx+dy+dz}{2(x+y+z)}.$$

Integrating, we obtain

$$-\log(x-y) = \frac{1}{2}[\log(x+y+z) - \log c_2]$$

$$\Rightarrow \qquad (x-y)^2(x+y+z) = c_2 \qquad (2)$$

Hence, the complete solution is given by Eqs. (1) and (2).

EXAMPLE 12.15

Solve $\dfrac{dx}{x^2-yz}=\dfrac{dy}{y^2-zx}=\dfrac{dz}{z^2-xy}$.

Solution: The given equation can be rewritten as

$$\frac{dx-dy}{(x-y)(x+y+z)}=\frac{dy-dz}{(y-z)(x+y+z)}=\frac{dz-dx}{(z-x)(x+y+z)}$$

$$\Rightarrow \qquad \frac{dx-dy}{x-y}=\frac{dy-dz}{y-z}=\frac{dz-dx}{z-x}$$

Consider, the first two members,

$$\frac{dx-dy}{x-y}=\frac{dy-dz}{y-z}.$$

Integrating, we obtain

$$\log(x-y) = \log(y-z) + \log c_1$$

$$\Rightarrow \qquad \frac{x-y}{y-z} = c_1 \qquad (1)$$

Again, consider the first and last members

$$\frac{dx-dy}{x-y}=\frac{dz-dx}{z-x}.$$

Integrating, we obtain

$$\log(x-y) = \log(z-x) + \log c_2$$

$$\Rightarrow \qquad \frac{x-y}{z-x} = c_2 \qquad (2)$$

Hence, the complete solution is given by, Eqs. (1) and (2).

EXAMPLE 12.16

Solve $\dfrac{dx}{x^2-y^2-yz}=\dfrac{dy}{x^2-y^2-xz}=\dfrac{dz}{z(x-y)}.$

Solution: From the given equation, we have

$$\frac{dx-dy}{-z(y-x)}=\frac{dz}{z(x-y)}$$

$$dx - dy = dz$$

Integrating, we obtain

$$-x + y + z = c_1 \qquad (1)$$

Again, from the given equation, we have

$$\frac{xdx-ydy}{(x^2-y^2)(x-y)}=\frac{dz}{z(x-y)}$$

$$\Rightarrow \qquad \frac{xdx-ydy}{x^2-y^2}=\frac{dz}{z}$$

Integrating, we obtain

$$\frac{1}{2}\log(x^2-y^2) = \log z + \log c_2$$

$$\Rightarrow \qquad \frac{x^2-y^2}{z^2} = c_2' \qquad (2)$$

The required complete solution is given by Eqs. (1) and (2).

EXAMPLE 12.17

Solve $\dfrac{dx}{1+y}=\dfrac{dy}{1+x}=\dfrac{dz}{z}.$

Solution: From the given equations, we have

$$\frac{dx-dy}{y-x}=\frac{dx+dy}{x+y+2}=\frac{dz}{z}$$

Considering the first and last members

$$\frac{dx-dy}{y-x}=\frac{dz}{z}.$$

Integrating, we obtain

$$-\log(x-y) = \log z - \log c_1$$

$$\therefore \qquad z = \frac{c_1}{x-y} \qquad (1)$$

Again, considering the last two members

$$\frac{dx+dy}{x+y+2}=\frac{dz}{z}.$$

Integrating, we obtain

$$\log(x+y+2) = \log z - \log c_2$$

$$\therefore \qquad z = (x+y+2)c_2 \qquad (2)$$

The complete solution is given by Eqs. (1) and (2).

EXAMPLE 12.18

Solve $\dfrac{dx}{y^3x-2x^4}=\dfrac{dy}{2y^4-x^3y}=\dfrac{dz}{9z(x^3-y^3)}$.

Solution: Considering the first two members

$$\frac{dx}{y^3x-2x^4}=\frac{dy}{2y^4-x^3y}$$

$$\Rightarrow \qquad (2y^4-x^3y)dx = (y^3x-2x^4)dy$$

Dividing by x^3y^3, we obtain

$$\left(\frac{2y}{x^3}-\frac{1}{y^2}\right)dx = \left(\frac{1}{x^2}-\frac{2x}{y^3}\right)dy$$

$$\Rightarrow \qquad \left(\frac{1}{x^2}dy-\frac{2y}{x^3}dx\right)+\left(\frac{1}{y^2}dx-\frac{2x}{y^3}dy\right) = 0$$

Integrating, we get

$$\frac{y}{x^2}+\frac{x}{y^2} = c_1 \qquad (1)$$

Again using $\frac{1}{x}, \frac{1}{y}, \frac{1}{3z}$ as multiples, we have

$$\text{Each fraction} = \frac{\frac{1}{x}dx + \frac{1}{y}dy + \frac{1}{3z}dz}{(y^3 - 2x^3) + (2y^3 - x^3) + 3(x^3 - y^3)}$$

$$= \frac{\frac{1}{x}dx + \frac{1}{y}dy + \frac{1}{3z}dz}{0}$$

$$\therefore \quad \frac{1}{x}dx + \frac{1}{y}dy + \frac{1}{3z}dz = 0$$

Integrating, we obtain

$$\log x + \log y + \frac{1}{3}\log z = \log c_2$$

$$\Rightarrow \quad xyz^{1/3} = c_2 \qquad (2)$$

Hence, Eqs. (1) and (2) together give the required complete solution.

EXAMPLE 12.19

Solve $\frac{dx}{\cos(x+y)} = \frac{dy}{\sin(x+y)} = \frac{dz}{z}$

Solution: From the given equation, we have

$$\frac{dx+dy}{\cos(x+y)+\sin(x+y)} = \frac{dx-dy}{\cos(x+y)-\sin(x+y)} = \frac{dz}{z}$$

Considering the first two members

$$\frac{\cos(x+y)-\sin(x+y)}{\cos(x+y)-\sin(x+y)}(dx+dy) = dx - dy$$

Integrating, we obtain

$$\log[\cos(x+y) + \sin(x+y)] = x - y + \log c_1$$

$$\Rightarrow \quad [\cos(x+y) + \sin(x+y)]e^{y-x} = c_1 \qquad (1)$$

Again, considering the first and last members

$$\frac{dx+dy}{\cos(x+y)+\sin(x+y)} = \frac{dz}{z}$$

$$\Rightarrow \quad \frac{1}{\sqrt{2}}\frac{dx+dy}{\sin\left(x+y+\frac{\pi}{4}\right)} = \frac{dz}{z}$$

$$\Rightarrow \qquad \operatorname{cosec}\left(x+y+\frac{\pi}{4}\right)(dx + dy) = \sqrt{2}\,\frac{dz}{z}$$

Integrating, we obtain

$$\log\tan\frac{1}{2}\left(x+y+\frac{\pi}{4}\right) = \sqrt{2}\log z - \log c_2$$

$$\Rightarrow \qquad z^{\sqrt{2}}\cot\frac{1}{2}\left(x+y+\frac{\pi}{4}\right) = c_2 \qquad (2)$$

Hence, Eqs. (1) and (2) give the required complete solution. ▲▲

EXERCISE 12.2

Solve the following simultaneous equations given in symmetrical form:

1. $\dfrac{dx}{x^2} = \dfrac{dy}{y^2} = \dfrac{dz}{nxy}$.

2. $\dfrac{dx}{yz} = \dfrac{dy}{zx} = \dfrac{dz}{xy}$.

3. $\dfrac{dx}{z} = \dfrac{dy}{-z} = \dfrac{dz}{z^2 + (x+y)^2}$.

4. $\dfrac{dx}{1} = \dfrac{dy}{-2} = \dfrac{dz}{3x^2\sin(y+2x)}$.

5. $\dfrac{dx}{x^2+y^2+yz} = \dfrac{dy}{x^2+y^2-xz} = \dfrac{dz}{z(x+y)}$.

6. $\dfrac{dx}{mz-ny} = \dfrac{dy}{nx-lz} = \dfrac{dz}{ly-mx}$.

7. $\dfrac{xdx}{z^2-2yz-y^2} = \dfrac{dy}{y+z} = \dfrac{dz}{y-z}$.

8. $\dfrac{dx}{y^2+z^2-x^2} = \dfrac{dy}{-2xy} = \dfrac{dz}{-2xz}$.

9. $\dfrac{dx}{z(x+y)} = \dfrac{dy}{z(x-y)} = \dfrac{dz}{x^2+y^2}$.

10. $\dfrac{dx}{x(y^2-z^2)} = \dfrac{dy}{y(z^2+x^2)} = \dfrac{dz}{z(x^2+y^2)}$.

11. $\dfrac{-dx}{z(x+y)} = \dfrac{dy}{y(x+y)} = \dfrac{dz}{(x-y)(2x+2y+z)}$.

12. $\dfrac{dx}{y(x+y)+az} = \dfrac{dy}{x(x+y)-az} = \dfrac{dz}{z(x+y)}$.

13. $\dfrac{yzdx}{y-z} = \dfrac{zxdy}{z-x} = \dfrac{xydz}{x-y}$.

14. $\dfrac{dx}{y^2(x-y)} = \dfrac{dy}{-x^2(x-y)} = \dfrac{dz}{z(x^2+y^2)}$.

15. $\dfrac{dx}{x+y-xy^2} = \dfrac{dy}{x^2y-x-y} = \dfrac{dz}{z(y^2-x^2)}$.

ANSWERS

Exercise 12.2

1. $y - x = c_1xy,\ z = \dfrac{nxy}{y-x}\log\dfrac{y}{x} + c_2$.
2. $x^2 - y^2 = c_1,\ x^2 - z^2 = c_2$.
3. $x + y = c_1,\ \log[z^2 + (x + y)^2] - 2x = c_2$.
4. $2x + y = c_1,\ z - x^3\sin(y + 2x) = c_2$.
5. $x - y - z = c_1,\ x^2 + y^2 = z^2c_2$.
6. $lx + my + nz = c_1,\ \dfrac{x^2}{2} + \dfrac{y^2}{2} + \dfrac{z^2}{2} = c_2$.
7. $\dfrac{y^2}{2} - yz - \dfrac{z^2}{2} = c_1,\ \dfrac{x^2}{2} + \dfrac{y^2}{2} + \dfrac{z^2}{2} = c_2$.
8. $y = c_1z,\ x^2 + y^2 + z^2 = c_2z$.
9. $x^2 - y^2 - z^2 = c_1,\ 2xy - z^2 = c_2$.
10. $yz = c_1x,\ x^2 + y^2 + z^2 = c_2$.
11. $xy = c_1,\ z(x + y) + x^2 + y^2 = c_2$.
12. $\dfrac{x+y}{z} = c_1,\ x^2 - y^2 - 2az = c_2$.
13. $xyz = c_1,\ x + y + z = c_2$.
14. $x^3 + y^3 = c_1,\ (x - y)/z = c_2$.
15. $z(1 - xy) = c_1,\ x^2 + y^2 - 2\log(1 - xy) = c_2$.

▲▲

Total Differential Equations

In this section, we propose to discuss single differential equations having one independent variable and more than one dependent variables. These equations are known as total or Pfaffian differential equations.

Definition: An ordinary differential equation of the from

$$Pdx + Qdy + Rdz = 0 \tag{1}$$

where P, Q, R are functions of x, y, z is called the total differential equation or single differential equation or Pfaffian differential equation in three variables x, y, z.

Equation (1) can be directly integrated if there exists a function $u(x, y, z)$ whose total differential du is equal to the left hand member of (1).

In other cases, Eq. (1) may or may not be integrable. We now proceed to find the condition which P, Q, R must satisfy, so that Eq. (1) may be integrable. This will be called the condition or criterion of integrability of the single differential Eq. (1).

Condition of integrability of $Pdx + Qdy + Rdz = 0$. A necessary and sufficient condition for $Pdx + Qdy + Rdz = 0$ may be integrable is

$$P\left(\frac{\partial Q}{\partial z} - \frac{\partial R}{\partial y}\right) + Q\left(\frac{\partial R}{\partial x} - \frac{\partial P}{\partial z}\right) + R\left(\frac{\partial P}{\partial y} - \frac{\partial Q}{\partial x}\right) = 0.$$

Proof – Necessary Condition

Consider $Pdx + Qdy + Rdz = 0$ (1)

Let $u(x, y, z,)$ be a function whose total differential du is equal to $Pdx + Qdy + Rdz$ or this expression multiplied by some factor which may be a function of x, y, z then $u(x, y, z)$ = constant is an integral of (1)

But, we know that

$$du = \frac{\partial u}{\partial x}dx + \frac{\partial u}{\partial y}dy + \frac{\partial u}{\partial z}dz$$

So $\frac{\partial u}{\partial x}, \frac{\partial u}{\partial y}$ and $\frac{\partial u}{\partial z}$ are proportional to P, Q and R respectively, which means

$$\frac{\frac{\partial u}{\partial x}}{P} = \frac{\frac{\partial u}{\partial y}}{Q} = \frac{\frac{\partial u}{\partial z}}{R} = \lambda \text{ (say)}$$

where, λ is a function of x, y, z. Then, we have

$$\frac{\partial u}{\partial x} = \lambda P; \ \frac{\partial u}{\partial y} = \lambda Q \text{ and } \frac{\partial u}{\partial z} = \lambda R$$

From the first two of these equations, we have

$$\frac{\partial}{\partial y}(\lambda P) = \frac{\partial}{\partial y}\left(\frac{\partial u}{\partial x}\right) = \frac{\partial^2 u}{\partial y \partial x} = \frac{\partial^2 u}{\partial x \partial y} = \frac{\partial}{\partial x}\left(\frac{\partial u}{\partial y}\right) = \frac{\partial}{\partial x}(\lambda Q)$$

$$\Rightarrow \quad \lambda\frac{\partial P}{\partial y} + P\frac{\partial \lambda}{\partial y} = \lambda\frac{\partial Q}{\partial x} + Q\frac{\partial \lambda}{\partial x}$$

Thus, we have

$$\lambda\left(\frac{\partial P}{\partial y}-\frac{\partial Q}{\partial x}\right)=Q\frac{\partial \lambda}{\partial x}-P\frac{\partial \lambda}{\partial y} \tag{2}$$

Similarly,

$$\lambda\left(\frac{\partial Q}{\partial z}-\frac{\partial R}{\partial y}\right)=R\frac{\partial \lambda}{\partial y}-Q\frac{\partial \lambda}{\partial z} \tag{3}$$

and

$$\lambda\left(\frac{\partial R}{\partial x}-\frac{\partial P}{\partial z}\right)=P\frac{\partial \lambda}{\partial z}-R\frac{\partial \lambda}{\partial x} \tag{4}$$

Multiplying Eqs. (2), (3) and (4) by R, P and Q respectively and adding, we get

$$P\left(\frac{\partial Q}{\partial z}-\frac{\partial R}{\partial y}\right)+Q\left(\frac{\partial R}{\partial x}-\frac{\partial P}{\partial z}\right)+R\left(\frac{\partial P}{\partial y}-\frac{\partial Q}{\partial x}\right)=0$$

$$\Rightarrow \qquad \begin{vmatrix} P & Q & R \\ P & Q & R \\ \frac{\partial}{\partial x} & \frac{\partial}{\partial y} & \frac{\partial}{\partial z} \end{vmatrix} = 0.$$

This is the necessary condition for the integrability of Eq. (1).

Sufficient Condition

If the relation

$$P\left(\frac{\partial Q}{\partial z}-\frac{\partial R}{\partial y}\right)+Q\left(\frac{\partial R}{\partial x}-\frac{\partial P}{\partial z}\right)+R\left(\frac{\partial P}{\partial y}-\frac{\partial Q}{\partial x}\right)=0 \tag{5}$$

is satisfied, the equation has a primitive of the from $u(x, y, z)$ = constant.

Now, let the relation (5) be satisfied for the coefficients P, Q, R of the Eq. (1), that is,

$$Pdx + Qdy + Rdz = 0.$$

Then the same condition is satisfied by

$$P_1 = \mu P,\ Q_1 = \mu Q \text{ and } R_1 = \mu R$$

where, μ is any function of x, y, z. So $\mu Pdx + \mu Qdy + \mu Rdz = 0$

∴ We have

$$\frac{\partial Q_1}{\partial z}-\frac{\partial R_1}{\partial y} = \mu\frac{\partial Q}{\partial z}+Q\frac{\partial \mu}{\partial z}-\left(\mu\frac{\partial P}{\partial y}+R\frac{\partial \mu}{\partial y}\right)$$

$$\Rightarrow \qquad \frac{\partial Q_1}{\partial z}-\frac{\partial R_1}{\partial y} = \mu\left(\frac{\partial Q}{\partial z}-\frac{\partial R}{\partial y}\right)+Q\frac{\partial \mu}{\partial z}-R\frac{\partial \mu}{\partial y} \tag{6}$$

Similarly,
$$\frac{\partial R_1}{\partial x}-\frac{\partial P_1}{\partial z} = \mu\left(\frac{\partial R}{\partial x}-\frac{\partial P}{\partial z}\right)+R\frac{\partial \mu}{\partial x}-P\frac{\partial \mu}{\partial z} \tag{7}$$

and
$$\frac{\partial P_1}{\partial y}-\frac{\partial Q_1}{\partial x} = \mu\left(\frac{\partial P}{\partial y}-\frac{\partial Q}{\partial x}\right)+P\frac{\partial \mu}{\partial y}-Q\frac{\partial \mu}{\partial z} \tag{8}$$

Multiplying Eqs. (6), (7) and (8) by P_1, Q, and R_1, respectively, adding and replacing P_1, Q_1, R_1, by μP, μQ, μR, respectively in the resulting right hand member, we obtain

$$P_1\left(\frac{\partial Q_1}{\partial z}-\frac{\partial R_1}{\partial y}\right)+Q_1\left(\frac{\partial R_1}{\partial x}-\frac{\partial P_1}{\partial z}\right)+R_1\left(\frac{\partial P_1}{\partial y}-\frac{\partial Q_1}{\partial x}\right)$$

$$= \mu\left[P\left(\frac{\partial Q}{\partial z}-\frac{\partial R}{\partial y}\right)+Q\left(\frac{\partial R}{\partial x}-\frac{\partial P}{\partial z}\right)+R\left(\frac{\partial P}{\partial y}-\frac{\partial Q}{\partial x}\right)\right] = 0$$

Now, $Pdx + Qdy$ may be regarded as an exact differential for if not so, then multiplying Eq. (1) by the integrating factor $\mu(x, y, z)$ we can make it so. Thus, there is no loss of generality in regarding $Pdx + Qdy$ as an exact differential. For this the condition is

$$\frac{\partial P}{\partial y}=\frac{\partial Q}{\partial x}$$

and Let
$$\int(Pdx + Qdy) = V$$

Then, $\frac{\partial V}{\partial x}=P$ and $\frac{\partial V}{\partial y}=Q$ and also

$$\frac{\partial P}{\partial z} = \frac{\partial^2 V}{\partial z\partial x} \text{ and } \frac{\partial Q}{\partial z}=\frac{\partial^2 V}{\partial z\partial y}$$

Substituting all these values in Eq. (5), we have

$$\frac{\partial V}{\partial x}\left(\frac{\partial^2 V}{\partial z\partial y}-\frac{\partial R}{\partial y}\right)+\frac{\partial V}{\partial y}\left(\frac{\partial R}{\partial x}-\frac{\partial^2 V}{\partial z\partial x}\right) = 0$$

$$\Rightarrow \qquad \frac{\partial V}{\partial x}\frac{\partial}{\partial y}\left(\frac{\partial V}{\partial z}-R\right)-\frac{\partial V}{\partial y}\frac{\partial}{\partial x}\left(\frac{\partial V}{\partial z}-R\right) = 0$$

$$\Rightarrow \qquad \begin{vmatrix} \frac{\partial V}{\partial x} & \frac{\partial}{\partial x}\left(\frac{\partial V}{\partial z}-R\right) \\ \frac{\partial V}{\partial y} & \frac{\partial}{\partial y}\left(\frac{\partial V}{\partial z}-R\right) \end{vmatrix} = 0$$

This shows that a relation independent of x and y exists between V and $\frac{\partial V}{\partial z} - R$.

Therefore, $\frac{\partial V}{\partial z} - R$ can be expressed as a function of z and V alone, that is, we can take

$$\frac{\partial V}{\partial z} - R = f(z, V)$$

Now,

$$Pdx + Qdy + Rdz = \frac{\partial V}{\partial x}dx + \frac{\partial V}{\partial y}dy + \left(\frac{\partial V}{\partial z} - f\right)dz$$

$$= \left(\frac{\partial V}{\partial x}dx + \frac{\partial V}{\partial y}dy + \frac{\partial V}{\partial z}dz\right) - fdz$$

$$= dV - f(z, V)\,dz$$

Thus, Eq. (1) can be written as

$$dV - f(z, V)\,dz = 0$$

which is an equation in two variables.

Hence, its integration will give an integral of the from $F(z, V) = 0$.

Hence, the condition is sufficient.

$\therefore$

$$P\left(\frac{\partial Q}{\partial z} - \frac{\partial R}{\partial y}\right) + Q\left(\frac{\partial R}{\partial x} - \frac{\partial P}{\partial z}\right) + R\left(\frac{\partial P}{\partial y} - \frac{\partial Q}{\partial x}\right) = 0$$

is the necessary and sufficient condition for the integrability of

$$Pdx + Qdy + Rdz = 0.$$

Note: The condition for the equation $Pdx + Qdy + Rdz = 0$ to be exact is

$$\frac{\partial P}{\partial y} = \frac{\partial Q}{\partial x},\ \frac{\partial Q}{\partial z} = \frac{\partial R}{\partial y} \text{ and } \frac{\partial R}{\partial x} = \frac{\partial P}{\partial z}.$$

Thus, every exact equation satisfies the condition of integrability (5).

Note: In vector form, the total differential Eq. (1) can be written as

$$A \cdot dr = 0 \quad \text{where} \quad A = P\hat{i} + Q\hat{j} + R\hat{k};\ r = x\hat{i} + y\hat{j} + z\hat{k}$$

and the condition of integrability is $A \cdot \text{Curl } A = 0$.

Geometrical interpretation of $Pdx + Qdy + Rdz = 0$.

Consider the surface $u(x, y, z) = \text{constant}$.

We know that $\frac{\partial u}{\partial x}, \frac{\partial u}{\partial y}$ and $\frac{\partial u}{\partial z}$ are proportional to the direction cosines of the normal at (x, y, z) on the surface.

But $$\frac{\partial u}{\partial x}dx + \frac{\partial u}{\partial y}dy + \frac{\partial u}{\partial z}dz = 0$$

Comparing this with $Pdx + Qdy + Rdz = 0$ (1)

We see that $\frac{\partial u}{\partial x}, \frac{\partial u}{\partial y}$ and $\frac{\partial u}{\partial z}$ are proportional to P, Q and R.

So Eq. (1) represents a single infinite system of surfaces such that the direction cosines of the normal at any point (x, y, z) on the surface belonging to these systems are proportional to the value of P, Q, R at that point.

Thus, it follows that a point moving in such a manner that its coordinates and the direction cosines of its path always satisfy Eq. (1) as part through any point in an infinity of directions and yet, while passing through any point must remain on the particular surface $u(x, y, z)$ = constant, while passes through the point.

Methods for solving $Pdx + Qdy + Rdz = 0$. First see that the condition of integrability is satisfied.

Method 1. Solution by inspection. When the condition of integrability is satisfied, it is possible in many cases that by reassigning the terms, it becomes exact and the solution is found readily. ▲▲

EXAMPLE 12.20

Solve $(yz + kx)dx + (zx - kz)dy + (xy - ky)dz = 0$.

Solution: Here, $P = yz + kx$, $Q = zx - kz$, $R = xy - ky$

$$\therefore\ P\left(\frac{\partial Q}{\partial z} - \frac{\partial R}{\partial y}\right) + Q\left(\frac{\partial R}{\partial x} - \frac{\partial P}{\partial z}\right) + R\left(\frac{\partial P}{\partial y} - \frac{\partial Q}{\partial x}\right)$$

$$= (yz + kx)(x - k - x + k) + (zx - kz)(y - y) + (xy - ky)(z - z) = 0$$

Hence, the condition of integrability is satisfied. Thus, the equation is exact and can be written as

$$yzdx + zxdy + xydy + kxdx - k(zdy + ydz) = 0$$

$$\Rightarrow \quad d(xyz) + kd\left(\frac{x^2}{2}\right) - kd(yz) = 0$$

Integrating, we get

$$xyz + \frac{kx^2}{2} - kyz = 0.$$

EXAMPLE 12.21

Solve the equation

$$(x^2z - y^3)dx + 3xy^2dy + x^3dz = 0.$$

Solution: First showing that it is integrable.

Here, $P = x^2z - y^3$, $Q = 3xy^2$, $R = x^3$

$$\therefore \quad \left(\frac{\partial Q}{\partial z} - \frac{\partial R}{\partial y}\right) + Q\left(\frac{\partial R}{\partial x} - \frac{\partial P}{\partial z}\right) + R\left(\frac{\partial P}{\partial y} - \frac{\partial Q}{\partial x}\right)$$

$$= (x^2z - y^3)(0 - 0) + 3xy^2(3x^2 - x^2) + x^3(-3y^2 - 3y^2) = 0$$

Hence, the given equation is integrable. Therefore, the given equation can be written as

$$x^2(zdx + xdz) - y^3dx + 3xy^2dy = 0$$

$$\Rightarrow \quad zdx + xdz - \frac{y^3}{x^2}dx + \frac{3y^2}{x}dy = 0$$

$$\Rightarrow \quad d(zx) + d(y^3/x) = 0$$

Integrating, we get

$$zx + \frac{y^3}{x} = c \quad \Rightarrow \quad x^2z + y^3 = cx.$$

EXAMPLE 12.22

Solve $(yz + xyz)dx + (zx + xyz)dy + (xy + xyz)dz = 0$.

Solution: Here, $P = yz + xyz$, $Q = zx + xyz$, $R = xy + xyz$

$$\therefore \quad P\left(\frac{\partial Q}{\partial z} - \frac{\partial R}{\partial y}\right) + Q\left(\frac{\partial R}{\partial x} - \frac{\partial P}{\partial z}\right) + R\left(\frac{\partial P}{\partial y} - \frac{\partial Q}{\partial x}\right)$$

$$= (yz + xyz)(x + xy - x - xz) + (zx + xyz)(y + yz - y - xy) + (xy + xyz)(z + xz - z - yz)$$
$$= 0$$

Hence, the condition of integrability is satisfied. Dividing the given equation by xyz, we have

$$\left(\frac{1}{x} + 1\right)dx + \left(\frac{1}{y} + 1\right)dy + \left(\frac{1}{z} + 1\right)dz = 0$$

Integrating, we get

$$\log x + x + \log y + y + \log z + z = c$$

$$\Rightarrow \quad \log xyz + x + y + z = c.$$

EXAMPLE 12.23

Verify that the equation

$$(y - z)(y + z - 2x)dx + (z - x)(z + x - 2y)dy + (x - y)(x + y - 2z)dz = 0$$

is exact and find the solution.

Solution: Here, $P = y^2 - z^2 - 2x(y - z)$, $Q = z^2 - x^2 - 2y(z - x)$, and $R = x^2 - y^2 - 2z(x - y)$

$$\frac{\partial P}{\partial y} = 2y - 2x, \frac{\partial Q}{\partial x} = -2x + 2y, \frac{\partial Q}{\partial z} = 2z - 2y,$$

$$\frac{\partial R}{\partial y} = -2y + 2z, \frac{\partial R}{\partial x} = 2x - 2z \text{ and } \frac{\partial P}{\partial z} = -2z + 2x$$

$$\therefore \quad P\left(\frac{\partial Q}{\partial z} - \frac{\partial R}{\partial y}\right) + Q\left(\frac{\partial R}{\partial x} - \frac{\partial P}{\partial z}\right) + R\left(\frac{\partial P}{\partial y} - \frac{\partial Q}{\partial x}\right) = 0$$

Hence, the condition of integrability is satisfied.

Also, $$\frac{\partial P}{\partial y} = \frac{\partial Q}{\partial x}, \frac{\partial Q}{\partial z} = \frac{\partial R}{\partial y} \text{ and } \frac{\partial R}{\partial x} = \frac{\partial P}{\partial z}$$

Therefore, the given equation is also exact. The given equation can be written as

$$(y^2dx + 2xydy) - (z^2dx + 2zx\,dx) + (z^2dy + 2zy\,dz) - (x^2dy + 2xydx) + (x^2dx + 2xz\,dz) - (y^2dz + 2yz\,dz) = 0$$

Integrating, we get

$$y^2x - z^2x + z^2y - x^2y + x^2z - y^2z = c$$

$$\Rightarrow \quad x(y^2 - z^2) + y(z^2 - x^2) + z(x^2 - y^2) = c$$

EXAMPLE 12.24

Find $f(y)$, such that

$$\left(\frac{yz + z}{x}\right)dx - zdy + f(y)dz = 0$$

is integrable and hence solve the equation.

Solution: The given equation can be written as

$$(yz + z)dx - zxdy + xf(y)dz = 0$$

Here, $P = yz + z$, $Q = -xz$, $R = xf(y)$

Now, $P\left(\frac{\partial Q}{\partial z} - \frac{\partial R}{\partial y}\right) + Q\left(\frac{\partial R}{\partial x} - \frac{\partial P}{\partial z}\right) + R\left(\frac{\partial P}{\partial y} - \frac{\partial Q}{\partial x}\right) = 0$

$\Rightarrow \quad (yz + z)(-x - xf'(y)) + (-xz)(f(y) - (y + 1)) + xf(y)(z - (-z)) = 0$ where $f'(y) = \frac{df}{dy}$

$$\Rightarrow \quad xz(1 + y)f' = xzf$$

$$\Rightarrow \quad \frac{df}{f} = \frac{dy}{1+y}$$

$\Rightarrow \quad f = K(1 + y)$ where, K is arbitrary constant.

Substituting in the given equation, we get

$$z(y+1)dx - zxdy + xk(1+y)dz$$

$$\Rightarrow \qquad \frac{dx}{x} - \frac{dy}{1+y} + \frac{k}{z}dz = 0$$

Integrating, we get

$$\log x - \log(l+y) + k\log z = \log c$$

$$\Rightarrow \qquad xz^k = c(l+y)$$

EXAMPLE 12.25

Solve $z(1+z)\cos x\,dx - z(1+z)dy + (1-z^2)(y-\sin x)dz$.

Solution: Here, $P = z(1+z)\cos x$, $Q = -z(1+z)$; $R = (1-z)^2(y-\sin x) = 0$

$$\therefore \quad P\left(\frac{\partial Q}{\partial z} - \frac{\partial R}{\partial y}\right) + Q\left(\frac{\partial R}{\partial x} - \frac{\partial P}{\partial z}\right) + R\left(\frac{\partial P}{\partial y} - \frac{\partial Q}{\partial x}\right)$$

$$= z(1+z)\cos x[-(1+2z) - (1-z^2)) - z(1+z)(1-z^2)$$

$$(-\cos x) - (1+2z)] + (1-z^2)(y-\sin x)(0-0) = 0$$

Hence, the condition of integrability is satisfied.

Now the given equation can be written as

$$z(1+z)(\cos x\,dx - dy) + (1-z^2)(y-\sin x)dz = 0$$

$$\Rightarrow \qquad \frac{\cos x\,dx - dy}{\sin x - y} + \frac{1-z^2}{z(1+z)}dz = 0$$

$$\Rightarrow \qquad \frac{dy - \cos x\,dx}{y - \sin x} - \left(\frac{1}{z} - 1\right)dz = 0$$

Integrating, we get

$$\log(y-\sin x) - \log z - z = +\log c$$

$$\Rightarrow \qquad (y-\sin x)kz = e^{-z}$$

$$\Rightarrow \qquad y - \sin z = cze^{-z}$$

which is the required solution. ▲▲

Method 2. One variable regarded as a constant.

First, the condition of integrability is verified. If any two terms like $Pdx + Qdy = 0$ can be solved easily, then we assume the third variable z as a constant that is $dz = 0$. Let $u = \phi(z)$ be the solution of $Pdx + Qdy = 0$, where, $\phi(z)$ is a function of z only, i.e., constant with respect to x and y.

Then to solve the equation

$$Pdx + Qdy + Rdz = 0 \tag{1}$$

Differentiate $u = \phi(z)$ with respect to x, y, z and compare with Eq. (1). If the coefficient of $d\phi$ or dz involve functions of x and y, it will be possible to eliminate them with the help of $u = \phi(z)$. Thus, we shall be able to get $d\phi/dz$, which is independent of x and y. Hence, ϕ is determined by integration from $d\phi/dz$ thus obtained, which is the complete solution of Eq. (1).

We shall illustrate the method by means of the following examples.

EXAMPLE 12.26

Solve $3x^2dx + 3y^2dy - (x^3 + y^3 + e^{2z})dz = 0$.

Solution: Here, $P = 3x^2$, $Q = 3y^2$; $R = -(x^3 + y^3 + e^{2z})$

$$\therefore \quad P\left(\frac{\partial Q}{\partial z} - \frac{\partial R}{\partial y}\right) + Q\left(\frac{\partial R}{\partial x} - \frac{\partial P}{\partial z}\right) + R\left(\frac{\partial P}{\partial y} - \frac{\partial Q}{\partial x}\right)$$

$$= 3x^2(0 + 3y^2) + 3y^2(-3x^2 - 0) - (x^3 + y^3 + e^{2z})(0 - 0) = 0$$

Hence, the condition of integrability is satisfied.

Taking z = constant, so that $dz = 0$, the given equation gives

$$3x^2dx + 3y^2dy = 0$$

Integrating, we get

$$x^3 + y^3 = \phi(z) \quad \text{(say)}$$

Now, differentiating this, we have

$$3x^2dx + 3y^2dy - \phi'(z)dz = 0$$

Comparing this with the given equation, we get

$$\phi'(z) = x^3 + y^3 + e^{2z} = \phi(z) + e^{2z}$$

$$\therefore \quad \frac{d\phi}{dz} - \phi = e^{2z}$$

which is the linear equation with IF $= e^{\int -1dx} = e^{-x}$

$\therefore$ its solution is

$$\phi e^{-z} = \int e^{2z}e^{-z}dz + c = e^z + c$$

$$\Rightarrow \quad \phi = e^{2z} + ce^z$$

$$\Rightarrow \quad x^3 + y^3 = e^{2z} + ce^z$$

which is the required solution.

EXAMPLE 12.27

Solve $(2x^2 + 2xy + 2xz^2 + 1)dx + dy + 2zdz = 0$.

Solution: Here, $P = 2x^2 + 2xy + 2xz^2 + 1$; $Q = 1$; $R = 2z$

$$\therefore \quad P\left(\frac{\partial Q}{\partial z} - \frac{\partial R}{\partial y}\right) + Q\left(\frac{\partial R}{\partial x} - \frac{\partial P}{\partial z}\right) + R\left(\frac{\partial P}{\partial y} - \frac{\partial Q}{\partial x}\right)$$

$$= (2x^2 + 2xy + 2xz^2 + 1)(0 - 0) + 1(0 - 4xz) + 2z(2x - 0) = 0$$

Hence, the condition of integrability is satisfied.

Taking x = constant so that $dx = 0$, the given equation gives

$$dy + 2zdz = 0$$

Integrating, we get

$$y + z^2 = \phi(x) \text{ (say)}$$

Differentiating this, we have

$$dy + 2zdz = \phi'(x)dx$$

Comparing with the given equation, we have

$$\begin{aligned} \phi'(x) &= -(2x^2 + 2xy + 2xz^2 + 1) \\ &= -2x^2 - 1 - 2x(y + z^2) \\ &= -2x^2 - 1 - 2x\phi(x) \end{aligned}$$

$$\Rightarrow \qquad \frac{d\phi}{dx} + 2x\phi = -2x^2 - 1$$

which is a linear equation with IF $= e^{\int 2x dx} = e^{x^2}$

$\therefore$ its solution is

$$\begin{aligned} \phi \cdot e^{x^2} &= -\int (2x^2 + 1)e^{x^2} dx + c \\ &= -\int x(e^{x^2} 2x)\, dx - \int e^{x^2} dx + c \\ &= -xe^{x^2} + \int e^{x^2} dx - \int e^{x^2} dx + c \end{aligned}$$

$$\Rightarrow \qquad \phi = -x + ce^{-x^2}$$

$$\Rightarrow \qquad (y + x + z^2)e^{x^2} = c$$

which is the required solution.

EXAMPLE 12.28

Find $f(y)$ if $f(y)dx - xzdy - xy \log y dz = 0$ is integrable. Find the corresponding integral.

Solution: Here $P = f(y)$; $Q = -zx$; $R = -xy \log y$.

If the condition of integrability is satisfied, then we have

$$P\left(\frac{\partial Q}{\partial z}-\frac{\partial R}{\partial y}\right)+Q\left(\frac{\partial R}{\partial x}-\frac{\partial P}{\partial z}\right)+R\left(\frac{\partial P}{\partial y}-\frac{\partial Q}{\partial x}\right) = 0$$

$$\Rightarrow \quad f(y)(-x + x\log y + x) - zx(-y\log y - 0) - xy\log y(f'(y) + z) = 0$$

$$\Rightarrow \quad xf(y)\log y - xy\log y\, f'(y) = 0$$

$$\Rightarrow \quad f(y) - yf'(y) = 0$$

$$\Rightarrow \quad \frac{f'(y)}{f(y)} = \frac{1}{y}$$

Integrating, we have

$$\log f(y) = \log y + \log k$$

$$\Rightarrow \quad f(y) = ky.$$

$\therefore$ The given equation reduces to

$$kydx - zxdy - xy\log y\,dz = 0$$

Taking z to be constant, so that $dz = 0$
From the given equation, we have

$$kydx - zxdy = 0$$

$$\Rightarrow \quad \frac{kdx}{x} - z\frac{dy}{y} = 0.$$

Integrating, we have

$$k\log x - z\log y = \phi(z) \text{ (say)}$$

Differentiating this, we have

$$\frac{k}{x}dx - \frac{z}{y}dy - \log y\,dz = \phi'(z)dz$$

$$\Rightarrow \quad kydx - zxdy - (xy\log y + xy\,\phi'(z))dz = 0$$

Comparing this with the given equation, we have

$$xy\log y + xy\,\phi'(z) = xy\log y$$

$$\Rightarrow \quad \phi'(z) = 0$$

$$\therefore \quad \phi(z) = \log c$$

$$\Rightarrow \quad k\log x - z\log y = \log c$$

$$\Rightarrow \quad x^k = cy^z$$

which is the required solution.

EXAMPLE 12.29

Solve $z(z - y)dx + z(z + x)dy + x(x + y)dz = 0$.

Solution: Here, $P = z(z - y)$, $Q = z(z + x)$; $R = x(x + y)$

$$\therefore \quad P\left(\frac{\partial Q}{\partial z} - \frac{\partial R}{\partial y}\right) + Q\left(\frac{\partial R}{\partial x} - \frac{\partial P}{\partial z}\right) + R\left(\frac{\partial P}{\partial y} - \frac{\partial Q}{\partial x}\right)$$

$$= z(z - y)(2z + x - x) + z(z + x)(2x + y - 2z + y) + x(x + y)(-z - x) = 0.$$

Hence, the condition of integrability is satisfied.

Taking x = constant, so that $dx = 0$, the given equation gives

$$z(z + x)dy + x(x + y)dz = 0$$

$$\Rightarrow \quad \frac{dy}{x+y} + \frac{xdz}{z(z+x)} = 0$$

$$\Rightarrow \quad \frac{dy}{x+y} + \left(\frac{1}{z} - \frac{1}{z+x}\right)dz = 0$$

Integrating, we have

$$\log(x + y) + \log z - \log(z + x) = \log \phi(x)$$

Differentiating this, we get

$$\Rightarrow \quad \frac{(x+y)z}{z+x} = \phi(x)$$

$$\frac{(z+x)[(dx+dy)z+(x+y)dz]-(x+y)\,z(dz+dx)}{(z+x)^2} = \phi'(x)dx$$

$$\Rightarrow \quad \frac{z(zd-y)dx+z(z+x)dy+x(x+y)dz}{(z+x)^2} = \phi'(x)dx$$

$$\Rightarrow \quad 0 = d\phi$$

$$\Rightarrow \quad \phi(x) = \text{constant}$$

$$\Rightarrow \quad \frac{(x+y)z}{z+x} = c$$

$$\Rightarrow \quad (x + y)z = c(x + z)$$

which is the required solution.

EXAMPLE 12.30

Solve $(z + z^3)\cos x dx - (z + z^3)dy + (1 - z^2)(y - \sin x)dz = 0$

Solution: Here, $P = (z + z^3)\cos x$; $Q = -(z + z^3)$; $R = (1 - z^2)(y - \sin x)$

$$\therefore \quad P\left(\frac{\partial Q}{\partial z} - \frac{\partial R}{\partial y}\right) + Q\left(\frac{\partial R}{\partial x} - \frac{\partial P}{\partial z}\right) + R\left(\frac{\partial P}{\partial y} - \frac{\partial Q}{\partial x}\right)$$

$$= (z + z^3)\cos x(-1 - 3z^2 - 1 + 2z) - (z + z^3)\,(-(1 - z^2)\cos x - (1 + 3z^2)\cos x) + (1 - z^2)\,(y - \sin x)(0 - 0) = 0$$

Hence, the condition of integrability is satisfied.

Taking z = constant, so that $dz = 0$, the given equation becomes

$$(z + z^3)\cos x\,dx - (z + z^3)dy = 0$$

$$\Rightarrow \qquad \cos x\,dx - dy = 0.$$

Integrating, we have

$$\sin x - y = \phi(z) \text{ (say)}.$$

Differentiating this, we have

$$\cos x\,dx - dy = \phi'(z)dz$$

$$\Rightarrow \qquad (z + z^3)\cos x\,dx - (z + z^3)\,dy - (z + z^3)\,\phi'(z)dz = 0$$

Comparing this with the given equation, we have

$$(1 - z^2)\,(y - \sin x) = -(z + z^3)\,\phi'(z)$$

$$\Rightarrow \qquad \phi(z) = \frac{z + z^3}{1 - z^2}\phi'(z)$$

$$\Rightarrow \qquad \frac{\phi'(z)}{\phi(z)} = \frac{1 - z^2}{z(1 + z^2)} = \left(\frac{1}{z} - \frac{2z}{z^2 + 1}\right).$$

Integrating, we have

$$\log \phi(z) = \log z - \log(z^2 + 1) + \log c$$

$$\Rightarrow \qquad (z^2 + 1)\phi = cz$$

$$\Rightarrow \qquad (\sin x - y)(z^2 + 1) = cz.$$

which is the required solution.

EXAMPLE 12.31

Solve $2yzdx + zxdy - xy(1 + z)dz = 0$.

Solution: Here, $P = 2yz$; $Q = zx$; $R = -xy(1 + z)$.

$$\therefore \quad P\left(\frac{\partial Q}{\partial z} - \frac{\partial R}{\partial y}\right) + Q\left(\frac{\partial R}{\partial x} - \frac{\partial P}{\partial z}\right) + R\left(\frac{\partial P}{\partial y} - \frac{\partial Q}{\partial x}\right)$$

$$= 2yz\,(x + x(1 + z) + zx(-y(1 + z) - 2y) - xy(1 + z)(2z - 2) = 0$$

Hence, the condition of integrability is satisfied.

Taking z = constant, we have $dz = 0$, so that the given equation becomes

$$2yzdx + zxdy = 0$$

$$\Rightarrow \qquad \frac{2dx}{x} + \frac{dy}{y} = 0.$$

Integrating, we have

$$2\log x + \log y = \log \phi(z)$$

$$\Rightarrow \qquad \phi(z) = x^2 y.$$

Differentiating this, we have

$$2xydx + x^2dy - \phi'(z)dz = 0$$

$$\Rightarrow \qquad 2yzdx + zxdy - \frac{z}{x}\phi'(z)dz = 0$$

Comparing this with the given equation, we get

$$\frac{z}{x}\phi'(z) = xy(1+z)$$

$$\Rightarrow \qquad \phi'(z) = \frac{x^2y}{z}(1+z) = \frac{\phi}{z}(1+z)$$

$$\Rightarrow \qquad \frac{d\phi}{\phi} = \left(\frac{1+z}{z}\right)dz$$

Integrating, we have

$$\log \phi = \log z + z + \log c$$

$$\Rightarrow \qquad \phi = cze^z$$

$$\Rightarrow \qquad x^2y = cze^z$$

which is the required solution.

EXAMPLE 12.32

Equation $y^2z(x\cos x - \sin x)\,dx + x^2z(y\cos y - \sin y)dy + xy(y\sin x + x\sin y + xy\cos z)$ $dz = 0$ satisfies the condition of integrability. Solve it.

Solution: Taking z = constant, so that $dz = 0$.

The given equation reduces to $y^2(x\cos x - \sin x)dx + x^2(y\cos y - \sin y)dy = 0$

$$\Rightarrow \qquad \frac{x\cos x - \sin x}{x^2}dx + \frac{y\cos y - \sin y}{y^2}dy = 0.$$

Integrating, we have

$$\frac{\sin x}{x} + \frac{\sin y}{y} = \phi(z) \quad \text{(say)}$$

where, $\phi(z)$ is a function of z.

Differentiating this, we have

$$\frac{x\cos x - \sin x}{x^2}dx + \frac{y\cos y - \sin y}{y^2}dy = \phi'(z)dz$$

$$\Rightarrow \quad zy^2(x\cos x - \sin x)dx + zx^2(y\cos y - \sin y)dy - x^2y^2z\,\phi'(z)dz = 0$$

Comparing this with the given equation, we have

$$-x^2y^2z\,\phi'(z) = xy(y\sin x + x\sin y + xy\cos z)$$

$$\Rightarrow \quad -z\,\phi'(z) = \frac{\sin x}{x} + \frac{\sin y}{y} + \cos z$$

$$\Rightarrow \quad -z\frac{d\phi}{dz} = \phi + \cos z$$

$$\Rightarrow \quad \frac{d\phi}{dz} + \frac{1}{z}\phi = \frac{-\cos z}{z}$$

which is a linear equation with IF $= e^{\int \frac{1}{z}dz} = e^{\log z} = z$
Therefore, its solution is

$$\phi \cdot z = -\int \frac{\cos z}{z} \cdot z dz + c = -\sin z + c$$

$$\Rightarrow \quad z\left(\frac{\sin x}{x} + \frac{\sin y}{y}\right) + \sin z = c$$

which is the required solution.

Method 3. Method of auxiliary equations ▲▲

If
$$Pdx + Qdy + Rdz = 0 \tag{1}$$
is integrable, then we have

$$P\left(\frac{\partial Q}{\partial z} - \frac{\partial R}{\partial y}\right) + Q\left(\frac{\partial R}{\partial x} - \frac{\partial P}{\partial z}\right) + R\left(\frac{\partial P}{\partial y} - \frac{\partial Q}{\partial x}\right) = 0 \tag{2}$$

Comparing Eqs. (1) and (2), we get simultaneous equations

$$\frac{dx}{\frac{\partial Q}{\partial z} - \frac{\partial R}{\partial y}} = \frac{dy}{\frac{\partial R}{\partial x} - \frac{\partial P}{\partial z}} = \frac{dz}{\frac{\partial P}{\partial y} - \frac{\partial Q}{\partial x}}$$

which are called auxiliary equations and can be solved like simultaneous equations.

Let $u = a$ and $v = b$ be its two integrals. Then, we find the value of $Adu + Bdv = 0$ and compare with Eq. (1) to find A and B.

Substituting these values of A and B in $Adu + Bdv = 0$ and solving, we get the required solution.

Note: This method fails if the equation is exact, that is,

when $$\frac{\partial P}{\partial y}=\frac{\partial Q}{\partial x};\frac{\partial Q}{\partial z}=\frac{\partial R}{\partial y};\frac{\partial R}{\partial x}=\frac{\partial P}{\partial z}$$

EXAMPLE 12.33

Solve $xz^3dx - zdy + 2ydz = 0$

Solution: Here, $P = xz^3$; $Q = -z$; $R = 2y$

$$\therefore\ P\left(\frac{\partial Q}{\partial z}-\frac{\partial R}{\partial y}\right)+Q\left(\frac{\partial R}{\partial x}-\frac{\partial P}{\partial z}\right)+R\left(\frac{\partial P}{\partial y}-\frac{\partial Q}{\partial x}\right) = xz^3(-1-2) - z(0-3xz^2) + 2y(0-0) = 0$$

Hence, the condition of integrability is satisfied. Now, the auxiliary equations are

$$\frac{dx}{-3}=\frac{dy}{-3xz^2}=\frac{dz}{0}$$

$$\Rightarrow\quad \frac{dx}{1}=\frac{dy}{xz^2}=\frac{dz}{0}$$

From the first and last members, we have

$$dz = 0 \Rightarrow z = u \text{ (say)}$$

From the first two members, we have

$$xz^2dx = dy \Rightarrow xz^2dx - dy = 0$$

Integrating, we have

$$\frac{x^2}{2}z^2 - y = \text{constant}$$

$$\Rightarrow\quad x^2z^2 - 2y = v \text{ (say)}$$

Hence $Adu + Bdv = 0$ gives

$$Adz + B(2xz^2dx - 2dy + 2x^2zdz) = 0$$

$$\Rightarrow\quad 2xz^2Bdx - 2Bdy + (A + 2x^2zB)dz = 0$$

Comparing the above equation with the given equation, we have

$$xz^3 = 2xz^2B;\quad -z = -2B;\quad 2y = A + 2x^2zB$$

$$\therefore\quad B = \frac{z}{2}=\frac{u}{2}$$

and $$A = 2y - 2x^2zB = 2y - x^2z^2 = -v$$

$\therefore$ $$Adu + Bdv = 0 \Rightarrow -vdu + \frac{u}{2}dv = 0$$

$\Rightarrow$ $$-\frac{2du}{u} + \frac{dv}{v} = 0$$

Integrating, we have

$$-2\log u + \log v = \log c$$

$\Rightarrow$ $$v = cu^2$$

$\Rightarrow$ $$x^2z^2 - 2y = cz^2$$

which is the required solution.

EXAMPLE 12.34

Solve $z(z-y)dx + (z+x)zdy + x(x+y)dz = 0$

Solution: Here, $P = z(z-y)$; $Q = (z+x)z$; $R = x(x+y)$

$$\therefore \quad P\left(\frac{\partial Q}{\partial z} - \frac{\partial R}{\partial y}\right) + Q\left(\frac{\partial R}{\partial x} - \frac{\partial P}{\partial z}\right) + R\left(\frac{\partial P}{\partial y} - \frac{\partial Q}{\partial x}\right)$$

$$= z(z-y)(2z + x - x) + (z+x)z\,(2x + y + 2z + y) + x(x+y)(-z-z) = 0$$

Hence, the condition of integrability is satisfied.
Hence, the auxiliary equations are

$$\frac{dx}{2z} = \frac{dy}{2x+2y+2z} = \frac{dz}{-2z}$$

$\Rightarrow$ $$\frac{dx}{z} = \frac{dy}{x+y+z} = \frac{dz}{-z}$$

From the first and last members

$$\frac{dx}{z} = \frac{dz}{-z} \Rightarrow dx + dz = 0$$

$\therefore$ $$x + z = u \text{ (say)}.$$

Again, we have

$$\frac{dx+dy}{x+y} = \frac{dz}{-z}$$

Integrating, we have

$$\log(x+y) + \log z = \log v$$

$\Rightarrow$ $$(x+y)z = v \text{ (say)}$$

Now, let the given equation be

$$
\begin{aligned}
& & Adu + Bdv &= 0 \\
&\Rightarrow & A(dx + dz) + B((x + y)dz + (dx + dy)z) &= 0 \\
&\Rightarrow & (A + zB)dx + Bzdy + (A + B(x + y))dz &= 0
\end{aligned}
$$

Comparing this with the given equation, we have

$$A + zB = z(z - y);\ Bz = (z + x)z;\ A + B(x + y) = x(x + y)$$

$$\therefore \quad B = x + z = u \text{ and } A + (x + y)(x - B) = -(x + y)z = -v$$

Hence, the given equation is

$$
\begin{aligned}
& & Adu + Bdv &= 0 \\
&\Rightarrow & -vdu + udv &= 0 \\
&\Rightarrow & \frac{-du}{u} + \frac{dv}{v} &= 0
\end{aligned}
$$

Integrating, we have

$$
\begin{aligned}
& & -\log u + \log v &= \log c \\
&\Rightarrow & v &= cu \\
&\Rightarrow & (x + y)z &= c(x + z)
\end{aligned}
$$

which is the required solution.

EXAMPLE 12.35

Solve $(2xz - yz)dx + (2yz - zx)dy - (x^2 - xy + y^2)dz = 0$.

Solution: Here $P = 2xz - yz;\ Q = 2yz - zx;\ R = x^2 + xy - y^2$

$$\therefore\ P\left(\frac{\partial Q}{\partial z} - \frac{\partial R}{\partial y}\right) + Q\left(\frac{\partial R}{\partial x} - \frac{\partial P}{\partial z}\right) + R\left(\frac{\partial P}{\partial y} - \frac{\partial Q}{\partial x}\right)$$

$$= (2xz - yz)(2y - x - x + 2y) + (2yz - zx)(-2x + y - 2x + y) - (x^2 - xy + y^2)(-z + z) = 0$$

Hence, the condition of integrability is satisfied.
Now, the auxiliary equations are

$$\frac{dx}{2(2y - x)} = \frac{dy}{2(y - 2x)} = \frac{dz}{0}$$

Taking the third member, we have

$$dz = 0 \quad \Rightarrow \quad z = u \text{ (say)}$$

Taking the first and second members, we have

$$(y - 2x)dx = (2y - x)dy$$

$$\Rightarrow \qquad ydx + xdy - 2xdx - 2ydy = 0$$

Integrating, we have

$$xy - x^2 - y^2 = v \text{ (say)}$$

Now, let the given equation be

$$Adu + Bdv = 0$$

$$\Rightarrow \qquad Adz + B(xdy + ydx - 2xdx - 2ydy) = 0$$

$$\Rightarrow \qquad B(y - 2x)dx + B(x - 2y)dy + Adz = 0$$

Comparing this with the given equation, we have

$$B(y - 2x) = z(2x - y);\ B(x - 2y) = z(2y - x) \quad \text{and} \quad A = -(x^2 - xy + y^2)$$

$$\Rightarrow \qquad B = z = -u \quad \text{and} \quad A = v$$

Hence, the given equation is

$$Adu + Bdv = 0$$

$$\Rightarrow \qquad vdu - udv = 0$$

$$\Rightarrow \qquad \frac{du}{u} - \frac{dv}{v} = 0$$

Integrating, we have

$$\log u - \log v = \log c$$

$$\Rightarrow \qquad u = cv$$

$$\Rightarrow \qquad z = c(xy - x^2 - y^2)$$

which is the required solution.

EXAMPLE 12.36

Solve $(y^2 + yz + z^2)dx + (z^2 + zx + x^2)dy + (x^2 + xy + y^2)dz = 0$.

Solution: Here, $P = y^2 + yz + z^2$; $Q = z^2 + zx + x^2$ and $R = x^2 + xy + y^2$

$$\therefore \quad P\left(\frac{\partial Q}{\partial z} - \frac{\partial R}{\partial y}\right) + Q\left(\frac{\partial R}{\partial x} - \frac{\partial P}{\partial z}\right) + R\left(\frac{\partial P}{\partial y} - \frac{\partial Q}{\partial x}\right)$$

$$= (y^2 + yz + z^2)(2z + x - x - 2y) + (z^2 + zx + x^2)(2x + y - y - 2z) + (x^2 + xy + y^2)(2y + z - z - 2x) = 0$$

Hence, the condition of integrability is satisfied. The auxiliary equations are

$$\frac{dx}{z - y} = \frac{dy}{x - z} = \frac{dz}{y - x}$$

$$\text{Each fraction} = \frac{dx + dy + dz}{0}$$

$$\therefore \qquad dx + dy + dz = 0$$

Integrating, we have

$$x + y + z = u \text{ (say).}$$

Again, using multipliers $z + y$, $x + z$, $y + x$

$$\text{Each fraction} = \frac{(z+y)dx + (x+z)dy + (y+x)dz}{0}$$

$$\Rightarrow \qquad (xdy + ydx) + (ydz + zdy) + (zdx + xdz) = 0$$

$$\Rightarrow \qquad d(xy) + d(yz) + d(zx) = 0$$

Integrating, we have

$$xy + yz + zx = v \text{ (say)}$$

The given equation is equal to

$$Adu + Bdv = 0$$

$$\Rightarrow \qquad A(dx + dy + dz) + B(ydx + xdy + ydz + zdy + zdx + xdz) = 0$$

$$\Rightarrow \qquad (A + B(y + z))dx + (A + B(x + z)dy + (A + B(y + x)) = 0$$

Comparing with the given equation, we have

$$A + B(y + z) = y^2 + yz + z^2 \qquad (1)$$

$$A + B(z + x) = z^2 + xz + x^2 \qquad (2)$$

$$A + B(x + y) = x^2 + xy + y^2 \qquad (3)$$

Subtracting Eq. (2) from Eq. (1), we get

$$B(y - x) = y^2 + z(y - x) - x^2$$

$$= (y - x)(x + y + z)$$

$$\therefore \qquad B = x + y + z = u$$

Therefore,

$$A = y^2 + yz + z^2 - B(y + z)$$

$$= y^2 + yz + z^2 - (x + y + z)(y + z)$$

$$= -(xy + xz + yz) = -v$$

Thus, the given equation becomes

$$Adu + Bdv = -Vdu + Udv = 0$$

$$\Rightarrow \qquad \frac{-du}{u} + \frac{dv}{v} = 0$$

Integrating, we have

$$-\log u + \log v = \log c$$

$$\Rightarrow \qquad v = cu$$

$$\Rightarrow \qquad xy + xz + yz = c(x + y + z)$$

which is the required solution.

EXAMPLE 12.37

Solve $(x^2y - y^3 - y^2z)dx + (xy^2 - x^2z - x^3)dy + (xy^2 + x^2y)dz = 0.$

Solution: Here $P = x^2y - y^3 - y^2$; $Q = xy^2 - x^2z - x^3$ and $R = xy^2 + x^2y$

$$\therefore\ P\left(\frac{\partial Q}{\partial z}-\frac{\partial R}{\partial y}\right)+Q\left(\frac{\partial R}{\partial x}-\frac{\partial P}{\partial z}\right)+R\left(\frac{\partial P}{\partial y}-\frac{\partial Q}{\partial x}\right)$$

$$= (x^2y - y^3 - y^2z)(-x^2 - 2xy - x^2) + (xy^2 - x^2z - x^3)$$
$$(y^2 + 2xy + y^2) + (xy^2 + x^2y)(x^2 - 3y^2 - 2yz - y^2 + 2xz + 3x^2) = 0$$

Hence, the condition of integrability is satisfied.
The auxiliary equations are

$$\frac{dx}{-x^2-2xy-x^2} = \frac{dy}{y^2+2xy+y^2} = \frac{dz}{x^2-3y^2-2yz-y^2} + 2xz + 3x^2$$

$$\Rightarrow \qquad \frac{dx}{-x(x+y)} = \frac{dy}{y(x+y)} = \frac{dz}{2x^2-2y^2-yz+xz}$$

Taking the first two members, we have

$$\frac{dx}{x}+\frac{dy}{y} = 0$$

Integrating, we have

$$\log x + \log y = \log u$$
$$\Rightarrow \qquad xy = u \text{ (say)}$$

Again, we have

$$\frac{dx+dy}{-(x+y)(x-y)} = \frac{dx+dy+dz}{x^2-y^2-yz+xz}$$

$$\Rightarrow \qquad -\frac{dx+dy}{x+y} = \frac{dx+dy+dz}{x+y+z}$$

Integrating, we have

$$\log(x+y) + \log(x+y+z) = \log v$$
$$\Rightarrow \qquad (x+y)(x+y+z) = v \text{ (say)}$$

Let the given equation can be written as

$$Adu + Bdv = 0$$
$$\Rightarrow \qquad A(ydx + xdy) + B[(x+y)(dx+dy+dz) + (x+y+z)(dx+dy)] = 0$$
$$\Rightarrow \qquad (Ay + B(2x+2y+z))dx + (Ax + B(2x+2y+dy) + B(x+y)dz = 0$$

Comparing this with the given equation, we have

$$Ay + B(2x + 2y + z) = x^2y - y^3 - y^2z$$

$$Ax + B(2x + 2y + z) = xy^2 - x^2z - x^3$$

and $$(x + y)B = xy^2 + x^2y \Rightarrow B = xy = u$$

and $$Ay + xy(2x + 2y + z) = x^2y - y^3 - y^2z$$

$$\Rightarrow \quad Ay = -x^2y - 2xy^2 - xyz - y^3 - y^2z$$

$$\Rightarrow \quad A = -x^2 - 2xy - xz - y^2 - yz$$

$$= -(x + y)^2 - (x + y)z$$

$$= -(x + y)(x + y + z) = -v$$

$$\therefore \quad Adu + Bdv = 0$$

$$\Rightarrow \quad -vdu + udv = 0 \Rightarrow \frac{du}{u} - \frac{dv}{v} = 0$$

Integrating, we have

$$\log u - \log v = -\log c$$

$$\Rightarrow \quad uc = v$$

$$\Rightarrow \quad cxy = (x + y)(x + y + z)$$

which is the required solution.

EXAMPLE 12.38

Solve $(y^2 + yz)dx + (xz + z^2)dy + (y^2 - xy)dz = 0$.

Solution: Here $P = y^2 + yz$; $Q = xz + z^2$ and $R = y^2 - xy$

$$\therefore \quad P\left(\frac{\partial Q}{\partial z} - \frac{\partial R}{\partial y}\right) + Q\left(\frac{\partial R}{\partial x} - \frac{\partial P}{\partial z}\right) + R\left(\frac{\partial P}{\partial y} - \frac{\partial Q}{\partial x}\right)$$

$$= (y^2 + yz)(x + 2z - 2y + x) + (xz + z^2)(-y - y) + (y^2 - xy)(2y + z - z) = 0$$

Hence, the condition of integrability is satisfied.

The auxiliary equations are

$$\frac{dx}{2(x - y + z)} = \frac{dy}{-2y} = \frac{dz}{2y}$$

$$\Rightarrow \quad \frac{dx}{x - y + z} = \frac{dy}{-y} = \frac{dz}{y}$$

Taking the last two members, we have

$$dy + dz = 0$$

Integrating, we have

$$y + z = u \text{ (say)}$$

Again, we have

$$\frac{dx + dz}{x + z} = \frac{dy}{-y}$$

Integrating, we have

$$\log(x + z) + \log y = \log v$$

Let the given equation be written as

$$Adu + Bdv = A(dy + dz) + B[(x + z)dy + y(dx + dz)] = 0$$

$$\Rightarrow \quad Bydx + [A + B(x + z)]dy + (A + By)dz = 0$$

Comparing this with the given equation, we have

$$By = y^2 + yz,\ A + B(x + z) = xz + z^2,\ A + By = y^2 - xy$$

$$\therefore \quad B = y + z = u \quad \text{and} \quad A = -By + y^2 - xy$$

$$= -(y + z)y + y^2 - xy$$

$$= -y(z + x) = -v$$

Hence, the given equation becomes

$$-vdu + udv = 0$$

$$\Rightarrow \quad \frac{du}{u} - \frac{dv}{v} = 0$$

Integrating, we have

$$\log u - \log v = -\log c$$

$$\Rightarrow \quad v = cu$$

$$\Rightarrow \quad y(z + x) = c(y + z)$$

which is the required solution. ▲▲

Method 4. Homogeneous equations

In case P, Q, R are homogeneous functions of x, y, z then one variable, say z, may be separated from the other two by the substitution

$$x = zu;\ y = zv$$

so that

$$dx = zdu + udz;\ dy = zdv + vdz$$

Thus, the given equation will be reduced to the from in which either the coefficient of dz is zero or not zero.

If the coefficient of dz is zero, we shall have an equation in only two variables u and v

If the coefficient of dz is not zero then we shall be able to separate z from u and v.

Therefore, in both the cases, the new equation may easily be integrated.

EXAMPLE 12.39

Solve $(x - y)dx - xdy + zdz = 0$.

Solution: The condition of integrability can be easily verified.

Here, P, Q, R are homogeneous functions

Let, $x = uz, y = vz$

so that $dx = zdu + udz, dy = zdv + vdz$

Substituting in the given equation, we have

$$(uz - vz)(zdu + udz) - uz(zdV + Vdz) + zdz = 0$$

$$\Rightarrow \quad z^2[(u - v)du - udv] + z[u(u - v) - uv + 1]dz = 0$$

$$\Rightarrow \quad \frac{(u-v)du - udv}{u^2 - 2uv + 1} + \frac{dz}{z} = 0$$

Integrating, we have

$$\frac{1}{2}\log(u^2 - 2uv + 1) + \log z = \frac{1}{2}\log c$$

$$\Rightarrow \quad (u^2 - 2av + 1)z^2 = c$$

$$\Rightarrow \quad u^2z^2 - 2uzvz + z^2 = c$$

$$\Rightarrow \quad x^2 - 2xy + z^2 = c$$

which is the required solution.

EXAMPLE 12.40

Solve $(z^2 - zy)dx + (z^2 + xz)dy + (x^2 + xy)dz = 0$.

Solution: The condition of integrability can easily be verified. Here P, Q, R are homogeneous functions.

Let $x = uz; y = vz$

So that $dx = zdu + udz; dy = zdv + vdz$

Substituting in the given equation, we have

$$z^2(1 - v)(zdu + udz) + z^2(1 + u)(zdv + vdz) + z^2(u^2 + uv)dz = 0$$

$$\Rightarrow \quad z[(1 - v)du + (1 + u)dv] + [u - uv + v + uv + u^2 + uv]dz = 0$$

$$\Rightarrow \quad z[(1 - v)du + (1 + u)dv] + (u + v)(u + 1)dz = 0$$

$$\Rightarrow \quad z[(1 + u)(du + dv) - (u + v)du] + (u + v)(u + 1)dz = 0$$

$$\Rightarrow \quad \frac{du + dv}{u + v} - \frac{du}{u + 1} + \frac{dz}{z} = 0$$

Integrating, we have

$$\log(u + v) - \log(u + 1) + \log z = \log c$$

$$\Rightarrow \quad (u + v)z = c(1 + u)$$

$$\Rightarrow \qquad (x+y) = c\left(1+\frac{x}{z}\right)$$

$$\Rightarrow \qquad z(x+y) = c(x+z)$$

which is the required solution.

EXAMPLE 12.41

Solve $(yz + z^2)dx - xz\,dy + xydz = 0$.

Solution: The condition of integrability can easily be verified. Here, P, Q, R are homogeneous functions.

Let $x = uz$, $y = vz$

So that $dx = zdu + udz$, $dy = zdv + vdz$

Substituting in the given equation, we have

$$(vz^2 + z^2)(zdu + udz) - uz^2(zdv + vdz) + uvz^2dz$$

$$\Rightarrow \qquad (v+1)(zdu + udz) - u(zdv + vdz + uvdz) = 0$$

$$\Rightarrow \qquad z[(v+1)du - udv] + [u(v+1) - uv + uv]dz = 0$$

$$\Rightarrow \qquad \frac{(v+1)du - udv}{u(v+1)} + \frac{dz}{z} = 0$$

$$\Rightarrow \qquad \frac{du}{u} - \frac{dv}{v+1} + \frac{dz}{z} = 0$$

Integrating, we have

$$\log u - \log(v+1) + \log z = \log c$$

$$\Rightarrow \qquad \frac{uz}{v+1} = c$$

$$\Rightarrow \qquad uz^2 = c(vz + z)$$

$$\Rightarrow \qquad xz = c(y+z)$$

which is the required solution.

EXAMPLE 12.42

Solve $z^2dx + (z^2 - 2yz)dy + (2y^2 - yz - xz)dz = 0$.

Solution: The condition of integrability can easily be verified. Here P, Q, R are homogeneous functions.

Let $x = uz$, $y = vz$

So that $dx = zdu + udz$, $dy = zdv + vdz$

Substituting in the given equation, we have

$$z^2(zdu + udz) + z^2(1-2v)(zdv + vdz) + (2v^2 - v - u)z^2dz = 0$$

$$\Rightarrow \qquad z^3du + z^3(1-2v)dv + [u + v(1-2v) + 2v^2 - v - u]z^2dz = 0$$

$$\Rightarrow \qquad z^3du + z^3(1 - 2v)dv = 0$$

$$\Rightarrow \qquad du + (1 - 2v)dv = 0$$

Integrating, we have

$$u + v - v^2 = c$$

$$\Rightarrow \qquad (x + y)z - y^2 = cz$$

which is the required solution.

EXAMPLE 12.43

Solve $(2xyz + y^2z + yz^2)dx + (x^2z + 2xyz + xz^2)dy + (x^2y + xy^2 + 2xyz)dz = 0$.

Solution: The given equation satisfies the integrability conditions

Here, P, Q, R are homogeneous functions of x, y, z.

Let $x = uz$, $y = vz$

so that $\qquad dx = zdu + udz,\ dy = zdv + vdz$

Substituting in the given equation, we have

$$z^3[(2uv + v^2 + v)(zdu + udz) + (u^2 + 2uv + u)(zdv + vdz) + (u^2v + uv^2 + 2uv)dz] = 0$$

$$\Rightarrow \quad z[(2uv + v^2 + v)du + (u^2 + 2uv + u)dv] + [(2uv + v^2 + v)u + (u^2 + 2uv + u)v + u^2v + uv^2 + 2uv]dz = 0$$

$$\Rightarrow \qquad (2uv + v^2 + v)du + (u^2 + 2uv + u)dv + 4(u^2v + uv^2 + uv)\frac{dz}{z} = 0$$

$$\Rightarrow \qquad \frac{(2uvdu + u^2dv) + (v^2du + 2uvdv) + (vdu + udv)}{u^2v + uv^2 + uv} + 4\frac{dz}{z} = 0$$

Integrating, we have

$$\log(u^2v + uv^2 + uv) + 4\log z = \log c$$

$$\Rightarrow \qquad z^4(u^2v + uv^2 + uv) = c$$

$$\Rightarrow \qquad xyz(x + y + z) = c$$

which is the required solution.

EXAMPLE 12.44

Solve $(y^2 + yz)dx + (xz + z^2)dy + (y^2 - xy)dz = 0$.

(a) by taking a variable constant,

(b) by homogeneous method, and

(c) by auxiliary equations.

Solution: Here, $P = y^2 + yz$, $Q = xz + z^2$, $R = y^2 - xy$

$$\therefore \quad P\left(\frac{\partial Q}{\partial z} - \frac{\partial R}{\partial y}\right) + Q\left(\frac{\partial R}{\partial x} - \frac{\partial P}{\partial z}\right) + R\left(\frac{\partial P}{\partial y} - \frac{\partial Q}{\partial x}\right)$$

$$= (y^2 + yz)(x + 2z - 2y + y) + (xz + z^2)(-y - y) + (y^2 - xy)(2y + z - z) = 0$$

Thus, the condition of integrability is satisfied.

(a) ***By taking a variable constant.*** Let z = constant so that $dz = 0$. Therefore, the given equation becomes

$$(y^2 + yz)dx + (xz + z^2)\,dy = 0$$

$$\Rightarrow \quad y(y + z)dx + z(x + z)\,dy = 0$$

$$\Rightarrow \quad \frac{dx}{x+z} + \frac{zdy}{y(y+z)} = 0$$

$$\Rightarrow \quad \frac{dx}{x+z} + \left(\frac{1}{y} - \frac{1}{y+z}\right)dy = 0$$

Integrating, we have

$$\log(x + z) + \log y - \log(y + z) = \text{constant.}$$

$$\Rightarrow \quad \frac{(x+z)y}{y+z} = \phi(z) \text{ (say)}$$

Now, differentiating this, we have

$$\frac{(y+z)\,[(dx+dz)y + (x+z)dy] - y(x+z)\,(dy+dz)}{(y+z)^2} = \phi'(z)dz$$

$$\Rightarrow \quad \frac{(y^2 + yz)dx + [(y+z)\,(x+z) - y(x+z)dy + [(y+z)y - y(x+z)]dz}{(y+z)^2} = \phi'(z)dz$$

$$\Rightarrow \quad \frac{(y^2 + yz)dx + (xz + z^2)dy + (y^2 - xy)dz}{(y+z)^2} = \phi'(z)dz$$

$$\Rightarrow \quad \phi'(z)dz = 0$$

$$\Rightarrow \quad \phi(z) = c \text{ (constant)}$$

$$\therefore \quad (x + z)y = c\,(y + z)$$

(b) ***Homogeneous equation***

Here P, Q, R are clearly homogeneous functions in x, y, z.

$\therefore$ Let $\quad x = uz,\ y = vz$

So that $\quad dx = zdu + udz,\ dy = zdv + vdz$

Substituting in the given equation, we have

$$z^2(v^2 + v)\,(zdu + udz) + z^2\,(u + 1)\,(zdv + vdz) + z^2(v^2 - uv)dz = 0$$

$$\Rightarrow \quad z[(v^2 + v)du + (u + 1)dv] + [u/v^2 + v) + v(u + 1) + v^2 - uv]dz = 0$$

$$\Rightarrow \quad z[(v^2 + v)du + (u + 1)dv] + [uv^2 + uv + v + v^2]dz = 0$$

$$\Rightarrow \quad z[(v^2 + v)du + (u + 1)dv] + [(v^2 + v)(u + 1)dz = 0$$

$$\Rightarrow \quad \frac{du}{u+1} + \frac{dv}{v^2+v} + \frac{dz}{z} = 0$$

$$\Rightarrow \quad \frac{du}{u+1} + \left(\frac{1}{v} - \frac{1}{v+1}\right)dv + \frac{dz}{z} = 0$$

Integrating, we have

$$\log(u + 1) + \log v - \log(v + 1) + \log z = \log c$$

$$\Rightarrow \quad \frac{(u+1)vz}{v+1} = c$$

$$\Rightarrow \quad (x + z)y = c(y + z).$$

(c) ***Forming auxiliary equations***

Comparing the given equation with the condition of integrability, we have

$$\frac{dx}{2(x-y+z)} = \frac{dy}{-2y} = \frac{dz}{2y}$$

$$\Rightarrow \quad \frac{dx}{x-y+z} = \frac{dy}{-y} = \frac{dz}{y}$$

Taking the last two members, we have

$$dy + dz = 0$$

$$\therefore \quad y + z = u \text{ (say)}$$

Again, we have

$$\frac{dx+dz}{x+z} = \frac{dy}{-y}$$

Integrating, we have

$$\log(x + z) + \log y = \log v \text{ (say)}$$

$$\Rightarrow \quad y(x + z) = v$$

Let the given equation be

$$Adu + Bdv = 0$$

$$\Rightarrow \quad A[dy + dz] + B(dy(x + z) + y(dx + dz)] = 0$$

$$\Rightarrow \quad Bydx + [A + B(x + z)]dy + (A + By)dz = 0$$

Comparing this with the given equations, we have

$$By = y^2 + yz,\ A + B(x + z) = xz + z^2,\ A + By = y^2 - xy$$

$$\therefore \quad B = y + z = u$$

and $$A = -By + y^2 - xy$$
$$= -(y+z)y + y^2 - xy$$
$$= -y(z+x) = -v$$

Hence, the given equation is

$$-v\,du + u\,dv = 0$$

$$\Rightarrow \quad \frac{du}{u} - \frac{dv}{v} = 0$$

Integrating, we have

$$\log u - \log v = -\log c$$

$$\Rightarrow \quad \frac{v}{u} = c$$

$$\Rightarrow \quad y(x+z) = c(x+z)$$

The non-integrable single equation.

Suppose that the condition of integrability is not satisfied by the equation

$$P\,dx + Q\,dy + R\,dz = 0 \tag{1}$$

Then Eq. (1) represents a family of curves orthogonal to the family represented by the equations

$$\frac{dx}{P} = \frac{dy}{Q} = \frac{dz}{R} \tag{2}$$

There exists no single relation between x, y, and z. However, we can find an infinite number of curves that lie on any given surface and satisfy Eq. (1). The following examples will illustrate the process.

EXAMPLE 12.45

Find the curves represented by the solution of $y\,dx + (z-y)\,dy + x\,dz = 0$, which lie on the plane $2x - y - z = 1$

Solution: Here $P = y$, $Q = z - y$, $R = x$

$$\therefore\ P\left(\frac{\partial Q}{\partial z} - \frac{\partial R}{\partial y}\right) + Q\left(\frac{\partial R}{\partial x} - \frac{\partial P}{\partial z}\right) + R\left(\frac{\partial P}{\partial y} - \frac{\partial Q}{\partial x}\right)$$
$$= y(1-0) + (z-y)(1-0) + x(1-0) = z + x \neq 0$$

This shows that the condition of integrability is not satisfied by the given differential equation.

Differentiating $2x - y + z = 1$, we have

$$2dx - dy - dz = 0$$

$$\Rightarrow \quad dz = 2dx - dy$$

Substituting dz in the given equation, we have

$$y\,dx + (z - y)\,dy + x(2dx - dy) = 0$$

$$\Rightarrow \quad (y + 2x)\,dx + (z - y - x)\,dy = 0$$

$$\Rightarrow \quad (y + 2x)\,dx + (2x - y - 1 - y - x)\,dy = 0$$

$$\Rightarrow \quad (y + 2x)\,dx + (2x - 2y - 1)\,dy = 0$$

$$\Rightarrow \quad (y\,dx + x\,dy) + 2x\,dx - 2y\,dy - dy = 0$$

Integrating, we have

$$xy + x^2 - y^2 - y = c$$

EXAMPLE 12.46

Show that the following equation is not integrable

$$3y\,dx + (z - 3y)\,dy + x\,dz = 0$$

Prove that the projection on the plane xy of the curves that satisfy the equation and lie in the plane $2x + y - z = a$ are the rectangular hyperbola: $x^2 + 3xy - y^2 + ay = c$

Solution: Here, $P = 3y$, $Q = z - 3y$, $R = x$

$$\therefore \quad P\left(\frac{\partial Q}{\partial z} - \frac{\partial R}{\partial y}\right) + Q\left(\frac{\partial R}{\partial x} - \frac{\partial P}{\partial z}\right) + R\left(\frac{\partial P}{\partial y} - \frac{\partial Q}{\partial x}\right)$$

$$= 3y(1 - 0) + (z - 3y)\,(1 - 0) + x(3 - 0) = 3x + z \neq 0$$

Thus, the condition of integrability is not satisfied. Therefore, the given equation is not integrable.

Now differentiating $2x + y - z = a$, we have

$$2dx + dy - dz = 0$$

Substituting dz in the given differential equation, we have

$$3y\,dx + (z - 3y)\,dy + x(2dx + dy) = 0$$

$$\Rightarrow \quad (3y + 2x)\,dx + (z - 3y + x)\,dy = 0$$

$$\Rightarrow \quad (3y + 2x)\,dx + (2x + y - a - 3y + x)\,dy = 0$$

$$\Rightarrow \quad (3y + 2x)\,dx + (3x - 2y - a)\,dy = 0$$

$$\Rightarrow \quad 2x\,dx + 3(y\,dx + x\,dy) - 2y\,dy - a\,dy = 0$$

Integrating, we have

$$x^2 + 3xy - y^2 - dy = c$$

which is the required result. ▲▲

EXERCISE 12.3

Solve the following total differential equations.

1. $(k - z)(ydx + xdy) + xydz = 0.$
2. $(yz + 2x)dx + (zx + 2z)dy + (xy - 2y)dz = 0.$
3. $yz\log zdx - zx\log zdy + xydz = 0.$
4. $yzdx + 2zxdy - 3xydz = 0.$
5. $(y^2 + z^2 - x^2)dx - 2xydy - 2xzdz = 0.$
6. $3x^2dx + 3y^2dy - (x^3 + y^3 + e^{2z})dz = 0.$
7. $2yzdx + zxdy - xy(1 + z)dz = 0.$
8. $(yz + xyz)dx + (zx + xyz)dy + (xy + xyz)dz = 0.$
9. $2yzdx + zxdy - xy(1 + z)dz = 0.$
10. $2zxdx + zdy - dz = 0.$
11. $\dfrac{yzdx}{x^2+y^2} - \dfrac{zxdy}{x^2+y^2} - \tan^{-1}\left(\dfrac{y}{x}\right)dz = 0.$
12. $xdx - ydy - \sqrt{a^2 - x^2 - y^2}\,dz = 0.$
13. $\dfrac{y+z-2x}{(y-x)(z-x)}dx + \dfrac{z+x-2y}{(z-y)(x-y)}dy + \dfrac{x+y-2z}{(x-z)(y-z)}dz.$
14. $(z + z^3)\cos x\,dx - (z + z^3)dy + (1 - z)^3(y - \sin x)dz.$
15. $(e^xy + e^z)dx + (e^yz + e^x)dy + (e^y - e^xy - e^yz)dz = 0.$
16. $3x^2dx + 3y^2dy - (x^3 + y^3 + e^{2z})dz = 0.$
17. $y^2z(x\cos x - \sin x)dx + x^2z(y\cos y - \sin y)dy + xy(y\sin x + x\sin y + xy\cos z)dz = 0.$
18. $xz^3dx - zdy + 2ydz = 0.$
19. $z(z - y)dx + (z + x)zdy + x(x + y)dy = 0.$
20. $(x^2 - y^2 - z^2 + 2xy + 2xz)dx + (y^2 - z^2 - x^2 + 2yz + 2yx)dy + (z^2 - x^2 - y^2 + 2zx + 2yz)$ $dz = 0.$
21. $(mz - ny)dx + (nx - lz)dy + (ly - mx)dz = 0.$
22. $(2xz - yz)dx + (2yz - xz)dy - (x^2 - xy + y^2)dz = 0.$
23. $yz(y + z)dx + zx(x + z)dy + xy(x + y)dy = 0.$
24. $2(y + z)dx - (x + z)dy + (2y - x + z)dz = 0.$
25. $2(yz + 3x)dx - (zx - 3z)dy + (xy - 3y)dz = 0.$
26. $(y + z)dx + dy + dz = 0.$

27. $y^2dx - zdy + ydz = 0.$

28. $(2x^3 - z)zdx + 2x^2yzdy + x(z + x)dz = 0.$

29. $zdz + (x - a)dx = \sqrt{b^2 - z^2 - (x-a)}dy.$

30. $(2x^2y + 2xy^2 + 2xyz + 1)dx + (x^3 + x^2y + x^2z + 2xyz + 2y^2z - 2yz^2 + 1)dy + (xy^2 + y^3 + y^2z + 1)dz = 0.$

31. $\dfrac{dx}{x+a} + \dfrac{dy}{y+b} + \dfrac{dz}{z+c} = 0.$

32. $(z + z^3)(\cos x dx - dy) + (1 - z^3)(y - \sin x)dz = 0.$

33. $(x - 3y - z)dx - (2y - 3x)dy + (z - x)dz = 0.$

34. $z^2dx + (z^2 - 2yz)dy + (2y^2 - yz - xz)dz = 0.$

35. Check the integrability of the equation

$$(x^2y - y^3 - y^2z)dx + (xy^2 - x^2z - x^3)dy + (xy^2 + x^2y)dz = 0.$$

and solve it (a) by the method of inspection, (b) by treating one variable constant, (c) by homogeneous method and (d) by using auxiliary equations.

36. Verify the equation $(y - z)(y + z - 2x)dx + (z - x)(z + x - 2y)dy + (x - y)(x + y - 2z)dz = 0$ is exact and find the solution.

37. Find $f(z)$, such that

$$\frac{y^2 + z^2 - x^2}{zx}dx - ydy + f(z)dz = 0 \text{ is integrable. Hence solve it.}$$

38. Find $f(y)$ if $f(y)dx - zxdy - xy\log y dz = 0$ is integrable. Find the corresponding integral.

39. Verify that the equation

$$x(y^2 - a^2)dx + y(x^2 - z^2)dy - z(y^2 - a^2)dz = 0 \text{ is integrable and solve it.}$$

40. Verify that the equation

$$z(z + y^2)dx + z(z + x^2)dy - xy(x + y)dz = 0 \text{ is integrable and find its primitive.}$$

41. Obtain the system of the curves lying on the system of $zx = K$ and satisfying the differential equation

$$yzdx + z^2dy + y(z + x)dz = 0.$$

42. Show that there is no single integral of

$$dz = 2ydx + xdy.$$

Prove that the curves of this equation that lie in the plane $z = x + y$ lie also on the surface of the family

$$(x - 1)^2(2y - 1) = c.$$

43. Find the orthogonal projection on the plane of xz of curves which lie on the paraboloid $3z = x^2 + y^2$ and satisfy the equation $2dz = (x + z)\,dx + ydy$.
44. Show that the curves of

$$xdx + ydy + c\sqrt{1-\frac{x^2}{a^2}-\frac{y^2}{b^2}}\,dz = 0.$$

that lie on the ellipsoid

$$\frac{x^2}{a^2}+\frac{y^2}{b^2}+\frac{z^2}{c^2} = 1.$$

lie also on the family of concentric spheres.
45. Solve $dz = mydx + ndy$.
46. Find the most general solution of the equation

$$ydx + (z - y)\,dy + xdz = 0.$$

which is consistent with the relation $2x - y - z = 1$.

ANSWERS

Exercise 12.3

1. $xy = c\,(k - z)$.
2. $xyz + x^2 - 2yz = c$.
3. $x \log z = cy$.
4. $xy^2 = cz^3$.
5. $x^2 + y^2 + z^2 = cx$.
6. $x^3 + y^3 = e^{2z} + ce^z$.
7. $x^2y = cze^z$.
8. $\log xyz + x + y + z = c$.
9. $\frac{y}{x} - z^2 = c$.
10. $x^2 + y - \log z = c$.
11. $y = x + \tan\,(c/z)$.
12. $x^2 + y^2 + (z + c)^2 = a^2$.
13. $(x - y)\,(y - z)\,(z - x) = c$.
14. $(y - \sin x) = cze^{-z}$.
15. $e^xy + e^yz + e^zx = ce^z$.
16. $x^3 + y^3 = e^{2z} + ce^z$.

17. $z\left(\frac{\sin x}{x}+\frac{\sin y}{y}\right)=c-\sin z.$

18. $x^2z^2 - 2y = cz^2.$

19. $(x + y)z = c(x + z).$

20. $(x^2 + y^2 + z^2) = c(x + y + z).$

21. $nx - lz = c(mz - nx).$

22. $x^2 + y^2 - xy = cz.$

23. $xyz = c(x + y + z).$

24. $(x + z)^2 = c(y + z).$

25. $xyz + x^2 - 3xyz = c.$

26. $y + z = ce^{-x}.$

27. $yx + z = yc.$

28. $x(x^2 + y^2 + \log x) + z = cx.$

29. $b^2 - z^2 - (x - a)^2 = (c - y)^2.$

30. $x^2y + y^2z + \log(x + y + z) = c.$

31. $(x + a)(y + b)(z + c) = k.$

32. $(\sin x - y)(x^2 + 1) = cz.$

33. $x^2 + 2y^2 - 6xy - 2xz + z^2 = c.$

34. $xz + yz - y^2 = cz^2.$

35. $x^2 + y^2 + z(x + y) = cxy.$

36. $x(y^2 - z^2) + y(z^2 - x^2) + z(x^2 - y^2) = c.$

37. $f(z) = -z;\ x^2 + y^2 + z^2 = xc.$

38. $f(y) = cy;\ xc = ky^z.$

39. $(x^2 - z^2)(y^2 - a^2) = c.$

40. $x(y^2 + z) = c(x + y)(1 - cy).$

41. $x = c_1y$ and $xz = c_2.$

42. $z = ce^{2x}.$

43. $x^2 + y^2 + z^2 = c^2.$

44. $z = m\int f(x)dx + nf(x) + c$ where $y = f(x).$

45. $xy + x^2 - y^2 - y = c.$

13

CHAPTER

Series Solution of Differential Equations

In the previous chapters, we have learnt that certain types of higher-order linear differential equations with constant coefficients and a special case of a variable coefficient differential equation (Euler–Cauchy type equation) have solutions that can be expressed as finite linear combination of known elementary functions. In general, however, higher order linear equations have no solutions that can be expressed in such a simple manner. Thus, we must seek other means of expression for the solutions of these equations. One such means of expression is furnished by infinite series representations and the present chapter is devoted to the methods of obtaining solutions in infinite series form. The series solution methods can be classified into two categories, power series method and general series solution method that is Frobenius method.

Ordinary and Singular Points of an Equation. Consider the variable coefficient second order homogeneous linear differential equation.

$$a_0(x)\frac{d^2y}{dx^2} + a_1(x)\frac{dy}{dx} + a_2(x)y = 0;\ a_0(x) \neq 0 \tag{1}$$

Dividing by $a_0(x)$, we can write it in the standard form, normal form or canonical form as

$$\frac{d^2y}{dx^2} + p(x)\frac{dy}{dx} + q(x)y = 0 \tag{2}$$

where, $p(x) = a_1(x)/a_0(x)$ and $q(x) = a_2(x)/a_0(x)$.

We define the following.

Definition: A function f is said to be analytic at x_0 if its Taylor series about x_0, that is, $\sum_{n=0}^{\infty}\frac{f^{(n)}(x_0)}{n!}(x-x_0)^n$ exists and converges to $f(x)$ for all x in same open interval I including x_0.

Definition: The point x_0 is called an ordinary point of the differential Eq. (1) if $a_0(x_0) \neq 0$. In other words, if at x_0 both of $p(x)$ and $q(x)$ in the equivalent normalized equation (2) are analytic then x_0 is said to be an ordinary point. Ordinary point is also called a regular point of the differential equation.

Definition: A point x_0 is called a singular point of the differential Eq. (1) if $a_0(x_0) = 0$. We can also say that if either one or both of $p(x)$ and $q(x)$ in the equivalent normalized Eq. (2) are not analytic at x_0, then x_0 is said to be a singular point of the differential Eq. (1).

We can further classify singular points as given below.

Definition: A singular point x_0 is said to be a regular singular point if the functions defined by the products $(x - x_0)\, p(x)$ and $(x - x_0)^2\, q(x)$ are both analytic at x_0.

Definition: A singular point x_0 is said to be an irregular singular point if either or both of the functions $(x - x_0)\, p(x)$ and $(x - x_0)^2\, q(x)$ are not analytic at x_0. ▲▲

EXAMPLE 13.1

Find the ordinary and singular points for $x^2y'' + xy' + (x^2 - n^2)y = 0$ (where n is a constant, x is any real number). Also classify the singular points.

Solution: We can write the given differential equation as $y'' + \frac{1}{x}y' + \left(\frac{x^2 - n^2}{x^2}\right)y = 0$ (1)

From the given equation on comparing with

$$a_0y'' + a_1y' + a_2y = 0 \text{ we obtain}$$

$$a_0 = x^2$$

Now, $$a_0 = 0 \Rightarrow x = 0$$

So, $x = 0$ is a singular point and $\mathbb{R} - \{0\}$ are ordinary points.

On comparing Eq. (1) with the canonical form $y'' + p(x)y' + q(x)y = 0$ we get

$$p(x) = \frac{1}{x} \quad q(x) = \frac{x^2 - n^2}{x^2}$$

Since $$\operatorname*{Lt}_{x\to 0} (x - 0)\, p(x) = \operatorname*{Lt}_{x\to 0} x \cdot \frac{1}{x} = 1$$

and $$\operatorname*{Lt}_{x\to 0} (x - 0)^2\, q(x) = \operatorname*{Lt}_{x\to 0} x^2\left(\frac{x^2 - n^2}{x^2}\right) = -n^2 \text{ exist.}$$

Therefore, the expressions $(x - 0)\, p(x)$ and $(x - 0)^2\, q(x)$ are analytic at $x = 0$.

Hence, $x = 0$ is a regular singular point of the given equation.

EXAMPLE 13.2

Classify the singular points of the equation $(1 - x^2)y'' - 2xy' + n(n + 1)y = 0$, n a constant.

Solution: We have $$a_0 = 1 - x^2$$

Now, $$a_0 = 0 \Rightarrow x = \pm 1$$

So, 1, –1 are singular points and $R - \{1, -1\}$ are ordinary points.

Also on comparing with the canonical form we obtain

$$p(x) = \frac{-2x}{1-x^2} \text{ and } q(x) = \frac{n(n+1)}{1-x^2}.$$

For $x = 1$, since $\underset{x\to 1}{\text{Lt}} (x - 1)p(x) = \underset{x\to 1}{\text{Lt}} (x-1)\frac{-2x}{1-x^2} = 1$

and $$\underset{x\to 1}{\text{Lt}} (x - 1)^2 q(x) = \underset{x\to 1}{\text{Lt}} (x - 1)^2 \frac{n(n+1)}{1-x^2} = 0 \text{ exist.}$$

Therefore, the expressions $(x - 1)\, p(x)$ and $(x - 1)^2\, q(x)$ are analytic at $x = 1$. Hence, $x = 1$ is a regular singular point of the given equation. Similarly, we can see $\underset{x\to 1}{\text{Lt}} (x + 1)\, p(x) = 1$ and $\underset{x\to -1}{\text{Lt}} (x + 1)^2\, q(x) = 0$ exist. Therefore, the expressions $(x + 1)\, p(x)$ and $(x + 1)^2\, q(x)$ are analytic at $x = -1$.

Hence, $x - -1$ is also a regular singular point of the given equation.

EXAMPLE 13.3

Classify the singular points of the equation

$$x^3y'' + 3xy' + 6y = 0$$

Solution: We have $$a_0 = x^3$$

Since, $$a_0 = 0 \Rightarrow x = 0$$

Therefore, $x = 0$ is singular point and $R - \{0\}$ are ordinary points.

Also on comparing with the general canonical form, we obtain $p(x) = \frac{3x}{x^3} = \frac{3}{x^2}$ and $q(x) = \frac{6}{x^3}$.

Since $\underset{x\to 0}{\text{Lt}} (x - 0)\, p(x) = \underset{x\to 0}{\text{Lt}}\, x \cdot \frac{3}{x^2} = \underset{x\to 0}{\text{Lt}} \frac{3}{x}$ and $\underset{x\to 0}{\text{Lt}} (x-0)^2 q(x) = \underset{x\to 0}{\text{Lt}}\, x^2 \cdot \frac{6}{x^3} = \underset{x\to 0}{\text{Lt}} \frac{6}{x}$ do not exist.

Therefore, the expressions $(x - 0)\, p(x)$ and $(x - 0)^2\, q(x)$ are not analytic at $x = 0$. Hence, $x = 0$ is an irregular singular point of the given equation.

Power Series Solution method about a given point

Consider the second order variable coefficient homogeneous linear differential equation

$$a_0(x)y'' + a_1(x)y' + a_2(x)y = 0 \tag{1}$$

The power series solution method about the given point x_0 involves the following steps:

1. Given point x_0 should be an ordinary/regular point.
2. Assume its solution to be of the form $y = \sum_{m=0}^{\infty} c_m(x - x_0)^m$
3. Calculate y', y'' and substitute these values in Eq. (1). Equate the coefficients of x^m to get the corresponding value of c_m.
4. Substituting these values of c_m's in $y = \sum_{m=0}^{\infty} c_m (x - x_0)^m$, we get the required power series solution with two arbitrary constants.

EXAMPLE 13.4

Find the power series solution about the origin of the following equation

$$(1 - x^2)y'' - 2xy' + 2y = 0.$$

Solution: Here the given point of origin is an ordinary point as

$$a_0 = 1 - x^2 \neq 0 \text{ at } x = 0.$$

Assume the solution of the given equation to be $y = \sum_{m=0}^{\infty} c_m(x - 0)^m = c_0 + c_1x + c_2x^2 + \ldots$

Then $\qquad y' = c_1 + 2c_2 x + 3c_3x^2 + \ldots$

and $\qquad y'' = 2c_2 + 6c_3x + 12c_4x^2 + \ldots$

Substituting these values in the given equation, we get

$(1 - x^2)(2c_2 + 6c_3x + 12c_4x^2 + \ldots) - 2x(c_1 + 2c_2x + 3c_3x^2 + \ldots) + 2(c_0 + c_1x + c_2x^2 + \ldots) = 0$

Equating to zero the coefficients of various powers of x, we get

$$c_2 = -c_0, \quad c_3 = 0, \quad c_4 = c_2/3 = -c_0/3, \quad c_5 = c_3/2 = 0, \ldots$$

Therefore, the solution becomes

$$y = c_0 + c_1x - c_0x^2 - \frac{c_0}{3}x^4 + \ldots$$

$$= c_0\left[1 - x^2 - \frac{x^4}{3} + \ldots\right] + c_1x.$$

where, c_0, c, are arbitrary constants.

EXAMPLE 13.5

Find the power series solution about $x = 2$ of the initial value problem

$$4y'' - 4y' + y = 0;\ y(2) = 0,\ y'(2) = 1/e.$$

Express the solution in the closed form.

Solution: As we know power series of a function about a given point is unique, we will use the Taylor series to determine the solution. Thus, we have the solution as

$$y(x) = y(2) + (x-2)\,y'(2) + \frac{(x-2)^2}{2!}\,y''(2) + \ldots$$

From the given equation, we have

$$y'' = \frac{1}{4}(4y' - y)$$

Differentiating $(m-2)$ times, we get

$$y^{(m)} = \frac{1}{4}(4y^{(m-1)} - y^{(m-2)})$$

Substituting $x = 2$, we obtain

$$y^{(m)}(2) = \frac{1}{4}(4y^{(m-1)}(2) - y^{(m-2)}(2));\ m = 1, 2, 3, \ldots$$

Using the values $y(2) = 0$ and $y'(2) = 1/e$, we obtain

$$y''(2) = \frac{1}{4}(4y'(2) - y(2)) = \frac{1}{4}\left(\frac{4}{e} - 0\right) = \frac{1}{e}$$

$$y'''(2) = \frac{1}{4}(4y''(2) - y'(2)) = \frac{1}{4}\left(\frac{4}{e} - \frac{1}{e}\right) = \frac{3}{4e}$$

$$y^{\text{iv}}(2) = \frac{1}{4}(4y'''(2) - y'''(2)) = \frac{1}{4}\left(4 - \frac{3}{4e} - \frac{1}{e}\right) = \frac{1}{2e}$$

$$y^{\text{v}}(2) = \frac{1}{4}(4y^{\text{iv}}(2) - y'''(2)) = \frac{1}{4}\left(4 - \frac{1}{2e} - \frac{3}{4e}\right) = \frac{5}{16e}, \text{ and so on.}$$

By substituting these values, the solution is given by

$$y(x) = 0 + (x-2)\left(\frac{1}{e}\right) + \frac{(x-2)^2}{2!}\left(\frac{1}{e}\right) + \frac{(x-2)^3}{3!}\left(\frac{3}{4e}\right) + \frac{(x-2)^4}{4!}\left(\frac{1}{2e}\right) + \ldots$$

$$= \frac{(x-2)}{e}\left\{1 + \frac{((n-2)/2)}{1!} + \frac{((x-2)/2)^2}{2!} + \frac{((x-2)/2)^3}{3!} + \ldots\right\}$$

$$= \frac{1}{e}(x-2)e^{(x-2)/2}.$$

EXAMPLE 13.6

Solve in series

$$(1 - x^2)y'' - xy' + 4y = 0.$$

Solution: Here $x = 0$ is an ordinary point since $a_0 = 1 - x^2 \neq 0$ at $x = 0$.

Write the power series as

$$y = \sum_{m=0}^{\infty} c_m(x - 0)^m = \sum_{m=0}^{\infty} c_m x^m.$$

Then, we have

$$y' = \sum_{m=1}^{\infty} mc_m x^{m-1} \text{ and } y'' = \sum_{m=2}^{\infty} m(m-1)c_m x^{m-2}.$$

Substituting the expressions for y, y', y'' in the given differential equation, we get

$$c_{m+2} = \frac{m^2 - 4}{(m+2)(m+1)} c_m;\ m \geq 2.$$

where, c_0 and c_1 are arbitrary constants.

We find that

$$c_4 = 0,\ c_5 = \frac{c_3}{4} = -\frac{c_1}{8},\ c_6 = 0,\ c_7 = \frac{3}{6}c_5 = +\frac{3}{6}\frac{c_1}{8},\ c_8 = 0, \text{ and so on.}$$

The power series solution is

$$y = c_0 + c_1 x - 2c_0 x^2 - \frac{c_1}{2}x^3 + 0 \cdot x^4 - \frac{c_1}{8}x^5 + 0 \cdot x^6 + \frac{3}{6}\frac{c_1}{8}x^7 + 0 \cdot x^8 + \ldots$$

$$= c_0(1 - 2x^2) + c_1 x\left(1 - \frac{x^2}{2} - \frac{x^4}{8} + \frac{3}{6}\frac{x^6}{8} - \ldots\right).$$

EXAMPLE 13.7

Find the power series solution about the point $x_0 = 2$ of the equation $y'' + (x-1)y' + y = 0$.

Solution: Write the power series as $y = \sum_{m=0}^{\infty} c_m(x-2)^m$.

Let $x - 2 = u \Rightarrow x = u + 2$

Thus, the given equation becomes

$$\frac{d^2y}{du^2} + (u+1)\frac{dy}{du} + y = 0 \tag{1}$$

and

$$y = \sum_{m=0}^{\infty} c_m u^m.$$

Substituting y, y', y'' in Eq. (1), we obtain

$$\sum_{m=2}^{\infty} m(m-1)c_m u^{m-2} + (u+1)\sum_{m=1}^{\infty} mc_m u^{m-1} + \sum_{m=0}^{\infty} c_m u^m = 0.$$

$$(1-x^2)\sum_{m=2}^{\infty} m(m-1)c_m x^{m-2} - x\sum_{m=1}^{\infty} mc_m x^{m-1} + 4\sum_{m=0}^{\infty} c_m x^m = 0.$$

$$\Rightarrow \quad \sum_{m=2}^{\infty} m(m-1)c_m x^{m-2} - \sum_{m=2}^{\infty} m(m-1)c_m x^m - \sum_{m=1}^{\infty} mc_m x^m + 4\sum_{m=0}^{\infty} c_m x^m = 0.$$

$$\Rightarrow \quad 2c_2 + 6c_3x + \sum_{m=4}^{\infty} m(m-1)c_m x^{m-2} - \sum_{m=2}^{\infty} m(m-1)c_m x^m - c_1x - \sum_{m=2}^{\infty} mc_m x^m + 4c_0$$
$$+ 4c_1x + 4\sum_{m=2}^{\infty} c_m x^m = 0.$$

For making power of x same in each term, in the third term change the index of summation from m to $m+2$.

$$2c_2 + 6c_3x + \sum_{m+2=4}^{\infty} (m+2)(\overline{m+2}-1)cx_{m+2}^{\overline{m+2}-2} - \sum_{m=2}^{\infty} m(m-1)c_m x^m - c_1x - \sum_{m=2}^{\infty} mc_m x^m$$
$$+ 4c_0 + 4c_1x + 4\sum_{m=2}^{\infty} c_m x^m = 0$$

$$\Rightarrow (2c_2 + 4c_0) + (6c_3 - c_1 + 4c_1)x + \sum_{m=2}^{\infty} [(m+2)(m+1)c_{m+2} - (m(m-1) + m - 4)c_m]x^m = 0$$

Setting the coefficients of successive powers of x to zero, we obtain

$$2c_2 + 4c_0 = 0,\ 6c_3 + 3c_1 = 0$$

$$(m+2)(m+1)c_{m+2} - (m(m-1) + m - 4)c_m = 0$$

$$\Rightarrow \quad c_2 = -2c_0,\ c_3 = -c_1/2,$$

$$\Rightarrow \quad \sum_{m=2}^{\infty} m(m-1)c_m u^{m-2} + \sum_{m=1}^{\infty} mc_m u^m + \sum_{m=1}^{\infty} mc_m u^{m+1} + \sum_{m=0}^{\infty} c_m u^m = 0$$

$$\Rightarrow \quad 2c_2 + \sum_{m=3}^{\infty} m(m-1)c_m u^{m-2} + \sum_{m=1}^{\infty} mc_m u^m + c_1 + \sum_{m=2}^{\infty} mc_m u^{m-1} + c_0 + \sum_{m=1}^{\infty} c_m u^{m-1} = 0$$

For making power of u same in each summation term, we change the summation index from m to $m+2$ in the first summation term and from m to $m+1$ in the third summation term.

$$2c_2 + \sum_{m+2=3}^{\infty} (m+2)(\overline{m+2}-1)c_{m+2}u^{\overline{m+2}-2} + \sum_{m=1}^{\infty} mc_m u^m + c_1 + \sum_{m+1=2}^{\infty} (m+2)c_{m+1}u^{\overline{m+1}-1} + c_0$$
$$+ \sum_{m=1}^{\infty} c_m u^m = 0$$

$$\Rightarrow 2c_2 + \sum_{m=1}^{\infty} (m+2)(m+1)c_{m+2}u^m + \sum_{m=1}^{\infty} mc_m u^m + c_1 + \sum_{m=1}^{\infty} (m+1)c_{m+1}u^m + c_0 + \sum_{m=1}^{\infty} c_m u^m = 0$$

$$\Rightarrow (2c_2 + c_1 + c_0) + \sum_{m=1}^{\infty} [(m+2)(m+1)c_{m+2} + mc_m + (m+1)c_{m+1} + c_m]u^m = 0.$$

Setting the coefficients of successive power of x to zero, we obtain

$$2c_2 + c_1 + c_0 = 0,\ c_{m+2} = -\frac{(m+1)(c_{m+1}+c_m)}{(m+2)(m+1)}$$

$$= -\frac{c_{m+1}+c_m}{m+2};\ m \geq 1.$$

where, c_0 and c_1 are arbitrary constants.

We obtain

$$c_2 = -\frac{1}{2}(c_0 + c_1),\ c_3 = -\frac{1}{3}(c_2 + c_1)$$

$$= -\frac{1}{3}\left\{-\frac{1}{2}(c_0 + c_1) + c_1\right\} = -\frac{1}{6}(-c_0 + c_1) \text{ and so on.}$$

The power series solution is

$$y = c_0 + c_1 u + \left\{-\frac{1}{2}(c_0 + c_1)\right\}u^2 + \left\{-\frac{1}{6}(-c_0 + c_1)\right\}u^3 + \ldots$$

$$= c_0\left\{1 - \frac{1}{2}(x+2)^2 + \frac{1}{6}(x-2)^3 + \ldots\right\} + c_1\left\{(x-2) - \frac{1}{2}(x-2)^2 - \frac{1}{6}(x-2)^3 + \ldots\right\}.$$

EXAMPLE 13.8

Using the power series method, solve the following initial value problem about the point $x_0 = 2$.

$$(1 + x)y'' - xy' - y = 0;\ y(2) = 1,\ y'(2) = 0.$$

Solution: Write the power series as $y = \sum_{m=0}^{\infty} c_m (x-2)^m$

Let $x - 2 = u \Rightarrow x = u + 2.$

Thus, the given equation becomes

$$(u+3)\frac{d^2y}{du^2}-(u+2)\frac{dy}{du}-y=0;\text{ and } y=\sum_{m=0}^{\infty}c_m u^m.$$

$$y(0)=1,\quad y'(0)=0 \tag{1}$$

Substituting y, y', y'' in Eq. (1), we obtain

$$(u+3)\sum_{m=2}^{\infty}m(m-1)c_m u^{m-2}-(u+2)\sum_{m=1}^{\infty}mc_m u^{m-1}-\sum_{m=0}^{\infty}c_m u^m=0$$

$$\Rightarrow \quad \sum_{m=2}^{\infty}m(m-1)c_m u^{m-1}+3\sum_{m=2}^{\infty}m(m-1)c_m u^{m-2}-\sum_{m=1}^{\infty}mc_m u^m-2\sum_{m=1}^{\infty}mc_m u^{m-1}$$
$$-\sum_{m=0}^{\infty}c_m u^m=0.$$

For making power of u same in each summation term, we change the summation index from m to $m+1$, $m+2$ and $m+1$ in first the second and fourth summation terms, respectively.

$$\sum_{m+1=2}^{\infty}(m+1)(\overline{m+1}-1)c_{m+1}u^{\overline{m+1}-1}+3\sum_{m+2=2}^{\infty}(m+2)(\overline{m+2}-1)c_{m+2}u^{\overline{m+2}-2}-\sum_{m=1}^{\infty}mc_m u^m$$
$$-2\sum_{m+1=1}^{\infty}(m+1)c_{m+1}u^{\overline{m+1}-1}-\sum_{m=0}^{\infty}c_m u^m=0$$

$$\Rightarrow \quad \sum_{m=1}^{\infty}(m+1)mc_{m+1}u^m+3\sum_{m=0}^{\infty}(m+2)(m+1)c_{m+2}u^m-\sum_{m=1}^{\infty}mc_m u^m$$
$$-2\sum_{m=0}^{\infty}(m+1)c_{m+1}u^m-\sum_{m=0}^{\infty}c_m u^m=0$$

$$\Rightarrow \quad \sum_{m=1}^{\infty}(m+1)mc_{m+1}u^m+6c_2+3\sum_{m=1}^{\infty}(m+2)(m+1)c_{m+2}u^m-\sum_{m=1}^{\infty}mc_m u^m-2c_1$$
$$-2\sum_{m=1}^{\infty}(m+1)c_{m+1}u^m-c_0-\sum_{m=1}^{\infty}c_m u^m=0.$$

$$\Rightarrow 6c_2-2c_1-c_0+\sum_{m=1}^{\infty}[(m+1)mc_{m+1}+3(m+2)(m+1)c_{m+2}-mc_m-2(m+1)c_{m+1}-c_m]u^m$$

Setting the coefficients of successive powers of u to zero, we obtain

$$6c_2-2c_1-c_0=0,\; c_{m+2}=\frac{(m+1)(m-2)c_{m+1}-(m+1)c_m}{3(m+2)(m+1)};\; m\ge 2.$$

where, c_0 and c_1 are arbitrary constants.

We obtain

$$c_2 = \frac{1}{6}(c_0 + 2c_1),\ c_3 = -\frac{-2c_2 - 2c_1}{3-3-2}$$

$$= \frac{1}{18}\left[\frac{2}{6}(c_0 + 2c_1) + 2c_1\right] = \frac{1}{54}(c_0 + 8c_1), \text{ and so on.}$$

The power series solution is

$$y = c_0 + c_1 u + \frac{1}{6}(c_0 + 2c_1)u^2 + \frac{1}{54}(c_0 + 8c_1)u^3 + \ldots \tag{2}$$

$\because$ $y(2) = 1$

$\Rightarrow$ $y = 1$ at $x = 2 \Rightarrow y = 1$ at $u = 0$

$\therefore$ Eq. (2) gives $1 = c_0$

Differentiating Eq. (2) with respect to u, we get

$$y' = c_1 + \frac{1}{3}(c_0 + 2c_1)u + \frac{1}{18}(c_0 + 8c_1)u^2 + \ldots \tag{3}$$

$\because$ $y'(2) = 0$

$\Rightarrow$ $y' = 0$ at $x = 2 \Rightarrow y' = 0$ at $u = 0$

$\therefore$ Equation (3) gives $0 = c_1$

On substituting these values, Eq. (2) becomes

$$y = 1 + \frac{1}{6}u^2 + \frac{1}{54}u^3 + \ldots$$

$$= 1 + \frac{1}{6}(x-2)^2 + \frac{1}{54}(x-2)^3 + \ldots$$

which is the required power series solution of the given differential equation.

EXAMPLE 13.9

Find the first five non-vanishing terms in the power series solution of the initial value problem

$$(1 - x^2)y'' + 2xy' + y = 0;\ y(0) = 1,\ y'(0) = 1.$$

Solution: Write the power series as $y = \sum_{m=0}^{\infty} c_m x^m$. Substituting the values of y, y', y'' in the given equation, we get

$$(1 - x^2)\sum_{m=2}^{\infty} m(m-1)c_m x^{m-2} + 2x\sum_{m=1}^{\infty} mc_m x^{m-1} + \sum_{m=0}^{\infty} c_m x^m = 0$$

$$\Rightarrow \quad \sum_{m=2}^{\infty} m(m-1)c_m x^{m-2} - \sum_{m=2}^{\infty} m(m-1)c_m x^m + 2\sum_{m=1}^{\infty} mc_m x^m + \sum_{m=0}^{\infty} c_m x^m = 0$$

$$\Rightarrow 2c_2 + 6c_3x + \sum_{m=4}^{\infty} m(m-1)c_m x^{m-2} - \sum_{m=2}^{\infty} m(m-1)c_m x^m + 2c_1x + 2\sum_{m=2}^{\infty} mc_m x^m + c_0 + c_1x + \sum_{m=2}^{\infty} c_m x^m = 0.$$

In the first term, changing the index of summation from m to $m + 2$, we obtain.

$$\Rightarrow (2c_2 + c_0) + (6c_3 + 2c_1 + c_1)x + \sum_{m+2=4}^{\infty} (m+2)(\overline{m+2}-1)c_{m+2}x^{\overline{m+2}-2} - \sum_{m=2}^{\infty} m(m-1)c_m x^m + 2\sum_{m=2}^{\infty} mc_m x^m + \sum_{m=2}^{\infty} c_m x^m = 0.$$

$$\Rightarrow (2c_2 + c_0) + 3(2c_3 + c_1)x + \sum_{m=2}^{\infty} [(m+2)(m+1)c_{m+2} - (m^2 - 3m - 1)c_m]x^m = 0.$$

Setting the coefficient of successive powers of x to zero, we obtain

$$2c_2 + c_0 - 0,\ 2c_3 + c_1 - 0,\ (m+2)(m+1)c_{m+2} - (m^2 - 3m - 1)c_m;\ m \geq 2.$$

We obtain

$$c_2 = -\frac{1}{2}c_0, \quad c_3 = -\frac{1}{2}c_1, \quad c_{m+2} = \frac{(m^2 - 3m - 1)c_m}{(m+2)(m+1)};\ m \geq 2.$$

where, c_0 and c_1 are arbitrary constants.

We have

$$c_4 = -\frac{3}{12}c_2 = \frac{1}{8}c_0, \text{ and so on.}$$

The power series solution is

$$y = c_0 + c_1x - \frac{1}{2}c_0x^2 - \frac{1}{2}c_1x^3 + \frac{1}{18}c_0x^4 + \ldots$$

$$= c_0\left\{1 - \frac{1}{2}x^2 + \frac{1}{8}x^4 - \ldots\right\} + c_1\left\{x - \frac{1}{2}x^3 + \ldots\right\}.$$

The initial condition $y(0) = 1$ gives $c_0 = 1$ and $y'(0) = 1$ gives $c_1 = 1$. The first five non-vanishing terms in the power series solution are

$$y = 1 + x - \frac{1}{2}x^2 - \frac{1}{2}x^3 + \frac{1}{8}x^4.$$

EXAMPLE 13.10

Solve the following initial value problem in series about the point $x_0 = 0$

$$y'' + 4y = 0;\ y(0) = 3,\ y'(0) = 4.$$

Solution: We write the power series as $y = \sum_{m=0}^{\infty} c_m x^m$.

Substituting $y, y,' y''$ in the given differential equation, we obtain

$$\sum_{m=2}^{\infty} m(m-1)c_m x^{m-2} + 4\sum_{m=0}^{\infty} c_m x^m = 0$$

Changing index of summation from m to $m + 2$ in the first summation term, we obtain

$$\sum_{m=0}^{\infty} (m+2)(m+1)c_{m+2}x^m + 4\sum_{m=0}^{\infty} c_m x^m = 0.$$

Equating the coefficient of x^m to zero, we get

$$c_{m+2} = \frac{-4c_m}{(m+2)(m+1)};\ m \geq 0$$

We have

$$c_2 = -2c_0,\ c_3 = -\frac{2}{3}c_1,\ c_4 = \frac{-4c_2}{12} = \frac{2}{3}c_0\ ,\ c_5 = \frac{-4c_3}{20} = \frac{2}{15}c_1, \text{ and so on.}$$

where, c_0 and c_1 are arbitrary constants.

The power series solution is

$$y = c_0 + c_1x - 2c_0x^2 - \frac{2}{3}c_1x^3 + \frac{2}{3}c_0x^4 + \frac{2}{15}c_1x^5 + \ldots$$

$$= c_0\left\{1 - 2x^2 + \frac{2}{3}x^4 - \ldots\right\} + c_1\left\{x - \frac{2}{3}x^3 + \frac{2}{15}x^5 + \ldots\right\}.$$

The initial condition $y(0) = 1$ given $c_0 = 3$ and $y'(0) = 4$ given $c_1 = 4$. Substituting these values, the required power series solution is

$$y = 3\left\{1 - 2x^2 + \frac{2}{3}x^4 + \ldots\right\} + 4\left\{x - \frac{2}{3}x^3 + \frac{2}{15}x^5 + \ldots\right\}.$$

EXAMPLE 13.11

Find a polynomial approximation of the fourth degree in the solution of the following initial value problem about the point $x_0 = 0$.

$$(1 + 2x)y'' - y' + y = 0;\ y(0) = 0;\ y'(0) = 1.$$

Solution: We write power series as $y = \sum_{m=0}^{\infty} c_m x^m$.

Substituting y, y', y'' in the given differential equation, we obtain

$$(1+2x)\sum_{m=2}^{\infty} m(m-1)c_m x^{m-2} - \sum_{m=1}^{\infty} mc_m x^{m-1} + \sum_{m=0}^{\infty} c_m x^m = 0$$

$$\Rightarrow \quad \sum_{m=2}^{\infty} m(m-1)c_m x^{m-2} + 2\sum_{m=2}^{\infty} m(m-1)c_m c^{m-1} - \sum_{m=1}^{\infty} mc_m x^{m-1} + \sum_{m=0}^{\infty} c_m x^m = 0.$$

Changing the index of summation from m to $m + 2$, $m + 1$ and $m + 1$ in the first, second and third summation terms, respectively, we obtain

$$\sum_{m=0}^{\infty} (m+2)(m+1)c_{m+2}x^m + 2\sum_{m=1}^{\infty} (m+1)mc_{m+1}x^m - \sum_{m=0}^{\infty} (m+1)c_{m+1}x^m + \sum_{m=0}^{\infty} c_m x^m = 0.$$

$$\Rightarrow 2c_2 + \sum_{m=1}^{\infty} (m+2)(m+1)c_{m+2}x^m + 2\sum_{m=1}^{\infty} (m+1)mc_{m+1}x^m - c_1 - \sum_{m=1}^{\infty} (m+1)c_{m+1}x^m$$

$$+ c_0 + \sum_{m=1}^{\infty} c_m x^m = 0$$

$$\Rightarrow (2c_2 + c_0) + \sum_{m=1}^{\infty} [(m+2)(m+1)c_{m+2} + 2(m+1)mc_{m+1} - (m+1)c_{m+1} + c_m]x^m = 0$$

Equating the successive powers of x to zero, we obtain $2c_2 - c_1 + c_0 = 0$, $(m + 2)(m + 1)$ $c_{m+2} + (m+1)(2m-1)c_{m+1} + c_m = 0$

We have

$$c_2 = -\frac{c_0 + c_1}{2},\ c_{m+2} = -\frac{c_m + (m+1)(2m-1)c_{m+1}}{(m+2)(m+1)};\ m \geq 0.$$

Now,

$$c_2 = -\frac{c_0 - c_1}{2} = -\frac{c_0 + c_1}{2}$$

$$c_3 = -\frac{c_1 + 2c_2}{6} = -\frac{1}{6}\{c_1 - c_0 + c_1\} = \frac{c_0 - 2c_1}{6}$$

$$c_4 = -\frac{c_2 + 9c_3}{12} = -\frac{1}{12}\left\{-\frac{c_0 + c_1}{2} + 3\frac{(c_0 - 2c_1)}{2}\right\}$$

$$= -\frac{1}{24}(2c_0 - 5c_1), \text{ and so on.}$$

where, c_0 and c_1 are arbitrary constants. The power series solution is

$$y = c_0 + c_1 x + \left(\frac{-c_0 + c_1}{2}\right)x^2 + \left(\frac{c_0 - 2c_1}{6}\right)x^3 - \frac{1}{24}(2c_0 - 5c_1)x^4 + \ldots$$

$$= c_0\left\{1 - \frac{x^2}{2} + \frac{x^3}{6} - \frac{x^4}{12} + \ldots\right\} + c_1\left\{x + \frac{x^2}{2} - \frac{x^3}{3} + \frac{5}{24}x^4 + \ldots\right\}.$$

The initial condition $y(0) = 0$ gives $c_0 = 0$ and $y'(0) = 1$ gives $c_1 = 1$.
Substituting these values, the required power series solution is

$$y = x + \frac{x^2}{2} - \frac{x^3}{3} + \frac{5}{24}x^4.$$

which is the required polynomial approximation of the fourth degree. ▲▲

EXERCISE 13.1

Find the singular points of the following differential equations and classify them.

1. $x^2y'' + axy' + by = 0$; a, b constants.
2. $x^3(x - 2)y'' + x^2y' + by = 0$.
3. $x^2y'' - 5y' + 3x^2y = 0$.
4. $x^2y'' + (\sin x)y' + (\cos x)y = 0$.
5. $(x^2 + x - 2)^2 y'' + 3(x + 2)y' + (x - 1)y = 0$.
6. $x^3(x^2 - 1)y'' - x(x + 1)y' - (x - 1)y = 0$.
7. Find the power series solution about $x = 0$ of the differential equation $y' - 2y = 0$.
8. Find the power series solution about the origin of the equation $(1 - x^2)y'' - 4xy' + 2y = 0$.
9. Using the power series method, solve the following initial value problem about the point $x_0 = 1$.
   $(1 + 2x)y'' + xy' + y = 0$; $y(1) = 1$, $y'(1) = -1$.
10. Find the power series solution of the following initial value problem about the point $x_0 = 0$.
    $$(1 - x)y'' + xy' + 2y = 0;\ y(0) = 2,\ y'(0) = 1.$$
11. Find a polynomial approximation of the fourth degree in the solution of the following initial value problem about the point $x_0 = 0$.
    $$(1 + x)y'' + y' + 3y = 0;\ y(0) = 1,\ y'(0) = 2.$$
12. Solve in power series about the point $x_0 = 0$ the differential equation
    $$(1 - x^2)y'' - 2xy' + 2y = 0.$$
13. Solve in series about the point $x_0 = 0$ $(x^2 + 1)y'' + xy' + 2xy = 0$, gives $y(0) = 4$, $y'(0) = 0$
14. Solve $(x^2 - 1)y'' + 4x + 2y = 0$ in power series about 0.
15. Solve the following differential equation in power series about the point $x_0 = 0$
    $(1 - x^2)y'' + 2xy' - 2y = 0$.

ANSWERS

Exercise 13.1

1. $x = 0$, regular singular point.
2. $x = 2$, regular singular point and $x = 0$, irregular singular point.

3. $x = 0$, irregular singular point.
4. $x = 0$, regular singular point.
5. $x = 1$, irregular singular point and $x = -2$, regular singular point.
6. $x = 0$, irregular singular point and $x = \pm 1$, regular singular point.
7. $y = c_0\left\{1+2x+\frac{(2x)^2}{2!}+\frac{(2x)^3}{3!}+\ldots\right\} = c_0 e^{2x}$.
8. $y = c_0\left\{1-x^2-\frac{2}{3}x^4+\ldots\right\}+c_1\left\{x+\frac{1}{3}x^3+\frac{4}{15}x^5+\ldots\right\}$.
9. $y = \left\{1-\frac{1}{6}(x-1)^2+\frac{1}{18}(x-1)^3-\frac{1}{108}(x-1)^4-\ldots\right\}$

$$-\left\{(x-1)-\frac{1}{6}(x-1)^2-\frac{1}{18}(x-1)^3+\frac{1}{27}(x-1)^4-\ldots\right\}.$$

10. $y = 2 + x - 2x^2 - \frac{7}{6}x^3 + \frac{1}{12}x^4$.
11. $y = 1 + 2x - \frac{5}{2}x^2 + \frac{2}{3}x^3 + \frac{1}{8}x^4$.
12. $y = c_0\left(1-x^2-\frac{x^4}{3}-\frac{x^6}{5}-\frac{x^8}{7}-\ldots\right)+c_1 x$.
13. $y = 4 + 6x - \frac{7}{3}x^3 - x^4 + \frac{21}{20}x^5$.
14. $y = c_0(1 + x^2 + x^4 + x^6 + \ldots) + c_1(x + x^3 + x^5 + x^2 + \ldots)$.
15. $y = c_0\left(1+\frac{x^4}{2!}+\frac{x^6}{4!}+\ldots\right)+c_1\left(x^2-\frac{x^6}{3!}+\frac{x^{10}}{5!}-\ldots\right)$. ▲▲

FROBENIUS METHOD

In the previous section, we have discussed method for finding series solution about an ordinary point, that is, power series solution of the equation. In this section, we shall study the Frobenius method for obtaining a series solution about a regular singular point of the equation

$$a_0(x)\frac{d^2y}{dx^2} + a_1(x)\frac{dy}{dx} + a_2(x)y = 0 \tag{1}$$

as given below:

1. Find regular singular point x_0 (if not given).
2. Assume the solution to be $y = (x - x_0)^r \sum_{m=0}^{\infty} c_m(x - x_0)^m$ where, r is a real number.
3. Substitute $y, \frac{dy}{dx}, \frac{d^2y}{dx^2}$ in Eq. (1) as above.
4. Equate to zero the coefficient of the lowest degree term in x (with $m = 0$) to get a quadratic equation known as the indicial equation and find indicial roots, say r_1 and r_2. The complete solution depends on the values of roots of the indicial equation.

Case I: When $r_1 \neq r_2$ and $r_1 - r_2$ is not an integer.

Find the coefficient c_m corresponding to $(x - x_0)^m$ term and put these in

$$y = \sum_{m=0}^{\infty} c_m(x - x_0)^{m+r}.$$

The complete solution is

$$y = Ay_1 + By_2$$

where,

$$y_1 = y \text{ at } r = r_1$$

$$y_2 = y \text{ at } r = r_2$$

and A, B are arbitrary constants.

Case II: When $r_1 = r_2$

The complete solution is

$$y = Ay_1 + By_2$$

where,

$$y_1 = y \text{ at } r = r_1$$

and

$$y_2 = \frac{\partial y}{\partial r} \text{ at } r = r_1$$

Case III: When $r_1 \neq r_2$ but $r_1 - r_2$ is an integer.

(a) If $c_1 = 0$ then the complete solution is

$$y = Ay_1 + By_2$$

where,

$$y_1 = y \text{ at } r = r_1$$

and

$$y_2 = y \text{ at } r = r_2$$

(b) If r_0 (which is either r_1 or r_2) makes c_1 undefined then put $c_0 = a_0\,(r - r_0)$ to make c_1 finite. The complete solution is

$$y = Ay_1 + By_2$$

where,

$$y_1 = y \text{ at } r = r_0$$

and $$y_2 = \frac{\partial y}{\partial r} \text{ at } r = r_0.$$

(c) If r_0 (which is either r_1 or r_2) makes c_1 arbitrary then find the complete solution by using $r = r_0$ only.

EXAMPLE 13.12

Solve in series the equation

$$2x^2y'' + xy' - (x^2 + 1)y = 0.$$

Solution: The point $x = 0$ is a regular singular point of the equation. Let the solution be written as

$$y = \sum_{m=0}^{\infty} c_m(x - 0)^{m+r}$$

Differentiating, we obtain

$$y' = \sum_{m=0}^{\infty} (m + r)c_m x^{m+r-1},\ y'' = \sum_{m=0}^{\infty} (m + r)(m + l - 1)c_m x^{m+l-2}.$$

Substituting in the given differential equation, we obtain

$$2x^2 \sum_{m=0}^{\infty} (m + r)(m + r - 1)c_m x^{m+r-2} + x \sum_{m=0}^{\infty} (m + r)c_m x^{m+r-1} - (x^2 + 1) \sum_{m=0}^{\infty} c_m x^{m+r} = 0$$

$$\Rightarrow 2 \sum_{m=0}^{\infty} (m + r)(m + r - 1)c_m x^{m+r} + \sum_{m=0}^{\infty} (m + r)c_m x^{m+r} - \sum_{m=0}^{\infty} c_m x^{m+r+2} - \sum_{m=0}^{\infty} c_m x^{m+r} = 0.$$

The lowest degree term (with $m = 0$) is the term containing x^r. Setting the coefficient of x^r to zero, we obtain

$$c_0[2r(l - 1) + r - 1] = 0$$

as $c_0 \neq 0$ we have $$2r^2 - r - 1 = 0$$

$$\Rightarrow \qquad r = 1, -1/2.$$

As $r_1 = 1$ and $r_2 = -1/2$ are not equal and $r_1 - r_2 = \frac{3}{2}$ is not an integer, we have Case I.

Since the lowest coefficient term in each summation term is x^{r+2}, so equating to zero the next power of x which is x^{r+1} (with $m = 1$), we obtain

$$c_1[2(l + 1)r + (l + 1) - 1] = 0$$

$$\Rightarrow \qquad c_1(2r^2 + 3r) = 0.$$

For $r = 1, -1/2,$ we obtain $c_1 = 0$ since $2r^2 + 3r \neq 0$ for these values of r.

For the remaining terms , we get

$$2\sum_{m=2}^{\infty}(m+r)(m+r-1)c_m x^{m+r}+\sum_{m=2}^{\infty}(m+r)c_m x^{m+r}-\sum_{m=0}^{\infty}c_m x^{m+r+2}-\sum_{m=2}^{\infty}c_m x^{m+r}=0.$$

Changing the index of summation from m to $m-2$ in the third summation term, we obtain

$$\sum_{m=2}^{\infty}[2(m+r)(m+r-1)+(m+r)-1]c_m x^{m+r}-\sum_{m-2=0}^{\infty}c_{m-2}x^{\overline{m-2}+l+2}=0$$

$$\sum_{m=2}^{\infty}[(m+r)(2m+2r-1)-1]c_m x^{m+r}-\sum_{m=2}^{\infty}c_{m-2}x^{m+r}=0.$$

Setting the coefficient of x^{m+r} to zero, we obtain the recurrence relation

$$[(m+r)(2m+2r-1)-1]c_m-c_{m-2}=0$$

$$\Rightarrow \qquad c_m=\frac{c_{m-2}}{(m+r)(2m+2r-1)-1};\ m\geq 2.$$

Since $c_1=0$, we obtain $\qquad c_3=0=c_5=\ldots$

For $r=1$, we get

$$c_m=\frac{c_{m-2}}{(m+1)(2m+1)-1}$$

$$c_2=\frac{c_0}{14},\ c_4=\frac{c_2}{44}=\frac{c_0}{616},\ \ldots$$

For $r=-\dfrac{1}{2}$, we get

$$c_m=\frac{c_{m-2}}{\left(m-\frac{1}{2}\right)(2m-2)-1}$$

$$c_2=\frac{c_0}{2},\ c_4=\frac{c_2}{20}=\frac{c_0}{40},\ \ldots$$

Now, according to case I, we have

$$y_1=x'[c_0+c_2x^2+c_4x^4+\ldots]$$

$$=x\left\{c_0+\frac{c_0}{14}x^2+\frac{c_0}{616}x^4+\ldots\right\}$$

$$=u(x),\text{ say}$$

and

$$y_2=x^{-1/2}\left\{c_0+\frac{c_0}{2}x^2+\frac{c_0}{40}x^4+\ldots\right\}=v(x),\text{ say}$$

The general solution is

$$y=Au(x)+Bv(x)$$

where, A and B are the arbitrary constants.

EXAMPLE 13.13

Find the series solution about $x = 0$, of the equation $xy'' + y' - xy = 0$ by the Frobenius method.

Solution: We find that $x = 0$ is a regular singular point of the given differential equation.

Let the solution be written as $y = \sum_{m=0}^{\infty} c_m(x - 0)^{m+r}$.

Substituting y, y', y'' in the given differential equation, we get

$$\sum_{m=0}^{\infty} (m + r)(m + r - 1)c_m x^{m+r-1} + \sum_{m=0}^{\infty} (m + r)c_m x^{(m+r+1)} - \sum_{m=0}^{\infty} c_m x^{m+r+1} = 0.$$

The lowest degree term (with $m = 0$) is the term containing x^{r-1}. Setting the coefficient of $x^{(r-1)}$ to zero, we obtain

$$c_0\,[r/r - 1) + r] = 0 \Rightarrow c_0 r^2 = 0$$

as $c_0 \neq 0$ we have $\qquad r^2 = 0$

$\Rightarrow \qquad r = 0, 0.$

Since $r_1 = 0 = r_2$, we have case II.

Since the lowest common term in each summation term is x^{r+1}, so equating to zero the next power of x which is x^r (with $m = 1$), we obtain

$$c_1(r + 1)^2 = 0$$

For $r = 0, 0$; we obtain $c_1 = 0$ since $(r + 1)^2 \neq 0$ for these values of r for the remaining terms, we get

$$\sum_{m=2}^{\infty} (m + r)^2 c_m x^{m+r-1} - \sum_{m=0}^{\infty} c_m x^{m+r+1} = 0.$$

Changing the index of summation from m to $m + 2$ in the first summation term, we get

$$\sum_{m=0}^{\infty} (m + r + 2)^2 c_{m+2} x^{m+r+1} - \sum_{m=0}^{\infty} c_m x^{m+r+1} = 0.$$

Setting the coefficient of x^{m+r+1} to zero, we obtain the recurrence relation

$$(m + r + 2)^2 c_{m+2} - c_m = 0$$

$$\Rightarrow \qquad c_{m+2} = \frac{c_m}{(m+r+2)^2};\ m \geq 0.$$

Since $c_1 = 0$; we obtain $c_3 = 0 = c_5 = \ldots$

Also $c_2 = \dfrac{c_0}{(r+2)^2}$, $c_4 = \dfrac{c_2}{(r+4)^2} = \dfrac{c_0}{(r+2)^2(r+4)^2}$, $c_6 = \dfrac{c_4}{(r+6)^2} = \dfrac{c_0}{(r+2)^2(r+4)^2(r+6)^2}$,

and so on.

Therefore,

$$y = c_0 x^r \left\{1 + \frac{x^2}{(r+2)^2} + \frac{x^4}{(r+2)^2(r+4)^2} + \frac{x^6}{(r+2)^2(r+4)^2(r+6)^2} + \ldots\right\} \tag{1}$$

For $r = 0$

$$y_1 = c_0\left[1 + \frac{x^2}{2^2} + \frac{x^4}{2^2 \cdot 4^2} + \frac{x^6}{2^2 \cdot 4^2 \cdot 6^2} + \ldots\right] = u(x), \text{ say.}$$

To get other linearly independent solution y_2, put y, y', y'' (as obtained in Eq. (1)) in the given differential equation, we obtain

$$xy'' + y' - xy = c_0 r^2 x^{r-1} \tag{2}$$

As for $r = 0$, right hand side becomes zero. So we have one linearly independent solution $y_1 = y$ at $r = 0$ as done above.

Consider now r as a parameter (variable) and differentiate Eq. (2) partially with respect to r, we obtain

$$\left(x\frac{d^2}{dx^2} + \frac{d}{dx} - x\right)\left(\frac{\partial y}{\partial r}\right) = c_0[2rx^{r-1} + r^2 x^{r-1} \log x]$$

Again, for $r = 0$ the right hand side becomes zero, so the second linearly independent solution is $y_2 = \dfrac{\partial y}{\partial r}$ at $r = 0$.

Now,

$$\frac{\partial y}{\partial r} = c_0 x^r (\log x)\left[1 + \frac{x^2}{(r+2)^2} + \frac{x^4}{(r+2)^2(r+4)^2}\right]$$

$$+ c_0 x^r\left[-\frac{2x^2}{(r+2)^3} - x^4\left[\frac{2}{(r+2)^3(r+4)^2} + \frac{2}{(r+2)^2(r+4)^3}\right] + \ldots\right].$$

Thus,

$$y_2 = \left.\frac{\partial y}{\partial r}\right|_{r=0} = c_0(\log x)\left[1 + \frac{x^2}{2^2} + \frac{x^4}{2^2 \cdot 4^2} + \ldots\right] + c_0\left[-\frac{x^2}{2^2} - \frac{3x^4}{128} + \ldots\right].$$

$$= (\log x)\, u(x) + v(x); \text{ say.}$$

The general solution is

$$y = Au(x) + B[(\log x)\, u(x) + v(x)]$$

$$= (A + B \log x)\, u(x) + Bv(x)$$

where, A and B are the arbitrary constants.

EXAMPLE 13.14

Find the series solution of the following differential equation by the Frobenius method

$$xy'' + (1 - 2x)y' + (x - 1)y = 0.$$

Solution: We find that $x = 0$ is a regular singular point of the given differential equation.

Let the solution be written as $y = \sum_{m=0}^{\infty} c_m(x - 0)^{m + r}$.

Substituting y, y', y'' in the given differential equation, we get

$$\sum_{m=0}^{\infty} (m + r)(m + r - 1)c_m x^{m + r - 1} + \sum_{m=0}^{\infty} (m + r)c_m x^{m + r - 1} - 2\sum_{m=0}^{\infty} (m + r)c_m x^{m + r}$$

$$+ \sum_{m=0}^{\infty} c_m x^{m + r + 1} - \sum_{m=0}^{\infty} c_m x^{m + r} = 0.$$

The lowest degree term (with $n = 0$) is the term containing x^{r-1}. Setting the coefficient of x^{r-1} to zero, we obtain

$$c_0[r(r - 1) + r] = 0 \Rightarrow c_0 r = 0$$

as $c_0 \neq 0$ we have $r_2 = 0$

$$\Rightarrow \qquad r = 0, 0$$

Since $r_1 = 0 = r_2$, we have case II.

Since the lowest common term in each summation term is x^{r+1}, so equating to zero the next power of x which is x^r; we obtain $c_1(l + 1)^2 = -2^r c_0 - c_0 = 0 \Rightarrow c_1 = \dfrac{2l+1}{(l+1)^2}\, c_0$

For $r = 0, 0$; we obtain $c_1 = c_0$. For the remaining terms, we get

$$\sum_{m=2}^{\infty} (m + r)^2\, c_m x^{(m + r + 1)} - 2\sum_{m=1}^{\infty} [2(m + r) + 1]c_m x^{mu} + \sum_{m=0}^{\infty} c_m x^{(m + r + 1)} = 0.$$

Changing the index of summation from m to $m + 2$ and $m + 1$ in the first and second summation terms, respectively, we get

$$\sum_{m=0}^{\infty} (m + r + 2)^2\, c_{m + 2} x^{m + l + 1} \sum_{m=0}^{\infty} [2(m + l + 1) + 1]\, c_{m + 1} x^{m + l + 1} + \sum_{m=0}^{\infty} c_m x^{m + r + 1} = 0$$

Setting the coefficient of $x^{m + r + 1}$ to zero, we obtain the recurrence relation

$$(m + r + 2)^2\, c_{m + 2} - (2m + 2r + 3)c_{m + 1} + c_m = 0$$

$$\Rightarrow \qquad c_{m + 2} = \frac{(2m+2r+3)c_{m+1} - c_m}{(m+r+2)^2};\ m \geq 0.$$

Thus,

$$c_2 = \frac{(2r+3)c_1 - c_0}{(r+2)^2}$$

$$= \frac{(2r+3)\dfrac{(2r+1)}{(r+1)^2}c_0 - c_0}{(r+2)^2} = \frac{\left[(2r+1)(2r+3)-(r+1)^2\right]c_0}{(r+1)^2(r+2)^2}$$

$$c_3 = \frac{(2r+5)c_2 - c_1}{(r+3)^2} = \frac{\{(2r+5)[2r+1)(2r+3)-(r+1)^2]-(2r+1)(r+2)^2\}}{(r+1)^2(r+2)^2(r+3)^2}c_0, \text{ and so on.}$$

Therefore, $y = c_0x^r\left[1+\dfrac{(2r+1)}{(r+1)^2}x+\dfrac{(2r+1)(2r+3)-(r+1)^2}{(r+1)^2(r+2)^2}x^2\right.$

$$\left.+\frac{(2r+5)[(2r+1)(2r+3)-(r+1)^2]-(2r+1)(r+1)^2}{(r+1)^2(r+2)^2(r+3)^2}x^3+\ldots\right]$$

Substituting $r = 0$, we obtain

$$y_1 = c_0\left[1+\frac{x}{1^2}+\frac{2}{1^2-2^2}x^2+\frac{6}{1^2-2^2-3^2}x^3+\ldots\right] = u(x); \text{ say.}$$

Now, $\dfrac{\partial y}{\partial r} = c_0x^r(\log x)\left[1+\dfrac{(2r+1)}{(r+1)^2}x+\ldots\right]+c_0x^r\left[0+\left(\dfrac{2}{(r+1)^2}-\dfrac{2(2r+1)}{(r+1)^3}\right)x+\ldots\right]$

Therefore, $y_2 = \left.\dfrac{\partial y}{\partial r}\right|_{r=0} = (\log x)\,u(x) + v(x)$

The general solution is

$$\begin{aligned} y &= Au(x) + B((\log x)\,u(x) + v(x)) \\ &= (A + B\log x)\,u(x) + Bv(x) \end{aligned}$$

where, A and B are the arbitrary constants.

EXAMPLE 13.15

Find a series solution of the equation $x^2y'' + x^3y' + (x^2 - 2)y = 0$ about the point $x = 0$.

Solution: We find that $x = 0$ is a regular singular point of the given differential equation.

Let the solution be written as $y = \sum\limits_{m=0}^{\infty} c_m(x-0)^{m+r}$.

Substituting y, y', y'' in the given differential equation, we obtain

$$\sum_{m=0}^{\infty}(m+r)(m+r-1)c_mx^{m+r} + \sum_{m=0}^{\infty}(m+r)c_mx^{m+r+2} + \sum_{m=0}^{\infty}c_mx^{m+r+2} - 2\sum_{m=0}^{\infty}c_mx^{m+r} = 0.$$

The lowest degree term (with $m = 0$) is the term containing x^r. Setting the coefficient of x^r to zero, we obtain

$$c_0[r(r-1)-2] = 0 \Rightarrow c_0(l^2 - l - 2) = 0$$

as $c_0 \neq 0$, we have $\quad r^2 - r - 2 = 0$

$\Rightarrow \quad r = -1, 2$

Since $r_1 = -1$ and $r_2 = 2$, so $r_1 \neq r_2$ and $r_1 - r_2 = -3$ is an integer. Therefore, we have case III.

Since the lowest common term in each summation term is x^{r+2}, so equating to zero, the next power of x which is x^{r+1} (with $m = 1$), we obtain

$$c_1[(r+1)r - 2] = 0 \Rightarrow c_1(r^2 + r - 2) = 0$$

For $r = -1, 2$ we have $r^2 + r - 2 \neq 0$. Therefore, $c_1 = 0$, for these values of r. So we have case III (a).

For the remaining terms, we get

$$\sum_{m=2}^{\infty} [(m+r)(m+r-1) - 2]c_m x^{m+r} + \sum_{m=0}^{\infty} [(m+r)+1]c_m x^{m+r+2} = 0.$$

Changing the index of summation from m to $m+2$ in the first summation term, we get

$$\sum_{m=0}^{\infty} [(m+r+2)(m+r+1) - 2]c_{m+2} x^{m+r+2} + \sum_{m=0}^{\infty} [m+r+1]c_m x^{m+r+2} = 0.$$

Setting the coefficient of x^{m+r+2} to zero, we obtain the recurrence relation

$$[(m+r+2)(m+r+1) - 2]c_{m+2} + (m+r+1)c_m = 0$$

$$\Rightarrow \quad c_{m+2} = -\frac{(m+r+1)}{(m+r+1)(m+r+2)-2} c_m;\ m \geq 0.$$

Thus, $c_3 = 0 = c_5 = \ldots$ since $c_1 = 0$

$$c_2 = -\frac{(r+1)c_0}{(r+1)(r+2)-2};\ c_4 = -\frac{(r+3)c_2}{(r+3)(r+4)-2},$$

and so on.

For $r = -1$

$$c_2 = 0 = c_4 = \ldots$$

$$\therefore \quad y_1 = x^{-1}[c_0 + 0 \cdot x + 0 \cdot x^2 + \ldots] = u(x);\ \text{say.}$$

For $r = 2$

$$c_2 = -\frac{3}{10}c_0;\ c_4 = -\frac{5}{28}c_2 = \frac{3}{56}c_0,\ \text{and so on.}$$

$$\therefore \quad y_2 = x^2\left[c_0 - \frac{3}{10}c_0 x^2 + \frac{3}{56}c_0 x^4 + \ldots\right]$$

The general solution is

$$y = \frac{A}{x} + B\left(x^3 - \frac{3}{10}x^5 + \frac{3}{56}x^6 - \ldots\right)$$

where, A and B are the arbitrary constants.

EXAMPLE 13.16

Find the series solution about $x = 0$ of the differential equation

$$x(1 + x)y'' + 3xy' + y = 0.$$

Solution: We find that $x = 0$ is a regular singular point of the given differential equation. Let the solution be written as $y = \sum_{m=0}^{\infty} c_m(x - 0)^{m + r}$.

Substituting y, y', y'' in the given differential equation, we get

$$\sum_{m=0}^{\infty} (m + r)(m + r - 1)c_m x^{m + r - 1} + \sum_{m=0}^{\infty} (m + r)(m + r - 1)c_m x^{m + r} + 3\sum_{m=0}^{\infty} (m + r)c_m x^{m + r} + \sum_{m=0}^{\infty} c_m x^{m + r} = 0.$$

The lowest degree term (with $m = 0$) is the term containing x^{r-1}. Setting the coefficient of x^{r-1} to zero, we obtain

$$c_0 r(r - 1) = 0$$

as $c_0 \neq 0$ we have $\quad r(r - 1) = 0 \Rightarrow r = 0, 1$

Since $r_1 = 0$ and $r_1 = 1$, we have $r_1 \neq r_2$ and $r_1 - r_2 = -1$ is an integer. Therefore, we have case III.

Since the lowest common term in each summation term is x^r, so we have

$$\sum_{m=1}^{\infty} (m + r)(m + l - 1)c_m x^{m + r - 1} + \sum_{m=0}^{\infty} [(m + r)(m + r + 2) + 1]c_m x^{m + r} = 0.$$

Changing the index of summation from m to $m + 1$ in the first summation term; we get

$$\sum_{m=0}^{\infty} (m + r + 1)(m + r)c_{m+1} x^{m + r} + \sum_{m=0}^{\infty} [(m + l)(m + r + 2) + 1]c_m x^{m + r} = 0.$$

Setting the coefficient of $x^{m + r}$ to zero; we obtain the recurrence relation

$$(m + r + 1)(m + r)c_{m+1} + [(m + r)(m + r + 2) + 1]c_m = 0$$

$$\Rightarrow \qquad c_{m+1} = -\frac{(m+r+1)^2}{(m+r+1)(m+r)}c_m = -\frac{m+r+1}{m+r}c_m\,;\ m \geq 0.$$

Thus, $\qquad c_1 = -\frac{r+1}{r}c_0$

We note that c_1 becomes infinite when $r = 0$ and hence all the remaining coefficients also become infinite. Therefore, according to case III (b), to make c_1, finite, let

$$c_0 = a_0(r - 0) = a_0 r$$

Clearly $c_1 = -(r + 1)a_0$ is finite for $r = 0$.

Now,
$$c_2 = -\frac{r+2}{r+1}c_1 = (r + 2)a_0$$

$$c_3 = -\frac{r+3}{r+2}c_2 = -(r + 3)a_0, \text{ and so on.}$$

Therefore,

$$y = a_0x^r\,[r - (r + 1)a_0^x + (r + 2)a_0x^2 - (r + 3)a_0x^3 + \ldots] \quad (1)$$

Put y, y', y'' (as obtained from Eq. (1)) in the given differential equation, we obtain

$$x(1 + x)y'' + 3xy' + y = a_0r^2(r - 1) \quad (2)$$

Clearly for $r = 0$, the right hand side of Eq. (2) becomes zero. Therefore, the one linearly independent solution of the given differential equation is y at $r = 0$

$$y_1 = y\big|_{r=0} = a_0[-x + 2x^2 - 3x^3 + \ldots]$$

$$= -a_0x(1 - 2x + 3x^2 \ldots] = -a_0x(1 + x)^{-2}$$

Differentiating Eq. (2) partially with respect to r, we get

$$\left[x(1+x)\frac{d^2}{dx^2} + 3x\frac{d}{dx} + 1\right]\left(\frac{\partial y}{\partial r}\right) = a_0(3r^2 - 2r) \quad (3)$$

Again, for $r = 0$, the right-hand side becomes zero. So, the second linearly independent solution of the given differential equation is $y_2 = \left.\frac{\partial y}{\partial r}\right|_{r=0}$

Now $\frac{\partial y}{\partial r} = a_0x^r(\log x)\,[r - (r + 1)x + (r + 2)x^2 - \ldots] + a_0x^r(1 - x + x^2 - \ldots]$

$\therefore$ $y_2 = (\log x)a_0\,(- x + 2x^2 - 3x^2 + \ldots) + a_0(1 - x + x^2 - \ldots)$

$= (\log x)y_1 + a_0(1 + x)^{-1}.$

The general solution is

$$y = Ax(1 + x)^{-2} + B[(\log x)x\,(1 + x)^{-2} + (1 + x)^{-1}]$$
$$= (A + B\log x)x\,(1 + x)^{-2} + B(1 + x)^{-1}.$$

where, A and B are the arbitrary constants.

Note: As we can note that for $r = 1$, the right hand side of (2) also becomes zero. So we can take one linearly independent solution $y_1 = y$ at $r = 1$ also. But since for $r = 1$, the right hand side of Eq. (3) is non-zero, so we cannot have second linearly independent solution of the given differential equation for $r = 1$. Therefore, only the value of r which makes c_1 infinite will give us the complete general solution of the given differential equation.

EXAMPLE 13.17

Find the series solution about the origin of the differential equation

$$x^2y'' + 6xy' + (6 + x^2)y = 0.$$

Solution: We find that $x = 0$ is a regular singular point of the given differential equation.

Let the solution be written as $y = \sum_{m=0}^{\infty} c_m(x - 0)^{m+r}$.

Substituting y, y', y'' in the given differential equation, we get

$$\sum_{m=0}^{\infty} (m + r)(m + r - 1)c_m x^{m+r} + 6\sum_{m=0}^{\infty} (m + r)c_m x^{m+r} + 6\sum_{m=0}^{\infty} c_m x^{m+r} + \sum_{m=0}^{\infty} c_m x^{m+l+2} = 0.$$

The lowest degree term (with $m = 0$) is the term containing x^r. Setting the coefficient of x^r to zero, we obtain

$$c_0[r(r - 1) + 6r + 6] = 0$$

$\Rightarrow$ $$c_0 (r^2 + 5r + 6) = 0$$

as $c_0 \neq 0$ we have $$r^2 + 5r + 6 = 0$$

$\Rightarrow$ $$r = -2, -3$$

As $r_1 = -2$ and $r_2 = -3$ $r_1 \neq r_2$ and $r_1 - r_2 = 1$ is an integer so we have case III.

Since the lowest term in each summation term is x^{r+2}; so equating to zero, the next power of x which is x^{r+1} (with $m = 1$); we obtain

$$c_1[(r + 1)r + 6(r + 1) + 6] = 0$$

$\Rightarrow$ $$c_1(r^2 + 7r + 12) = 0$$

For $r = -2$ $\quad r^2 + 7r + 12 \neq 0 \quad \Rightarrow \quad c_1 = 0$

For $r = -3$ $\quad r^2 + 7r + 12 = 0 \quad \Rightarrow \quad c_1$ is arbitrary.

So, we have case III (c).

Therefore, the indicial root $r = -3$; given the complete solution as the corresponding solution contains two arbitrary constants. The remaining terms are

$$\sum_{m=2}^{\infty} [(m + r)(m + r + 5) + 6]c_m x^{m+r} + \sum_{m=0}^{\infty} c_m x^{m+r+2} = 0.$$

Changing the index of summation from m to $m + 2$ in the first summation term, we get

$$\sum_{m=0}^{\infty} [(m + r + 2)(m + r + 7) + 6]c_{m+2} x^{m+r+2} + \sum_{m=0}^{\infty} c_m x^{m+r+2} = 0.$$

Setting the coefficient of x^{m+r+2} to zero, we obtain the recurrence relation

$$[(m + r + 2)(m + r + 7) + 6]c_{m+2} + c_m = 0$$

$$\Rightarrow \qquad c_{m+2} = -\frac{c_m}{(m+r+2)(m+r+7)+6}; \; m \geq 0$$

Thus,
$$c_2 = -\frac{c_0}{(r+2)(r+7)+6},\; c_3 = -\frac{c_1}{(r+3)(r+8)+6},\; c_4 = -\frac{c_2}{(r+4)(r+9)+6},$$

$$c_5 = -\frac{c_3}{(r+5)(r+10)+6}, \text{ and so on.}$$

For $r = -3$

$$c_2 = -\frac{c_0}{2},\; c_3 = -\frac{c_1}{6},\; c_4 = -\frac{c_2}{11} = \frac{c_0}{24},\; c_5 = -\frac{c_3}{20} = \frac{c_1}{120}, \text{ and so on.}$$

Therefore, the general solution is

$$y = x^{-3}\left[c_0 + c_1 x - \frac{c_0}{2}x^2 - \frac{c_1}{6}x^3 + \frac{c_0}{24}x^4 + \frac{c_1}{120}x^5 - \ldots\right]$$

$$= x^{-3}\left[c_0\left(1 - \frac{x^2}{2!} + \frac{x^4}{4!} - \ldots\right) + c_1\left(x - \frac{x^3}{3!} + \frac{x^5}{5!} - \ldots\right)\right]$$

$$= x^{-3}[c_0 \cos x + c_1 \sin x]$$

$$= c_0\left(\frac{\cos x}{x^3}\right) + c_1\left(\frac{\sin x}{x^3}\right)$$

where, c_0 and c_1 are the arbitrary constants.

Note: If we consider $r = -2$, then we have

$$c_3 = 0 = c_5 = \ldots \text{ since } c_1 = 0$$

$$c_2 = \frac{c_0}{6},\; c_4 = -\frac{c_2}{20} = \frac{c_0}{120}, \text{ and so on}$$

Therefore,

$$y = x^{-2}\left[c_0 - \frac{c_0}{6}x^2 + \frac{c_0}{120}x^4 - \ldots\right]$$

$$= x^{-3}c_0\left(x - \frac{x^3}{3!} + \frac{x^5}{5!} - \ldots\right)$$

$$= c_0\left(\frac{\sin x}{x^3}\right)$$

So, in this case we are able to find only one linearly independent solution of the given differential equation, which is already contained in the solution obtained for $r = -3$.

EXAMPLE 13.18

Find the Frobenius series solution about $x = 0$, of the equation

$$(1 + x^2)y'' - 2xy' + 6y = 0.$$

Solution: The point $x = 0$ is a regular (ordinary) point of the given differential equation. So, we can also use the power series method here. But we will find its solution by using the Frobenius method. Let the solution be written as $y = \sum_{m=0}^{\infty} c_m(x - 0)^{m+r}$.

Substituting y, y', y'' in the given differential equation, we get

$$\sum_{m=0}^{\infty} (m + r)(m + r - 1)c_m x^{m+r-2} + \sum_{m=0}^{\infty} (m + r)(m + r - 1)c_m x^{m+r} - 2\sum_{m=0}^{\infty} (m + r)c_m x^{m+r}$$

$$+ 6\sum_{m=0}^{\infty} c_m x^{m+r} = 0.$$

The lowest degree term (with $m = 0$) is the term containing x^{r-2}. Setting the coefficient of x^{r-2} to zero, we obtain

$$c_0{}^r(r - 1) = 0$$

as $c_0 \neq 0$ we have $r(r - 1) = 0$

$\Rightarrow$ $r = 0, 1$

As $r_1 = 0$ and $r_1 = 1$ we have $r_1 \neq r_2$ and $r_1 - r_2 = -1$ is an integer, so we have case III.

Since the lowest common term in each summation is x^r, so equation to zero the next power of x which is x^{r-1} (with $m = 1$), we obtain $c_1(r + 1)r = 0$.

For $r = 0$, $(r + 1)r = 0 \Rightarrow c_1$ is arbitrary.

So, we have case III (c).

For $r = 1$ $(r + 1)r \neq 0 \Rightarrow c_1 = 0$.

Therefore, the indicial root $r = 0$ gives the complete solution as the corresponding solution contains two arbitrary constants. The remaining terms are

$$\sum_{m=2}^{\infty} (m + r)(m + r - 1)c_m x^{m+r-2} - \sum_{m=0}^{\infty} [(m + r)(m + r + 1) - 6]c_m x^{m+r} = 0.$$

Changing the index of summation from m to $m + 2$ in the first summation term, we get

$$\sum_{m=0}^{\infty} (m + r + 2)(m + r + 1)c_{m+2} x^{m+r} - \sum_{m=0}^{\infty} [(m + r)(m + r + 1) - 6]c_m x^{m+r} = 0.$$

Setting the coefficient of x^{m+r} to zero, we obtain the recurrence relation

$$(m + r + 2)(m + r + 1)c_{m+2} - [(m + r)(m + r + 1) - 6]c_m = 0$$

$$\Rightarrow \qquad c_{m+2} = \frac{(m+r)(m+r+1)-6}{(m+r+1)(m+r+2)}c_m;\ m \geq 0.$$

Thus, $$c_2 = \frac{r(r+1)-6}{(r+1)(r+2)} c_0$$

$$c_3 = \frac{(r+1)(r+2)-6}{(r+2)(r+3)} c_1$$

$$c_4 = \frac{(r+2)(r+3)-6}{(r+3)(r+4)} c_2 \text{ and}$$

$$c_5 = \frac{(r+3)(r+4)-6}{(r+4)(r+5)} \text{ and so on.}$$

For $r = 0$

$$c_2 = -3\, c_0,\ c_3 = -\frac{4}{6} c_1 = -\frac{2}{3} c_1$$

$$c_4 - 0,\ c_5 = \frac{6}{20} c_3 = -\frac{c_1}{5}, \text{ and so on.}$$

Therefore, the general solution is

$$y = x^0 \left[c_0 + c_1 x - 3c_0 x^2 - \frac{2}{3} c_1 x^3 + 0 \cdot x^4 - \frac{c_1}{5} x^5 + \ldots \right]$$

$$= c_0(1 - 3x^2) + c_1 \left(x - \frac{2}{3} x^3 - \frac{x^5}{5} + \ldots \right)$$

where, c_0 and c_1 are the arbitrary constants.

Note: If we consider $r = -1$, then we have

$$c_3 = 0 = c_5 = \ldots \text{ since } c_1 = 0$$

$$c_2 = -\frac{2}{3} c_0,\ c_4 = -\frac{6}{20} c_2 = +\frac{c_0}{5}, \text{ and so on.}$$

Therefore,

$$y = x\left[c_0 - \frac{2}{3} c_0 x^2 + \frac{c_0}{5} x^4 - \ldots \right]$$

$$= c_0 \left[x - \frac{2}{3} x^3 + \frac{1}{5} x^5 - \ldots \right]$$

So, in this case, we are able to find only one linearly independent solution of the given differential equation which is already contained in the solution obtained for $r = 0$.

EXAMPLE 13.19

Find the series solution about $x = 0$ of the differential equation $x^2y'' + xy' + (x^2 - 1)y = 0$.

Solution: We find that $x = 0$ is a regular singular point of the given differential equation.

Let the solution be written as $y = \sum_{m=0}^{\infty} c_m(x - 0)^{m+r}$.

Substituting y, y', y'' in the given differential equation, we get

$$\sum_{m=0}^{\infty} (m+r)(m+r-1)c_m x^{m+r} + \sum_{m=0}^{\infty} (m+r)c_m x^{m+r} + \sum_{m=0}^{\infty} c_m x^{m+r+2} - \sum_{m=0}^{\infty} c_m x^{m+r} = 0.$$

The lowest degree term (with $m = 0$) is the term containing x^r. Setting the coefficient of x^r to zero, we obtain $c_0[r(r-1) + r - 1] = 0$

$$\Rightarrow \qquad c_0(r^2 - 1) = 0$$

as $c_0 \neq 0$ we have $\qquad r^2 - 1 = 0 \Rightarrow r = \pm 1$

As $r_1 = 1$ and $r_2 = -1$ we have $r_1 \neq r_2$ and $r_1 - r_2 = 2$ is an integer. So, we have case III.

Since the lowest common term in each summation term is x^{r+2}, so equating to zero the next power of x which is x^{r+1} (with $m = 1$), we obtain

$$c_1[(r+1)r + (r+1) - 1] = 0$$

$$\Rightarrow \qquad c_1(r^2 + 2r) = 0$$

As $r^2 + 2r \neq 0$ for $r = 1, -1$; we obtain $c_1 = 0$.
For the remaining terms, we get

$$\sum_{m=2}^{\infty} [(m+r)^2 - 1]c_m x^{m+r} + \sum_{m=0}^{\infty} c_m x^{m+r+2} = 0.$$

Changing the index of summation term m to $m + 2$ in the first summation term, we get

$$\sum_{m=0}^{\infty} [(m+r+2)^2 - 1]c_{m+2} x^{m+r+2} + \sum_{m=0}^{\infty} c_m x^{m+r+2} = 0.$$

Setting the coefficient of x^{m+r+2} to zero, we obtain the recurrence relation

$$[(m+r+2)^2 - 1]c_{m+2} + c_m = 0$$

$$\Rightarrow \qquad c_{m+2} = -\frac{c_m}{(m+r+2)^2 - 1}; \; m \geq 0$$

$$\Rightarrow \qquad c_{m+2} = -\frac{c_m}{(m+r+1)(m+r+3)}; \; m \geq 0$$

Thus, $\qquad c_2 = -\dfrac{c_0}{(r+1)(r+3)}$

We note that c_2 becomes infinite when $r = -1$ and hence all the remaining coefficients also becomes infinite. Therefore, according to case III (b) to make c_2 finite let

$$c_0 = a_0(r - (-1)) = a_0(r + 1)$$

Clearly, $c_2 = -\dfrac{a_0}{(l+3)}$ is finite for $r = -1$.

Now, $c_3 = 0 = c_5 = \ldots$ since $c_1 = 0$

$$c_4 = -\frac{c_2}{(r+3)(r+5)} = \frac{1}{(r+3)^2(r+5)}, \text{ and so on.}$$

Therefore,

$$y = a_0x^r\left[(r+1) - \frac{1}{r+3}x^2 + \frac{1}{(r+3)^2(r+5)}x^4 - \ldots\right]$$

Thus,
$$y_1 = y\big|_{r=-1} = a_0x^{-1}\left[0 - \frac{1}{2}x^2 + \frac{1}{2^2 \cdot 4}x^4 - \ldots\right]$$

$$y_1 = -\frac{a_0}{2}\left[x - \frac{x^3}{2 \cdot 4} + \ldots\right]$$

Also,
$$\frac{\partial y}{\partial r} = a_0x^r(\log x)\left((r+1) - \frac{1}{r+3}x^2 + \frac{1}{(r+3)^2(r+5)}x^4 - \ldots\right)$$

$$+ a_0x^r\left(1 + \frac{1}{(r+3)^2}x^2 + \left(\frac{-2}{(r+3)^2(r+5)} - \frac{1}{(r+3)^2(r+5)^2}\right) - \ldots\right)$$

Hence, $y_2 = \dfrac{\partial y}{\partial r}\bigg|_{r=-1} = (\log x)y_1 + a_0x^{-1}\left(1 + \dfrac{x^2}{2^2} - \left(\dfrac{2}{2^2 \cdot 4} + \dfrac{1}{2^2 \cdot 4^2}x^4\right) - \ldots\right)$

Therefore, the general solution is

$$y = Ay_1 + By_2$$

$$= (A + B\log x)\frac{1}{2}\left(x - \frac{x^3}{2 \cdot 4} + \ldots\right) + \frac{B}{x}\left(1 + \frac{x^2}{2^2} - \left(\frac{2}{2^3 \cdot 4} + \frac{1}{2^2 \cdot 4^2}\right)x^4 - \ldots\right).$$

Note: For $r = 1$, we have

$$c_2 = -\frac{c_0}{2 \cdot 4}, \text{ and so on.}$$

Therefore,

$$y = x^1\left[c_0 - \frac{c_0}{2\cdot 4}x^2 + \ldots\right]$$

$$= c_0\left(x - \frac{x^3}{2\cdot 4} + \ldots\right)$$

which is already contained in the general solution obtained above by using only $r = -1$.

EXAMPLE 13.20

Find the series solution about $x = 1$ of the following differential equation

$$2(1 - x)y'' - xy' + y = 0.$$

Solution: The point $x = 1$ is a regular singular point of the given differential equation.

Let the solution be written as $y = \sum_{m=0}^{\infty} c_m(x - 1)^{m+r}$.

Let $x - 1 = u \Rightarrow x = u + 1$

So, we obtain

$$-2u\frac{d^2y}{du^2} - (u+1)\frac{dy}{du} + y = 0 \text{ about } u = 0$$

and

$$y = \sum_{m=0}^{\infty} c_m u^{m+r}.$$

Substituting y, dy/du, d^2y/du^2 in the given differential equation, we get

$$-2\sum_{m=0}^{\infty}(m + r)(m + r - 1)c_m u^{m+r-1} - \sum_{m=0}^{\infty}(m + r)c_m u^{m+r} - \sum_{m=0}^{\infty}(m + r)c_m u^{m+r-1} + \sum_{m=0}^{\infty} c_m u^{m+r} = 0.$$

The lowest degree term (with $m = 0$) is the term containing u^{r-1}. Setting the coefficient of u^{r-1} to zero, we obtain

$$c_0[-2r(r - 1) - r] = 0 \Rightarrow c_0(2r^2 + Br) = 0$$

as $c_0 \neq 0$ we have $\quad r(2r + 1) = 0$

$\Rightarrow \quad r = 0, 1/2$

As $r_1 = 0$ and $r_2 = 1/2$, we have $r_1 \neq r_2$ and $r_1 - r_2 = -1/2$ is an integer, so we have case III.

Since the lowest common term in each summation term is u^r, so the remaining terms are

$$\sum_{m=1}^{\infty}(m + r)(-2m - 2r + 1)c_m x^{m+r-1} - \sum_{m=0}^{\infty}(m + r - 1)c_m x^{m+r} = 0.$$

Changing the index of summation from m to $m + 1$ in the first summation term, we get

$$\sum_{m=0}^{\infty} (m + r + 1)(-2m - 2r - 1)c_{m+1}u^{m+r} - \sum_{m=0}^{\infty} (m + r - 1)c_m u^{m+r} = 0.$$

Setting the coefficient of u^{m+r} to zero, we obtain the recurrence relation

$$(m + r + 1)(-2m - 2r - 1)c_{m+1} - (m + r - 1)c_m = 0$$

$$\Rightarrow \quad c_{m+1} = -\frac{(m+r-1)}{(m+r+1)(2m+2r+1)}c_m; \; m \geq 0$$

Thus,

$$c_1 = -\frac{r-1}{(r+1)(2r+1)}c_0$$

$$c_2 = -\frac{r}{(r+2)(2l+3)}c_1$$

$$c_3 = -\frac{(r+1)}{(r+3)(2r+5)}c_2, \text{ and so on.}$$

For $r = 0$

$$c_1 = c_0$$

$$c_2 = 0 = c_3 = \ldots$$

Therefore,

$$y_1 = u^0[c_0 + c_0 u] = c_0[1 + (x - 1)]$$

Similarly, for $r = 1/2$

$$c_1 = -\frac{-1/2}{\frac{3}{2}\cdot 2}c_0 = \frac{1}{6}c_0$$

$$c_2 = -\frac{1/2}{\frac{5}{2}\cdot 4}c_1 = -\frac{c_1}{20} = -\frac{c_0}{120}$$

$$c_3 = -\frac{3/2}{\frac{7}{2}\cdot 6}c_2 = -\frac{c_2}{14} = \frac{c_0}{1680}, \text{ and so on.}$$

Therefore,

$$y_2 = u^{1/2}\left[c_0 + \frac{1}{6}c_0 u - \frac{c_0}{120}u^2 + \frac{c_0}{1680}u^3 - \ldots\right]$$

$$= c_0(u - 1)^{1/2}\left[1 + \frac{(u-1)}{6} - \frac{(u-1)^2}{120} + \frac{(u-1)^3}{1680} - \ldots\right]$$

Hence, the general solution is

$$y = A[1 + (u - 1)] + B(u - 1)^{1/2}\left[1+\frac{(u-1)}{6}-\frac{(u-1)^2}{120}+\frac{(u-1)^3}{1680}-\ldots\right]$$

where, A and B are not the arbitrary constants.

EXAMPLE 13.21

Find the series solution about the point $x = 2$ of the following differential equation $x(x - 2)y'' + 4y' + 3y = 0$.

Solution: We find that $x = 2$ is a regular singular point of the given differential equation.

Let the solution be written as $y = \sum_{m=0}^{\infty} c_m(x - 2)^{m+r}$.

Let $$u = x - 2 \Rightarrow x = u + 2$$

So, we have

$$(u + 2)u\frac{d^2y}{du^2}+4\frac{dy}{du}+ 3y = 0 \text{ about } u = 0 \text{ and } y = \sum_{m=0}^{\infty} c_m u^{m+r}.$$

Substituting y, dy/du, d^2y/du^2 in the given differential equation, we get

$$\sum_{m=0}^{\infty}(m + r)(m + r - 1)c_m u^{m+l} + 2\sum_{m=0}^{\infty}(m + r)(m + r - 1)c_m u^{m+r-1}$$
$$+ 4\sum_{m=0}^{\infty}(m + r)c_m u^{m+r-1} + 3\sum_{m=0}^{\infty} c_m u^{m+r} = 0.$$

The lowest degree term (with $m = 0$) is the term containing u^{r-1}. Setting the coefficient of u^{r-1} to zero, we obtain

$$c_0[2r(r - 1) + 4r] = 0$$

$$\Rightarrow \quad c_0(2r^2 + 2r) = 0$$

as $c_0 \neq 0$ we have $$2r(r + 1) = 0$$

$$\Rightarrow \quad r = 0, -1$$

As $r_1 = 0$ and $r_2 = -1$ we have $r_1 \neq r_2$ and $r_1 - r_2 = 1$ is an integer. So, we have case III. Since the lowest common term in each summation term is x^r, so the remaining terms are

$$\sum_{m=1}^{\infty}(m + r)(2m + 2r + 2)c_m u^{r-1} + \sum_{m=0}^{\infty}[(m + r)(m + r - 1) + 3]c_m u^r = 0.$$

Changing the index of summation from m to $m + 1$ in the first summation terms, we get

$$\sum_{m=0}^{\infty}(m + r + 1)(2m + 2r + 4)c_{m+1}u^r + \sum_{m=0}^{\infty}[(m + r)(m + r - 1) + 3)c_m u^r = 0.$$

Setting the coefficient of u^r equal to zero, we obtain the recurrence relation

$$(m + r + 1)(2m + 2r + 4)c_{m+1} + [(m + r)(m + r - 1) + 3]c_m = 0$$

$$\Rightarrow \qquad c_{m+1} = -\frac{(m+r)(m+r-1)+3}{2(m+r+1)(m+r+2)} c_m; \; m \geq 0$$

Thus, $$c_1 = -\frac{r(r-1)+3}{2(r+1)(r+2)} c_0$$

So, if we take $r = -1$, c_1 is not defined.
Therefore, to make c_1 finite, we put
$c_0 = a_0(r - (-1)) = a_0(r + 1)$ and use case III (b) to find the general solution

Thus, $$c_1 = -\frac{r(r-1)+3}{2(r+2)} a_0$$

Also for $r = 0$, $$c_1 = -\frac{3}{4} c_0.$$

Similarly, $$c_2 = -\frac{(r+1)r+3}{2(r+2)(r+3)} c_1$$

$$\Rightarrow \qquad c_2 = \frac{(r(r+1)+3)(r(r-1)+3)}{4(r+2)^2(r+3)} a_0, \text{ and so on}$$

Therefore,

$$y = u^r a_0 \left[(r+1) - \frac{r(r-1)+3}{2(r+2)} u + \frac{(r^2+r+3)(r^2-r+3)}{4(r+2)^2(r+3)} u^2 - \ldots \right] \tag{1}$$

The first linearly independent solution is

$$y_1 = y|_{r=-1} = u^{-1} a_0 \left\{ 0 - \frac{5}{2} u + \frac{15}{8} u^2 - \frac{15}{32} u^3 + \ldots \right\}$$

$$= -\frac{5}{2} a_0 \left\{ 1 - \frac{3}{4}(x-2) + \frac{3}{10}(x-2)^2 - \ldots \right\}.$$

For the second linearly independent solution differentiate Eq. (1) partially with respect to r, we get

$$\frac{\partial y}{\partial r} = u^r (\log u) a_0 \left[(r+1) - \frac{r^2-r+3}{2(r+2)} u + \frac{(r^2+(r+3)(r^2-r+3)}{4(r+2)^2(r+3)} u^2 - \ldots \right]$$

$$+ \; u^r a_0 \left[1 - \left(\frac{2r-1}{2(r+2)} - \frac{r^2-r+3}{2(r+2)^2} \right) u + \left(\frac{(2r+1)(r^2-r+3)}{4(r+2)^2(r+3)} + \frac{(r^2+r+3)(2r-1)}{4(r+2)^2(r+3)} \right. \right.$$

$$\left. \left. + \frac{-2(r^2+r+3)(r^2-r+3)}{4(r+2)^3(r+2)} - \frac{(r^2+r+3)(r^2-r+3)}{4(r+2)^2(r+3)^2} \right) u^2 - \ldots \right]$$

Therefore,

$$y_2 = \left.\frac{\partial y}{\partial r}\right|_{r=-1} = (\log u)y_1 + u^{-1}a_0\left[1 - \left(\frac{-3}{2} - \frac{5}{2}\right)u + \left(\frac{-5}{8} + \frac{-9}{8} - \frac{30}{8} - \frac{15}{16}\right)u^2 - \ldots\right]$$

$$= (\log(x-2))y_1 + (x-2)^{-1}a_0\left[1 + 4(x-2) - \frac{103}{16}(x-2)^2 - \ldots\right]$$

Therefore, the required general solution is

$$y = (A + B\log(x-2))\left[1 - \frac{3}{4}(x-2) + \frac{3}{16}(x-2)^2 - \ldots\right]$$

$$+ B(x-2)^{-1}\left[1 + 4(x-2) - \frac{103}{16}(x-2)^2 - \ldots\right]$$

where, A and B are the arbitrary constants.

Note: The second indicial root will give us the one linearly independent solution only which is already contained in the general solution found above

For $r = 0$, $$c_1 = -\frac{3}{4}c_0$$

$$c_2 = \frac{3}{16}c_0, \text{ and so on.}$$

Hence, $$y = u^0\left\{c_0 - \frac{3}{4}c_0 u + \frac{3}{16}c_0 u^2 - \ldots\right\}$$

$\Rightarrow$ $$y = c_0\left\{1 - \frac{3}{4}(x-2) + \frac{3}{16}(x-2)^2 - \ldots\right\}.$$

EXAMPLE 13.22

Solve in series the differential equation $(1 - x^2)y'' - 2xy' + n(n+1)y = 0$ (where, n is a positive integer) about $x = 0$.

Solution: The given point $x = 0$ is an ordinary point of the given differential equation, so we can also use the power series method here. But we will find its solution by using the Frobenius method. Let the solution be written as $y = \sum_{m=0}^{\infty} c_m(x-0)^{m+r}$.

Substituting y, y', y'' in the given differential equation, we get

$$\sum_{m=0}^{\infty}(m+r)(m+r-1)c_m x^{m+r-2} - \sum_{m=0}^{\infty}(m+r)(m+r-1)c_m x^{m+r}$$

$$-2\sum_{m=0}^{\infty}(m+r)c_m x^{m+r} + n(n+1)\sum_{m=0}^{\infty}c_m x^{m+r} = 0.$$

The lowest degree term (with $m = 0$) is the term containing x^{r-2}. Setting the coefficient of x^{r-2} zero, we obtain

$$c_0[r(r-1)] = 0$$

as $c_0 \neq 0$ we have
$$r(r-1) = 0$$

$\Rightarrow$
$$r = 0, 1$$

As $r_1 = 0$ and $r_2 = 1$ we have $r_1 \neq r_2$ and $r_1 - r_2 = -1$ is an integer so we have case III. Since the lowest common term in each summation term is x^r, so equating to zero the next power of x which is x^{r-1} (with $m = 1$), we obtain

$$c_1(r+1)r = 0.$$

For $r = 1$ $(r + 1)r = 0 \Rightarrow c_1$ is arbitrary. So, we have case III (c).

For $r = 1$ $(r + 1)r \neq 0 \Rightarrow c_1 = 0$.

Therefore, the indicial root $r = 0$, gives the complete solution as the correspanding solution contains two arbitrary constants. The remaining terms are

$$\sum_{m=2}^{\infty} (m+r)(m+r-1)c_m x^{m+l-2} - \sum_{m=0}^{\infty} [(m+r)(m+r+1) - n(n+1)]c_m x^{m+r} = 0.$$

Changing the index of summation from m to $m + 2$ in the first summation term, we get

$$\sum_{m=0}^{\infty} (m+r+2)(m+r+1)c_{m+2} x^{m+r} - \sum_{m=0}^{\infty} [(m+r)(m+r+1) - n(n+1)c_m x^{m+r} = 0.$$

Setting the coefficient of x^{m+r} equal to zero, we obtain the recurrence relation

$$(m+r+2)(m+r+1)c_{m+2} - [(m+r)(m+r+1) - n(n+1)]c_m = 0$$

$\Rightarrow$
$$c_{m+2} = \frac{(m+r)(m+r+1) - n(n+1)}{(m+r+1)(m+r+2)} c_m;\ m \geq 0.$$

Thus,
$$c_2 = \frac{r(r+1) - n(n+1)}{(r+1)(r+2)} c_0$$

$$c_3 = \frac{(r+1)(r+2) - n(n+1)}{(r+2)(r+3)} c_1$$

$$c_4 = \frac{(r+2)(r+3) - n(n+1)}{(r+3)(r+4)} c_2$$

$$c_5 = \frac{(r+3)(r+4) - n(n+1)}{(r+4)(r+5)} c_3, \text{ and so on.}$$

For $r = 0$

$$c_2 = -\frac{n(n+1)}{2!} c_0$$

$$c_3 = \frac{2 - n^2 - n}{6} c_1 = -\frac{(n+2)(n-1)}{3!} c_1$$

$$c_4 = \frac{6-n^2-n}{3-4}c_2 = \frac{(n+3)(n-2)}{3-4}c_2$$

$$= \frac{(n+3)(n+1)(n-2)n}{4!}c_0$$

$$c_5 = \frac{12-n^2-n}{4-5}c_3 = \frac{(n+4)(n-3)}{4-5}c_3$$

$$= \frac{(n+4)(n+2)(n-1)(n-3)}{5!}c_1, \text{ and so on.}$$

Therefore, the general solution is

$$y = x^0\left[c_0 + c_1x - c_0\frac{n(n+1)}{2!}x^2 - c_1\frac{(n+2)(n-1)}{3!}x^3\right.$$

$$\left.+ c_0\frac{(n+3)(n+1)(n-2)n}{4!}x^4 + \frac{(n+4)(n+2)(n-3)(n-1)}{5!}x^5 - \ldots\right]$$

$$\Rightarrow \qquad y = c_0\left[1 - \frac{n(n+1)}{2!}x^2 - \frac{(n+3)(n+1)(n-2)}{4!}x^4 + \ldots\right]$$

$$+ c_1\left[x - \frac{(n+2)(n-1)}{3!}x^3 + \frac{(n+4)(n+2)(n-1)(n-3)}{5!}x^5 - \ldots\right]$$

where, c_0 and c_1 are the arbitrary constants.

Note: The second indicial root $r = 1$, will give us one linearly independent solution only which is already contained in the general solution found above

For $r = 1$ we have $c_1 = 0 = c_3 = c_5 = \ldots$

$$c_2 = \frac{2-n(n+1)}{2\cdot 3}c_0 = -\frac{n^2+n-2}{3!}c_0$$

$$= -\frac{(n+2)(n-1)}{3!}c_0$$

$$c_4 = \frac{12-n(n+1)}{4\cdot 5}c_2 = -\frac{n^2+n-12}{4\cdot 5}c_2$$

$$= -\frac{(n+4)(n-3)}{4\cdot 5}c_2 = \frac{(n+4)(n+2)(n-1)(n-3)}{5!}c_0$$

Hence,

$$y = x^1\left[c_0 - c_0\frac{(n+2)(n-1)}{3!}x^2 + \frac{(n+4)(n+2)(n-1)(n-3)}{5!}x^4 - \ldots\right]$$

$$= c_0\left[x - \frac{(n+2)(n-1)}{3!}x^3 + \frac{(n+4)(n+2)(n-1)(n-3)}{5!}x^5 - \ldots\right]$$

▲▲

EXERCISE 13.2

Find the series solution of the following differential equation by the Frobenius method about $x = 0$.

1. $x^2y'' + xy' + (x^2 - n^2)y = 0$ where n is a fraction.
2. $9x(1 + x)y'' - 6y' + 2y = 0$.
3. $3x(1 - x)y'' + 2(1 - x)y' + 4y = 0$.
4. $2x(1 - x)y'' + (1 - x)y' + 3y = 0$.
5. $2x^2(1 + x^2)y'' - 3x(1 + x^2)y' + 2y = 0$.
6. $xy'' + (1 - x)y' + 3y = 0$.
7. $(x + x^2)y'' + (1 + x)y' - y = 0$.
8. $4(x^2 + x^4)y'' + 16x^3y' + y = 0$.
9. $xy'' - y = 0$.
10. $y'' + x^2y = 0$.
11. $x^2y'' + 6xy + (6 - 4x^2)y = 0$.
12. $x^2y'' + 4xy' + (x^2 + 2)y = 0$.
13. $x^2y'' + xy' + (x^2 - 1)y = 0$.
14. $2x^2y'' + xy' - 3y = 0$.
15. $y'' - 2xy' + 2xy = 0$.
16. Find the series solution about $x = -1$ of the following differential equation
$$9x(1 + x)y'' - 6y' + 2y = 0.$$
17. Solve the following different equation using Frobenius method
$$xy'' + 3y' - x^2y = 0.$$

ANSWERS

Exercise 13.2

1. $r = \pm n,\ c_1 = c_3 = \ldots = 0$

$$c_{m+2} = -\frac{c_m}{(m+r+2+n)(m+r+2-n)};\ m \geq 0.$$

$$y = x^r\left[1 - \frac{x^2}{(r+2+n)(r+2-n)} + \frac{x^4}{(r+4+n)(r+4-n)(r+2+n)(r+2-n)} - \ldots\right]$$

$y_1 = y$ at $r = n$ and $y^2 = y$ at $r = -n$.

2. $r = 0, 5/3; c_{m+1} = -\dfrac{[9(m+r)(m+r-1)+2]}{3(m+r+1)(3m+3r-2)} c_m; \ (m \geq 0)$

$$y_1 = 1 + \frac{x}{3} - \frac{1}{9}x^3 + \frac{5}{81}x^5 - \ldots = (1+x)^{1/3} \text{ and } y_2 = x^{5/3}\left[1 - \frac{1}{2}x + \frac{7}{22}x^2 - \ldots\right].$$

3. $r = 0, 1/3; c_{m+1} = \dfrac{[(m+r)(3m+3r-1)-4}{(m+r+1)(3m+3r+2)} c_m; \ (m \geq 0)$

$$y_1 = 1 - 2x + \frac{2}{5}x^2 + \frac{1}{10}x^3 + \ldots \text{ and } y_2 = x^{1/3}(1-x).$$

4. $r = 0, 1/2; c_{m+1} = \dfrac{(m+r)(2m+2r-1)-3}{(m+r+1)(2m+2r+1)} c_m; \ (m \geq 0)$

$$y_1 = 1 - 3x + x^2 + \frac{1}{5}x^3 + \ldots \text{ and } y^2 = x^{1/2}(1-x).$$

5. $r = 2, 1/2; c_{m+1} = -\dfrac{(m+r)(2m+2r-5)c_m}{(m+r+2)(2m+2r-1)+2}; \ (m \geq 0)$

$$y_1 = x^2\left[1 + \frac{1}{7}x^2 - \frac{5}{77}x^4 + \ldots\right] \text{ and } y_2 = x^{1/2}(1+x^2).$$

6. $r = 0, 0; c_{m+1} = -\dfrac{(m+r-3)}{(m+r+1)^2} c_m; \ (m \geq 0)$

$$y_1 = \left[1 - \frac{3}{1^2}x + \frac{2\cdot 3}{1^2\cdot 2^2}x^2 - \frac{1\cdot 2\cdot 3}{1^2\cdot 2^2\cdot 3^2}x^3 + \ldots\right] \text{ and } y_2 = (\log x)y_1 + \left[7x - \frac{23}{4}x^2 + \ldots\right].$$

7. $r = 0, 0; c_{m+1} = -\dfrac{(m+r-1)}{(m+r+1)} c_m; \ (m \geq 0)$

$$y_1 = 1 + x \text{ and } y_2 = (\log x)y_1 + \left[-2x - \frac{x^2}{2} + \frac{x^3}{6} + \ldots\right].$$

8. $r = 1/2, 1/2; c_{m+1} = -\dfrac{4(m+r)(m+r+3)}{[2(m+r)+3]^2} c_m; \ (m \geq 0)$

$$y_1 = x^{1/2}\left[1 - \frac{7}{4^2}x + \frac{5\cdot 7\cdot 11}{4^2\cdot 8^2}x^4 - \ldots\right]$$

and $$y_2 = (\log x)y_1 + x^{1/2}\left[-\frac{9}{16}x^3 + \frac{1053}{2048}x^4 - \ldots\right].$$

9. $r = 0, 1;\ c_{m+1} = \dfrac{c_m}{(m+r)(m+r+1)};\ (m \geq 0)$

Set $c_0 = a_0(r - 0) = a_0 r$ where $a_0 \neq 0$

$$y_1 = x + \frac{x^2}{1^2 \cdot 2} + \frac{x^3}{1^2 \cdot 2^2 \cdot 3} + \ldots$$

and $y_2 = y_1(\log x) + \left[1 - \dfrac{x}{1^2} - \left(\dfrac{2}{1^3 \cdot 2} + \dfrac{1}{1^2 \cdot 2^2}\right)x^2\right.$

$$\left. - \left(\frac{2}{1^3 \cdot 2^2 \cdot 3} + \frac{2}{1^2 \cdot 2^3 \cdot 3} + \frac{1}{1^2 \cdot 2^2 \cdot 3^2}\right)x^3 - \ldots\right].$$

10. $r = 0$; $r = 0$ gives complete solution $c_{m+1} = -\dfrac{c_m}{(m+r+3)(m+r+4)};\ (m \geq 0)$

$$y_1 = 1 - \frac{x^4}{3 \cdot 4} + \frac{x^8}{3 \cdot 4 \cdot 7 \cdot 8} - \ldots$$

and
$$y_2 = x - \frac{x^5}{4 \cdot 5} + \frac{x^9}{4 \cdot 5 \cdot 8 \cdot 9} - \ldots$$

11. $r = -2, -3;\ c_{m+2} = \dfrac{4c_m}{(m+r+2)(m+r+7)+6};\ (m \geq 0)$

$r = -3$ gives complete solution

$$y = c_0 x^{-3}\left[1 + 2x^2 + \frac{2}{3}x^4 + \ldots\right] + c_1 x^{-3}\left[x + \frac{2}{3}x^3 + \frac{2}{15}x^5 + \ldots\right].$$

12. $r = -1, -2$; $r = -2$ gives complete solution $c_{m+2} = -\dfrac{c_m}{(m+l+2)(m+l+5)+2};\ (m \geq 0)$
$y_1 = (\cos x)/x^2$ and $y_2 = (\sin x)/x^2$.

13. $r = \pm 1;\ c_{m+2} = -\dfrac{c_m}{(m+r+1)(m+r+3)};\ (m \geq 0)$

Set $c_0 = a_0(r + 1)$ where $a_0 \neq 0$

$$y_1 = -\frac{1}{2}\left(x - \frac{x^3}{2 \cdot 4} + \ldots\right)$$

and
$$y_2 = \left\{(\log x)y_1 + \frac{1}{x}\left\{1 + \frac{x^2}{2^2} - \left(\frac{2}{2^3 \cdot 4} + \frac{1}{2^2 \cdot 4^2}\right)x^4 + \ldots\right\}\right\}.$$

14. $r = -1, 3/2$; $y = 1/x$ and $y_2 = x^{3/2}$.

15. $r = 0, 1$ where, $a_0 \neq 0$ $r = 0$ gives complete solution

$$y_1 = 1 - \frac{2n}{2!}x^2 + \frac{2^2 n(n-2)}{4!}x^4 - \ldots$$

and
$$y_2 = x\left\{1 - \frac{2(n-1)}{3!}x^2 + \frac{2^2(n-1)(n-3)}{5!}x^4 - \ldots\right\}.$$

16. $r = 0, 1/3;\ c_{m+1} = \dfrac{9(m+r)(m+r-1)+2}{(m+r+1)(9m+9r+6)}c_m;\ (m \geq 0)$

$$y_1 = 1 + \frac{1}{3}(x+1) + \frac{1}{45}(x+1)^2 + \frac{1}{162}(x+1)^3 + \ldots \text{ and } y_2 = (x+1)^{1/3}.$$

17. $y_1 = \dfrac{1}{2!} + \dfrac{1-4}{5!}x^3 + \dfrac{1-4-7}{8!}x^6 + \ldots$ and $y_2 = x^{-2}\left\{1 + \dfrac{2}{3!}x^3 + \dfrac{2-5}{6!}x^6 + \ldots\right\}.$

14

CHAPTER

Legendre Polynomials and Bessel Functions

In many problems of mathematical physics and engineering, we come across second order linear differential equations, whose solutions give rise to special functions different from standard functions like sine, cosine, exponential, logarithmic functions, etc. In this chapter, we will discuss two important second order linear differential equations and obtain their solutions giving two special functions.

Legendre Differential Equation and Legendre Polynomials

An equation of the form

$$(1 - x^2)y'' - 2xy' + n(n + 1)y = 0 \tag{1}$$

where, n is a real constant, is known as Legendre differential equation. This equation has a number of applications in the physical problems involving spherical geometry and gravitation.

Equation (1) can also be written as

$$[(1 - x^2)y']' + n(n + 1)y = 0 \tag{2}$$

In most applications, n takes positive integral values. So in this chapter, we assume n to be a positive real number.

First, we will find a series solution of Eq. (1) using power series method about $x = 0$.

Therefore, let $y = \sum_{m=0}^{\infty} c_m x^m$.

Then $y' = \sum_{m=1}^{\infty} mc_m x^{m-1}$ and $y'' = \sum_{m=2}^{\infty} m(m - 1)c_m x^{m-2}$

Substituting y, y', y'' in Eq. (1), we obtain

$$\sum_{m=2}^{\infty} m(m - 1)c_m x^{m-2} - \sum_{m=2}^{\infty} m(m - 1)c_m x^m - 2\sum_{m=1}^{\infty} mc_m x^m + n(n + 1)\sum_{m=0}^{\infty} c_m x^m = 0.$$

By changing index of summation from m to $m + 2$ in the first term, we obtain

$$\sum_{m+2=2}^{\infty} (m+2)(m+1)c_{m+2}x^m - \sum_{m=2}^{\infty} m(m-1)c_m x^m - 2\sum_{m=1}^{\infty} mc_m x^m + n(n+1)\sum_{m=0}^{\infty} c_m x^m = 0$$

$$\Rightarrow \sum_{m=0}^{\infty} (m+2)(m+1)c_{m+2}x^m - \sum_{m=2}^{\infty} m(m-1)c_m x^m - 2\sum_{m=1}^{\infty} c_m x^m + n(n+1)\sum_{m=0}^{\infty} c_m x^m = 0$$

$$\Rightarrow 2c_2 + 6c_3 x + \sum_{m=2}^{\infty} (m+2)(m+1)c_{m+2}x^m - \sum_{m=2}^{\infty} m(m-1)c_m x^m - 2c_1 x - 2\sum_{m=2}^{\infty} mc_m x^m$$

$$+ n(n+1)\left[c_0 + c_1 x + \sum_{m=2}^{\infty} c_m x^m\right] = 0.$$

$$\Rightarrow \quad 2c_2 + 6c_3 x - 2c_1 x + n(n+1)(c_0 + c_1 x) + \sum_{m=2}^{\infty} [(m+2)(m+1)c_{m+2}$$

$$- [m(m-1) + 2m - n(n+1)]c_m]x^m = 0.$$

Equating the coefficients of various powers of x to zero, we get

$$2c_2 + n(n+1)c_0 = 0,\ 6c_3 - 2c_1 + n(n+1)c_1 = 0$$

and
$$(m+2)(m+1)c_{m+2} - (m^2 + m - n^2 - n)c_m = 0.$$

Therefore, $c_2 = -\dfrac{1}{2!} n(n+1)c_0,\ c_3 = \dfrac{1}{6}(2 - n^2 - n)c_1 = -\dfrac{1}{3!}(n-1)(n+2)c_1$

$$c_{m+2} = -\frac{(n-m)(m+n+1)}{(m+2)(m+1)} c_m;\ m \geq 2.$$

We have

$$c_4 = -\frac{(n-2)(n+3)}{4\cdot 3} c_2 = \frac{(n-2)\,n(n+1)(n+3)}{4\cdot 3\cdot 2} c_0$$

$$= \frac{1}{4!}(n-2)\,n(n+1)(n+3)c_0.$$

$$c_5 = -\frac{(n-3)(n+4)}{5\cdot 4} c_3 = \frac{1}{5!}(n-3)(n-1)(n+2)(n+4)c_1 \text{ and so on.}$$

Substituting in the power series solution, we obtain

$$y = c_0\left[1 - \frac{1}{2!} n(n+1)x^2 + \frac{1}{4!}(n-2)\,n(n+1)(n+3)x^4 - \dots\right]$$

$$+ c_1\left[x - \frac{1}{3!}(n-1)(n+2)x^3 + \frac{1}{5!}(n-3)(n-1)(n+2)(n+4)x^5 - \dots\right].$$

Thus, we see the solution is expressed in the ascending powers of x.

We now solve Eq. (1) using Frobenius method in series of descending powers of x as this form of solution finds applications to many physical problems. For this, let the series solution of Eq. (1) be

$$y = \sum_{m=0}^{\infty} c_m x^{k-m}$$

Then $\quad y' = \sum_{m=0}^{\infty} c_m(k-m)x^{k-m-1}$ and $y'' = \sum_{m=0}^{\infty} c_m(k-m)(k-m-1)x^{k-m-2}$.

Substituting y, y', y'' in Eq. (1), we obtain

$$\sum_{m=0}^{\infty} c_m(k-m)(k-m-1)x^{k-m-2} - \sum_{m=0}^{\infty} c_m(k-m)(k-m-1)x^{k-m} - 2\sum_{m=0}^{\infty} c_m(k-m)x^{k-m}$$

$$+ n(n+1)\sum_{m=0}^{\infty} c_m x^{k-m} = 0.$$

$$\Rightarrow \sum_{m=0}^{\infty} c_m(k-m)(k-m-1)x^{k-m-2} - \sum_{m=0}^{\infty} c_m[(k-m)(k-m+1) - n(n+1)]x^{k-m} = 0.$$

$$\Rightarrow \sum_{m=0}^{\infty} c_m(k-m)(k-m-1)x^{k-m-2} - \sum_{m=0}^{\infty} c_m(k-m-n)(k-m+n+1)x^{k-m} = 0.$$

Equating the coefficient of highest power of x obtained by putting $m = 0$ (since the series solution is assumed to be in the descending power of x), we get the indicial equation as

$$c_0(k-n)(k+n+1) = 0$$

Since $c_0 \neq 0$ we obtain

$$k = n, -(n+1)$$

Now equating the coefficient of next power of x (obtained by putting $m = 1$), we obtain

$$c_1(k-1-n)(k+n) = 0$$

Since $\quad (k-1-n)(k+n) \neq 0$ for $k = n, -(n+1)$,

We have $\quad c_1 = 0.$

To get a recurrence relation, we have

$$\sum_{m=0}^{\infty} c_m(k-m)(k-m-1)x^{k-m-2} - \sum_{m=2}^{\infty} c_m(k-m-n)(k-m+n+1)x^{k-m} = 0.$$

Changing the index of summation from m to $m - 2$ in the first term, we obtain

$$\sum_{m=2}^{\infty} c_{m-2}(k - m + 2)(k - m + 1)x^{k-m} - \sum_{m=2}^{\infty} c_m(k - m - n)(k - m + n + 1)x^{k-m} = 0.$$

So equating coefficient of x^{k-m} to zero, we obtain

$$c_m = \frac{(k-m+2)(k-m+1)}{(k-m-n)(k-m+n+1)} c_{m-2} \tag{3}$$

As $c_1 = 0$, we have $c_3 = c_5 = c_7 = \ldots = 0$.

To obtain $c_2, c_4, c_6, \ldots$ we consider two cases.

Case I: When $k = n$

From Eq. (3), we have

$$c_m = \frac{(n-m+2)(n-m+1)}{(-m)(-m+2n+1)} c_{m-2}$$

So that

$$c_2 = \frac{n(n-1)}{(-2)(2n-1)} c_0$$

$$c_4 = \frac{(n-2)(n-3)}{(-4)(2n-3)} c_2 = \frac{n(n-1)(n-2)(n-3)}{4 \cdot 2 \cdot (2n-1)(2n-3)} c_0, \text{ and so on.}$$

Thus, for $k = n$

$$y = c_6\left[x^n - \frac{n(n-1)}{2(2n-1)}x^{n-2} + \frac{n(n-1)(n-2)(n-3)}{2 \cdot 4 \cdot (2n-1)(2n-3)}x^{n-4} - \ldots\right] \tag{4}$$

Case II: When $k = -(n + 1)$

From Eq. (3), we have

$$c_m = \frac{(-n-m+1)(-n-m)}{(-m-2n-1)(-m)} c_{m-2}$$

So that

$$c_2 = \frac{(-n-1)(-n-2)}{(-2n-3)(-2)} c_0 = \frac{(n+1)(n+2)}{2(2n+3)} c_0$$

$$c_4 = \frac{(-n-3)(-n-4)}{(-2n-5)(-4)} c_2 = \frac{(n+1)(n+2)(n+3)(n+4)}{2 \cdot 4 \cdot (2n+3)(2n+5)} c_0, \text{ and so on.}$$

Thus, for $k = -(n+1)$

$$y = c_0^{x^{-n-1}}\left[+\frac{(n+1)(n+2)}{2(2n+3)}x^{-n-3} + \frac{(n+1)(n+2)(n+3)(n+4)}{2 \cdot 4 \cdot (2n+3)(2n+5)}x^{-n-5} + \ldots\right]. \tag{5}$$

If n is a positive integer and $c_0 = \dfrac{1\cdot 3\cdot 5\ldots(2n-1)}{n!}$ then solution (4) can be expressed as

$$y = \frac{1\cdot 3\cdot 5\ldots(2n-1)}{n!}\left[x^n - \frac{n(n-1)}{2(2n-1)}x^{n-2} + \frac{n(n-1)(n-2)(n-3)}{2\cdot 4\cdot(2n-1)(2n-3)}x^{n-4} - \ldots\right] \tag{6}$$

which is a terminating series. If we denote Eq. (6) by $P_n(x)$, then

$$P_n(x) = \sum_{m=0}^{[n/2]} (-1)^n \frac{(2n-2m)!}{2^n n!(n-2m)!(n-m)!} x^{n-2m}$$

where, $$\left[\frac{n}{2}\right] = \begin{cases} \dfrac{n}{2} & \text{if } n \text{ is even} \\ \dfrac{n-1}{2} & \text{if } n \text{ is odd} \end{cases}$$

$P_n(x)$ is called Legendre Polynomial of the first kind.

Further, if n is a positive integer and

$$c_0 = \frac{n!}{1\cdot 3\cdot 5\ldots(2n+1)}$$

then solution (5) can be expressed as

$$y = \frac{n!}{1\cdot 3\cdot 5\ldots(2n+1)}$$

$$\left\{x^{-n-1} + \frac{(n+1)(n+2)}{2(2n+3)}x^{-n-3} + \frac{(n+1)(n+2)(n+3)(n+4)}{2\cdot 4\cdot(2n+3)(2n+5)}x^{-n-5} + \ldots\right\} \tag{7}$$

We denote the solution (7) by $Q_n(x)$.

Here, $Q_n(x)$ is a non-terminating infinite series as n is positive and is known as Legendre polynomial of the second kind.

Hence, the complete solution of Eq. (1) is

$$y = AP_n(x) + BQ_n(x)$$

where A and B are arbitrary constants. We will now discuss an important relation known as Rodrigue's formula which is used to find the expressions for the Legendre polynomials.

Rodrigue's Formula

$$P_n(x) = \frac{1}{n!\,2^n}\frac{d^n}{dx^n}(x^2-1)^n$$

Proof: Let $u = (x^2 - 1)^n$

Differentiating, we get

$$u_1 = \frac{du}{dx} = 2nx(x^2 - 1)^{n-1} = \frac{2nxu}{x^2 - 1}$$

$$\Rightarrow \qquad (1 - x^2)u_1, + 2nxu = 0$$

Differentiating this equation $(n + 1)$ times by using the Leibnitz theorem, we obtain

$$^{n+1}C_0(1-x^2)\frac{d^{n+1}}{dx^{n+1}}(u_1) + ^{n+1}C_1\frac{d}{dx}(1-x^2)\frac{d^n}{dx^n}(u_1) + ^{n+1}C_2\frac{d^2}{dx^2}(1-x^2)\frac{d^{n-1}}{dx^{n-1}}(u_1)$$

$$+ 2n\left[^{n+1}C_0 x\frac{d^{n+1}}{dx^{n+1}}u + ^{n+1}C_1\frac{d}{dx}(x)\frac{d^n}{dx^n}u\right] = 0$$

Using $u_n = \dfrac{d^n u}{dx^n}$ we have

$$(1 - x^2)u_{n+2} + (n + 1)(-2x)u_{n+1} + \frac{(n+1)n}{2}(-2)u_n + 2n[xu_{n+1}\ (n + 1)u_n] = 0$$

$$\Rightarrow (1 - x^2)u_{n+2} - 2xu_{n+1} + n(n + 1)u_n = 0$$

$$\Rightarrow (1 - x^2)(u_n)'' - 2x(u_n)' + n(n + 1)u_n = 0.$$

which is the Legendre differential equation for $y = u_n$. Also, its finite series solution is $P_n(x)$.

Therefore, $P_n(x) = au_n$ where, a is a constant

$$= a\frac{d^n}{dx^n}(x^2 - 1)^n$$

To determine the constant a, we will use $P_n(1) = 1$ (which we will prove later on). Now let $x = 1$

$$P_n(1) = a\left[\frac{d^n}{dx^n}(x-1)^n(x+1)^n\right]_{x=1}$$

$$= a\left[\sum_{j=0}^{n} {}^nc_j \frac{d^j}{dx^j}(x-1)^n \frac{d^{n-j}}{dx^{n-j}}(x+1)^n\right]$$

$= a[n!\ (x + 1)^n +$ terms containing product of various powers of $(x - 1)$ and $(x + 1)]_{x=1}$

$= a[n!\ (1 + 1)^n + 0 + \ldots]$

$= a(n!\ 2^n).$

Therefore, $a = \dfrac{1}{n!\,2^n}$

and hence, $$P_n(x) = \frac{1}{n!\,2^n}\frac{d^n}{dx^n}(x^2-1)^n.$$

which is known as the Rodrigue's formula. Using Rodrigue's formula, we can find the values of $P_0(x)$, $P_1(x)$, $P_2(x)$, $P_3(x)$ and so on as given below.

For $n = 0$ $$P_0(x) = \frac{1}{0!\,2^0}\frac{d^0}{dx^0}(x^2-1)^0 = 1$$

For $n = 1$ $$P_1(x) = \frac{1}{1!\,2^1}\frac{d^1}{dx^1}(x^2-1)^1 = \frac{1}{2}(2x) = x$$

For $n = 2$ $$P_2(x) = \frac{1}{2!\,2^2}\frac{d^2}{dx^2}(x^2-1)^2$$

$$= \frac{1}{2}\frac{d}{dx}(x^3-x) = \frac{1}{2}(3x^2-1)$$

For $n = 3$ $$P_3(x) = \frac{1}{3!\,2^3}\frac{d^3}{dx^3}(x^2-1)^3$$

$$= \frac{1}{8}\frac{d^2}{dx^2}(x(x^2-1)^2) = \frac{1}{8}\frac{d}{dx}(5x^4-6x^2+1)$$

$$= \frac{5x^3-3x}{2}$$

Similarly, $P_4(x) = \dfrac{35x^4-30x^2+3}{8}$, and so on. ▲▲

EXAMPLE 14.1

Express $x^4 + 2x^3 - 6x^2 + 5x - 3$ in terms of Legendre polynomials.

Solution: Writing various powers of x in terms of Legendre polynomials using the above relations.

We get

$$1 = P_0(x) \quad x = P_1(x)$$

$$x^2 = \frac{1}{3}(2P_2 + 1) = \frac{1}{3}(2P_2 + P_0)$$

$$x^3 = \frac{1}{5}(2P_3 + 3x) = \frac{1}{5}(2P_3 + 3P_1)$$

$$x^4 = \frac{1}{35}(8P_4 + 30x^2 - 3) = \frac{1}{35}\left(8P_4 + 30 - \frac{1}{3}(2P_2 + P_0) - 3P_0\right) = \frac{1}{35}(8P_4 + 20P_2 + 7P_0)$$

Therefore,

$$x^4 + 2x^3 - 6x^2 + 5x - 3$$

$$= \frac{1}{35}(8P_4 + 20P_2 + 7P_0) + 2 - \frac{1}{5}(2P_3 + 3P_1) - 6 - \frac{1}{3}(2P_2 + P_0) + 5P_1 - 3P_0$$

$$= \frac{1}{35}(8P_4 + 28P_3 - 120P_2 + 217P_1 - 168P_0).$$

Generating function for Legendre polynomials

We shall prove that

$$(1 - 2xt + t^2)^{-1/2} = \sum_{n=0}^{\infty} t^n P_n(x);\ t \neq 1$$

The function on the left hand side is called the generating function of the Legendre polynomials. Which means $P_n(x)$ is the coefficient of t^n in the expansion of $(1 - 2xt + t^2)^{-1/2}$ in ascending powers of x.

Expanding $(1 - t(2x - t))^{-1/2}$ in binomial series, we get

$$(1 - t(2x - t)^{-1/2} = 1 + \frac{1}{2}t(2x - t) + \frac{1\cdot 3}{2\cdot 4}t^2(2x - t)^2$$

$$+\ldots+\frac{1\cdot 3\ldots(2(n-r)-1)}{2\cdot 4\ldots(2n-2r)}t^{n-1}(2x-t)^{n-1}+\ldots+\frac{1\cdot 3\ldots(2n-1)}{2\cdot 4\ldots(2n)}t^n(2n-t)^{n-1}+\ldots$$

$$= 1+\frac{2!}{(1!)^2\, 2^2}t(2x-t)+\frac{4!}{(2!)^2\, 2^2}t^2(2x-t)^2+\ldots+$$

$$\frac{(2n-2r)!}{[(n-r)!]^2 2^{2n-2r}}t^{n-r}(2x-t)^{n-r}+\ldots+\frac{(2n)!}{(n!)^2\, 2^{2n}}t^n(2x-t)^n+\ldots$$

The term in t^n from the term containing $t^{n-r}(2x - t)^{n-r}$

$$= \frac{(2n-2r)!}{[(n-r)!]^2 2^{2n-2r}}\, t^{n-r}\, {}^{n-r}c_r(-t)^r\,(2x)^{n-2r}$$

$$= \frac{(2n-2r)!}{[(n-r)!]^2\, 2^{2n-2r}}\times\frac{(n-r)!}{r!\,(n-2r)!}(-1)^r t^n (2x)^{n-2r}$$

$$= \frac{(-1)^r(2n-2r)!}{2^n\, r!\,(n-r)!(n-2r)!}x^{n-2r}t^n$$

Collecting all terms in t^n which will occur in the term containing $t^n(2x - t)^n$ and the preceding terms we see that terms in t^n

$$= \sum_{r=0}^{[n/2]} \frac{(-1)^r (2n-2r)!}{2^n r!(n-r)!(n-2r)!} x^{n-2r} t^n = P_n(x)t^n$$

where,
$$[n/2] = \begin{cases} n/2 & \text{if } n \text{ is even} \\ \dfrac{n-1}{2} & \text{if } n \text{ is odd} \end{cases}$$

Therefore, $(1 - 2xt + t^2)^{-1/2} = \sum_{n=0}^{\infty} P_n(x)t^n;\ t \neq 1.$

This proves that $P_n(x)$ is the coefficient of t^n in the expansion of $(1 - 2xt + t^2)^{-1/2}$.

EXAMPLE 14.2

Show that

(i) $P_n(1) = 1$

(ii) $P_n(-1) = (-1)^n$

(iii) $P_n'(1) = \frac{1}{2} n(n + 1)$

(iv) $P_n'(-1) = \frac{(-1)^{n-1}}{2} n(n + 1)$

(v) $P_n(-x) = (-1)^n P_n(x)$

Solution:

(i) Substituting $x = 1$ in the generating function,
We get

$$(1 - 2t + t^2)^{-1/2} = \sum_{n=0}^{\infty} P_n(1)t^n$$

$$\therefore \quad \sum_{n=0}^{\infty} P_n(1)t^n = (1 - t)^{-1} = 1 + t + t^2 + \dots + t^n + \dots$$

$$= \sum_{n=0}^{\infty} t^n$$

Comparing the coefficient of t^n on both sides, we get

$$P_n(1) = 1$$

(ii) Substituting $x = -1$ in the generating function,

We get

$$(1 + 2t + t^2)^{-1/2} = \sum_{n=0}^{\infty} P_n(-1)t^n$$

$$\therefore \qquad \sum_{n=0}^{\infty} P_n(-1)t^n = (1 + t)^{-1} = 1 - t + t^2 + \dots + (-1)^n t^n + \dots$$

$$= \sum_{n=0}^{\infty} (-1)^n t^n.$$

Comparing the coefficient of t^n on both sides, we get

$$P_n(-1) = (-1)^n$$

(iii) Since $P_n(x)$ satisfies Legendre differential equation $(1 - x^2)y'' - 2xy' + n(n + 1)y = 0$,

we get $$(1 - x^2)P_n''(x) - 2xP_n'(x) + n(n + 1) P_n(x) = 0 \tag{1}$$

Substituting $x = 1$, in Eq. (1) we get

$$0 - 2P_n'(1) + n(n + 1) P_n(1) = 0$$

$$\therefore \qquad P_n'(1) = \frac{n(n+1)}{2} \quad \text{as } P_n(1) = 1 \text{ by (i)}$$

(iv) Substituting $x = -1$ in Eq. (1), we get

$$0 - 2(-1) P_n'(-1) + n(n + 1)P_n(-1) = 0$$

$$\therefore \qquad P_n'(-1) = -\frac{n(n+1)}{2} P_n(-1)$$

$$= -\frac{(-1)^{n-1}}{2} n(n + 1) \text{ as } P_n(-1) = (-1)^n \text{ by (ii).}$$

(v) Replacing x by $-x$ in the generating function, we get

$$(1 + 2xt + t^2)^{-1/2} = \sum_{n=0}^{\infty} P_n(-x)t^n$$

Now, replacing t by $-t$, we get

$$(1 - 2xt + t^2)^{-1/2} = \sum_{n=0}^{\infty} P_n(-x)(-t)^n$$

$$\therefore \qquad \sum_{n=0}^{\infty} P_n(x)t^n = \sum_{n=0}^{\infty} P_n(-x)(-t)^n$$

Comparing the coefficient of t^n on both sides, we get

$$P_n(x) = P_n(-x)\,(-1)^n$$

or
$$P_n(-x) = (-1)^n P_n(x).$$

Substituting $x = 1$ in (v), we get

$$\begin{aligned} P_n(-1) &= (-1)^n P_n(1) \\ &= (-1)^n \text{ by using (i).} \end{aligned}$$

RECURRENCE RELATIONS FOR LEGENDRE POLYNOMIALS

The following recurrence relations can be easily derived from the generating function for $P_n(x)$.

(i) $(n + 1)P_{n+1}(x) = (2n + 1)xP_n(x) - nP_{n-1}(x)$.

(ii) $nP_n(x) = xP_n'(x) - P_{n-1}'(x)$

(iii) $(2n + 1)P_n(x) = P_{n+1}'(x) - P_{n-1}'(x)$

(iv) $P_{n+1}'(x) - xP_n'(x) = (n + 1)P_n(x)$

(v) $(1 - x^2)P_n'(x) = n[P_{n-1}(x) - xP_n(x)]$

(vi) $(1 - x^2)P_n'(x) = (n + 1)[xP_n(x) - P_{n+1}(x)]$

Proof:

(i) Differentiating the generating function

$$(1 - 2xt + t^2)^{-1/2} = \sum_{n=0}^{\infty} P_n(x)t^n$$

Partially with respect to t, we get

$$-\frac{1}{2}(1 - 2xt + t^2)^{-3/2}(-2x + 2t) = \sum_{n=0}^{\infty} nP_n(x)t^{n-1}$$

$$\Rightarrow \qquad (x - t)\frac{(1 - 2xt + t^2)^{1/2}}{1 - 2xt + t^2} = \sum_{n=0}^{\infty} nP_n(x)t^{n-1} \qquad (1)$$

$$\Rightarrow \qquad (x - t)\sum_{n=0}^{\infty} P_n(x)t^n = (1 - 2xt + t^2)\sum_{n=0}^{\infty} nP_n(x)t^{n-1}$$

Comparing the coefficient of t^n on both sides, we obtain

$$xP_n(x) - P_{n-1}(x) = (n + 1)P_{n+1}(x) - 2nxP_n(x) + (n - 1)P_{n-1}(x)$$

$$\Rightarrow \qquad (2n + 1)xP_n(x) - nP_{n-1}(x) = (n + 1)P_{n+1}(x)$$

(ii) Differentiating the generating function partially with respect to x, we obtain

$$-\frac{1}{2}(1 - 2xt + t^2)^{-3/2}(-2t) = \sum_{n=0}^{\infty} P_n'(x)t^n$$

$$t\frac{(1-2xt+t^2)^{1/2}}{1-2xt+t^2} = \sum_{n=0}^{\infty} P_n'(x)t^n \tag{2}$$

Dividing the expressions in Eqs (1) and (2), we get

$$\frac{x-t}{t} = \frac{\sum_{n=0}^{\infty} nP_n(x)t^{n-1}}{\sum_{n=0}^{\infty} P_n'(x)t^n}$$

$$\Rightarrow \qquad (x - t)\sum_{n=0}^{\infty} P_n'(x)t^n = \sum_{n=0}^{\infty} nP_n(x)t^n$$

Comparing the coefficient of t^n on both sides, we get

$$xP_n'(x) - P_{n-1}'(x) = nP_n(x).$$

(iii) Differentiating recurrence relation (i) with respect to x, we get

$$(n + 1)P_{n+1}'(x) = (2n + 1)\cdot 1 \cdot P_n(x) + (2n + 1)xP_n'(x) - nP_{n-1}'(x) \tag{3}$$

Substituting for $xP_n'(x)$ from recurrence relation (ii), we obtain

$$(n + 1)P_{n+1}'(x) = (2n + 1)\, P_n(x) + (2n + 1)[nP_n/x) + P_{n-1}'(x)] - nP_{n-1}'(x)$$

$$\Rightarrow (n + 1)P_{n+1}'(x) = (n + 1)\,(2n + 1)P_n(x) + (n + 1)\, P_{n-1}'(x)$$

$$\Rightarrow \quad (2n + 1)P_n(x) = P_{n+1}'(x) - P_{n-1}'(x).$$

(iv) Rewriting Eq. (3) again

$$(n + 1)P_{n+1}'(x) = (2n + 1)P_n(x) + (2n + 1)xP_n'(x) - nP_{n-1}'(x).$$

Substituting for $P_{n-1}'(x)$ from recurrence relation (ii), we obtain

$$(n + 1)P_{n+1}'(x) = (2n + 1)P_n(x) + (2n + 1)xP_n'(x) - n[xP_n'(x) - nP_n(x)]$$

$$\Rightarrow \quad (n + 1)P_{n+1}'(x) = (2n + 1 + n^2)P_n(x) + (n + 1)xP_n'(x)$$

$$\Rightarrow \qquad P_{n+1}'(x) = (n + 1)P_n(x) + xP_n'(x).$$

(v) From recurrence relations (ii) and (iv), we get

$$\text{(ii) } nP_n(x) = xP_n'(x) - P_{n-1}'(x)$$

and $$\text{(iv) } (n + 1)P_n(x) = P_{n+1}'(x) - xP_n'(x)$$

Multiplying (iii) by x, we get

$$nxP_n(x) = x^2P_n'(x) - xP_{n-1}'(x) \quad \text{(a)}$$

Replacing n by $n - 1$ in (iv), we get

$$ {}^nP_{n-1}(x) = P_n'(x) - {}^xP_{n-1}'(x) \quad \text{(b)}$$

Subtracting Eq. (a) from Eq. (b), we have

$$nP_{n-1}(x) - nxP_n(x) = (1 - x^2)P_n'(x)$$

$$\Rightarrow \quad n[P_{n-1}(x) - xP_n(x)] = (1 - x^2)P_n'(x).$$

(vi) From recurrence relation (i), we have

$$(n + 1)P_{n+1}(x) = (2n + 1)^xP_n(x) - {}^nP_{n-1}(x)$$

$$\Rightarrow \quad (n + 1)P_{n+1}(x) = (n + 1)xP_n(x) + nxP_n(x) - nP_{n-1}(x)$$

$$\Rightarrow \quad (n + 1)[P_{n+1}(x) - xP_n(x)] = n[xP_n(x) - P_{n-1}(x)]$$

Substituting in recurrence relation (v), we obtain

$$(n + 1)\,[-P_{n+1}(x) + xP_n(x)] = (1 - x^2)P_n'(x).$$

EXAMPLE 14.3

Prove that

$$P_{n+1}'(x) + P_n'(x) = P_0(x) + 3P_1(x) + 5P_2(x) + \ldots + (2n + 1)P_n(x).$$

Solution: Writing recurrence relation (iii),

$$P_{n+1}'(x) - P_{n-1}'(x) = (2n + 1)P_n(x)$$

For $n = 1, 2, \ldots n - 1, n$ we have

$$P_2'(x) - P_0'(x) = 3P_1(x)$$

$$P_3'(x) - P_1'(x) = 5P_2(x)$$

$$\ldots \ldots \ldots \ldots \ldots \ldots \ldots \ldots \ldots \ldots \ldots$$

$$P_n'(x) - P_{n-2}'(x) = (2n - 1)P_{n-1}(x)$$

$$P_{n+1}'(x) - P_{n-1}'(x) = (2n + 1)P_n(x)$$

On adding these, we obtain

$$P_{n+1}'(x) + P_n'(x) - P_0'(x) - P_1'(x) = 3P_1(x) + 5P_2(x) + \ldots + (2n + 1)P_n(x)$$

$$\Rightarrow \quad P_{n+1}'(x) + P_n'(x) = P_0'(x) + P_1'(x) + 3P_1(x) + 5P_2(x) + \ldots + (2n + 1)P_n(x)$$

Using $P_0(x) = 1$, $P_1(x) = x$

$$P_0'(x) = 0, \; P_1'(x) = 1 = P_0(x)$$

We obtain

$$P_{n+1}'(x) + P_n'(x) = P_0(x) + 3P_1(x) + 5P_2(x) + \dots + (2n+1)P_n(x).$$

Orthogonal property of Legendre polynomials

We shall prove that

$$\int_{-1}^{1} P_m(x)P_n(x)dx = \begin{cases} 0 & m \neq n \\ \dfrac{2}{2n+1} & m = n \end{cases}$$

Since $P_n(x)$ satisfies the Legendre differential equation $(1 - x^2)y'' - 2xy' + n(n+1)y = 0$, we have

$$(1 - x^2)P_m'' - 2xP'_m + m(m+1)P_m = 0 \quad \text{(a)}$$

and $$(1 - x^2)P_n'' - 2xP_n' + n(n+1)P_n = 0 \quad \text{(b)}$$

Multiply Eq. (a) by $P_n(x)$ and Eq. (b) by $P_m(x)$ and subtract. We obtain

$$(1 - x^2)(P_m'' P_n - P_mP_n'') - 2x(P_m' P_n - P_m P_n') + [m(m+1) - n(n+1)]P_mP_n = 0$$

Combining the first two terms, we get

$$\frac{d}{dx}[(1 - x^2)(P_m'P_n - P_mP_n')] + [m(m+1) - n(n+1)]P_mP_n = 0$$

Integrating over [–1, 1], we obtain

$$[(1 - x^2)(P_m'P_n - P_mP_n')]_{-1}^{1} + [m(m+1) - n(n+1) \times \int_{-1}^{1} P_mP_n dx = 0$$

$$\Rightarrow \quad 0 + [m(m+1) - n(n+1)] \int_{-1}^{1} P_m P_n dx = 0$$

Since $m \neq n \Rightarrow m(m+1) - n(n+1) \neq 0$

Therefore, $$\int_{-1}^{1} P_m(x)P_n(x)dx = 0 \quad \text{(c)}$$

when $m = n$, start with generating function

$$(1 - 2xt + t^2)^{-1/2} = \sum_{m=0}^{\infty} P_m(x)t^m \quad \text{(d)}$$

and $$(1 - 2xt + t^2)^{-1/2} = \sum_{n=0}^{\infty} P_n(x)t^n \quad \text{(e)}$$

Multiplying the corresponding sides of Eqs. (d) and (e), we get

$$\sum_{m=0}^{\infty}\sum_{n=0}^{\infty} P_m(x)\, P_n(x) t^{m+n} = (1 - 2xt + t^2)^{-1}$$

Integrating both sides with respect to x, we get

$$\sum_{m=0}^{\infty}\sum_{n=0}^{\infty}\left(\int_{-1}^{1} P_m(x)P_n(x)dx\right)t^{m+n} = \int_{-1}^{1}(1-2xt+t^2)^{-1}dx$$

Using Eq. (c), we obtain

$$\sum_{n=0}^{\infty}\left(\int_{-1}^{1}(P_n(x))^2 dx\right)t^{2n} = \int_{-1}^{1}\frac{dx}{1-2xt+t^2} = \left[\frac{\log(1-2xt+t^2)}{-2t}\right]_{-1}^{1}$$

$$= -\frac{1}{2t}[\log(1 - 2t + t^2) - \log(1 + 2t + t^2)]$$

$$= \frac{1}{2t}[\log(1 + t)^2 - \log(1 - t)^2]$$

$$= \frac{1}{t}[\log(1 + t) - \log(1 - t)]$$

$$= \frac{1}{t}\left\{\left(t - \frac{t^2}{2} + \frac{t^3}{3} - \dots\right) - \left(-t - \frac{t^2}{2} - \frac{t^2}{3} - \dots\right)\right\}$$

$$= 2\left\{1 + \frac{t^2}{3} + \frac{t^5}{5} + \dots\right\} = 2\sum_{n=0}^{\infty}\frac{t^{2n}}{2n+1}$$

Comparing the coefficient of t^{2n} from both sides, we obtain

$$\int_{-1}^{1} P_n^2(x)dx = \frac{2}{2n+1}.$$

EXAMPLE 14.4

Show that

$$\int_{-1}^{1} xP_n(x)\, P_{n-1}(x)dx = \frac{2n}{4n^2-1};\ n = 1, 2 \dots.$$

Solution: Using the recurrence relation (i), we have

$$(2n + 1)xP_n = (n + 1)P_{n+1} + nP_{n-1}$$

$$\therefore \qquad xP_n = \frac{n+1}{2n+1}P_{n+1} + \frac{n}{2n+1}P_{n-1}$$

Multiplying by P_{n-1}, we get

$$xP_nP_{n-1} = \frac{n+1}{2n+1}P_{n+1}P_{n-1} + \frac{n}{2n+1}P_{n-1}^2$$

Integrating both sides with respect to x, from $x = -1$ to $x = 1$, we get

$$\int_{-1}^{1} xP_n(x)P_{n-1}(x)dx = \frac{n+1}{2n+1}\int_{-1}^{1} P_{n+1}(x)dx + \frac{n}{2n+1}\int_{-1}^{1} P_{n-1}^2(x)dx$$

$$= \frac{n+1}{2n+1}(0) + \frac{n}{2n+1} \times \frac{2}{2(n-1)+1} \text{ (by orthogonal property)}$$

Hence, $\int_{-1}^{1} xP_n(x)P_{n-1}(x)dx = \frac{2n}{4n^2-1}$.

EXAMPLE 14.5

Show that $\int_{-1}^{1} (1 - x^2)P_m'(x)P_n'(x)dx = \begin{cases} 0 & m \neq n \\ \frac{2n(n+1)}{2n+1} & m = n \end{cases}$

Solution: Integrating by parts, we obtain

$$\int_{-1}^{1} (1 - x^2)P_m'(x)P_n'(x)dx = [(1 - x^2)P_m'(x)P_n(x)]_{-1}^{1} - \int_{-1}^{1} \frac{d}{dx}((1 - x^2)P_m'(x))P_n(x)dx$$

$$= \int_{-1}^{1} P_n(x)\,[(1 - x^2)P_m''(x) - 2xP_m'(x)]dx$$

Since $P_m(x)$ satisfies Legendre differential equation $(1 - x^2)y'' - 2xy' + m(m + 1)y = 0$ we have

$$(1 - x^2)P_m''(x) - 2xP_m'(x) = -m(m + 1)P_m(x)$$

Substituting this, we obtain,

$$\int_{-1}^{1} (1 - x^2)P_m'(x)P_n'(x)dx = m(m + 1)\int_{-1}^{1} P_m(x)P_n(x)dx$$

when $m \neq n$ $\int_{-1}^{1} P_m(x)P_n(x)dx = 0$ (by orthogonal property)

$$\therefore \quad \int_{-1}^{1} (1 - x^2)P_m'(x)P_n'(x)dx = 0$$

when, $m = n$ $\int_{-1}^{1} P_m(x)P_n(x)dx = \frac{2}{2n+1}$ (by orthogonal property)

$$\therefore \quad \int_{-1}^{1} (1 - x^2)P_m'(x)P_n'(x)dx = n(n+1) \times \frac{2}{2n+1} = \frac{2n(n+1)}{2n+1}.$$

EXAMPLE 14.6

Using the generating function of Legendre polynomials, prove that

$$P_{2n+1}(0) = 0,\ P_{2n}(0) = (-1)^n \frac{1\cdot 3\cdot 5\ldots(2n-1)}{2\cdot 4\cdot 6\ldots(2n)}$$

Solution: Let $x = 0$ in generating function of Legendre polynomials, we have

$$(1 + t^2)^{-1/2} = \sum_{n=0}^{\infty} P_n(0)t^n$$

$$\therefore \quad \sum_{n=0}^{\infty} P_n(0)t^n 1 - \left(\frac{1}{2}\right)t^2 + \frac{(-1/2)(-1/2-1)}{2!}(t^2)^2$$

$$+\frac{(-1/2)(-1/2-1)(-1/2-2)}{3!}(t^2)^3 + \ldots + \frac{(-1/2)(-1/2-1)\ldots(-1/2-(n-1))}{n!} \times (t^2)^n + \ldots$$

$$= 1 - \frac{1}{2}t^2 + \frac{1\cdot 3}{2\cdot 4}t^4 - \frac{1\cdot 3\cdot 5}{2\cdot 4\cdot 6}t^6 + \ldots + \frac{(-1)^n 1\cdot 3\ldots(2n-1)}{2^n\ n!}t^{2n} + \ldots$$

$$= 1 - \frac{1}{2}t^2 - 1\frac{1\cdot 3}{2\cdot 4}t^4 - \frac{1\cdot 3\cdot 5}{2\cdot 4\cdot 6}t^6 + \ldots + \frac{(-1)^n 1\cdot 3\ldots(2n-1)}{2\cdot 4\ldots(2n)}t^{2n} + \ldots$$

$$= \sum_{n=0}^{\infty} \frac{(-1)^n 1\cdot 3\ldots(2n-1)}{2\cdot 4\ldots(2n)} t^{2n}$$

Comparing coefficient of t^{2n+1}, t^{2n} on both sides, we get

$$P_{2n+1}(0) = 0$$

and

$$P_{2n} = \frac{(-1)^n 1\cdot 3\ldots(2n-1)}{2\cdot 4\ldots(2n)}.$$

EXAMPLE 14.7

Prove that (i) $\int_{-1}^{1} P_n(x)dx = \begin{cases} 2 & \text{if } n = 0 \\ 0 & \text{if } n \geq 1 \end{cases}.$

Solution: When $n = 0\ P_0(x) = 1$

$$\therefore \quad \int_{-1}^{1} P_0(x)dx = \int_{-1}^{1} 1 \cdot dx = [x]_{-1}^{1} = 2$$

when $n \geq 1$, using Rodrigue's formula, we have

$$\int_{-1}^{1} P_n(x)dx = \frac{1}{2^n\, n!}\int_{-1}^{1} \frac{d^n}{dx^n}(x^2-1)^n dx$$

$$= \frac{1}{2^n\, n!}\left[\frac{d^{n-1}}{dx^{n-1}}(x^2-1)^n\right]_{-1}^{1}$$

$$= \frac{1}{2^n\, n!}\left[\frac{d^{n-1}}{dx^{n-1}}(x-1)^n(x+1)^n\right]_{-1}^{1}$$

$$= \frac{1}{2^n\, n!}\left[{}^{n-1}c_0\left(\frac{d^{n-1}}{dx^{n-1}}(x-1)^n\right)(x+1)^n + {}^{n-1}c_1\left(\frac{d^{n-2}}{dx^{n-2}}(x-1)^n\right)\frac{d}{dx}(x+1)^n\right.$$

$$\left. + \ldots + {}^{n-1}c_{n-1}(x-1)^n\frac{d^{n-1}}{dx^{n-1}}(x+1)^n\right]_{-1}^{1} \quad \text{(by using Leibnitz Theorem)}$$

$$= \frac{1}{2^n\, n!}[n!\,(x-1)\,(x+1)^n + \ldots + n!\,(x+1)(x-1)^n]_{-1}^{1} = 0.$$

$$\left\{\because \frac{d^n}{dx^n}(ax+b)^m = a^n\frac{m!}{(m-n)!}(ax+b)^{m-n}\right\}$$

EXAMPLE 14.8

Using the Rodrigue's formula, prove that

$$\int_{-1}^{1} x^m P_n(x)dx = 0,\ m < n.$$

Solution: Using Rodrigue's formula, we have

$$\int_{-1}^{1} x^m P_n(x)dx = \frac{1}{2^n\, n!}\int_{-1}^{1} x^m\frac{d^n}{dx^n}(x^2-1)^n dx$$

$$= \frac{1}{2^n\, n!}\left\{\left[x^m\frac{d^{n-1}}{dx^{n-1}}(x^2-1)^n\right]_{-1}^{1} \int_{-1}^{1} mx^{m-1}\frac{d^{n-1}}{dx^{n-1}}(x^2-1)^n dx\right\}$$

$$= \frac{1}{2^n\ n!}\left\{0 - m\int_{-1}^{1} x^{m-1}\frac{d^{n-1}}{dx^{n-1}}(x^2-1)^n\,dx\right\}$$

{First term is zero using Example 14.7}

$$= \frac{(-1)^1 m}{2^n\ n!}\left\{\left[x^{m-1}\frac{d^{n-2}}{dx^{n-2}}(x^2-1)^n\right]_{-1}^{1} - \int_{-1}^{1}(m-1)x^{m-2}\times\frac{d^{n-2}}{dx^{n-2}}(x^2-1)^n\,dx\right\}$$

$$= \frac{(-1)^2 m(m-1)}{2^n\ n!}\int_{-1}^{1} x^{m-2}\frac{d^{n-2}}{dx^{n-2}}(x^2-1)^n\,dx$$

{First term is again zero using Example 14.7}

$$= \frac{(-1)^m m(m-1)\ldots 3\cdot 2\cdot 1}{2^n\ n!}\int_{-1}^{1} x^{m-m}\frac{d^{n-m}}{dx^{n-m}}\ (x^2-1)^n dx$$

(On continuing the similar steps)

$$= \frac{(-1)^m m!}{2^n\ n!}\left[\frac{d^{n-m-1}}{dx^{n-m-1}}(x^2-1)\right]_{-1}^{1}$$

$$= \frac{(-1)^n\ m!}{2^n\ n!}\times 0 = 0.$$

EXAMPLE 14.9

Using the Rodrigue's formula, show that

$$\int_{-1}^{1} P_n^2(x)dx = \frac{2}{2n+1}.$$

Solution: Using Rodrigue's formula, we obtain

$$\int_{-1}^{1} P_n^2(x)dx = \frac{1}{(2^n\ n!)^2}\int_{-1}^{1} D^n(x^2-1)^n D^n(x^2-1)^n\,dx \ \left\{\text{Using } D = \frac{d}{dx}\right\}$$

$$= \frac{1}{(2^n\ n!)^2}\left\{[D^n(x^2-1)^n D^{n-1}(x^2-1)^n]_{-1}^{1} - \int_{-1}^{1} D^{n+1}(x^2-1)^n \times D^{n-1}(x^2-1)^n\,dx\right\}$$

Since $[D^m(n^2-1)^n]_{-1}^{1} = 0$ for $m < n$, we have

$$\int_{-1}^{1} P_n^2(x)dx = \frac{-1}{(2^n\, n!)^2}\int_{-1}^{1} D^{n+1}(x^2-1)^n\, D^{n-1}(x^2-1)^n\, dx$$

[Integrating by parts $(n-1)$ times].

$$= \frac{(-1)^n}{(2^n\, n!)^2}\int_{-1}^{1} D^{n+m}(x^2-1)^n\, D^{n-n}(x^2-1)^n\, dx$$

$$= \frac{(-1)^n}{(2^n\, n!)^2}\int_{-1}^{1} D^{2n}(x^2-1)^n\,(x^2-1)^n\, dx$$

$$= \frac{(-1)^n}{(2^n\, n!)^2}\int_{-1}^{1} (2n)!\,(x^2-1)^n\, dx$$

$$= \frac{(2n)!}{(2^n\, n!)^2}\int_{-1}^{1} (1-x^2)^n\, dx$$

$$= \frac{2(2n)!}{(2^n\, n!)^2}\int_{0}^{1} (1-x^2)dx \qquad \{\text{Let } x = \sin\theta\}$$

$$= \frac{2(2n)!}{(2^n\, n!)^2}\int_{0}^{\pi/2} \cos^{2n+1}\theta\, d\theta$$

$$= \frac{2(2n)!}{(2^n\, n!)^2}\,\frac{2n(2n-2)\ldots 4\cdot 2}{(2n+1)(2n-1)\ldots 3\cdot 1}$$

$$= \frac{2(2n)!}{(2^n\, n!)^2}\,\frac{[2n(2n-2)\ldots 4\cdot 2]^2}{(2n+1)2n(2n-1)\ldots 3\cdot 2\cdot 1}$$

$$= \frac{2(2n)!}{(2^n\, n!)^2}\,\frac{(2^n\, n!)^2}{(2n+1)!} = \frac{2}{2n+1}$$

EXAMPLE 14.10

Show that

$$\int_{-1}^{1} x^2 P_{n-1}(x)P_{n+1}(x)dx = \frac{2n(n+1)}{(2n-1)(2n+1)(2n+3)}$$

Solution: We have from the recurrence relation (i)

$$(2n + 1)xP_n = (n + 1)P_{n+1} + nP_{n-1}$$

$$\Rightarrow \qquad xP_n = \frac{n+1}{2n+1}P_{n+1} + \frac{n}{2n+1}P_{n-1}$$

$$\therefore \qquad xP_{n-1} = \frac{n}{2n-1}P_n + \frac{(n-1)}{2n-1}P_{n-2}$$

and

$$xP_{n+1} = \frac{n+2}{2n+3}P_{n+2} + \frac{n+1}{2n+3}P_n$$

$$\therefore \qquad x^2P_{n-1}P_{n+1} = \frac{1}{(2n-1)(2n+3)}[n(n+2)P_nP_{n+2} + n(n+1)P_n^2$$

$$+ (n-1)(n+2)P_{n-2}P_{n+2} + (n^2-1)P_{n-2}P_n]$$

Integrating both sides from 1 to 1 and using arthogonality of Legendre polynomials, we get

$$\int_{-1}^{1} x^2P_{n-1}(x)P_{n+1}(x)dx = \frac{n(n+1)}{(2n-1)(2n+3)}\int_{-1}^{1} P_n^2(x)dx$$

$$= \frac{2n(n+1)}{(2n-1)(2n+1)(2n+3)}$$

EXAMPLE 14.11

Prove that $\int_{-1}^{1} xP_n(x)P_m'(x)dx = 0$ or 2 or $\frac{2n}{2n+1}$.

Solution: From the recurrence relation (ii), we have

$$nP_n = xP_n' - P_{n-1} \tag{1}$$

Again from recurrence relation (iii), we have

$$P_{n+1}' - P_{n-1}' = (2n+1)P_n \tag{2}$$

Replacing n by $n - 2$, $n - 4$, $n - 6$, succesively in Eq. (2), we get

$$\left.\begin{aligned} P_{n-1}' - P_{n-3}' &= (2n-3)P_{n-2} \\ P_{n-3}' - P_{n-5}' &= (2n-7)P_{n-4} \end{aligned}\right\} \tag{3}$$

Adding Eqs. (1) and (3), we get

$$xP_n' = nP_n + (2n-3)P_{n-2} + (2n-7)P_{n-4} + \ldots$$

Replacing n by m, we get

$$xP_m' = mP_m + (2m-3)P_{m-2} + (2m-7)P_{m-4} + \dots$$

Multiplying both sides by P_n, we get

$$xP_nP_m' = mP_nP_m + (2m-3)P_nP_{m-2} + (2m-7)P_nP_{m-4} + \dots$$

Integrating both sides with respect to x from -1 to 1, we get

$$\int_{-1}^{1} xP_n(x)P_m'(x)dx = m\int_{-1}^{1} P_n(x)P_m(x)dx + (2m-3)\int_{-1}^{1} P_n(x)P(x)_{m-2} \times dx + (2m-7)\cdot \int_{-1}^{1} P_n(x)P_{m-4}(x)dx + \dots$$

Three cases arise

Case I: When n is different from m, $m-2$, $m-4$, and so on. Then

$$\int_{-1}^{1} xP_n(x)P_m'(x)dx = 0 + 0 + 0 + \dots = 0 \text{ \{using orthogonal property\}}.$$

Case II: When $n = m$. Then

$$\int_{-1}^{1} xP_n(x)P_m'(x)dx = n \cdot \frac{2}{2n+1} + 0 + 0 + \dots = \frac{2n}{2n+1} \text{ \{using orthogonal property\}}.$$

Case III: When $n = m-2$. Then $n \neq m$, $n \neq m-4$, ... and so on. So, we obtain

$$\int_{-1}^{1} xP_n(x)P_m'(x)dx = 0 + (2m-3) \times \frac{2}{2(m-2)+1} + 0 + \dots \text{ \{using orthogonal property\}}$$

$$= (2m-3) \times \frac{2}{2m-3} = 2$$

Similarly, we can prove that

$$\int_{-1}^{1} xP_n(x)P_m'(x)dx = 2 \text{ when } n = m-4 \text{ or } n = m-6 \text{ etc.}$$

Thus, combining all the three cases, we have

$$\int_{-1}^{1} xP_n(x)P_m'(x)dx = 0 \text{ or } 2 \text{ or } \frac{2n}{2n+1}.$$

▲▲

EXERCISE 14.1

1. Express $2 - 3x + 4x^2$ in terms of Legendre polynomials.

2. Show that $x^4 = \frac{8}{35}P_4(x) + \frac{4}{7}P_2(x) + \frac{1}{5}P_0(x)$.

3. Using the Rodrigue's formula, prove that

$$\int_{-1}^{1} f(x)P_n(x)dx = \frac{(-1)^n}{2^n\, n!}\int_{-1}^{1}(x^2-1)^n f^{(n)}(n)dx.$$

4. Show that

$$\int P_n(x)dx = \frac{1}{2n+1}[P_{n+1}(x) - P_{n-1}(x)].$$

5. Express $5P_4(x) + 10P_3(x) + 2P_2 + P_1(x)$ in terms of powers of x.

6. Prove that $(2n + 1)(1 - x^2)P_n'(x) = n(n + 1)(P_{n-1}(x) - P_{n+1}(x)$.

7. Express x^5 in terms of Legendre polynomials.

8. Show that $\sum_{n=0}^{\infty}\frac{P_n(x)}{n+1} = \log\left(\frac{\sqrt{1-x}+\sqrt{2}}{\sqrt{1-x}}\right)$; $-1 < x < 1$.

   [*Hint:* Integrate generating function with respect to t from 0 to 1]

9. Show that $\sum_{n=0}^{\infty}(2n + 1)P_n(x)t^n = (1 - t^2)(1 - 2xt + t^2)^{-3/2}$

   [*Hint:* Differentiate generating function with respect to t and then multiply by $2t$].

10. Prove that

    (i) $P_{2n}'(0) = 0$

    (ii) $P_{2n+1}'(0) = \frac{(-1)^n(2n+1)!}{2^{2n}(n!)^2}$.

11. Prove that

    (i) $\int_0^1 P_{2n}(x)dx = 0$

    (ii) $\int_0^1 P_{2n+1}(x)dx = (-1)^n$ [*Hint:* Use recurrence relation III].

12. Show that

$$\int_{-1}^{1} (x^2 - 1)P_{n+1}P_n'dx = \frac{2n(n+1)}{(2n+1)(2n+3)}$$

{*Hint:* Use $(2n + 1)(1 - x^2)P_n'(x) = n(n + 1)[P_{n-1}(x) - P_{n+1}(x)]$}.

13. Prove that $\int_{-1}^{1} P_n(x)(1 - 2xt + t^2)^{-1/2}dx = \frac{2t^n}{2n+1}$.

{*Hint:* Multiply generating function by $P_n(x)$}

14. Show that $\int_{-1}^{1} x^n P_n(x)dx = \frac{2^{n+1}(n!)^2}{(2n+1)!}$.

15. Show that $P_5'(x) = 9P_4(x) + 5P_2(x) + P_0(x)$.

ANSWERS

Exercise 14.1

1. $\frac{10}{3}P_0(x) - 3P_1(x) + \frac{8}{3}P_2(x)$.

5. $(175x^4 + 200x^3 - 126x^2 - 112x + 7)/8$.

7. $\frac{8}{63}\left[P_5(x) + \frac{7}{2}P_3(x) + \frac{69}{8}P_1(x)\right]$.

Bessel's differential equation and Bessel's function

An equation of the form

$$x^2y'' + xy' + (x^2 - \alpha^2)y = 0 \tag{1}$$

is known as Bessel equation of order α, where α is a real number. The solution of Eq. (1) is known as Bessel function. Bessel equations have a number of applications in physical and engineering science, like problems involving vibrations or heat conduction in cylindrical regions. The point $x = 0$ is a regular singular point of Eq. (1) and we find solution of Eq. (1) by using Frobenius method. Consider

$$y = \sum_{m=0}^{\infty} c_m x^{m+r}$$

Substituting $y' = \sum_{m=0}^{\infty} (m + r)c_m x^{m+r-1}$ and $y' = \sum_{m=0}^{\infty} (m + r)(m + r - 1)c_m x^{m+r-2}$ in Eq. (1), we obtain

$$\sum_{m=0}^{\infty}(m+r)(m+r-1)c_m x^{m+r}+\sum_{m=0}^{\infty}c_m(m+r)x^{m+r}+\sum_{m=0}^{\infty}c_m x^{m+r+2}-\alpha^2\sum_{m=0}^{\infty}c_m x^{m+r}=0.$$

$$\Rightarrow \sum_{m=0}^{\infty}[(m+r)^2-\alpha^2]c_m x^{m+r}+\sum_{m=0}^{\infty}c_m x^{m+r+2}=0$$

Equating the coefficient of lowest power of x which is x^r (obtained by taking $m = 0$), to zero, we have

$$c_0(r^2-\alpha^2)=0 \Rightarrow r=\alpha, -\alpha \text{ as } c_0 \neq 0.$$

Equating the coefficient of next power of x that is x^{r+1} to zero, we obtain

$$[(r+1)^2-\alpha^2]c_1=0$$

Since for $r = \alpha, -\alpha$ $\quad (r+1)^2-\alpha^2 \neq 0$

We have $c_1 = 0$.

The remaining terms of the summation are

$$\sum_{m=2}^{\infty}[(m+r)^2-\alpha^2]c_m x^{m+r}+\sum_{m=0}^{\infty}c_m x^{m+r+2}=0.$$

Changing the index of summation from m to $m + 2$ in the first summation term, we have

$$\sum_{m=0}^{\infty}[(m+r+2)^2-\alpha^2]c_{m+2}x^{m+r+2}+\sum_{m=0}^{\infty}c_m x^{m+r+2}=0.$$

Equating the coefficient of x^{m+r+2} to zero, we get

$$c_{m+2}=\frac{-c_m}{(m+r+2)^2-\alpha^2};\ m \geq 0$$

Since $c_1 = 0$, we obtain $c_3 = 0 = c_5 = \ldots$

For $m = 0$
$$c_2=\frac{-c_0}{(r+2)^2-\alpha^2}$$

For $m = 2$
$$c_4=\frac{-c_2}{(r+4)^2-\alpha^2}$$

$$=\frac{c_0}{[(r+2)^2-\alpha^2][(r+4)^2-\alpha^2]},\ \text{and so on}$$

So we obtain

$$y=c_0x^r\left[1-\frac{1}{(r+2)^2-\alpha^2}x^2\frac{1}{[(r+2)^2-\alpha^2][(r+4)^2-\alpha^2]}-\ldots\right]$$

For $r = r$, we get

$$y_1 = c_0 x^r \left[1 - \frac{x^2}{2^2(1+r)} + \frac{x^4}{2^4(2!)(1+r)(2+r)} - \dots\right] \tag{2}$$

and for $r = -r$ we get

$$y = c_0 x^{-r} \left[1 - \frac{x^2}{2^2(1-r)} + \frac{x^4}{2^4(2!)(1-r)(2-r)} - \dots\right] \tag{3}$$

If we make a suitable choice of c_0, we obtain an important function called the Bessel's function of first kind of order 2 denoted by $J_r(x)$. Let the value of c_0 be chosen as

$$c_0 = \frac{1}{2^r \Gamma(r+1)}$$

where Γ is the common function defined by (see Appendix) $\Gamma(r) = \int_0^\infty x^{r-1} e^{-x} dx$

We have $\Gamma(r + 1) = r\Gamma(r)$
Using the above value of c_0, we obtain

$$y_1 = \left(\frac{x}{2}\right)^r \left[\frac{1}{\Gamma(r+1)} - \frac{1}{1!\,\Gamma(r+2)}\left(\frac{x}{2}\right)^2 + \frac{1}{2!\,\Gamma(r+2)}\left(\frac{x}{2}\right)^4 - \dots\right]$$

$$= \sum_{m=0}^{\infty} \frac{(-1)^r}{m!\,\Gamma(m+r+1)}\left(\frac{x}{2}\right)^{2m+r} = J_r(x)$$

Similarly, $$y_2 = \sum_{m=0}^{\infty} \frac{(-1)^{-r}}{m!\,\Gamma(m-r+1)}\left(\frac{x}{2}\right)^{2m-r} = J_{-r}(x)$$

Case I: When α is not an integer . In this case $J_r(x)$ and $J_{-r}(x)$ are linearly independent, hence the complete solution of Eq. (1) is

$$y = AJ_r(x) + BJ_{-r}(x).$$

Case II: When α is an integer.

In this case $J_{-r}(x) = (-1)^r J_r(x)$ (which we will prove later on):

Clearly, $J_r(x)$ and $J_{-r}(x)$ are not linearly independent, therefore $y_2(x)$ is not the second linearly independent solution. To find the second linearly independent solution, we proceed as follows.

Let $y = u(x)J_r(x)$ be solution of Eq. (1).

Substituting y, $y' = u'J_r + uJ_r'$ and $y'' = u''J_r + 2u'J_r' + uJ_r''$ in Eq. (1), we obtain

$$x^2(u''J_r + 2u'J_r' + uJ_r'') + x(u'J_r + uJ_r') + (x^2 - \alpha^2)uJ_r = 0$$

$$\Rightarrow \quad u[x^2J_r'' + xJ_r' + (x^2 - r^2)J_r] + x^2u''J_r + 2x^2u'J_r' + xu'J_r = 0 \tag{4}$$

Now, since $y_1 = J_\alpha(x)$ is the first linearly independent solution of Eq. (1), therefore

$$x^2J_r'' + xJ_r' + (x^2 - a^2)J_r = 0.$$

$\therefore$ Equation (4) reduces to

$$x^2u''J_r + 2x^2u'J_r' + xu'J_r = 0$$

Dividing by $x^2u'J_\omega$, we obtain

$$\frac{u''}{u'} + 2\frac{J_r'}{J_r} + \frac{1}{x} = 0.$$

Integrating with respect to x, we have

$$\log u' + 2\log J_r + \log x = \log B$$

$$\Rightarrow \quad \log(u'J_r^2x) = \log B$$

$$\Rightarrow \quad u' = \frac{B}{xJ_r^2} \text{ or } u = B\int\frac{dx}{xJ_r^2} + A$$

Thus,
$$y = AJ_r(x) + BJ_r(x)\int\frac{dx}{x[J_r(x)]^2}$$

Hence, the complete solution of Eq. (1) when α is an integer is

$$y = AJ_r(x) + B1/r(x)$$

where
$$1/r(x) = J_r(x)\int\frac{dx}{x[J_r(x)]^2}$$

is called the Bessel function of the second kind of order r.

Note: When α is an integer, we have $\Gamma(r + 1) = r!$

The Bessel function becomes

$$J_\alpha(x) = \sum_{m=0}^{\infty}\frac{(-1)^m}{m!\,(m+r)!}\left(\frac{x}{2}\right)^{2m+r}$$

EXAMPLE 14.12

Find the solution of the following differential equations using Bessel's function

(a) $x^2y'' + xy' + \left(x^2 - \frac{1}{16}\right)y = 0$

Solution: The given equation is the Bessel's differential equation with $\alpha = 1/4$ which is not an integer. Therefore, the complete solution is

$$y = AJ_{1/4}(x) + BJ_{-1/4}(x).$$

(b) $x^2y'' + xy' + (x^2 - 9)y = 0$

Solution: The given equation is the Bessel's differential equation with $\alpha = 3$ which is an integer.

Therefore, the complete solution is

$$y = AJ_3(x) + By_3(x)$$

where, $1/3(x)$ is Bessel's function of second kind of order 3.

(c) $x^2y'' + xy' + (9x^2 - 4)y = 0$

Solution: Let $z = 3x$

$$\therefore \qquad y' = \frac{dy}{dx} = \frac{dy}{dz}\frac{dz}{dx} = 3\frac{dy}{dz} \text{ and } y'' = \frac{d^2y}{dz^2} = \frac{d}{dz}\left(3\frac{dy}{dz}\right)\frac{dz}{dx} = 9\frac{d^2y}{dz^2}$$

On substituting, the given equation becomes

$$z^2\frac{d^2y}{dz^2} + z\frac{dy}{dz} + (z^2 - 4)y = 0$$

which is the Bessel's differential equation with $\alpha = 2$ which is an integer. Therefore, the complete solution of given equation is

$$y = AJ_2(z) + B1/2(z) = AJ_2(3x) + B1/2(3x).$$

EXAMPLE 14.13

Reduce the equation

$$4x^2y''+ + 4xy' + (64x^2 - 9)y = 0$$

Solution: Using the substitution $4x = z$ and hence, find the general solution of the equation.

When $4x = z$, we have

$$y' = \frac{dy}{dx} = \frac{dy}{dz}\frac{dz}{dx} = 4\frac{dy}{dz} \text{ and } y'' = \frac{d^2y}{dx^2} = \frac{d}{dz}\left(4\frac{dy}{dz}\right)\frac{dz}{dx} = 16\frac{d^2y}{dz^2}$$

The given equation reduces to

$$4z^2\frac{d^2y}{dx^2} = 4z\frac{dy}{dz} + (4z^2 - 9)y = 0$$

or

$$z^2\frac{d^2y}{dz^2} = z\frac{dy}{dz} + \left(z^2 - \frac{9}{4}\right)y = 0.$$

This equation is the Bessel's differential equation with $\alpha = 3/2$ which is not an integer. Hence, the general solution is

$$y = AJ_{3/2}(z) + BJ_{-3/2}(z) = AJ_{3/2}(4x) + BJ_{-3/2}(4x).$$

EXAMPLE 14.14

Find the solution of the differential equation

$$xy'' + y' + y = 0$$

by using the substitution $z = 2\sqrt{x}$.

Solution: When $z = 2\sqrt{x}$, we have

$$y' = \frac{dy}{dx} = \frac{dy}{dz}\frac{dz}{dx} = \frac{1}{\sqrt{x}}\frac{dy}{dz} \text{ and } y'' = \frac{d^2y}{dx^2} = \frac{1}{dx}\left(\frac{1}{\sqrt{x}}\frac{dy}{dz}\right)$$

$$= -\frac{1}{2x\sqrt{x}}\frac{dy}{dz} + \frac{1}{\sqrt{x}}\frac{d}{dz}\left(\frac{dy}{dz}\right)\frac{dz}{dx}$$

$$= -\frac{1}{2x\sqrt{x}}\frac{dy}{dz} + \frac{1}{x}\frac{d^2y}{dz^2}$$

Substituting in the given differential equation, we get

$$x\left(-\frac{1}{2x\sqrt{x}}\frac{dy}{dz} + \frac{1}{x}\frac{d^2y}{dz^2}\right) + \frac{1}{\sqrt{x}}\frac{dy}{dz} + y = 0$$

$$\Rightarrow \quad \frac{d^2y}{dz^2} + \frac{1}{2\sqrt{x}}\frac{dy}{dz} + y = 0$$

$$\Rightarrow \quad \frac{d^2y}{dz^2} + \frac{1}{z}\frac{dy}{dz} + y = 0 \qquad \left\{\text{Using } z = 2\sqrt{x}\right\}$$

$$\Rightarrow \quad z^2\frac{d^2y}{dz^2} + z\frac{dy}{dz} + (z^2 - 0)y = 0.$$

This equation is Bessel's equation with $\alpha = 0$ which is an integer. Therefore, the general solution is

$$y = AJ_0(z) + B1/0(z) = AJ_0\ (2\sqrt{x}) + B1/0\ (2\sqrt{x}).$$

EXAMPLE 14.15

Prove that

(a) $J_{1/2}(x) = \sqrt{\frac{2}{\pi x}}\sin x$

(b) $J_{-1/2}(x) = \sqrt{\frac{2}{\pi x}}\cos x$

(c) $[J_{1/2}(x)]^2 + [J_{-1/2}(x)]^2 = \dfrac{2}{\pi x}$

Solution: We have

$$J_{1/2}(x) = \sum_{m=0}^{\infty} \frac{(-1)^m}{m!\,\Gamma\left(m+\frac{1}{2}+1\right)}\left(\frac{x}{2}\right)^{2m+\frac{1}{2}}$$

$$= \left(\frac{x}{2}\right)^{1/2} \sum_{m=0}^{\infty} \frac{(-1)^m}{m!\left(m+\frac{3}{2}\right)}\left(\frac{x}{2}\right)^{2m}$$

$$= \left(\frac{x}{2}\right)^{1/2}\left[\frac{1}{\Gamma(3/2)} - \frac{1}{1!\,\Gamma(5/2)}\left(\frac{x}{2}\right)^2 + \frac{1}{2!\,\Gamma(7/2)}\left(\frac{x}{2}\right)^4 - \ldots\right]$$

$$= \left(\frac{x}{2}\right)^{1/2}\left[\frac{1}{\Gamma(3/2)} - \frac{1}{1!\frac{3}{2}\Gamma\left(\frac{3}{2}\right)}\left(\frac{x}{2}\right)^2 + \frac{1}{2!-\frac{5}{2}\cdot\frac{3}{2}\Gamma\left(\frac{3}{2}\right)}\left(\frac{x}{2}\right)^4 - \ldots\right]$$

$$= \left(\frac{x}{2}\right)^{1/2}\frac{1}{\Gamma(3/2)}\left[1-\frac{x^2}{3!}+\frac{x^4}{5!}-\ldots\right]$$

$$= \left(\frac{x}{2}\right)^{1/2}\frac{1}{\frac{1}{2}\Gamma\left(\frac{1}{2}\right)}\frac{1}{x}\left[x-\frac{x^3}{3!}+\frac{x}{5!}-\ldots\right]$$

$$= \sqrt{\frac{2}{\pi x}}\ \sin x \text{ which is (a).}$$

Similarly,

$$J_{-1/2}(x) = \sum_{m=0}^{\infty} \frac{(-1)^m}{m!\,\Gamma\left(m-\frac{1}{2}+1\right)}\left(\frac{x}{2}\right)^{2m-\frac{1}{2}}$$

$$= \left(\frac{x}{2}\right)^{1/2} \sum_{m=0}^{\infty} \frac{(-1)^m}{m!\,\Gamma\left(m+\frac{1}{2}\right)}\left(\frac{x}{2}\right)^{2m}$$

$$= \left(\frac{2}{x}\right)^{1/2}\left[\frac{1}{\Gamma(1/2)} - \frac{1}{1!\,\Gamma(3/2)}\left(\frac{x}{2}\right)^2 + \frac{1}{2!\,\Gamma(5/2)}\left(\frac{x}{2}\right)^4 - \ldots\right]$$

$$= \left(\frac{2}{x}\right)^{1/2}\left[\frac{1}{\Gamma(1/2)} - \frac{1}{1!\frac{1}{2}\Gamma\left(\frac{1}{2}\right)}\left(\frac{x}{2}\right)^2 + \frac{1}{2!\frac{3}{2}\cdot\frac{1}{2}\Gamma\left(\frac{1}{2}\right)}\left(\frac{x}{2}\right)^4 - \ldots\right]$$

$$= \left(\frac{2}{x}\right)^{1/2}\frac{1}{\Gamma(1/2)}\left[1 - \frac{x^2}{2!} + \frac{x^4}{4!} - \ldots\right]$$

$$= \sqrt{\frac{2}{\pi x}}\cos x \text{ which is (b).}$$

Squaring and adding (a) and (b), we obtain $[J_{1/2}(x)]^2 + [J_{-1/2}(x)]^2 = \frac{2}{\pi x}$ which is (c).

EXAMPLE 14.16

It α is an integer, show that $J_{-\alpha}(x) = (-1)^\alpha J_\alpha(x)$.

We have

$$J_{-\alpha}(x) = \sum_{m=0}^{\infty}\frac{(-1)^m}{m!\Gamma(m-\alpha+1)}\left(\frac{x}{2}\right)^{2m-\alpha}$$

$$= \sum_{m=0}^{\infty}\frac{(-1)^m}{m!(m-\alpha)!}\left(\frac{x}{2}\right)^{2m-\alpha}$$

$$= \sum_{m=\alpha}^{\infty}\frac{(-1)^m}{m!(m-\alpha)!}\left(\frac{x}{2}\right)^{2m-\alpha} \quad \{\text{using definition of } (m-\alpha)!\}$$

Changing the index of summation from m to $m + \alpha$, we obtain

$$J_{-\alpha}(x) = \sum_{m=0}^{\infty}\frac{(-1)^{m+\alpha}}{(m-\alpha)!\,m!}\left(\frac{x}{2}\right)^{2m+\alpha}$$

$$= (-1)^\alpha\sum_{m=0}^{\infty}\frac{(-1)^m}{m!\Gamma(m+\alpha+1)}\left(\frac{x}{2}\right)^{2m+\alpha}$$

$$= (-1)^\alpha J_\alpha(x).$$

which is the required result.

Recurrence relations for Bessel's function $J_\alpha(x)$

(i) $(x^\alpha J_\alpha(x))' = x^\alpha J_{\alpha-1}(x)$

(ii) $\int x^\alpha J_{\alpha-1}(x)dx = x^\alpha J_\alpha(x) + c$

(iii) $(x^{-\alpha}J_\alpha(x))' = -x^{-\alpha}J_{\alpha+1}(x)$

(iv) $\int x^{-\alpha} J_{\alpha+1}(x)dx = -x^{-\alpha} J_{\alpha}(x) + c$

(v) $xJ_{\alpha}'(x) = xJ_{\alpha-1}(x) - \alpha J_{\alpha}(x)$

(vi) $xJ_{\alpha}'(x) = \alpha J_{\alpha}(x) - xJ_{\alpha+1}(x)$

(vii) $2J_{\alpha}'(x) = J_{\alpha-1}(x) - J_{\alpha+1}(x)$

(viii) $2\alpha J_{\alpha}(x) = x[J_{\alpha-1}(x) + J_{\alpha+1}(x)]$

Solution: We have

$$J_{\alpha}(x) = \sum_{m=0}^{\infty} \frac{(-1)^m}{m!\,\Gamma(m+\alpha+1)} \left(\frac{x}{2}\right)^{2m+\alpha} \tag{1}$$

Multiplying Eq. (1) both sides by x^{α}, we get

$$x^{\alpha} J_{\alpha}(x) = \sum_{m=0}^{\infty} \frac{(-1)^m}{m!\,\Gamma(m+\alpha+1)} \left(\frac{x}{2}\right)^{2m+2\alpha}_{2m+\alpha}$$

Differentiating this equation with respect to x, we obtain

$$(x^{\alpha} J_{\alpha}(x))' = \sum_{m=0}^{\infty} \frac{(-1)^m}{m!\,\Gamma(m+\alpha+1)} (2m+2\alpha) \frac{x^{2m+2\alpha-1}}{2^{2m+\alpha}}$$

$$= x^{\alpha} \sum_{m=0}^{\infty} \frac{(-1)^m}{m!\,\Gamma(m+\alpha)} \left(\frac{x}{2}\right)^{2m+\alpha-1}$$

$$= x^{\alpha} J_{\alpha-1}(x)$$

Therefore, $(x^{\alpha} J_{\alpha}(x))' = x^{\alpha} J_{\alpha-1}(x)$ which is (i).

Integrating, we obtain (ii) $\int x^{\alpha} J_{\alpha-1}(x)dx = x^{\alpha} J_{\alpha}(x) + c$.

Multiplying Eq. (1) both sides by $x^{-\alpha}$, we get

$$x^{-\alpha} J_{\alpha}(x) = \sum_{m=0}^{\infty} \frac{(-1)^m}{m!\,\Gamma(m+\alpha+1)} \frac{x^{2m}}{2^{2m+\alpha}}$$

Differentiating this equation with respect to x, we obtain

$$(x^{-\alpha} J_{\alpha}(x))' = \sum_{m=0}^{\infty} \frac{(-1)^m}{m!\,\Gamma(m+\alpha+1)} 2m \frac{x^{2m-1}}{2^{2m+\alpha}}$$

$$= \sum_{m=1}^{\infty} \frac{(-1)^m}{(m-1)!\,\Gamma(m+\alpha+1)} \frac{x^{2m+1}}{2^{2m+\alpha-1}}$$

Changing the index of summation from m to $m + 1$, we have

$$(x^{-\alpha}J_\alpha(x))' = \sum_{m=0}^{\infty} \frac{(-1)^{m+1}}{m!\,\Gamma(m+\alpha+2)} \frac{x^{2m+1}}{2^{2m+\alpha-1}}$$

$$= -x^{-n} \sum_{m=0}^{\infty} \frac{(-1)^m}{m!\,\Gamma(m+\overline{\alpha+1}+1} \left(\frac{x}{2}\right)^{2m+\overline{\alpha+1}}$$

$$= -x^{-\alpha}J_{\alpha+1}(x)$$

Therefore, $(x^{-\alpha}J_\alpha(x))' = -x^{-\alpha}J_{\alpha+1}(x)$ which is (iii).
Integrating, we obtain (iv)

$$\int x^{-\alpha}J_{\alpha+1}(x)dx = -x^{-\alpha}J_\alpha(x) + c.$$

Simplifying (i), we get

$$x^{\alpha}J_\alpha'(x) + \alpha x^{\alpha-1}J_\alpha(x) = x^{\alpha}J_{\alpha-1}(x)$$

Dividing by $x^{\alpha-1}$, we get

$$xJ_\alpha'(x) + \alpha J_\alpha(x) = xJ_{\alpha-1}(x)$$

$$\Rightarrow \quad xJ_\alpha(x) = xJ_{\alpha-1}(x) - \alpha J_\alpha(x) \text{ which is (v).}$$

Similarly, simplifying (iii), we get

$$x^{-\alpha}J_\alpha'(x) - \alpha x^{-\alpha-1}J_\alpha(x) = -x^{-\alpha}J_{\alpha+1}(x)$$

Dividing by $x^{-\alpha-1}$, we obtain

$$xJ_\alpha'(x) - \alpha J_\alpha(x) = -xJ_{\alpha+1}(x)$$

$$\Rightarrow \quad xJ_\alpha'(x) = nJ_\alpha(x) - xJ_{\alpha+1}(x) \text{ which is (vi).}$$

Adding (v) and (vi), we get

$$2xJ_\alpha'(x) = x[J_{\alpha-1}(x) - J_{\alpha+1}(x)]$$

$$\Rightarrow \quad 2J_\alpha'(x) = J_{\alpha-1}(x) - J_{\alpha+1}(x) \text{ which is (vii).}$$

Subtracting (v) from (vi), we get

$$2\alpha J_\alpha(x) - xJ_{\alpha+1}(x) - xJ_{\alpha-1}(x) = 0$$

$$\Rightarrow \quad 2\alpha J_\alpha(x) = x[J_{\alpha+1}(x) + J_{\alpha+1}(x)] \text{ which is (iii).}$$

EXAMPLE 14.17

Express $J_5(x)$ in terms of $J_0(x)$ and $J_1(x)$.

Solution: We know that

$$2\alpha J_\alpha(x) = x[J_{\alpha-1}(x) + J_{\alpha+1}(x)]$$

$$\Rightarrow \qquad J_{\alpha+1}(x) = \frac{2\alpha}{x} J_\alpha(x) - J_{\alpha-1}(x)$$

Putting $\alpha = 1, 2, 3, 4$ successively, we get

$$J_2(x) = \frac{2}{x} J_1(x) - J_0(x)$$

$$J_3(x) = \frac{4}{x} J_2(x) - J_1(x)$$

$$J_4(x) = \frac{6}{x} J_3(x) - J_2(x)$$

$$J_5(x) = \frac{8}{x} J_4(x) - J_3(x)$$

Now,

$$J_3(x) = \frac{4}{x}\left[\frac{2}{x} J_1(x) - J_0(x)\right] - J_1(x)$$

$$= \left(\frac{8}{x^2} - 1\right) J_1(x) - \frac{4}{x} J_0(x)$$

$$J_4(x) = \frac{6}{x}\left[\left(\frac{8}{x^2} - 1\right) J_1(x) - \frac{4}{x} J_0(x)\right] - \left[\frac{2}{x} J_1(x) - J_0(x)\right]$$

$$= \left(\frac{48}{x^3} - \frac{8}{x}\right) J_1(x) + \left(1 - \frac{24}{x^2}\right) J_0(x)$$

$$J_5(x) = \frac{8}{x}\left[\left(\frac{48}{x^3} - \frac{8}{x}\right) J_1(x) + \left(1 - \frac{24}{x^2}\right) J_0(x)\right] - \left[\left(\frac{8}{x^2} - 1\right) J_1(x) - \frac{4}{x} J_0(x)\right]$$

$$= \left(\frac{384}{x^4} - \frac{72}{x^2} - 1\right) J_1(x) + \left(\frac{12}{x} - \frac{192}{x^3}\right) J_0(x)$$

EXAMPLE 14.18

Show that

$$J_{5/2}(x) = \sqrt{\frac{2}{\pi x}}\left[\frac{1}{x^2}(3 - x^2)\sin x - \frac{3}{x}\cos x\right]$$

and

$$J_{-5/2}(x) = \sqrt{\frac{2}{\pi x}}\left[\frac{1}{x^2}(3 - x^2)\cos x + \frac{3}{x}\sin x\right]$$

Solution: We know $$J_{\alpha+1}(x) = \frac{2\alpha}{x} J_\alpha(x) - J_{\alpha-1}(x)$$

when $\alpha = \frac{1}{2}$, we get

$$J_{3/2}(x) = \frac{1}{x} J_{1/2}(x) - J_{-1/2}(x)$$

when $\alpha = \frac{3}{2}$, we get

$$J_{5/2}(x) = \frac{3}{x} J_{3/2}(x) - J_{1/2}(x)$$

$$= \frac{3}{x}\left[\frac{1}{x}J_{1/2}(x) - J_{-1/2}(x)\right] - J_{1/2}(x)$$

$$= \left(\frac{3}{x^2} - 1\right)J_{1/2}(x) - \frac{3}{x}J_{-1/2}(x)$$

But $J_{1/2}(x) = \sqrt{\frac{2}{\pi x}} \sin x$ and $J_{-1/2}(x) = \sqrt{\frac{2}{\pi x}} \cos x$

Therefore,

$$J_{5/2}(x) = \sqrt{\frac{2}{\pi x}}\left[\frac{1}{x^2}(3 - x^2)\sin x - \frac{3}{x}\cos x\right]$$

Also, we know

$$J_{\alpha-1}(x) = \frac{2\alpha}{x} J_\alpha(x) - J_{\alpha+1}(x)$$

when $\alpha = -\frac{1}{2}$, we get $J_{-3/2}(x) = -\frac{1}{x}J_{-1/2}(x) - J_{1/2}(x)$

when $\alpha = -\frac{3}{2}$, we get $J_{-5/2}(x) = -\frac{3}{x}J_{-3/2}(x) - J_{-1/2}(x)$

$$= -\frac{3}{x}\left[-\frac{1}{x}J_{-1/2}(x) - J_{1/2}(x)\right] - J_{-1/2}(x)$$

$$= \left(\frac{3}{x^2} - 1\right) J_{-1/2}(x) + \frac{3}{x}J_{1/2}(x)$$

Therefore,

$$J_{-5/2}(x) = \sqrt{\frac{2}{\pi x}}\left[\frac{1}{x^2}(3-x^2)\cos x + \frac{3}{x}\sin x\right].$$

EXAMPLE 14.19

Prove that $\quad 4J_0''' + 3J_0' + J_3 = 0.$

Solution: We know $\quad (x^{-\alpha}J_\alpha)' = -x^{-\alpha}J_{\alpha+1}$

let $\alpha = 0$, we have

$$J_0' = -J_1$$

Differentiating with respect to x, we get

$$J_0'' = -J_1' = -\frac{1}{2}[J_0 - J_2] \qquad \{\because 2J_\alpha' = J_{\alpha-1} - J_{\alpha+1}\}$$

Differentiating again with respect to x, we get

$$J_0''' = -\frac{1}{2}(J_0' - J_2')$$

$$= -\frac{1}{2}\left(J_0' - \frac{1}{2}(J_1 - J_3)\right)$$

$$\Rightarrow \quad 4J_0''' + 2J_0' - J_1 + J_3 = 0$$

$$\Rightarrow \quad 4J_0''' + 2J_0' - (-J_0') + J_3 = 0$$

$$\Rightarrow \quad 4J_0''' + 3J_0' + J_3 = 0$$

EXAMPLE 14.20

Prove that

(i) $[xJ_\alpha(x)J_{\alpha+1}(x)]' = x[J_\alpha^2(x) - J_{\alpha+1}^2(x)]$

(ii) $[J_\alpha^2(x) + J_{\alpha+1}^2(x)]' = 2\left[\frac{\alpha}{x}J_\alpha^2(x) - \frac{\alpha+1}{x}J_{\alpha+1}^2(x)\right]$

Solution: We have

$$\frac{d}{dx}[xJ_\alpha(x)J_{\alpha+1}(x) = \frac{d}{dx}[x^{-\alpha}J_\alpha(x)x^{\alpha+1})J_{\alpha+1}(x)]$$

$$= x^{-\alpha}J_\alpha(x)\frac{d}{dx}[x^{\alpha+1}J_{\alpha+1}(x)] + x^{\alpha+1}J_{\alpha+1}(x)\frac{d}{dx}[x^{-\alpha}J_\alpha(x)]$$

$$= x^{-\alpha}J_\alpha(x)[x^{\alpha+1}J_\alpha(x)] + x^{\alpha+1}J_{\alpha+1}(x)\,(-x^{-\alpha}J_{\alpha+1}(x)] \quad \{\text{using (i), (iii)}\}$$

$$= x[J_\alpha^2(x) - J_{\alpha+1}^2(x)] \text{ which proves (i).}$$

Now $\dfrac{d}{dx}[J_\alpha^2(x) + J_{\alpha+1}^2(x)] = 2J_\alpha(x)J_\alpha'(x) + 2J_{\alpha+1}(x)J_{\alpha+1}'(x)$

We know

$$J_\alpha'(x) = \frac{\alpha}{x}J_\alpha(x) - J_{\alpha+1}(x)$$

and

$$J_\alpha'(x) = -\frac{\alpha}{x}J_\alpha(x) + J_{\alpha-1}(x)$$

$\Rightarrow$

$$J_{\alpha+1}'(x) = -\frac{(\alpha+1)}{x}J_{\alpha+1}(x) + J_\alpha(x)$$

Substituting these, we obtain

$$\frac{d}{dx}[J_\alpha^2(x) + J_{\alpha+1}^2(x)] = 2J_\alpha(x)\left[\frac{\alpha}{x}J_\alpha(x) - J_{\alpha+1}(x)\right]$$

$$+ 2J_{\alpha+1}(x)\left[-\frac{(\alpha+1)}{x}J_{\alpha+1}(x) + J_\alpha(x)\right]$$

$$= 2\left[\frac{\alpha}{x}J_\alpha^2(x) - \frac{\alpha+1}{x}J_{\alpha+1}^2(x)\right] \text{ which proves (ii).}$$

EXAMPLE 14.21

Show that

$$J_0^2 + 2(J_1^2 + J_2^2 + J_3^2 + \dots) = 1.$$

Solution: From the previous example, we have

$$[J_\alpha^2(x) + J_{\alpha+1}^2(x)]' = 2\left[\frac{\alpha}{x}J_\alpha^2(x) - \frac{\alpha+1}{x}J_{\alpha+1}^2(x)\right]$$

Substituting $\alpha = 0, 1, 2, \dots$ successively, we get

$$[J_0^2 + J_1^2]' = 2\left[0 - \frac{1}{x}J_1^2\right]$$

$$[J_1^2 + J_2^2]' = 2\left[\frac{1}{x}J_1^2 - \frac{2}{x}J_2^2\right]$$

$$[J_2^2 + J_3^2]' = 2\left[\frac{2}{x}J_2^2 - \frac{3}{x}J_3^2\right]$$

Adding, we obtain

$$[J_0^2 + 2(J_1^2 + J_2^2 + J_3^2 + \ldots)]' = 0.$$

Since $J_\alpha \to 0$ as $\alpha \to \infty$.

Integrating, we obtain

$$J_0^2 + 2(J_1^2 + J_2^2 + J_3^2 + \ldots) = c$$

when $x = 0$, we have

$$J_0(0) = 1$$

and $$J_\alpha(0) = 0 \text{ for } \alpha > 0.$$

Thus, $$1 = c$$

Therefore, we have

$$J_0^2 + 2(J_1^2 + J_2^2 + J_3^2 + \ldots) = 1.$$

EXAMPLE 14.22

Prove that

(i) $\int x^3 J_0(x)dx = x^3 J_1(x) - 2x^2 J_2(x) + c$

(ii) $\int \frac{1}{x} J_2(x)dx = -\frac{1}{x} J_1(x) + c$

(iii) $\int x J_0^2(x)dx = \frac{1}{2} x^2 [J_0^2(x) + J_1^2(x)]$

Solution: We have $\int x^\alpha J_{\alpha-1}(x)dx = x^\alpha J_\alpha(x) + c$

For $\alpha = 1 \int xJ_0 dx = xJ_1 + c$

Therefore, $\int x^2(xJ_0)dx = x^2(xJ_1) - \int (2x)\,(xJ_1)dx$

$$= x^3 J_1 - 2\int x^2 J_1 dx$$

$$= x^3 J_1 - 2x^2 J_2 + c \qquad \left\{\because \text{ for } \alpha = 2 \int x^2 J_1 dx = x^2 J_2 + c\right\}$$

which is (i)

Also, we have $\int x^{-\alpha}J_{\alpha+1}dx = -x^{-\alpha}J\alpha + c$

For $n = 1$ $\int x^{-1}J_2dx = -x^{-1}J_1 + c$

Therefore, $\int x^{-1}J_2dx = -x^{-1}J_1 + c$ which is (ii).

$$\int xJ_0^2dx = \frac{x^2}{2}J_0^2 - \int(2J_0J_0')\frac{x^2}{2}dx \qquad \{\text{Integrating by parts}\}$$

$$= \frac{x^2}{2}J_0^2 + \int x^2J_0\,J_1dx \qquad \{\because \text{ for } n = 0\ (x^{\alpha}J_{\alpha}) = x^{-\alpha}J_{\alpha+1} \text{ gives } J_0' = -J_1\}$$

$$= \frac{x^2}{2}J_0^2 + \int(xJ_1)\frac{d}{dx}(xJ_1)dx \qquad \{\because \text{ for, } n = 1\ (x^{\alpha}J_{\alpha}) = x^{\alpha}J_{\alpha-1} \text{ we have } (xJ_1)' = xJ_0\}$$

$$= \frac{x^2}{2}J_0^2 + \frac{(xJ_1)^2}{2} = \frac{1}{2}\,x^2[J_0^2 + J_1^2].$$

EXAMPLE 14.23

Evaluate the following integrals in terms of the Bessel's functions

(i) $\int x^{-1}J_4(x)dx$

(ii) $\int x^5J_0(x)dx$

Solution: We have $\int x^{-\alpha}J_{\alpha+1}dx = -x^{-\alpha}J_{\alpha} + c$

$$\int x^2\,(x^{-3}J_4)dx = x^2(-x^{-3}J_3) - \int(2x)\,(x^{-3}J_3)dx$$

$$= -x^{-1}J_4 + \int 2x^{-2}J_3dx$$

$$= -x^{-1}J_4 - 2x^{-2}J_2 + c \qquad \left\{\because \int x^{\alpha}J_{\alpha-1}dx = x^{\alpha}J_{\alpha} + c \Rightarrow \int xJ_0dx = xJ_1 + c\right.$$

$$= \int x^4(xJ_0)dx$$

$$= x^4(xJ_1) - \int 4x^3(xJ_1)dx$$

$$= x^5 J, - 4 \int x^4 J, dx$$

$$= x^J J_1 - 4 \int x^2 (x^2 J_1) dx \qquad \left\{\because \int x^2 J_1 dx = x^2 J_2 + c\right\}$$

$$= x^J J_1 - 4\left[x^2(x^2 J_2) - \int 2x(x^2 J_2) dx\right]$$

$$= x^5 J_1 - 4x^4 J_2 + 8 \int x^3 J_2 dx$$

$$= x^5 J_1 - 4x^4 J_2 + 8x^3 J_3 + c.$$

EXAMPLE 14.24

Show that

(i) $\int J_0(x) \cos x dx = xJ_0(x) \cos x + xJ_1(x) \sin x + c$

(ii) $\int J_1(x) \sin x dx = xJ_1(x) \sin x + J_0(x)(x \cos x - \sin x) + c$

Solution: Integrating by parts taking 1 as the first function

$$\int J_0 \cos x dx = xJ_0 \cos x - \int (J_0' \cos x - J_0 \sin x) x dx$$

$$= xJ_0 \cos x + \int x(J_1 \cos x + J_0 \sin x) dx \qquad \{\because J_0' = -J_1\}$$

$$= xJ_0 \cos x + \int (xJ_1) \cos x dx + \int xJ_0 \sin x dx$$

$$= xJ_0 \cos x + xJ_1 \sin x - \int [J_1 + xJ_1^1] \sin x dx + \int xJ_0 \sin x dx$$

$$= xJ_0 \cos x + xJ_1 \sin x - \int \sin x (xJ_1^1 + J_1 - xJ_0) dx$$

$$\{\because xJ_\alpha^1 = xJ_{\alpha-1} - \alpha J_\alpha \therefore xJ_1^1 = xJ_0 - J_1\}$$

$$= xJ_0 \cos x + xJ_1 \sin x + c.$$

Also, $\int J_1 \sin x dx = x(J_1 \sin x) - \int (J_1^1 \sin x + J_1 \cos x) x dx$

$$= xJ_1 \sin x - \int (xJ_1^1) \sin x dx - \int xJ_1 \cos x dx$$

$$\{\because xJ_\alpha^1 = xJ_{\alpha-1} - \alpha J_\alpha \Rightarrow xJ_1^1 = xJ_0 - J_1\}$$

$= xJ_1 \sin x - \int (xJ_0 - J_1) \sin xdx - \int xJ_1 \cos xdx$

$= xJ_1 \sin x - \int xJ_0 \sin xdx + \int J_1 \sin xdx - \int xJ_1 \cos xdx$

$= xJ_1 \sin x - xJ_0 (- \cos x) + \int (xJ_0' + J_0)(- \cos x)dx + \int J_1 \sin xdx - \int xJ_1 \cos xdx$

$= xJ_1 \sin x + xJ_0 \cos x + \int (-xJ_1 + J_0) (- \cos x)dx + \int J_1 \sin xdx - \int xJ_1 \cos xdx \; \{\because J_0' = -J_1\}$

$= xJ_1 \sin x + xJ_0 \cos x - \int J_0 \cos xdx + \int J_1 \sin xdx$

$= xJ_1 \sin + xJ_0 \cos x - \cos x - \left[J_0 \sin x - \int J_0' \sin xdx\right] + \int J_1 \sin xdx$

$= xJ_1 \sin x + xJ_0 \cos x - J_0 \sin x - \int J_1 \sin xdx + \int J_1 \sin xdx \qquad \{\because J0' = -J_1\}$

$= xJ_1 \sin x + xJ_0 (x \cos x - \sin x) + c.$

Generating function for Bessel's function $J_\alpha(x)$ we shall prove that

$$e^{\frac{1}{2}x\left(t-\frac{1}{t}\right)} - \sum_{n=-\infty}^{\infty} J_\alpha(x)t^\alpha$$

The function on the left hand side is called the generating functions of the Bessel's function of the first kind $J_\alpha(x)$.

We have

$$e^{\frac{1}{2}x\left(t-\frac{1}{t}\right)} = e^{xt/2} \cdot e^{-x/2t}$$

$$= \left\{1 - \left(\frac{xt}{2}\right) + \frac{1}{2!}\left(\frac{xt}{2}\right)^2 + \ldots + \frac{1}{n!}\left(\frac{xt}{2}\right)^n + \ldots\right\}$$

$$\left\{1 - \left(\frac{x}{2t}\right) + \frac{1}{2!}\left(\frac{x}{2t}\right)^2 + \ldots + \frac{(-1)^n}{n!}\left(\frac{x}{2t}\right)^n + \ldots\right\}$$

The coefficient t^n in this product

$$= \frac{1}{n!}\left(\frac{x}{2}\right)^n - \frac{1}{(n+1)!}\left(\frac{x}{2}\right)^{n+2} + \frac{1}{2!\,(n+2)!}\left(\frac{x}{2}\right)^{n+4} - \ldots \Rightarrow J_n(x) = \sum_{m=0}^{\infty} \frac{(-1)^m}{m!\,(n+m)!}\left(\frac{x}{2}\right)^{n+2m}$$

As all integral powers (positive as well as negative) of t occur, we have

$$e^{1/2x\left(t-\frac{1}{t}\right)} = J_0(x) + tJ_1(x) + t^2J^2(x) + \ldots + t^{-1}J_{-1}(x) + t^{-2}J_{-2}(x) + \ldots$$

$$= \sum_{n=-\infty}^{\infty} t^n J_n(x).$$

EXAMPLE 14.25

Show that

$$\frac{1}{\pi}\int_0^{\pi} \cos(n\theta - x\sin\theta)d\theta = J_n(x)$$

for all integral values of n.

Solution: We have

$$e^{\frac{1}{2}x\left(t-\frac{1}{t}\right)} = J_0(x) + \left(t-\frac{1}{t}\right)J_1(x) + \left(t^2+\frac{1}{t^2}\right)J_2(x) + \ldots + \quad (1)$$

Let $\qquad t = e^{i\theta} = \cos\theta + i\sin\theta$

Then $\qquad \dfrac{1}{t} = e^{-i\theta} = \cos\theta - i\sin\theta$

Also, $\qquad t^n = \cos n\theta + i\sin n\theta$

$$\frac{1}{t^n} = \cos n\theta - i\sin n\theta$$

Thus, $\qquad t^n + \dfrac{1}{t^n} = 2\cos n\theta \quad t^n - \dfrac{1}{t^n} = 2i\sin n\theta$

So Eq. (1) becomes

$$e^{\frac{1}{2}x}(2i\sin\theta) = e^{i(x\sin\theta)} = J_0 + (2i\sin\theta)J_1 + (2\cos\cdot 2\theta)J_2 + (2i\sin 3\theta)J^3 + \ldots$$

$$\Rightarrow \cos(x\sin\theta) + i\sin(x\sin\theta) = J_0 + 2[J_2\cos 2\theta + J_4\cos 4\theta + \ldots] + i2[J_1\sin\theta + J_3\sin 3\theta + \ldots]$$

Comparing the real and imaginary parts, we obtain

$$\cos(x\sin\theta) = J_0 + 2[J_2\cos 2\theta + J_4\cos 4\theta + \ldots] \quad (2)$$

and

$$\sin(x\sin\theta) = 2[J_1\sin\theta + J_3\sin 3\theta + \ldots] \quad (3)$$

These series are usually called the Jacobi series.

Now multiplying both sides of Eq. (2) by $\cos n\theta$ and both sides of Eq. (3) by $\sin n\theta$ and integrating over the interval $[0, \pi]$, we obtain

$$\frac{1}{\pi}\int_0^{\pi} \cos(x \sin \theta)\cos n\theta d\theta = \begin{cases} J_n(x) & n \text{ is even} \\ 0 & n \text{ is odd} \end{cases} \tag{4}$$

and $$\frac{1}{\pi}\int_0^{\pi} \sin(x \sin \theta)\sin n\theta d\theta = \begin{cases} 0 & n \text{ is even} \\ J_n(x) & n \text{ is odd} \end{cases} \tag{5}$$

Adding Eqs. (4) and (5), we obtain

$$\frac{1}{\pi}\int_0^{\pi} [(\cos n\theta \cos(x \sin \theta) + \sin n\theta \sin (x \sin \theta)d\theta = Jn(x)$$

$\Rightarrow$ $$\frac{1}{\pi}\int_0^{\pi} \cos(n\theta - x \sin \theta)d\theta = Jn/x) \text{ for all integral values of } n.$$

EXAMPLE 14.26

Show that $J_0(x) = \dfrac{1}{\pi}\displaystyle\int_0^{\pi} \cos(x \cos \theta)d\theta$.

Solution: We have

$$e^{\frac{1}{2}x\left(t-\frac{1}{t}\right)} = J_0 + J_1\left(t - \frac{1}{t}\right) + J_2\left(t^2 + \frac{1}{t^2}\right) + \ldots$$

Let $$t = ie^{i\theta} = i(\cos \theta + i \sin \theta)$$

$$\frac{1}{t} = ie^{-i\theta} = -i(\cos \theta - i \sin \theta)$$

Then $$t - \frac{1}{t^2} = 2i \cos \theta$$

Also, $$t^2 = -(\cos 2\theta + i \sin 2\theta)$$

$$\frac{1}{t^2} = -(\cos 2\theta - i \sin 2\theta)$$

Then $$t^2 + \frac{1}{t^2} = -2 \cos 2\theta$$

Therefore,

$$e^{\frac{1}{2}x\left(t-\frac{1}{t}\right)} = e^{\frac{1}{2}x(2i\cos\theta)} = e^{ix\cos\theta}$$

$$= J_0 + J_1(2i \cos \theta) + J_2(-2 \cos 2\theta) + J_3(-2i \cos 3\theta) + \ldots$$

$\Rightarrow \cos(x \cos \theta) + i \sin(x \sin \theta)$

$$= J_0 - 2[J_2 \cos 2\theta - J_4 \cos 4\theta + \ldots] + i2[J_1 \cos \theta - J_3 \cos 3\theta + \ldots]$$

Comparing real and imaginany parts, we obtain

$$\cos(x \cos \theta) = J_0 - 2[J_2 \cos 2\theta - J_4 \cos 4\theta + \ldots]$$

and $$\sin(x \cos \theta) = 2[J_1 \cos \theta - J_3 \cos 3\theta + \ldots]$$

Integrating over the interval $[0, \pi]$, we obtain

$$\int_0^\pi \cos(x \cos \theta) d\theta = \int_0^\pi [J_0 - 2(J_2 \cos 2\theta - J_4 \cos 4\theta + \ldots)] d\theta$$

$$= \left[J_0 - 2\left(J_2 \frac{\sin 2\theta}{2} - J_4 \frac{\sin 4\theta}{4} + \ldots \right) \right]_0^\pi$$

$$= \pi J_0$$

Hence, $$\frac{1}{\pi}\int_0^\pi \cos(x \cos \theta) d\theta = J_0(x).$$

Orthogonality of Bessel's function

If a and b are roots of the equation

$$J_\alpha(x) = 0, \text{ then}$$

$$\int_0^1 x J_\alpha(ax) J_\alpha(bx) dx = \begin{cases} 0 & a \neq b \\ \frac{1}{2}[J_{\alpha+1}(a)]^2 & a = b \end{cases}$$

Proof: We can see easily that the solution of the equation

$$x^2 u'' + xu' + (a^2 x^2 - \alpha^2) u = 0 \quad (1)$$

and $$x^2 v'' + xv' + (b^2 x^2 - \alpha^2) v = 0 \quad (2)$$

are $$u = J_\alpha(ax) \text{ and } v = J_\alpha(bx), \text{ respectively.}$$

Multiplying Eq. (1) by v and Eq. (2) by u and subtracting, we get

$$x^2(u''v - uv'') + x(u'v - uv') + (a^2 - b^2)x^2 uv = 0$$

$$\Rightarrow \quad x(u''v - uv'') + 1 \cdot (u'v - uv') + (a^2 - b^2)xuv = 0$$

$$\Rightarrow \quad \frac{d}{dx}[x(u'v - uv')] = (b^2 - a^2)xuv$$

Now, integrating both sides from 0 to 1, we obtain

$$(b^2 - a^2)\int_0^1 xuv dx = [x(u'v - uv')]_0^1$$

$$= [u'v - uv']_{x=1}$$

Since $$u = J_\alpha(ax)$$

$$\therefore \quad u' = \frac{du}{dx} = J_\alpha'(ax)a$$

Similarly, $$v' = J_\alpha'(bx)b$$

So, we have

$$\int_0^1 xJ_\alpha(ax)J_\alpha(bx)dx = \frac{aJ_\alpha'(a)J_\alpha(b) - bJ_\alpha(a)J_\alpha'(b)}{b^2 - a^2} \tag{3}$$

If a and b are distinct roots of $J_\alpha(x) = 0$ then $J_\alpha(a) = J_\alpha(b) = 0$ and Eq. (3) reduces to

$$\int_0^1 xJ_\alpha(ax)J_2(bx)dx = 0 \text{ as } a \neq b$$

This is known as the orthogonality relation of Bessel functions. When $a = b$ then treating b as a variable approaching a, we obtain

$$\underset{b \to a}{\text{Lt}} \int_0^1 xJ_\alpha(ax)J_\alpha(bx)dx = \underset{b \to a}{\text{Lt}} \frac{aJ_\alpha'(a)J_\alpha(b) - bJ_\alpha(a)J_\alpha'(b)}{b^2 - a^2}$$

$$= \underset{b \to a}{\text{Lt}} \frac{aJ_\alpha'(a)J_\alpha(b)}{b^2 - a^2} \qquad \left(\frac{0}{0} \text{ form}\right)$$

$$= \underset{b \to a}{\text{Lt}} \frac{aJ_\alpha'(a)J_\alpha'(b)}{2b} \qquad \text{[using L' Hospital Rule]}$$

$$= \frac{1}{2}\,[J_\alpha'(a)]^2$$

$$= \frac{1}{2}\,[J_{\alpha+1}(a)]^2 \qquad \{\because xJ_\alpha \because = \alpha J_\alpha - xJ_{\alpha+1}\}$$

$$\Rightarrow \quad J_\alpha' = \frac{\alpha}{x} J_\alpha - J_{\alpha+1}$$

when $$x = a$$

$$J_\alpha'(a) = \frac{\alpha}{a} J_\alpha(a) - J_{\alpha+1}(a)\; J_\alpha'(a) = J_{\alpha+1}(a), \text{ since } J_\alpha(a) = 0\}$$

▲▲

EXERCISE 14.2

Find the general solution of the following differential equations:

1. $x^2y'' + xy' + \left(x^2 - \frac{1}{4}\right)y = 0.$
2. $36x^2y'' + 36xy' + (36x^2 - 4)y = 0.$
3. $x^2y'' + xy' + (x^2 - 25)y = 0.$ Reduce the given differential equations to Bessel's equation, using the given substitution and find the general solution also.
4. $4x^2y'' + 4xy' + (x^2 - 1)4 = 0,\ x = 2z.$
5. $9x^2y'' + 9xy' + (81x^2 - 1)y = 0,\ z = 3x.$
6. $x^2y'' + xy' + 4(x^4 - \alpha^2)y = 0,\ x^2 = z,\ \alpha$ is a real fraction.
7. $x^2y'' + xy' + (9x^2 - 4)y = 0,\ z = 3x.$
8. $xy'' + 5y' + xy = 0,\ y = z/x^2.$
9. $y'' + \frac{y'}{x} + \left(8 - \frac{1}{x^2}\right) y = 0,\ z = 2\sqrt{2x}.$
10. $4y'' + 9xy = 0,\ z = x^{3/2}.$
11. $xy'' + xy' + \frac{1}{4} y = 0,\ z = z = \sqrt{x}.$

Prove the following:

12. $J_2'(x) = \left(1 - \frac{4}{x^2}\right)J_1(x) + \frac{2}{x}J_0(x).$
13. $J_3'(x) = \left(\frac{12}{x^2} - 1\right)J_0(x) - \left(\frac{24}{x^3} - \frac{5}{x}\right)J_1 x.$
14. $J_1''(x) = \frac{1}{x} J_2(x) - J_1(x).$
15. $J_n''(x) = \frac{1}{4} [J_{n-2}(x) - 2J_n(x) + J_{n+2}(x)].$
16. $\frac{d}{dx} [J_n 2(x)] = \frac{x}{2n} [J_{n-1}^2(x) - J_{n+1}^2(x)].$

Prove the following:

17. $\int J_3(x)dx = -J_2(x) - \frac{2}{x} J_1(x) + c.$

18. $\int J_5(x)dx = -[2J_4(x) + 2J_2(x) + J_0(x)] + c.$

19. $\int x^3 J_0(x)dx = x(x^2 - 4)J_1(x) + 2x^2 J_0(x)] + c.$

20. $\int_0^1 J_1(2x)dx = \frac{1}{\alpha}$ where α is a root of $J_0(x) = 0$ that is $J_0(\alpha) = 0$.

[*Hint:* Let $\alpha x = t$ and integrate using $J_0' = -J_1$].

21. $\int_0^1 (x - x^3)J_0(x)dx = 4J_1(1) - 2J_0(1).$

22. $\int_0^1 x^2 J_0(x)dx = J_1(1) + J_0(1) - \int_0^1 J_0(x)dx.$

23. $\int_0^k xJ_0(\alpha x)dx = \frac{k}{\alpha} J_1(\alpha k).$

24. $J_{n-1}(x) = x \int_0^1 J_n(xy)y^{n-1}dy.$ [*Hint:* Let $xy = t$].

25. Prove that $\cos n = J_0(x) - 2J_2(x) + 2J_4(x) - \ldots = J_0(x) + 2\sum_{n=1}^{\infty}(-1)^n J_{2n}(x)$ and $\sin x = 2J_1(x) - 2J_3(x) + 2J_5(x) - \ldots = 2\sum_{n=1}^{\infty}(-1)^n J_{2n+1}(x).$

[*Hint:* Let $\theta = \pi/2$ in Example – 14.25]

26. Show that

$$J_0^2(x) + 2J_1^2(x) + 2J_2^2(x) + \ldots = 1.$$

Using generating function for Bessel's function.
[*Hint:* Squaring Eqs. (2) and (3) of Example 14.25 and integrating from 0 to π].

ANSWERS

Exercise 14.2

1. $y = AJ_{1/2}(x) + BJ_{-1/2}(x) = \sqrt{\dfrac{2}{\pi x}}\,(A \sin x + B \cos x)$.
2. $y = AJ_{1/3}(x) + BJ_{-1/3}(x)$.
3. $4 = AJ_5(x) + B1/5(x)$.
4. $y = AJ_{1/2}(x/2) + BJ_{-1/2}(x/2)$.
5. $y = AJ_{1/3}(3x) + BJ_{-1/3}(3x)$.
6. $y = AJ_{\alpha}(x^2) + BJ_{-\alpha}(x^2)$.
7. $y = AJ_2(3x) + B1/2(3x)$.
8. $y = [AJ_2(x) + B1/2(x)]/x^2$.
9. $y = AJ_1\left(2\sqrt{2x}\right) + By,\left(2\sqrt{2x}\right)$.
10. $y = \sqrt{x}\,[AJ_{1/3}(x^{3/2}) + BJ_{-1/3}/x^{3/2})]$.
11. $y = AA_0\left(\sqrt{x}\right) + B1/0\left(\sqrt{x}\right)$.

15

CHAPTER

Applications of Ordinary Differential Equations

In chapter 1, we pointed out that differential equations originate from the mathematical formulation of a great variety of problems in science and engineering. In this chapter, we shall consider only such practical problems which give rise to ordinary differential equation. First, we formulate the problem mathematically, thereby obtaining a differential equation. Then, we solve the equation by using methods discussed earlier. It may be noted that we have already studied geometrical applications in chapter 7. ▲▲

EXAMPLE 15.1

Resisted Vertical Motion

Suppose a particle falls under gravity in a resisting medium whose resistance varies with velocity. Find the relation between distance (s) and velocity (v) of the particle.

Solution: Initially, particle starts from rest and let after falling a distance s in time t, the velocity of the particle be v. Now, the forces acting on the particle are its weight mg downwards and resistance $m\lambda v$ upwards where, λ is the coefficient of resistance per unit mass of the medium.

Therefore, the equation of motion is

$$m\frac{dv}{dt} = mg - m\lambda v$$

$$\Rightarrow \quad \frac{dv}{dt} = g - \lambda v$$

$$\Rightarrow \quad \frac{dv}{g-\lambda v} = dt$$

Integrating, we obtain

$$-\frac{1}{\lambda}\log(g - \lambda v) = t + c$$

Since $v = 0$ when $t = 0$

$$\therefore \quad c = -\frac{1}{\lambda}\log g$$

Thus,

$$\frac{1}{\lambda}\log\left[\frac{g}{g-\lambda v}\right] = t \tag{1}$$

$$\Rightarrow \quad \frac{g-\lambda v}{g} = e^{-\lambda t}$$

$$\therefore \quad v = \frac{g}{\lambda}(1 - e^{-\lambda t})$$

$$\Rightarrow \quad \frac{ds}{dt} = \frac{g}{\lambda}(1 - e^{-\lambda t})$$

which represents relation between velocity v and time t.

Integrating, we have

$$s = \frac{g}{\lambda}\left(t + \frac{1}{\lambda}e^{-\lambda t}\right) + c'$$

Since, $s = 0$ when $t = 0$

$$\therefore \quad c' = -\frac{g}{\lambda^2}$$

Thus,

$$s = \frac{g}{\lambda}t + \frac{g}{\lambda^2}(e^{-\lambda t} - 1)$$

which gives us the distance s after time t from the rest position.

Using Eq. (1), we obtain

$$s = \frac{2}{\lambda^2}\log\left[\frac{g}{g-\lambda v}\right] - \frac{v}{\lambda}$$

which gives us the relation between distance s and velocity v of the particle.

EXAMPLE 15.2

Velocity of Escape from the Earth

In this case, we have to find the minimum initial velocity of a particle which is fired in the radial direction from the earth and is supposed to escape from the earth. Neglect the air resistance and the gravitational pull of other celestial bodies.

Solution: Since the particle is acted upon only by the gravitational attraction of the earth, therefore according to Newton's law of gravitation, the acceleration a of the particle is proportional to $1/r^2$ where, r is the distance from the centre of the earth of the particle.

Thus,

$$a = \frac{dv}{dt} = \frac{\mu}{r^2}$$

where, v is the velocity and t is the time.

The acceleration is negative because v is decreasing and thus $\mu < 0$.
when $r = R$ (the earth's radius) and $a = -g$ (the acceleration due to gravity at the surface of the earth), we get

$$-g = \frac{\mu}{R^2} \quad \Rightarrow \quad \mu = -gR^2.$$

Thus,
$$a = \frac{dv}{dt} = \frac{dv}{dr}\frac{dr}{dt} = \frac{dv}{dr}v = -\frac{gR^2}{r^2}$$

Separating the variables, we get

$$vdv = -gR^2 \frac{dr}{r^2}$$

Integrating, we obtain

$$\frac{v^2}{2} = \frac{gR^2}{r} + c \tag{1}$$

On the surface of the earth $r = R$ and $v = v_0$ (the initial velocity). Then

$$c = \frac{v_0^2}{g} - gR$$

So, Eq. (1) becomes

$$v^2 = \frac{2gR^2}{r} + V_0^2 - 2gR$$

when v vanishes, the particle stops and the velocity will change from positive to negative and the particle will return to the earth. Thus, the velocity will remain positive, if and only if $v_0^2 \geq 2gR$ and thus the particle projected from the earth with this velocity will escape from the earth. Hence, the minimum such velocity of projection $v_0 = \sqrt{2gR}$ is called the velocity of escape from the earth.

Since
$$R = 6372 \text{ km and } g = 9.8 \text{ m/s}^2$$
$$= 0.0098 \text{ km/s}^2$$

we have

$$v_0 = \sqrt{2gR} \cong 11.2 \text{ km/s}.$$

EXAMPLE 15.3

Atmospheric Pressure

In this problem, we will find the atmospheric pressure p at a height x above the sea level, when pressure is assumed to be constant and variable both.

Solution: Take a vertical column of air of unit cross-section. Let an element of this column be bounded by two horizontal planes at height x and $x + \delta x$ above the sea level with corresponding pressures p and $p + \delta p$ respectively. Let ρ be the average density of the element.

Now, since the thin column δx of air is being pressed upwards with pressure p and downwards with $p + \delta p$ and the weight $\rho g \delta x$, we get by considering its equilibrium

$$p = p + \delta p + \rho g \delta x$$

$$\Rightarrow \quad \delta p + \rho g \delta x = 0$$

$$\therefore \quad \frac{\delta p}{\delta x} = -\rho g$$

Taking the limit as $\delta h \to 0$, we have

$$\frac{dp}{dx} = -\rho g \tag{1}$$

which is the differential equation giving the atmospheric pressure p at height x. Now, we obtain its solution by assuming the following assumptions.

(i) When the temperature is constant by Boyle's law $p = h\rho \Rightarrow \rho = \dfrac{h}{p}$

Substituting in Eq. (1), we obtain

$$\frac{dp}{dx} = -\frac{pg}{h} \Rightarrow \frac{dp}{p} = -\frac{g}{h}dx$$

Integrating, we have $\log p = -\dfrac{g}{h}x + c$

At the sea level, $x = 0$, $p = p_0$ (say)

$$\therefore \quad \log p_0 = c$$

$$\Rightarrow \quad \log \frac{p}{p_0} = -\frac{gx}{h}$$

$$\therefore \quad p = p_0 e^{-gx/h}$$

(ii) When the temperature varies, we have

$$p = h\rho^n$$

Proceeding as above, we shall find p as given by

$$\frac{n}{n-1}[p_0^{1-1/n} - p^{1-1/n}] = gh^{-1/n}x.$$

EXAMPLE 15.4

Newton's Law of Cooling

According to this law the time rate at which a body cools is proportional to the difference between the temperature of the body and the temperature of the surrounding medium. If $T(t)$ denotes the temperature of the body at any time t and T_0 is the constant temperature of the outside medium, then

$$\frac{dT}{dt} = k(T - T_0)$$

where, k is the usual constant of proportionality.

On separating the variables, we obtain

$$\frac{dT}{T - T_0} = kdt$$

Integrating, we have

$$\log(T - T_0) = kt + \log c$$

$$\Rightarrow \qquad T = T_0 + ce^{kt} \qquad (1)$$

We now consider the following problem.

Water is heated to the boiling point temperature 100°C. It is then removed from heat and kept in a room which is at a constant temperature of 60°C. After 3 minutes, the temperature of the water is 90°C.

(a) Find the temperature of water after 6 minutes.

(b) When will the temperature of water be 75°C and 61°C.

Solution: Here at $t = 0$ $T = 100$ and $T_0 = 60$

So, Eq. (1) gives $\qquad c = 40$

Also at $t = 3$ $\qquad T = 90$

Therefore, $\qquad 90 = 60 + 40e^{k \times 3}$

$$\Rightarrow \qquad k = \frac{1}{3}\ln\left(\frac{3}{4}\right) = -0.095894024$$

(a) When $t = 6$

$$T = 60 + 40\, e^{-0.095894024 \times 6} = 82.5°\text{C}$$

(b) When $T = 75°$C

$$75 = 60 + 40e^{-0.095894024 \times t}$$

$\Rightarrow \qquad t = 10.2$ minutes.

When $T = 61°C$

$$61 = 60 + 40e^{-0.095894024 \times t}$$

$$\Rightarrow \quad t = 38.5 \text{ minutes.}$$

EXAMPLE 15.5

Growth of Bacteria

In biology, it is often observed that the rate at which certain bacteria grow is proportional to the number of bacteria present at any time. This gives rise to the following initial value problem

$$\frac{dN}{dt} = \lambda N \text{ with } N(t_0) = N_0.$$

where, N_0 is number of bacteria present initially, that is, at time t_0 and N is the number of bacteria after time t. We shall illustrate this problem by considering the next example.

A culture initially has N_0 number of bacteria. At $t = 1$ hour the number of bacteria is measured to be $\frac{3N_0}{2}$. If the rate of growth is proportional to the number of bacteria present, determine the time necessary for the number of bacteria to triple.

Solution: We have

$$\frac{dN}{dt} = \lambda N \text{ with } N(0) = N_0$$

Separating the variables, we have

$$\frac{dN}{N} = \lambda dt$$

Integrating, we obtain

$$\log N = \lambda t + \log c$$

$$\Rightarrow \quad N = ce^{\lambda t}$$

At $t = 0$ we have $\quad N_0 = c$

$$\therefore \quad N = N_0 e^{\lambda t}$$

At $t = 1$, we have $\quad N = \frac{3N_0}{2}$

$$\therefore \quad \frac{3N_0}{2} = N_0 e^{\lambda \times 1}$$

$$\Rightarrow \qquad k = \ln\left(\frac{3}{2}\right) = 0.4055$$

Thus,

$$N = N_0\, e^{0.4055\, t}$$

When $N = 3N_0$ we have

$$3N_0 = N_0\, e^{0.4055\, t}$$

$$\Rightarrow \qquad t = \frac{\ln 3}{0.4055} = 2.71 \text{ hours.}$$

EXAMPLE 15.6

Decay of Radioactive Materials

The law of radioactive decay states that disintegration at any instant is proportional to the amount of material present.

If A is the amount of material at any time t, then $\frac{dA}{dt} = -\mu A$

where, μ is the decay constant.

We illustrate this by the following problem. Uranium disintegrates at a rate proportional to the amount present at any instant. If M_1, and M_2 grams of uranium are present at times T_1 and T_2 respectively, find the half-life of uranium.

Solution: Let M be the mass of uranium at time T. Then the equation of disintegration of uranium is

$$\frac{dM}{dT} = -\mu M$$

where, μ is decay constant

Separating the variables, we have

$$\frac{dM}{M} = -\mu T$$

Integrating, we obtain

$$\ln M = -\mu T + c$$

at $T = T_1$, $M = M_1$, so $\ln M_1 = -\mu T_1 + c$

at $T = T_2$, $M = M_2$ so $\ln M_2 = -\mu T_2 + c$

Subtracting, we get

$$\ln M_2 - \ln M_1, = -\mu(T_2 - T_1)$$

$$\Rightarrow \qquad \mu = \frac{\ln(M_1 / M_2)}{T_2 - T_1}$$

Let the mass reduce to half its initial value M_0 in time $T_{1/2}$, then

$$\ln M_{0/2} - \ln M_0 = -\mu(T_{1/2} - 0)$$

$$\Rightarrow \quad T_{1/2} = \frac{1}{\mu}\ln 2 = \frac{(T_2 - T_1)\ln 2}{\ln(M_1 / M_2)}$$

EXAMPLE 15.7

Mixture Problem

A tank contains 100 litres of fresh water. Two litres of brine, each containing 1 gm of dissolved salt run into the tank per minute and the solution kept uniform by stirring runs out at the rate of 1 litre per minute. Find the amount of salt present when the tank contains 150 litres of brine.

Solution: Let q gm be the quantity of salt present in the brine at time t, then dq/dt is the rate at which the salt content is changing. The rate at which the salt content increases due to the inflow = $2 \times 1 = 2$ gm/min. Let C gm be the concentration of brine at time t.

The rate at which the salt content decreases due to the outflow = $C \times 1 = C$ gm/min.

Now according to the equation of continuity $\frac{dq}{dt}$ = rate of inflow – rate of outflow.

$$\frac{dq}{dt} = 2 - c \qquad (1)$$

Now, the initial volume of liquid is 100 litres. In one minute, 2 litres of brine enter the tank and 1 litre of brine leaves the tank, so that the volume of liquid in the tank increases at the rate = 2 – 1 = 1 litre/min.

∴ The volume of liquid at time t is $(100 + t)$ litres containing q gm of salt.

$$\therefore \quad C = \frac{q}{100 + t}$$

Therefore, Eq. (1) becomes

$$\frac{dq}{dt} = 2 - \frac{q}{100 + t}$$

$$\Rightarrow \quad \frac{dq}{dt} + \frac{q}{100 + t} = 2$$

which is a linear equation in q and t with

$$\text{IF} = e^{\int dt/(100+t)} = 100 + t$$

∴ The solution is

$$q(100 + t) = \int 2(100 + t)dt + a$$

$$\Rightarrow \quad (100 + t)q = 2\left(100t + \frac{t^2}{2}\right) + a$$

Initially, when $t = 0$, $q = 0$ so that $a = 0$

$$\therefore \qquad q = \frac{2\left(100t + \frac{t^2}{2}\right)}{100 + t}$$

When volume of the liquid is 150 litres

then $$150 = 100 + t$$

$$\therefore \qquad t = 50 \text{ minutes}$$

and salt content

$$q = \frac{2\left[100 \times 5 + \frac{(50)^2}{2}\right]}{100 + 50} = 83.3 \text{ gm.}$$

EXAMPLE 15.8

Simple Harmonic Motion

Experimentally, it is found that a 6 lb weight stretches a certain spring 6 inches. If the weight is pulled 4 inches below the equilibrium position and released:

(a) set up the differential equation and associated conditions describing the motion;

(b) find the position of the weight as a function of time;

(c) find the amplitude, period and frequency of motion; and

(d) determine the position, velocity and acceleration of the weight 1/2 sec after it has been released.

Solution: By Hooke's law

$$F = kx = mg$$

$$k\left(\frac{6}{12}\right) = 6$$

$$\cdot\ k = 12$$

where, k is the spring constant.

So the differential equation representing the motion is

$$m\frac{d^2x}{dt^2} = -kx$$

$$\Rightarrow \qquad \frac{d^2x}{dt^2} + \frac{kg}{mg}x = 0$$

$$\Rightarrow \qquad \frac{d^2x}{dt^2} + \frac{12 \times 32}{6} x = 0$$

$$\Rightarrow \qquad \frac{d^2x}{dt^2} + 64x = 0$$

Auxiliary equation is $\qquad D^2 + 64 = 0$

$$D = \pm 8i$$

Therefore, the general solution is

$$x = A \cos 8t + B \sin 8t$$

At $t = 0$ $\qquad x = \frac{4}{12} = \frac{1}{3}$ ft

Thus, $\qquad \frac{1}{3} = A$

Therefore, $\qquad x = \frac{1}{3} \cos 8t + B \sin 8t$

Also, since the weight is released from 4 inches below the equilibrium position, we have

$$v = \frac{dx}{dt} = 0 \text{ at } t = 0$$

Now, $\qquad \frac{dx}{dt} = -\frac{8}{3} \sin 3t + 8B \cos 8t$

$\therefore \qquad 0 = -\frac{8}{3} \sin 3(0) + 8B \cos 8(0)$

$\Rightarrow \qquad 0 = B$

Hence, the required solution is

$$x = \frac{1}{3} \cos 8t$$

Clearly, amplitude is $\frac{1}{3}$ ft and time period $T = 2\pi\sqrt{\frac{mg}{kg}} = 2\pi\sqrt{\frac{6}{12 \times 32}} = \frac{\pi}{4}$ sec.

Also, the frequency $v = \frac{1}{7} = \frac{4}{\pi}$ $\sec^{-1}$.

$$v = \frac{dx}{dt} = -\frac{8}{3} \sin 8t$$

$$a = \frac{d^2x}{dt^2} = -\frac{64}{3}\cos 8t$$

So, when $$t = \frac{1}{2}\text{ sec}$$

$$x = \frac{1}{3}\cos 8\left(\frac{1}{2}\right) = \frac{1}{3}\cos 4 = \frac{1}{3}\cos 229° = -0.219\text{ ft.}$$

$$v = -\frac{8}{3}\sin 8\left(\frac{1}{2}\right) = -\frac{8}{3}\cos 229° = 2.01\text{ ft/sec}$$

$$a = -\frac{64}{3}\cos\left(\frac{1}{2}\right) = -\frac{64}{3}\cos 229° = 14.0\text{ ft/sec}^2.$$

EXAMPLE 15.9

Simple Pendulum

A simple pendulum of length l is oscillating through a small angle θ in a medium for which the resistance is proportional to the velocity. Obtain the differential equation of its motion and discuss the motion.

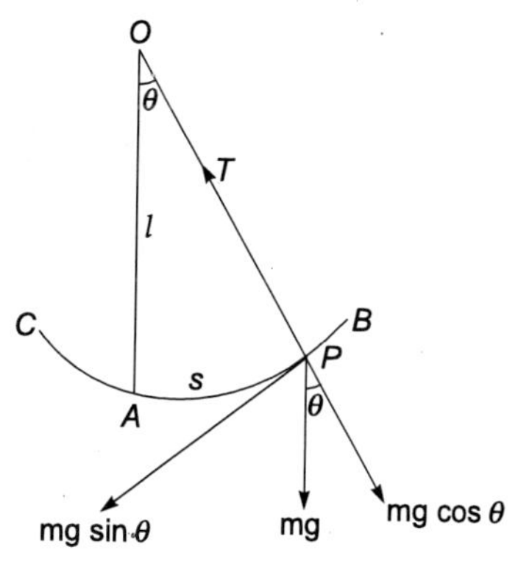

Solution: Let O be the fixed point and P be the position of the bob of mass m at any time t and A be the mean position.

Let arc $AP = s$ and $\angle AOP = \theta$

The equation of motion along the tangent is

$$m\frac{d^2s}{dt^2} = -mg\sin\theta - \lambda\frac{ds}{dt}$$

where λ is a constant.

$$\therefore\ m\frac{d^2(l\theta)}{dt^2} + mg\sin\theta + \lambda\frac{d}{dt}(l\theta) = 0$$

Replace $\sin\theta$ by θ (as θ is small) and put $w = g/l$ and $\lambda/m = 2k$, to obtain

$$\frac{d^2\theta}{dt^2} + 2k\frac{d\theta}{dt} + \frac{g\theta}{l} = 0 \Rightarrow \frac{d^2\theta}{dt^2} + 2k\frac{d\theta}{dt} + w\theta = 0.$$

which is the required differential equational of motion.

The auxiliary equation is

$$D^2 + 2kD + W = 0$$

$$\therefore\quad D = -k \pm \sqrt{W^2 - K^2}$$

The oscillatory motion of the bob is possible only when $K < W$. The solution of the differential equation is

$$\theta = e^{-kt}\left[c_1 \cos\sqrt{w^2 - k^2}\,t + c_2 \sin\sqrt{w^2 - k^2}\,t\right]$$

with period of the vibratory motion $\dfrac{2\pi}{\sqrt{w^2 - k^2}}$.

EXAMPLE 15.10

Chemical Reactions

Let us consider a first order reaction $A \rightarrow$ Products. From the differential equation representing this reaction and hence find the corresponding solution also.

Solution: Suppose that at the beginning of the reaction ($t = 0$), the concentration of A is a moles/litre. If after time t, x moles of A have changed, the concentration of A is $a - x$. We know that for a first order reaction, the rate of reaction dx/dt is directly proportional to the concentration of the reactants. Thus,

$$\frac{dx}{dt} = k(a - x)$$

$$\Rightarrow \quad \frac{dx}{a - x} = kdt$$

Integrating, we have

$$-\log(a - x) = kt + c$$

Initially, when $t = 0$,

$$x = 0$$

$$\therefore \quad c = -\log a$$

$$\log \frac{a}{a - x} = kt$$

$$\Rightarrow \quad k = \frac{1}{t}\log\frac{a}{a - x}.$$

EXAMPLE 15.11

Projectile with Resistance

Find the path of a particle projected with a velocity v at an angle α to the horizon in a medium whose resistance apart from gravity, varies as velocity. Also, find the greatest height attained.

Solution: Let x and y be horizontal and vertical axes respectively with origin O as the point of projection

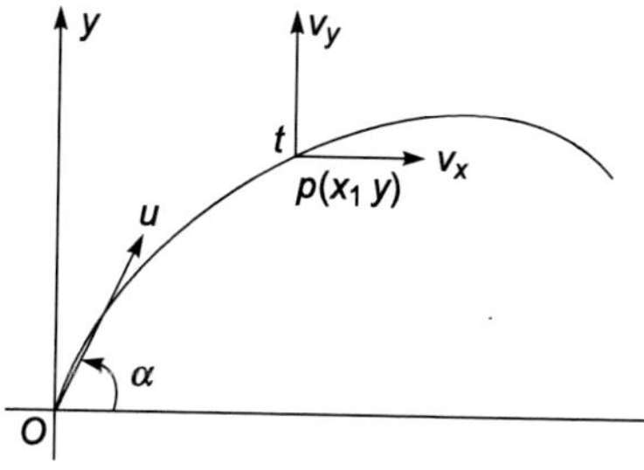

Let $P(x, y)$ be the position of the projectile at the time t when projected with initial velocity u.

$\therefore$ the equations of motion parallel to x and y axes are

$$m\frac{dv_x}{dt} = -mkv_x \text{ and } m\frac{dv_y}{dt} = -mg - mkv_y$$

$$\Rightarrow \quad \frac{dv_x}{dt} = -kv_x \text{ and } \frac{dv_y}{dt} = -(g + kv_y)$$

Separating the variables, we obtain

$$\frac{dv_x}{v_x} = -kdt \text{ and } \frac{dv_y}{g + kv_y} = -dt$$

Integrating, we have

$$\log v_x = -kt + c_1 \text{ and } \frac{1}{k}\log(g + kv_y) = -t + c_2$$

Initially, when $t = 0$, $v_x = u \cos \alpha$, $v_y = u \sin \alpha$

Therefore, we have

$$\log u \cos \alpha = c_1 \text{ and } \frac{1}{k}\log(g + ku \sin \alpha) = c_2, \text{ we get}$$

So, on substituting c_1 and c_2, we get

$$\log\left(\frac{v_x}{u\cos\alpha}\right) = -kt \, \text{and} \, d\frac{1}{k}\log\left(\frac{g + kv_y}{g + ku\sin\alpha}\right) = -t$$

$$\Rightarrow \quad \frac{dx}{dt} = v_x = u \cos \alpha e^{-xt} \text{ and } \frac{dy}{dt} = v_y = \frac{1}{k}[(g + xu \sin \alpha)e^{-kt} - g]$$

Again, integrating we get

$$x = \frac{u\cos\alpha e^{-kt}}{-k} + c_3 \text{ and } y = -\frac{1}{k}\left(\frac{g}{k} + u\sin\alpha\right)e^{-kt} - \frac{g}{k}t + c_4$$

Initially, when $t = 0$, $x = 0$ and $y = 0$

Therefore,

$$0 = -\frac{u\cos\alpha}{k} + c_3 \text{ and } 0 = -\frac{1}{k}\left(\frac{g}{k} + u\sin\alpha\right) + c_4$$

On substation c_3 and c_4, we obtain

$$x = \frac{u\cos\alpha}{k}(1 - e^{-kt}) \text{ and } y = \frac{1}{k}\left(\frac{g}{k} + u\sin\alpha\right)(1 - e^{-kt})$$

Eliminating t, we obtain

$$y = \left(\frac{g}{k} + u\sin\alpha\right)\frac{x}{u\cos\alpha} + \frac{g}{k^2}\log\left(1 - \frac{kx}{u\cos\alpha}\right)$$

which is the required equation of trajectory.

The projectile will attain the greatest height

when $$\frac{dy}{dt} = 0$$

$$\Rightarrow \quad \frac{1}{k}\left(\frac{g}{h} + u\sin\alpha\right)(ke^{-kt}) - \frac{g}{k} = 0$$

$$\Rightarrow \quad t = \frac{1}{k}\log\left(1 + \frac{ku\sin\alpha}{g}\right)$$

Thus, the greatest height $= \dfrac{u\sin\alpha}{k} - \dfrac{g}{k^2}\log\left(1 + \dfrac{ku\sin\alpha}{g}\right)$. ▲▲

EXERCISE 15.1

1. A particle of mass m moves under gravity in a medium whose resistance is K times its velocity where k is a constant. If the particle is projected vertically upwards with a velocity v; show that the time to reach the highest point is $\left(1 + \dfrac{kv}{mg}\right)$.
2. It is found that 0.5 per cent of radium disappears in 12 years. Find:
   (a) what percentage will disappear in 1000 years; and
   (b) what is the half-life of radium?
3. When a cable is removed from an oven, its temperature is 300°F. Three minutes later, its temperature is 200°F. How long will it take to cool to room temperature of 70°F?
4. Bacteria in a certain culture increases at a rate proportional to the number present. If the number increases from 1000 to 2000 in 1 hour, how many are present at the end of 1.5 hours?

5. Bacteria in a certain culture increase at a rate proportional to the number present. If the number doubles in one hour, how long does it take for the number to triple?
6. A body of mas m falls from rest under gravity and air resistance is proportional to the square of its velocity. Find velocity as a function of time.
7. Show that a particle projected from the earth's surface with a velocity of 7 miles/sec, will not return to the earth.
   (Take earth's radius = 3960 miles and $g = 32.17$ ft/sec^2).
8. The rate at which water flows from a hole at the bottom of a tank is proportional to the square root of the depth of the water. If half the water flows from a cylindrical tank (with vertical axis) in 5 minutes, find the time required to empty the tank.
9. Up to a certain height in the atmosphere, it is found that the pressure p and density ρ are connected by the relation $p = k\rho^n (n > 1)$. If this relation continues to hold upto any height, show that density would vanish at a finite height.
10. Under certain conditions, cane-sugar in water is converted into dextrose at a rate which is proportional to the amount unconverted at any time. If 75 gm, at time t – 0, 8 gm are converted during the first 30 minutes, find the amount converted in 1.5 hours.
11. A tank is initially filled with 100 gallons of salt solution containing 1 lb of salt per gallon. Fresh brine containing 2 lb of salt per gallon runs into the tank, at the rate of 5 gallons per minute and the mixture assumed to be uniform by stirring, runs out at the same rate. Find the amount of salt in the tank at any time and determine how long, it will take for this amount to reach 150 lb.
12. At the end of three successive seconds, the distance of a point moving with simple harmonic motion from its mean position, measured in the same direction are 1, 3, 5. Show that the period of complete oscillation is $2\pi/\theta$, where, $\cos\theta = \frac{3}{5}$.
13. An elastic string of natural length $2a$ and modulus λ is stretched between two points A and B distant $4a$ apart on a smooth horizontal table. A particle of mass m is attached to the middle of the string. Show that it can vibrate in line AB with period $2\pi/w$ where, $w^2 = \frac{2\lambda}{am}$.
14. Find how many seconds a clock would lose per day, if the length of its pendulum were increased in the ratio 900 : 901.
15. A pendulum of length (l) hangs against a wall inclined at an angle (θ) to the horizontal. Show that the time of complete oscillation is $2\pi\sqrt{\frac{l}{g\sin\theta}}$.
16. A particle starts from rest, a distance 10 cm from a fixed point O. It moves along a horizontal straight line towards O under the influence of an attractive force

at O. This force at any time varies as the distance of the particle from O. If the acceleration of the particle is 9 cm/sec^2 directed towards O, when the particle is 1 cm from O, describe the motion.

17. A body weighing 10 kg is hung from a spring. A pull of 20 kg weight will stretch the spring to 10 cm. The body is pulled down to 20 cm below the static equilibrium position and then released. Find the displacement of the body from its equilibrium position at 1 sec, the maximum velocity and the period of oscillation.
18. A spring of negligible weight which stretches 1 inch under tension of 2 lb is fixed at one end and is attached to a weight of w lb at the other. It is found that resonance occurs when an axial periodic force 2 cos $2t$ lb acts on the weight. Show that when the free vibrations have died out, the forced vibrations are given by $x = ct \sin 2t$ and find the values of w and c.
19. The equations of motion of an electron under certain conditions are

$$m\frac{d^2x}{dt^2} + eH\frac{dy}{dt} = CE$$

and
$$m\frac{d^2y}{dt^2} - eH\frac{dx}{dt} = 0$$

with conditions $x = \frac{dx}{dt} = y = \frac{dy}{dt} = 0$, when $t = 0$; find the path of the electron.

20. A particle is projected with velocity u making an angle α with the horizontal. Neglecting air resistance, show that the equation of its path is the parabola

$$y = x + \tan\alpha - \frac{gx^2}{2u^2\cos^2\alpha}.$$

Find the time of flight, the greatest height attained and the range on the horizontal plane.

ANSWERS

Exercise 15.1

2. (a) 34.2 per cent (b) 1672.1770 years.
3. About half an hour.
4. 2828.4271.
5. 1.5849 hours.
6. $v = \sqrt{\frac{mg}{k}}\tanh\left(\frac{9k}{m}t + c\right)$.
8. 17 min 4 sec.

10. 21.5 gm.
11. $100(2 - e^{-t/20})$; 13 – 9 min.
14. 48 seconds per day.
17. 0.2 cos 14 m, 2.8 m/sec, 0.45 sec.
18. 6 gm, 1/12.
19. $x = \dfrac{\varepsilon}{HW}(1 - \cos wt), y = \dfrac{E}{HW}(wt - \sin wt)\ (wt - \sin wt)$
20. $\dfrac{2u \sin \alpha}{g}; \dfrac{u^2 \sin^2 \alpha}{2g}; \dfrac{u^2 \sin 2\alpha}{g}.$

SECTION-B

Partial Differential Equations

1
CHAPTER

Introduction to Partial Differential Equations

So far the differential equations involving only one independent variable were discussed. Now, the equations involving more than one independent variables will be introduced.

Definition. An equation containing one or more partial derivatives of an unknown function of two or more independent variables is known as a partial differential equation.

For example,

$$x\frac{\partial z}{\partial x} - y\frac{\partial z}{\partial y} = (z - xy)^2$$

$$\frac{\partial^2 z}{\partial x^2} + xy\frac{\partial^2 z}{\partial y^2} = 2x\frac{\partial^2 z}{\partial x \partial y}$$

are partial differential equations with z as dependent and x, y as independent variables. Most of the problems of solid mechanics, fluid dynamics, quantum mechanics, heat flow, wave propagation, engineering sciences and mathematical physics are partial differential equations. Therefore, it is important for the students of science and engineering to know how the partial differential equations are formed and their solution methods.

Some Important Terms

We have already stated that an equation involving the partial derivatives of an unknown function of two or more independent variables is known as a partial differential equation. Just as in the case of ordinary differential equations, we define the order and degree of a partial differential equation to be the order and power of the derivative of highest order occurring in the equation, respectively.

Definition. A partial differential equation is said to be linear if the dependent variable and its partial derivatives occur only in the first degree and are not multiplied.

Otherwise, it is called non-linear partial differential equation.

We enlist some partial differential equations with their order, degree, and linear/non-linear characteristics:

Partial differential equation	Order	Degree	Linear/Nonlinear
$\frac{\partial z}{\partial x}+\frac{\partial z}{\partial y} = z + xy$	1	1	Linear
$(x^2 - z^2)\frac{\partial z}{\partial x}+yz\frac{\partial z}{\partial y} = -3xy$	1	1	Nonlinear
$\left(\frac{\partial z}{\partial x}\right)^2+2\frac{\partial z}{\partial y} - 2z = 0$	1	2	Nonlinear
$x\frac{\partial^2 z}{\partial x^3}+y\left(\frac{\partial z}{\partial y}\right)^3 + kz^2 = 0$	2	1	Nonlinear
$z\left(\frac{\partial z}{\partial x}\right)^2 + x^2 = 3$	1	2	Nonlinear
$\left(\frac{\partial z}{\partial x}\right)^2+\left(\frac{\partial z}{\partial y}\right)^2 = 1$	1	2	Nonlinear

Formation of Partial Differential Equations

In earlier chapters, we have seen how an ordinary differential equation was obtained by eliminating arbitrary constants but unlike ordinary differential equations, a partial differential equation can be formed either by eliminating arbitrary constants or eliminating arbitrary functions depending upon the relation given. We now discuss each method with illustration:

(a) ***Formation of partial differential equations by the elimination of arbitrary constants.*** Consider an equation

$$f(x, y, z, a, b) = 0 \tag{1}$$

where, a and b are arbitrary constants. Differentiating Eq. (1) with respect to x and y partially, we get

$$\frac{\partial f}{\partial x}+\frac{\partial f}{\partial z}\frac{\partial z}{\partial x} = 0 \text{ or } \frac{\partial f}{\partial x}+\frac{\partial f}{\partial z}p = 0 \tag{2}$$

and

$$\frac{\partial f}{\partial y}+\frac{\partial f}{\partial z}\frac{\partial z}{\partial y} = 0 \text{ or } \frac{\partial f}{\partial y}+\frac{\partial f}{\partial z}q = 0 \tag{3}$$

Eliminating arbitrary constants a and b between Eqs. (1), (2) and (3), we shall obtain a partial differential equation of the form

$$\phi(x, y, z, p, q) = 0$$

which is a partial differential equation of the first order.

Note:

1. If in the function f, there are more arbitrary constants then the number of independent variables the above procedure of eliminating arbitrary constants will give rise to a partial differential equation of higher order than the first order.
2. If there are n arbitrary constants, we need $n + 1$ equations to eliminate them.

▲▲

EXAMPLE 1.1

Form a partial differential equation by eliminating a, b from

$$z = axe^y + \frac{1}{2}a^2e^{2y} + b.$$

Solution: Differentiating the given equation partially with respect to x and y, we get

$$\frac{\partial z}{\partial x} = ae^y \tag{1}$$

and

$$\frac{\partial z}{\partial y} = axe^y + a^2e^{2y}$$

$$- x\left(\frac{\partial z}{\partial x}\right)+\left(\frac{\partial z}{\partial x}\right)^2 \quad \text{\{by using Eq. (1)\}}$$

$\therefore$

$$q = xp + p^2$$

which is the required partial differential equation.

EXAMPLE 1.2

Form a partial differential equation by eliminating arbitrary constant a, b, c from $\frac{x^2}{a^2}+\frac{y^2}{b^2}+\frac{z^2}{c^2} = 1.$

Solution: Differentiating the given equation partially with respect to x and y, we get

$$\frac{2x}{a^2}+\frac{2z}{c^2}\frac{\partial z}{\partial x} = 0$$

$\Rightarrow$

$$c^2x + a^2zp = 0 \tag{1}$$

and

$$\frac{2y}{b^2}+\frac{2z}{c^2}\frac{\partial z}{\partial y} = 0$$

$\Rightarrow$

$$c^2y + b^2zq = 0 \tag{2}$$

Now, differentiating Eq. (1) partially with respect to x, we have

$$c^2 + a^2\left\{z\frac{\partial p}{\partial x} + p\frac{\partial z}{\partial x}\right\} = 0$$

$$\Rightarrow \qquad \frac{c^2}{a^2} + z\frac{\partial p}{\partial x} + p^2 = 0$$

From Eq. (1) $\qquad \frac{c^2}{a^2} = -\frac{z}{x}p$

Substituting this, we get

$$-\frac{z}{x}p + z\frac{\partial p}{\partial x} + p^2 = 0$$

$$\Rightarrow \qquad xz\frac{\partial p}{\partial x} - zp + xp^2 = 0$$

$$\Rightarrow \qquad xz\cdot r + xp^2 - zp = 0 \qquad \left\{\because \frac{\partial p}{\partial x} = \frac{\partial^2 z}{\partial x^2} = r\right\}$$

which is the required partial differential equation.

We can also differentiate Eq. (2) partially with respect to y to get

$$c^2 + b^2\left\{z\frac{\partial q}{\partial y} + q\frac{\partial z}{\partial y}\right\} = 0$$

$$\Rightarrow \qquad \frac{c^2}{b^2} + z\frac{\partial q}{\partial y} + q\frac{\partial z}{\partial y} = 0$$

From Eq. (2) $\qquad \frac{c^2}{b^2} = -\frac{zq}{y}$

Substituting this, we get

$$-\frac{z}{y}q + z\frac{\partial q}{\partial y} + q - q = 0$$

$$\Rightarrow \qquad yz\frac{\partial q}{\partial y} + yq^2 - zq = 0$$

$$\Rightarrow \qquad yzt + yq^2 - zq = 0 \qquad \left\{\because \frac{\partial q}{\partial y} = \frac{\partial^2 z}{\partial y^2} = t\right\}$$

The reason to get two differential equations is the number of arbitrary constants are more than the number of independent variables.

EXAMPLE 1.3

Find the partial differential equation of all spheres of radius r having centres in the xy plane.

Solution: Let $(a, b, 0)$ be the centre in xy plane of the sphere of radius r. The equation of sphere is

$$(x - a)^2 + (y - b)^2 + z^2 = r^2$$

Here a and b are arbitrary constants.

Differentiating partially with respect to x and y, we get

$$2(x - a) + 2z\frac{\partial z}{\partial x} = 0 \Rightarrow x - a = -zp$$

and $$2(y - b) + 2z\frac{\partial z}{\partial y} = 0 \Rightarrow z - b = -zq$$

Substituting these, we obtain

$$(-zp)^2 + (-zq)^2 + z^2 = r^2$$

$\Rightarrow$ $$z^2(p^2 + q^2 + 1) = r^2$$

which is the required partial differential equation.

EXAMPLE 1.4

Find the partial differential equation of planes having equal x, y intercepts.

Solution: The equation of plane is

$$\frac{x}{a} + \frac{y}{a} + \frac{z}{b} = 1$$

$\Rightarrow$ $$z = b - \frac{b}{a}(x + y)$$

Differentiating partially with respect to x and y, we get

$$\frac{\partial z}{\partial x} = -\frac{b}{a} \text{ and } \frac{\partial z}{\partial y} = -\frac{b}{a}$$

so that $$\frac{\partial z}{\partial x} = \frac{\partial z}{\partial y} \Rightarrow p = q$$

which is the required partial differential equation.

EXAMPLE 1.5

Find the partial differential equation of all planes which are at a constant distance K from the origin.

Solution: The equation of the plane in normal from is

$$lx + my + nz = K$$

where, l, m, n are the direction cosines of the normal form the origin to the plane.

We know that $l^2 + m^2 + n^2 = 1$

$$\therefore \qquad n = \sqrt{1-l^2-m^2}$$

Thus, we obtain

$$lx + my + \left(\sqrt{1-l^2-m^2}\right)z = K.$$

where l and m are two arbitrary constants to be eliminated.

Differentiating this equation, partially with respect to x and y, we get

$$l+\left(\sqrt{1-l^2-m^2}\right)\frac{\partial z}{\partial x}=0$$

$$\Rightarrow \qquad l = -p\sqrt{1-l^2-m^2}$$

and
$$m+\left(\sqrt{1-l^2-m^2}\right)\frac{\partial z}{\partial y}=0$$

$$\Rightarrow \qquad m = -q\sqrt{1-l^2-m^2}$$

Now,
$$l^2 + m^2 = (p^2+q^2)(1-l^2-m^2)$$

$$\Rightarrow \qquad l^2 + m^2 = \frac{p^2+q^2}{1+p^2+q^2}$$

and
$$1-l^2-m^2 = \frac{1}{1+p^2+q^2}$$

Therefore, we obtain

$$l=-\frac{p}{\sqrt{1+p^2+q^2}} \text{ and } m=-\frac{q}{\sqrt{1+p^2+q^2}}$$

Now on substituting these values in

$$lx + my + \left(\sqrt{1-l^2-m^2}\right)z = K, \text{ we get}$$

$$\frac{-px}{\sqrt{1+p^2+q^2}}-\frac{qy}{\sqrt{1+p^2+q^2}}+\frac{z}{\sqrt{1+p^2+q^2}} = K$$

$$\Rightarrow \qquad z = px + qy + K\sqrt{1+p^2+q^2},$$

which is the required partial differential equation.

EXAMPLE 1.6

Find the differential equation of the set of all right circular cones whose axis coincides with z-axis.

Solution: The equation of a right circular cone whose axis coincide with z-axis, having vertical angle α and vertex at $(0, 0, c)$ is given by

$$x^2 + y^2 = (z - c)^2 \tan^2 \alpha$$

Here, c and α are arbitrary constants.

Differentiating partially with respect to x and y, we get

$$2x = 2(z - c)\frac{\partial z}{\partial x} \tan^2 \alpha$$

$$\Rightarrow \qquad x = (z - c)p \tan^2 \alpha$$

$$\Rightarrow \qquad xy = (z - c)yp \tan^2 \alpha$$

and

$$2y = 2(z - c)\frac{\partial z}{\partial y} \tan^2 \alpha$$

$$\Rightarrow \qquad y = (z - c)q \tan^2 \alpha$$

$$\Rightarrow \qquad xy = (z - c)xq \tan^2 \alpha.$$

Equating, we get

$$(z - c)yp = (z - c)xq$$

$$\Rightarrow \qquad yp - xq = 0$$

which is the required differential equation.

EXAMPLE 1.7

Show that the differential equation of all cones which have their vertex at the origin is $px + qy = z$. Verify that $yz + zx + xy = 0$ is a surface satisfying the above equation.

Solution: The equation of any cone with vertex at origin is

$$ax^2 + by^2 + cz^2 + 2fgz + 2gzx + 2hxy = 0$$

where, a, b, c, f, g, h are parameters.

Differentiating partially with respect to x and y, we obtain

$$2ax + 2cz\frac{\partial z}{\partial x} + 2fy\frac{\partial z}{\partial x} + 2g\frac{\partial z}{\partial x}x + 2gz + 2hy = 0$$

$$\Rightarrow \qquad ax + gz + hy + p(cz + gx + fy) = 0 \tag{1}$$

and $$2by + 2cz\frac{\partial z}{\partial y} + 2fz + 2fy\frac{\partial z}{\partial y} + 2g\frac{\partial z}{\partial y}x + 2hx = 0$$

$\Rightarrow$ $$by + fz + hx + q(cz + fy + gx) = 0 \quad (2)$$

Multiplying Eq. (1) by x and Eq. (2) by y and adding, we have

$$(ax^2 + by^2 + gzx + fyz + 2hxy) + (px + qy)(cz + fy + gz) = 0$$

$\Rightarrow$ $$-(cz^2 + fyz + gzx) + (px + qy)(cz + fy + gx) = 0$$

$\Rightarrow$ $$(cz + fy + gx)(px + qy - z) = 0$$

$\Rightarrow$ $$px + qy - z = 0$$

which is the required partial differential equation

Given equation is

$$yz + zx + xy = 0$$

Differentiating partially with respect to x and y, we get

$$yp + px + z + y = 0 \text{ and } z + yq + xq + x = 0$$

Solving for p and q, we obtain

$$p = -\frac{z+y}{x+y} \text{ and } q = -\frac{z+x}{x+y}$$

$\therefore$ $$px + qy - z = -\frac{x(z+y)}{x+y} - \frac{y(z+x)}{x+y} - z = 0.$$

Hence, $yz + zx + xy = 0$ is a surface satisfying $px + qy - z = 0$.

▲▲

EXERCISE 1.1

Obtain the partial differential equations by eliminating the arbitrary constants in the following cases.

1. $z = (x + a)(y + b)$.
2. $ax^2 + by^2 + z^2 = 1$.
3. $2z = \sqrt{x+a} + \sqrt{y-a} + 2b$.
4. $z^2 \cot^2 \alpha = (x - a)^2 + (y - b)^2$.
5. $x^2 + y^2 = (z - c)^2 \tan^2 \alpha$ where, α is a parameter.
6. $z = a(x + \log y) - \frac{x^2}{2} - b^2$.
7. From differential equation by eliminating constants A and p from $z = Ae^{pt} \sin px$.

8. Show that the partial differential equation obtained by eliminating the arbitrary constant a and b from $z = ax + by + f(a, b)$ is given by $z = px + qy + f(p, q)$.
9. Find the differential equation of the family of spheres of radius 4 with centres on the xy-plane.
10. Find the partial differential equation of the family of spheres of radius 7 with centres on the plane $x + y = 0$.
11. Form the partial differential equation by eliminating arbitrary constants a and b in
    (i) $z = ax + by + a^2 + b^2$
    (ii) $z = ax + a^2y^2 + b$.
12. From the partial differential equation by eliminating the arbitrary constants a and b from
    (i) $z = ax^3 + by^3$
    (ii) $z = (x^2 + a)(y^2 + b)$
    (iii) $\log(az - 1) = x + ay + b$
    (iv) $z = xy + y\sqrt{x^2 - a^2} + b$.
13. Find the partial differential equation of all spheres whose centres lie on the x-axis.
14. Find the partial differential equation by eliminating a and b from the equation

$$z = a(x + y) + b.$$

15. Eliminate the arbitrary constants from $z = ax + by + cxy$ to form a partial differential equation.

ANSWERS

Exercise 1.1

1. $pq = z$.
2. $z(px + qy) = z^2 - 1$.
3. $p^2 + q^2 = x + y$.
4. $p^2 + q^2 = \tan^2 \alpha$.
5. $yp = xq$.
6. $p + x = qy$.
7. $\partial^2 z/\partial x^2 + \partial^2 z/\partial t^2 = 0$.
9. $(x - y)^2 (p^2 + q^2 + 1) = 16(p - q)^2$.
10. $(p^2 + q^2 + 1)(x - y)^2 = 49(p - q)$.
11. (i) $z = px + qy + p^2 + q^2$; (ii) $q = 2p^2y$.
12. (i) $3z = px + py$; (ii) $4xyz = pq$; (iii) $zq = p(q + 1)$; (iv) $pq = px + qy$.

13. $xq = py$.
14. $p = q$.
15. $px + qy - xys = z$ or $r = 0$ or $t = 0$.

Formation of partial differential equation by the elimination of arbitrary function: Let u and v be two functions of x, y and z connected by the relation $f(u, v) = 0$.

Differentiating partially with respect to x and y, we get

$$\frac{\partial f}{\partial u}\left(\frac{\partial u}{\partial x}+\frac{\partial u}{\partial z}\frac{\partial z}{\partial x}\right)+\frac{\partial f}{\partial v}\left(\frac{\partial v}{\partial x}+\frac{\partial v}{\partial z}\frac{\partial z}{\partial x}\right)=0$$

$$\Rightarrow \quad \frac{\partial f}{\partial u}\left(\frac{\partial u}{\partial x}+\frac{\partial u}{\partial z}p\right)+\frac{\partial f}{\partial v}\left(\frac{\partial v}{\partial x}+\frac{\partial v}{\partial z}p\right)=0$$

and

$$\frac{\partial f}{\partial u}\left(\frac{\partial u}{\partial y}+\frac{\partial u}{\partial z}\frac{\partial z}{\partial y}\right)+\frac{\partial f}{\partial v}\left(\frac{\partial v}{\partial y}+\frac{\partial v}{\partial z}\frac{\partial z}{\partial y}\right)=0$$

$$\Rightarrow \quad \frac{\partial f}{\partial u}\left(\frac{\partial u}{\partial y}+\frac{\partial u}{\partial z}q\right)+\frac{\partial f}{\partial v}\left(\frac{\partial v}{\partial y}+\frac{\partial v}{\partial z}q\right)=0$$

Eliminating $\frac{\partial f}{\partial u}$ and $\frac{\partial f}{\partial v}$, we get

$$\left(\frac{\partial u}{\partial x}+\frac{\partial u}{\partial z}p\right)\left(\frac{\partial v}{\partial y}+\frac{\partial v}{\partial z}q\right)=\left(\frac{\partial u}{\partial y}+\frac{\partial u}{\partial z}q\right)\left(\frac{\partial v}{\partial x}+\frac{\partial v}{\partial z}p\right)$$

$$\Rightarrow \quad \left(\frac{\partial u}{\partial x}+\frac{\partial u}{\partial z}p\right)\left(\frac{\partial v}{\partial y}+\frac{\partial v}{\partial z}q\right)-\left(\frac{\partial u}{\partial y}+\frac{\partial u}{\partial z}q\right)\left(\frac{\partial v}{\partial x}+\frac{\partial v}{\partial z}p\right)=0$$

$$\Rightarrow \quad \left(\frac{\partial u}{\partial y}\frac{\partial v}{\partial z}-\frac{\partial u}{\partial z}\frac{\partial v}{\partial y}\right)p+\left(\frac{\partial u}{\partial z}\frac{\partial v}{\partial x}-\frac{\partial u}{\partial u}\frac{\partial v}{\partial z}\right)q$$

$$=\frac{\partial u}{\partial x}\frac{\partial v}{\partial y}-\frac{\partial u}{\partial y}\frac{\partial v}{\partial x}$$

$\Rightarrow$ $Pp + Qq = R$ which is the required equation.

where,

$$P=\frac{\partial u}{\partial y}\frac{\partial v}{\partial z}-\frac{\partial u}{\partial z}\frac{\partial v}{\partial y}=\begin{vmatrix}\frac{\partial u}{\partial y} & \frac{\partial u}{\partial z}\\ \frac{\partial v}{\partial y} & \frac{\partial v}{\partial z}\end{vmatrix}=\frac{\partial(u,v)}{\partial(y,z)}$$

$$Q = \frac{\partial u}{\partial z}\frac{\partial v}{\partial x} - \frac{\partial u}{\partial x}\frac{\partial v}{\partial z} = \begin{vmatrix} \frac{\partial u}{\partial z} & \frac{\partial u}{\partial x} \\ \frac{\partial v}{\partial z} & \frac{\partial v}{\partial x} \end{vmatrix} = \frac{\partial(u,v)}{\partial(z,x)}$$

and
$$R = \frac{\partial u}{\partial x}\frac{\partial v}{\partial y} - \frac{\partial u}{\partial y}\frac{\partial v}{\partial x} = \begin{vmatrix} \frac{\partial u}{\partial x} & \frac{\partial u}{\partial y} \\ \frac{\partial v}{\partial x} & \frac{\partial v}{\partial y} \end{vmatrix} = \frac{\partial(u,v)}{\partial(x,y)}$$

Note:

If the given relation between x, y and z contains two arbitrary functions, then in general, their elimination will give rise to a partial differential equation of the order higher than one.

EXAMPLE 1.8

Form a differential equation by eliminating the function f from

$$z = y^2 + 2f\left(\frac{1}{x} + \log y\right)$$

Solution: Differentiating partially with respect to x and y, we have

$$\frac{\partial z}{\partial x} = 2f'\left(\frac{1}{x} + \log y\right)\left(-\frac{1}{x^2}\right)$$

$$\Rightarrow \qquad -px^2 = 2f'\left(\frac{1}{x} + \log y\right)$$

and
$$\frac{\partial z}{\partial y} = 2y + 2f'\left(\frac{1}{x} + \log y\right)\left(\frac{1}{y}\right)$$

$$\Rightarrow \qquad qy - 2y^2 = 2f'\left(\frac{1}{x} + \log y\right)$$

Thus eliminating $2f\left(\frac{1}{x} + \log y\right)$, we obtain

$$-px^2 = qy - 2y^2$$

$$\Rightarrow \qquad x^2p + yq = 2y^2$$

which is the required equation.

EXAMPLE 1.9

Form a partial differential equation by eliminating functions f and g from

$$z = f(x + iy) + g(x - iy)$$

Solution: Differentiating, we have

$$\frac{\partial z}{\partial x} = f'(x + iy) + g'(x - iy)$$

$$\frac{\partial z}{\partial y} = if'(x + iy) - ig'(x - iy)$$

$$\frac{\partial^2 z}{\partial x^2} = f''(x + iy) + g''(x + y)$$

$$\frac{\partial^2 z}{\partial y^2} = i^2 f''(x + iy) + (-i)^2 g'(x - iy)$$

$$\therefore \quad \frac{\partial^2 z}{\partial y^2} + \frac{\partial^2 z}{\partial y^2} = 0, \text{ which is the required equation.}$$

EXAMPLE 1.10

Form a partial differential equation by eliminating the arbitrary function ϕ from $\phi(x + y + z, x^2 + y^2 - z^2) = 0$. What is the order of the partial differential equation?

Solution: Let $x + y + z = u$ and $x^2 + y^2 - z^2 = v$

Thus, we have $\phi(u, v) = 0$

Differentiating with respect to x and y partially, we get

$$\frac{\partial \phi}{\partial u}\left(\frac{\partial u}{\partial x} + \frac{\partial u}{\partial z}\frac{\partial z}{\partial x}\right) + \frac{\partial \phi}{\partial v}\left(\frac{\partial v}{\partial x} + \frac{\partial v}{\partial z}\frac{\partial z}{\partial x}\right) = 0$$

$$\Rightarrow \quad \frac{\partial \phi}{\partial u}(1 + p) + \frac{\partial \phi}{\partial v}(2x - 2zp) = 0$$

and

$$\frac{\partial \phi}{\partial u}\left(\frac{\partial u}{\partial y} + \frac{\partial u}{\partial z}\frac{\partial z}{\partial y}\right) + \frac{\partial \phi}{\partial v}\left(\frac{\partial v}{\partial y} + \frac{\partial v}{\partial z}\frac{\partial z}{\partial y}\right) = 0$$

$$\Rightarrow \quad \frac{\partial \phi}{\partial u}(1 + q) + \frac{\partial \phi}{\partial v}(2y - 2zq) = 0$$

Eliminating $\frac{\partial \phi}{\partial u}$ and $\frac{\partial \phi}{\partial v}$, we get

$$\frac{2x - 2zp}{1 + p} = \frac{2y - 2zq}{1 + q}$$

$$\Rightarrow \quad (1 + q)(x - zp) - (1 + p)(y - zq) = 0$$

$$\Rightarrow \quad (y + z)p - (x + z)q = x - y$$

which is the required partial differential equation, whose order is one.

EXAMPLE 1.11

Form a partial differential equation by eliminating the arbitrary functions f and ϕ from $z = yf(x) + x\phi(y)$.

Solution: Differentiating partially with respect to x and y, we get

$$\frac{\partial z}{\partial x} = yf'(x) + \phi(y)$$

and
$$\frac{\partial z}{\partial y} = f(x) + x\phi'(y)$$

Again,
$$\frac{\partial^2 z}{\partial x \partial y} = f'(x) + \phi'(y)$$

$$\frac{\partial^2 z}{\partial x \partial y} = \frac{1}{y}\left(\frac{\partial z}{\partial x} - \phi(y)\right) + \frac{1}{x}\left(\frac{\partial z}{\partial y} - f(x)\right)$$

$$\Rightarrow \qquad xy\frac{\partial^2 y}{\partial x \partial y} = x\frac{\partial z}{\partial x} + y\frac{\partial z}{\partial y} - (x\phi(y) + yf(x))$$

$$\Rightarrow \qquad xy\frac{\partial^2 z}{\partial x \partial y} = x\frac{\partial z}{\partial x} + y\frac{\partial z}{\partial y} - z$$

$$\Rightarrow \qquad xys = xp + yq - z$$

which is the required partial differential equation, whose order is 2.

EXAMPLE 1.12

Form a partial differential equation by eliminating the arbitrary function f from the equation $x + y + z = f(x^2 + y^2 + z^2)$.

Solution: Differentiating partially with respect to x and y, we get

$$1 + \frac{\partial z}{\partial x} = f'(x^2 + y^2 + z^2)\left(2x + 2z\frac{\partial z}{\partial x}\right)$$

$$\Rightarrow \qquad 1 + p = f'(x^2 + y^2 + z^2)(2x + 2zp)$$

and
$$1 + \frac{\partial z}{\partial y} = f'(x^2 + y^2 + z^2)\left(2y + 2z\frac{\partial z}{\partial y}\right)$$

$$\Rightarrow \qquad 1 + q = f'(x^2 + y^2 + z^2)(2y + 2zq).$$

Eliminating $f'(x^2 + y^2 + z^2)$, we obtain

$$\frac{(1+p)}{2x+2zp} = \frac{1+q}{2y+2zq}$$

$$\Rightarrow \quad (1 + p)(y + zq) = (1 + q)(x + zp)$$

$$\Rightarrow \quad (y - z)p + (z - x)q = x - y.$$

which is the required partial differential equation.

EXAMPLE 1.13

Equation of any cone with vertex at $P(a, b, c)$ is of the from $f\left(\frac{x-a}{z-c}, \frac{y-b}{z-c}\right) = 0$. Find the differential equation of the cone.

Solution: Let $\frac{x-a}{z-c} = u$ and $\frac{y-b}{z-c} = v$

Then, the equation of the given cone becomes $f(u, v) = 0$.
Differentiating partially with respect to x and y, we have

$$\frac{\partial f}{\partial u}\left(\frac{\partial u}{\partial x} + \frac{\partial u}{\partial z}\frac{\partial z}{\partial x}\right) + \frac{\partial f}{\partial v}\left(\frac{\partial v}{\partial x} + \frac{\partial v}{\partial z}\frac{\partial z}{\partial x}\right) = 0$$

$$\Rightarrow \quad \frac{\partial f}{\partial u}\left(\frac{1}{z-c} - \frac{x-a}{(z-c)^2}p\right) + \frac{\partial f}{\partial v}\left(0 - \frac{y-b}{(z-c)^2}p\right) = 0$$

and

$$\frac{\partial f}{\partial u}\left(\frac{\partial u}{\partial y} + \frac{\partial u}{\partial z}\frac{\partial z}{\partial y}\right) + \frac{\partial f}{\partial v}\left(\frac{\partial v}{\partial y} + \frac{\partial v}{\partial z}\frac{\partial z}{\partial y}\right) = 0$$

$$\Rightarrow \quad \frac{\partial f}{\partial u}\left(0 - \frac{x-a}{(z-c)^2}q\right) + \frac{\partial f}{\partial v}\left(\frac{1}{z-c} - \frac{y-b}{(z-c)^2}q\right) = 0$$

Eliminating $\frac{\partial f}{\partial u}$ and $\frac{\partial f}{\partial v}$, we get

$$\frac{-\frac{y-b}{(z-c)^2}p}{\frac{1}{z-c} - \frac{x-a}{(z-c)^2}p} = \frac{\frac{1}{z-c} - \frac{y-b}{(z-c)^2}q}{-\frac{x-a}{(z-c)^2}q}$$

$$\frac{(y-b)p}{z-c-(x-a)p} = \frac{z-c-(y-b)q}{(x-a)q}$$

$$\Rightarrow \quad (z - c)^2 + (z - c)[-(y - b)q - (x - a)p] + (x - a)(y - b)pq = (x - a)(y - b)pq$$

$$\Rightarrow \quad z - c = (x - a)p + (y - b)q$$

which is the required differential equation of the given cone.

EXAMPLE 1.14

Form a partial differential equation by eliminating arbitrary functions f and g from $z = f(x^2 - y) + g(x^2 + y)$.

Solution: Differentiating partially with respect to x and y, we get

$$\frac{\partial z}{\partial x} = 2xf'(x^2 - y) + 2xg'(x^2 + y)$$

and
$$\frac{\partial z}{\partial y} = -f'(x^2 - y) + g'(x^2 + y)$$

Again, differentiating with respect to x and y respectively, we get

$$\frac{\partial^2 z}{\partial x^2} = 2[f'(x^2 - y) + g'(x^2 + y)] + 4x^2[f''(x^2 - y) + g''(x^2 + y)]$$

and
$$\frac{\partial^2 z}{\partial y^2} = f''(x^2 - y) + g''(x^2 + y)$$

Therefore,

$$\frac{\partial^2 z}{\partial x^2} = 2\frac{1}{2x}\frac{\partial z}{\partial x} + 4x^2\frac{\partial^2 z}{\partial y^2}$$

$\Rightarrow$
$$xr = p + 4x^3 t.$$

which is the required partial differential equation of order 2.

EXAMPLE 1.15

Form the partial differential equation by eliminating the arbitrary function ϕ where, $\phi(xy + z^2, x + y + z) = 0$.

Solution: Let $u = xy + z^2$ and $v = x + y + z$

$\therefore$ Given equation becomes $\phi(u, v) = 0$

Differentiating partially with respect to x and y, we get

$$\frac{\partial \phi}{\partial u}\left(\frac{\partial u}{\partial x} + \frac{\partial u}{\partial z}\frac{\partial z}{\partial x}\right) + \frac{\partial \phi}{\partial v}\left(\frac{\partial v}{\partial x} + \frac{\partial v}{\partial z}\frac{\partial z}{\partial x}\right) = 0$$

$\Rightarrow$
$$\frac{\partial \phi}{\partial u}(y + 2zp) + \frac{\partial \phi}{\partial v}(1 + 1 \cdot p) = 0$$

and
$$\frac{\partial \phi}{\partial u}\left(\frac{\partial u}{\partial y} + \frac{\partial u}{\partial z}\frac{\partial z}{\partial y}\right) + \frac{\partial \phi}{\partial v}\left(\frac{\partial v}{\partial y} + \frac{\partial v}{\partial z}\frac{\partial z}{\partial y}\right) = 0$$

$\Rightarrow$
$$\frac{\partial \phi}{\partial u}(x + 2zq) + \frac{\partial \phi}{\partial v}(1 + 1 \cdot q) = 0$$

Eliminating $\frac{\partial \phi}{\partial u}$ and $\frac{\partial \phi}{\partial v}$, we obtain

$$\frac{1+p}{y+2zp} - \frac{1+q}{x+2zq}$$

$\Rightarrow$ $$(y + 2zp)(1 + q) - (x + 2zq)(1 + p) = 0$$

$\Rightarrow$ $$(2z - x)p + (y - 2z)q = x - y.$$

which is the required differential equation.

EXERCISE 1.2

Form the partial differential equation by eliminating the arbitrary function in

1. $xyz = \phi(x^2 + y^2 - z^2)$.
2. $xy + yz + zx = f\left(\frac{z}{x+y}\right)$.
3. $x^2 + y^2 + z^2 = f(zy + z - x)$.
4. $ax + by + cz = \phi(x^2 + y^2 + z^2)$.
5. $z = yf(x) + xg(y)$.
6. $z = xf(ax + by) + g(ax + by)$.
7. $z = f(2x + y) + g(3x - y)$.

Form the partial differential equations by eliminating the arbitrary function ϕ from the relation.

8. $\phi(x^2 + y^2 + z^2, xyz) = 0$.
9. $\phi\left(\frac{y}{x}, x^2 + y^2 + z^2\right) = 0$.
10. $\phi(x^2 - y^2, x^2 - z^2) = 0$.
11. $\phi(x^2 + y^2, z - xy) = 0$.
12. Form a partial differential equation by eliminating the function f from $z = e^{ax+by} f(ax + by)$.
13. Form a partial differential equation by eliminating the arbitrary functions f and g from $z = f(x^2 - y) + g(x^2 + y)$.
14. Form a partial differential equation by eliminating the arbitrary function from $z = e^{ny} \phi(x - y)$.
15. Form a partial differential equation by eliminating the arbitrary function from $z = f(xy) + g(x/y)$.

ANSWERS

Exercise 1.2

1. $x(y^2 + z^2)p - y(z^2 - x^2)q = z(x^2 - y^2)$.
2. $(x + y)(x + 2z)p - (x + y)(y + 2z)q = z(x - y)$.
3. $(y^2 + y - z^2)p - (xy + x + z)q = (xz + y)$.
4. $(cy - bz)p + (az - cx)q = (bx - ay)$.
5. $z = xp + yq - xys$.
6. $2abs = b^2r + a^2t$.
7. $6t = r + s$
8. $x(z^2 - y^2)p + y(x^2 - z^2)q = (y^2 - x^2)z$.
9. $xzp + yzq = -(x^2 + y^2)$.
10. $yzp + zxp = xy$.
11. $py - qx = y^2 - x^2$.
12. $b(\partial z/\partial x) + a(\partial x/\partial y) = 2abz$.
13. $x\dfrac{\partial^2 z}{\partial x^2} = \dfrac{\partial z}{\partial x} + 4x^3\dfrac{\partial^2 z}{\partial y^2}$.
14. $p + q = nz$.
15. $x^2r - y^2t + xp - q = 0$.

▲▲

2

CHAPTER

Solution of Linear Partial Differential Equations of the First Order

A solution of a partial differential equation is a non-differential relation between the independent and dependent variables which satisfies the given partial differential equation. This is also known as integral of the partial differential equation.

For example, the solutions of $px + qy = z$ are $z = Ax + By$ and $z = x\phi\left(\frac{y}{x}\right)$ where, A, B are arbitrary constants and ϕ is an arbitrary function.

Complete Integral. The solution $\phi(x, y, z, a, b) = 0$ of a first order partial differential equation $f(x, y, z, p, q) = 0$, which contains two arbitrary constants a and b is called a complete integral of the partial differential equation.

For example, $z = ax + by + ab$ is a complete solution of $z = px + qy + pq$.

Particular Integral. The solution obtained from a complete integral by assigning particular values to the arbitrary constants is called a particular solution.

For example, $z = x + 2y + 2$ is a particular solution of $z = px + qy + pq$.

General Integral

Let
$$\phi(x, y, z, a, b) = 0 \tag{1}$$
be the complete integral of the partial differential equation
$$f(x, y, z, p, q) = 0 \tag{2}$$

If one of the constants, say b in Eq. (1) is a function of the other that is $b = h(a)$ then Eq. (1) becomes
$$\phi(x, y, z, a, h(a)) = 0 \tag{3}$$

Then, the envelope of the family of surface (3), obtained by eliminating 'a' between Eq. (3) and $\frac{\partial \phi}{\partial a} = 0$, is called the general integral of the partial differential Eq. (2).

For example, $z = f(x^2 - y^2)$ is the general solution of $py + qx = 0$.

Singular Integral. Let $\phi(x, y, z, a, b) = 0$ (1)

be the complete integral of $f(x, y, z, p, q) = 0$ (2)

where, a and b are arbitrary constants.

Then, the envelope of the family of surface (1), obtained by eliminating a and b between Eq. (1), $\frac{\partial \phi}{\partial a} = 0$ and $\frac{\partial \phi}{\partial b} = 0$ is called the singular integral of the partial differential Eq. (2).

For example, $xy + z = 0$ is a singular integral of $z = px + qy + pq$.

Note: One most important difference between the solution of partial differential equation and those of ordinary differential equation is the general solution of a linear ordinary differential equation contains arbitrary constants of integration whereas the general solution of a linear partial differential equation contains arbitrary functions.

Equations Easily Integrable

We now consider such partial differential equations which can be solved by direct integration. Here, the usual constant of integration consists of an arbitrary function of the variable considered constant during the integration. ▲▲

EXAMPLE 2.1

Solve $\frac{\partial^3 z}{\partial x^2 \partial y} + 18xy^2 + \sin(2x - y) = 0.$

Solution: Keep y fixed and integrate the given equation twice to obtain

$$\frac{\partial^2 z}{\partial x \partial y} + 9x^2y^2 - \frac{1}{2}\cos(2x - y) = f(y)$$

$$\frac{\partial z}{\partial y} + 3x^3y^2 - \frac{1}{4}\sin(2x - y) = xf(y) + g(y)$$

Now keep x fixed and integrate with respect to y, we get

$$z + x^3y^3 - \frac{1}{4}\cos(2x - y) = x\int f(y)dy + \int g(y)dy + h(x)$$

The result may be simplified by writing $\int f(y)dy = u(y)$, $\int g(y)dy = v(g)$.

Thus, $$z = \frac{1}{4}\cos(2x - y) - x^3y^3 + xu(y) + v(y) + h(x)$$

where, u, v and h are arbitrary functions.

EXAMPLE 2.2

Solve $\frac{\partial^2 z}{\partial x^2} + z = 0$ when $x = 0$, $z = e^y$ and $\frac{\partial z}{\partial x} = 1.$

Solution: If z were a function of x alone, the solution would have been $z = A \sin x + B \cos x$, where, A and B are constants. Since z is a function of x and y, A and B are arbitrary functions of y. Hence, the solution of the given equation is

$$z = f(y) \sin x + g(y) \cos x$$

$$\frac{\partial z}{\partial x} = f(y) \cos x - g(y) \sin x$$

When $\quad x = 0 \; z = e^y$

$\therefore \quad e^y = g(y)$

and $\quad x = 0, \; \frac{\partial z}{\partial x} = 1$

$\therefore \quad 1 = f(y)$

Hence, the required solution is

$$z = \sin x + e^y \cos x.$$

EXAMPLE 2.3

Solve $y\frac{\partial^2 z}{\partial x \partial y} + \frac{\partial z}{\partial x} = 4xy.$

Solution: The given equation can be written as

$$y\frac{\partial p}{\partial y} + p = 4xy$$

which is a linear differential equation of the first order if x is taken as a constant. Its solution is

$$py = 2xy^2 + u(x)$$

$$\Rightarrow \quad y\frac{\partial z}{\partial x} = 2xy^2 + u(x)$$

Integrating it (keeping y constant); we get

$$z = x^2 y + \frac{1}{y} u_1(x) + v(x)$$

where $u_1(x) = \int u(x)dx$ and $v(x)$ are arbitrary functions.

EXAMPLE 2.4

Solve $\frac{\partial^2 z}{\partial x \partial y} = \sin x \sin y$, for which $\frac{\partial z}{\partial y} = -2 \sin y$ when $x = 0$ and $z = 0$, when y is an odd multiple of $\pi/2$.

Solution: Integrating with respect to x (keeping y constant), we get

$$\frac{\partial z}{\partial y} = -\cos x \sin y + f(y)$$

when $$x = 0, \frac{\partial z}{\partial y} = -2 \sin y$$

$\therefore$ $$-2 \sin y = -\sin y + f(y)$$

$\Rightarrow$ $$f(y) = -\sin y$$

Thus, $$\frac{\partial z}{\partial y} = -\cos x \sin y - \sin y$$

Now, integrating with respect to y (keeping x constant), we get

$$z = \cos x \cos y + \cos y + g(x)$$

when y is an odd multiple of $\pi/2$, $$z = 0$$

$\therefore$ $$0 = 0 + 0 + g(x) \Rightarrow g(x) = 0$$

Hence, the complete solution is

$$z = (1 + \cos x) \cos y.$$

EXAMPLE 2.5

Solve the partial differential equation $\frac{\partial^2 z}{\partial x \partial y} = \frac{1}{x+y}$ given that $z = y \log y$ and $\frac{\partial z}{\partial y} = 1 + \log y$, when $x = 0$.

Solution: Integrating partially with respect to x, we get

$$\frac{\partial z}{\partial y} = \int \frac{1}{x+y} dx + f(y)$$

$\Rightarrow$ $$\frac{\partial z}{\partial y} = \log(x + y) + f(y)$$

when $$x = 0, \frac{\partial z}{\partial y} = 1 + \log y$$

$\therefore$ $$1 + \log y = \log(0 + y) + f(y)$$

$\Rightarrow$ $$f(y) = 1$$

Thus, $$\frac{\partial z}{\partial y} = \log(x + y) + 1$$

Now, integrating partially with respect to y, we get

$$z = \int \log(x + y)dy + y + g(x)$$

$$= y \log(x + y) - \int \frac{y}{x+y} dy + y + g(x)$$

$$= y \log(x + y) - \int \left(1 - \frac{x}{x+y}\right) dy + y + g(x)$$

$$= y \log(x + y) - y + x \log(x + y) + y + g(x)$$

$$= (y + x) \log (x + y) + g(x)$$

when $x = 0, z = y \log y$

$\therefore$ $y \log y = y \log y + g(x) \Rightarrow g(x) = 0$

Therefore, $z = (y + x) \log(x + y).$

EXAMPLE 2.6

Solve $\dfrac{\partial^2 z}{\partial x^2} = a^2 z$ given that when $x = 0$, $\dfrac{\partial z}{\partial x} = a \sin y$ and $\dfrac{\partial z}{\partial y} = 0.$

Solution: If z is a function of only x, then the solution will be

$$z = Ae^{ax} + Be^{-ax}$$

where A and B are arbitrary constants.

But, z is a function of x and y.

$\therefore$ Considering A and B as arbitrary functions of y, we get

$$z = f(y)e^{ax} + g(y)e^{-ax}$$

Now, $$\frac{\partial z}{\partial x} = af(y)e^{ax} - ag(y)e^{-ax}$$

$$\frac{\partial z}{\partial y} = f'(y)e^{ax} + g'(x)e^{-ax}$$

when $x = 0$, $\dfrac{\partial z}{\partial x} = a \sin y$

$\therefore$ $a \sin y = af(y) - ag(y)$

$\Rightarrow$ $\sin y = f(y) - g(y)$ (1)

when $x = 0$, $\dfrac{\partial z}{\partial y} = 0.$

$$\therefore \qquad 0 = f'(y) + g'(y) \qquad (2)$$

Differentiating Eq. (1), with respect to y, we get

$$\cos y = f'(y) - g'(y)$$

Thus, $$f'(y) = \frac{1}{2}\cos y \text{ and } g'(y) = -\frac{\cos y}{2}$$

$$\Rightarrow \qquad f(y) = \frac{1}{2}\sin y + c_1 \text{ and } g(y) = -\frac{1}{2}\sin y + c_2$$

Therefore, $$z = \left(\frac{1}{2}\sin y + c_1\right)e^{ax} + \left(-\frac{1}{2}\sin y + c_2\right)e^{-ax}$$

$$= \sin y\left(\frac{e^{ax} - e^{-ax}}{2}\right) + c_1 e^{ax} + c_2 e^{-ax}$$

$$= \sin y \sinh x + \text{constant.}$$

EXAMPLE 2.7

Solve $\frac{\partial z}{\partial x} = 3x - y$ and $\frac{\partial z}{\partial y} = -x + \cos y$.

Solution: Integrating $\frac{\partial z}{\partial x} = 3x - y$ with respect to x, we get

$$z = \frac{3x^2}{2} - yx + f(y)$$

Now, differentiating partially with respect to y, we get

$$\frac{\partial z}{\partial y} = -x + f'(y)$$

We are also given $$\frac{\partial z}{\partial y} = -x + \cos y$$

Therefore, $$-x + \cos y = -x + f'(y)$$

$$\Rightarrow \qquad \cos y = f'(y)$$

$$\Rightarrow \qquad \sin y + c = f(y)$$

Thus, we get

$$z = \frac{3x^2}{2} - yx + \sin x + c.$$

▲▲

EXERCISE 2.1

Solve the following partial differential equations:

1. $\dfrac{\partial^3 z}{\partial x^2 \partial y} = 0.$

2. $\dfrac{\partial^2 z}{\partial x^2} = \sin x.$

3. $\dfrac{\partial^2 z}{\partial x \partial y} = e^{-y} \cos x$ given that $z = 0$ when $y = 0$ and $\dfrac{\partial z}{\partial y} = 0$ when $x = 0.$

4. $\dfrac{\partial^2 z}{\partial y^2} = \sin(xy).$

5. $\dfrac{\partial^2 z}{\partial x \partial y} = \dfrac{1}{xy}.$

6. $\dfrac{\partial^2 z}{\partial x \partial y} = 6 - \dfrac{x}{y}.$

7. $x\dfrac{\partial z}{\partial x} = 2x + y + 3z.$

8. $\dfrac{\partial^3 z}{\partial x^2 \partial y} = \cos(ax + by).$

9. $\dfrac{\partial^2 z}{\partial y \partial x} = 4x \sin(3xy).$

10. $\dfrac{\partial^2 z}{\partial y^2} = z$ when $y = 0$; $z = e^x$ and $\dfrac{\partial z}{\partial y} = e^{-y}.$

11. Solve $\dfrac{\partial^2 z}{\partial x^2} + z = 0$, given that when $x = 0$, $z = e^y$ and $\dfrac{\partial z}{\partial x} = 1.$

12. Solve $\dfrac{\partial^2 z}{\partial y^2} = z$ given that $z = e^x$ and $\dfrac{\partial z}{\partial y} = e^{-x}$ when $y = 0.$

13. Solve $\dfrac{\partial z}{\partial x} = 6x + 3y$; $\dfrac{\partial z}{\partial y} = 3x - 4y.$

14. Solve $\log\left[\dfrac{\partial^2 z}{\partial x \partial y}\right] = x + y.$

15. Solve $\frac{\partial^2 z}{\partial x^2} = a^2 z$ given that $x = 0$, $\frac{\partial z}{\partial x} = a \sin y$ and $\frac{\partial z}{\partial y} = 0$.

ANSWERS

Exercise 2.1

1. $z = F(x) + xg(y) + H(y)$ where $F(x) = \iint f(x)dxdx$.
2. $z = -\sin x + xf(y) + g(y)$.
3. $z(x, y) = \sin x(1 - e^{-y})$.
4. $z = \frac{1}{x^2} \sin (xy) + yf(x) + g(x)$.
5. $z = \log x \log y + g(y) + f(x)$.
6. $z = 6xy - \frac{x^2}{2} \log y + f(y) + g(x)$.
7. $z = -x - \frac{y}{3} + x^3 f(y)$.
8. $z = -\frac{1}{a^2 b} \sin(ax + by) + xF(y) + f(y) + \phi(x)$.
9. $z = 24x \cos 3xy - 36x^2 \sin 3xy + f(x) + \phi(y)$.
10. $z = e^y \cosh x + e^{-y} \sinh x$.
11. $z = e^y \cos x + \sin x$.
12. $z = e^y \cosh x + e^{-y} \sinh x$.
13. $z = 3x^2 + 3xy - 2y^2 + c$.
14. $z = e^{x+y} + g(y) + \phi(x)$.
15. $z = A \cosh ax + \sin y \sinh ax$.

▲▲

Lagrange's Equation

The linear first order partial differential equation of the form $P(x, y, z)p + Q(x, y, z)q = R(x, y, z)$ is called the Lagrange's equation in two independent variables x, y.

Lagrange's method of solving the linear partial differential equation of order one

$$Pp + Qq + R$$

where, P, Q and R are functions of x, y, z.

The general solution of the equation $Pp + Qq = R$ is given by $\phi(u, v) = 0$ where, ϕ is an arbitrary function and $u(x, y, z) = c_1$, $v(x, y, z) = c_2$ are two linearly independent solutions of the equations

$$\frac{dx}{P} = \frac{dy}{Q} = \frac{dz}{R}.$$

Note: These equations are called the auxiliary or subsidiary equations.

Proof: Differentiating $\phi(u, v) = 0$ partially with respect to x and y, we get

$$\frac{\partial \phi}{\partial u}\left(\frac{\partial u}{\partial x}+\frac{\partial u}{\partial z}\frac{\partial z}{\partial x}\right)+\frac{\partial \phi}{\partial v}\left(\frac{\partial v}{\partial x}+\frac{\partial v}{\partial z}\frac{\partial z}{\partial x}\right) = 0$$

and
$$\frac{\partial \phi}{\partial u}\left(\frac{\partial u}{\partial y}+\frac{\partial u}{\partial z}\frac{\partial z}{\partial y}\right)+\frac{\partial \phi}{\partial v}\left(\frac{\partial v}{\partial x}+\frac{\partial v}{\partial z}\frac{\partial z}{\partial y}\right) = 0$$

$\Rightarrow$
$$\frac{\partial \phi}{\partial u}\left(\frac{\partial u}{\partial x}+p\frac{\partial u}{\partial z}\right)+\frac{\partial \phi}{\partial v}\left(\frac{\partial v}{\partial x}+p\frac{\partial z}{\partial z}\right) = 0$$

and
$$\frac{\partial \phi}{\partial u}\left(\frac{\partial u}{\partial y}+q\frac{\partial u}{\partial z}\right)+\frac{\partial \phi}{\partial v}\left(\frac{\partial v}{\partial x}+q\frac{\partial v}{\partial z}\right) = 0$$

Eliminating $\frac{\partial \phi}{\partial u}$ and $\frac{\partial \phi}{\partial v}$, we have

$$\frac{\frac{\partial u}{\partial x}+p\frac{\partial u}{\partial z}}{\frac{\partial v}{\partial x}+p\frac{\partial v}{\partial z}} = \frac{\frac{\partial u}{\partial y}+q\frac{\partial u}{\partial z}}{\frac{\partial v}{\partial y}+q\frac{\partial v}{\partial z}}$$

$\Rightarrow$
$$\left(\frac{\partial u}{\partial x}+p\frac{\partial u}{\partial z}\right)\left(\frac{\partial v}{\partial y}+q\frac{\partial v}{\partial z}\right)-\left(\frac{\partial u}{\partial y}+q\frac{\partial u}{\partial z}\right)\left(\frac{\partial v}{\partial x}+p\frac{\partial v}{\partial z}\right)=0$$

$\Rightarrow$
$$\left(\frac{\partial u}{\partial z}\frac{\partial v}{\partial y}-\frac{\partial u}{\partial y}\frac{\partial v}{\partial z}\right)p+\left(-\frac{\partial u}{\partial x}\frac{\partial v}{\partial z}+\frac{\partial u}{\partial z}\frac{\partial v}{\partial x}\right)q+\left(\frac{\partial u}{\partial x}\frac{\partial v}{\partial y}-\frac{\partial u}{\partial y}\frac{\partial v}{\partial x}\right)=0$$

$\therefore$
$$\left(\frac{\partial u}{\partial y}\frac{\partial v}{\partial z}-\frac{\partial u}{\partial z}\frac{\partial v}{\partial y}\right)p+\left(\frac{\partial u}{\partial x}\frac{\partial v}{\partial z}-\frac{\partial u}{\partial z}\frac{\partial v}{\partial x}\right)q = \frac{\partial u}{\partial x}\frac{\partial v}{\partial y}-\frac{\partial u}{\partial y}\frac{\partial v}{\partial x} \tag{1}$$

Hence, $\phi(u, v) = 0$ is a solution of Eq. (1).

Taking the total differential of $u(x, y, z) = c_1$ and $v(x, y, z) = c_2$, we get

$$\frac{\partial u}{\partial x}dx+\frac{\partial u}{\partial y}dy+\frac{\partial u}{\partial z}dz=0 \tag{2}$$

and $$\frac{\partial v}{\partial x}dx+\frac{\partial v}{\partial y}dy+\frac{\partial v}{\partial z}dz=0 \tag{3}$$

As u and v are independent functions, solving Eqs. (2) and (3), we obtain

$$\frac{dx}{\frac{\partial u}{\partial y}\frac{\partial v}{\partial z}-\frac{\partial u}{\partial z}\frac{\partial v}{\partial y}}=\frac{dy}{\frac{\partial u}{\partial z}\frac{\partial v}{\partial x}-\frac{\partial u}{\partial x}\frac{\partial v}{\partial z}}=\frac{dz}{\frac{\partial u}{\partial x}\frac{\partial v}{\partial y}-\frac{\partial u}{\partial y}\frac{\partial v}{\partial x}}$$

Comparing with $\frac{dx}{P}=\frac{dy}{Q}=\frac{dz}{R}$, we obtain

$$\frac{\frac{\partial u}{\partial y}\frac{\partial v}{\partial z}-\frac{\partial u}{\partial z}\frac{\partial v}{\partial y}}{P}=\frac{\frac{\partial u}{\partial z}\frac{\partial v}{\partial x}-\frac{\partial u}{\partial x}\frac{\partial v}{\partial z}}{Q}=\frac{\frac{\partial u}{\partial x}\frac{\partial v}{\partial y}-\frac{\partial u}{\partial y}\frac{\partial v}{\partial x}}{R}=\lambda\,(\text{say})$$

$$\Rightarrow \quad \frac{\partial u}{\partial y}\frac{\partial v}{\partial z}-\frac{\partial u}{\partial z}\frac{\partial v}{\partial y}=\lambda P,\quad \frac{\partial u}{\partial z}\frac{\partial v}{\partial x}-\frac{\partial u}{\partial x}\frac{\partial v}{\partial z}=\lambda Q$$

and $$\frac{\partial u}{\partial x}\frac{\partial v}{\partial y}-\frac{\partial u}{\partial y}\frac{\partial v}{\partial x}=\lambda R$$

Substituting these values in Eq. (1), we get

$$(\lambda P)p+(\lambda Q)q=\lambda R$$

$$\Rightarrow \quad Pp+Qq=R$$

which is the given equation.

Therefore, if $u(x, y, z) = c_1$ and $v(x, y, z) = c_2$ are two independent solutions of the system of differential equations $\frac{dx}{P}=\frac{dy}{Q}=\frac{dz}{R}$, then $\phi(u, v) = 0$ is a solution of $Pp + Qq = R$ where ϕ is an arbitrary function.

Working method to solve $Pp + Qq = R$

- Write the Lagrange's auxiliary or subsidiary equations $\frac{dx}{P}=\frac{dy}{Q}=\frac{dz}{R}$.
- Find two independent solutions $u(x, y, z) = c_1$ and $v(x, y, z) = c_2$ of the auxiliary equations.
- The general solution of the given partial differential equation $Pp + Qq = R$ is obtained by substituting the values of u and v (obtained above) in any one of the following equivalent forms

$$\phi(u, v) = 0,\ v = \phi(u) \text{ or } u = \phi(v)$$

where ϕ is an arbitrary function.

Now we will discuss various rules for finding u and v.

Rule I

If one of the variables is either absent or cancels out from any two terms then an integral can be found by usual methods. Same can be done for another set of fractions.

▲▲

EXAMPLE 2.8

Find the general solution of the Largrange's equation $2yzp + zxq = 3xy$.

Solution: The auxiliary equations are

$$\frac{dx}{2yz} = \frac{dy}{zx} = \frac{dz}{3xy}$$

Taking the first pair of terms, we get

$$xdx = 2ydy$$

Integrating, we get

$$\frac{x^2}{2} = y^2 + c \Rightarrow x^2 - 2y^2 = c_1$$

where $c_1 = 2c$.

Taking the second and third terms, we get

$$3ydy = zdz$$

Integrating, we get

$$\frac{3}{2}y^2 = \frac{z^2}{2} + k$$

$\Rightarrow$ $3y^2 - z^2 = c_2$

where $c_2 = 2k$

The general solution is given by $\phi(c_1, c_2) = 0$

$\Rightarrow$ $\phi(x^2 - 2y^2, 3y^2 - z^2) = 0$

or $3y^2 - z^2 = f(x^2 - 2y^2)$.

EXAMPLE 2.9

Find the general solution of the partial differential equation $xy^2p + y^3q = (zxy^2 - 4x^3)$

Solution: The auxiliary equations are

$$\frac{dx}{xy^2} = \frac{dy}{y^3} = \frac{dz}{zxy^2 - 4x^3}$$

Taking the first pair of terms, we get

$$\frac{dx}{x} = \frac{dy}{y}$$

Integrating, we obtain $\log x = \log y - \log a \Rightarrow ax = y$
Using the first and last terms, we get

$$\frac{dx}{a^2x^3} = \frac{dz}{a^2x^3z - 4x^3}$$

$$\Rightarrow \qquad dx = \frac{dz}{z - \frac{4}{a^2}}$$

Integrating, we get

$$x = \log\left[z - \frac{4}{a^2}\right] + b \Rightarrow x = \log\left[z - \frac{4x^2}{y^2}\right] + b$$

The general solution is given by

$$\phi(a, b) = 0$$

$$\Rightarrow \qquad \phi\left(\frac{y}{x}, x - \log\left[z - \frac{4x^2}{y^2}\right]\right) = 0$$

EXAMPLE 2.10

Find the general solution of $\frac{y^2z}{x}p + zxq = y^2$.

Solution: The auxiliary equations are

$$\frac{dx}{\frac{y^2z}{x}} = \frac{dy}{zx} = \frac{dz}{y^2}$$

$$\Rightarrow \qquad \frac{dx}{y^2z} = \frac{dy}{zx^2} = \frac{dz}{xy^2}$$

Taking the first two terms, we get

$$x^2dx = y^2dy$$

Integrating, we obtain

$$\frac{x^3}{3} = \frac{y^3}{3} + a \Rightarrow x^3 - y^3 = c_1$$

where $c_1 = 3a$

Again, taking the first and last terms, we get

$$xdx = zdz$$

Integrating, we obtain $\frac{x^2}{2} = \frac{z^2}{2} + b$

$\Rightarrow$ $x^2 - z^2 = c_2$, where $c_2 = 2b$.

Hence, the general solution is $\phi(c_1, c_2) = 0$

$\Rightarrow$ $\phi(x^3 - y^3, x^2 - z^2) = 0.$

EXAMPLE 2.11

Solve the following partial differential equation for general solution

$$y^2p + x^2q = x^2y^2z^2.$$

Solution: The auxiliary equations are

$$\frac{dx}{y^2} = \frac{dy}{x^2} = \frac{dz}{x^2y^2z^2}$$

Taking the first two terms, we get

$$x^2dx = y^2dy$$

Integrating, we obtain $x^3 - y^3 = c_1$

Taking the first and last term, we get

$$x^2dx = \frac{dz}{z^2}$$

Integrating, we obtain

$$\frac{x^3}{3} = -\frac{1}{z} + k$$

$\Rightarrow$ $x^3 + \frac{3}{z} = c_2$ where $c_2 = 3k$

$\therefore$ The general solution is given by $\phi(c_1, c_2) = 0$

$\Rightarrow$ $\phi\left(x^3 - y^3, x^3 + \frac{3}{z}\right) = 0$

or $x^3 + \frac{3}{z} = f(x^3 - y^3).$

EXAMPLE 2.12

Solve for general solution the partial differential equation $p \tan x + q \tan y = \tan z$.

Solution: The auxiliary equations are

$$\frac{dx}{\tan x} = \frac{dy}{\tan y} = \frac{dz}{\tan z}.$$

Taking first two terms, we get

$$\frac{dx}{\tan x} = \frac{dy}{\tan y}$$

Integrating, we obtain

$$\log \sin x = \log \sin y + \log a$$

$$\Rightarrow \qquad \frac{\sin x}{\sin y} = a$$

Taking the last two terms, we get

$$\frac{dy}{\tan y} = \frac{dz}{\tan z}$$

Integrating, we obtain

$$\log \sin y = \log \sin z + \log b$$

$$\Rightarrow \qquad \frac{\sin y}{\sin z} = b$$

$\therefore$ The general solution is given by

$$\phi(a, b) = 0 \Rightarrow \phi\left(\frac{\sin x}{\sin y}, \frac{\sin y}{\sin z}\right) = 0$$

EXAMPLE 2.13

Solve $y^2p - xyp = x(z - 2y)$.

Solution: The auxiliary equations are

$$\frac{dx}{y^2} = -\frac{dy}{xy} = \frac{dz}{x(z-2y)}$$

Taking the first two terms, we get

$$xdx = -ydy$$

Integrating, we obtain $x^2 + y^2 = c_1$

Now, taking the last two terms, we get

$$-\frac{dy}{y} = \frac{dz}{z-2y} \Rightarrow \frac{dz}{dy} = -\frac{z-2y}{y}$$

$$\Rightarrow \qquad \frac{dz}{dy} + \frac{1}{y}z = 2$$

which is linear in y and z with

$$\text{IF} = e^{\int \frac{1}{y} dy} = e^{\log y} = y$$

Hence, its solution is

$$z - y = \int 2y dy + c_2$$

$$\Rightarrow \qquad yz - y^2 = c_2$$

Hence, the general solution is $\phi(c_1, c_2) = 0$

$$\Rightarrow \qquad \phi(x^2 + y^2, yz = y^2) = 0.$$

EXAMPLE 2.14

Solve $(x^2 + 2y^2)p - xyq = xz$.

Solution: The auxiliary equations are

$$\frac{dx}{x^2 + 2y^2} = -\frac{dy}{xy} = \frac{dz}{xz}$$

Taking the last two terms, we get

$$-\frac{dy}{y} = \frac{dz}{z}$$

Integrating, we obtain $-\log y + \log a = \log z \Rightarrow yz = a$

Taking first two terms, we get

$$\frac{dx}{dy} = -\frac{x^2 + 2y^2}{xy}$$

$$\Rightarrow \qquad x\frac{dx}{dy} + \frac{1}{y}x^2 = -2y$$

Let $\qquad x^2 = t \Rightarrow x\dfrac{dx}{dy} = \dfrac{1}{2}\dfrac{dt}{dy}$

Therefore, $\qquad \dfrac{dt}{dy} + \dfrac{2}{y}t = -4y$

which is a linear differential equation with

$$\text{IF} = e^{\int \frac{2}{y} dy} = e^{2\log y} = y^2$$

and hence, its solution is

$$t \cdot y^2 = \int -4y \cdot y^2 dy + b$$

$$\Rightarrow \qquad x^2y^2 + y^4 = b$$

Hence, the general solution is given by

$$\phi(a, b) = 0$$

$$\Rightarrow \qquad \phi(yz, y^2x^2 + y^4) = 0.$$

EXAMPLE 2.15

Solve the following equation

$$p\sqrt{x} + q\sqrt{y} = \sqrt{z}.$$

Solution: The auxiliary equations are

$$\frac{dx}{\sqrt{x}} = \frac{dy}{\sqrt{y}} = \frac{dz}{\sqrt{z}}$$

Taking the first two terms, we get

$$\frac{dx}{\sqrt{x}} = \frac{dy}{\sqrt{y}}$$

Integrating, we obtain $\dfrac{\sqrt{x}}{1/2} = \dfrac{\sqrt{y}}{1/2} + a$

$$\Rightarrow \qquad \sqrt{x} - \sqrt{y} = c_1 \text{ where } c_1 = a/2$$

Taking first and last terms, we get

$$\frac{dx}{\sqrt{x}} = \frac{dz}{\sqrt{z}}$$

Integrating, we obtain $\dfrac{\sqrt{x}}{1/2} = \dfrac{\sqrt{z}}{1/2} + b$

$\Rightarrow \sqrt{x} - \sqrt{z} = c_2$ where $c_2 = b/2$

Hence, the general solution is

$$\phi(c_1, c_2) = 0$$

$$\Rightarrow \qquad \phi(\sqrt{x} - \sqrt{y}, \sqrt{x} - \sqrt{z}) = 0$$

or

$$\sqrt{x} - \sqrt{y} = f(\sqrt{x} - \sqrt{z}).$$

Rule II

Solution of $Pp + Qq = R$ is obtained by taking two members of the auxiliary equation and integrating to get one independent solution which is used to find another independent solution.

EXAMPLE 2.16

Solve $p + 3q = 5z + \tan(y - 3x)$.

Solution: The auxiliary equations are

$$\frac{dx}{1} = \frac{dy}{3} = \frac{dz}{5z + \tan(y - 3x)}$$

Taking the first two terms, we get

$$3dx = dy$$

Integrating, we obtain $3x + c_1 = y \Rightarrow y - 3x = c_1$

Taking first and third fractions, we get

$$\frac{dx}{1} = \frac{dz}{5z + \tan c_1}$$

Integrating, we obtain

$$x = \frac{\log(5z + \tan c_1)}{5} + c_2$$

$\Rightarrow$ $$5x - \log(5z + \tan c_1) - c_2'$$

where $$c_2' = 5c_2$$

Hence, the required general solution is

$$f(y - 3x, 5x - \log(5z + \tan(y - 3x))) = 0$$

EXAMPLE 2.17

Solve $py + qx = xyz^2(x^2 - y^2)$

Solution: The auxiliary equations are

$$\frac{dx}{y} = \frac{dy}{x} = \frac{dz}{xyz^2(x^2 - y^2)}$$

Taking the first two terms, we get

$$xdx = ydy$$

Integrating, we obtain $x^2 - y^2 = c_1$

Taking second and third fractions, we get

$$\frac{dy}{x} = \frac{dz}{xyz^2 c_1}$$

$$\Rightarrow \qquad c_1 y\,dy = \frac{dz}{z^2}$$

Integrating, we obtain $\quad c_1 y^2 = -\frac{2}{z} + c_2$

Hence, the required general solution is

$$\phi\left(x^2 - y^2,\ y^2(x^2 + y^2) + \frac{2}{z}\right) = 0$$

EXAMPLE 2.18

Solve $(p - q)(x + y) = z$.

Solution: The auxiliary equations are

$$\frac{dx}{x+y} = -\frac{dy}{(x+y)} = \frac{dz}{z}$$

Taking first two members, we have

$$dx = -dy$$

Integrating, we obtain $\quad x + y = c_1$

From the first and third terms, we get

$$\frac{dx}{c_1} = \frac{dz}{z}$$

Integrating, we obtain $x = c_1 \log z + c_2$
Hence, the general solution is given by

$$\phi(x + y, x - (x + y) \log z) = 0$$

EXAMPLE 2.19

Solve $zp - zq = z^2 + (x + y)^2$.

Solution: The auxiliary equations are

$$\frac{dx}{z} = -\frac{dy}{z} = \frac{dz}{z^2 + (x+y)^2}$$

Taking the first two terms, we get

$$dx = -dz$$

Integrating, we obtain $\quad x + z = c_1$

Taking the first and third terms, we have

$$\frac{dzc}{z} = \frac{dz}{z^2 + c_1^2} \Rightarrow dx = \frac{zdz}{z^2 + c_1^2}$$

Integrating, we obtain

$$2x = \log(z^2 + c_1^2) + c_2$$

Hence, the required general solution is

$$\phi(x + y, 2x - \log(z^2 + (x + y)^2)) = 0.$$

EXAMPLE 2.20

Solve $5p - 6q = 25x^4 \cos(6x + 5y)$.

Solution: The auxiliary equations are

$$\frac{dx}{5} = -\frac{dy}{6} = \frac{dz}{25x^4 \cos(6x+5y)}$$

Taking the first two terms, we get

$$6dx = -5dy$$

Integrating, we obtain $6x + 5y = c_1$

Taking the first and third fractions, we have

$$\frac{dx}{5} = \frac{dz}{25x^4 \cos c_1}$$

$\Rightarrow$ $$5x^4 \cos c_1 dx = dz$$

Integrating, we obtain

$$x^5 \cos c_1 = z + c_2$$

Hence, the required general solution is

$$\phi(6x + 5y, x^5 \cos(6x + 5y) - z) = 0.$$

EXAMPLE 2.21

Solve $xy^2p + y^3q = zxy^2 - 4x^3$.

Solution: The auxiliary equations are

$$\frac{dx}{xy^2} = \frac{dy}{y^3} = \frac{dz}{zxy^2 - 4x^3}$$

Taking the first pair of terms, we get

$$\frac{dx}{x} = \frac{dy}{y}$$

Integrating, we have $\log x + \log c_1 = \log y$

$\Rightarrow$ $$c_1x = y.$$

Using this relation, we get

$$\frac{dx}{c_1^2 x^3} = \frac{dz}{x(c_1^2 x^2 z - 4x^2)}$$

$$\Rightarrow \qquad dx = \frac{dz}{z - \dfrac{4}{c_1^2}}$$

Integrating, we get

$$x = \log\left[z - \frac{4}{c_1^2}\right] + c_2$$

Hence, the general solution is given by

$$\phi\left[\frac{y}{x^2},\ x - \log\left(z - \frac{4x^2}{y^2}\right)\right] = 0$$

Rule III

By using simple algebra, we have

$$\frac{dx}{P} = \frac{dy}{Q} = \frac{dz}{R} = \frac{Sdx + Tdy + Udz}{PS + QT + RU}$$

where S, T, U are functions of x, y, z.

If $PS + QT + RU = 0$, then we have

$Sdx + Tdy + Udz = 0$

which can be integrated to give $u(x, y, z) = c_1$ and the other independent solution can be obtained either by selecting another set of multipliers S, T, U or by taking two members of auxiliary equations.

EXAMPLE 2.22

Find the general solution of $\left(\dfrac{y-z}{yz}\right)dx + \left(\dfrac{z-x}{zx}\right)dy = \dfrac{x-y}{xy}$.

Solution: The auxiliary equations are

$$\frac{dx}{\dfrac{1}{z} - \dfrac{1}{y}} = \frac{dy}{\dfrac{1}{x} - \dfrac{1}{z}} = \frac{dz}{\dfrac{1}{y} - \dfrac{1}{x}}$$

Using 1, 1, 1 as multipliers, we get $\dfrac{dx + dy + dz}{\dfrac{1}{z} - \dfrac{1}{y} + \dfrac{1}{x} - \dfrac{1}{z} + \dfrac{1}{y} - \dfrac{1}{x}}$

$$= \frac{dx + dy + dz}{0}$$

$\therefore$ $$dx + dy + dz = 0$$

which on integrating gives $\quad x + y + z = c_1$

Again, using $\frac{1}{x}, \frac{1}{y}, \frac{1}{z}$ as multipliers, we get

$$\frac{\frac{1}{x}dx + \frac{1}{y}dy + \frac{1}{z}dz}{\frac{1}{x}\left(\frac{1}{z} - \frac{1}{y}\right) + \frac{1}{y}\left(\frac{1}{x} - \frac{1}{z}\right) + \frac{1}{z}\left(\frac{1}{y} - \frac{1}{x}\right)} = \frac{\frac{1}{x}dx + \frac{1}{y}dy + \frac{1}{z}dz}{0}$$

$\therefore$ $$\frac{1}{x}dx + \frac{1}{y}dy + \frac{1}{z}dz = 0$$

which on integrating gives $\log x + \log y + \log z = \log c_2$

or $$xyz = c_2$$

Hence, the general solution is $\phi(c_1, c_2) = 0$

$\therefore$ $$\phi(x + y + z, xyz) = 0.$$

EXAMPLE 2.23

Solve the partial differential equation $(z^2 - 2yz - y^2)p + (xy + zx)q = (xy - zx)$

Solution: The auxiliary equations are

$$\frac{dx}{z^2 - 2yz - y^2} = \frac{dy}{xy + zx} = \frac{dz}{xy - zx}$$

Using x, y, z as multipliers, we get

$$\frac{xdx + ydy + zdz}{x(z^2 - 2yz - y^2) + y(xy + zx) + z(xy - zx)} = \frac{xdx + ydy + zdz}{0}$$

$\therefore$ $$xdx + ydy + zdz = 0$$

which on integration gives $x^2 + y^2 + z^2 = c_1$

From the last two members, we have

$$\frac{dy}{xy + zx} = \frac{dz}{xy - zx}$$

$$\Rightarrow \qquad \frac{dy}{y+z} = \frac{dz}{y-z}$$

$$\Rightarrow \qquad (y - z)dy = (y + z)dz$$

$$\Rightarrow \qquad ydy - (zdy + ydz) - zdz = 0$$

which on integration gives $y^2 - 2yz - z^2 = c_2$

Hence, the general solution is $\phi(c_1, c_2) = 0$

$$\therefore \qquad \phi(x^2 + y^2 + z^2, y^2 - 2yz - z^2) = 0.$$

EXAMPLE 2.24

Solve $(mz - ny)p + (nx - lz)q = ly - mx$.

Solution: The auxiliary equations are

$$\frac{dx}{mz-ny} = \frac{dy}{nx-lz} = \frac{dz}{ly-mx}$$

Using x, y, z as multipliers, we get

$$\frac{xdx+ydy+zdz}{x(mz-ny)+y(nx-lz)+z(ly-mx)} = \frac{xdx+ydy+zdz}{0}$$

$$\therefore \qquad xdx + ydy + zdz = 0$$

which on integration gives $x^2 + y^2 + z^2 = c_1$

Again, using l, m, n as multipliers, we get

$$\frac{ldx+mdy+ndz}{l(mz-ny)+m(nx-lz)+n(ly-mx)} = \frac{ldx+mdy+ndz}{0}$$

$$\therefore \qquad ldx + mdy + ndz = 0$$

which on integration gives $lx + my + nz = c_2$

Hence, the required general solution is given by

$$\phi(c_1, c_2) = \phi(x^2 + y^2 + z^2, lx + my + nz) = 0$$

EXAMPLE 2.25

Solve $x(y^2 + z)p - y(x^2 + z)q = z(x^2 - y^2)$

Solution: The auxiliary equations are

$$\frac{dx}{x(y^2+z)} = -\frac{dy}{y(x^2+z)} = \frac{dz}{z(x^2-y^2)}$$

Using $\frac{1}{x}, \frac{1}{y}, \frac{1}{z}$ as multipliers, we have

$$\frac{\frac{1}{x}dx + \frac{1}{y}dy + \frac{1}{z}dz}{(y^2+z)-(x^2+z)+(x^2-y^2)} = \frac{\frac{1}{x}dx + \frac{1}{y}dy + \frac{1}{z}dz}{0}$$

$$\therefore \qquad \frac{1}{x}dx + \frac{1}{y}dy + \frac{1}{z}dz = 0$$

which on integration gives $\log x + \log y + \log z = \log c_1$

$$\Rightarrow \qquad xyz = c_1$$

Again, using $x, y, -1$ as multipliers, we get

$$\frac{xdx + ydy - dz}{x^2(y^2+z) - y^2(x^2+z) - z(x^2-y^2)} = \frac{xdx + ydy - dz}{0}$$

$$\therefore \qquad xdx + ydy - dz = 0$$

which on integration gives $\qquad x^2 + y^2 - 2z = c_2$

Hence, the required general solution is given by

$$\phi(c_1, c_2) = \phi(xyz, x^2 + y^2 - 2z) = 0$$

EXAMPLE 2.26

Solve $(x + 2z)p + (4zx - y)q = 2x^2 + y$.

Solution: The auxiliary equations are

$$\frac{dx}{x+2z} = \frac{dy}{4zx-y} = \frac{dz}{2x^2+y}$$

Using $y, x, -2z$ as multipliers, each fraction is equal to

$$\frac{ydx + xdy - 2zdz}{y(x+2z) + x(4zx-y) - 2z(2x^2+y)} = \frac{ydx + xdy - 2zdz}{0}$$

$$\therefore \qquad ydx + xdy - 2zdz = 0$$

which on integration gives $xy - z^2 = c_1$

Again, using $2x, -1, -1$ as multipliers, each fraction of auxiliary equation is equal to

$$\frac{2xdx - dy - dz}{2x(x+2z) - (4zx-y) - (2x^2+y)} = \frac{2xdx - dy - dz}{0}$$

$$\therefore \qquad 2xdx - dy - dz = 0$$

which on integration gives $x^2 - y - z = c_2$.

Hence, the required general solution is given by

$$\phi(c_1, c_2) = \phi(xy - z^2, x^2 - y - z) = 0$$

EXAMPLE 2.27

Solve $z(x + 2y)p - z(y + 2x)q = y^2 - x^2$.

Solution: The auxiliary equation are given by

$$\frac{dx}{z(x+2y)} = \frac{dy}{-z(y+2x)} = \frac{dz}{y^2 - x^2}$$

Using x, y, z as multipliers, each fraction is equal to

$$\frac{xdx + ydy + zdz}{xz(x+2y) - yz(y+2x) + z(y^2 - x^2)} = \frac{xdx + ydy + zdz}{0}$$

$\therefore$ $xdx + ydy + zdz = 0$ which on integration yields $x^2 + y^2 + z^2 = c_1$

Taking the first two members, we have

$$\frac{dx}{x+2y} = -\frac{dy}{y+2x}$$

$$\Rightarrow \qquad (y + 2x)dx + (x + 2y)dy = 0$$

$$\Rightarrow \qquad 2xdx + (ydx + xdy) + 2ydy = 0$$

which on integration gives $x^2 + xy + y^2 = c_2$

Hence, the required general solution is given by

$$\phi(c_1, c_2) = \phi(x^2 + y^2 + z^2, x^2 + xy + y^2) = 0$$

EXAMPLE 2.28

Solve $x(x^2 + 3y^2)p - y(3x^2 + y^2)q = 2z(y^2 - x^2)$.

Solution: The auxiliary equations are given by

$$\frac{dx}{x(x^2+3y^2)} = -\frac{dy}{y(3x^2+y^2)} = \frac{dz}{2z(y^2-x^2)}$$

Using $\frac{1}{x}, \frac{1}{y}, -\frac{1}{z}$ as multipliers, each fraction is equal to

$$\frac{\frac{1}{x}dx+\frac{1}{y}dy-\frac{1}{z}dz}{(x^2+3y^2)-(3x^2+y^2)-2(y^2-x^2)}=\frac{\frac{1}{x}dx+\frac{1}{y}dy-\frac{1}{z}dz}{0}$$

$\therefore$
$$\frac{1}{x}dx+\frac{1}{y}dy-\frac{1}{z}dz=0$$

which on integration gives $\log x + \log y - \log z = \log c_1$

$\Rightarrow$
$$\frac{xy}{z}=c_1$$

Taking the first two members, we get

$$\frac{dx}{x(x^2+3y^2)}=-\frac{dy}{y(3x^2+y^2)}$$

$\Rightarrow$
$$\frac{dy}{dx}=-\left(\frac{y}{x}\right)\frac{3x^2+y^2}{x^2+3y^2}$$

Put $y = vx$ so that $\dfrac{dy}{dx} = v + x\dfrac{dv}{dx}$

$\therefore$
$$v+x\frac{dv}{dx}=-v\frac{3+v^2}{1+3v^2}$$

$\Rightarrow$
$$x\frac{dv}{dx}=-\frac{4(1+v^2)v}{1+3v^2}$$

$\Rightarrow$
$$\frac{4}{x}dx+\left(\frac{1}{v}+\frac{2v}{1+v^2}\right)dv=0$$

which on integration gives

$$4\log x+\log v+\log(1+v^2)=\log c_2$$

$\Rightarrow$
$$x^4 v(1+v^2)=c_2$$

$\Rightarrow$
$$x^4\left(\frac{y}{x}\right)\left[1+\left(\frac{y}{x}\right)^2\right]=c_2$$

$\Rightarrow$
$$xy(x^2+y^2)=c_2$$

Using $\dfrac{xy}{z} = c_1$, we obtain

$z(x^2 + y^2) = c_2'$ where $c_2' = c_2/c_1$

Hence, the required general solution is given by

$$\phi(c_1, c_2') = \phi\left(\frac{xy}{z}, z(x^2 + y^2)\right) = 0$$

EXAMPLE 2.29

Solve $x(y - z)p + y(z - x)q = z(x - y)$.

Solution: The auxiliary equations are given by

$$\frac{dx}{x(y-z)} = \frac{dy}{y(z-x)} = \frac{dz}{z(x-y)}$$

Using $\frac{1}{x}, \frac{1}{y}, \frac{1}{z}$ as multipliers, each fraction is equal to

$$\frac{\frac{1}{x}dx + \frac{1}{y}dy + \frac{1}{z}dz}{(y-z)+(z-x)+(x-y)} = \frac{\frac{1}{x}dx + \frac{1}{y}dy + \frac{1}{z}dz}{0}$$

$\therefore$ $\frac{1}{x}dx + \frac{1}{y}dy + \frac{1}{z}dz = 0$ which on integration yields $xyz = c_1$

Again, using 1, 1, 1 as multipliers, each fraction is equal to

$$\frac{dx + dy + dz}{x(y-z) + y(z-x) + z(x-y)} = \frac{dx + dy + dz}{0}$$

$$\therefore \qquad dx + dy + dz = 0$$

which on integration gives $x + y + z = c_2$.

Hence, the required general solution is given by

$$\phi(c_1, c_2) = \phi(xyz, x + y + z) = 0$$

Rule IV

As discussed earlier, we can choose S, T, U functions of x, y, z such that

$$\frac{dx}{P} = \frac{dy}{Q} = \frac{dz}{R} = \frac{Sdx + Tdy + Udz}{PS + QT + RU}$$

Suppose S, T, U are selected in such a way that $d(PS + QT + RU) = Sdx + Tdy + Udz)$

Then $\frac{Sdx + Tdy + Udz}{PS + QT + RU}$ can be combined with a suitable member of auxiliary equations to get one independent solution and the other independent solution can be obtained either by selecting another set of multipliers or by talking two members of auxiliary equations.

EXAMPLE 2.30

Solve the partial differential equation $(x^2 - y^2 - z^2)p + 2xyq = 2xz$.

Solution: The auxiliary equations are

$$\frac{dx}{x^2 - y^2 - z^2} = \frac{dy}{2xy} = \frac{dz}{2xz}$$

From the last two members, we have

$$\frac{dx}{x} = \frac{dy}{y}$$

which on integration gives

$$\log y = \log z + \log c_1 \Rightarrow \frac{y}{z} = c_1$$

Using x, y, z as multipliers, we find each fraction is equal to

$$\frac{xdx + ydy + zdz}{x(x^2 - y^2 - z^2) + y(2xy) + z(2xz)} = \frac{xdx + ydy + zdz}{x(x^2 + y^2 + z^2)}$$

On combining this with the third member of auxiliary equations, we get

$$\frac{dz}{2zx} = \frac{xdx + ydy + zdz}{x(x^2 + y^2 + z^2)}$$

which on integration gives

$$\log z + \log c_2 = \log(x^2 + y^2 + z^2)$$

$$\Rightarrow \quad \frac{x^2 + y^2 + z^2}{z} = c_2$$

Hence, the general solution is

$$\phi(c_1, c_2) = \phi\left(\frac{y}{z}, \frac{x^2 + y^2 + z^2}{z}\right) = 0$$

EXAMPLE 2.31

Solve $p \cos(x + y) + q \sin(x + y) = z$.

Solution: The auxiliary equations are given by

$$\frac{dx}{\cos(x + y)} = \frac{dy}{\sin(x + y)} = \frac{dz}{z}$$

Consider $\dfrac{dx + dy}{\cos(x + y) + \sin(x + y)} = \dfrac{dz}{z}$

$$\Rightarrow \qquad \frac{dx+dy}{\sqrt{2}\left\{\frac{1}{\sqrt{2}}\cos(x+y)+\frac{1}{\sqrt{2}}\sin(x+y)\right\}}=\frac{dz}{z}$$

$$\Rightarrow \qquad \frac{1}{\sqrt{2}}\ \frac{dx+dy}{\cos\frac{\pi}{4}\cos(x+y)\sin\frac{\pi}{4}\sin(x+y)}=\frac{dz}{z}$$

$$\frac{dx+dy}{\cos\left(x+y-\frac{\pi}{4}\right)}=\sqrt{2}\ \frac{dz}{z}$$

$$\Rightarrow \qquad \sec\left(x+y-\frac{\pi}{4}\right)d(x+y)=\sqrt{2}\frac{dz}{z}$$

which on integration gives

$$\log\left|\sec\left(x+y-\frac{\pi}{4}\right)+\tan\left(x+y-\frac{\pi}{4}\right)\right|=\sqrt{2}\,[\log z+\log c_1]$$

$$\Rightarrow \qquad \sec\left(x+y-\frac{\pi}{4}\right)+\tan\left(x+y-\frac{\pi}{4}\right)=\pm\, c_1^{\sqrt{2}}\, z^{\sqrt{2}}$$

$$\Rightarrow \qquad z^{-\sqrt{2}}\left\{\sec\left(x+y-\frac{\pi}{4}\right)+\tan\left(x+y-\frac{\pi}{4}\right)\right\}=c_1'$$

Again, using simple algebra, we obtain

$$\frac{dx+dy}{\cos(x+y)+\sin(x+y)}=\frac{dx-dy}{\cos(x+y)-\sin(x+y)}$$

$$\Rightarrow \qquad \frac{\cos(x+y)-\sin(x+y)}{\cos(x+y)+\sin(x+y)}(dx+dy)=dx-dy$$

$$\Rightarrow \qquad \frac{\cos(x+y)-\sin(x+y)}{\sin(x+y)+\cos(x+y)}\ d(x+y)=dx-dy$$

Integrating, we get

$$\log(\sin(x+y)+\cos(x+y))=x-y+\log c_2$$

$$\Rightarrow \qquad \sin(x+y)+\cos(x+y)=c_2e^{x-y}$$

$$\Rightarrow \qquad e^{-(x-y)}\,[\sin(x+y)+\cos(x+y)]=c_2$$

Hence, the required general solution is given by

$$\phi(c_1', c_2) = \phi\left(z^{-\sqrt{2}}\left[\sec\left(x+y-\frac{\pi}{4}\right)+\tan\left(x+y-\frac{\pi}{4}\right)\right], e^{y-x}[\sin(x+y)+\cos(x+y)]\right) = 0$$

EXAMPLE 2.32

Solve $(x^2 - y^2 - yz)p + (x^2 - y^2 - zx)q = z(x - y)$.

Solution: The auxiliary equation arc

$$\frac{dx}{x^2-y^2-z^2} = \frac{dy}{x^2-y^2-zx} = \frac{dz}{z(x-y)}$$

Using 1, –1, 0 as multipliers, we have

$$\frac{dx-dy}{(x^2-y^2-yz)-(x^2-y^2-zx)} = \frac{dx-dy}{z(x-y)}$$

Using x, $-y$, 0 as multipliers, we have

$$\frac{xdx-ydy}{x(x^2-y^2-z^2)-y(x^2-y^2-zx)} = \frac{xdx-ydy}{(x-y)(x^2-y^2)}$$

$$\therefore \qquad \frac{dz}{z(x-y)} = \frac{dx-dy}{z(x-y)} = \frac{xdx-ydy}{(x-y)(x^2-y^2)}$$

$$\Rightarrow \qquad \frac{dz}{z} = \frac{dx-dy}{z} = \frac{xdx-ydy}{x^2-y^2}$$

Taking the first two members, we have

$$dz = dx - dy$$

which on integration yields $z - x + y = c_1$

Again, taking the first and third members, we obtain

$$\frac{dz}{z} = \frac{xdx-ydy}{x^2-y^2}$$

which on integration gives

$$2\log z + c_2 = \log(x^2 - y^2)$$

$$\Rightarrow \qquad \frac{x^2-y^2}{z^2} = c_2$$

Hence, the required general solution is given by

$$\phi(c_1, c_2) = \phi\left(z - x + y,\ \frac{x^2 - y^2}{z^2}\right) = 0.$$

EXAMPLE 2.33

Solve $xp + yq = z - a\sqrt{(x^2 + y^2 + z^2)}$.

Solution: The auxiliary equations are

$$\frac{dx}{x} = \frac{dy}{y} = \frac{dz}{z - a\sqrt{(x^2 + y^2 + z^2)}}$$

Taking the first two members, we have

$$\frac{dx}{x} = \frac{dy}{y}$$

which on integration yields $x/y = c_1$

Using x, y, z as multipliers, we get

$$\frac{xdx + ydy + zdz}{x^2 + y^2 + z^2 - az\sqrt{(x^2 + y^2 + z^2)}}$$

Combining with the first and last members, we have

$$\frac{dx}{x} = \frac{dz}{z - a\sqrt{(x^2 + y^2 + z^2)}} = \frac{xdx + ydy + zdz}{x^2 + y^2 + z^2 - az\sqrt{(x^2 + y^2 + z^2)}}$$

Put $x^2 + y^2 + z^2 = t^2$ so that $xdx + ydy + zdz = tdt$

$$\therefore \quad \frac{dx}{x} = \frac{dz}{z - at} = \frac{tdt}{t^2 - azt}$$

$$\Rightarrow \quad \frac{dx}{x} - \frac{dz}{z - at} = \frac{dt}{t - az}$$

$$\therefore \quad \frac{dx}{x} = \frac{dz + dt}{z + t - a(z + t)} = \frac{dz + dt}{(1 - a)(z + t)}$$

$$\Rightarrow \quad (1 - a)\frac{dx}{x} = \frac{d(z + t)}{z + t}$$

Integrating, we have

$$(1 - a)\log x - \log(z + t) = \log c_2$$

$$\Rightarrow \qquad \frac{x^{1-a}}{z+\sqrt{(x^2+y^2+z^2)}} = c_2$$

Hence, the required general solution is

$$\phi(c_1, c_2) = \phi\left(\frac{x}{y}, \frac{x^{1-a}}{z+\sqrt{(x^2+y^2+z^2)}}\right) = 0$$

EXAMPLE 2.34

Solve $px(z - 2y^2) = (z - qy)(z - y^2 - 2x^2)$.

Solution: The given equation can be written as

$$x(z - 2y^2)p + y(z - y^2 - 2x^2)q = z(z - y^2 - 2x^2)$$

The auxiliary equations are given by

$$\frac{dx}{x(z-2y^2)} = \frac{dy}{y(z-y^2-2x^2)} = \frac{dz}{z(z-y^2-2x^2)}$$

Taking the second and third members, we have

$$\frac{dx}{x} = \frac{dy}{y}$$

which on integration yields $\frac{y}{x} = c_1$.

Using 0, $-2y$, 1 as multipliers, we get

$$\frac{-2ydy+dz}{-2y^2(z-y^2-2x^2)+z(z-y^2-2x^2)} = \frac{d(z-y^2)}{(z-2y^2)(z-y^2-2x^2)}$$

Combining this with the first member of auxiliary equations, we obtain

$$\frac{dx}{x(z-2y^2)} = \frac{d(z-y^2)}{(z-2y^2)(z-y^2-2x^2)}$$

$$\Rightarrow \qquad \frac{d(z-y^2)}{dx} = \frac{z-y^2-2x^2}{x}$$

Put $z - y^2 = u$

$$\therefore \qquad \frac{du}{dx} = \frac{u-2x^2}{x}$$

$\Rightarrow \quad \dfrac{du}{dx} - \dfrac{u}{x} = -2x$ which is an ordinary linear differential equation with

$$\text{IF} = e^{-\int \frac{1}{x} dx} = \frac{1}{x}$$

$$\therefore \quad u \cdot \frac{1}{x} = \int -2x\left(\frac{1}{x}\right) dx + c_2$$

$$\Rightarrow \quad \frac{z - y^2}{x} = -2x + c_2$$

$$\Rightarrow \quad \frac{z - y^2 + 2x^2}{x} = c_2$$

Hence, the required general solution is given by

$$\phi(c_1, c_2) = \phi\left(\frac{y}{z}, \frac{z - y^2 + 2x^2}{x}\right) = 0.$$

EXAMPLE 2.35

Find the general solution of the partial differential equation $2xzp + 2yzq = z^2 - x^2 - y^2$.

Solution: The auxiliary equations are given by

$$\frac{dx}{2xz} = \frac{dy}{2yz} = \frac{dz}{z^2 - x^2 - y^2}$$

From the first two members, we get

$$\frac{dx}{x} = \frac{dy}{y}$$

which on integration yields $\qquad y/x = c_1$

Using x, y, z as multipliers, we get

$$\frac{xdx + ydy + zdz}{2x^2z + 2y^2z + z(z^2 - x^2 - y^2)} = \frac{xdx + ydy + zdz}{z(x^2 + y^2 + z^2)}$$

Combining with the first member, we have

$$\frac{dx}{2xz} = \frac{xdx + ydy + zdz}{z(x^2 + y^2 + z^2)}$$

$$\Rightarrow \quad \frac{dx}{x} = \frac{2(xdx + ydy + zdz)}{x^2 + y^2 + z^2}$$

Integrating, we get

$$\log x + \log c_2 = \log(x^2 + y^2 + z^2)$$

$$\Rightarrow \qquad \frac{x^2 + y^2 + z^2}{x} = c_2$$

Hence, the required general solution is given by

$$\phi(c_1, c_2) = \phi\left(\frac{y}{x}, \frac{x^2 + y^2 + z^2}{x}\right) = 0$$

EXAMPLE 2.36

Solve $(y^2 + yz + z^2)p + (z^2 + zx + x^2)q = x^2 + xy + y^2$.

Solution: The auxiliary equations are given by

$$\frac{dx}{y^2 + yz + z^2} = \frac{dy}{z^2 + zx + x^2} = \frac{dz}{x^2 + xy + y^2}$$

$$\Rightarrow \qquad \frac{dx - dy}{(y^2 - x^2) + z(y - x)} = \frac{dy - dz}{(z^2 - y^2) + x(z - y)} = \frac{dz - dx}{(x^2 - z^2) + y(x - z)}$$

$$\Rightarrow \qquad \frac{dx - dy}{(y - x)(y + x + z)} = \frac{dy - dz}{(z - y)(z + y + z)} = \frac{dz - dx}{(x - z)(x + z + y)}$$

$$\frac{dx - dy}{y - x} = \frac{dy - dz}{z - y} = \frac{dz - dx}{x - z}$$

Taking the first two members, we have

$$\frac{dx - dy}{x - y} = \frac{dy - dz}{y - z}$$

Integrating, we get

$$\log(x - y) = \log(y - z) + \log c_1$$

$$\Rightarrow \qquad \frac{x - y}{y - z} = c_1$$

Again, taking the second and third terms, we obtain $\frac{dy - dz}{y - z} = \frac{dz - dx}{z - x}$

Integrating, we get

$$\log(y - z) = \log(z - x) + \log c_2$$

$$\Rightarrow \qquad \frac{y-z}{z-x} = c_2$$

Hence, the required general solution is given by $\phi(c_1, c_2) = \phi\left(\frac{x-y}{y-z}, \frac{y-z}{z-x}\right) = 0$

EXAMPLE 2.37

Find the general solution of the partial differential equation

$$px(x + y) - qy(x + y) + (x - y)(2x + 2y + z) = 0.$$

Solution: The auxiliary equations are given by

$$\frac{dx}{x(x+y)} = \frac{dy}{-y(x+y)} = \frac{dz}{-(x-y)(2x+2y+z)}$$

Taking the first two members, we get

$$\frac{dx}{x} = -\frac{dy}{y}$$

Integrating, we get $xy = c_1$

Again, $$\frac{dx+dy}{(x-y)(x+y)} = \frac{dx+dy+dz}{(x+y)(x-y)-(x-y)(2x+2y+z)}$$

$$\Rightarrow \qquad \frac{dx+dy}{x+y} = \frac{dx+dy+dz}{-x-y-z}$$

$$\Rightarrow \qquad \frac{d(x+y)}{x+y} + \frac{d(x+y+z)}{x+y+z} = 0$$

which on integrating yields

$$(x + y)(x + y + z) = c_2$$

Hence, the required solution is given by

$$\phi(c_1, c_2) = \phi(xy, (x + y)(x + y + z)) = 0.$$

Integral surface passing through a given curve

Here, we present methods to find the integral surface of the partial differential equation

$$Pp + Qq = R$$

which passes through a given curve.

(a) Let $Pp + Qq = R$ be the given equation and its auxiliary equations give $\phi(x, y, z) = c_1$ and $\psi(x, y, z) = c_2$ as two independent solutions. Suppose we wish to obtain the integral surface which passes through the curve whose equation in parametric form is given by $x = x(t)$, $y = y(t)$, $z = z(t)$ where, t is a parameter.

$$\therefore \qquad \phi(x(t), y(t), z(t)) = c_1 \text{ and } \psi(x(t), y(t), z(t)) = c_2.$$

We eliminate parameter t to get a relation involving c_1 and c_2 and finally, substitute the values of c_1 and c_2 to obtain the required integral surface.

(b) In this case, suppose we wish to obtain the integral surface passing through the curve which is determined by the two equations

$$f(x, y, z) = 0 \text{ and } g(x, y, z) = 0$$

Eliminate x, y, z from these equations and $\phi(x, y, z) = c_1$; $\psi(x, y, z) = c_2$ to get a relation between c_1 and c_2. Then, required integral surface is obtained by replacing c_1 by $\phi(x, y, z)$ and c_2 by $\psi(x, y, z)$.

EXAMPLE 2.38

Find the equation of integral surface of $(y - z)p + (z - x)q = x - y$, which passes through $y = 2x$, $z = 0$.

Solution: The auxiliary equations are

$$\frac{dx}{y-z} = \frac{dy}{z-x} = \frac{dz}{x-y}$$

Using 1, 1, 1 as multipliers, each member is equal to

$$\frac{dx+dy+dz}{(y-z)+(z-x)+(x-y)} = \frac{dx+dy+dz}{0}$$

$\therefore$ $dx + dy + dz = 0$ which on integration gives $x + y + z = c_1$

Again, using x, y, z as multipliers, each member is equal to

$$\frac{xdx+ydy+zdz}{x(y-z)+y(z-x)+z(x-y)} = \frac{xdx+ydy+zdz}{0}$$

$\therefore$ $xdx + ydy + zdz = 0$ which on integration yields $x^2 + y^2 + z^2 = c_2$.

Hence, the general solution is given by

$$x + y + z = c_1 \text{ and } x^2 + y^2 + z^2 = c_2.$$

Using the parametric equations of the curve, $x = t$, $y = 2t$, $z = 0$ we get

$$t + 2t = c_1 \quad \text{and} \quad t^2 + 4t^2 = c_2$$

$$3t = c_1 \quad \text{and} \quad 5t^2 = c_2$$

$$t = \frac{c_1}{3} \quad \text{and} \quad 5\left(\frac{c_1}{3}\right)^2 = c_2$$

$$\therefore \qquad 5c_1^2 = 9c_2$$

Substituting values of c_1 and c_2, we get

$$5(x + y + z)^2 = 9(x^2 + y^2 + z^2)$$

which is the required integral surface.

EXAMPLE 2.39

Find the equation of the integral surface of the differential equation $2y(z - 3)p + (2x - z)q = y(2x - 3)$, which passes through the circle $z = 0$, $x^2 + y^2 = 2x$.

Solution: The auxiliary equations are

$$\frac{dx}{2y(z-3)} = \frac{dy}{2x-z} = \frac{dz}{y(2x-3)}$$

Using $\frac{1}{2}$, y, -1 as multipliers, each member is equal to

$$\frac{\frac{1}{2}dx + ydy - dz}{y(z-3) + y(2x-z) - y(2x-3)} = \frac{\frac{1}{2}dx + ydy - dz}{0}$$

$\therefore\ \frac{1}{2}dx + ydy - dz = 0$ which on integration yields $\frac{1}{2}x + \frac{y^2}{2} - z = c_1$ or $x + y^2 - 2z = c_1$

Taking the first and third members, we obtain

$$\frac{dx}{2(z-3)} = \frac{dz}{2x-3}$$

$$\Rightarrow \qquad (2x - 3)dx = 2(z - 3)dz$$

which on integration gives $x^2 - 3x - z^2 + 6z = c_2$.

Now, the parametric equations of the given circle are $x = t$, $y = (2t - t^2)^{1/2}$, $z = 0$. Substituting these, we get

$$t^2 - 3t = c_2 \quad \text{and} \quad 3t - t^2 = c_1$$

$$\therefore \qquad c_1 + c_2 = 0$$

Using values of c_1 and c_2, we get

$$x + y^2 - 2z + x^2 - 3x - z^2 + 6z = 0$$

$$\Rightarrow \qquad x^2 + y^2 - z^2 - 2x + 4z = 0$$

which is the required integral surface.

EXAMPLE 2.40

Find the integral surface of $x(y^2 + z)p - y(x^2 + z)q = (x^2 - y^2)z$ which passes through the straight line $x = t$, $y = -t$, $z = 1$.

Solution: The auxiliary equations are

$$\frac{dx}{x(y^2+z)} = \frac{dy}{-y(x^2+z)} = \frac{dz}{(x^2-y^2)z}$$

Using $\frac{1}{x}, \frac{1}{y}, \frac{1}{z}$ as multipliers, each member is equal to

$$\frac{\frac{1}{x}dx + \frac{1}{y}dy + \frac{1}{z}dz}{(y^2+z)-(x^2+z)+(x^2-y^2)} = \frac{\frac{1}{x}dx + \frac{1}{y}dy + \frac{1}{z}dz}{0}$$

$\therefore\ \frac{1}{x}dx + \frac{1}{y}dy + \frac{1}{z}dz = 0$ which on integration yields $xyz = c_1$

Again, using $x, y, -1$ as multipliers, each member is equal to

$$\frac{xdx + ydy - dz}{x^2(y^2+z) - y^2(x^2+z) - (x^2-y^2)z^2}$$

$$= \frac{xdx + ydy - dz}{0}$$

$\therefore\ xdx + ydy - dz = 0$ which on integration gives $x^2 + y^2 - 2z = c_2$.

Using the parametric equations of the given straight line, we get

$$(t)(-t)(1) = c_1 \quad \text{and} \quad t^2 + t^2 - 2 = c_2$$

$$\Rightarrow \quad -t^2 = c_1 \quad \text{and} \quad 2t^2 - 2 = c_2$$

$$\therefore \quad -2c_1 - 2 = c_2$$

$$\Rightarrow \quad 2c_1 + c_2 - 2 = 0.$$

Substituting the values of c_1 and c_2; we get

$$2xyz + x^2 + y^2 - 2z + 2 = 0$$

which is the required integral surface.

EXAMPLE 2.41

Find the equation of the integral surface satisfying $4yzp + q + 2y = 0$ and passing through $y^2 + z^2 = 1$ and $x + z = 2$.

Solution: The auxiliary equations are

$$\frac{dx}{4yz} = \frac{dy}{1} = \frac{dz}{-2y}$$

Taking first and third members, we obtain

$$\frac{dx}{2z} = \frac{dz}{-2y} \Rightarrow dx + 2zdz = 0$$

which on integration yields $x + z^2 = c_1$

Again, using the second and third members, we get

$$\frac{dy}{1} = \frac{dz}{-2y} \Rightarrow 2ydy + dz = 0$$

which on integration gives $y^2 + z = c_2$

On adding, we get $y^2 + z^2 + x + z = c_1 + c_2$

Using $y^2 + z^2 = 1$ and $x + z = 2$, we obtain

$$c_1 + c_2 = 3$$

Substituting the values of c_1 and c_2, the required integral surface is given by

$$x + z^2 + y^2 + z = 3$$

$$\Rightarrow \qquad y^2 + z^2 + x + z - 3 = 0.$$

EXAMPLE 2.42

Find the integral surface of the equation $(x - y)y^2p + (y - x)x^2q = (x^2 + y^2)z$; through the curves $xz = a^3$, $y = 0$.

Solution: The auxiliary equations are

$$\frac{dx}{(x-y)y^2} = \frac{dy}{(y-x)x^2} = \frac{dz}{(x^2+y^2)z}$$

Taking the first two members, we get

$$\frac{dx}{y^2} = \frac{dy}{-x^2} \Rightarrow x^2dx + y^2dy = 0$$

which on integration yields $x^3 + y^3 = c$,

Again, $$\frac{dx - dy}{(x-y)(x^2+y^2)} = \frac{dz}{(x^2+y^2)z}$$

$$\Rightarrow \qquad \frac{dx - dy}{x - y} = \frac{dz}{z}$$

which on integration gives $\dfrac{x-y}{z} = c_2$

Using $xz = a^3$ and $y = 0$, we have

$$x^3 = c_1 \quad \text{and} \quad x/z = c_2$$

$$\therefore \qquad x^2 = a^3 c_2 \;\Rightarrow\; c_1^2 = a^9 c_2^3.$$

Substituting values of c_1 and c_2, the equation of the required integral surface is

$$(x^3 + y^3)^2 = a^a \left[\frac{x-y}{z}\right]^3$$

$$\Rightarrow \qquad z^3(x^3 + y^3)^2 = a^9(x - y)^3.$$

EXAMPLE 2.43

Find the integral surface of the linear first order partial differential equation $yp + xq = z - 1$, which passes through the curve $z = x^2 + y^2 + 1;\ y = 2x$.

Solution: The auxiliary equations are

$$\frac{dx}{y} = \frac{dy}{x} = \frac{dz}{z-1}$$

Taking the first two members, we obtain

$$\frac{dx}{y} = \frac{dy}{x} \Rightarrow xdx - ydy = 0$$

which on integration yields $x^2 - y^2 = c_1$

Again, $\dfrac{dx+dy}{y+x} = \dfrac{dz}{z-1} \Rightarrow \dfrac{d(x+y)}{x+y} = \dfrac{dz}{z-1}$

Integrating, we get $\dfrac{x+y}{z-1} = c_2$

The parametric form of the given curve is

$$x = t,\ y = 2t,\ z = 5t^2 + 1.$$

$$\therefore \qquad t^2 - 4t^2 = c_1 \text{ and } \frac{t+2t}{(5t^2+1)-1} = c_2$$

$$\Rightarrow \qquad -3t^2 = c_1 \text{ and } \frac{3}{5t} = c_2$$

$$\Rightarrow \qquad 3t^2 = c_1', \text{ and } \frac{3}{5t} = c_2 \text{ where } c_1' = -c_1$$

$$\therefore \qquad \frac{3}{5\sqrt{\dfrac{c_1'}{3}}} = c_2 \Rightarrow \frac{3\sqrt{3}}{5\sqrt{c_1'}} = c_2$$

Substituting values of c_1' and c_2, we have

$$\frac{3\sqrt{3}}{5\sqrt{y^2 - x^2}} = \frac{x+y}{z-1}$$

$\Rightarrow$ $$5(y^2 - x^2)^{1/2} / 3\sqrt{3} = (z-1)/(x+y)$$

which is the required integral surface.

EXAMPLE 2.44

Find the integral surface of $x^2p + y^2q = -z^2$; which passes through the hyperbola $x + y = xy$; $z = 1$.

Solution: The auxiliary equations are

$$\frac{dx}{x^2} = \frac{dy}{y^2} = \frac{dz}{z^2}$$

Taking the first and third members, we have

$$\frac{dx}{x^2} = \frac{dz}{-z^2}$$

which on integration gives $-\frac{1}{x} = \frac{1}{z} - c_1$

$\Rightarrow$ $$\frac{1}{z} + \frac{1}{x} = c_1.$$

Again, taking the second and third members, we obtain

$$\frac{dy}{y^2} = \frac{dz}{-z^2}$$

which on integration gives $\frac{1}{y} + \frac{1}{z} = c_2.$

On adding these, we get $$\frac{2}{z} + \frac{1}{x} + \frac{1}{y} = c_1 + c_2$$

$\Rightarrow$ $$\frac{2}{z} + \frac{x+y}{xy} = c_1 + c_2$$

$\Rightarrow$ $$\frac{2}{1} + \frac{xy}{xy} = c_1 + c_2$$

$\therefore$ $$c_1 + c_2 = 3$$

Substituting the values of c_1 and c_2, the required equation of the integral surface is

$$\frac{1}{z}+\frac{1}{x}+\frac{1}{z}+\frac{1}{y} = 3$$

$$\Rightarrow \qquad yz + zx + 2xy = 3xyz.$$

EXAMPLE 2.45

Find the integral surface of the partial differential equation $(x - y)p + (y - x - z)q = 2$ through the circle $z = 1$, $x^2 + y^2 = 1$.

Solution: The auxiliary equations are

$$\frac{dx}{x-y} = \frac{dy}{y-x-z} = \frac{dz}{z}$$

Using 1, 1, 1 as multipliers, each member is equal to

$$\frac{dx+dy+dz}{(x-y)+(y-x-z)+z} = \frac{dx+dy+dz}{0}$$

$\therefore$ $dx + dy + dz = 0$ which on integration yields $x + y + z = c_1$.

Again, $\dfrac{dx-dy}{(x-y)-(y-x-z)} = \dfrac{dz}{z}$

$$\Rightarrow \qquad dx - dy = dz$$

which on integration gives $x - y - z = c_2$.

Using $z = 1$, $x^2 + y^2 = 1$; we have

$$x + y = c_1 - 1 \text{ and } x - y = c_2 + 1$$

$$\Rightarrow \qquad (x + y)^2 = (c_1 - 1)^2 \text{ and } (x - y)^2 = (c_2 + 1)^2$$

$$\Rightarrow \qquad x^2 + y^2 + 2xy = (c_1 - 1)^2 \text{ and } x^2 + y^2 - 2xy = (c_2 + 1)^2$$

On adding, we get

$$2(x^2 + y^2) = (c_1 - 1)^2 + (c_2 + 1)^2$$

$$\Rightarrow \qquad (c_1 - 1)^2 + (c_2 + 1)^2 = 2$$

$$\Rightarrow \qquad c_1^2 + c_2^2 + 2c_2 = 0$$

$$\Rightarrow \qquad (x + y + z)^2 - 2(x + y + z) + (x - y - z)^2 + 2(x - y - z) = 0$$

which is the required equation of the integral surface.

EXAMPLE 2.46

Find the integral surface of $(x^2 - yz)p + (y^2 - zx)q = z^2 - xy$.

which passes through the straight line $x = 1; y = 0$.

Solution: The auxiliary equations are

$$\frac{dx}{x^2 - yz} = \frac{dy}{y^2 - zx} = \frac{dz}{z^2 - xy}$$

Now,
$$\frac{dx - dy}{x^2 - yz - y^2 + zx} = \frac{dy - dz}{y^2 - zx - z^2 + xy}$$

$$\Rightarrow \quad \frac{dx - dy}{(x-y)(x+y+z)} = \frac{dy - dz}{(y-z)(x+y+z)}$$

$$\Rightarrow \quad \frac{d(x-y)}{x-y} = \frac{d(y-z)}{y-z}$$

which on integration yields $\dfrac{x-y}{y-z} = c_1$

Again,
$$\frac{dx + dy + dz}{x^2 - yz + y^2 - zx + z^2 - xy} = \frac{xdx + ydy + zdz}{x(x^2 - yz) + y(y^2 - zx) + z(z^2 - xy)}$$

$$\Rightarrow \quad \frac{d(x+y+z)}{x^2 + y^2 + z^2 - xy - yz - zx} = \frac{xdx + ydy + zdz}{x^3 + y^3 + z^3 - 3xyz}$$

$$\Rightarrow \quad \frac{d(x+y+z)}{1} = \frac{xdx + ydy + zdz}{x+y+z}$$

$$\Rightarrow \quad (x + y + z)\, d(x + y + z) = xdx + ydy + zdz.$$

which on integration gives

$$(x + y + z)^2 = x^2 + y^2 + z^2 + 2c_2$$

$$\Rightarrow \quad xy + yz + zx = c_2.$$

Using $x = 1; y = 0$; we have

$$-\frac{1}{z} = c_1 \text{ and } + z = c_2$$

$$\therefore \quad c_1 c_2 = -1$$

Substituting values of c_1 and c_2, we have

$$\left(\frac{x-y}{y-z}\right)(xy + yz + zx) + 1 = 0$$

$$\Rightarrow \quad (x - y)(xy + yz + zx) + (y - z) = 0$$

which is the required integral surface.

EXAMPLE 2.47

Solve $xp + yq = z$. Find a solution representing a surface meeting the parabola $y^2 = 4x$; $z = 1$.

Solution: The auxiliary equations are

$$\frac{dx}{x} = \frac{dy}{y} = \frac{dz}{z}$$

Taking the first and third members, we have

$$\frac{dx}{x} = \frac{dz}{z}$$

which on integration yields $\frac{x}{z} = c_1$.

Again, taking the second and third members, we have

$$\frac{dx}{x} = \frac{dy}{y}$$

which on integration gives $\frac{y}{z} = c_2$.

∴ the required general solution is given by

$$\phi(c_1, c_2) = \phi\left(\frac{x}{z}, \frac{y}{z}\right) = 0.$$

Using $y^2 = 4x$; $z = 1$; we get

$$x = c_1 \text{ and } y = c_2$$

∴ $$c_2^2 = 4c_1.$$

Substituting values of c_1 and c_2; we have

$$\left(\frac{y}{z}\right)^2 = 4\left(\frac{x}{z}\right)$$

⇒ $$y^2 = 4xz$$

which is the required equation of the integral surface.

Surface orthogonal to a given system of surfaces

An interesting application of the theory of linear partial differential equations of the first order is the determination of the system of surface orthogonal to a given system of surface.

Let $$f(x, y, z) = c \tag{1}$$

represents a system of surface, where c is parameter. Suppose, we wish to obtain a system of surfaces, which cut each of Eq. (1) at right angles. Then, the direction ratios of the normal at the point (x, y, z) to (1); which passes through that point are $\partial f/\partial x$, $\partial f/\partial y$, $\partial f/\partial z$.

Let the surface $z = \phi(x, y)$ (2)

cuts each surface of Eq. (1) at right angles. Then, the normal at (x, y, z) to Eq. (2) has direction ratios $\partial z/\partial x, \partial z/\partial y, -1$ or $p, q, -1$.

Since normals at (x, y, z) to Eqs. (1) and (2) are at right angles, we have

$$p(\partial f/\partial x) + q(\partial f/\partial y) - 1\,(\partial f/\partial z) = 0$$

$$\Rightarrow \qquad Pp + Qq = R.$$

Conversely, any solution of the linear partial differential equation $Pp + Qq = R$ is orthogonal to every surface of the system characterized by Eq. (1), for $Pp + Qq = R$ simply states that the normal to any solution of Eq. (4) is perpendicular to the normal to that member of the system (1), which passes through the same point. The linear equation $Pp + Qq = R$ is therefore the general partial differential equation determining the surface orthogonal to the members of the system (1), that is, the surfaces generated by the integral curves of the equations $\dfrac{dx}{\partial f/\partial x} = \dfrac{dy}{\partial f/\partial y} = \dfrac{dz}{\partial f/\partial z}$.

EXAMPLE 2.48

Find the surface which is orthogonal to the one parameter system $z = cxy(x^2 + y^2)$ which passes through the hyperbola $x^2 - y^2 = a^2$; $z = 0$.

Solution: The given system of surfaces is

$$f(x, y, z) = z/(x^3y + xy^3) = c.$$

$$\therefore \qquad \frac{\partial f}{\partial x} = -\frac{z(3x^2y + y^3)}{(x^3y + xy^3)^2};\ \frac{\partial f}{\partial y} = -\frac{z(x^3 + 3xy^2)}{(x^3y + xy^3)^2};\ \frac{\partial f}{\partial z} = \frac{1}{x^3y + xy^3}$$

The required orthogonal surface is solution of $p(\partial f/\partial x) + q(\partial f/\partial y) = (\partial f/\partial z)$

$$\Rightarrow \qquad -(3x^2y + y^3)p - (x^3 + 3xy^2)q = \frac{x^3y + xy^3}{z}$$

$$\Rightarrow \qquad \left(\frac{3x^2 + y^2}{x}\right)p + \left(\frac{x^2 + 3y^2}{y}\right)q = -\left(\frac{x^2 + y^2}{z}\right)$$

The auxiliary equations are

$$\frac{dx}{(3x^2 + y^2)/x} = \frac{dy}{(x^2 + 3y^2)/y} = \frac{dz}{-(x^2 + y^2)/z}$$

Taking the first two members, we have

$$xdx = ydy$$

which on integration yields $x^2 - y^2 = c_1$

Using x, y, z as multipliers, each member is equal to

$$\frac{xdx + ydy + 4zdz}{(3x^2 + y^2) + (x^2 + 3y^2) - 4(x^2 + y^2)}$$

$$= \frac{xdx + ydy + 4zdz}{0}$$

$\therefore$ $$xdx + ydy + 4zdz = 0$$

which on integration gives $x^2 + y^2 + 4z^2 = c_2$.

Hence, any surface which is orthogonal to the given family of surface is of the form $c_2 = \phi(c_1)$

$\Rightarrow$ $$x^2 + y^2 + 4z^2 = \phi(x^2 - y^2)$$

where ϕ is an arbitrary function.

For the particular surface passing through the hyperbola $x^2 - y^2 = a^2$, $z = 0$; we must take $\phi(x^2 - y^2) = a^4(x^2 + y^2)/(x^2 - y^2)^2$.

Hence, the required surface is given by $(x^2 + y^2 + 4z^2)(x^2 - y^2)^2 = a^4(x^2 + y^2)$.

EXAMPLE 2.49

Find the family orthogonal to $\phi(z(x + y)^2, x^2 - y^2) - 0$.

Solution: Let $u = z(x + y)^2$ and $v = x^2 - y^2$.

$\therefore$ $$\phi(u, v) = 0$$

Differentiating w.r.t. x and y partially by turn, we get

$$\frac{\partial \phi}{\partial u}\left(\frac{\partial u}{\partial x} + p\frac{\partial u}{\partial z}\right) + \frac{\partial \phi}{\partial v}\left(\frac{\partial v}{\partial x} + p\frac{\partial v}{\partial z}\right) = 0$$

and $$\frac{\partial \phi}{\partial u}\left(\frac{\partial u}{\partial y} + q\frac{\partial v}{\partial z}\right) + \frac{\partial \phi}{\partial v}\left(\frac{\partial v}{\partial y} + q\frac{\partial v}{\partial z}\right) = 0$$

Putting $\frac{\partial u}{\partial x} = 2z(x+y), \frac{\partial u}{\partial y} = 2z(x+y), \frac{\partial u}{\partial z} = (x+y)^2$ and $\frac{\partial v}{\partial x} = 2x, \frac{\partial v}{\partial y} = -2y, \frac{\partial v}{\partial z} = 0;$

we get

$$\frac{\partial \phi}{\partial u}[2z(x + y) + p(x + y)^2]\frac{\partial \phi}{\partial v}[2x + q(0) = 0$$

and $$\frac{\partial \phi}{\partial u}[2z(x + y) + q(x + y)^2] + \frac{\partial \phi}{\partial v}[-2y + q(0)] = 0.$$

$\therefore$ $$\frac{-\frac{\partial \phi}{\partial u}}{\frac{\partial \phi}{\partial v}} = \frac{2x}{2z(x+y)+p(x+y)^2} = \frac{-2y}{2z(x+y)+q(x+y)^2}$$

$$\Rightarrow \qquad x(x + y)\,[2z + q(x + y)] = -y(x + y)\,[2z + p(x + y)]$$

$$\Rightarrow \qquad 2xz + qx(x + y) = -2yz - py(x + y)$$

$$\Rightarrow \qquad py(x + y) + qx(x + y) = -2z(y + x)$$

$$\Rightarrow \qquad py + qx = -2z \tag{1}$$

which is the differential equation of the family of surfaces given by $\phi(u, v) = 0$. So the differential equation of the family of surfaces orthogonal to Eq. (1) is given by

$$ydx + xdy - 2zdz = 0 \Rightarrow d(xy) - 2zdz = 0.$$

which on integration yields $xy - z^2 = c$.

which is the required family of orthogonal surface, c being a parameter.

EXAMPLE 2.50

Find the family of surfaces orthogonal to the family of surface given by the differential equation $(y + z)p + (z + x)q = x + y$.

Solution: The differential equation of the family of surface orthogonal to the given family which is represented by

$$(y + z)p + (z + x)q = x + y \text{ is given by}$$

$$(y + z)dx + (z + x)dy + (x + y)dz = 0$$

$$\Rightarrow (ydx + xdy) + (ydz + zdy) + (zdx + xdz) = 0.$$

which on integration gives

$$xy + yz + zx = c$$

which is the required family, c being a parameter. ▲▲

EXERCISE 2.2

Solve the following partial differential equations

1. $z(yp + xq) = xy$.
2. $\dfrac{y^2zp}{x} + zxq = y^2$.
3. $(mx - ny)p + (nx - lz)q = (ly - mx)$.
4. $x(y^2 - z^2)p + y(z^2 - x^2)q = z(x^2 - y^2)$.
5. $x(y^2 + z)p - y(x^2 + z)q = z(x^2 - y^2)$.
6. $z(xp - yq) = y^2 - x^2$.
7. $yp - xq = y^2 - x^2$.

8. $x^2p + y^2q = (x + y)z$.
9. $(x^2 - yz)p + (y^2 - zx)q = (z^2 - xy)$.
10. $p\sqrt{x} + q\sqrt{y} = \sqrt{z}$.
11. $px^2 + qy^2 = z^2$.
12. $p \tan x + q \tan y = \tan z$.
13. $p \cot x + q \cot y = \cot z$.
14. $\frac{y^2 z}{x} p - xyq = y^2$.
15. $p - q = \log(x + y)$.
16. $(y + z)p + (z + x)q = x + y$.
17. $pz - qz = z^2 + (x + y)^2$.
18. $(x + y)zp + (x - y)zq = x^2 + y^2$.
19. $(y^2 + z^2)p - xyq + xz = 0$.
20. $(x - 2z)p + (2z - y)q = y - x$.
21. $x(y^2 + z^2)p + y(z^2 + x^2)q = z(y^2 - x^2)$.
22. $(3x - 4y)p + (4x - 2z)q = 2y - 3x$.
23. $xzp + yzq = xy$.
24. $py + qx = xyz^2(x^2 - y^2)$.
25. $(2x^2 + y^2 + z^2 - 2yz - zx - xy)p + (x^2 + 2y^2 + z^2 - yz - 2zx - xy)q = x^2 + y^2 + 2z^2 - yz - zx - 2xy$.
26. $z - xp - yq = a\sqrt{(x^2 + y^2 + z^2)}$.
27. $(y^3x - 2x^4)p + (2y^4 - x^3y)q = 9z(x^3 - y^3)$.
28. $\left(\frac{y - z}{yz}\right)p + \left(\frac{z - x}{zx}\right)q = \frac{x - y}{xy}$.
29. $(3x + y - z)p + (x + y - z)q = 2(z - y)$.
30. $x^2p + y^2q = mxy$.
31. $2p + 3q = \tan z$.
32. $p \cos x + q \cos y = \cos z$.
33. $(y + z)p + (z + x)q = (x + y)$.
34. $(x^2 + y^2)p + 2xyq = (x + y)z$.
35. $xp + zq = -y$.
36. $(1 + y)p + (1 - x)q = z$.

37. $(y^2 + z^2)p - xyq = -zx.$
38. $z(xp - qy) = y^2 - x^2.$
39. $(z - y)p + (x - z)q = y - x.$
40. $px + qz = -y.$
41. $5p + 6q = 25x^4 \cos(6x + 5y).$
42. $p + 5q = z - \cot(5x - y).$
43. $zp - zq = z^2 + (x + y)^2.$
44. $p - 2q = 3x^2 \sin(2x + y).$
45. $xy^2p - y^3q = -axz.$
46. $zp - zq = x + y.$
47. $(p - q)(x + y) = z.$
48. $p + q = \sin x.$
49. $p + q = \frac{z}{a}.$
50. $p + q = \cos x.$
51. $y^2p + x^2q = x^2y^2z^2.$
52. $py + qx = xyz^2(x^2 - y^2).$
53. $px(z - 2y^2) = (z - qy)(z - y^2 - 2x^3).$
54. $xyp + y^2q + 2x^2 - xyz = 0.$
55. $(y^2 + yz + z^2)p + (z^2 + zx + x^2)q = x^2 + xy + y^2.$
56. $(x + y - z)(p - q) + a(px - qy + x - y) = 0.$ {*Hint*: Substitute $u = x + y$ and $v = x - y$}.
57. $p - qy \log y = z \log y.$
58. $x^2p + y^2q = x + y.$
59. $px - qy = (z - xy)^2.$
60. $zp + zq = z^2 + (x - y)^2.$
61. $x(y^n - z^n)p + y(z^n - x^n)q = z(x^n - y^n).$
62. $(xz + y^2)p + (yz - 2x^2)q + 2xy + z^2 = 0.$
63. $xyp + y(2x - y)q = 2xz.$
64. $(x^2 - y^2 - z^2)p + 2xy = 2xz.$
65. $(p - q)(e^{x+y} + xyz) + (yq - xp)(z + xye^{-x-y}) + (x - y)(1 - z^2) = 0.$

{*Hint:* Let $u = e^{x+y}$; $v = xy$}.

66. $(3x + y - z)p + (x + y - z)q = 2(z - y).$
67. $px - qy = (z - xy)^2.$

68. $\cos(x + y)p + \sin(x + y)q = z + \dfrac{1}{x}$.

69. $z(z^2 - xy)(px - qy) = x^4$.

70. $xyp + y(2x - y)q = 2xz$.

71. Find the equation of the integral surface of the differential equation

$$2y(z - 3)p + (2x - z)q = y(2x - 3)$$

which passes through the circle $z = 0$, $x^2 + y^2 = 2x$.

72. Find the integral surface of $x^2p + y^2q + z^2 = 0$; which passes through the hyperbola $xy = x + y$; $z = 0$.

73. Find the integral surface of $p + q = 1$; which passes through $x = 0$; $y^2 = z$.

74. Find the equation of the integral surface of the differential equation $xp + yq = z$; which passes through the curve $x + y = 1$; $yz = 1$.

75. Find the particular integral of $x(y - z)p + y(z - x)q = z(x - y)$; to represent surface passing through the curve $x = y = z$

76. Find the equation of surfaces orthogonal to $f\left(\dfrac{x}{z}, \dfrac{y}{z}\right) = 0$.

77. Find the orthogonal surfaces of $\dfrac{\sin z}{\sin y} = f\left(\dfrac{\sin x}{\sin y}\right)$.

78. Find the orthogonal surfaces of $\dfrac{1}{x} - \dfrac{1}{y} = f\left(\dfrac{1}{x} - \dfrac{1}{z}\right)$.

79. Find the equation of the family of surfaces; which cut orthogonally the curves of family $x^2 + y^2 + z^2 = \lambda xy$; λ being a parameter.

80. Find the integral surface of the partial differential equation $(x - y)p + (y - x - z)q = z$; through the circle $z = 1$; $x^2 + y^2 = 1$.

81. Find the equation of surface satisfying $4yzp + q + 2y = 0$ and passing through $y^2 + z^2 = 1$, $x + z = 2$.

82. Find the general integral of the partial differential equation $(2xy - 1)p + (z - 2x^2)q = 2(x - yz)$; and also the particular integral which passes through the line $x = 1$; $y = 0$.

83. Find the general solution of the equation $2x(y + z^2)p + y(2y + z^2)q = z^2$; and deduce that $yz(z^2 + yz - 2y) = x^2$ is a solution.

84. Find the general solution of $x(z + 2a)p + (xz + 2yz + 2ay)q = z(z + a)$. Find also the integral surfaces; which pass through the curves

(a) $y = 0$; $z^2 = 4ax$

(b) $y = 0; z^3 + x(z + a)^2 = 0.$

85. Solve $xp + yq = z$. Find a solution representing a surface meeting the parabola $y^2 = 4x, z = 1$.

86. Find the surface; which intersects the surfaces of the system $z(x + y) = c(3z + 1)$ orthogonally and which passes through the circle $x^2 + y^2 = 1; z = 1$.

87. Write down the system of equations for obtaining the general equation of surfaces orthogonal to the family given by $x(x^2 + y^2 + z^2) = cy^2$.

88. Find the general solution of the partial differential equation

$$(3 - 2yz)p + x(2z - 1)p = 2x(y - 3).$$

Hence, obtain the particular solution; which passes through the curve $z = 0; x^2 + y^2 = 4$.

89. Find the equation of the surface which cuts orthogonally the system of surfaces $2xy + 3yz = a(z + 2)$; where, a is an arbitrary constant and passes through the circle $z = 0; x^2 + y^2 = 9$.

90. Find the equation of the system of surfaces; which cut orthogonally the family of conics $z^2 = c(x^2 + y^2)$. Obtain the particular surface; which passes through the circle $z = 3; x^2 + y^2 = 9$.

ANSWERS

Exercise 2.2

1. $\phi(x^2 - z^2, y^2 - z^2) = 0.$
2. $\phi(x^3 - y^3, x^2 - z^2) = 0.$
3. $\phi(x^2 + y^2 + z^2, lx + my + nz) = 0.$
4. $\phi(x^2 + y^2 + z^2, xyz) = 0.$
5. $\phi(x^2 + y^2 - 2z, xyz) = 0.$
6. $\phi(xy, x^2 + y^2 + z^2) = 0.$
7. $\phi(x^2 + y^2, xy - z) = 0.$
8. $\phi\left(\frac{1}{x} - \frac{1}{y}, \frac{x - y}{z}\right) = 0.$
9. $\phi\left(\frac{x - y}{y - z}, xy + yz + zx\right) = 0.$
10. $\phi(\sqrt{x} - \sqrt{y}, \sqrt{y} - \sqrt{z}) = 0.$
11. $\phi\left(\frac{1}{x} - \frac{1}{y}, \frac{1}{y} - \frac{1}{z}\right) = 0.$

12. $\phi\left(\dfrac{\sin x}{\sin y}, \dfrac{\sin y}{\sin z}\right) = 0.$

13. $\phi\left(\dfrac{\cos z}{\cos y}, \dfrac{\cos y}{\cos x}\right) = 0.$

14. $\phi(x^2 - z^2, x^3 - y^3) = 0.$

15. $\phi(x \log(x + y) - z, (x + y) = 0.$

16. $\phi\left(\dfrac{x-y}{y-z}, (x-y)^2(x+y+z)\right) = 0.$

17. $\phi[(x + y, \log(z^2 + x^2 + y^2 + 2xy) - 2x] = 0.$

18. $\phi(x^2 - y^2 - z^2, 2xy - z^2) = 0.$

19. $\phi\left(\dfrac{y}{x}, x^2 + y^2 + z^2\right).$

20. $\phi(xy + z^2, x + y + z) = 0.$

21. $\phi\left[x^2 - y^2 - z^2, \dfrac{x}{yz}\right] = 0.$

22. $\phi[x^2 + y^2 + z^2; 2x + 3y + 4z] = 0.$

23. $\phi\left[\dfrac{x}{y}, xy - z^2\right] = 0.$

24. $y^2(x^2 - y^2) + \dfrac{2}{z} = \phi(x^2 - y^2).$

25. $\phi\left(\dfrac{x-y}{y-z}, \dfrac{y-z}{z-x}\right) = 0.$

26. $x^{1-a} = \left[z + \sqrt{(x^2 + y^2 + z^2)}\right]\phi(x/y).$

27. $\phi\left(\dfrac{y}{x^2} + \dfrac{x}{y^2}, xyz^{1/3}\right) = 0.$

28. $\phi(x + y + z, xyz) = 0.$

29. $\phi\left(x - 3y - z, \dfrac{(x-y+z)^2}{x+y-z}\right) = 0.$

30. $\phi\left(\frac{x-y}{xy},\left(\frac{z-mxy}{x-y}\right)\log\left(\frac{x}{y}\right)\right)=0.$

31. $\phi(3x - 2y, \log(\sec x) - 2z) = 0.$

32. $\phi\left(\frac{\sec x+\tan x}{\sec y+\tan y},\frac{\sec x+\tan x}{\sec z+\tan z}\right)=0.$

33. $\phi\left[\frac{x-y}{y-z},(x+y+z)(x-y)^2\right]=0.$

34. $\phi\left(\frac{x+y}{z},\frac{2y}{x^2-y^2}\right)=0.$

35. $\phi\left[y^2+z^2,\frac{z}{e^{\tan-1(y/z)}}\right]=0.$

36. $\phi\left[(1+x)^2-(1+y)^2,\frac{x+y+2}{z}\right]=0.$

37. $\phi\left(x^2+y^2+z^2,\frac{y}{z}\right)=0.$

38. $\phi(x^2 + y^2 + z^2, xy) = 0.$

39. $\phi(x + y + z, x^2 + y^2 + z^2) = 0.$

40. $\phi\left[y^2+z^2,\log x-\sin^{-1}\frac{y}{\sqrt{y^2+z^2}}\right]=0.$

41. $\phi[6x + 5y, x^5\cos(6x + 5y)- z] = 0.$

42. $\phi(5x - y, x - \log(z - \cot(5x - y))) = 0.$

43. $\phi(x + y, 2x - \log(z^2 + (x + y)^2)) = 0.$

44. $\phi(2x + y, x^3 \sin(2x + y) = 0.$

45. $\phi\left(xy,\frac{ax}{3y^2}+\log z\right)=0.$

46. $\phi\left(x+y,(x+y)z-\frac{z^2}{2}\right)=0.$

47. $\phi(x + y, x - (x + y)\log z) = 0.$

48. $z + \cos x = \phi(x - y).$

49. $e^{y/a} = z\phi(x - y)$.

50. $\sin x - z = \phi(x - y)$.

51. $\phi(x^3 - y^3, y^3 + 3z^{-1}) = 0$.

52. $y^2(x^2 - y^2) + (2/z) = \phi(x^2 - y^2)$.

53. $(y^3 - ax - x^3)/x = \phi(z/y)$.

54. $x - \log(z - (2x/y)) = \phi(x/y)$.

55. $\phi\left(\dfrac{y-z}{x-y}, \dfrac{x-z}{x-y}\right) = 0$.

56. $\phi(x + y + z, a(x - y)^2 + 4(x + y)z - a(x + y)^2) = 0$.

57. $\phi(yz, e^x \log y) = 0$.

58. $\phi\left(\dfrac{1}{y} - \dfrac{1}{z}, e^{-z}(x - y)\right) = 0$.

59. $\phi(xy, xe^{1/(z - xy)}) = 0$.

60. $\log[z^2 + (x - y)^2] - 2x = \phi(x - y)$.

61. $x^n + y^n + z^n = \phi(xyz)$.

62. $\phi(yz + x^2, 2xz - y^2) = 0$.

63. $\phi(xy - x^2, z/xy) = 0$.

64. $\phi\left[\dfrac{y}{z}, \dfrac{x^2 + y^2 + z^2}{z}\right] = 0$

65. $\phi[xy - ze^{x + y}, e^{x + y} - xyz] = 0$.

66. $\phi\left[\dfrac{x - y + z}{\sqrt{x + y - z}}, x - 3y - z\right] = 0$

67. $\phi(xy, xe^{1/(z - xy)}) = 0$.

68. $\phi(z^2 + 1)^{1/\sqrt{2}} \tan\left(\dfrac{3\pi}{8} - \dfrac{x + y}{2}\right) = \phi(e^{y - x}[\cos(x + y) + \sin(x + y)])$.

69. $\phi(xy, (z^2 - xy)^2 - x^4) = 0$.

70. $\phi\left(xy - x^2, \dfrac{z}{xy}\right) = 0$.

71. $x^2 + y^2 - z^2 - 2x + 4z = 0$.

72. $xz + yz + 2xy = 3xyz$.

73. $(x - y)^2 = z - x.$

74. $(x + y)^2 = yz.$

75. $(x + y + z)^3 = 27xyz.$

76. $x^2 + y^2 + z^2 = k.$

77. $\cos x \cos y \cos z = k.$

78. $x^3 + y^3 + z^3 = k.$

79. $x^2 + y^2 + z^2 = \phi\left(\dfrac{x^2 - y^2}{z^2}\right)$

80. $z^4(x + y + z)^2 + (y - x - z)^2 - 2z^4(x + y + z) + 2z^2(y - x - z) = 0.$

81. $y^2 + z^2 + x + z - 3 = 0.$

82. $x^2 + y^2 + z - xz - y = 1.$

85. $\phi(x/2, y/2) = 0; y^2 = 4xz.$

86. $x^2 + y^2 + 2z^3 - z^2 = -2.$

87. $x^2 + y^2 + z^2 = z\phi((2x^2 + y^2)/z^2).$

88. $\phi(y^2 - 6y - z^2 + z, x^2 + z^2 + 6y) = 0; x^2 + y^2 + z = 4.$

89. $3(x^2 + y^2 + z^2) - z^3 = 27.$

90. $\phi\left(\dfrac{y}{x}, x^2 + y^2 + z^2\right) = 0\,; x^2 + y^2 + z^2 = 18.$

▲▲

3

CHAPTER

Nonlinear Partial Differential Equations of the First Order

Partial differential equations in which $\frac{\partial z}{\partial x}(=p)$ and $\frac{\partial z}{\partial y}(=q)$ occur other than in the first degree, are called nonlinear partial differential equations of the first order. In the previous chapter, we have discussed linear partial differential equations of the first order. Here, in this chapter, we shall discuss some standard methods for the solution of nonlinear partial differential equations of the first order. In general, a nonlinear partial differential equation of the first order is written as

$$f(x, y, z, p, q) = 0 \tag{1}$$

where f is nonlinear in p and q.

The solution of this equation can be categorized into the following three types:

(a) We have seen that $f(x, y, z, p, q) = 0$ can be derived by eliminating a and b from $f(x, y, z, a, b) = 0$, $\partial f/\partial a = 0$ and $\partial f/\partial b = 0$. Therefore, the solutions of Eq. (1) are the two parameter family of surfaces $f(x, y, z, a, b) = 0$. This solution is called the complete integral or complete solution.

(b) Under a given condition on the solution by choosing a particular case $a = g(b)$ or $b = g(a)$, the complete integral reduces to a one parameter family $f(x, y, z, a, g(a)) = 0$. Now we find the envelope of this equations by eliminating constant a from $f(x, y, a, g(a)) = 0$ and $\partial f/\partial a = 0$. The eliminate or the envelope is a surface $\phi(x, y, z) = 0$. This envelope is also a solution of Eq. (1). When $g(a)$ is arbitrary, this solution is called the general solution or general integral of Eq. (1). If a particular value is used for $g(a)$, then the solution is called a particular integral.

(c) We try to obtain the envelope of the two parameter family of surfaces defined by $f(x, y, z, a, b) = 0$ by eliminating the arbitrary constants a, b from the equation $f(x, y, z, a, b) = 0$ and the equations $\partial f/\partial a = 0$ and $\partial f/\partial b = 0$. If this envelope exists, then it is called the singular solution or singular integral of Eq. (1). Obviously, not all first order equations may have singular solutions.

Now, we shall discuss four standard types of nonlinear partial differential equations of first order.

Type I Equations containing *p* and *q* only.

Let the equation of this form be $f(p, q) = 0$. Complete integral is given by

$$z = ax + by + c$$

where, a and b are connected by the relation $f(a, b) = 0$.

Therefore, $b = \phi(a)$ and the complete integral becomes

$$z = ax + \phi(a)y + c.$$

which contains two arbitrary constants a and c, which is equal to the number of independent variables x and y.

General integral is obtained by eliminating a from

$$z = ax + \phi(a)y + \psi(a) \quad \text{where} \quad c = \psi(a)$$

and $0 = x + \phi'(a)y + \psi'(a)$ (obtained by differential partially with respect to 'a' the above equation).

The singular integral of $f(p, q) = 0$ is obtained by eliminating a and c between the complete integral $z = ax + \phi(a)y + c$ and the equations obtained by differentiating this partially with respect to a and c, that is, $0 = x + y\phi'(a)$ and $0 = 1$.

But the last equation $0 = 1$ is inadmissible. So, we conclude that the equations of standard form (I) given by $f(p, q) = 0$ have no singular integral or singular solution.

▲▲

EXAMPLE 3.1

Solve $pq = p + q$.

Solution: The given equation is of the from $f(p, q) = 0$.

$\therefore$ its complete integral is

$$z = ax + by + c$$

where $f(a, b) = 0 \quad \Rightarrow \quad ab = a + b \quad \Rightarrow \quad b = a/(a - 1)$

Thus, $z = ax + \left(\dfrac{a}{a-1}\right)y + c$

which contains two arbitrary constants a and c.

General integral. Let $c = \psi(a)$ in complete integral, we have

$$z = ax + \frac{a}{a-1}y + \psi(a) \tag{1}$$

Differentiating partially with respect to a, we get

$$0 = x - \frac{1}{(a-1)^2}y + \psi'(a) \tag{2}$$

On eliminating 'a' from Eqs. (1) and (2), we get general integral of the given equation. As discussed earlier, the given equation has no singular solution.

EXAMPLE 3.2

Solve $p^2q^2 = 1$.

Solution: The given equation is of the form $f(p, q) = 0$.
$\therefore$ its complete integral is

$$z = ax + by + c$$

where $f(a, b) = 0 = a^2b^2 - 1$

$$\therefore \qquad b = \pm\frac{1}{a}$$

Take $b = \frac{1}{a}$ $(a \neq 0)$.

$\therefore$ The complete integral is

$$z = ax + \frac{1}{a}y + c \ (a \neq 0).$$

General integral. Let $c = \psi(a)$.

$$\therefore \qquad z = ax + \frac{1}{a}y + \psi(a) \qquad (1)$$

Differentiating partially with respect to a, we get

$$a = x - \frac{1}{a^2}y + \psi'(a) \qquad (2)$$

Eliminating a from Eqs. (1) and (2), we get general integral of the given equation. As discussed earlier, the given equation has no singular integral.

EXAMPLE 3.3

Solve $p = 4q^2 + 2$.

Solution: The given equation is of the form $f(p, q) = 0$
$\therefore$ its complete integral is

$$z = ax + by + c$$

where $f(a, b) = 0 = a - 4b^2 - 2$

$$\therefore \qquad b = \pm\frac{\sqrt{a-2}}{2}$$

Let
$$b = \frac{\sqrt{a-2}}{2} \quad (a \geq 2)$$

Therefore, $z = ax + \frac{\sqrt{a-2}}{2}y + c$.

As discussed earlier, let $c = \psi(a)$

$$\therefore \qquad z = ax + \frac{\sqrt{a-2}}{2}y + \psi(a) \tag{1}$$

Differentiating Eq. (1) partially with respect to a, we get

$$0 = x + \frac{y}{4\sqrt{a-2}} + \psi'(a) \tag{2}$$

Eliminating a from Eqs. (1) and (2), we obtain general integral of the given equation. As discussed earlier, it has no singular integral.

EXAMPLE 3.4

Solve $x^2p^2 + y^2q^2 = z^2$.

Solution: The given equation can be written as

$$\left(\frac{x}{z}\frac{\partial z}{\partial x}\right)^2 + \left(\frac{y}{z}\frac{\partial z}{\partial y}\right)^2 = 1 \tag{1}$$

Let $x = e^X$, $y = e^Y$ and $z = e^Z$, we obtain

$$X = \log x,\ Y = \log y \quad \text{and} \quad Z = \log z$$

$$dX = \frac{1}{x}dx\ ,\ dY = \frac{1}{y}dy \quad \text{and} \quad dZ = \frac{1}{z}dz$$

$$\therefore \qquad \frac{\partial z}{\partial X} = \frac{x}{z}\frac{\partial z}{\partial x} \quad \text{and} \quad \frac{\partial Z}{\partial Y} = \frac{y}{z}\frac{\partial z}{\partial y}$$

$$\Rightarrow \qquad P = \frac{x}{z}p \quad \text{and} \quad Q = \frac{y}{z}q$$

So, Eq. (1) becomes

$$P^2 + Q^2 = 1$$

which is of the standard for $f(P, Q) = 0$.

$\therefore$ its solution is

$$Z = aX + bY + 0$$

where $$f(a, b) = 0 \Rightarrow a^2 + b^2 = 1$$

$$\Rightarrow \qquad b = \sqrt{1-a^2}$$

Thus, the complete integral of Eq. (1) is given by

$$Z = aX + \sqrt{1-a^2}\ Y + c$$

or $$\log z = a \log x + \sqrt{1-a^2}\ \log y + c.$$

Taking

$$a = \cos\alpha \quad \text{and} \quad c = \log c'$$

$$\log z = \cos\alpha \log x + \sin\alpha \log y + \log c'$$

or
$$z = c' x^{\cos\alpha} y^{\sin\alpha}$$

which contains two arbitrary constants α and c'.

General integral. Let $c' = \psi(\alpha)$

$\therefore$
$$z = \psi(a)\, x^{\cos\alpha} y^{\sin\alpha} \tag{2}$$

Differentiating partially with respect to α, we get

$$0 = \psi'(\alpha)\, x^{\cos\alpha} y^{\sin\alpha} + \psi(\alpha)\, x^{\cos\alpha} \log_e x\,(-\sin\alpha) y^{\sin\alpha} + \psi(\alpha)\, x^{\cos\alpha} y^{\sin\alpha} \log_e y(\cos\alpha) \tag{3}$$

The general integral of Eq. (1) is obtained by eliminating α between Eqs. (2) and (3).

Singular integral. Differentiating the complete integral

$$z = c'\, x^{\cos\alpha} y^{\sin\alpha} \tag{4}$$

with respect to α and c', we get

$$0 = c' x^{\cos\alpha} (\log_e \cos\alpha)(-\sin\alpha) y^{\sin\alpha} + c' x^{\cos\alpha} y^{\sin\alpha} (\log_e \sin\alpha)(\cos\alpha) \tag{5}$$

and
$$0 = x^{\cos\alpha} y^{\sin\alpha} \tag{6}$$

Eliminating α and c', between Eqs. (4), (5) and (6); the singular integral is $z = 0$, which satisfy the given Eq. (1).

EXAMPLE 3.5

Solve for complete integral, the equation $(x + y)(p + q)^2 + (x - y)(p - q)^2 = 1$.

Solution: Let $x + y = X^2$ and $x - y = Y^2$

$\therefore$
$$p = \frac{\partial z}{\partial x} = \frac{\partial z}{\partial X}\frac{\partial X}{\partial x} + \frac{\partial z}{\partial Y}\frac{\partial Y}{\partial x} = \frac{1}{2X}\frac{\partial z}{\partial X} + \frac{1}{2Y}\frac{\partial z}{\partial Y}$$

and
$$q = \frac{\partial z}{\partial y} = \frac{\partial z}{\partial Y}\frac{\partial Y}{\partial y} + \frac{\partial z}{\partial Y}\frac{\partial Y}{\partial y} = \frac{1}{2X}\frac{\partial z}{\partial X} - \frac{1}{2Y}\frac{\partial z}{\partial Y}$$

so that $p + q = \dfrac{1}{X}\dfrac{\partial z}{\partial X}$ and $p - q = \dfrac{1}{Y}\dfrac{\partial z}{\partial Y}$.

On substituting these, the given equation reduces to

$$\left(\frac{\partial z}{\partial X}\right)^2 + \left(\frac{\partial z}{\partial Y}\right)^2 = 0 \quad \text{or} \quad P^2 + Q^2 = 1$$

which is of the standard form $f(P, Q) = 0$.

∴ its solution is

$$z = aX + bY + c.$$

where $$f(a, b) = 0 = a^2 + b^2 \Rightarrow b = \sqrt{1-a^2}$$

∴ the complete integral is

$$z = aX + \sqrt{1-a^2}\, Y + c$$

or $$z = a\sqrt{x+y} + \sqrt{1-a^2}\,\sqrt{x-y} + c$$

which contains two arbitrary constants a and c and is the complete integral of the given equation.

EXAMPLE 3.6

Find the complete integral of $pq = x^m y^n z^{2l}$.

Solution: The equation can be written as

$$\frac{z^{-l}}{x^m}\frac{\partial z}{\partial x}\frac{z^{-l}}{y^n}\frac{\partial z}{\partial y} = 1$$

Let $$x^m dx = dX,\ y^n dy = dY \text{ and } z^l dz = dZ$$

so that $$X = \frac{x^{m+1}}{m+1}, Y = \frac{y^{n+1}}{n+1} \text{ and } Z = \frac{z^{-l+1}}{-l+1}$$

Therefore, we obtain

$$\frac{\partial Z}{\partial X}\frac{\partial Z}{\partial Y} = 1 \quad \text{or} \quad PQ = 1$$

which is of the standard form $f(P, Q) = 0$.
∴ its solution is

$$Z = aX + bY + c$$

where $f(a, b) = 0 = ab - 1 \Rightarrow b = \dfrac{1}{a}\ (a \neq 0)$.

∴ the complete integral is $Z = aX + \left(\dfrac{1}{a}\right)Y + c$

or $$\frac{z^{-l+1}}{-l+1} = a\frac{x^{m+1}}{m+1} + \frac{1}{a}\frac{y^{n+1}}{n+1} + c$$

which contains two arbitrary constants a and c.

EXAMPLE 3.7

Solve for complete integral the equation $(x^2 + y^2)(p^2 + q) = 1$.

Solution: Let $x = r\cos\theta$, $y = r\sin\theta$ so that $x^2 + y^2 = r^2$, $\theta = \tan^{-1}(y/x)$

$\therefore$
$$\frac{\partial r}{\partial x} = \frac{x}{r} = \cos\theta, \frac{\partial r}{\partial y} = \frac{y}{r} = \sin\theta$$

$$\frac{\partial \theta}{\partial x} = \frac{1}{1+(y/x)^2}\left(\frac{-y}{x^2}\right) = \frac{-y}{x^2+y^2} = -\frac{1}{r}\sin\theta$$

$$\frac{\partial \theta}{\partial y} = \frac{1}{1+(y/x)^2}\left(\frac{1}{x}\right) = \frac{x}{x^2+y^2} = \frac{1}{r}\cos\theta$$

$\therefore$
$$p = \frac{\partial z}{\partial x} = \frac{\partial z}{\partial r}\frac{\partial r}{\partial x} + \frac{\partial z}{\partial \theta}\frac{\partial \theta}{\partial x} = \frac{\partial z}{\partial r}\cos\theta - \frac{\sin\theta}{r}\frac{\partial z}{\partial \theta}$$

and
$$q = \frac{\partial z}{\partial y} = \frac{\partial z}{\partial r}\frac{\partial r}{\partial y} + \frac{\partial z}{\partial \theta}\frac{\partial \theta}{\partial y} = \frac{\partial z}{\partial r}\sin\theta + \frac{\cos\theta}{r}\frac{\partial z}{\partial \theta}.$$

Substituting in the given equation, we have

$$r^2\left\{\left(\frac{\partial z}{\partial r}\right)^2 + \frac{1}{r^2}\left(\frac{\partial z}{\partial \theta}\right)^2\right\} = 1$$

$\Rightarrow$
$$\left(r\frac{\partial z}{\partial r}\right)^2 + \left(\frac{\partial z}{\partial \theta}\right)^2 = 1$$

Again, let $\frac{1}{r}dr = dR$ so that $R = \log r$

$\therefore$
$$\left(\frac{\partial z}{R}\right)^2 + \left(\frac{\partial z}{\partial \theta}\right)^2 = 1 \text{ or } P^2 + Q^2 = 1$$

which is of the standard for $f(P, Q) = 0$.

$\therefore$ its solution is

$$z = aR + bQ + c$$

where $f(a, b) = 0 = a^2 + b^2 - 1 \Rightarrow b = \sqrt{1-a^2}$

$\therefore$
$$z = aR + \sqrt{1-a^2}\,\theta + c$$

or
$$z = a\log r + \sqrt{1-a^2}\tan^{-1}(y/x) + c$$

or
$$z = \frac{a}{2}\log(x^2 + y^2) + \sqrt{1-a^2}\tan^{-1}(y/x) + c$$

which contains two arbitrary constants a and c and is the complete integral of the given equation.

EXAMPLE 3.8

Find complete integral of $x^2p^2 + y^2q^2 = z$.

Solution: The given equation can be written as $\left(\frac{x}{\sqrt{z}}\frac{\partial z}{\partial x}\right)^2 + \left(\frac{y}{\sqrt{z}}\frac{\partial z}{\partial y}\right)^2 = 1$.

Let $$\frac{1}{\sqrt{z}}\,dz = dZ,\ \frac{1}{x}\,dx = dX \quad \text{and} \quad \frac{1}{y}\,dy = dY$$

so that $$2\sqrt{z} = Z,\ \log x = X \quad \text{and} \quad \log y = Y$$

$\therefore$ $$P = \frac{\partial z}{\partial X} = \frac{x}{\sqrt{x}}\frac{\partial z}{\partial x} \quad \text{and} \quad Q = \frac{\partial Z}{\partial Y} = \frac{y}{\sqrt{z}}\frac{\partial z}{\partial y}$$

$\therefore$ $P^2 + Q^2 = 1$ which is of the standard form $f(P, Q) = 0$ and its solution is

$$Z = aX + bY + c$$

where $f(a, b) = 0 = a^2 + b^2 = 1 \Rightarrow b = \sqrt{1-a^2}$.

$\therefore$ $$2\sqrt{z} = a \log x + \sqrt{1-a^2}\,\log y + c$$

which is complete integral of the given equation.

EXAMPLE 3.9

Find the complete integral of $(1 - x^2)yp^2 + x^2q = 0$.

Solution: The given equations can be written as $\left(\frac{\sqrt{1-x^2}}{x}\frac{\partial z}{\partial x}\right)^2 + \left(\frac{1}{y}\frac{\partial z}{\partial y}\right) = 0$.

Let $$\frac{x}{\sqrt{1-x^2}}\,dx = dX \quad \text{and} \quad ydy = dY$$

so that $$-\sqrt{1-x^2} = X \quad \text{and} \quad y^2/2 = Y$$

$\therefore$ $$\left(\frac{\partial z}{\partial X}\right)^2 + \left(\frac{\partial z}{\partial Y}\right) = 0 \quad \text{or} \quad P^2 + Q = 0.$$

which is of the standard for $f(P, Q) = 0$

$\therefore$ its solution is

$$z = aX + bY + c$$

where $$f(a, b) = 0 = a^2 + b \Rightarrow b = -a^2$$

$\therefore$ $$z = -a\sqrt{1-x^2} - \frac{a^2y^2}{2} + c$$

which is the required complete integral.

EXAMPLE 3.10

Find the complete integral of $(y - x)(qy - px) = (p - q)^2$.

Solution: Let $$X = x + y \text{ and } Y = xy$$

Now, $$p = \frac{\partial z}{\partial x} = \frac{\partial z}{\partial X}\frac{\partial X}{\partial x} + \frac{\partial z}{\partial Y}\frac{\partial Y}{\partial x} = \frac{\partial z}{\partial X} + y\frac{\partial z}{\partial Y}$$

and $$q = \frac{\partial z}{\partial y} = \frac{\partial z}{\partial X}\frac{\partial X}{\partial y} + \frac{\partial z}{\partial Y}\frac{\partial Y}{\partial y} = \frac{\partial z}{\partial X} + x\frac{\partial z}{\partial Y}$$

$\therefore$ $$(y - x)\left[y\left(\frac{\partial z}{\partial X} + x\frac{\partial z}{\partial Y}\right) - x\left(\frac{\partial z}{\partial X} + y\frac{\partial z}{\partial Y}\right)\right] = \left[(y - x)\frac{\partial z}{\partial Y}\right]^2$$

or $$\frac{\partial z}{\partial X} = \left(\frac{\partial z}{\partial Y}\right)^2 \quad \text{or} \quad P = Q^2$$

which is of the standard form $f(P, Q) = 0$

$\therefore$ its solution is

$$z = aX + bY + c$$

where $$f(a, b) = 0 = a - b^2 \Rightarrow a = b^2.$$

$\therefore$ the required complete integral is $z = b^2(x + y) + bxy + c$.

EXAMPLE 3.11

Find the complete integral of $yp + xq = pq$.

Solution: The given equation can be written as

$$\frac{1}{x}\frac{\partial z}{\partial x} + \frac{1}{y}\frac{\partial z}{\partial y} = \left(\frac{1}{x}\frac{\partial z}{\partial x}\right)\left(\frac{1}{y}\frac{\partial z}{\partial y}\right)$$

Let $$xdx = dX \text{ and } ydy = dY$$

so that $$x^2/2 = X \text{ and } y^2/2 = Y$$

$\therefore$ $$\frac{\partial z}{\partial X} + \frac{\partial z}{\partial Y} = \left(\frac{\partial z}{\partial X}\right)\left(\frac{\partial z}{\partial Y}\right) \quad \text{or} \quad P + Q = PQ$$

which is of the standard form $f(P, Q) = 0$.

∴ its solution is $$z = aX + bY + c$$

where, $$f(a, b) = 0 = a + b - ab \Rightarrow b = \frac{a}{a-1}$$

∴ the required complete integral is

$$z = a\left(\frac{x^2}{2}\right) + \frac{a}{a-1}\left(\frac{y^2}{2}\right) + c.$$

where a and c are arbitrary constants.

Type II Equations not containing x and y

Let the equation of this form be

$$f(p, q, z) = 0 \tag{1}$$

Assume $z = F(u)$ where, $u = x + ay$; a is an arbitrary constant, be a solution of Eq. (1).

then $$p = \frac{\partial z}{\partial x} = \frac{\partial z}{\partial u}\frac{\partial u}{\partial x} = \frac{\partial z}{\partial u}$$

and $$q = \frac{\partial z}{\partial y} = \frac{\partial z}{\partial u}\frac{\partial u}{\partial y} = a\frac{\partial z}{\partial u}$$

∴ Equation (1) reduces to the form

$$f\left(\frac{dz}{du}, a\frac{dz}{du}, z\right) = 0$$

which is an ordinary differential equation of the first order and can be integrated and thus complete integral will be known.

Put $b = \phi(a)$ in the complete integral

$$f(x, y, z, a, b) = 0$$

we get $$f(x, y, z, a, \phi(a)) = 0$$

∴ The general integral is given by eliminating a from

$$f(x, y, z, a, \phi(a)) = 0 \quad \text{and} \quad \partial f/\partial a = 0.$$

Singular integral is obtained by eliminating a and b from

$$f(x, y, z, a, b) = 0, \partial f/\partial a = 0 \quad \text{and} \quad \partial f/\partial b = 0$$

EXAMPLE 3.12

Solve $z = p^2 - q^2$.

Solution: The given partial differential equation is of the form $f(p, q, z) = 0$

Let $z = F(u)$ where $u = x + ay$; a is an arbitrary constant.

∴ $$p = \frac{\partial z}{\partial x} = \frac{\partial z}{\partial x}\frac{\partial x}{\partial x} = \frac{dz}{du} \quad \text{and} \quad q = \frac{\partial z}{\partial y} = \frac{\partial z}{\partial u}\frac{\partial u}{\partial y} = a\frac{dz}{du}$$

Substituting in the given equation, we have

$$z = (1 - a^2)\left(\frac{dz}{du}\right)^2 \quad \text{or} \quad \sqrt{z} = \sqrt{1-a^2}\,\frac{dz}{du}$$

or $$\sqrt{1-a^2}\, z^{-1/2} dz = du$$

Integrating, we get

$$\sqrt{1-a^2}\; 2\sqrt{z} = u + b$$

or $$4(1 - a^2)z = (x + ay + b)^2 \tag{1}$$

which contains two arbitrary constants a and b and is the complete integral of the given differential equation.

Let $$b = \phi(a)$$

$\therefore$ $$4(1 - a^2)z = (x + ay + \phi(a))^2 \tag{2}$$

Differentiating partially with respect to a, we get

$$-8az = 2(x + ay + \phi(a))\,(y + \phi'(a)) \tag{3}$$

The general integral can be obtained by eliminating a between Eqs. (2) and (3).
Differentiating the complete integral Eq. (1), partially with respect to a and b, we get

$$8az = 2(x + ay + b)y \tag{4}$$

and $$0 = 2(x + ay + b) \tag{5}$$

Eliminating a and b between Eqs. (1), (4) and (5), the singular integral of given equation is $z = 0$; as it satisfies the given differential equation $\{\because z = 0 \Rightarrow p = 0; q = 0\}$.

EXAMPLE 3.13

Solve $z^2(1 + p^2 + q^2) = k^2$.

Solution: The given equation is of the form $f(p, q, z) = 0$.

Let $$z = F(u) \text{ and } u = x + ay$$

so that $$p = \frac{dz}{du} \text{ and } q = a\frac{dz}{du}$$

$\therefore$ $$z^2\left[1+\left(\frac{dz}{du}\right)^2 + a^2\left(\frac{dz}{du}\right)^2\right] = k^2$$

$\Rightarrow$ $$(1 + a^2)\left(\frac{dz}{du}\right)^2 = \frac{k^2 - z^2}{z^2}$$

$\Rightarrow$ $$\frac{dz}{du} = \pm\frac{\sqrt{k^2 - z^2}}{\sqrt{1+a^2}\,z}$$

$$\Rightarrow \qquad \pm\sqrt{1+a^2}\,\frac{z}{\sqrt{k^2-z^2}}\,dz = du$$

Integrating, we get

$$\pm\sqrt{1+a^2}\times\sqrt{k^2-z^2} = u + b$$

$$\Rightarrow \qquad \mp\sqrt{1+a^2}\,\sqrt{k^2-z^2} = x + ay + b$$

$$\Rightarrow \qquad (1 + a^2)\,(k^2 - z^2) = (x + ay + b)^2 \qquad (1)$$

which is the required complete integral

Let $\qquad b = \phi(a)$

$$\therefore \qquad (1 + a^2)\,(k^2 - z^2) = (x + ay + \phi(a))^2 \qquad (2)$$

Differentiating partially with respect to a, we get

$$2a(k^2 - z^2) = 2(x + ay + \phi(a))\,(y + \phi'(a)) \qquad (3)$$

The general integral is obtained by eliminating a from Eqs. (2) and (3).
Differentiating Eq. (1) partially with respect to a and b respectively, we get

$$2a(k^2 - z^2) = 2(x + ay + b)y$$

and $\qquad 0 = 2(x + ay + b)$

On eliminating a and b, we get

$$z = \pm k$$

which is the required singular integral.

EXAMPLE 3.14

Solve $p(p + q) = 8z$.

Solution: The given equation is of the from $f(p, q, z) = 0$.
Let $z = F(u)$ and $u = x + ay$ so that

$$p = \frac{dz}{du} \quad \text{and} \quad q = a\frac{dz}{du}$$

$$\therefore \qquad \frac{dz}{du}\left(\frac{dz}{du} + a\frac{dz}{du}\right) = 8z$$

$$\Rightarrow \qquad \left(\frac{dz}{du}\right)^2 (1+a) = 8z$$

$$\Rightarrow \qquad \sqrt{1+a}\,\frac{dz}{du} = \pm 2\sqrt{2}$$

$$\Rightarrow \qquad \pm\sqrt{1+a}\,\frac{dz}{2\sqrt{z}} = \sqrt{2}du$$

Integrating, we have

$$\pm\sqrt{1+a}\,\sqrt{z}=\sqrt{2}\,(u+b)$$

$$\Rightarrow \qquad (1+a)z = 2(u+b)^2 = 2(x+ay+b)^2 \tag{1}$$

which is the required complete integral.

Let $$b = \phi(a)$$

$$\therefore \qquad (1+a)z = 2(x+ay+\phi(a))^2 \tag{2}$$

Differentiating partially with respect to a, we get

$$z = 4(x+ay+\phi(a))\,(y+\phi'(a)) \tag{3}$$

On eliminating a from Eqs. (2) and (3), we get the general integral.
Differentiating Eq. (1) with respect to a and b respectively we get

$$z = 4(x+ay+b)y$$

and $$0 = 4(x+ay+b)$$

Eliminating a and b, we get

$$z = 0$$

which is the required singular integral.

EXAMPLE 3.15

Find a complete integral of $pq = x^m y^n z^l$.

Solution: The given equation can be written as

$$\left(\frac{1}{x^m}\frac{\partial z}{\partial x}\right)\left(\frac{1}{y^n}\frac{\partial z}{\partial y}\right) = z^l$$

Let $$x^m dx = dX \quad \text{and} \quad y^n dy = dY$$

so that $$\frac{x^{m+1}}{m+1} = X \quad \text{and} \quad \frac{y^{n+1}}{n+1} = Y$$

$$\therefore \qquad \frac{\partial z}{\partial X}\frac{\partial z}{\partial Y} = z^l \quad \text{or} \quad PQ = z^l$$

which is of the form $f(P, Q, Z) = 0$.

Let $$z = F(u) \text{ and } u = X + aY$$

so that $$P = \frac{dz}{du} \text{ and } Q = a\frac{dz}{du}$$

$$\therefore \qquad \left(\frac{dz}{du}\right)\left(a\frac{dz}{du}\right) = z^l \text{ or } z^{-l/2}dz = \frac{du}{\sqrt{a}}$$

Integrating, we have $\dfrac{z^{-l/2+1}}{-l/2+1} = \dfrac{u}{\sqrt{a}} + b$

or $$\frac{z^{-l/2+1}}{-l/2+1} = \frac{1}{\sqrt{a}}\left(\frac{x^{m+1}}{m+1} + a\frac{y^{n+1}}{n+1}\right) + b$$

which is the required complete integral with a and b as arbitrary constants.

EXAMPLE 3.16

Solve $q^2y^2 = z(z - px)$.

Solution: The given equation can be written as

$$\left(y\frac{\partial z}{\partial y}\right)^2 = z\left(z - x\frac{\partial z}{\partial x}\right)$$

Let $$\frac{1}{x}\,dx = dX \quad \text{and} \quad \frac{1}{y}\,dy = dY$$

so that $$\log x = X \quad \text{and} \quad \log y = Y$$

$\therefore$ $$\left(\frac{\partial z}{\partial Y}\right)^2 = z\left(z - \frac{\partial z}{\partial X}\right) \quad \text{or} \quad Q^2 = z(z - P)$$

which is of the form $f(P, Q, z) = 0$.

Let $$z = f(u) \quad \text{where} \quad u = X + aY$$

so that $$P = \frac{\partial z}{\partial X} = \frac{dz}{du} \text{ and } Q = \frac{\partial z}{\partial y} = a\frac{dz}{du}$$

$\therefore$ $$a^2\left(\frac{dz}{du}\right)^2 = z\left(z - \frac{dz}{du}\right)$$

or $$a^2\left(\frac{dz}{du}\right)^2 = z\frac{dz}{du} - z^2 = 0$$

$\therefore$ $$\frac{dz}{du} = \frac{z}{2a^2}\left(-1 \pm \sqrt{1+4a^2}\right)$$

or $$2a^2\frac{dz}{z} = \left(-1 \pm \sqrt{1+4a^2}\right)du$$

Integrating, we have

$$2a^2 \log z = \left(-1 \pm \sqrt{1+4a^2}\right)(u + \log b)$$

or $$2a^2 \log z = \left(-1 \pm \sqrt{1+4a^2}\right)(\log x + a \log y + \log b)$$

or $$z^{2a^2/\left(-1\pm\sqrt{1+4a^2}\right)} = bxy^a$$

which is the required complete integral.

EXAMPLE 3.17

Find complete integral of $z^2(p^2z^2 + q^2) = 1$. Find also the singular solution if it exists.

Solution: The given differential equation is of the form $f(p, q, z) = 0$.

Let $z = F(u)$ where $u = x + ay$

$\therefore$ $$p = \frac{\partial z}{\partial x} = \frac{dz}{du} \text{ and } q = \frac{\partial z}{\partial y} = a\frac{dz}{du}$$

Thus, $$z^2(z^2 + a^2)\left(\frac{dz}{du}\right)^2 = 1$$

or $$z\sqrt{z^2 + a^2}\, dz = du$$

Integrating, we have

$$\frac{1}{3}(z^2 + a^2)^{3/2} = u + b \quad \text{or} \quad (z^2 + a^2)^3 = 9(x + ay + b)^2.$$

which is the required complete integral with a and b being the arbitrary constants.

Differentiating it partially with respect to a and b, we get

$$6a(z^2 + a^2)^2 = 18y(x + ay + b)$$

and $$0 = 18(x + ay + b) \quad \Rightarrow \quad a = 0 \text{ so } z = 0$$

But when $z = 0$, $p = 0 = q$ which does not satisfy the given differential equation, so $z = 0$ is not the singular integral. Hence, the problem does not have singular solution.

EXAMPLE 3.18

Find the complete integral of $p^2x^2 = z(z - qy)$.

Solution: The given equation can be written as

$$\left(x\frac{\partial z}{\partial x}\right)^2 = z\left(z - y\frac{\partial z}{\partial y}\right)$$

Let $$\frac{1}{x}dx = dX \quad \text{and} \quad \frac{1}{y}dy = dY$$

so that $$\log x = X \text{ and } \log y = Y$$

$\therefore$ $$\left(\frac{\partial z}{\partial X}\right)^2 = z\left(z - \frac{\partial z}{\partial Y}\right) \quad \text{or } P^2 = z(Z - Q)$$

which is of the form $f(P, Q, Z) = 0$.

Let $$z = F(u) \quad \text{where} \quad u = X + aY$$

so that $$P = \frac{\partial z}{\partial X} = \frac{\partial z}{\partial u} \quad \text{and} \quad Q = \frac{\partial z}{\partial Y} = a\frac{\partial z}{\partial u}$$

$\therefore$ $$\left(\frac{dz}{du}\right)^2 = z\left(z - a\frac{dz}{du}\right)$$

or $$\left(\frac{dz}{du}\right)^2 + az\frac{dz}{du} - z^2 = 0$$

$\Rightarrow$ $$\frac{dz}{du} = \frac{-az \pm \sqrt{a^2z^2 + 4z^2}}{2a}$$

$\Rightarrow$ $$\frac{dz}{du} = \left(\frac{-a \pm \sqrt{a^2 + 4}}{2a}\right)z$$

$\Rightarrow$ $$\frac{dz}{du} = kz \text{ or } \frac{dz}{z} = kdu$$

Integrating, we have

$$\frac{1}{k}\log z = u + \log b \quad \Rightarrow \quad z^{1/k} = xby^a$$

which is the required complete integral with a and b as arbitrary constants.

EXAMPLE 3.19

Solve $z^2(p^2 + q^2 + 1) = 1$.

Solution: The given equation is of the from $f(p, q, z) = 0$

Let $$z = F(u) \quad \text{where} \quad u = x + ay$$

so that $$p = \frac{\partial z}{\partial x} = \frac{dz}{du} \quad \text{and} \quad q = \frac{\partial z}{\partial y} = a\frac{dz}{du}$$

$\therefore$ $$z^2\left[(1+a^2)\left(\frac{dz}{du}\right)^2 + 1\right] = 1$$

$\Rightarrow$ $$(1 + a^2)\left(\frac{dz}{du}\right)^2 = \frac{1-z^2}{z^2}$$

$\Rightarrow$ $$\frac{z}{\sqrt{1-z^2}}dz = \frac{du}{\sqrt{1+a^2}}$$

$\Rightarrow$ $$\sqrt{1+a^2}\,\frac{zdz}{\sqrt{1-z^2}} = du$$

Integrating, we have

$$\sqrt{1+a^2}\,\frac{(1-z^2)^{1/2}}{-1} = u + b \quad \text{or} \quad (1 + a^2)(1 - z^2) = (x + ay + b)^2$$

which is the required complete integral with a and b as arbitrary constants.

EXAMPLE 3.20

Solve $p^3 + q^3 = 3pqz$ $(z > 0)$.

Solution: The given equation is of the form $f(p, q, z) = 0$.

Let $\quad z = F(u) \quad \text{where} \quad u = x + ay$

so that $\quad p = \dfrac{\partial z}{\partial x} = \dfrac{dz}{du} \quad \text{and} \quad q = \dfrac{\partial z}{\partial y} = a\dfrac{dz}{du}$

$$\therefore \quad (1+a^3)\left(\frac{dz}{du}\right)^3 = 3az\left(\frac{dz}{du}\right)^2$$

$$\Rightarrow \quad (1 + a^3)\frac{dz}{du} = 3az$$

$$\Rightarrow \quad (1 + a^3)\frac{dz}{z} = 3adu$$

Integrating, we have

$$(1 + a^3)\log z = 3au + b$$

$$\Rightarrow \quad (1 + a^3)\log z = 3a(x + ay) + b.$$

which is the complete integral with a and b as arbitrary constants.

Differentiating it partially with respect to a and b, we get

$$3a^2 \log z = 3(x + 2ay)$$

and $\quad 0 = 1$ which is absurd.

Hence, singular integral does not exist for the given equation.

EXAMPLE 3.21

Solve $z^2(p^2 + q^2 + 2) = 1$.

Solution: The given equation is of the form $f(p, q, z) = 0$.

Let $\quad z = F(u) \quad \text{where} \quad u = x + ay$

so that $\quad p = \dfrac{\partial z}{\partial x} = \dfrac{dz}{du} \quad \text{and} \quad q = \dfrac{\partial z}{\partial y} = a\dfrac{dz}{du}$

$$\therefore \qquad z^2\left\{(1+a^2)\left(\frac{dz}{du}\right)^2+2\right\}=1$$

$$\Rightarrow \qquad (1+a^2)\left(\frac{dz}{dy}\right)^2 = \frac{1-2z^2}{z^2}$$

$$\Rightarrow \qquad \sqrt{1+a^2}\,\frac{z}{\sqrt{1-2z^2}}\,dz = du$$

Integrating, we have

$$\sqrt{1+a^2}\,\frac{(1-2z^2)^{1/2}}{-2} = u + b$$

$$\Rightarrow \qquad (1 + a^2)(1 - 2z^2) = 4(x + ay + b)^2$$

which is the required complete integral with a and b as arbitrary constants.

Differentiating it partially with respect to a and b, we have

$$2a(1 - 2z^2) = 8(x + ay + b)y$$

and

$$0 = 8(x + ay + b)$$

$$\therefore \qquad a = 0$$

Thus,

$$1 - 2z^2 = 0$$

which is the required singular integral.

EXAMPLE 3.22

Solve $p^2 + q^2 = 4z$.

Solution: The given equation is of the from $f(p, q, z) = 0$.

Let $\qquad z = F(u) \quad \text{where} \quad u = x + ay$

so that $\qquad p = \dfrac{\partial z}{\partial x} = \dfrac{dz}{du} \quad \text{and} \quad q = \dfrac{\partial z}{\partial y} = a\dfrac{dz}{du}$

$$\therefore \qquad (1+a^2)\left(\frac{dz}{du}\right)^2 = 4z \;\Rightarrow\; \sqrt{1+a^2}\,\frac{dz}{\sqrt{z}} = 2du.$$

$$\Rightarrow \qquad \frac{\sqrt{1+a^2}}{2}\,\frac{dz}{\sqrt{z}} = du$$

Integrating, we have $\qquad \dfrac{\sqrt{1+a^2}}{2}\,\dfrac{z^{1/2}}{\frac{1}{2}} = u + b$

$$\Rightarrow \qquad (1 + a^2)z = (x + ay + b)^2$$

which is the required complete integral with a and b as arbitrary constants.

Differentiating it with respect to a and b partially, we have

$$2az = 2(x + ay + b)y \text{ and } 0 = 2(x + ay + b)$$

$$\therefore \quad a = 0$$

Thus, $z = 0$ is the required singular integral.

Type III Equations of the type $f(x, p) = g(y, q)$.

In this type of equations z is absent and the terms containing x and p can be separated from the terms containing y and q.

Let
$$f(x, p) = a = g(y, q) \tag{1}$$

where, a is an arbitrary constant.

Solving Eq. (1) for p and q, we get

$$p = F(x, a) \text{ and } q = G(y, a)$$

$$\therefore \quad dz = F(x, a)dx + G(y, a)dy$$

Integrating, we obtain

$$z = \int F(x, a)dx + \int G(y, a)dy + b \tag{2}$$

which contains two arbitrary constants a and b and is the complete integral of the given equation.

The general integral can be obtained as in other cases.

Differentiating Eq. (2) partially with respect to b, we get $0 = 1$, which is inadmissible, so there is no singular integral in this case.

EXAMPLE 3.23

Solve $py + qx + pq = 0$.

Solution: The given equation can be written as

$$p(y + q) = -qx \text{ or } \frac{p}{x} = -\frac{q}{(y+q)}$$

which is of the from $f(x, p) = g(y, q)$.

Let
$$\frac{p}{x} = a = -\frac{q}{y+q}$$

$$\Rightarrow \quad p = ax \text{ and } q = -\frac{ay}{(1+a)}$$

Thus, $dz = pdx + qdy = axdx - \dfrac{ay}{1+a}dy$.

Integrating, we have

$$z=\frac{1}{2}ax^2-\frac{a}{1+a}\frac{1}{2}y^2+b$$

which is the required complete integral with a and b as arbitrary constants.

Let $$b = \phi(a)$$

$$\therefore \quad z = \frac{1}{2}ax^2 - \frac{1}{2}\frac{a}{1+a}y^2 + \phi(a)$$

Differentiating partially with respect to a, we get

$$0=\frac{1}{2}x^2-\frac{1}{2}\frac{1}{(1+a)^2}y^2+\phi'(a)$$

By eliminating a from these two equations, we get general integral.

For singular integral differentiate $z=\frac{1}{2}ax^2-\frac{1}{2}\frac{a}{1+a}y^2+b$ partially with respect to a and b, we have

$$0=\frac{1}{2}x^2-\frac{1}{2}\frac{1}{(1+a^2)}y^2$$

and $0 = 1$, which is inadmissible.

Hence, the given equation has no singular integral.

EXAMPLE 3.24

Solve $q = xyp^2$.

Solution: The given equation can be written as $xp^2 = q/y$ which is of the from $f(x, p) = g(y, q)$.

Let $xp^2 = a = q/y$.

$$\Rightarrow \quad p=\sqrt{a/x}, \; q = ay$$

where a is an arbitrary constants.

$$\therefore \quad dz = pdx + qdy = \sqrt{\frac{a}{x}}\,dx + aydy$$

Integrating, we have

$$z=2\sqrt{ax}+\frac{1}{2}ay^2+b$$

which is the complete integral with a and b arbitrary constants.

Let $$b = \phi(a)$$

$$\therefore \quad z = 2\sqrt{ax} + \frac{1}{2}ay^2 + \phi(a)$$

Differentiate partially with respect to a, we get

$$0 = \sqrt{x/a} + \frac{1}{2}y^2 + \phi'(a)$$

Elimination of a from these two equations, will give us general integral.

Differentiate $z = 2\sqrt{ax} + \frac{1}{2}ay^2 + b$ partially with respect to a and b, we have

$$0 = \sqrt{x/a} + \frac{1}{2}y^2 \quad \text{and} \quad 0 = 1, \text{ which is inadmissible.}$$

Hence, there is no singular integral of the given equation.

EXAMPLE 3.25

Find a complete integral of $yp = 2yx + \log q$.

Solution: The given equation can be written as

$$p - 2x = \frac{\log q}{y}$$

which is of the form $f(x, p) = g(y, q)$.

Let $$p - 2x = a = \frac{\log q}{y}$$

$$\therefore \quad p = 2x + a \quad \text{and} \quad q = e^{ay}$$

Thus, $$dz = pdx + qdy = (2x + a)dx + e^{ay}dy$$

Integrating, we have

$$z = x^2 + ax + \frac{e^{ay}}{a} + b.$$

which is the required complete integral with a and b as arbitrary constants.

EXAMPLE 3.26

Find a complete integral of $z(p^2 - q^2) = x - y$.

Solution: The given equation can be written as

$$\left(\sqrt{z}\frac{\partial z}{\partial x}\right)^2 - \left(\sqrt{z}\frac{\partial z}{\partial y}\right)^2 = x - y$$

Let $$\sqrt{z}\,dz = dZ \text{ so that } \frac{2}{3}z^{3/2} = Z$$

$$\therefore \quad \left(\frac{\partial z}{\partial x}\right)^2 - \left(\frac{\partial Z}{\partial y}\right)^2 = x - y \quad \text{or} \quad P^2 - Q^2 = x - y$$

or $$P^2 - x = Q^2 - y$$

which is of the form $f(x, P) = g(y, Q)$.

Let $P^2 - x = a = Q^2 - y$ where a is an arbitrary constant.

$$\Rightarrow \quad P = \sqrt{x+a} \quad \text{and} \quad Q = \sqrt{y+a}$$

$$\therefore \quad dZ = Pdx + Qdy = \sqrt{x+a}\,dx + \sqrt{y+a}\,dy$$

Integrating, we have

$$Z = \frac{2}{3}(x + a)^{3/2} + \frac{2}{3}(y + a)^{3/2} + b$$

or $$z^{3/2} = (x + a)^{3/2} + (y + a)^{3/2} + c$$

where $c = 3b/2$ is an arbitrary constant which is the complete integral of the given differential equation.

EXAMPLE 3.27

Find a complete integral of $z^2(p^2 + q^2) = x^2 + y^2$.

Solution: The given equation can be written as

$$\left(z\frac{\partial z}{\partial x}\right)^2 + \left(z\frac{\partial z}{\partial y}\right)^2 = x^2 + y^2$$

Let $$zdx = dZ \text{ so that } \frac{1}{2}z^2 = Z$$

$$\therefore \quad \left(\frac{\partial Z}{\partial x}\right)^2 + \left(\frac{\partial Z}{\partial y}\right)^2 = x^2 + y^2$$

or $$P^2 + Q^2 = x^2 + y^2 \text{ or } P^2 - x^2 = y^2 - Q^2.$$

which is of the form $f(x, p) = g(y, q)$

$$\Rightarrow \quad P^2 - x^2 = a = y^2 - Q^2$$

$$\Rightarrow \quad P = \sqrt{x^2 + a} \quad \text{and} \quad Q = \sqrt{y^2 - a}$$

$\therefore$ $$dZ = Pdx + Qdy = \sqrt{x^2+a}\,dx + \sqrt{y^2-a}\,dy$$

Integrating, we have

$$Z = \frac{1}{2}x\sqrt{x^2+a} + \frac{a}{2}\log\left(x+\sqrt{x^2+a}\right) + \frac{1}{2}y\sqrt{y^2-a} - \frac{a}{2}\log\left(y+\sqrt{y^2-a}\right)$$

$\Rightarrow$ $$z^2 = x\sqrt{x^2+a} + y\sqrt{y^2-a} + a\left[\log\left(x+\sqrt{x^2+a}\right) - \log\left(y+\sqrt{y^2-a}\right)\right] + c$$

where $c = 26$ is an arbitrary constant.

EXAMPLE 3.28

Find a complete integral of $z(xp - yq) = y^2 - x^2$.

Solution: The given equation can be written as $x\left(z\frac{\partial z}{\partial x}\right) - y\left(z\frac{\partial z}{\partial y}\right) = y^2 - x^2$.

Let $$zdz = dZ \text{ so that } z^2/2 = Z$$

$\therefore$ $$x\left(\frac{\partial z}{\partial x}\right) - y\left(z\frac{\partial z}{\partial y}\right) = y^2 - x^2$$

or $$xP - yQ = y^2 - x^2 \text{ or } xP + x^2 = yQ + y^2.$$

which is of the form $f(x, p) = g(y, Q)$.

$\therefore$ $$xP + x^2 = a = yQ + y^2$$

$\therefore$ $$P = \frac{a}{x} - x \quad \text{and} \quad Q = \frac{a}{y} - y$$

$\therefore$ $$dZ = Pdx + Qdy = \left(\frac{a}{x} - x\right)dx + \left(\frac{a}{y} - y\right)dy$$

Integrating, we have

$$Z = a\log x - \frac{x^2}{2} + a\log y - \frac{y^2}{2} + b$$

or $$z^2 = 2a\log(xy) - x^2 - y^2 + b$$

where a and b are arbitrary constants.

EXAMPLE 3.29

Find a complete integral of $\frac{p^2}{x} - \frac{q^2}{y} = \frac{1}{z}\left(\frac{1}{x} + \frac{1}{y}\right)$.

Solution: The given equation can be written as

$$\frac{1}{x}\left(\sqrt{z}\frac{\partial z}{\partial x}\right)^2 - \frac{1}{y}\left(\sqrt{z}\frac{\partial z}{\partial y}\right)^2 = \frac{1}{x}+\frac{1}{y}$$

Let $\sqrt{z}\,dz = dZ$ so that $\frac{2}{3}z^{3/2} = Z$

$$\therefore \quad \frac{1}{x}\left(\frac{\partial Z}{\partial x}\right)^2 - \frac{1}{y}\left(\frac{\partial Z}{\partial y}\right)^2 = \frac{1}{x}+\frac{1}{y}$$

or $$\frac{P^2}{x} - \frac{Q^2}{y} = \frac{1}{x}+\frac{1}{y}$$

or $$\frac{P^2-1}{x} = \frac{Q^2+1}{y}$$

which is of the form $f(x, P) = g(y, Q)$.

$$\therefore \quad \frac{P^2-1}{x} = a = \frac{Q^2+1}{y}$$

$$\Rightarrow \quad P = \sqrt{1+ax} \quad \text{and} \quad Q = \sqrt{ay-1}$$

$$\therefore \quad dZ = Pdx + Qdy = \sqrt{1+ax}\,dx + \sqrt{ay-1}\,dy$$

Integrating, we have

$$Z = \frac{2}{3}\frac{(1+ax)^{3/2}}{a} + \frac{2}{3}\frac{(ay-1)^{3/2}}{a} + b$$

$$\Rightarrow \quad az^{3/2} = (1+ax)^{3/2} + (ay-1)^{3/2} + b$$

which is the required complete integral with a and b as arbitrary constants.

EXAMPLE 3.30

Find the complete integral of $2x(z^2q^2 + 1) = pz$.

Solution: The given equation can be written as

$$2x\left[\left(z\frac{\partial z}{\partial y}\right)^2 + 1\right] = z\frac{\partial z}{\partial x}$$

Let $zdx = dZ$ so that $z^{2/2} = Z$

$$\therefore \quad 2x\left[\left(\frac{\partial Z}{\partial y}\right)^2 + 1\right] = \frac{\partial Z}{\partial x}$$

or $$2x[Q^2 + 1] = P \text{ or } P/2x = 1 = Q^2$$

which is of the from $f(x, P) = g(y, Q)$.

$\therefore$ $$\frac{P}{2x} - 1 = a = Q^2 \Rightarrow P = 2x(1 + a) \quad \text{and} \quad Q = \sqrt{a}$$

Thus, $$dZ = Pdx + Qdy = 2x(1 + a)dx + \sqrt{a}dy$$

Integrating, we have

$$Z = x^2(1 + a) + \sqrt{ay} + b$$

or $$z^{2/2} = x^2(1 + a) + \sqrt{a}\, y + b$$

where a and b are arbitrary constants.

EXAMPLE 3.31

Solve $x^2p^2 = q^2y$.

Solution: The given equation is of the form $f(x, p) = g(y, q)$

$\therefore$ $$x^2p^2 = a = q^2y$$

$\Rightarrow$ $$p = \sqrt{a}/x \quad \text{and} \quad q = \sqrt{a/y}$$

Therefore, $dz = pdx + qdy = \dfrac{\sqrt{a}}{x}dx + \sqrt{\dfrac{a}{y}}\, dy$.

Integrating, we have

$$z = \sqrt{a}\log x + 2\sqrt{ay} + b$$

which is the complete integral with a and b as arbitrary constants.

Differentiate this with respect to a and b partially, we get

$$0 = \frac{1}{2\sqrt{a}}\log x + \frac{\sqrt{y}}{a}$$

and $0 = 1$, which is inadmissible.

So singular integral does not exist for the given differential equation.

EXAMPLE 3.32

Solve for complete integral $pe^y = qe^x$.

Solution: The given equation can be written as $pe^{-x} = qe^{-y}$ which is of the form $f(x, p) = g(y, q)$

$\therefore$ $$pe^{-x} = a = qe^{-y}$$

$\Rightarrow$ $$p = ae^x \text{ and } q = ae^y$$

Thus, $$dz = pdx + qdy = ae^x\, dx + ae^y\, dy$$

Integrating, we have

$$z = a(e^x + e^y) + b$$

which is the required complete integral.

EXAMPLE 3.33

Solve $\sqrt{p}+\sqrt{q}=2x$.

Solution: The given equation can be written as

$$\sqrt{p}-2x=-\sqrt{q}$$

which is of the form $f(x, p) = g(y, q)$.

Let $$\sqrt{p}-2x=a=-\sqrt{q}$$

$\Rightarrow$ $$p = (a + 2x)^2 \text{ and } q = a^2$$

Thus, $dz = pdx + qdy = (a + 2x)^2\, dx + a^2\, dy$.
Integrating, we have

$$z = \frac{(a+2x)^3}{b}+a^2y+b$$

which is the complete integral of the given equation. Further, it can be easily shown that a singular integral does not exist for the given differential equation.

Type IV Equations of the type $z = px + qy + f(p, q)$.

This type of equation may be considered analogous to Clairaut's form $y = px + f(p)$ where, $p=\frac{dy}{dx}$ in ordinary differential equations.

The complete integral of the equation in this form is given by

$$z = ax + by + f(a, b)$$

General integral is obtained as in other cases.
Differentiating the complete integral partially with respect to a and b, we get

$$0 = a + \frac{\partial f}{\partial a} \quad \text{and} \quad 0 = b + \frac{\partial f}{\partial b}$$

Singular integral is obtained by eliminating a and b between

$$z = ax + by + f(a, b),\ 0 = a + \frac{\partial f}{\partial a} \text{ and } 0 = b + \frac{\partial f}{\partial b}.$$

EXAMPLE 3.34

Solve $z = px + qy + \log(pq)$.

Solution: The given differential equation is of Clairout's form.
$\therefore$ its complete integral is

$$z = ax + by + \log(ab)$$

where a and b are arbitrary constants.

Let $\quad b = \phi(a)$

$$\therefore \quad z = ax + y\phi(a) + \log(a\phi(a))$$

$$\Rightarrow \quad z = ax + y\phi(a) + \log a + \log \phi(a).$$

Differentiating partially with respect to a, we get

$$0 = x + y\phi'(a) + \frac{1}{a} + \frac{\phi'(a)}{\phi(a)}$$

The general integral is obtained by eliminating a from these two equations.
Differentiating $z = ax + by + \log(ab)$ partially with respect to a and b, we have

$$0 = x + \frac{1}{a}$$

and

$$0 = y + \frac{1}{b}.$$

$$\Rightarrow \quad a = -\frac{1}{x} \text{ and } b = -\frac{1}{y}$$

$$\therefore \quad z = \left(-\frac{1}{x}\right)x + \left(-\frac{1}{y}\right)y + \log\left(-\frac{1}{x} - \frac{-1}{y}\right)$$

or

$$z = -2 - \log xy$$

which satisfy the given equation and hence, it is the singular integral.

EXAMPLE 3.35

Solve $z = px + qy + c\sqrt{1+p^2+q^2}$.

Solution: The given differential equation is of Clairout's form.

$\therefore$ Its complete integral is

$$z = ax + by + c\sqrt{1+a^2+b^2}$$

where a and b are arbitrary constants.

Let $\quad b = \phi(a)$

$$\therefore \quad z = ax + y\phi(a) + c\sqrt{1+a^2+\phi^2(a)}$$

Differentiating partially with respect to a, we get

$$0 = x + y\phi'(a) + \frac{c(a+\phi(a)\phi'(a))}{\sqrt{1+a^2+\phi^2(a)}}.$$

General integral is obtained by eliminating a between these two equations.

Differentiate $z = ax + by + c\sqrt{1+a^2+b^2}$ partially with respect to a and b, we have

$$0 = x + ca/\sqrt{1+a^2+b^2}$$

and $$0 = y + cb/\sqrt{1+a^2+b^2}$$

$\therefore$ $$x^2 + y^2 = c^2(a^2 + b^2)/(1 + a^2 + b^2)$$

$\Rightarrow$ $$x^2 + y^2 - c^2 = -c^2/(1 + a^2 + b^2)$$

$\Rightarrow$ $$1 + a^2 + b^2 = c^2/(c^2 - x^2 - y^2)$$

Thus, $$a = -\frac{x\sqrt{1+a^2+b^2}}{c} = \frac{-x}{\sqrt{c^2-x^2-y^2}}$$

and $$b = -\frac{y\sqrt{1+a^2+b^2}}{c} = \frac{-y}{\sqrt{c^2-x^2-y^2}}$$

$\therefore$ $$z = \frac{-x^2-y^2}{\sqrt{(c-x^2-y^2)}} + \frac{c^2}{\sqrt{c^2-x^2-y^2}} = \sqrt{c^2-x^2-y^2}$$

or $$x^2 + y^2 + z^2 = c^2.$$

which satisfy the given equation and hence, it is the singular integral of the given differential equation.

EXAMPLE 3.36

Find the singular integral of the partial differential equation

$$z = px + qy + 4(pq)^{1/4}$$

Solution: The given equation is of Clairout's form.

$\therefore$ Its complete integral is

$$z = ax + by + 4(ab)^{1/4}$$

Differentiate it partially with respect to a and b, we get

$$0 = x + (ab)^{-3/4}(b)$$

$\Rightarrow$ $$x = -a^{-3/4}b^{1/4}$$

and $$0 = y + (ab)^{-3/4}\,(a)$$

$\Rightarrow$ $$y = -a^{1/4}\, b^{-3/4}$$

$\therefore$ $$xy = (ab)^{-1/2}$$

$\Rightarrow$ $$ab = \frac{1}{x^2y^2}$$

Thus, $$a = -\frac{1}{x^{3/2}y^{1/2}} \quad \text{and} \quad b = -\frac{1}{x^{1/2}y^{3/2}}$$

$\therefore$ $$z = -\frac{1}{x^{1/2}y^{1/2}} + -\frac{1}{x^{1/2}y^{1/2}} + 4\frac{1}{x^{1/2}y^{1/2}}$$

$\Rightarrow$ $$z = -\frac{2}{x^{1/2}y^{1/2}} \Rightarrow z^2 - \frac{4}{xy}$$

$\Rightarrow$ $$xyz^2 - 4 = 0$$

is the required singular integral of the given differential equation.

EXAMPLE 3.37

Find a complete and singular integral of $4xyz = pq + 2px^2y + 2qxy^2$.

Solution: The given equation can be written as

$$z = \left(\frac{1}{2x}\frac{\partial z}{\partial x}\right)\left(\frac{1}{2y}\frac{\partial z}{\partial y}\right) + \left(\frac{1}{2x}\frac{\partial z}{\partial x}\right)x^2 + \left(\frac{1}{2y}\frac{\partial z}{\partial y}\right)y^2$$

Let $2xdx = dX$ and $2ydy = dY$; so that

$x^2 = X$ and $y^2 = Y$

$\therefore$ $$z = \left(\frac{\partial z}{\partial X}\right)\left(\frac{\partial z}{\partial Y}\right) + \left(\frac{\partial z}{\partial X}\right)X + \left(\frac{\partial z}{\partial y}\right)Y$$

or $$z = PX + QY + PQ$$

which is of Clairout's form. So, its complete integral is

$$z = aX + bY + ab$$

or $$z = ax^2 + by^2 + ab$$

where a and b are arbitrary constants.

Differentiating partially with respect to a and b, we get

$$0 = x^2 + b \text{ and } 0 = y^2 + a$$

$\Rightarrow$ $$a = -y^2 \text{ and } b = -x^2$$

$$\therefore \qquad z = -y^2x^2 - x^2y^2 + x^2y^2$$

or

$$z = -x^2y^2$$

which is the required singular solution.

EXAMPLE 3.38

Prove that the complete integral of $z = px + qy - 2p - 3q$ represents all possible planes through the point (2, 3, 0). Also, find the envelope of all planes represented by the complete integral.

Solution: The given equation is of Clairout's form.

$\therefore$ Its complete integral is

$$z = ax + by - 2a - 3b \tag{1}$$

where a and b are arbitrary constants. Since it is linear in x, y, z, so it represents a plane. Also, the point (2, 3, 0) satisfy it. Hence, the complete integral of the given equation represents all possible planes through the point (2, 3, 0).

Differentiating it with respect to a and b partially, we get

$$0 = x - 2 \text{ and } 0 = y - 3$$

$$\Rightarrow \qquad x = 2 \text{ and } y = 3$$

$$\therefore \qquad z = 0$$

which satisfy the given equation and hence, it is the envelope of all planes represented by the complete Eq. (1).

EXAMPLE 3.39

Show that the complete solution of $z = px + qy + \sqrt{1+p^2+q^2}$ represents a family of planes, which are at a distance one unit from origin.

Solution: The given equation is of Clairout's form.

$\therefore$ Its complete integral is

$$z = ax + by + z = \sqrt{1+a^2+b^2} \tag{1}$$

where a and b are arbitrary constants.

$$\Rightarrow \qquad -ax - by + z = \sqrt{1+a^2+b^2} \tag{2}$$

which is of type $Ax + By + Cz = D$.

$\therefore$ It represents family of planes, whose normals have direction ratios $-a, -b, 1$. Here the

$$\text{perpendicular distance from (0, 0, 0) to plane} = \frac{\left|0+0+0-\sqrt{1+a^2+b^2}\right|}{\sqrt{(-a)^2+(-b)^2+(1)^2}} = 1.$$

So, Eq. (1) represents family of planes which are at a distance one unit from origin.

EXAMPLE 3.40

Show that the complete solution of $z = px + qy + \dfrac{pq}{pq - p - q}$ represents a family of planes whose algebraic sum of intercepts made on co-ordinate axes is one.

Solution: The given equation is of Clairout's form.

$\therefore$ Its complete integral is

$$z = ax + by + \frac{ab}{ab - a - b}$$

where a and b are arbitrary constants.

$$\Rightarrow \qquad -ax - by + z = \frac{ab}{ab - a - b}$$

which represents family of planes.

$$\therefore \qquad \frac{\dfrac{-ax}{ab}}{\dfrac{ab}{ab-a-b}} + \frac{\dfrac{-by}{ab}}{\dfrac{ab}{ab-a-b}} + \frac{\dfrac{z}{ab}}{\dfrac{ab}{ab-a-b}} = 1$$

$$\Rightarrow \qquad \frac{x}{\dfrac{-b}{ab-a-b}} + \frac{y}{\dfrac{-a}{ab-a-b}} + \frac{z}{\dfrac{ab}{ab-a-b}} = 1.$$

which is the equation of plane in intercept form.

$\therefore$ Intercepts made on co-ordinate axes are $\dfrac{b}{ab-a-b}, \dfrac{a}{ab-a-b}, \dfrac{ab}{ab-a-b}$, respectively.

So their algebraic sum

$$= \frac{-b}{ab-a-b} + \frac{-a}{ab-a-b} + \frac{ab}{ab-a-b} = 1.$$

Hence, the result.

EXAMPLE 3.41

Solve $z = px + qy + 2\sqrt{pq}$.

Solution: The given equation is of Clairout's form.

So its complete integral is

$$z = ax + by + 2\sqrt{ab}$$

Differentiate it partially with respect to a and b, we get

$$0 = x + \sqrt{\frac{b}{a}} \text{ and } 0 = y + \sqrt{\frac{a}{b}}$$

$$\Rightarrow \qquad x = -\sqrt{\frac{b}{a}} \text{ and } y = -\sqrt{\frac{a}{b}}.$$

Now $$z - x = (ax + by + 2\sqrt{ab}) - x = \sqrt{\frac{b}{a}}$$

and $$z - y = (ax + by + 2\sqrt{ab}) - y = \sqrt{\frac{a}{b}}$$

∴ $$(z - x)(z - y) = 1.$$

which is the required singular solution.

EXAMPLE 3.42

Find the complete integral of $p^2x + q^2y = (z - 2px - 2qy)^2$.

Solution: Taking positive root, the given equation reduces to

$$z = 2px + 2qy + (p^2x + q^2y)^{1/2}$$

or $$z = \sqrt{x}\left(2\sqrt{x}\frac{\partial z}{\partial x}\right) + \sqrt{y}\left(2\sqrt{y}\frac{\partial x}{\partial y}\right) + \frac{1}{2}\left[\left(2\sqrt{x}\frac{\partial z}{\partial x}\right)^2 + \left(2\sqrt{y}\frac{\partial z}{\partial y}\right)^2\right]^{1/2}$$

Let $\frac{1}{2\sqrt{x}}dx = dX$ and $\frac{1}{2\sqrt{y}}dy = dY$ so that

$$\sqrt{x} = X \text{ and } \sqrt{y} = Y$$

∴ $$z = X\frac{\partial z}{\partial X} + Y\frac{\partial y}{\partial Y} + \frac{1}{2}\left[\left(\frac{\partial z}{\partial X}\right)^2 + \left(\frac{\partial z}{\partial Y}\right)^2\right]^{1/2}$$

or $$z = XP + YQ + \frac{1}{2}[P^2 + Q^2]^{1/2}$$

which is of Clairout's form and its complete integral is

$$z = Xa + Yb + \frac{1}{2}[a^2 + b^2]^{1/2}$$

or $$z = a\sqrt{x} + b\sqrt{y} + \frac{1}{2}[a^2 + b^2]^{1/2}.$$

EXAMPLE 3.43

Find the complete and singular integral of $z = px + qy + p^2q^2$.

Solution: The given equation is of Clairout's form. So, its complete integral is

$$z = ax + by + a^2b^2$$

where a, b are arbitrary constants.

Differentiate it partially with respect to a and b, we get

$$0 = x + 2ab^2 \text{ and } 0 = y + 2a^2b$$

$\therefore$ $$a = -(y^2/2x)^{1/3} \text{ and } b = -(x^2/2y)^{1/3}$$

Thus $$z = -x(y^2/2x)^{1/3} - y(x^2/2y)^{1/3} + (x^2y^2/16)^{1/3}$$

or $$z = -\frac{3}{4}xy^{1/3}\,x^{2/3}\,y^{2/3}$$

which is the required singular solution.

EXAMPLE 3.44

Find the complete integral and singular integral of the differential equation.

$$z = px + qy + p^2 - q^2$$

Solution: The given equation is of Clairout's form. So, its complete integral is given by

$$z = ax + by + a^2 - b_2 \text{ (}a, b \text{ being arbitrary constants).}$$

Differentiate it partially with respect to a and b, we get

$$0 = x + 2a \text{ and } 0 = y - 2b$$

$\therefore$ $$a = -x/2 \text{ and } b = y/2$$

Thus, $$z = \left(-\frac{x}{2}\right)x + \left(\frac{y}{2}\right)y + \frac{x^2}{4} - \frac{y^2}{4}$$

or $$z = -\frac{x^2}{4} + \frac{y^2}{4}$$

or $$4z = y^2 - x^2$$

which is the required singular solution.

EXERCISE 3.1

Find the complete integral of the following partial differential equations:

1. $p = e^q$.
2. $q = pq + p^2$.
3. $xy^2p = z(y^2 + z^2\,q^2)$.
4. $p^m \sec^{2m} x + z^l q^n \operatorname{cosec} 2^n y = z^{lm/(m-n)}$.
5. $z^2p^2y + 6zpxy + 2zqx^2 + 4x^2y = 0$.
6. $(y - x)(qy - px) = (p - q)^2$.
7. $p(1 + q^2) = q(z - a)$.
8. $p^2z^2 + q^2 = 1$.
9. $p^2z^2 = z(z - qy)$.

10. $x(1 + y)p = y(1 + x)q$.
11. $p^2q^2 + x^2y^2 = x^2q^2 (x^2 + y^2)$.
12. $pqz = p^2 (xq + p^2) + q^2(yp + q^2)$.
13. $(p + q) (z - px - qy) = 1$.
14. $yp = 2yx \log q$.
15. $z = px + qy + p^m q^n$.
16. $p + q = \sin x + \sin y$.
17. $pq = xy$.
18. $(y - x) (qy - px) = (p - q)^2$.
19. $(x + y) (p + q)^2 + (x - y) (p - q)^2$. {*Hint:* let $x + y = X^2$ and $x - y = Y^2$}.
20. $z(p^2 - q^2) = x - y$.
21. $(x^2 + y^2) (p^2 + q^2) = 1$.
22. $z = px + qy + \sin(pq)$.
23. $z = px + qy + \log(p^3 + q^3)$.
24. $q^2y^2 = z(z - px)$.
25. $p^2x^2 + q^2y^2 = z$.

Find the complete and singular solutions of the following partial differential equations:

26. $p^3 + q^3 = 2 > z$.
27. $z^2(p^2 + q^2 + 1) = c^2$.
28. $p = 4q^2 + 2$.
29. $z = px + qy + p^2 + pq + q^2$.
30. $z^2(p^2 + q^2 + 4) = 1$.
31. $\sqrt{p} + \sqrt{q} = 5y$.
32. $py + qx + pq = 0$.
33. $z = px + qy + p/q$.
34. $z = px + qy + (\alpha p^2 + \beta q^2 + 1)^{1/2}$.
35. $z^2(p^2 + q^2 + 2) = 1$.
36. $q(p - \cos x) = \cos y$.
37. $x^2p^2 = q^2y$.
38. $p - q = x^2 + y^2$.
39. $p^2 - x = q^2 - y$.
40. $px + q = p^2$.
41. $p^2 + 5p + 3q + 4 = 0$.
42. $p^2q^3 = 1$.
43. $p^2 = zq$.
44. $z = p^2 + q^2$.

45. $4(1 + z^3) = 9z^4pq$.
46. $q^2 = z^2p^2(1 - p^2)$.
47. $z^2(p^2z^2 + q^2) = 1$.
48. Show that the complete solution of $z = px + qy - 5p + 4q$ represents all planes through $(5, -4, 0)$. Also, find the envelope of all planes represented by complete solution.
49. Find the complete integral, general integral and singular integral of $pq = 4xy$. Show that the equation is satisfied by $z = 2x + y + c$; c being an arbitrary constant. What is thc character of this integral?
50. Find the complete integral of $z = p^2 - q^2$.
51. Find the complete integral of the partial differential equation

$$2p^2q^2 + 3x^2y^2 = 8x^2q^2(x^2 + y^2).$$

52. Find the complete integral of the partial differential equation

$$p^2q^2 + x^2y^2 = x^2q^2(x^2 + y^2).$$

ANSWERS

Exercise 3.1

1. $z = ax + y \log a + c$.
2. $z = ax + \dfrac{a^2}{1-a} y + c$.
3. $z^2 = ax^2 + y^2\sqrt{a-1} + c$.
4. $\dfrac{m-n}{m-n-l} z^{(m-n-l)/(m-n)} = \dfrac{1}{2}a\left(x+\dfrac{1}{2}\sin 2x\right)+\dfrac{1}{2}(1-a^m)^{1/n}\left(y-\dfrac{1}{2}\sin 2y\right)+c$.
5. $z^2 = ax^2 - [(a^2 + 6a + 4)/2]y^2 + c$.
6. $z = a^2(x + y) + axy + b$.
7. $4a(z - a) = 4 + (x + ay + b)^2$.
8. $z\sqrt{z^2+a^2} + a^2 \log [z+\sqrt{z^2+a^2}] = 2(x + ay + b)$.
9. $x^by^a = z^{1/k}$ where $k = (-1 \pm \sqrt{a^2+4})/2a^2$.
10. $z = a(\log xy + x + y) + b$.
11. $z = \dfrac{2}{3}(x^2 + a^2)^{3/2} + (y^2 - a^2)^{1/2} + b$.
12. $abz = a^2(xb + a^2) + b^2(ya + b^2)$.
13. $z = ax + by + 1/(a + b)$.

14. $z = x^2 + ax + \frac{1}{a} e^{ay} + b.$

15. $z = ax + by + a^m b^n.$

16. $z = a(x - y) - (\cos x + \cos y) + b.$

17. $2z = ax^2 + \frac{y^2}{a} + b.$

18. $z = a(x + y) + \sqrt{a}\, xy + b.$

19. $z = a\sqrt{x+y} + \sqrt{(1-a^2)(x-y)} + b.$

20. $z^{3/2} = (x + a)^{3/2} + (y + a)^{3/2} + b.$

21. $z = \frac{a}{2} \log (x^2 + y^2) \pm \sqrt{1-a^2} \tan^{-1} \frac{y}{x} + b.$

22. $z = ax + by + \sin(ab).$

23. $z = ax + by + \log (a^3 + b^3).$

24. $z^k = bxy^a$ where $k = 2a^2 / (-1 \pm \sqrt{1+4a^2})$.

25. $x^a y^{\sqrt{1-a^2}} = b e^{2\sqrt{z}}.$

26. $(z + a^3)z^2 = 8(x + ay + b)^3; z = 0.$

27. $(1 + a^2)(c^2 - z^2) = (x + ay + b)^2 ; z^2 = c^2.$

28. $z = ax + \frac{\sqrt{a-2}}{2} y + b$; no singular solution.

29. $z = ax + by + a^2 + ab + b^2; 3z + x^2 + y^2 - xy = 0.$

30. $(1 + a^2)(1 - 4z^2) = (4x + 4ay + b)^2; 1 - 4z^2 = 0.$

31. $z = a^2x + \frac{1}{15} (5y - a)^3 + b$; no singular solution.

32. $2z = ax^2 - \frac{a}{a+1} y^2 + b$; no singular solution.

33. $z = ax + by + a/b; xz + 4 = 0.$

34. $z = ax + by + (\alpha a^2 + \beta b^2 + 1)^{1/2}; \frac{x^2}{\alpha} + \frac{y^2}{\beta} + z^2 = 1.$

35. $(1 + a^2)(1 - 2z^2) = 4(x + ay + b)^2; 2z^2 - 1 = 0.$

36. $z = ax + \sin x + \frac{1}{a} \sin y + b$; no singular solution.

37. $z=\sqrt{a}\log x+2\sqrt{ay}+b$; no singular solution.

38. $z = (x^3 - y^3)/3 + a(x + y) + b$; no singular solution.

39. $3z = 2(x + a)^{3/2} + 2(y + a)^{3/2} + b$; no singular solution.

40. $z = \frac{1}{4}[x^2 + x(x^2 + 4a)^{1/2}] + a \log[x + (x^2 + 4a)^{1/2} + ay$; no singular solution.

41. $z = ax - \frac{a^2+5a+4}{3}y + b$; no singular solution.

42. $z = ax + \frac{1}{a^{2/3}}y + b$; no singular solution.

43. $z = be^{a(x + ay)}$; no singular solution.

44. $4(1 + a^2)z = (x + ay + b)^2$; $z = 0$.

45. $a(1 + z^3) = (x + ay + b)^2$; $1 + z^3 = 0$.

46. $(x + ay + b)^2 = z^2 - a^2$; $z - 0$.

47. $9(x + ay + b)^2 = (z^2 + a^2)^3$; no singular solution.

48. $z = 0$.

49. $z = ax^2 + \frac{1}{a}y^2 + b$; $z = ax^2 + \frac{1}{a}y^2 + \phi(a)$ and $0 = x^2 - \frac{1}{a^2}y^2 + \phi'(a)$; no singular integral; particular case of the complete integral.

50. $2z^{1/2} = ax \pm (a^2 - 1)^{1/2}y + c$.

51. $z = \frac{2}{3}(x^2 + a^2)^{3/2} + \left(\frac{3}{2}\right)^{1/2}\left(\frac{1}{2}\right)(y^2 - a^2)^{1/2} + b$.

52. $z = a(x^2 - 1)^{1/2} + \frac{a^2y^2}{2} + b$.

▲▲

4

CHAPTER

General Methods of Solution of Nonlinear Partial Differential Equations of the First Order

In this chapter, we will discuss the general methods for the solution of nonlinear partial differential equations. For nonlinear partial differential equations of any degree but with only two independent variables Charpit's method will be used and for equations involving more than two independent variables Jacobs's method is given.

Charpit's Method

As discussed earlier, Charpit's method is the general method for solving nonlinear partial differential equations of order one but of any degree with two independent variables. Let the given equation be

$$f(x, y, z, p, q) = 0 \tag{1}$$

Our purpose is to find a second partial differential equation

$$\phi(x, y, z, p, q, a) = 0 \tag{2}$$

with a as arbitrary constant such that Eqs. (1) and (2) can be solved to obtain

$$p = p(x, y, z, a) \text{ and } q = q(x, y, z, a)$$

and $\quad dz = pdx + qdy$ becomes integrable. (3)

On integration Eq. (3), we will get complete integral of Eq. (1). The general and singular integrals can be obtained in the usual manner.

To obtain relation (2), partially differentiate Eqs. (1) and (2) with respect to x and y to get

$$\frac{\partial f}{\partial x} + \frac{\partial f}{\partial z}p + \frac{\partial f}{\partial p}\frac{\partial p}{\partial x} + \frac{\partial f}{\partial q}\frac{\partial q}{\partial x} = 0 \tag{4}$$

$$\frac{\partial \phi}{\partial x} + \frac{\partial \phi}{\partial z}p + \frac{\partial \phi}{\partial p}\frac{\partial p}{\partial x} + \frac{\partial \phi}{\partial q}\frac{\partial q}{\partial x} = 0 \tag{5}$$

$$\frac{\partial f}{\partial y}+\frac{\partial f}{\partial z}q+\frac{\partial f}{\partial p}\frac{\partial p}{\partial y}+\frac{\partial f}{\partial q}\frac{\partial q}{\partial y}=0 \tag{6}$$

$$\frac{\partial \phi}{\partial y}+\frac{\partial \phi}{\partial z}q+\frac{\partial \phi}{\partial p}\frac{\partial p}{\partial y}+\frac{\partial \phi}{\partial q}\frac{\partial q}{\partial y}=0 \tag{7}$$

Eliminating $\frac{\partial p}{\partial x}$ from Eqs. (4) and (5), we have

$$\left(\frac{\partial f}{\partial x}+\frac{\partial f}{\partial z}p+\frac{\partial f}{\partial q}\frac{\partial q}{\partial x}\right)\frac{\partial \phi}{\partial p}-\left(\frac{\partial \phi}{\partial x}+\frac{\partial \phi}{\partial z}p+\frac{\partial \phi}{\partial q}\frac{\partial q}{\partial x}\right)\frac{\partial f}{\partial p}=0$$

or
$$\left(\frac{\partial f}{\partial x}\frac{\partial \phi}{\partial p}-\frac{\partial \phi}{\partial x}\frac{\partial f}{\partial p}\right)+\left(\frac{\partial f}{\partial z}\frac{\partial \phi}{\partial p}-\frac{\partial \phi}{\partial z}\frac{\partial f}{\partial p}\right)p+\left(\frac{\partial f}{\partial q}\frac{\partial \phi}{\partial p}-\frac{\partial \phi}{\partial q}\frac{\partial f}{\partial p}\right)\frac{\partial q}{\partial x}=0 \tag{8}$$

Similarly, eliminating $\frac{\partial q}{\partial y}$ from Eqs. (6) and (7), we have

$$\left(\frac{\partial f}{\partial y}\frac{\partial \phi}{\partial q}-\frac{\partial \phi}{\partial y}\frac{\partial f}{\partial q}\right)+\left(\frac{\partial f}{\partial z}\frac{\partial \phi}{\partial q}-\frac{\partial \phi}{\partial z}\frac{\partial f}{\partial q}\right)q+\left(\frac{\partial f}{\partial p}\frac{\partial \phi}{\partial q}-\frac{\partial \phi}{\partial p}\frac{\partial f}{\partial q}\right)\frac{\partial p}{\partial y}=0 \tag{9}$$

Since
$$\frac{\partial q}{\partial x}=\frac{\partial^2 z}{\partial x\partial y}=\frac{\partial^2 z}{\partial y\partial x}=\frac{\partial p}{\partial y}$$

Adding Eqs. (8) and (9), we have

$$\left(\frac{\partial f}{\partial x}+p\frac{\partial f}{\partial z}\right)\frac{\partial \phi}{\partial p}+\left(\frac{\partial f}{\partial y}+q\frac{\partial f}{\partial z}\right)\frac{\partial \phi}{\partial q}+\left(-p\frac{\partial f}{\partial p}-q\frac{\partial f}{\partial q}\right)\frac{\partial \phi}{\partial z}+\left(-\frac{\partial f}{\partial p}\right)\frac{\partial \phi}{\partial x}+\left(-\frac{\partial f}{\partial q}\right)\frac{\partial \phi}{\partial y}=0 \tag{10}$$

Equation (10) is Lagrange's equation with x, y, z, p, q as independent variables and ϕ as the dependent variables. Its solution will depend on the auxiliary equations.

$$\frac{dp}{\frac{\partial f}{\partial x}+p\frac{\partial f}{\partial z}}=\frac{dq}{\frac{\partial f}{\partial y}+q\frac{\partial f}{\partial z}}=\frac{dz}{-p\frac{\partial f}{\partial p}-q\frac{\partial f}{\partial q}}=\frac{dx}{-\frac{\partial f}{\partial p}}=\frac{dy}{-\frac{\partial f}{\partial q}}=\frac{d\phi}{0} \tag{11}$$

The equations given by Eq. (11) are known as Charpit's auxiliary equations. Any integral of Eq. (11) will satisfy Eq. (10). Now, we find a solution of Eq. (11) in which either p or q occur for $\phi \equiv 0$. Then from $f \equiv 0$ given by Eq. (1) and $\phi \equiv 0$; find the values of p and q and substitute in $dz = pdx + qdy$. Integrating this, we obtain the complete integral of Eq. (1).

Working Method

To find the complete integral of $f(x, y, x, p, q) = 0$ by Charpit's method, proceed as follows.

- Write the given differential equation in the from

$$f(x, y, z, p, q) = 0 \tag{1}$$

if not given.

- Write the Charpit's auxiliary equations

$$\frac{dp}{\frac{\partial f}{\partial x}+p\frac{\partial f}{\partial z}}=\frac{dq}{\frac{\partial f}{\partial y}+q\frac{\partial f}{\partial z}}=\frac{dz}{-p\frac{\partial f}{\partial p}-q\frac{\partial f}{\partial q}}=\frac{dx}{\frac{-\partial f}{\partial p}}=\frac{dy}{\frac{-\partial f}{\partial q}} \tag{2}$$

- Select any two factors in Eq. (2), which on integration give the simplest relation involving at least one of p and q.

  Let this relation be

$$\phi(x, y, z, p, q, a) = 0 \tag{3}$$

where a is an arbitrary constant.

- Solve Eqs. (1) and (3) to get the values of p and q in terms of x, y, and z.
- Substitute the values of p and q in $dz = pdx + qdy$.
- Integrate it to get the complete integral of the given partial differential Eq. (1). The general and singular integrals can be obtained in the usual manner.

▲▲

EXAMPLE 4.1

Solve $p = (qy + z)^2$

Solution: Let $f(x, y, z, p, q) = p - (qy + z)^2$.

The Charpit's auxiliary equations are

$$\frac{dp}{\frac{\partial f}{\partial x}+p\frac{\partial f}{\partial z}}=\frac{dq}{\frac{\partial f}{\partial y}+q\frac{\partial f}{\partial z}}=\frac{dz}{-p\frac{\partial f}{\partial p}-q\frac{\partial f}{\partial q}}=\frac{dx}{\frac{-\partial f}{\partial p}}=\frac{dy}{\frac{-\partial f}{\partial q}}$$

or

$$\frac{dp}{-2p(qy+z)}=\frac{dq}{-2q(qy+z)}=\frac{dz}{-p(1)-q(-2y)(qy+z)}=\frac{dx}{-1}=\frac{dy}{2y(qy+z)}$$

Taking the first and fifth members, we have

$$\frac{dp}{p}=\frac{dy}{-y}$$

Integrating, we obtain $py = a$ or $p = a/y$.

Using Eq. (1), we get $\frac{a}{y} = (qy + z)^2$

$$\Rightarrow \qquad q=\frac{\sqrt{a}}{y^{3/2}}-\frac{z}{y}$$

Hence,

$$dz = pdx + qdy = \frac{a}{y}dx+\left(\frac{\sqrt{a}}{y^{3/2}}-\frac{z}{y}\right)dy$$

or $$ydz + zdy = adx + \frac{\sqrt{a}}{\sqrt{y}}dy$$

Integrating, we have

$$yz = ax + 2\sqrt{ay} + b$$

which is the complete integral.

Differentiate it partially with respect to a and b; we get

$0 = x + \sqrt{\frac{y}{a}}$ and $0 = 1$, which is inadmissible, hence there is no singular integral.

EXAMPLE 4.2

Solve $z = p(x + p) + q(y + q)$ by Charpit's method.

Solution: Let $f(x, y, z, p, q) = z - p(x + p) - q(y + q) = 0$.

The Charpit's auxiliary equations are

$$\frac{dp}{-p+p} = \frac{dq}{-q+q} = \frac{dz}{-p(-x-2p)-q(-y-2q)} = \frac{dx}{x+2p} = \frac{dy}{y+2q}$$

$\therefore$ $$dp = 0 = dq.$$

Integrating, we have $p = a$ and $q = b$

$\therefore$ $$z = ax + by + a^2 + b^2$$

which is the required complete integral.

Differentiate it partially with respect to a and b; we have

$$o = x + 2a \text{ and } o = y + 2b$$

$\therefore$ $$a = -x/2 \text{ and } b = -y/2$$

Hence, $$z = \left(-\frac{x}{2}\right)x + \left(-\frac{y}{2}\right)y + \frac{x^2}{4} + \frac{y^2}{4} = \frac{-x^2 - y^2}{4}$$

or $$4z + x^2 + y^2 = 0$$

which is the required singular integral.

EXAMPLE 4.3

Solve $px + qy = pq$.

Solution: Let $f(x, y, z, p, q) = px + qy - pq = 0$

The Charpit's auxiliary equations are

$$\frac{dp}{p+0} = \frac{dq}{q+0} = \frac{dz}{-p(x-q)-q(y-p)} = \frac{dx}{-(x-q)} = \frac{dy}{-(y-p)}$$

Taking the first two members, we have

$$\frac{dp}{p} = \frac{dq}{q}$$

Integrating, we obtain $p = aq$

From the given equation, we get

$$aqx + qy = aq^2$$

$\therefore$ $$q = \frac{y+ax}{a}$$

and $$p = aq = y + ax.$$

Hence, $$dz = pdx + qdy = (y + ax)dx + \frac{y+ax}{a}dy$$

or $$adz = (y + ax)\,(dy + adx).$$

Integrating, we have

$$az = \frac{1}{2}(y + ax)^2 + b$$

which is the complete integral.

Differentiate it with respect to a and b; we get

$$z = x(y + ax) \text{ and } 0 = 1;$$

which is inadmissible. Hence, these is no singular integral.

EXAMPLE 4.4

Solve $(p^2 + q^2)y = qz$.

Solution: Let $f(x, y, z, p, q) = (p^2 + q^2)y - qz = 0$

The Charpit's auxiliary equations are

$$\frac{dp}{-pq} = \frac{dq}{p^2+q^2-q^2} = \frac{dz}{-2p^2y-2q^2y+qz} = \frac{dx}{-2py} = \frac{dy}{-2qy+z}$$

Taking the first two members, we get

$$\frac{dp}{-pq} = \frac{dq}{p^2} \text{ or } pdp + qdq = 0$$

$\therefore$ $$p^2 + q^2 = a$$

From the given equation $ay = qz$ or $q \neq z = \dfrac{ay}{z}$

and $$p=\sqrt{a-q^2}=\sqrt{a-\frac{a^2y^2}{z^2}}=\sqrt{\frac{a}{z}}\sqrt{z^2-ay^2}$$

Hence, $$dz = pdx + qdy = \frac{\sqrt{a}}{z}\sqrt{z^2-ay^2}\,dx+\frac{ay}{z}dy$$

or $$\frac{zdz-aydy}{\sqrt{z^2-ay^2}}=\sqrt{a}\,dx$$

Integrating, we obtain $(z^2 - ay^2)^{1/2} = \sqrt{a}\,x + b$ or $z^2 - ay^2 = (\sqrt{a}\,x+b)^2$ which is the required complete integral (a, b are arbitrary constants).

For simplification, let $a = c^2$

$\therefore$ $$z^2 - c^2y^2 = (cx + b)^2$$

Differentiate it with respect to c and b, we get $0 = 2cy^2 + 2(cx + b)x$

and $$0 = 2(cx + b)$$

$\therefore$ $y = 0$ and hence, $z = 0$ is the singular integral.

EXAMPLE 4.5

Solve $2(z + px + qy) = yp^2$.

Solution: Let $f(x, y, z, p, q) = 2(z + px + qy) - yp^2 = 0$.

The Charpit's auxiliary equations are

$$\frac{dp}{2p+2p}=\frac{dq}{2q-p^2+2q}=\frac{dz}{-p(2x-2yp)-q(2y)}=\frac{dx}{-(2x-2yp)}=\frac{dy}{-2y}$$

Taking the first and last members, we have

$$\frac{dp}{4p}=\frac{dy}{-2y} \text{ or } \frac{dp}{p}+2\frac{dy}{y}=0$$

Integrating, we get $\log p + 2\log y = \log a$ or $py^2 = a$

$\therefore$ $$p = \frac{a}{y^2} \text{ and } q=-\frac{z}{y}-\frac{ax}{y^3}+\frac{a^2}{2y^4}$$

Hence, $$dz = pdx + qdy = \frac{a}{y^2}dx+\left[-\frac{z}{y}+\frac{ax}{y^3}+\frac{a^2}{2y^4}\right]=0$$

or $$ydz + zdy - a\left(\frac{ydx-xdy}{y^2}\right)-\frac{a^2}{2y^3}dy=0$$

Integrating, we have

$$yz - a\left(\frac{x}{y}\right) + \frac{a^2}{4y^2} = b$$

which is the complete integral with a, b as arbitrary constants.

Differentiate it with respect to a and b, we get, $\left(\frac{x}{y}\right) + \frac{a}{2y^2} = 0$ and $0 = 1$ which is absurd and hence, there is no singular solution of the given equation.

EXAMPLE 4.6

Solve $yzp^2 = q$.

Solution: Let $\qquad f(x, y, z, p, q) = yzp^2 - q = 0.$

The Charpit's auxiliary equations are

$$\frac{dp}{0 + p(yp^2)} = \frac{dq}{zp^2 + q(yp^2)} = \frac{dz}{-p(2yzp) - q(-1)} = \frac{dx}{-2yzp} = \frac{dy}{+1}$$

Taking the first and last members, we get

$$\frac{dp}{yp^3} = \frac{dy}{1} \text{ or } \frac{dp}{p^3} = ydy$$

Integrating, we have $-\frac{1}{p^2} + a = y^2$

or
$$p^2 = \frac{1}{a - y^2}$$

From the given equation $\quad q = yzp^2 = \frac{yz}{a - y^2}$

Hence,
$$dz = pdx + qdy = \frac{1}{\sqrt{a - y^2}}dx + \frac{yz}{a - y^2}dy$$

$$\Rightarrow \qquad \sqrt{a - y^2}\, dz - \frac{yz}{\sqrt{a - y^2}}\, dy = dx$$

Integrating, we have $\qquad z\sqrt{a - y^2} = x + b$

or
$$z^2(a - y^2) = (x + b)^2$$

which is complete integral with a, b as arbitrary constants.

Differentiate it with respect to a and b, we get $z^2 = 0$ and $0 = 2(x + b)$
Hence, $z = 0$ is singular solution.

EXAMPLE 4.7

Find the complete integral of the equation $p^2 + q^2 - 2px - 2qy + 1 = 0$.

Solution: Let $f(x, y, z, p, q) = p^2 + q^2 - 2px - 2qy + 1 = 0$.

The Charpit's auxiliary equations are

$$\frac{dp}{-2p} = \frac{dq}{-2q} = \frac{dz}{-p(2p-2x)-q(2q-2y)} = \frac{dx}{-(2p-2x)} = \frac{dy}{-(2q-2y)}$$

Taking the first two members, we have

$$\frac{dp}{p} = \frac{dq}{q}$$

$$\therefore \quad p = aq$$

From the given equation

$$a^2q^2 + q^2 - 2aqx - 2qy + 1 = 0$$

or $$(a^2 + 1)q^2 - 2(ax + y)q + 1 = 0$$

$$\therefore \quad q = \frac{2(ax+y) \pm \sqrt{4(ax+y)^2 - 4(a^2+1)}}{2(a^2+1)}$$

Taking positive value only, we get

$$q = \frac{ax+y+\sqrt{(ax+y)^2-(a^2+1)}}{a^2+1}$$

and $$p = aq = \frac{a\,[ax+y+\sqrt{(ax+y)^2-(a^2+1)}]}{a^2+1}$$

Hence, $$dz = pdx + qdy$$

$$\Rightarrow \quad (a^2 + 1)dz = [ax+y+\sqrt{(ax+y)^2-(a^2+1)}]\,(adx+dy)$$

Integrating, we have

$$(a^2 + 1)z = \frac{(ax+y)^2}{2} + \frac{ax+y}{2}\sqrt{(ax+y)^2-(a^2+1)} - \frac{a^2+1}{2}\log(ax+y+\sqrt{(ax+y)^2-(a^2+1)}\,)+6$$

which is the complete integral.

EXAMPLE 4.8

Find the complete integral of the equation $(p^2 + q^2)^n (qx - py) = 1$

Solution: Let $$f(x, y, z, p, q) = (p^2 + q^2)^n (qx - py) - 1 = 0.$$

The Charpit's auxiliary equations are

$$\frac{dp}{q(p^2+q^2)^n} = \frac{dq}{-p(p^2+q^2)^n} = \ldots.$$

$$\Rightarrow \qquad 2pdp + 2qdq - 0$$

$$\Rightarrow \qquad p^2 + q^2 = \frac{1}{a^2}$$

From the given equation $\dfrac{(qx - py)}{2n} = 1$

$$\Rightarrow \qquad q = \frac{a^{2n} + py}{x}$$

$$\therefore \qquad p^2 + \left(\frac{a^{2n} + py}{x}\right)^2 = \frac{1}{a^2}$$

or $$(x^2 + y^2)p^2 + 2a^{2n}yp + a^{4n} - \frac{x^2}{a^2} = 0$$

$$\therefore \qquad p = -\frac{2a^{2ny} \pm \sqrt{4a^{4n}y^2 - 4(x^2+y^2)\left(a^{4n} - \dfrac{x^2}{a^2}\right)}}{2(x^2+y^2)}$$

Taking positive value only, we get

$$p = -\frac{a^{2n}y + x\sqrt{-a^{4n} + \dfrac{x^2+y^2}{a^2}}}{x^2+y^2}$$

and $$q = \frac{a^{2n} + py}{x} = \frac{xa^{2n} + y\sqrt{-a^{4n} + \dfrac{x^2+y^2}{a^2}}}{x^2+y^2}$$

Hence, $$dz = pdx + qdy$$

$$= a^{2n}\frac{(-ydx + xdy)}{x^2+y^2} + \sqrt{\frac{x^2+y^2}{a^2} - a^{4n}}\left(\frac{xdx + ydy}{x^2+y^2}\right)$$

$$\Rightarrow \qquad dz = a^{2n} d\left(\tan^{-1}\frac{y}{x}\right) + \frac{1}{a}\sqrt{t^2 - a^{4n+2}}\,\frac{dt}{t}$$

[Putting $x^2 + y^2 = t^2 \Rightarrow xdy + ydy = dt$].

Integrating, we get

$$z = a^{2n}\tan^{-1}\left(\frac{y}{x}\right) + \frac{1}{a}\int\left[\frac{t}{\sqrt{t^2 - a^{4n+2}}} - \frac{a^{4n+2}}{t\sqrt{t^2 - a^{4n}}}dt\right]$$

or $\quad z = a^{2n}\tan^{-1}\left(\frac{y}{x}\right) + \frac{1}{a}[\sqrt{t^2 - a^{4n+2}} - a^{2n+1}\sec^{-1}]\left(\frac{t}{a^{2n+1}} + b\right)$ where $t = \sqrt{x^2 + y^2}$.

EXAMPLE 4.9

Solve for complete integral

$$z^2(p^2z^2 + q^2) = 1.$$

Solution: Let, $\quad f(x, y, z, p, q) = p^2z^4 + z^2q^2 - 1 = 0.$

The Charpit's auxiliary equations are

$$\frac{dp}{p(4p^2z^3 + 2zq^2)} = \frac{dq}{q(4p^2z^3 + 2zq^2)} = \frac{dz}{-2p^2z^4 - 2q^2z^2} = \frac{dx}{-2pz^4} = \frac{dy}{-2qz^2}$$

Taking the first two members, we get

$$\frac{dp}{p} = \frac{dq}{q} \text{ or } p = aq.$$

From the given equation

$$z^2(a^2q^2z^2 + q^2) = 1$$

$$\Rightarrow \qquad q = \pm\frac{1}{z\sqrt{a^2z^2 + 1}}$$

Then, $\qquad p = aq = \pm\frac{a}{z\sqrt{a^2z^2 + 1}}$

Hence, $\qquad dz = pdx + qdy \pm\frac{a}{z\sqrt{a^2z^2 + 1}}dx \pm \frac{1}{z\sqrt{a^2z^2 + 1}}dy$

$$\Rightarrow \qquad z\sqrt{a^2z^2 + 1}\,dz = \pm\,(adx + dy)$$

Integrating, we get

$$\frac{(a^2z^2+1)^{3/2}}{3a^2} = \pm (ax + y + b)$$

or $$(a^2z^2 + 1)^3 = 9a^4(ax + y + b)$$

which is the complete integral with a, b as arbitrary constants.

EXAMPLE 4.10

Find the complete integral of

$$p^2 + q^2 - 2px - 2qy + 2xy = 0.$$

Solution: Let $f(x, y, z, p, q) = p^2 + q^2 - 2px - 2qy + 2xy = 0.$

The Charpit's auxiliary equations are

$$\frac{dp}{-2p+2y} = \frac{dq}{-2q+2x} = \frac{dz}{-p(2p-2x)-q(2q-2y)} = \frac{dx}{-(2p-2x)} = \frac{dy}{-(2q-2y)}$$

which gives $$\frac{dp+dq}{2(x+y-p-q)} = \frac{dx+dy}{2(x+y-p-q)}$$

or $$dp + dq = dx + dy$$

Integrating $(p - x) + (q - y) = a.$

We can rewrite the given equation as

$$(p - x)^2 + (q - y)^2 = (x - y)^2$$

$$\Rightarrow \quad (p - x)^2 + [a + (p - x)]^2 = (x - y)^2$$

$$\Rightarrow \quad (p - x)^2 - 2a(p - x) + a^2 - (x - y)^2 = 0$$

$$\therefore \quad p - x = \frac{2a \pm \sqrt{4a^2 - 8(a^2 - (x-y)^2)}}{4}$$

$$\Rightarrow \quad p = x + \frac{1}{2}(a \pm \sqrt{2(x-y)^2 - a^2})$$

and $$q = a + y - p + x$$

$$= y + \frac{1}{2}(a \mp \sqrt{2(x-y)^2 - a^2})$$

Hence, $$dz = pdx + qdy$$

or $$dz = xdy + ydy + \frac{a}{2}(dx+dy) \pm \frac{1}{\sqrt{2}}\sqrt{(x-y)^2 - \frac{a^2}{2}}\,(dx-dy)$$

Integrating, we have

$$z = \frac{x^2+y^2}{2} + \frac{a(x+y)}{2} \pm \frac{1}{\sqrt{2}}\left[\frac{x-y}{2}\sqrt{(x-y)^2 \frac{-a^2}{2}} - \frac{a^2}{4}\log(x-y) + \sqrt{(x-y)^2 - \frac{a^2}{2}}\right] + b$$

which is the desired complete integral with a, b as arbitrary constants.

EXAMPLE 4.11

Solve by Charpit's method the partial differential equation

$$p^2x(x-1) + 2pqxy + q^2y(y-1) - 2pxz - 2qyz + z^2 = 0.$$

Solution: Let $f(x, y, z, p, q) = p^2x(x-1) + 2pqxy + q^2y(y-1) - 2pxz - 2qyz + z^2 = 0$.

The Charpit's auxiliary equations are

$$\frac{dp}{p^2(2x-1) + p(-2px - 2qy + 2z) + 2pqy - 2pz} = \frac{dq}{q^2(2y-1) + 2pqx - 2qz + q(-2px - 2qy + 2z)}$$

$$= \frac{dz}{-p(2px(x-1) + 2qxy - 2xz) - q(2pxy + 2qy(y-1) - 2yz)}$$

$$= \frac{dx}{-2px(x-1) - 2qxy + 2xz} = \frac{dy}{-2pxy - 2qy(y-1) + 2yz}$$

Taking the first two members, we have

$$\frac{dp}{-p^2} = \frac{dq}{-q^2} = \frac{\frac{1}{p}dp - \frac{1}{q}dq}{-p+q}$$

Taking the last two members, we have

$$\frac{dx}{x(-2px + 2p - 2qy + 2z)} = \frac{dy}{y(-2px - 2qy + 2q + 2z)}$$

$$= \frac{\frac{1}{x}dx - \frac{1}{y}dy}{2(p-q)}$$

Hence, $$\frac{\frac{1}{p}dp - \frac{1}{q}dq}{-(p-q)} = \frac{\frac{1}{x}dx - \frac{1}{y}dy}{2(p-q)}$$

$$\Rightarrow \qquad -\frac{1}{p}dp+\frac{1}{q}dq=\frac{1}{2}\left[\frac{1}{x}dx-\frac{1}{y}dy\right]$$

Integrating, we have

$$-\log p + \log q + \log a = \frac{1}{2}[\log x - \log y]$$

$$\Rightarrow \qquad \frac{aq}{p} = (x/y)^{1/2}$$

$$\Rightarrow \qquad p = aq(y/x)^{1/2}$$

where a is an arbitrary constant.

From the given equation

$$(px + qy - z)^2 = p^2x + q^2y$$

$$\Rightarrow \qquad px + qy - z = (p^2x + q^2y)^{1/2}$$

$$\Rightarrow \qquad aq(xy)^{1/2} + qy - z = (a^2q^2y + q^2y)^{1/2}$$

$$\Rightarrow \qquad q[y + a(xy)^{1/2} - (1 + a^2)^{1/2}\, y^{1/2}] = z$$

so that

$$q=\frac{z}{y^{1/2}[y^{1/2}+ax^{1/2}-(1+a^2)^{1/2}]}$$

and

$$p=\frac{az}{x^{1/2}[y^{1/2}+ax^{1/2}-(1+a^2)^{1/2}]}$$

Hence,

$$dz = pdx + qdy$$

or

$$dz = \left[\frac{azdx}{x^{1/2}}+\frac{zdy}{y^{1/2}}\right]\frac{1}{(y^{1/2}+ax^{1/2}-(1+a^2)^{1/2})}$$

or

$$\frac{dz}{z}=\frac{ay^{1/2}dx+x^{1/2}dy}{(xy)^{1/2}(y^{1/2}+ax^{1/2}-(1+a^2)^{1/2})}$$

Integrating, we have

$\log z = 2\log(y^{1/2} + ax^{1/2} - (1 + a^2)^{1/2}) + \log b$ or $z = b(y^{1/2} + ax^{1/2} - (1 + a^2)^{1/2})^2$

where, a and b are arbitrary constants.

▲▲

EXERCISE 4.1

Solve for the complete integral the following equations using Charpit's method:

1. $p^3 + q^3 = 3pqz$.
2. $2(z + xp + yq) = yp^2$.
3. $(x^2 - y^2)pq - xy(p^2 - q^2) = 0$.
4. $2xz + pq - px^2 - 2qxy = 0$.
5. $2z + p^2 + qy + 2y^2 = 0$.
6. $px + qy = z\sqrt{1+pq}$.
7. $(p + q)(px + qy) = 1$.
8. $2z + p^2 + qy + 2y^2 = 0$.
9. $pq + x(2y + 1)p + (y^2 + y)q - (2y + 1)z = 0$.
10. $yp = 2xy + \log q$.
11. $p^2q(x^2 + y^2) = p^2 + q$.
12. $xp - yq = xq\, f(z - px - qy)$.
13. $px + qy = z(1 + pq)^{1/2}$.
14. $2(z + xp + qy) - yp^2 = 0$.
15. $2xz - px^2 - 2qxy + pq = 0$.

ANSWERS

Exercise 4.1

1. $3a(x + ay + b) = (1 + a^3)\log z$.
2. $4zy^3 = 4axy - a^2 + 4by^2$.
3. $z = \frac{a}{2}\log(x^2 + y^2) + \frac{1}{a}\tan^{-1}\frac{y}{x} + b$.
4. $z = ay + b(x^2 - a)$.
5. $2z = \frac{2ax}{y^2} - \left(y^2 - \frac{a^2}{y^4}\right) + \frac{2b}{y^2}$.
6. $\log z = -\left(\frac{(ax+y)^2}{2az^2} \pm \left(\frac{ax+y}{2\sqrt{a}\,z}\sqrt{\frac{(ax+y)^2}{az^2} - 1}\right) - \frac{1}{2}\log\left(\frac{ax+y}{\sqrt{a}\,z} + \sqrt{\frac{(ax+y)^2}{az^2} - 1}\right)\right) + b$.
7. $z\sqrt{a+1} = 2\sqrt{ax+y} + b$.

8. $2y^2z + y^2(a - x)^2 + y^4 = b$.

9. $z = ax + b(a + y + y^2)$.

10. $z = (a + 2x)^{2/4} + \left(\frac{1}{a}\right)e^{ay} + b$.

11. $z = \log\,(x + \sqrt{x^2 + a}) + \frac{1}{2\sqrt{a}}\log\frac{y - \sqrt{a}}{y + \sqrt{a}} + b$.

12. $(z - a)^2 = b(2xy + x^2f(a))$.

13. $\log z = -\left[\frac{v}{2}(v^2 - 1)^{1/2} - \frac{1}{2}\log v + (v^2 - 1)^{1/2}\right] - \frac{v^2}{2} + b$ where $v = \frac{ax + y}{z\sqrt{a}}$.

14. $yz = \frac{ax}{y} - \frac{a^2}{4y^2} + b$.

15. $z = ay + b(x^2 - a)$. ▲▲

Jacobi's Method

This method is used for the solution of nonlinear partial differential equations of order one with three or more independent variables.

Let us consider the nonlinear partial differential equation with three independent variables x_1, x_2, x_3 as $f(x_1, x_2, x_3, p_1, p_2, p_3) = 0$ (1)
z is the dependent variable not occurring in the except in its partial derivatives p_1, p_2, p_3 with respect to x_1, x_2, x_3 respectively.

Our purpose is to find two additional equations

$$\phi_1(x_1, x_2, x_3, p_1, p_2, p_3) = a_1 \tag{2}$$

and

$$\phi_2(x_1, x_2, x_3, p_1, p_2, p_3) = a_2 \tag{3}$$

where a_1, a_2 are arbitrary constants such that Eqs. (1), (2), (3) may be solved for p_1, p_2, p_3 as functions of x_1, x_2, x_3 which when substituted in $dz = p_1dx_1 + p_2dx_2 + p_3dx_3$ makes dz integrable, for which conditions are

$$\frac{\partial p_2}{\partial x_1} = \frac{\partial^2 z}{\partial x_1 \partial x_2} = \frac{\partial p_1}{\partial x_2}, \frac{\partial p_3}{\partial x_2} = \frac{\partial p_2}{\partial x_3}, \frac{\partial p_1}{\partial x_3} = \frac{\partial p_3}{\partial x_1}$$

Now differentiate Eqs. (1) and (2) partially with respect to x_1; we have

$$\frac{\partial f}{\partial x_1}+\frac{\partial f}{\partial p_1}\frac{\partial p_1}{\partial x_1}+\frac{\partial f}{\partial p_2}\frac{\partial p_2}{\partial x_1}+\frac{\partial f}{\partial p_3}\frac{\partial p_3}{\partial x_1}=0 \tag{4}$$

$$\frac{\partial \phi_1}{\partial x_1}+\frac{\partial \phi_1}{\partial p_1}\frac{\partial p_1}{\partial x_1}+\frac{\partial \phi_1}{\partial p_2}\frac{\partial p_2}{\partial x_1}+\frac{\partial \phi_1}{\partial p_3}\frac{\partial p_3}{\partial x_1}=0 \tag{5}$$

Eliminating $\frac{\partial p_1}{\partial x_1}$ from Eqs. (4) and (5); we have

$$\left(\frac{\partial f}{\partial x_1}\frac{\partial \phi_1}{\partial p_1}-\frac{\partial f}{\partial p_1}\frac{\partial \phi_1}{\partial x_1}\right)+\left(\frac{\partial f}{\partial p_2}\frac{\partial \phi_1}{\partial p_1}-\frac{\partial f}{\partial p_1}\frac{\partial \phi_1}{\partial p_2}\right)\frac{\partial p_2}{\partial x_1}+\left(\frac{\partial f}{\partial p_3}\frac{\partial \phi_1}{\partial p_1}-\frac{\partial f}{\partial p_1}\frac{\partial \phi_1}{\partial p_3}\right)\frac{\partial p_3}{\partial x_1}=0 \tag{6}$$

Similarly, we have

$$\left(\frac{\partial f}{\partial x_2}\frac{\partial \phi_1}{\partial p_2}-\frac{\partial f}{\partial p_2}\frac{\partial \phi_1}{\partial x_2}\right)+\left(\frac{\partial f}{\partial p_1}\frac{\partial \phi_1}{\partial p_2}-\frac{\partial f}{\partial p_2}\frac{\partial \phi_1}{\partial p_1}\right)\frac{\partial p_1}{\partial x_2}+\left(\frac{\partial f}{\partial p_3}\frac{\partial \phi_1}{\partial p_2}-\frac{\partial f}{\partial p_2}\frac{\partial \phi_1}{\partial p_3}\right)\frac{\partial p_3}{\partial x_2}=0 \tag{7}$$

and

$$\left(\frac{\partial f}{\partial x_3}\frac{\partial \phi_1}{\partial p_3}-\frac{\partial f}{\partial p_3}\frac{\partial \phi_1}{\partial x_3}\right)+\left(\frac{\partial f}{\partial p_1}\frac{\partial \phi_1}{\partial p_3}-\frac{\partial f}{\partial p_3}\frac{\partial \phi_1}{\partial p_1}\right)\frac{\partial p_1}{\partial x_3}+\left(\frac{\partial f}{\partial p_2}\frac{\partial \phi_1}{\partial p_3}-\frac{\partial f}{\partial p_3}\frac{\partial \phi_1}{\partial p_2}\right)\frac{\partial p_2}{\partial x_3}=0 \tag{8}$$

Adding Eqs. (6), (7), (8) and using $\frac{\partial p_2}{\partial x_1}=\frac{\partial p_1}{\partial x_2}, \frac{\partial p_3}{\partial x_2}=\frac{\partial p_2}{\partial x_3}$ and $\frac{\partial p_1}{\partial x_3}=\frac{\partial p_3}{\partial x_1}$; we have

$$\left(\frac{\partial f}{\partial x_1}\frac{\partial \phi_1}{\partial p_1}-\frac{\partial f}{\partial p_1}\frac{\partial \phi_1}{\partial x_1}\right)+\left(\frac{\partial f}{\partial x_2}\frac{\partial \phi_1}{\partial p_2}-\frac{\partial f}{\partial p_2}\frac{\partial \phi_1}{\partial x_2}\right)+\left(\frac{\partial f}{\partial x_3}\frac{\partial \phi_1}{\partial p_3}-\frac{\partial f}{\partial p_3}\frac{\partial \phi_1}{\partial x_3}\right)=0 \tag{9}$$

or $$(f,\phi_1)=\sum_{r=1}^{3}\left(\frac{\partial f}{\partial x_r}\frac{\partial \phi_1}{\partial p_r}-\frac{\partial f}{\partial p_r}\frac{\partial \phi_1}{\partial x_r}\right)=0$$

Similarly, we have

$$(f,\phi_2)=\sum_{r=1}^{3}\left(\frac{\partial f}{\partial x_r}\frac{\partial \phi_2}{\partial p_r}-\frac{\partial f}{\partial p_r}\frac{\partial \phi_2}{\partial x_r}\right) \tag{10}$$

and $$(\phi_1,\phi_2)=\sum_{r=1}^{3}\left(\frac{\partial \phi_1}{\partial x_r}\frac{\partial \phi_2}{\partial p_r}-\frac{\partial \phi_1}{\partial p_r}\frac{\partial \phi_2}{\partial x_r}\right) \tag{11}$$

Since Eq. (9) is a linear equation of the first order with $x_1, x_2, x_3, p_1, p_2, p_3$ as independent variables and ϕ, as dependent variable.

$\therefore$ The auxiliary equations are

$$\frac{\partial p_1}{\frac{\partial f}{\partial x_1}} = \frac{\partial x_1}{-\frac{\partial f}{\partial p_1}} = \frac{\partial p_2}{\frac{\partial f}{\partial x_2}} = \frac{\partial x_2}{-\frac{\partial f}{\partial p_2}} = \frac{\partial p_3}{\frac{\partial f}{\partial x_3}} = \frac{\partial x_3}{\frac{\partial f}{\partial p_3}} \tag{12}$$

which are known as Jacobi's auxiliary equations. Any integral of Eq. (12) will satisfy Eq. (9). We try to find two independent integrals $\phi_1(x_1, x_2, x_3, p_1, p_2, p_3) = a_1$, and $\phi_2(x_1, x_2, x_3, p_1, p_2, p_3) = a_2$ with the help of Eq. (12). If these relations satisfy Eq. (11), these are the required two additional relations, viz. Eqs. (2) and (3). We now solve Eqs. (1), (2) and (3) for p_1, p_2, p_3 in terms of x_1, x_2, x_3. Substituting these values in $dz = p_1dx + p_2dx_2 + p_3dx_3$ and then integrating the resulting equation, we shall obtain a complete integral of the given equation containing three arbitrary constants of integration.

Working Method

For the nonlinear partial differential equation with three independent variables proceed as follows:

- Write the given differential equation in the form

$$f(x_1, x_2, x_3, p_1, p_2, p_3) = 0 \tag{1}$$

- Write the Jacobi's auxiliary equations

$$\frac{\partial p_1}{\frac{\partial f}{\partial x_1}} = \frac{\partial x_1}{-\frac{\partial f}{\partial p_1}} = \frac{\partial p_2}{\frac{\partial f}{\partial x_2}} = \frac{\partial x_2}{-\frac{\partial f}{\partial p_2}} = \frac{\partial p_3}{\frac{\partial f}{\partial x_3}} = \frac{\partial x_3}{-\frac{\partial f}{\partial p_3}}$$

  Solving these equations, obtain two additional equations

$$\phi_1(x_1, x_2, x_3, p_1, p_2, p_3) = a_1 \tag{2}$$

  and

$$\phi_2(x_1, x_2, x_3, p_1, p_2, p_3) = a_2 \tag{3}$$

  where a_1, a_2 are arbitrary constants.

  Try to find the simplest relations (2) and (3); so that further solution may be as simple as possible.

- Verify that relations (2) and (3) satisfy the condition

$$(\phi_1, \phi_2) = \sum_{r=1}^{3}\left(\frac{\partial \phi_1}{\partial x_r}\frac{\partial \phi_2}{\partial p_r} - \frac{\partial \phi_1}{\partial p_r}\frac{\partial \phi_2}{\partial x_r}\right) = 0 \tag{4}$$

- If Eq. (4) is satisfied, then salve Eqs. (1), (2) and (3) to get the values of p_1, p_2, p_3 in terms of x_1, x_2, x_3.
- Substitute the values of p_1, p_2, p_3 in $dz = p_1dx_1 + p_2dx_2 + p_3dx_3$ (5)
- Integrate Eq. (5) to get the complete integral of Eq. (1).

EXAMPLE 4.12

Solve $p_1^3 + p_2^2 + p_3 = 1$, using Jacobi's method.

Solution: Let $$f(x_1, x_2, x_3, p_1, p_2, p_3) = p_1^3 + p_2^2 + p^3 - 1 = 0.$$

The Jacobi's auxiliary equations are

$$\frac{dx_1}{-3p_1^2} = \frac{dp_1}{0} = \frac{dx_2}{-2p_2} = \frac{dp_2}{0} = \frac{dx_3}{-1} = \frac{dp_3}{0}$$

$$\left\{ \text{By using } \frac{dx_1}{\frac{-\partial f}{\partial p_1}} = \frac{dp_1}{\frac{\partial f}{\partial x_1}} = \frac{dx_2}{\frac{-\partial f}{\partial p_2}} = \frac{dp_2}{\frac{\partial f}{\partial x_2}} = \frac{dx_3}{\frac{-\partial f}{\partial p_3}} = \frac{dp_3}{\frac{\partial f}{\partial x_3}} \right\}$$

$$\therefore \qquad dp_1 = 0 = dp_2$$

Integrating, we have $F_1 = p_1 = c_1$ and $F_2 = p_2 = c_2$

Now $$(F_1, F_2) = \sum_{r=1}^{3} \left(\frac{\partial F_1}{\partial x_r} \frac{\partial F_2}{\partial p_r} - \frac{\partial F_1}{\partial p_r} \frac{\partial F_2}{\partial x_r} \right)$$

$$= \left(\frac{\partial F_1}{\partial x_1} \frac{\partial F_2}{\partial p_1} - \frac{\partial F_1}{\partial p_1} \frac{\partial F_2}{\partial x_1} \right) + \left(\frac{\partial F_1}{\partial x_2} \frac{\partial F_2}{\partial p_2} - \frac{\partial F_1}{\partial p_2} \frac{\partial F_2}{\partial x_2} \right) + \left(\frac{\partial F_1}{\partial x_3} \frac{\partial F_2}{\partial p_3} - \frac{\partial F_1}{\partial p_3} \frac{\partial F_2}{\partial x_3} \right)$$

$$= 0$$

$\therefore$ These may be taken as additional equations. From the given equation $p_3 = 1 - c_1^3 - c_2^2$.

Hence, $$dz = p_1 dx_1 + p_2 dx_2 + p_3 dx_3$$
$$= c_1 dx_1 + c_2 dx_2 + (1 - c_1^3 - c_2^2) dx_3.$$

Integrating, we obtain

$$z = c_1 x_1 + c_2 x_2 + (1 - c_1^3 - c_2^3) x_3 + a_3.$$

which is the required complete integral.

EXAMPLE 4.13

Find the complete integral of

$$2p_1 x_1 x_3 + 3p_2 x_3^2 + p_2^2 p_3 = a.$$

Solution: Let $f(x_1, x_2, x_3, p_1, p_2, p_3) = 2p_1 x_1 x_3 + 3p_2 x_3^2 + p_2^2 p_3 - a = 0.$

The Jacobi's auxiliary equations are

$$\frac{dx_1}{-2x_1 x_3} = \frac{dp_1}{2p_1 x_3} = \frac{dx_2}{-3x_3^2 - 2p_2 p_3} = \frac{dp_2}{0} = \frac{dx_3}{-p_2^2} = \frac{dp_3}{2p_1 x_1 + 6p_2 x_3}$$

Taking the first and second members, we have

$$\frac{dx_1}{-x_1} = \frac{dp_1}{p_1} \text{ or } \frac{dx_1}{x_1} + \frac{dp_1}{p_1} = 0$$

$\therefore$ $$F_1 = p_1 x_1 = a_1$$

Again, taking the first and fourth members, we have

$$dp_2 = 0$$

$\therefore$ $$F_2 = p_2 = a_2$$

Now, $$(F_1, F_2) = \sum_{r=1}^{3}\left(\frac{\partial F_1}{\partial x_r}\frac{\partial F_2}{\partial p_r} - \frac{\partial F_1}{\partial p_r}\frac{\partial F_2}{\partial x_r}\right) = 0$$

$\therefore$ These may be taken as additional equations.

Now, $p_1 = \dfrac{a_1}{x_1}$, $p_2 = a_2$ and from the given equation $p_3 = -\dfrac{1}{a_2^2}(2a_1x_3 + 3a_2x_3^2)$

Hence, $dz = p_1dx_1 + p_2dx_2 + p_3dx_3 = \dfrac{a_1}{x_1}dx_1 + a_2dx_2 - \dfrac{1}{a_2^2}(2a_1x_3 + 3a_2x_3^2)dx_3$

Integrating, we have

$$z = a_1 \log x_1 + a_2x_2 - \frac{1}{a_2^2}(a_1x_3^2 + a_2x_3^3) + a_3$$

which is the required complete integral.

EXAMPLE 4.14

Find the complete integral of

$$x_3^2p_1^2p_2^2p_3^2 + p_1^2p_2^2 - p_3^2 = 0.$$

Solution: Let $f(x_1, x_2, x_3, p_1, p_2, p_3) = x_3^2p_1^2p_2^2p_3^2 + p_1^2p_2^2 - p_3^2 = 0.$

The Jacobi's auxiliary equations are

$$\frac{dx}{-2(x_3^2p_1p_2^2p_3^2 + p_1p_2^2} = \frac{dp_1}{0} = \frac{dx_2}{-2(x_3^2p_1^2p_2p_3 + p_1^2p_2)}$$

$$= \frac{dp_2}{0} = \frac{dx_3}{-2(x_3^2p_1^2p_2^2p_3 - p_3)} = \frac{dp_3}{2x_3p_1^2p_2^2p_3^2}$$

Hence, we have $$dp_1 = 0 = dp_2$$

$\Rightarrow$ $$F_1 = p_1 = a_1 \text{ and } F_2 = p_2 = a_2.$$

Also, $$(F_1, F_2) = \sum_{r=1}^{3}\left(\frac{\partial F_1}{\partial x_r}\frac{\partial F_2}{\partial p_r} - \frac{\partial F_1}{\partial p_r}\frac{\partial F_2}{\partial x_r}\right) = 0$$

∴ These may be taken as additional equations

From the given equation $p_3 = \pm a_1 a_2 / \sqrt{(1 - x_3^2 a_1^2 a_2^2)}$

Hence, $$dz = p_1 dx_1 + p_2 dx_2 + p_3 dx_3$$

$$= a_1 dx_1 + a_2 dx_2 \frac{\pm a_1 a_2}{\sqrt{1 - a_1^2 a_2^2 x_3^2}} dx_3$$

Integrating, we obtain

$$z = a_1 x_1 + a_2 x_2 \pm \sin^{-1}(a_1\, a_2\, x_3) + a_3.$$

which is the required complete integral.

EXAMPLE 4.15

Apply Jacobi's method to find the complete integral of

$$(p_1 + x_1)^2 + (p_2 + x_2)^2 + (p_3 + x_3)^2 = 3(x_1 + x_2 + x_3).$$

Solution: Let $f(x_1, x_2, x_3, p_1, p_2, p_3) = (p_1 + x_1)^2 + (p_2 + x_2)^2 + (p_3 + x_3)^2 - 3(x_1 + x_2 + x_3) = 0.$

The Jacobi's auxiliary equations are

$$\frac{dx_1}{-2(p_1 + x_1)} = \frac{dp_1}{2(p_1 + x_1) - 3} = \frac{dx_2}{-2(p_2 + x_2)} = \frac{dp_2}{2(p_2 + x_2) - 3} = \frac{dx_3}{-2(p_3 + x_3)} = \frac{dp_3}{2(p_3 + x_3) - 3}$$

Now $\frac{dx_1 + dp_1}{-3} = \frac{dx_2 + dp_2}{-3} = \frac{dx_3 + dp_3}{-3}$

∴ $dx_1 + dp_1 = dx_2 + dp_2$ and $dx_1 + dp_1 = dx_3 + dp_3$.

Integrating, we have

$$x_1 + p_1 = x_2 + p_2 + a_1 \text{ and } x_1 + p_1 = x_3 + p_3 + a_2$$

Let $$F_1 = p_1 - p_2 + x_1 - x_2 = a_1$$

and $$F_2 = p_1 - p_3 + x_1 - x_3 = a_2.$$

Also, $$(F_1, F_2) = \sum_{r=1}^{n}\left(\frac{\partial F_1}{\partial x_r}\frac{\partial F_2}{\partial p_r} - \frac{\partial F_1}{\partial p_r}\frac{\partial F_2}{\partial x_r}\right) = 0$$

∴ These may be taken as additional equations.

Now, $$p_2 = p_1 + x_1 - x_2 - a_1 \text{ and } p_3 = x_1 + p_1 - x_3 - a_2$$

From the given equation, we have

$$(p_1 + x_1)^2 + (p_1 + x_1 - a_1)^2 + (p_1 + x_1 - a_2)^2 - 3(x_1 + x_2 + x_3) = 0$$

or $$3(p_1 + x_1)^2 - 2(a_1 + a_2)\,(p_1 + x_1) + (a_1^2 + a_2^2 - 3(x_1 + x_2 + x_3)) = 0$$

$$\therefore \qquad p_1 + x_1 = \frac{2(a_1 + a_2) \pm \sqrt{4(a_1 + a_2)^2 - 12(a_1^2 + a_2^2 - 3(x_1 + x_2 + x_3)}}{6}$$

$$p_1 = -x_1 + \frac{1}{3}[(a_1 + a_2) \pm \sqrt{9(x_1 + x_2 + x_3) - 2(a_1^2 + a_2^2 - a_1 a_2)}]$$

$$\therefore \qquad p_2 = -x_2 + \frac{1}{3}[(2a_1 + a_2) \pm \sqrt{9(x_1 + x_2 + x_3) - 2(a_1^2 + a_2^2 - a_1 a_2)}]$$

and $$p_3 = -x_3 + \frac{1}{3}[(a_1 - 2a_2) \pm \sqrt{9(x_1 + x_2 + x_3) - 2(a_1^2 + a_2^2 - a_1 a_2)}]$$

Hence, $dz = p_1 dx_1 + p_2 dx_2 + p_3 dx_3 = -(x_1 dx_1 + x_2 dx_2 + x_3 dx_3) + \frac{1}{3}[(a_1 + a_2)dx_1 + (-2a_1 + a_2)dx_2 + (a_1 - 2a_2)dx_3] \pm \frac{1}{3}[\sqrt{9(x_1 + x_2 + x_3) - 2(a_1^2 + a_2^2 - a_1 a_2)}(dx_1 + dx_2 + dx_3)]$.

Integrating, we obtain

$$z = -\frac{1}{2}(x_1^2 + x_2^2 + x_3^2) + \frac{1}{3}[(a_1 + a_2)x_1 + (-2a_1 + a_2)x_2 + (a_1 - 2a_2)x_3 \pm \frac{2}{81}[9(x_1 + x_2 + x_3) - 2(a_1^2 + a_2^2 - a_1 a_2)]^{3/2} + a_3$$

which is the required complete integral of the given equation with a_1, a_2, a_3 as arbitrary constants.

EXAMPLE 4.16

Find the complete integral of

$$p_1 p_2 p_3 = z^3 x_1 x_2 x_3.$$

Solution: The given equation is not in the standard form as it involves z. So first, we reduce the given equation to the standard form as

$$\left(\frac{1}{z}\frac{\partial z}{\partial x_1}\right)\left(\frac{1}{z}\frac{\partial z}{\partial x_2}\right)\left(\frac{1}{z}\frac{\partial z}{\partial x_3}\right) = x_1 x_2 x_3$$

Let $\frac{1}{z}\,dz = dZ$ so that $\log z = Z$

$$\therefore \qquad \frac{\partial Z}{\partial x_1}\frac{\partial Z}{\partial x_2}\frac{\partial Z}{\partial x_3} = x_1x_2x_3 \text{ or } P_1P_2P_3 = x_1x_2x_3.$$

Let $\quad f(x_1, x_2, x_3, P_1, P_2, P_3) = P_1P_2P_3 - x_1x_2x_3 = 0.$

The Jacobi's auxiliary equations are

$$\frac{dx_1}{-P_2P_3} = \frac{dP_1}{-x_2x_3} = \frac{dx_2}{-P_1P_3} = \frac{dP_2}{-x_1x_3} = \frac{dx_3}{-P_1P_2} = \frac{dP_3}{-x_1x_2}$$

Taking the first and second members, we get

$$\frac{P_1dx_1}{-P_1P_2P_3} = \frac{x_1dP_1}{-x_1x_2x_3} \Rightarrow \frac{dx_1}{x_1} = \frac{dP_1}{P_1}$$

Integrating, we obtain $\quad P_1 = a_1x_1$

Let $\quad F_1 = P_1 - a_1x_1 = 0$

Similarly, taking the third and fourth members, we get

$$F_2 = P_2 - a_2x_2 = 0.$$

$\therefore$ These may be taken as additional equations as $(F_1, F_2) = 0$.

From the given equation $P_3 = x_3/a_1a_2$

Hence, $$dZ = P_1dx_1 + P_2dx_2 + P_3dx_3 = a_1x_1dx_1 + a_2x_2dx_2 + \frac{x_3}{a_1a_2}dx_3$$

Integrating, we obtain

$$2Z = a_1x_1^2 + a_2x_2^2 + \frac{x_3^2}{a_1a_2} + a_3$$

or $$2\log z = a_1x_1^2 + a_2x_2^2 + \frac{x_3^2}{a_1a_2} + a_3$$

which is the required complete integral.

EXAMPLE 4.17

Find the complete integral of

$$p_3x_3(p_1 + p_2) + x_1 + x_2 = 0.$$

Solution: Let $f(x_1, x_2, x_3, p_1, p_2, p_3) = p_3x_3(p_1 + p_2) + x_1 + x_2 = 0.$

The Jacobi's auxiliary equations are

$$\frac{dx_1}{-p_3x_3} = \frac{dp_1}{1} = \frac{dx_2}{-p_3x_3} = \frac{dp_2}{1} = \frac{dx_3}{-x_3(p_1+p_2)} = \frac{dp_3}{p_3(p_1+p_2)}$$

Taking the second and fourth members, we have

$$dp_1 = dp_2$$

Integrating, we get $\qquad F_1 = p_1 - p_2 = a_1$

Taking the last two members, we have

$$\frac{dx_3}{-x_3} = \frac{dp_3}{p_3}$$

Integrating, we get

$$-\log x_3 + \log a_2 = \log p_3$$

$$\Rightarrow \qquad F_2 = p_3 x_3 = a_2.$$

Now,
$$(F_1, F_2) = \sum_{r=1}^{3}\left(\frac{\partial F_1}{\partial x_r}\frac{\partial F_2}{\partial p_r} - \frac{\partial F_1}{\partial p_r}\frac{\partial F_2}{\partial x_r}\right) = 0.$$

∴ These may be taken as additional equations.
From the given equation, we have

$$p_3 = a_2/x_3,\ p_1 = p_2 + a_1$$

and
$$a_2(2p_2 + a_1) + x_1 + x_2 = 0$$

∴
$$p_2 = \frac{a_1}{2} = \frac{x_1 + x_2}{2a_2}$$

and
$$p_1 = \frac{a_1}{2} - \frac{x_1 + x_2}{2a_2}$$

Hence,
$$dz = p_1 dx_1 + p_2 dx_2 + p_3 dx_3$$

$$= \left(\frac{a_1}{2} - \frac{x_1 + x_2}{2a_2}\right)dx_1 + \left(\frac{-a_1}{2} - \frac{x_1 + x_2}{2a_2}\right)dx_2 + \frac{a_2}{x_3}dx_3$$

$$= \frac{a_1}{2}(dx_1 - dx_2) - \frac{1}{2a_2}(x_1 + x_2)(dx_1 + dx_2) + \frac{a_2}{x_3}dx_3$$

Integrating, we obtain

$$z = \frac{a_1}{2}(x_1 - x_2) - \frac{1}{4a_2}(x_1 + x_2)^2 + a_2 \log x_3 + a_3$$

which is the required complete integral with a_1, a_2, a_3 as arbitrary constants.

EXAMPLE 4.18

Find the complete integral of $(x_2 + x_3)(p_2 + p_3)^2 + zp_1 = 0$.

Solution: The given equation can be written as

$$(x_2 + x_3)\left(\frac{1}{z}\frac{\partial z}{\partial x_2} + \frac{1}{z}\frac{\partial z}{\partial x_3}\right)^2 + \left(\frac{1}{z}\frac{\partial z}{\partial x_1}\right) = 0$$

Let $\frac{1}{2}\, dz = dZ$ so that $\log z = Z$

$$\therefore \qquad (x_2 + x_3)\left(\frac{\partial Z}{\partial x_2} + \frac{\partial Z}{\partial x_3}\right)^2 + \left(\frac{\partial Z}{\partial x_1}\right) = 0$$

or

$$(x_2 + x_3)\,(P_2 + P_3)^2 + P_1 = 0.$$

The Jacobi's auxiliary equations are

$$\frac{dx_1}{-1} = \frac{dP_1}{0} = \frac{dx_2}{-2(x_2 x_3)(P_2 + P_3)} = \frac{dP_2}{(P_2 + P_3)^2}$$

$$= \frac{dx_3}{-2(x_2 + x_3)(P_2 + P_3)} = \frac{dP_3}{(P_2 + P_3)^2}$$

Taking the first and second members, we have

$$dP_1 = 0 \Rightarrow P_1 = -a_1 = F_1$$

Again, taking the third and fifth members, we have

$$dP_2 = dP_3 \Rightarrow P_2 - P_3 = a_2 = F_2.$$

$\therefore$ These may be taken as additional equations as $(F_1, F_2) = 0$.

From the given equation, we have

$$(x_2 + x_3)^2\,(2P_3 + a_2)^2 - a_1 = 0$$

$$\Rightarrow \qquad P_3 = -\frac{a_2}{2} \pm \sqrt{\frac{a_1}{x_2 + x_3}}$$

$$\therefore \qquad P_2 = P_3 + a_2 = \frac{a_2}{2} \pm \sqrt{\frac{a_1}{x_2 + x_3}}$$

Hence,

$$dZ = P_1 dx_1 + P_2 dx_2 + P_3 dx_3$$

$$= -a_1 dx_1 + \left[\frac{a_2}{2} \pm \sqrt{\frac{a_1}{x_2 + x_3}}\right] dx_2 + \left[-\frac{a_2}{2} \pm \sqrt{\frac{a_1}{x_2 + x_3}}\right] dx_3$$

$$= -a_1 dx_1 + \frac{a_2}{2}(dx_2 + dx_3) \pm \sqrt{\frac{a_1}{x_2 + x_3}}\,(dx_2 + dx_3)$$

Integrating, we obtain

$$Z = -a_1x_1 + \frac{a_2}{2}(x_2 + x_3) \pm 2\sqrt{a_1(x_2 + x_3)} + a_3$$

or
$$\log z = -a_1x_1 + \frac{a_2}{2}(x_2 + x_3) \pm 2\sqrt{a_1(x_2 + x_3)} + a_3$$

which is the required complete integral with a_1, a_2, a_3 as arbitrary constants.

EXAMPLE 4.19

Find the complete integral of

$$2x_1x_3zp_1p_3 + x_2p_2 = 0.$$

Solution: The given equation can be written as

$$2x_1x_3\left(z\frac{\partial z}{\partial x_1}\right)\left(z\frac{\partial z}{\partial x_3}\right) + x_2\left(z\frac{\partial z}{\partial x_2}\right) = 0$$

Let $zdz = dZ$ so that $\frac{z^2}{2} = Z$

$$\therefore\quad 2x_1x_3\left(\frac{\partial Z}{\partial x_1}\right)\left(\frac{Z}{\partial x_3}\right) + x_2\left(\frac{\partial Z}{\partial x_2}\right) = 0$$

or
$$2x_1x_3P_1P_3 + x_2P_2 = 0.$$

Let $f(x_1, x_2, x_3, P_1, P_2, P_3) = 2x_1x_3P_1P_3 + x_2P_2 = 0.$

The Jacobi's auxiliary equations are

$$\frac{dx_1}{-2x_1x_3P_3} = \frac{dP_1}{2x_3P_1P_3} = \frac{dx_2}{-x_2} = \frac{dP_2}{P_2} = \frac{dx}{-2x_1x_3P_1} = \frac{dP_3}{2x_1P_1P_3}$$

Taking the first and second members, we get

$$\frac{dx_1}{-x_1} = \frac{dP_1}{P_1}$$

Integrating, we have

$$P_1x_1 = a_1 = F_1$$

Taking the fifth and sixth members, we get

$$\frac{dx_3}{-x_3} = \frac{dP_3}{P_3}$$

Integrating, we have

$$P_3x_3 = a_2 = F_2$$

Now, $(F_1, F_2) = 0$

∴ These may be taken as additional equations.

From the given equation, we have

$$P_1 = \frac{a_1}{x_1}, P_3 = \frac{a_2}{x_3} \text{ and } P_2 = -\frac{2a_1a_2}{x_2}$$

Hence, $$dZ = P_1dx + P_2dx_2 + P_3dx_3$$

$$= \frac{a_1}{x_1}dx_1 - \frac{2a_1a_2}{x_2}dx_2 + \frac{a_2}{x_3}dx_3$$

Integrating, we obtain

$$Z = a_1 \log x_1 - 2a_1a_2 \log x_2 + a_2 \log x_3 + a_3$$

or $$\frac{z^2}{2} = a_1 \log x_1 - 2a_1a_2 \log x_2 + a_2 \log x_3 + a_3$$

which is the required complete integral with a_1, a_2, a_3 as arbitrary constants.

EXAMPLE 4.20

Solve $p^2 + q^2 = c^2$ by using Jacobi's method.

Solution: Let the solution of the given equation be

$$u(x, y, z) = 0 \tag{1}$$

So we can assume u as dependent variable and x, y, z as three independent variables. Differentiate Eq. (1) partially with respect to x and y respectively, we have

$$\frac{\partial u}{\partial x} + \frac{\partial u}{\partial z}\frac{\partial z}{\partial x} = 0 \text{ and } \frac{\partial u}{\partial y} + \frac{\partial u}{\partial z}\frac{\partial z}{\partial y} = 0$$

or $$p_1 + p_3p = 0 \text{ and } p_2 + p_3q = 0$$

where, $$p_1 = \frac{\partial u}{\partial x}, p_2 = \frac{\partial u}{\partial y}, p_3 = \frac{\partial u}{\partial z}, p = \frac{\partial z}{\partial x} \text{ and } q = \frac{\partial z}{\partial y}.$$

So, the given equation becomes

$$\frac{p_1^2}{p_3^2} + \frac{p_2^2}{p_3^2} = c^2 \text{ or } p_1^2 + p_2^2 = c^2p_3^2$$

Let $$f(x, y, z, p_1, p_2, p_3) = p_1^2 + p_2^2 - c^2p_3^2 = 0.$$

Now, the Jacobi's auxiliary equations are given by

$$\frac{dx}{-2p_1} = \frac{dp_1}{0} = \frac{dy}{-2p_2} = \frac{dp_2}{0} = \frac{dz}{2c^2p_3} = \frac{dp_3}{0}$$

From the second and fourth members, we have

$$dp_1 = 0 = dp_2$$

Integrating, we get $\qquad p_1 = a_1 = F_1$ and $p_2 = a_2 = F_2$

As $\qquad (F_1, F_2) = 0$

∴ These may be taken as additional equations

From the given equation, we have

$$p_3 = \frac{(a_1^2 + a_2^2)^{1/2}}{c}$$

Hence,

$$du = p_1dx + p_2dy + p_3dz$$

$$= a_1dx + a_2dy + \frac{(a_1^2 + a_2^2)^{1/2}}{c}dz$$

Integrating, we obtain

$$u = a_1x + a_2y + \frac{(a_1^2 + a_2^2)^{1/2}}{c}z + a_3$$

Taking $a_2 = 1$; the required solution $u = 0$ is given by $a_1x + y + \dfrac{(a_1^2 + 1)^{1/2}}{c}z + a_3 = 0$, which is the required complete integral of the given equation with a_1 and a_3 as arbitrary constants.

EXAMPLE 4.21

Solve by using Jacobi's method

$$p^2x + q^2y = z$$

Solution: Let the solution of the given equation be

$$u(x, y, z) = 0 \tag{1}$$

Assuming u as dependent variable and x, y, z as three independent variables, differentiate Eq. (1) partially with respect to x and y respectively, we get

$$\frac{\partial u}{\partial x} + \frac{\partial u}{\partial z}\frac{\partial z}{\partial x} = 0 \text{ and } \frac{\partial u}{\partial y} + \frac{\partial u}{\partial z}\frac{\partial z}{\partial y} = 0$$

or

$$p_1 + p_3p = 0 \text{ and } p_2 + p_3q = 0$$

∴

$$p = -\frac{p_1}{p_3} \text{ and } q = -\frac{p_2}{p_3}$$

where $p_1 = \dfrac{\partial u}{\partial x}, p_2 = \dfrac{\partial u}{\partial y}, p_3 = \dfrac{\partial u}{\partial z}, p = \dfrac{\partial z}{\partial x}$ and $q = \dfrac{\partial z}{\partial y}$

Thus, the given equation reduces to

$$p_1^2 x + p_2^2 y = z p_3^2$$

Let $f(x, y, z, p_1, p_2, p_3) = xp_1^2 + yp_2^2 - zp_3^2 = 0$.

Now, the Jacobi's auxiliary equation are given by

$$\frac{dx}{-2x_1 p_1} = \frac{dp_1}{p_1^2} = \frac{dy}{-2yp_2} = \frac{dp_2}{p_2^2} = \frac{dz}{2zp_3} = \frac{dp_3}{-p_3^2}$$

Taking the first two members, we get

$$\frac{dx}{x} + \frac{2dp_1}{p_1} = 0$$

Integrating, we obtain

$$\log x + 2 \log p_1 = \log a_1 \text{ or } xp_1^2 = a_1 \text{ or } p_1 = \left(\frac{a_1}{x}\right)^{1/2}$$

Similarly, the third and fourth members give

$$p_2 = \left(\frac{a_2}{y}\right)^{1/2}$$

From $\quad p_1^2 x + p_2^2 y = zp_3^2$; we have

$$p_3 = \left[\frac{(a_1 + a_2)}{z}\right]^{1/2}$$

Hence, $\quad du = p_1 dx + p_2 dy + p_3 dz$

$$= \left(\frac{a_1}{x}\right)^{1/2} dx + \left(\frac{a_2}{y}\right)^{1/2} dy + \left(\frac{a_1 + a_2}{z}\right)^{1/2} dz$$

Integrating, we obtain

$$u = 2[(a_1 x)^{1/2} + (a_2 y)^{1/2} + [(a_1 + a_2)z]^{1/2}] + a_3$$

Taking $a_2 = 1$; we have the required solution

$$u = 2(a_1 x)^{1/2} + 2y^{1/2} + 2(a_1 + 1)^{1/2} + a_3 = 0$$

which is the required complete integral containing two arbitrary constants a_1 and a_3.

EXAMPLE 4.22

Find the complete integral of

$$p_1 p_2 p_3 + p_4^3 x_1 x_2 x_3 x_4^3 = 0.$$

Solution: Let $f(x_1, x_2, x_3, x_4, p_1, p_2, p_3, p_4) = p_1 p_2 p_3 + p_4^3 x_1 x_2 x_3 x_4^3 = 0$.

The Jacobi's auxiliary equations are

$$\frac{dx_1}{-p_2p_3} = \frac{dp_1}{p_4^3x_2x_3x_4^3} = \frac{dx_2}{-p_1p_3} = \frac{dp_2}{p_4^3x_1x_3x_4^3}$$

$$= \frac{dx_3}{-p_1p_2} = \frac{dp_3}{p_4^3x_1x_2x_4^3} = \frac{dx_4}{-3p_4^2x_1x_2x_3x_4^3}$$

$$= \frac{dp_4}{3p_4^3x_1x_2x_3x_4^2}$$

Taking the first two members, we have

$$\frac{p_1dx_1}{-p_1p_2p_3} = \frac{xdp_1}{p_4^3x_1x_2x_3x_4^3}$$

or
$$\frac{dp_1}{p_1} = \frac{dx_1}{x_1}$$

Integrating, we obtain $\log p_1 = \log x_1 + \log a_1$

$\therefore$
$$F_1 = p_1 - a_1x_1 = 0$$

Similarly,
$$F_2 = p_2 - a_2x_2 = 0 \text{ and } F_3 = p_3 - a_3x_3 = 0$$

Now,
$$(F_1, F_2) = \sum_{r=1}^{4}\left(\frac{\partial F_1}{\partial x_r}\frac{\partial F_2}{\partial p_r} - \frac{\partial F_1}{\partial p_r}\frac{\partial F_2}{\partial x_r}\right) = 0$$

Similarly,
$$(F_2, F_3) = 0 \text{ and } (F_1, F_3) = 0$$

$\therefore$ These may be taken as three additional equations.

From the given equation, we have

$$p_1 = a_1x_1,\ p_2 = a_2x_2,\ p_3 = a_3x_3 \text{ and } p_4 = -\frac{(a_1a_2a_3)^{1/3}}{x_4}.$$

Hence, $dz = p_1dx_1 + p_2dx_2 + p_3dx_3 + p_4dx_4 = a_1x_1dx_1 + a_2x_2dx_2 + a_3x_3dx_3 - \dfrac{(a_1a_2a_3)^{1/3}}{x_4}dx_4$

Integrating, we obtain

$$2z = a_1x_1^2 + a_2x_2^2 + a_3x_3^2 - 2(a_1a_2a_3)^{1/3}\log x_4 + a_4.$$

which is the required complete integral containing arbitrary constants a_1, a_2, a_3 and a_4.

EXERCISE 4.2

Solve the following equations by using Jacobi's method:

1. $p_1x_1 + p_2x_2 = p_3^2$.
2. $2p_1x_1x_3 + 3p_2x_3^3 + p_2^2p_3 = 0$.
3. $p_1^2 + p_2p_3 - z(p_2 + p_3) = 0$.
4. $2x_1^2x_2p_1^2p_3 = x_1^2p_2 + 2x_2p_1^2$.
5. $p_1^2 + p_2p_3 - z(p_2 + p_3) = 0$.
6. $p_1x_1 + p_2x_2 - p_3^2 = 0$.
7. $(x_2 + x_3)(p_2 + p_3)^2 + 2p_1 = 0$.
8. $zpq = p + q$.
9. $2x_1x_3zp_1p_3 + x_2p_2 = 0$.
10. $p_1^2 + p_2p_3 - z(p_2 + p_3) = 0$.
11. $q = px + p^2$.
12. $z = pq$.
13. $p_1 + p_2 + p_3 - p_1p_2p_3 = 0$.
14. $x_1p_1^2 + x_2p_2^2 - x_3p_3^2 = 0$.
15. $p_1x_1^2 - p_2^2 - ap_3^2 = 0$.

ANSWERS

Exercise 4.2

1. $z = a_1 \log x_1 + a_2 \log x_2 + (a_1 + a_2)^{1/2} x_3 + a_3$.
2. $z = a_1 \log x_1 + a_2x_2 - \dfrac{(a_1x_3^2 + a_2x_3^3)}{a_2^2} + a_3$.
3. $\log z = (a_1 + a_2 - a_1a_2)^{1/2} x_1 + a_1x_2 + a_2x_3 + a_3$.
4. $z = \dfrac{a_2}{a_1}(a_1x_1^2 - 1)^{1/2} + a_1^2x_2^2 + a_1x_3 + a_3$.
5. $(1 + a_1a_2) \log z = (a_1 + a_2)(x_1 + c_1x_2 + c_2x_3 + c_3)$.
6. $z = a \log x_1 + b \log x_2 + x_3\sqrt{a+b} + c$.
7. $\log z + ax_1 + b(x_2 - x_3) \pm \sqrt{a(x_2 + x_3)} + c = 0$.
8. $z^2 = 2(a + 1)(x + y(a) + b$.

9. $z^2 = a_1 \log x_1 - a_1 a_2 \log x_2 + a_2 \log x_3 + a_3$.

10. $(1 + a_1 a_2) \log z = (a_1 + a_2)(x_1 + a_2 x_2 + a_2 x_3 + a_3)$.

11. $z = -\dfrac{x^2}{4} \pm \dfrac{1}{2}\left\{\dfrac{x}{2}\sqrt{x^2+4a} + 2a\log\left(x+\sqrt{x^2+4a}\right)\right\} + ay + b$.

12. $2\sqrt{z} = \sqrt{ax} + \dfrac{1}{\sqrt{a}}y + b$.

13. $z = a_1 x_1 + a_2 x_2 + a_3 x_3 + a_4$ where $a_1 + a_2 + a_3 - a_1 a_2 a_3 = 0$.

14. $z = 2(a_1 x_1)^{1/2} + 2(a_2 x_2)^{1/2} + 2\left[\dfrac{(a_1+a_2)}{2}x_3\right]^{1/2} + a_3$.

15. $z = -(a_1^2 - a a_2^2)x_1^{-1} + a_1 x_2 + a_2 a_1 x_3 + a_3$.

▲▲

5

CHAPTER

Homogeneous Linear Partial Differential Equations with Constant Coefficients

If the independent variable and its partial derivatives of various order occur linearly in an equation, then the equation is called the linear partial differential equation. If the coefficient of various terms are constant quantities, then it is called the linear partial differential equation with constant coefficients. The standard general form of a linear partial differential equation with constant coefficient is

$$\left(a_0 \frac{\partial^n z}{\partial x^n} + a_1 \frac{\partial^{n-1} z}{\partial x^{n-1} \partial y} + \ldots + a_n \frac{\partial^n z}{\partial y^n}\right) + \left(b_0 \frac{\partial^{n-1} z}{\partial x^{n-1}} + b_1 \frac{\partial^{n-1} z}{\partial^{n-2} z \partial y} + \ldots + b_{n-1} \frac{\partial^{n-1} z}{\partial y^{n-1}}\right)$$

$$+ \ldots + \left(k_0 \frac{\partial z}{\partial x} + k_1 \frac{\partial z}{\partial y}\right) + lz = f(x, y) \quad (1)$$

where $a_0, a_1 \ldots a_n, b_0, b_1 \ldots b_n, \ldots k_0, k_1, l$ are all constants.

We usually write $\frac{\partial}{\partial x} = D_x$ and $\frac{\partial}{\partial y} = D_y$, then Eq. (1) becomes

$$[(a_0 D_x^n + a_1 D_x^{n-1} D_y + \ldots + a_n D_y^n) + (b_0 D_x^{n-1} + b_1 D_x^{n-2} D_y + \ldots + b_{n-1} D_y^{n-1})$$

$$+ \ldots + (k_0 D_x + k_1 D_y) + l]z = f(x, y) \quad (2)$$

When all the partial derivatives appearing in Eq. (1) are of the same order, then the resulting equation is called a linear homogeneous partial differential equation with constant coefficients and is of the form

$$(a_0 D_x^n + a_1 D_x^{n-1} D_y + \ldots + a_n D_y^{n-1})z = f(x, y) \quad (3)$$

On the other hand, when all the partial derivatives are not of the same order, then it is called a non-homogeneous linear partial differential equation with constant coefficients.

In this chapter, we will study the various methods of solving homogeneous linear partial differential equations with constant coefficients.

Note: As in the case of ordinary differential equations, the complete solution can be written as

$$z = \text{Complementary Function (CF)} + \text{Particular Integral (PI)}$$

where the complementary function (CF) is the solution of

$$(a_0D_x^n + a_1D_x^{n-1}D_y + \dots + a_nD_y^n)z = 0$$

and the particular integral (PI) is a solution satisfying

$$(a_0D_x^n + a_1D_x^{n-1}D_y + \dots + a_nD_y^n)z = f(x, y)$$

and does not contain any arbitrary constant.

Now, we will discuss the method to find the complementary function.

We can rewrite the equation

$$(a_0D_x^n + a_1D_x^{n-1}D_y + \dots + a_nD_y^n)z = 0$$

as

$$F(D_x, D_y)z = 0$$

If $F(D_x, D_y)$ can be factorized into linear factors of the form $(a_iD_x + b_iD_y + c_i)$ then $F(D_x, D_y)$ is said to be reducible. Otherwise, it is said to be irreducible. $F(D_x, D_y)$ may contain some reducible and some irreducible factors.

For example,

$$(D_x^2 - D_y^2)z = (D_x + D_y)(D_x - D_y)z = 0$$

$$(D_x^3 - 6D_x^2D_y + 11D_xD_y^2 - 6D_y^3)z = (D_x - D_y)(D_x - 2D_y)(D_x - 3D_y)z = 0$$

are the reducible forms and

$$(2D_x^2 + D_y^2)z = 0$$

$$(D_x^3 + D_x^2D_y + D_y^3)z = 0$$

are irreducible forms while

$$(D_x^3 + DD_y^2)z = D_x(D_x^2 + D_y^2)z = 0$$

$$(D_x^3 - D_x^2D_y + D_xD_y^2 - D_y^3)z = (D_x - D_y)(D_x^2 + D_y^2)z = 0$$

have one reducible factor and one irreducible factor.

Note: We will discuss only the equations which are in reducible form.

As discussed earlier, complementary function is solution of $F(D_x, D_y)z = 0$

Let $F(D_x, D_y) = (D_x - \alpha_1D_y)(D_x - \alpha_2D_y) \dots (D_x - \alpha_nD_y)z = 0$

where $\alpha_1, \alpha_2, \dots \alpha_n$ are constants.

In the above equation, we have written $F(D_x, D_y)$ as a product of n linear factors. As usual, $a_0D_x^n + a_1D_x^{n-1}D_y + \dots + a_nD_y^n = 0$ is called the auxiliary equation. Writing $D_x/D_y = m$, we obtain $a_0m^n + a_1m^{n-1} + \dots + a_n = 0$. Now we factorize this equation and find its roots. Depending on the nature of roots, complementary function (CF) has different forms which we will illustrate by using the following examples. ▲▲

EXAMPLE 5.1

Find the solution of $(D_x^2 - D_y^2)z = 0$ by reducing it to linear factors.

Solution: The given equation can be written as

$$(D_x + D_y)(D_x - D_y)z = 0$$

Let $$(D_x - D_y)z = u$$

$\therefore$ $$(D_x + D_y)u = 0$$

or $$\frac{\partial u}{\partial x} + \frac{\partial u}{\partial y} = 0$$

or $$p' + q' = 0$$

which is of Lagrange's form.

Therefore, the Lagrange's auxiliary equations are

$$\frac{dx}{1} = \frac{dy}{1} = \frac{du}{0}$$

From the first two members, we get $dx = dy$ which on integration gives $x - y = c_1$.
Taking the second and third members, we get $du = 0$ which gives on integration $u = c_2$

Hence, $$u = \phi_1(x - y)$$

Thus, $$(D_x - D_y)z = \phi_1(x - y)$$

or $$\frac{\partial z}{\partial x} - \frac{\partial z}{\partial y} = \phi_1(x - y)$$

or $$p - q = \phi_1(x - y)$$

which is again of Lagrange's form. So its Lagrange's auxiliary equations are

$$\frac{dx}{1} = \frac{dy}{-1} = \frac{dz}{\phi_1(x-y)}$$

Taking the first two members, we obtain

$$dx + dy = 0$$

Integrating, we get $$x + y = c_1'$$

Again, $$\frac{dx - dy}{1-(-1)} = \frac{dz}{\phi_1(x-y)}$$

$$\phi_1(x - y)(dx - dy) = 2dz$$

Integrating, we get

$$\phi_1'(x - y) + c_2' = 2z$$

or $$c_2'' = z - \phi_1''(x - y)$$

The complete solution is

$$\psi(c_1', c_2'') = 0$$

or $$c_2'' = \psi_1(c_1')$$

or $$z - \phi_1''(x - y) = \psi_1(x + y)$$

or $$z = \psi_1(x + y) + \psi_2(y - x)$$

where $$\psi_2 = \phi_1''.$$

Alternately, on writing $D_x/D_y' = m$, the auxiliary equation is $m^2 - 1 = 0 \Rightarrow m = 1, -1$
So, the complete solution is

$$z = \psi_1(y + (1)x) + \psi_2(y + (-1)x)$$

or $$z = \psi_1(y + x) + \psi_2(y - x)$$

EXAMPLE 5.2

Solve $(D_x^2 - 4D_xD_y + 4D_y^2)z = 0$.

Solution: We can rewrite the given equation as

$$(D_x - 2D_y)^2 z = 0$$

On writing $D_x/D_y = m$, the auxiliary equation is $(m - 2)^2 = 0$ or $m = 2, 2$

Let $$(D_x - 2D_y)z = u$$

$\therefore$ $$(D_x - 2D_y)u = 0$$

or $$\frac{\partial u}{\partial x} - 2\frac{\partial u}{\partial y} = 0$$

or $$p_1 - 2q_1 = 0$$

which is of Lagrange's form. So, its Lagrange's auxiliary equation is

$$\frac{dx}{1} = \frac{dy}{-2} = \frac{du}{0}$$

Taking the first two members, we get $2dx + dy = 0$.
Integrating, we get $2x + y = c_1$.
Taking the second and third members, we get $du = 0 \Rightarrow u = c_2$
Hence, we have

$$u = \phi_1(2x + y)$$

$\therefore$ $$(D_x - 2D_y)z = \phi_1(2x + y)$$

or $$\frac{\partial z}{\partial x} - 2\frac{\partial z}{\partial y} = \phi_1(2x + y)$$

or
$$p - 2q = \phi_1(2x + y)$$

which is again of Lagrange's form. So, the Lagrange's auxiliary equations are

$$\frac{dx}{1} = \frac{dy}{-2} = \frac{dz}{\phi(x+y)}$$

Taking the first two members, we get

$$2x + y = 0$$

Integrating, we obtain

$$2x + y = c_1$$

Again, taking the first and third members, we get

$$\frac{dx}{1} = \frac{dz}{\phi_1(c_1)}$$

or
$$\phi_1(c_1)dx = dz$$

Integrating, we obtain

$$x\phi_1(c_1) + c_2' = z$$

or
$$x\phi_1(2x + y) + c_2' = z$$

or
$$c_2' = z - x\phi_1(2x + y)$$

Hence, the complete integral is

$$\psi(c_1, c_2') = 0 \text{ or } c_2' = \psi_1(c_1)$$

or
$$z - x\phi_1(2x + y) = \psi_1(2x + y)$$

or
$$z = \psi_1(2x + y) + x\phi_1(2x + y)$$

or
$$z = \psi_1(2x + y) + x\psi_2(2x + y)$$

where
$$\psi_2 = \phi_1.$$

Alternately, on writing $D_x/D_y' = m$, the auxiliary equation is $(m - 2)^2 = 0$ or $m = 2; 2$
So, the complete solution is

$$z = \psi_1(y + 2x) + x\psi_2(y + 2x).$$

Note: From the above examples, we can say when auxiliary roots $m_1, m_2, m_3, \ldots m_n$ are distinct, the complete integral is

$$z = \psi_1(y + m_1x) + \psi_2(y + m_2x) + \psi_3(y + m_3x) + \ldots + \psi_n(y + m_nx)$$
$$= m_n$$

and when $m_1 = m_2 = m_3 = \ldots = m_n = m$ (say) the complete integral is

$$z = \psi_1(y + mx) + x\psi_2(y + mx) + x^2\psi_3(y + mx) + \ldots + x^{n-1}\psi_n(y + mx)$$

EXAMPLE 5.3

Solve $(D_x^3 - D_x^2 D_y)z = 0$.

Solution: The given equation can be written as

$$(D_x - D_y)D_x^2 z = 0$$

Let $$D_x^2 z = u$$

$\therefore$ $$(D_x - D_y)u = 0$$

or $$\frac{\partial u}{\partial x} - \frac{\partial u}{\partial y} = 0$$

or $$p' - q' = 0$$

which is of Lagrange's form. So, Lagrange's auxiliary equations are

$$\frac{dx}{1} = \frac{dy}{-1} = \frac{du}{0}$$

Taking first two members, we get $dx + dy = 0$

Integrating, we obtain $$x + y = c_1$$

Again, $$du = 0 \Rightarrow u = c_2$$

Hence, its solution is $$u = \phi(x + y)$$

Thus, $$D_x^2 z = \phi(x + y)$$

Integrating, we obtain

$$z = \psi(x + y) + x\phi_1(y) + \phi_2(y)$$

where $$\psi(x + y) = \iint \phi(x + y)dx\, dx.$$

EXAMPLE 5.4

Solve $(D_x D_y^2 - D_y^3)z = 0$.

Solution: The given equation can be written as

$$(D_x - D_y)D_y^2 z = 0$$

Let $$D_y^2 z = u$$

$\therefore$ $$(D_x - D_y)u = 0$$

or $$\frac{\partial u}{\partial x} - \frac{\partial u}{\partial y} = 0$$

or $$p - q = 0$$

which is of Lagrange's form. So, Lagrange's auxiliary equation is

$$\frac{dx}{1} = \frac{dy}{-1} = \frac{du}{0}$$

From the previous example, its solution, $u = \phi(x + y)$

Thus, $$D_y^2 z = \phi(x + y).$$

Integrating, we obtain

$$z = \psi(x + y) + y\phi_1(x) + \phi_2(x)$$

where, $$\psi(x + y) = \iint \phi(x + y)dy\, dy.$$

Note: From the above two examples, we can say, if D_x's are common, multiply by successive powers of x and if D_y's are common, multiply by successive powers of y.

EXAMPLE 5.5

Solve $(D_x^3 - 3D_x^2 D_y + 2D_x D_y^2)z = 0$.

Solution: The auxiliary equation is

$$m^3 - 3m^2 + 2m = 0$$

or $$m(m - 1)(m + 2) = 0$$

or $$m = 0, 1, 2$$

Hence, the complete solution is

$$z = \phi_1(y + 0.x) + \phi_2(y + x) + \phi_3(y + 2x) = \phi_1(y) + \phi_2(y + x) + \phi_3(y + 2x).$$

EXAMPLE 5.6

Solve $(D_x^3 - 6D_x^2 D_y + 11D_x D_y^2 - 6D_y^3)z = 0$.

Solution: The auxiliary equation is

$$m^3 - 6m^2 + 11m - 6 = 0$$

or $$(m - 1)(m - 2)(m - 3) = 0$$

or $$m = 1, 2, 3$$

Hence, the complete solution is given by

$$z = \phi_1(y + x) + \phi_2(y + 2x) + \phi_3(y + 3x).$$

EXAMPLE 5.7

Solve $2r + 5s + 2t = 0$.

Solution: Since $r = \frac{\partial^2 z}{\partial x^2} = D_x^2 z$, $s = \frac{\partial^2 z}{\partial x \partial y} = D_x D_y z$ and $t = \frac{\partial^2 z}{\partial y^2} = D_y^2 z$ the given equation is

$$(2D_x^2 + 5D_xD_y + 2D_y^2)z = 0$$

The auxiliary equation is

$$2m^2 + 5m + 2 = 0$$

or $$(2m + 1)(m + 2) = 0$$

or $$m = -\frac{1}{2}, -2$$

Hence, the complete solution is

$$z = \phi_1\left(y - \frac{1}{2}x\right) + \phi_2(y - 2x)$$

or $$z = \psi_1(2y - x) + \psi_2(y - 2x).$$

EXAMPLE 5.8

Solve $\frac{\partial^2 z}{\partial x^2} + a^2 \frac{\partial^2 z}{\partial y^2} = 0.$

Solution: The given equation is $(D_x^2 + a^2 D_y^2)z = 0$

The auxiliary equation is $$m^2 + a^2 = 0$$

or $$(m + ia)(m - ia) = 0$$

or $$m = -ia, ia$$

Hence, the complete solution is

$$z = \phi_1(y + iax) + \phi_2(y - iax)$$

EXAMPLE 5.9

Solve $\frac{\partial^4 z}{\partial x^4} - \frac{\partial^2 z}{\partial y^4} = 0.$

Solution: The given equation is $(D_x^4 - D_y^4)z = 0$

The auxiliary equation is $$m^4 - 1 = 0$$

or $$(m^2 - 1)(m^2 + 1) = 0$$

or $$m = \pm 1, \pm i$$

Hence, the complete solution is

$$z = \phi_1(y + x) + \phi_2(y - x) + \phi_3(y + ix) + \phi_4(y - jx).$$

EXAMPLE 5.10

Solve $(4D_x^2 + 12D_xD_y + 9D_y^2)z = 0.$

Solution: The auxiliary equation is

$$4m^2 + 12m + 9 = 0$$

or
$$(2m + 3)^2 = 0$$

or
$$m = -3/2, -3/2$$

Hence, the solution is

$$z = \phi_1\left(y - \frac{3}{2}x\right) + x\phi_2\left(y - \frac{3}{2}x\right)$$

or
$$z = \psi_1(2y - 3x) + x\psi_2(2y - 3x).$$

EXAMPLE 5.11

Solve $(D_x^3 - 3D_x^2D_y + 3D_xD_y^2 - D_y^3)z = 0.$

Solution: The auxiliary equation is

$$m^3 - 3m^2 + 3m - 1 = 0$$

or
$$(m - 1)^3 = 0$$

or
$$m = 1, 1, 1$$

Hence, the solution is

$$z = \phi_1(y + x) + x\phi_2(y + x) + x^2\phi_3(y + x).$$

EXAMPLE 5.12

Solve $(D_x^4 - 2D_x^3D_y + 2D_xD_y^3 - D_y^4)z = 0$

Solution: The auxiliary equation is

$$m^4 - 2m^3 + 2m - 1 = 0$$

or
$$(m - 1)^3 (m + 1) = 0$$

or
$$m = 1, 1, 1, -1$$

Hence, the solution is

$$z = \phi_1(y + x) + x\phi_2(y + x) + x^2\phi_3(y + x) + \phi_4(y - x).$$

EXAMPLE 5.13

Find the solution of biharmonic equation

$$(D_x^4 - 2D_x^2D_y^2 + D_y^4)z = 0.$$

Solution: The auxiliary equation is

$$m^4 - 2m^2 + 1 = 0$$

or $$(m^2 - 1)^2 = 0$$

or $$m^2 = 1, 1$$

or $$m = \pm 1, \pm 1$$

Hence, the complete solution is

$$z = \phi_1(y + x) + x\phi_2(y + x) + \phi_3(y - x) + x\phi_4(y - x).$$

EXAMPLE 5.14

Solve $(D_x^2 D_y - DxD_y^2)z = 0$.

Solution: The auxiliary equation is

$$m^2 - m = 0$$

or $$m(m - 1) = 0$$

or $$m = 0, 1$$

Hence, the complete solution is

$$z = \phi_1(y) + \phi_2(y + x)$$

EXAMPLE 5.15

Solve $(D_x^3 - 2D_x^2 D_y)z = 0$.

Solution: The auxiliary equation is

$$m^3 - 2m^2 = 0$$

or $$m^2(m - 2) = 0$$

or $$m = 0, 0, 2$$

Hence, the complete solution is

$$z = \phi_1(y) + x\phi_2(y) + \phi_3(y + 2x).$$

▲▲

EXERCISE 5.1

Find the solution of the following homogeneous linear partial differential equations:

1. $(4D_x^3 - 3D_x D_y^2 + D_y^3)z = 0$.
2. $r = a^2 t$.
3. $\dfrac{\partial^3 z}{\partial x^3} - 7\dfrac{\partial^3 z}{\partial x \partial y^2} + 6\dfrac{\partial^3 z}{\partial y^3} = 0$.

4. $(D_x^2 - 3aD_xD_y + 2a^2D_y^2)z = 0.$
5. $25r - 40s + 16t = 0.$
6. $\frac{\partial^4 z}{\partial x^4} + \frac{\partial^4 z}{\partial y^4} - 2\frac{\partial^4 z}{\partial x^2 \partial y^2} = 0.$
7. $\frac{\partial^3 z}{\partial x^3} - 2\frac{\partial^3 z}{\partial x^2 \partial y} + \frac{\partial^3 z}{\partial x \partial y^2} = 0.$
8. $r - 4s + 4t = 0.$
9. $(D_x^4 - D_y^4)z = 0.$
10. $(D_x^4 + D_y^4)z = 0.$
11. $\frac{\partial^4 z}{\partial y^3 \partial x} + \frac{\partial^4 z}{\partial y^4} = 0.$
12. $(D_x^3 - D_x^2D_y - 6D_xD_y^2)z = 0.$
13. $\frac{\partial^5 z}{\partial x^3 \partial y^2} + \frac{\partial^5 z}{\partial x^2 \partial y^3} = 0.$
14. $(D_x^2D_y - 3D_xD_y^2 + 2D_y^3)z = 0.$
15. $(D_x^2 - 2D_xD_y)z = 0.$
16. $\frac{\partial^3 z}{\partial x^3} - 7\frac{\partial^3 z}{\partial x^2 \partial y} + 12\frac{\partial^3 z}{\partial x \partial y^2} = 0.$
17. $(D_x^2D_y - 3D_xD_y^2 + 2D_y^3)z = 0.$
18. $\frac{\partial^4 z}{\partial y^3 \partial x} + \frac{\partial^4 z}{\partial y^4} = 0.$
19. $(D_x^3 - D_x^2D_y - 6D_xD_y^2)z = 0.$
20. $\frac{\partial^4 z}{\partial x^4} - 2\frac{\partial^4 z}{\partial x^2 \partial y^2} + \frac{\partial^4 z}{\partial y^4} = 0.$
21. $(D_x^4 - 2D_x^3D_y + 2D_xD_y^3 - D_y^4)z = 0.$
22. $(D_x^2 - 2D_xD_y - D_y^2) = 0.$
23. $(4D_x^2 + 12D_xD_y + 9D_y^2)z = 0.$

24. $2\frac{\partial^2 z}{\partial x^2} - 3\frac{\partial^2 y}{\partial x \partial y} - 2\frac{\partial^2 z}{\partial y^2} = 0$.

25. $r - 4s + 4t = 0$.

26. $\frac{\partial^4 z}{\partial x^4} + \frac{\partial^4 z}{\partial y^4} + 2\frac{\partial^4 z}{\partial x^2 \partial y^2} = 0$.

ANSWERS

Exercise 5.1

1. $z = \phi_1(y - x) + \phi_2(2y + x) + x\phi_2(2y + x)$.
2. $z = \phi_1(y + ax) + \phi_2(y - ax)$.
3. $z = \phi_1(y + x) + \phi_2(y + 2x) + \phi_3(y - 3x)$.
4. $z = \phi_1(y + ax) + \phi_2(y + 2ax)$.
5. $z = \phi_1(5y + 4x) + x\phi_2(5y + 4x)$.
6. $z = \phi_1(y + x) + x\phi_2(y + x) + \phi_3(y - x) + x\phi_4(y - x)$.
7. $z = \phi_1(y) + \phi_2(y + x) + x\phi_3(y + x)$.
8. $z = \phi_1(y + 2x) + x\phi_2(y + 2x)$.
9. $z = \phi_1(y + x) + \phi_2(y - x) + \phi_3(y + ix) + \phi_4(y - jx)$.
10. $z = \phi_1(y + \alpha x) + \phi_2(y + \overline{\alpha x}) + \phi_3(y + \beta x) + \phi_4(y + \overline{\beta x})$ where $\alpha = \frac{1+i}{\sqrt{2}}$ and $\beta = \frac{-1+j}{\sqrt{2}}$.
11. $z = \phi_1(x) + y\phi_2(x) + y^2\phi_3(x) + \phi_4(y - x)$.
12. $z = \phi_1(y) + \phi_2(y - 3x) + \phi_3(y + 2x)$.
13. $z = \phi_1(x) + y\phi_2(x) + \phi_3(y) + x\phi_4(y) + \phi_5(y - x)$.
14. $z = \phi_1(x) + \phi_2(y + x) + \phi_3(y + 2x)$
15. $z = \phi_1(y) + \phi_2(y + 2x)$.
16. $z = \phi_1(y) + \phi_2(y + 3x) + \phi_3(y + 4x)$.
17. $z = \phi_1(x) + \phi_2(y + x) + \phi_3(y + 2x)$.
18. $z = \phi_1(y - x) + \phi_2(x) + y\phi_3(x) + y^2\phi_4(x)$.
19. $z = \phi_1(y - 3x) + \phi_2(y) + \phi_3(y + 2x)$.
20. $z = \phi_1(y + x) + \phi_2(y - x) + x\phi_3(y + x) + x\phi_4(y - x)$.
21. $z = \phi_1(y - x) + \phi_2(y + x) + \phi_3(y + x) + x^2\phi_4(y + x)$.
22. $z = \phi_1(y + (1 + \sqrt{2})x) + \phi_2(y + (1 - \sqrt{2})x)$.

23. $z = \phi_1(2y - 3x) + x\phi_2(2y - 3x)$.

24. $z = \phi_1(2y - x) + \phi_2(y + 2x)$.

25. $z = \phi_1(y + 2x) + x\phi_2(y + 2x)$.

26. $z = \phi_1(y + ix) + x\phi_2(y + jx) + \phi_3(y - ix) + x\phi_4(y - jx)$.

▲▲

Particular Integral

Consider $F(D_x, D_y)z = f(x, y)$ then particular integral is $\dfrac{1}{F(D_x, D_y)} f(x, y)$.

For finding particular integral, we give three rules depending on the nature of $f(x, y)$ and taking $F(D_x, D_y) = D_x^2 + c_1 D_x D_y + c_2 D_y^2$.

Rule I

When $f(x, y) = e^{\alpha x + \beta y}$ then $\dfrac{1}{F(D_x, D_y)} e^{\alpha x + \beta y} = \dfrac{1}{F(\alpha, \beta)} e^{\alpha x + \beta y}$ provided $F(\alpha, \beta) \neq 0$.

We have $D_x(e^{\alpha x + \beta y}) = \alpha e^{\alpha x + \beta y}$

$$D_y(e^{\alpha x + \beta y}) = \beta e^{\alpha x + \beta y}$$

$$D_x^2(e^{\alpha x + \beta y}) = \alpha^2 e^{\alpha x + \beta y}$$

$$D_y^2(e^{\alpha x + \beta y}) = \beta^2 e^{\alpha x + \beta y}$$

and
$$D_x D_y(e^{\alpha x + \beta y}) = D_x(\beta e^{\alpha x + \beta y}) = \alpha\beta e^{\alpha x + \beta y}.$$

So that

$$(D_x^2 + c_1 D_x D_y + c_2 D_y^2)(e^{\alpha x + \beta y}) = (\alpha^2 + c_1\alpha\beta + c_2\beta^2)(e^{\alpha x + \beta y})$$

$$\Rightarrow \quad F(D_x, D_y)e^{\alpha x + \beta y} = F(\alpha, \beta)e^{\alpha x + \beta y}$$

$$\Rightarrow \quad e^{\alpha x + \beta y} = F(\alpha, \beta)\frac{1}{F(D_x, D_y)} e^{\alpha x + \beta y}$$

$$\Rightarrow \quad \frac{1}{F(\alpha, \beta)} e^{\alpha x + \beta y} = \frac{1}{F(D_x, D_y)} e^{\alpha x + \beta y} \text{ provided } F(\alpha, \beta) \neq 0.$$

Rule II

When $f(x, y) = \sin(\alpha x + \beta y)$ or $\cos(\alpha x + \beta y)$ then we write $D_x^2 = -\alpha^2$, $D_x D_y = -\alpha\beta$ and $D_y^2 = -\beta^2$ in the denominator.

We shall illustrate this by taking $\dfrac{1}{F(D_x, D_y)} \cos(\alpha x + \beta y)$

Now, $D_x(\cos(\alpha x + \beta y)) = -\alpha \sin(\alpha x + \beta y)$

$$D_x^2(\cos(\alpha x + \beta y)) = -\alpha^2 \cos(\alpha x + \beta y)$$

$$D_y(\cos(\alpha x + \beta y)) = -\beta \sin(\alpha x + \beta y)$$

$$D_y^2(\cos(\alpha x + \beta y)) = -\beta^2 \cos(\alpha x + \beta y) \text{ and } D_x D_y = -\alpha\beta \cos(\alpha x + \beta y)$$

$\therefore$
$$F(D_x, D_y) \cos(\alpha x + \beta y)$$
$$= (D_x^2 + c_1 D_x D_y + c_2 D_y^2) \cos(\alpha x + \beta y)$$
$$= (-\alpha^2 + c_1(-\alpha\beta) + c_2(-\beta^2)) \cos(\alpha x + \beta y)$$

$\Rightarrow$
$$\cos(\alpha x + \beta y) = (-\alpha^2 + c_1(-\alpha\beta) + c_2(-\beta^2)) \frac{1}{F(D_x, D_y)} \cos(\alpha x + \beta y)$$

$\Rightarrow$
$$\frac{1}{-\alpha^2 + c_1(-\alpha\beta) + c_2(-\beta^2)} \cos(\alpha x + \beta y) = \frac{1}{F(D_x, D_y)} \cos(\alpha x + \beta y)$$

provided $-\alpha^2 + c_1(-\alpha\beta) + c_2(-\beta^2) \neq 0$.

Similarly, we can see

$$\frac{1}{F(D_x, D_y)} \sin(\alpha x + \beta y) = \frac{1}{-\alpha^2 + c_1(-\alpha\beta) + c_2(-\beta^2)} \sin(\alpha x + \beta y)$$

provided $-\alpha^2 + c_1(-\alpha\beta) + c_2(-\beta^2) \neq 0$.

Rule III

When $f(x, y)$ = sum or difference of terms of the form $x^m y^n$ where, m and n are constants then write

$$\frac{1}{F(D_x, D_y)} \phi(x^m, y^n) = [F(D_x, D_y)]^{-1} \phi(x^m, y^n)$$

To evaluate it, expand $[F(D_x, D_y)]^{-1}$ in ascending power of D_x if $m < n$ or D_y if $m > n$ by using Binomial theorem and then operate on $x^m y^n$ term by term. ▲▲

EXAMPLE 5.16

Solve $\dfrac{\partial^3 z}{\partial x^3} - 3\dfrac{\partial^3 z}{\partial x^2 \partial y} + 4\dfrac{\partial^3 z}{\partial y^3} = e^{x+2y}$.

Solution: The given equation is

$$(D_x^3 - 3D_x^2 D_y + 4D_y^3)z = e^{x+2y}$$

Here, auxiliary equation is

$$m^3 - 3m^2 + 4 = 0$$

or $(m + 1)(m - 2)^2 = 0$

or $m = -1, 2, 2$

$\therefore$ CF is

$$z_c = \phi_1(y - x) + \phi_2(y + 2x) + x\phi_3(y + 2x)$$

and PI is

$$z_p = \frac{1}{Dx^3 - 3Dx^2D_y + 4D_y^3}e^{x+2y}$$

$$= \frac{1}{(1)^3 - 3(1)^2(2) + 4(2)^3}e^{x+2y} \qquad \text{\{using Rule I\}}$$

$$= \frac{1}{27}e^{x+2y}$$

Hence, the complete solution is

$$z = z_c + z_p$$

$$z = \phi_1(y - x) + \phi_2(y + 2x) + x\phi_3(y + 2x) + \frac{1}{27}e^{x+2y}.$$

EXAMPLE 5.17

Solve $(D_x^3 - 6D_x^2D_y + 11D_xD_y^2 - 6D_y^3)z = 0.$

Solution: Here auxiliary equation is

$$m^3 - 6m^2 + 11m - 6 = 0$$

or $(m - 1)(m - 2)(m - 3) = 0$

or $m = 1, 2, 3$

$\therefore$ CF is

$$z_c = \phi_1(y + x) + \phi_2(y + 2x) + \phi_3(y + 3x)$$

and PI is

$$z_p = \frac{1}{Dx^3 - 6Dx^2D_y + 11D_xD_y^2 - 6Dy^3}e^{2x+3y}$$

$$= \frac{1}{(2)^3 - 6(2)^2(3) + 11(2)(3)^2 - 6(3)^3}e^{2x+3y}$$

{using Rule II}

$$= -\frac{1}{28}e^{2x+3y}.$$

Hence, the complete solution is

$$z = z_c + z_p$$

$$z = \phi_1(y + x) + \phi_2(y + 2x) + \phi_3(y + 3x) + \frac{1}{28}e^{2x+3y}.$$

EXAMPLE 5.18

Solve $\dfrac{\partial^2 z}{\partial x \partial y} = e^{x+y}$.

Solution: For finding CF

$$\frac{\partial^2 z}{\partial x \partial y} = 0$$

$$\Rightarrow \quad \frac{\partial}{\partial x}\left(\frac{\partial z}{\partial y}\right) = 0$$

$$\Rightarrow \quad \frac{\partial z}{\partial y} = \phi(y)$$

$$\therefore \quad z = \int \phi(y)dy + \phi_2(x)$$

Thus,

$$z_c = \phi_1(y) + \phi_2(x)$$

and PI is

$$z_p = \frac{1}{D_x D_y} e^{x+y} = \frac{1}{(1)(1)} e^{x+y} \quad \text{\{using Rule I\}}$$

$$= e^{x+y}$$

Hence, the complete solution is

$$z = z_c + z_p$$

$$z = \phi_1(y) + \phi_2(x) + e^{x+y}.$$

EXAMPLE 5.19

Solve $r - 2s + t = \sin(2x + 3y)$.

Solution: The given equation can be written as

$$(D_x^2 - 2D_x Dy + D_y^2)z = \sin(2x + 3y).$$

Here, the complete solution is

$$m^2 - 2m + 1 = 0$$

or

$$(m - 1)^2 = 0$$

or

$$m = 1, 1$$

$\therefore$ CF is

$$z_c = \phi_1(y + x) + x\phi_2(y + x)$$

and PI is

$$z_p = \frac{1}{D_x^2 - 2D_x D_y + D_y^2} \sin(2x + 3y)$$

$$= \frac{1}{-2^2 - 2(-2\cdot3) - 3^2} \sin(2x + 3y)$$

{using Rule II write $D_x^2 = -2^2$, $D_xD_y = -2\cdot3$ and $D_y^2 = -3^2$}

$$= -\sin(2x + 3y)$$

Hence, the complete solution is

$$z = z_c + z_p$$

$$z = \phi_1(y + x) + x\phi_2(y + x) - \sin(2x + 3y).$$

EXAMPLE 5.20

Solve $r + t = \cos mx \cos ny$.

Solution: The given equation is

$$(D_x^2 + D_y^2)z = \cos mx \cos ny$$

Here, auxiliary equation is $m^2 + 1 = 0$

or $m = \pm i$

$\therefore$ CF is

$$z_c = \phi_1(y + ix) + \phi_2(y - ix)$$

and PI is $z_p = \dfrac{1}{D_x^2 + D_y^2} \cos mx \cos ny$

$$= \frac{1}{D_x^2 + D_y^2}\left[\frac{1}{2}(\cos(mx+ny)+\cos(mx-ny))\right]$$

$$= \frac{1}{2}\left[\frac{1}{(-m^2)+(-n^2)}\cos(mx+ny)+\frac{1}{(-m^2)+(-n)^2}\cos(mx+ny)\right] \text{ \{using Rule II]}$$

$$= -\frac{\cos mx \cos ny}{m^2+n^2}$$

Hence, the complete solution is

$$z = z_c + z_p$$

$$z = \phi_1(y + ix) + \phi_2(y - ix) - \frac{\cos mx \cos ny}{m^2+n^2}$$

EXAMPLE 5.21

Solve $(D_x^3 - 4D_x^2D_y + 4D_xD_y^2)z = 8\cos(3y + 2x)$.

Solution: Here, the auxiliary equation is

$$m^3 - 4m^2 + 4m = 0$$

or
$$m(m-2)^2 = 0$$

or
$$m = 0, 2, 2$$

∴ CF is

$$z_c = \phi_1(y) + \phi_2(y+2x) + x\phi_3(y+2x)$$

and PI is
$$z_p = \frac{1}{D_x^3 - 4D_x^2 D_y + 4D_x D_y^2} 8\cos(3y+2x)$$

$$= \frac{1}{D_x^2 - 4D_x D_y + 4D_y^2}\frac{1}{D_x} 8\cos(3y+2x)$$

$$= \frac{1}{D_x^2 - 4D_x D_y + 4D_y^2} 8\int \cos(3y+2x)dx$$

$$= 8\frac{1}{D_x^2 - 4D_x D_y + 4D_y^2}\frac{\sin(3y+2x)}{+2}$$

$$= 4\frac{1}{(-2^2) - 4(-2^3) + 4(-3^2)} \sin(3y+2x)$$

$$= -\frac{2}{13}\sin(3y+2x)$$

Hence, the complete solution is

$$z = z_c + z_p$$

$$z = \phi_1(y) + \phi_2(y+2x) + x\phi_3(y+2x) - \frac{2}{13}\sin(3y+2x).$$

EXAMPLE 5.22

Solve $(D_x^2 - 6D_x D_y + 9D_y^2)z = 12x^2 + 36xy$.

Solution: Here, the auxiliary equation is

$$m^2 - 6m + 9 = 0$$

or
$$(m-3)^2 = 0$$

or
$$m = 3, 3.$$

∴ CF is

$$z_c = \phi_1(y+3x) + x\phi_2(y+3x)$$

and PI is
$$z_p = \frac{1}{D_x^2 - 6D_x D_y + 9D_y^2}(12x^2 + 36xy)$$

$$= \frac{1}{(D_x - 3D_y)^2}(12x^2 + 36xy)$$

$$= \frac{1}{D_x^2}\left(1 - \frac{3D_y}{D_x}\right)^{-2}(12x^2 + 36xy)$$

$$= \frac{1}{D_x^2}\left(1 + 6\frac{D_y}{D_x} + 27\frac{D_y^2}{D_x^2} + \ldots\right)(12x^2 + 36xy)$$

$$= \frac{1}{D_x^2}\left(12x^2 + 36xy + \frac{6}{D_x}(36x)\right)$$

$$= x^4 + 6x^2y + 9x^4 = 10x^4 + 6x^2y.$$

Hence, the complete solution is

$$z = z_c + z_p$$

$$z = \phi_1(y + 3x) + x\phi_2(y + 3x) + 10x^4 + 6x^2y.$$

EXAMPLE 5.23

Solve $(D_x^2 + 2D_xD_y + D_y^2)z = x^2 + xy + y^2$.

Solution: Here, the auxiliary equation is

$$m^2 + 2m + 1 = 0$$

or $$(m + 1)^2 = 0$$

or $$m = -1, -1$$

$\therefore$ CF is

$$z_c = \phi_1(y - x) + x\phi_2(y - x)$$

and PI is $$z_p = \frac{1}{(D_x + D_y)^2}(x^2 + xy + y^2)$$

$$= \frac{1}{D_x^2}\left(1 + \frac{D_y}{D_x}\right)^{-2}(x^2 + xy + y^2).$$

$$= \frac{1}{D_x^2}\left(1 - 2\frac{D_y}{D_x} + 3\frac{D_y^2}{D_x^2} - \ldots\right)(x^2 + xy + y^2)$$

$$= \frac{1}{D_x^2}\left(x^2 + xy + y^2 - \frac{2}{D_x}(x + 2y) + \frac{3}{D_x^2}(2)\right)$$

$$= \left\{\left(\frac{x^4}{12}+\frac{yx^3}{6}+\frac{x^2y^2}{2}\right)-2\left(\frac{x^4}{24}+\frac{yx^3}{3}\right)+6\left(\frac{x^4}{24}\right)\right\}$$

$$= \frac{1}{4}(x^4 - 2x^3y + 2x^2y^2).$$

Hence, the complete solution is

$$z = z_c + z_p$$

$$z = \phi_1(y - x) + x\phi_2(y - x) + \frac{1}{4}(x^4 - 2x^3y + 2x^2y^2).$$

EXAMPLE 5.24

Find the complete solution of

$$\frac{\partial^2 z}{\partial x^2}+(\alpha+\beta)\frac{\partial^2 z}{\partial x\partial y}+\alpha\beta\frac{\partial^2 z}{\partial y^2} = xy.$$

Solution: The given equation is

$$(D_x^2 + (\alpha + \beta)D_xD_y + \alpha\beta\, D_y{}^2)z = xy.$$

Here, the auxiliary equation is

$$m^2 + (\alpha + \beta)m + \alpha\beta = 0$$

or $$(m + \alpha)(m + \beta) = 0$$

or $$m = -\alpha, -\beta$$

∴ CF is

$$z_c = \phi_1(y - ax) + \phi_2(y - \beta x)$$

and PI is $$z_p = \frac{1}{D_x^2+(\alpha+\beta)D_xD_y+\alpha\beta D_y^2}(xy)$$

$$= \frac{1}{D_x^2}\left\{1+\left((\alpha+b)\frac{D_y}{D_x}+\alpha\beta\frac{D_y^2}{D_x^2}\right)\right\}^{-1}(xy)$$

$$= \frac{1}{D_x^2}\left\{1-\left((\alpha+\beta)\frac{D_y}{D_x}+\alpha\beta\frac{D_y^2}{D_x^2}\right)\right\}(xy)$$

$$= \frac{1}{D_x^2}\left[xy-\frac{(\alpha+\beta)}{D_x}(x)+\frac{\alpha\beta}{D_x^2}(0)\right]$$

$$= \frac{x^3y}{6}-(\alpha+\beta)\frac{x^4}{24}.$$

Hence, the complete solution is

$$z = z_c + z_p$$

$$z = \phi_1(y - \alpha x) + \phi_2(y - \beta x) + \frac{x^3 y}{6} - \frac{(\alpha+\beta)}{24}x^4.$$

EXAMPLE 5.25

Solve $\dfrac{\partial^2 z}{\partial x^2} - 3\dfrac{\partial^2 z}{\partial x \partial y} + 2\dfrac{\partial^2 z}{\partial y^2} = e^{2x - y} + \cos(x + 2y)$.

Solution: The given equation is

$$(D_x^2 - 3D_xD_y + 2D_y^2)z = e^{2x - y} + \cos(x + 2y)$$

Here, the auxiliary equation is

$$m^2 - 3m + 2 = 0$$

or $$(m - 2)(m - 1) = 0$$

or $$m = 1, 2$$

∴ CF is

$$z_c = \phi_1(y + x) + \phi_2(y + 2x)$$

and PI is $$y_p = \frac{1}{D_x^2 - 3D_xD_y + 2D_y^2}e^{2x-y} + \frac{1}{D_x^2 - 3D_xD_y + 2D_y^2}\cos(x + 2y)$$

$$= \frac{1}{(2)^2 - 3(2)(-1) + 2(-1)^2}e^{2x-y} + \frac{1}{-1^2 - 3(-1\cdot 2) + 2(-2^2)}\cos(x + 2y)$$

$$= \frac{1}{12}e^{2x-y} - \frac{1}{3}\cos(x + 2y)$$

Hence, the complete solution is

$$z = z_c + z_p$$

$$z = \phi_1(y + x) + \phi_2(y + 2x) + \frac{1}{12}e^{2x - y} - \frac{1}{3}\cos(x + 2y).$$

EXAMPLE 5.26

Solve $\dfrac{\partial^3 z}{\partial x^3} - 2\dfrac{\partial^3 z}{\partial x^2 \partial y} = 2e^{2x} + 3x^2 y$.

Solution: The given equation is $(D_x^3 - 2D_x^2D_y)z = 2e^{2x} + 3x^2y$.

Here, the auxiliary equation is

$$m^3 - 2m^2 = 0$$

or $$m^2(m - 2) = 0$$

or $$m = 0, 0, 2.$$

$\therefore$ CF is

$$z_c = \phi_1(y) + x\phi_2(y) + \phi_3(y + 2x)$$

and PI is

$$z_p = 2\frac{1}{D_x^3 - 2D_x^2 D_y}e^{2x} + 3\frac{1}{D_x^3 - 2D_x^2 D_y}x^2 y$$

$$= 2\frac{1}{(2)^3 - 2(2)^2(0)}e^{2x} + \frac{3}{D_x^3}\left(1 - 2\frac{D_y}{D_x}\right)^{-1}(x^2 y)$$

$$= \frac{1}{4}e^{2x} + \frac{3}{D_x^3}\left(1 + 2\frac{D_y}{D_x} + 4\frac{D_y^2}{D_x^2} + \ldots\right)(x^2 y)$$

$$= \frac{1}{4}e^{2x} + \frac{3}{D_x^3}\left(x^2 y + \frac{2}{D_x}x^2\right)$$

$$= \frac{1}{4}e^{2x} + 3\left(\frac{x^5 y}{160} + 2\frac{x^6}{360}\right)$$

$$= \frac{1}{4}e^{2x} + \frac{x^5 y}{20} + \frac{x^6}{60}$$

Hence, the complete solution is

$$z = z_c + z_p$$

$$z = \phi_1(y) + x\phi_2(y) + \phi_3(y + 2x) + \frac{1}{4}e^{2x} + \frac{x^5 y}{20} + \frac{x^6}{60}.$$

or $z = \phi_1(y) + x\phi_2(y) + \phi_3(y + 2x) + \dfrac{1}{60}(15e^{2x} + 3x^{5y} + x^6).$ ▲▲

THEOREM: If $F(D_x, D_y)$ be homogeneous function of D_x and D_y of degree n then

$$\frac{1}{F(D_x, D_y)}f(\alpha x + \beta y) = \frac{1}{F(\alpha, \beta)}\underbrace{\iiint \ldots \int}_{n=\text{times}} f(u)du du \ldots du$$

where $u = \alpha x + \beta y$ provided $F(\alpha, \beta) \neq 0$.

Proof: We know

$$D_x f(\alpha x + \beta y) = \alpha f'(\alpha x + \beta y)$$

$$D_x^2 f(\alpha x + \beta y) = \alpha^2 f''(\alpha x + \beta y)$$

$$\ldots \ldots \ldots \ldots \ldots \ldots \ldots \ldots \ldots \ldots \ldots \ldots$$

$$D^p f(\alpha x + \beta y) = \alpha^p f^{(p)}(\alpha x + \beta y)$$

Similarly,

$$D^q f(\alpha x + \beta y) = \beta^q f^{(q)}(\alpha x + \beta y)$$

and $$D_x^r D_y^t f(\alpha x+\beta y) = a^r \beta^t f^{(r+t)}(\alpha x+\beta y).$$

As $F(D_x, D_y)$ is a homogeneous function of degree n.

$\therefore$ $$F(D_x, D_y) f(\alpha x+\beta y) = F(\alpha, \beta) f^{(n)}(\alpha x+\beta y)$$

$\Rightarrow$ $$f(\alpha x+\beta y) = \frac{1}{F(D_x, D_y)} F(\alpha, \beta) f^{(n)}(\alpha x+\beta y)$$

$\Rightarrow$ $$\frac{1}{F(\alpha,\beta)} f(\alpha x+\beta y) = \frac{1}{F(D_x, D_y)} f^{(n)}(\alpha x+\beta y)$$

Let $$\alpha x+\beta y = u$$

$\Rightarrow$ $$\frac{1}{F(\alpha,\beta)} f(u) = \frac{1}{F(D_x, D_y)} f^{(n)}(u)$$

$\therefore$ $$\frac{1}{F(\alpha,\beta)} \underbrace{\iiint \ldots \int}_{n=\text{times}} f(u)du du \ldots du = \frac{1}{F(D_x, D_y)} f(u)$$

Exceptional case. When $F(\alpha, \beta) = 0$, then the above result does not hold good. We shall find partial integral as follows.

If $F(\alpha, \beta) = 0$ then $\beta D_x - \alpha D_y$ must be a factor of $F(D_x, D_y)$. $\{\because \beta D_x - \alpha D_y = \beta\alpha - \alpha\beta = 0\}$

Let this factor be repeated m times ($m \geq 1$).

Then $$\frac{1}{(\beta D_x - \alpha D_y)^m} f(\alpha x+\beta y) = \frac{x^m}{\beta^m (m!)} f(\alpha x+\beta y).$$

Consider the equation $(\beta D_x - \alpha D_y)z = x^r f(\alpha x+\beta y)$ (1)

or $$\beta p - \alpha q = x^r f(\alpha x+\beta y)$$

which is of Lagrange's form. So, its auxiliary equations are

$$\frac{dx}{\beta} = \frac{dy}{-\alpha} = \frac{dz}{x^r f(\alpha x+\beta y)}$$

Taking the first two members, we have

$$\alpha dx + \beta dy = 0$$

so that $$\alpha x+\beta y = c_1$$

Taking the first and third members, we get

$$\frac{dx}{\beta} = \frac{dz}{x^r f(c_1)} \text{ or } dz = \frac{x^r f(c_1)}{\beta} dx$$

Integrating, we get

$$z = \frac{x^{r+1}}{\beta(r+1)} f(c_1) = \frac{x^{r+1} f(\alpha x + \beta y)}{\beta(r+1)} \tag{2}$$

Then Eq. (1) can be written as

$$z = \frac{1}{(\beta D_x - \alpha D_y)} \; x^r f(\alpha x + \beta y) \tag{3}$$

From Eqs. (2) and (3), we have

$$\frac{1}{(\beta D_x - \alpha D_y)} x^r f(\alpha x + \beta y) = \frac{x^{r+1}}{\beta(r+1)} f(\alpha x + \beta y). \tag{4}$$

Hence, if $z = \dfrac{1}{(\beta D_x - \alpha D_y)^m} f(\alpha x + \beta y)$; then we have

$$z = \frac{1}{(\beta D_x - \alpha D_y)^{m-1}} \left[\frac{1}{(\beta D_x - \alpha D_y)} x^0 f(\alpha x + \beta y) \right]$$

$$= \frac{1}{(\beta D_x - \alpha D_y)^{m-1}} \left\{ \frac{x}{\beta} f(\alpha x + \beta y) \right\} \qquad \text{\{Using Eq. (4)\}}$$

$$= \frac{1}{\beta} \frac{1}{(\beta D_x - \alpha D_y)^{m-2}} \left[\frac{1}{(\beta D_x - \alpha D_y)} x f(\alpha x + \beta y) \right]$$

$$= \frac{1}{\beta} \frac{1}{(\beta D_x - \alpha D_y)^{m-2}} \left[\frac{x^2}{\beta \cdot 2} f(\alpha x + \beta y) \right]$$

$$= \frac{1}{2\beta^2} \frac{1}{(\beta D_x - \alpha D_y)^{m-2}} x^2 f(\alpha x + \beta y) \text{ and continuing}$$

$$= \frac{1}{m!\beta^m} \frac{1}{(\beta D_x - \alpha D_y)^{m-m}} x^m f(\alpha x + \beta y)$$

$$= \frac{x^m}{\beta^m (m!)} f(\alpha x + \beta y).$$

▲▲

EXAMPLE 5.27

Solve $(D_x^3 - 6D_x^2 D_y + 12 D_x D_y^2 - 8D_y^3)z = \cos(2x + y)$.

Solution: Here, the auxiliary equation is

$$m^3 - 6m^2 + 12m - 8 = 0.$$

or $$(m-2)^3 = 0$$

or $$m = 2, 2, 2.$$

∴ CF is

$$z_c = \phi_1(y + 2x) + x\phi_2(y + 2x) + x^2\phi_3(y + 2x).$$

and PI is $$z_p = \frac{1}{(D_x - 2D_y)^3} \cos(2x + y) = \frac{1}{0} \cos(2x + y) \quad \text{\{Rule fails\}}$$

$$= \frac{x^3}{(1)^3(3!)} \cos(2x + y) = \frac{x^3}{6} \cos(2x + y).$$

Hence, the complete solution is

$$z = z_c + z_p$$

$$z = \phi_1(y + 2x) + x\phi_2(y + 2x) + x^2\phi_3(y + 2x) + \frac{x^3}{6} \cos(2x + y).$$

EXAMPLE 5.28

Solve $\dfrac{\partial^2 z}{\partial x^2} - 5\dfrac{\partial^2 z}{\partial x \partial y} + 4\dfrac{\partial^2 z}{\partial y^2} = \sin(4x + y)$.

Solution: The given equation is

$$(D_x^2 - 5D_xD_y + 4D_y^2)z = \sin(4x + y)$$

Here, the auxiliary equation is

$$m^2 - 5m + 4 = 0$$

or $$(m-1)(m-4) = 0$$

or $$m = 1, 4.$$

∴ CF is

$$z_c = \phi_1(y + x) + \phi_2(y + 4x).$$

and PI is $$z_p = \frac{1}{D_x^2 - 5D_xD_y + 4D_y^2} \sin(4x + y)$$

$$= \frac{1}{-4^2 - 5(-4\cdot 1) + 4(-1^2)} \sin(4x + y)$$

$$= \frac{1}{0} \sin(4x + y) \quad \text{\{Rule fails\}}$$

$$= \frac{1}{(D_x - 4D_y)} \frac{1}{(D_x - D_y)} \sin(4x + y)$$

$$= \frac{1}{(D_x - 4D_y)} \left[\frac{D_x + D_y}{D_x^2 - D_y^2} \right] \sin(4x + y)$$

$$= \frac{1}{(D_x - 4D_y)} \left[\frac{4\cos(4x + y) + \cos(4x + y)}{(-4^2) - (-1^2)} \right]$$

$$= \frac{1}{(D_x - 4D_y)} \left(-\frac{1}{3} \right) \cos(4x + y) = -\frac{1}{3} \frac{x}{1 \cdot 1!} \cos(4x + y) \text{ (Rule fails)}$$

$$= -\frac{x}{3} \cos(4x + y).$$

Hence, the complete integral is

$$z = z_c + z_p$$

$$z = \phi_1(y + x) + \phi_2(y + 4x) - \frac{x}{3} \cos(4x + y).$$

EXAMPLE 5.29

Solve $r + 3s + 2t = 2x + 3y$.

Solution: The given equation is

$$(D_x^2 + 3D_xD_y + 2D_y^2)z = 2x + 3y.$$

Here, the auxiliary equation is

$$m^2 + 3m + 2 = 0$$

or

$$(m + 1)(m + 2) = 0$$

or

$$m = -1, -2.$$

$\therefore$ CF is

$$z_c = \phi_1(y - x) + \phi_2(y - 2x)$$

and PI is

$$z_p = \frac{1}{D_x^2 + 3D_xD_y + 2D_y^2} (2x + 3y)$$

$$= \frac{1}{(2)^2 + 3(2)(3) + 2(3)^2} \iint u\,du\,du \quad \text{where } a = 2x + 3y$$

$$= \frac{1}{40} \frac{u^3}{6} = \frac{(2x + 3y)^3}{240}$$

Hence, the complete solution is

$$z = z_c + z_p$$

$$z = \phi_1(y - x) + \phi_2(y - 2x) + \frac{(2x+3y)^3}{240}.$$

EXAMPLE 5.30

Solve $(4D_x^2 - 4D_xD_y + D_y^2)z = 16 \log (x + 2y)$.

Solution: Here, the auxiliary equation is

$$4m^2 - 4m + 1 = 0$$

or $$(2m - 1)^2 = 0$$

or $$m = \frac{1}{2}, \frac{1}{2}$$

$\therefore$ CF is

$$z_c = \phi_1\left(y + \frac{1}{2}x\right) + x\phi_2\left(y + \frac{1}{2}x\right)$$

$$= \psi_1(2y + x) + x\psi_2(2y + x)$$

and PI is $$z_p = 16\frac{1}{(2D_x - D_y)^2}\log (x + 2y)$$

$$= 16\frac{x^2}{2^2(2!)}\log (x + 2y) = 2x^2 \log (x + 2y).$$

Hence, the complete solution is

$$z = z_c + z_p$$

$$z = \psi_1(2y + x) + x\psi_2(2y + x) + 2x^2 \log (x + 2y).$$

EXAMPLE 5.31

Solve $(D_x^2 + D_xD_y - 2D_y^2)z = \sqrt{x+2y}$.

Solution: Here, the auxiliary equation is

$$m^2 + m - 2 = 0$$

or $$(m + 2)(m - 1) = 0$$

or $$m = 1, -2$$

$\therefore$ CF is

$$z_c = \phi_1(y + x) + \phi_2(y - 2x).$$

and PI is $$z_p = \frac{1}{D_x^2 + D_xD_y - 2D_y^2}\sqrt{x+2y}$$

$$= \frac{1}{(1)^2 + (1)(2) - 2(2)^2} \iint \sqrt{u}\, dudu \text{ where } u = x + 2y$$

$$= \frac{1}{-1} \frac{u^{5/2}}{\frac{3}{2} \cdot \frac{5}{2}} = -\frac{4}{15} = (x + 2y)^{5/2}.$$

Hence, the complete integral is

$$z = z_c + z_p$$

$$z = \phi_1(y + x) + \phi_2(y - 2x) - \frac{4}{15}(x + 2y)^{5/2}.$$

EXAMPLE 5.32

Solve the following partial differential equation

$$\frac{\partial^3 z}{\partial x^3} - 2\frac{\partial^3 z}{\partial x^2 \partial y} - \frac{\partial^3 z}{\partial x \partial y^2} + 2\frac{\partial^3 z}{\partial y^3} = e^{x + y}.$$

Solution: The given equation is

$$(D_x^3 - 2D_xD^2D_y - D_xD_y^2 + 2D_y^3)z = e^{x + y}.$$

Here, the auxiliary equation is

$$m^3 - 2m^2 - m + 2 = 0$$

$\Rightarrow$ $$m^2(m - 2) - 1\,(m - 2) = 0$$

$\Rightarrow$ $$(m^2 - 1)(m - 2) = 0$$

$\Rightarrow$ $$m = 1, -1, 2$$

∴ CF is

$$z_c = \phi_1(y + x) + \phi_2(y - x) + \phi_3(y + 2x).$$

and PI is $$z_p = \frac{1}{(D_x^2 - D_y^2)(D_x - 2D_y)} e^{x + y}$$

$$= \frac{1}{(D_x - D_y)} \left[\frac{1}{(D_x + D_y)(D_x - 2D_y)} e^{x+y} \right]$$

$$= \frac{1}{(D_x - D_y)} \left[\frac{1}{(1+1)(1-2)} e^{x+y} \right]$$

$$= -\frac{1}{2} \frac{1}{D_x - D_y} e^{x + y}$$

$$= -\frac{1}{2}\frac{x}{1(1!)}e^{x+y}$$

$$= -\frac{x}{2}e^{x+y}.$$

Hence, the complete solution is

$$z = z_c + z_p$$

$$z = \phi_1(y + x) + \phi_2(y - x) + \phi_3(y + 2x) - \frac{x}{2}e^{x+y}.$$

EXAMPLE 5.33

Find the complete solution of the following partial differential equation

$$\frac{\partial^2 z}{\partial x^2} - 3\frac{\partial^2 z}{\partial x \partial y} + 2\frac{\partial^2 z}{\partial y^2} = e^{2x-y} + e^{x+y} - \cos(x + 2y).$$

Solution: The given equation is

$$(D_x^2 - 3D_xD_y + 2D_y^2)z = e^{2x-y} + e^{x+y} - \cos(x + 2y)$$

Here, the auxiliary equation is

$$m^2 - 3m + 2 = 0$$

or $$(m - 2)(m - 1) = 0$$

or $$m = 1, 2$$

$\therefore$ CF is

$$z_c = \phi_1(y + x) + \phi_2(y + 2x)$$

and PI is $$z_p = \frac{1}{D_x^2 - 3D_xD_y + 2D_y^2}(e^{2x-y} + e^{x+y} - \cos(x + 2y)).$$

Now $$\frac{1}{D_x^2 - 3D_xD_y^2}e^{2x-y} = \frac{1}{(2)^2 - 3(2)(-1) + 2(-1)^2}\iint e^u du du$$

where $u = 2x - y$

$$= \frac{1}{12}e^u = \frac{1}{12}e^{2x-y}.$$

$$\frac{1}{(D_x - 2D_y)(D_x - d_y)}e^{x+y} = \frac{1}{0}e^{x+y}$$ {Rule fails}

$$= \frac{1}{D_x - D_y}\left[\frac{1}{D_x - 2D_y}e^{x+y}\right]$$

$$= \frac{1}{D_x - D_y}\left[\frac{1}{1-2(1)}e^{x+y}\right]$$

$$= -\frac{1}{D_x - D_y}e^{x+y}$$

$$= -\frac{x}{1(1!)}e^{x+y} = -xe^{x+y}$$

and $$\frac{1}{D_x^2 - 3D_xD_y + 2D_y^2}\cos(x+2y) = \frac{1}{(1)^2 - 3(1)(2) + 2(2)}\iint \cos u\,du\,du$$

where $u = x + 2y$

$$= \frac{1}{3}\times - \cos u = -\frac{1}{3}\cos(x+2y)$$

Hence, the complete solution is

$$z = z_c + z_p$$

$$z = \phi_1(y+x) + \phi_2(y+2x) + \frac{1}{12}e^{2x-y} - xe^{x+y} - \frac{1}{3}\cos(x+2y).$$

EXAMPLE 5.34

Solve $(D_x^2 - 6D_xD_y + 9D_y^2)z = \tan(y+3x)$.

Solution: Here, the auxiliary equation is

$$m^2 - 6m + 9 = 0$$

or $$(m-3)^2 = 0$$

or $$m = 3, 3.$$

$\therefore$ CF is

$$z_c = \phi_1(y+3x) + x\phi_2(y+3x).$$

and PI is $$z_p = \frac{1}{(D_x - 3D_y)^2}\tan(y+3x) = \frac{1}{0}\tan(y+3x)$$ {Rule fails}

$$= \frac{x^2}{(1)^2(2!)}\tan(y+3x) = \frac{x^2}{2}\tan(y+3x).$$

Hence, the complete solution is

$$z = z_c + z_p$$

$$z = \phi_1(y+3x) + x\phi_2(y+3x) + \frac{x^2}{2}\tan(y+3x).$$

General Method to Find Particular Integral

Let the given equation be $F(D_x, D_y)z = f(x, y)$ where $F(D_x, D_y)$ is a homogeneous function of D_x and D_y of degree n.

Let $\alpha_1, \alpha_2 \ldots \alpha_n$ be roots of auxiliary equation so that

$$F(D_x, D_y) = (D_x - \alpha_1 D_y)(D_x - \alpha_2 D_y) \ldots (D_x - \alpha_n D_y)$$

$$\therefore \quad z_p = \frac{1}{F(D_x, D_y)} f(x, y) = \frac{1}{(D_x - \alpha_1 D_y)(D_x - \alpha_2 D_y)\ldots(D_x - \alpha_n D_y)} f(x, y)$$

and $\dfrac{1}{D_x - \alpha D_y} f(x, y) = \int f(x, c - \alpha x)dx$ for some constant c.

Proof: Consider the equation

$$(D_x - \alpha D_y)z = f(x, y) \text{ or } p - \alpha q = f(x, y)$$

which is of Lagrange's form. So, its auxiliary equations are

$$\frac{dx}{1} = \frac{dy}{-\alpha} = \frac{dz}{f(x,y)}$$

Taking the first two members, we get

$$\alpha dx + dy = 0 \Rightarrow \alpha x + y = c$$

Again, taking the first and third members, we have

$$dx = \frac{dz}{f(x,y)} = \frac{dz}{f(x, c - \alpha x)}$$

$$\Rightarrow \quad dx = f(x, c - \alpha x)dx$$

Integrating, we get $\quad z = \int f(x, c - \alpha x)dx$

$$\Rightarrow \quad \frac{1}{D_x - \alpha D_y} f(x, y) = \int f(x, c - \alpha x)dx$$

Hence, the particular integral (z_p) can be obtained by applying this operation on the factors in succession, starting from the right and after integration constant c is replaced by $\alpha x + y$.

EXAMPLE 5.35

Solve $r + s - 6t = y \cos x$.

Solution: The equation can be written as

$$(D_x^2 + D_xD_y - 6D_y^2)z = y \cos x$$

Here, the auxiliary equation is

$$m^2 + m - 6 = 0$$

or
$$(m - 2)(m + 3) = 0$$

or
$$m = 2, -3.$$

$\therefore$ CF is

$$z_c = \phi_1(y + 2x) + \phi_2(y - 3x)$$

and PI is

$$z_p = \frac{1}{(D_x - 2D_y)(D_x + 3D_y)} y \cos x$$

$$= \frac{1}{D_x - 2D_y} \int (c + 3x) \cos x dx \qquad \{\because \alpha = -3\}$$

$$= \frac{1}{D_x - 2D_y} [c \sin x + 3x \sin x + 3 \cos x]$$

$$= \frac{1}{D_x - 2D_y} [(y - 3x) \sin x + 3x \sin x + 3 \cos x]$$

$$= \frac{1}{D_x - 2D_y} [y \sin x + 3 \cos x]$$

$$= \int ((c' - 2x) \sin x + 3 \cos x) dx \qquad \{\because \alpha = 2\}$$

$$= - c' \cos x - 2(-x \cos x + \sin x) + 3 \sin x$$

$$= - (y + 2x) \cos x + 2x \cos x + \sin x$$

$$= -y \cos x + \sin x.$$

Hence, the complete solution is

$$z = z_c + z_p$$
$$z = \phi_1(y + 2x) + \phi_2(y - 3x) - y \cos x + \sin x.$$

EXAMPLE 5.36

Solve $(D_x^2 + 5D_xD_y + 6D_y^2)z = \dfrac{1}{y - 2x}$.

Solution: Here, the auxiliary equation is

$$m^2 + 5m + 6 = 0$$

or $$(m + 2)(m + 3) = 0$$

or $$m = -2, -3$$

$\therefore$ CF is

$$z_c = \phi_1(y - 2x) + \phi_2(y - 3x).$$

and PI is $$z_p = \frac{1}{(D_x + 3D_y)(D_x + 2D_y)}\left(\frac{1}{y - 2x}\right)$$

$$= \frac{1}{D_x + 3D_y}\int\frac{1}{c + 2x - 2x}dx \qquad \{\because \alpha = -2\}$$

$$= \frac{1}{D_x + 3D_y}\left(\frac{x}{c}\right) = \frac{1}{D_x + 3D_y}\left(\frac{x}{y - 2x}\right)$$

$$= \int\frac{x}{c' + 3x - 2x}dx \qquad \{\because \alpha = -3\}$$

$$= \int\frac{x}{c' + x}dx = \int\left(1 - \frac{c'}{c' + x}\right)dx$$

$$= x - c' \log (c' + x)$$

$$= x - (y - 3x) \log (y - 2x).$$

Hence, the complete solution is

$$z = z_c + z_p$$

$$z = \phi_1(y - 2x) + \phi_2(y - 3x) + x - (y - 3x) \log (y - 2x).$$

EXAMPLE 5.37

Solve $r - t = \tan^3 x \tan y - \tan x + \tan^3 y$.

Solution: The given equation can be written as

$$(D_x^2 - D_y^2)z = \tan^3 x \tan y - \tan x + \tan^3 y$$

Here, the auxiliary equation is

$$m^2 - 1 = 0$$

or $$m = 1, -1$$

$\therefore$ CF is

$$z_c = \phi_1(y + x) + \phi_2(y - x).$$

and PI is $$z_p = \frac{1}{(D_x + D_y)(D_x - D_y)}\tan x \tan y\,(\sec^2 x - \sec^2 y)$$

$$= \frac{1}{(D_x + D_y)} \int (\tan x \sec^2 x \tan (c - x) - \tan x \tan (c - x) \sec^2 (c - x)) dx \qquad \{\because \alpha = +1\}$$

$$= \frac{1}{(D_x + D_y)} \int \left[\frac{1}{2} \tan^2 x \tan(c - x) + \frac{1}{2} \int \tan^2 x \sec^2 (c - x) dx + \frac{1}{2} \tan x \tan^2 (c - x) - \frac{1}{2} \int \tan^2 (c - x) \sec^2 x dx \right]$$

$$= \frac{1}{2} \frac{1}{D_x + D_y} \left[\tan^2 x \tan(c - x) + \tan x \tan^2 (c - x) + \int [\sec^2 x - \sec^2 (c - x)] dx \right]$$

$$= \frac{1}{2} \frac{1}{D_x + D_y} [\tan^2 x \tan y + \tan x \tan^2 y + \tan x + \tan y]$$

$$= \frac{1}{2} \frac{1}{D_x + D_y} [\sec^2 x \tan y + \tan x \sec^2 y]$$

$$= \frac{1}{2} \int [\sec^2 x \tan (c' + x) + \tan x \sec^2 (c' + x] dx \qquad \{\because \alpha = -1\}$$

$$= \frac{1}{2} \tan x \tan (c' + x)$$

$$= \frac{1}{2} \tan x \tan y.$$

Hence, the complete solution is

$$z = z_c + z_p$$

$$z = \phi_1(y + x) + \phi_2(y - x) + \frac{1}{2} \tan x \tan y.$$

EXAMPLE 5.38

Solve the partial differential equation

$$(D_x^2 + 3D_x D_y + 2D_y^2) z = \sin x + e^{-3x} + x^2 + y.$$

Solution: Here, the auxiliary equation is

$$m^2 + 3m + 2 = 0$$

or $$(m + 1)(m + 2) = 0$$

or $$m = -1, -2$$

$\therefore$ CF is

$$z_c = \phi_1(y - x) + \phi_2(y - 2x)$$

and PI is

$$z_p = \frac{1}{D_x^2 - 3D_xD_y + 2D_y^2}(\sin x) + \frac{1}{D_x^2 + 3D_xD_y + 2D_y^2}(e^{-3x}) + \frac{1}{D_x^2 + 3D_xD_y + 2D_y^2}(x^2 + y)$$

Now, $$\frac{1}{D_x^2 - 3D_xD_y + 2D_y^2}\sin x = \frac{1}{(-1^2) + 3(-1\cdot 0) + 2(-0^2)}\sin x$$

$$= -\sin x$$

$$\frac{1}{D_x^2 - 3D_xD_y + 2D_y^2}e^{-3x} = \frac{1}{(-3)^2 + 3(-3\cdot 0) + 2(-0^2)}\cdot e^{-3x}$$

$$= \frac{e^{-3x}}{9}$$

and $$\frac{1}{D_x^2 + 3D_xD_y + 2D_y^2}(x^2 + y) = \frac{1}{D_x^2}\left[1 + \left(3\frac{D_y}{D_x} + 2\frac{D_y^2}{D_x^2}\right)\right]^{-1}(x^2 + y)$$

$$= \frac{1}{D_x^2}\left[1 - \left(3\frac{D_y}{D_x} + 2\frac{D_y^2}{D_x^2}\right) + \ldots\right](x^2 + y)$$

$$= \frac{1}{D_x^2}\left\{x^2 + y - \left(\frac{3}{D}(1) + \frac{2}{D_x^2}(0)\right)\right\}$$

$$= \frac{x^4}{12} + \frac{yx^2}{2} - \frac{3x^3}{6} = \frac{x^4}{12} + \frac{yx^2}{2} - \frac{x^3}{2}$$

$\therefore$ $$z_p = -\sin x + \frac{e^{-3x}}{9} + \frac{x^4}{12} + \frac{yx^2}{2} - \frac{x^3}{2}$$

Hence, the complete solution is

$$z = z_c + z_p$$

$$z = \phi_1(y - x) + \phi_2(y - 2x) - \sin x + \frac{e^{-3x}}{9} + \frac{x^4}{12} + \frac{yx^2}{2} - \frac{x^3}{2}$$

EXAMPLE 5.39

Solve the partial differential equation $(D_x^2 - D_xD_y)z = \sin x \sin 2y + e^{x+y} - xy + e^x \cos 2y$.

Solution: Hence, the auxiliary equation is

$$m^2 - m = 0$$

or $$m(m - 1) = 0$$

or $$m = 0, 1$$

$\therefore$ CF is

$$z_c = \phi_1(y) + \phi_2(y + x)$$

and PI is

$$z_p = \frac{1}{D_x^2 - D_x D_y}(\sin x \sin 2y + e^{x+y} - xy + e^x \cos 2y)$$

$$= \frac{1}{D_x^2 - D_x D_y}(\sin x \sin 2y) + \frac{1}{D_x^2 - D_x D_y}(e^{x+y}) - \frac{1}{D_x^2 - D_x D_y}(xy) + \frac{1}{D_x^2 - D_x D_y}(e^x \cos 2y)$$

Now $\frac{1}{D_x^2 - D_x D_y}(\sin x \sin 2y) = \frac{1}{D_x^2 - D_x D_y}\frac{1}{2}[\cos(x - 2y) - \cos(x + 2y)]$

$$= \frac{1}{2}\left\{\frac{1}{-(1)^2 - (-(1)(-2))}\cos(x-2y) - \frac{1}{-(1)^2 - (-(1)(2))}\cos(x+2y)\right\}$$

$$= \frac{1}{2}\left[-\frac{1}{3}\cos(x-2y) - \cos(x+2y)\right]$$

$$\frac{1}{D_x^2 - D_x D_y}e^{x+y} = \frac{1}{D_x - D_y}\frac{1}{D_x}e^{x+y} = \frac{1}{D_x - D_y}e^{x+y} = \frac{x}{1(1!)}e^{x+y} = xe^{x+y}$$

$$\frac{1}{D_x^2 - D_x D_y}(xy) = \frac{1}{D_x^2}\left\{1 - \frac{D_y}{D_x}\right\}^{-1}(xy)$$

$$= \frac{1}{D_x^2}\left\{1 + \frac{D_y}{D_x} + \frac{D_y^2}{D_x^2} + \ldots\right\}(xy)$$

$$= \frac{1}{D_x^2}\left\{xy + \frac{x}{D_x} + 0 + \ldots\right\}$$

$$= \frac{x^3 y}{6} + \frac{x^4}{24}$$

and $\frac{1}{D_x^2 - D_x D_y}e^x \cos 2y$ = Real part of $\frac{1}{D_x^2 - D_x D_y}(e^x e^{2iy})$

= Real part of $\frac{1}{D_x^2 - D_x D_y}e^{x+2iy}$

$$= \text{Real part of } \frac{1}{(1)^2-(1)(2i)} e^{x+2iy}$$

$$= \text{Real part of } \frac{1+2i}{1+4}(e^{x+2iy})$$

$$= \frac{1}{5}e^x(\cos 2y - 2\sin 2y)$$

$$\therefore \quad z_p = -\frac{1}{6}\cos(x-2y) - \frac{1}{2}\cos(x+2y) + xe^{x+y} + \frac{x^3y}{6} + \frac{x^4}{24} + \frac{1}{5}e^x(\cos 2y - 2\sin 2y)$$

Hence, the complete solution is

$$z = z_c + z_p$$

$$z = \phi_1(y) + \phi_2(y+x) - \frac{1}{6}\cos(x-2y) - \frac{1}{2}\cos(x+2y) + xe^{x+y} + \frac{x^3y}{6} + \frac{x^4}{24} + \frac{1}{5}e^x(\cos 2y - 2\sin 2y).$$

EXAMPLE 5.40

Solve $(D_x^2 + D_xD_y - 6D_y^2)z = x^2 \sin(x+y)$.

Solution: Here, the auxiliary equation is

$$m^2 + m - 6 = 0$$

or

$$(m+3)(m-2) = 0$$

or

$$m = -3, 2$$

$\therefore$ CF is

$$z_c = \phi_1(y-3x) + \phi_2(y+2x)$$

and PI is

$$z_p = \frac{1}{D_x^2 + D_xD_y - 6D_y^2} x^2 \sin(x+y)$$

$$= \frac{1}{(D_x+3D_y)(D_x-2D_y)} x^2 \sin(x+y)$$

$$= \frac{1}{D_x+3D_y}\int x^2 \sin(x+c-2x)dx \qquad \{\because \alpha = 2\}$$

$$= \frac{1}{D_x+3D_y}\int x^2 \sin(c-x)dx$$

$$= \frac{1}{D_x+3D_y}\left[x^2\cos(c-x) - \int 2x\cos(c-x)dx\right]$$

$$= \frac{1}{D_x + 3D_y}\left[x^2\cos(c-x) - \left(-2x\sin(c-x) + \int 2\sin(c-x)dx\right)\right]$$

$$= \frac{1}{D_x + 3D_y}\,[x^2\cos(c-x) + 2x\sin(c-x) - 2\cos(c-x)]$$

$$= \frac{1}{D_x + 3D_y}\,[(x^2-2)\cos(x+y) + 2x\sin(x+y)]$$

$$= \int [(x^2-2)\cos(x+c'+4x) + 2x\sin(x+c'+4x)]dx \qquad \{\because \alpha = -3\}$$

$$= (x^2-2)\,\frac{\sin(c'+4x)}{4} - \int 2x\frac{\sin(c'+4x)}{4}dx + 2\int x\sin(c'+4x)dx$$

$$= \frac{1}{4}(x^2-2)\sin(c'+4x) + \frac{3}{2}\int x\sin(c'+4x)dx$$

$$= \frac{1}{4}(x^2-2)\sin(c'+4x) + \frac{3}{2}\left\{-x\frac{\cos(c'+4x)}{4} + \int\frac{\cos(c'+4x)}{4}dx\right\}$$

$$= \frac{1}{4}(x^2-2)\sin(c'+4x) - \frac{3}{8}x\cos(c'+4x) + \frac{3}{32}\sin(c'+4x)$$

$$= \frac{1}{4}(x^2-2)\sin(x+y) - \frac{3}{8}\cos(x+y) + \frac{3}{32}\sin(x+y)$$

$$= \frac{1}{4}\left(x^2 - \frac{13}{8}\right)\sin(x+y) - \frac{3}{8}x\cos(x+y)$$

Hence, the complete integral is

$$z = z_c + z_p$$

$$z = \phi_1(y-3x) + \phi_2(y+2x) + \frac{1}{4}\left(x^2 - \frac{13}{8}\right)\sin(x+y) - \frac{3}{8}x\cos(x+y).$$

EXAMPLE 5.41

Solve $(D_x^2 - D_xD_y - 2D_y^2)z = (y-1)e^x$.

Solution: Here, the auxiliary equation is

$$m^2 - m - 2 = 0$$

or $$(m-2)(m+1) = 0$$

or $$m = 2, -1.$$

$\therefore$ CF is

$$z_c = \phi_1(y + 2x) + \phi_2(y - x).$$

and PI is

$$z_p = \frac{1}{(D_x - 2D_y)(D_x + D_y)}(y - 1)e^x$$

$$= \frac{1}{D_x - 2D_y}\int (c + x - 1)e^x dx \qquad \{\because \alpha = -1\}$$

$$- \frac{1}{D_x - 2D_y}[(c + x - 1)e^x - e^x]$$

$$= \frac{1}{D_x - 2D_y}(y - 2)e^x$$

$$= \int (c - 2x - 2)e^x + dx \qquad \{\because \alpha = 2\}$$

$$= (c - 2x - 2)e^x + 2e^x$$

$$= (c - 2x)e^x = ye^x.$$

Hence, the complete integral is

$$z = z_c + z_p$$

$$z = \phi_1(y + 2x) + \phi_2(y - x) + ye^x.$$

EXAMPLE 5.42

Solve the following partial differential equation

$$\frac{\partial^3 z}{\partial x^3} - 3\frac{\partial^3 z}{\partial x \partial y^2} - 2\frac{\partial^3 z}{\partial y^3} = \sin(x + 2y) - e^y(2x + 3).$$

Solution: The given equation is

$$(D_x^3 - 3D_xD_y^2 - 2D_y^3)z = \sin(x + 2y) - e^y(2x + 3)$$

Here, the auxiliary equation is

$$m^3 - 3m - 2 = 0$$

or

$$(m - 2)(m^2 + 2m + 1) = 0$$

or

$$(m - 2)(m + 1)^2 = 0$$

or

$$m = 2, -1, -1$$

$\therefore$ CF is

$$z_c = \phi_1(y + 2x) + \phi_2(y - x) + x\phi_3(y - x).$$

and PI is $$z_p = \frac{1}{D_x^3 - 3D_xD_y{}^2 - 2D_y^3}\sin(x+2y) - \frac{1e^y}{D_x^3 - 3D_xD_y^2 - 2D_y^3}(2x+3)$$

Now, $$\frac{1}{D_x^3 - 3D_xD_y{}^2 - 2D_y^3}\sin(x+2y) = \frac{1}{(1)^3 - 3(1)(2)^2 - 2(2)^3}\iiint \sin u\,du\,du\,du$$

where $u = x + 2y$

$$= -\frac{1}{27}\cos u = -\frac{1}{27}\cos(x+2y)$$

and $$\frac{1}{D_x^3 - 3D_xD_y^2 - 2D_y^3}e^y(2x+3) = \frac{1}{(D_x+D_y)^2(D_x-2D_y)}e^y(2x+3)$$

$$= \frac{1}{(D_x+D_y)^2}\int e^{x-2x}(2x+3)dx \qquad \{\because \alpha = 2\}$$

$$= \frac{1}{(D_x+D_y)^2}\left[\frac{e^{c-2x}}{-2}(2x+3) - 2\frac{e^{c-2x}}{4}\right]$$

$$= \frac{1}{(D_x+D_y)^2}e^y\left(-x-\frac{3}{2}-\frac{1}{2}\right)$$

$$= \frac{1}{(D_x+D_y)^2}\frac{-1}{D_x+D_y}e^y(x+2)$$

$$= \frac{1}{D_x+D_y}\int e^{c+x}(x+2)dx \qquad \{\because \alpha = -1\}$$

$$= -\frac{1}{D_x+D_y}[(x+2)e^{c+x} - e^{c+x}]$$

$$= -\frac{1}{D_x+D_y}(x+1)e^y$$

$$= -\int e^{c+x}(x+1)dx \qquad \{\because \alpha = -1\}$$

$$= -[e^{c+x}(x+1) - e^{c+x}]$$

$$= -(e^y(x+1) - e^y = -xe^y$$

$$\therefore \qquad z_p = -\frac{1}{27}\cos(x+2y) + xe^y$$

Hence, the complete solution is $z = z_c + z_p$

$$z = \phi_1(y + 2x) + \phi_2(y - x) + x\phi_3(y - x) - \frac{1}{27}\cos(x + 2y) + xe^y.$$

EXAMPLE 5.43

Solve $(D_x^3 + D_x^2 D_y - D_x D_y^2 - D_y^3)z = e^y \cos 2x$.

Solution: Here, the auxiliary equation is

$$m^3 + m^2 - m - 1 = 0$$

or
$$(m + 1)^2 (m - 1) = 0$$

or
$$m = -1, -1, 1$$

$\therefore$ CF is

$$z_c = \phi_1(y + x) + \phi_2(y - x) + x\phi_3(y - x)$$

and PI is
$$z_p = \frac{1}{(D_x - D_y)(D_x + D_y)}\frac{1}{D_x + D_y} e^y \cos 2x$$

$$= \frac{1}{(D_x - D_y)(D_x + D_y)}\int e^{c + x} \cos 2x dx \qquad \{\because \alpha = -1\}$$

$$= \frac{1}{(D_x - D_y)(D_x + D_y)}\frac{e^{c+x}}{1^2 + 2^2}(\cos 2x + 2\sin 2x)$$

$$= \frac{1}{5}\frac{1}{D_x - D_y}\frac{1}{D_x + D_y} e^y (\cos 2x + 2\sin 2x)$$

$$= \frac{1}{5}\frac{1}{D_x - D_y}\int e^{c + x}(\cos 2x + 2\sin 2x)dx \qquad \{\because \alpha = -1\}$$

$$= \frac{1}{5}\frac{1}{D_x - D_y}\left[\frac{e^{c+x}}{1^2 + 2^2}(\cos 2x + 2\sin 2x) + 2\frac{e^{c+x}}{1^2 + 2^2}(\sin 2x - 2\cos 2x)\right]$$

$$= \frac{1}{25}\frac{1}{D_x - D_y} e^y(4\sin 2x - 3\cos 2x)$$

$$= \frac{1}{25}\int e^{c - x}(4\sin 2x - 3\cos 2x)dx \qquad \{\because \alpha = 1\}$$

$$= \frac{1}{25}\left[\frac{4e^{c-x}}{1^2 + 2^2}(-\sin 2x - 2\cos 2x) - \frac{3e^{c-x}}{1^2 + 2^2}(-\cos 2x + 2\sin 2x\right]$$

$$= -\frac{1}{25} e^y (\cos 2x + 2 \sin 2x).$$

Hence, the complete solution is

$$z = \phi_1(y + x) + \phi_2(y - x) + x\phi_2(y - x) - \frac{1}{25} e^y (\cos 2x + 2 \sin 2x).$$

▲▲

EXERCISE 5.2

Solve the following partial differential equations:

1. $(D_x^2 - 2D_xD_y + D_y^2)z = e^{x+y}$.
2. $(D_x^2 + 4D_xD_y + 3D_y^2)z = 3 \sin (3x + 4y) + 4 \cos (x + 2y)$.
3. $(D_x^3 - 2D_x2D_y)z = 4x^3y^3$.
4. $(D_x^3 - 2D_x^2D_y - D_xD_y^2 + 2D_y^3)z = e^{x+y}$.
5. $(D_x^2 + 3D_xD_y + 2D_y^2)z = x + y$.
6. $(D_x^2 + 2D_xD_y + D_y^2)z = 2 \cos y - x \sin y$.
7. $(D_x^2 + DxD_y - 2D_y^2)z = (y - 1)e^x$.
8. $r - s - 2t = (2x^2 + xy - y^2) \sin (xy) - \cos (xy)$.
9. $(D_x^2 - 4D_y^2)z = \frac{4x}{y^2} - \frac{y}{x^2}$.
10. $(D_x^2 - 3D_xD_y + D_y^2)z = \sin x \cos y$.
11. $(D_x^2 + 3D_xD_y + 2D_y^2)z = x + y$.
12. $(D_x^2 + 3D_xD_y + 2D_y^2)z = 6(x + y)$.
13. $\frac{\partial^2 z}{\partial x^2} - 6\frac{\partial^2 z}{\partial x \partial y} + 9\frac{\partial^2 z}{\partial y^2} = 12x^2 + 4xy$.
14. $2\frac{\partial^2 z}{\partial x^2} - 5\frac{\partial^2 z}{\partial x \partial y} + 2\frac{\partial^2 z}{\partial y^2} = 24(y - x)$.
15. $(D_x^3 - 3D_x^2D_y + 4D_y^3)z = e^{x+2y}$.
16. $(r - 2s + t) = e^{x+2y} + x^2$.
17. $(D_x^2 - 3D_xD_y)z = 2 \sin x \cos 2y$.
18. $(D_x^3 - 4D_x^2D_y + 4D_xD_y^2)z = 6 \cos (2x + 3y)$.

19. $\dfrac{\partial^2 z}{\partial x^2} - \dfrac{\partial^2 z}{\partial y^2} = 5 \sin 2x \sin 3y.$

20. $3\dfrac{\partial^2 z}{\partial x^2} - 10\dfrac{\partial^2 z}{\partial x \partial y} + 3\dfrac{\partial^2 z}{\partial y^2} = 10 \sin (2x + y).$

21. $r - 2s + t = \sin (2x + 3y).$

22. $r + t = \cos \alpha x \cos \beta y.$

23. $\dfrac{\partial^2 z}{\partial x^2} - 6\dfrac{\partial^2 z}{\partial x \partial y} + 9\dfrac{\partial^2 z}{\partial y^2} = 6x + 2y.$

24. $2r - 5s + 2t = 54(y - x).$

25. $2\dfrac{\partial^2 z}{\partial x^2} - \dfrac{\partial^2 z}{\partial x \partial y} - 3\dfrac{\partial^2 z}{\partial y^2} = 10e^{x - y}.$

26. $\dfrac{\partial^2 z}{\partial x^2} - 2\alpha\dfrac{\partial^2 z}{\partial x \partial y} + \alpha^2\dfrac{\partial^2 z}{\partial y^2} = g(y + \alpha x).$

27. $2r - s - 3t = 5e^{x - y}.$

28. $\dfrac{\partial^3 z}{\partial x^3} - 7\dfrac{\partial^3 z}{\partial x \partial y^2} - 6\dfrac{\partial^3 z}{\partial y^3} = \cos (x + 2y) + e^{y - 2x}.$

29. $(D_x^2 - 4D_xD_y + 4_y^2)z = \tan (y + 2x).$

30. $(D_x^2 + D_xD_y - 6D_y^2)z = y \sin x.$

31. $(D_x^2 - 2D_xD_y - 15D_y^2)z = 12xy.$

32. $\dfrac{\partial^3 z}{\partial x^3} + 2\dfrac{\partial^3 z}{\partial x^2 \partial y} - \dfrac{\partial^3 z}{\partial x \partial y^2} - 2\dfrac{\partial^3 z}{\partial y^3} = (y + 2)e^x.$

33. $r + 5s + 6t = \dfrac{1}{y - 2x}.$

34. $\dfrac{\partial^2 z}{\partial x^2} + \dfrac{\partial^2 z}{\partial x \partial y} - 6\dfrac{\partial^2 z}{\partial y^2} = y \cos x.$

35. $4\dfrac{\partial^2 z}{\partial x^2} - 4\dfrac{\partial^2 z}{\partial x \partial y} + \dfrac{\partial^2 z}{\partial y^2} \; 16 \log (x + 2y).$

36. $r - t = x - y.$

37. $(D_x^2 + 5D_xD_y + 6D_y^2)z = (y - 2x)^{-1}.$

38. $(D_x^2 - 3D_xD_y + 2D_y^2)z = e^{2x - y} + e^{x + y} + \cos(x + 2y)$.

39. $(D_x^3 - 4D_x^2D_y + 5D_xD_y^2 - 2D_y^3)z = e^{2x + y} + (y + x)^{1/2}$.

40. $(D_x^3 - 7D_xD_y^2 - 6D_y^3)z = \sin(x + 2y) + e^{3x + y}$.

41. $(D_x^3 - 3D_xD_y^2 - 2D_y^3)z = \cos(x + 2y) - e^y(3 + 2x)$.

42. $(D_x^3 - D_x^2D_y - 2D_xD_y^2)z = e^{x + 2y}(x^2 + 4y^2)$.

43. $(D_x^3 + 2D_xD_y - D_xD_y^2 - 2D_y^3)z = (y + 2)e^y$.

44. Find the solution of the equation $\dfrac{\partial^2 z}{\partial x^2} + \dfrac{\partial^2 z}{\partial y^2} = e^{-x}\cos y$ which approaches to ∞ as $x \to \infty$ and has the value $\cos y$ when $x = 0$.

45. Find a surface through two lines $x = z = 0$ and $x - y = z - 1 = 0$ which satisfies $(D_x^2 - 4D_xD_y + 4D_y^2)z = 0$.

ANSWERS

Exercise 5.2

1. $z = \phi_1(y + x) + x\phi_2(y + x) + \dfrac{x^2}{2}e^{x+y}$.

2. $z = \phi_1(y - x) + \phi_2(y - 3x) - \dfrac{1}{35}\sin(3x + 4y) - \dfrac{4}{21}\cos(x + 2y)$.

3. $z = \phi_1(y) + x\phi_2(y) + \phi_3(y + 2x) + \dfrac{x^6y^3}{30} + \dfrac{x^7y^2}{35} + \dfrac{x^8y}{70} + \dfrac{x^9}{315}$.

4. $z = \phi_1(y + x) + \phi_2(y - x) + \phi_3(y + 2x) - \dfrac{1}{2}xe^{x + y}$.

5. $z = \phi_1(y - x) + \phi_2(y - 2x) + \dfrac{(x+y)^3}{36}$.

6. $z = \phi_1(y - x) + x\phi_2(y - x) + x\sin y$.

7. $z = \phi_1(y + 2x) + \phi_2(y - x) + ye^x$.

8. $z = \phi_1(y + 2x) + \phi_2(y - x) + \sin xy$.

9. $z = \phi_1(y + 2x) + \phi_2(y - 2x) + x\log y + y\log x + 3x$.

10. $z = \phi_1\left(y + \left(\dfrac{3+\sqrt{5}}{2}\right)x\right) + \phi_2\left(y + \left(\dfrac{3-\sqrt{5}}{2}\right)x\right) + \dfrac{1}{2}\sin(x + y) - \dfrac{1}{10}\sin(x - y)$.

11. $z = \phi_1(y - x) + \phi_2(y - 2x) - \frac{x^3}{3} + \frac{x^2 y}{2}.$

12. $z = \phi_1(y - 2x) + \phi_2(y - x) - 2x^3 + 3x^2 y.$

13. $z = \phi_1(y + 3x) + x\phi_2(y + 3x) + 2x^4 + \frac{2}{3}x^3 y.$

14. $z = \phi_1(2y + x) + \phi_2(y + 2x) + 6x^2 y + 3x^3.$

15. $z = \phi_1(y - x) + \phi_2(y + 2x) + x\phi_3(y + 2x) + \frac{1}{27}e^{x + 2y}.$

16. $z = \phi_1(y + x) + x\phi_2(y + x) + e^{x + 2y} + \frac{x^5}{20}.$

17. $z = \phi_1(y) + \phi_2(y + 3x) - \frac{1}{7}\sin(x + 2y) + \frac{1}{5}\sin(x - 2y).$

18. $z = \phi_1(y) + \phi_2(y + 2x) + x\phi_3(y + 2x) - \frac{3}{16}\sin(2x + 3y).$

19. $z = \phi_1(y + x) + \phi_2(y - x) + \sin 2x \sin 3y.$

20. $z = \phi_1(3y + x) + \phi_2(y + 3x) + 2\sin(2x + y).$

21. $z = \phi_1(y + x) + x\phi_2(y + x) - \sin(2x + 3y).$

22. $z = \phi_1(y + ix) + \phi_2(y - jx) - \frac{1}{\alpha^2 + \beta^2}\cos \alpha x \cos \beta y.$

23. $z = \phi_1(y + 3x) + x\phi_2(y + 3x) + x^2(3x + y).$

24. $z = \phi_1(2y + x) + \phi_2(y + 2x) + (y - x)^3.$

25. $z = \phi_1(y - x) + \phi_2(2y + 3x) - 2xe^{x - y}.$

26. $z = \phi_1(y + \alpha x) + x\phi_2(y + \alpha x) + \frac{x^2}{2}g(y + \alpha x).$

27. $z = \phi_1(y - x) + \phi_2(2y + 3x) - xe^{x - y}.$

28. $z = \phi_1(y - x) + \phi_2(y - 2x) + \phi_3(y + 3x) + \frac{1}{75}\sin(x + 2y) + \frac{x}{5}e^{y - 2x}.$

29. $z = \phi_1(y + 2x) + x\phi_2(y + 2x) + x^2\tan(2x + y).$

30. $z = \phi_1(y - 3x) + \phi_2(y + 2x) - y\sin x - \cos x.$

31. $z = \phi_1(y + 5x) + \phi_2(y - 3x) + x^4 + 2x^3y$.

32. $z = \phi_1(y - 2x) + \phi_2(y - x) + \phi_3(y + x) + ye^x$.

33. $z = \phi_1(y - 3x) + \phi_2(y - 2x) + x \log (y - 2x)$.

34. $z = \phi_1(y - 3x) + \phi_2(y + 2x) + \sin x - y \cos x$.

35. $z = \phi_1(2y + x) + x\phi_2(2y + x) + 2x^2 \log (x + 2y)$.

36. $z = \phi_1(y + x) + \phi_2(y - x) + \frac{x}{4}(x - y)^2$.

37. $z = \phi_1(y - 2x) + \phi_2(y - 3x) + x \log (y - 2x)$.

38. $z = \phi_1(y + x) + \phi_2(y + 2x) + \frac{1}{12} e^{2x - y} - xe^{x + y} - \frac{1}{3} \cos (x + 2y)$.

39. $z = \phi_1(y + x) + x\phi_2(y + x) + \phi_3(y + 2x) + xe^{2x + y} - \frac{x^2}{3}(y + x)^{1/2}$.

40. $z = \phi_1(y - x) + \phi_2(y - 2x) + \phi_3(y + 3x) - \frac{1}{75} \cos (x + 2y) + \frac{x}{20} e^{3x + y}$.

41. $z = \phi_1(y - x) + x\phi_2(y - x) + \phi_3(y + 2x) + \frac{1}{27} \sin (y + 2x) + xe^y$.

42. $z = \phi_1(y) + \phi_2(y + 2x) + \phi_3(y - x) - \frac{1}{81}(9x^2 + 36y^2 - 18x - 72y + 76)e^{x + 2y}$.

43. $z = \phi_1(y + x) + \phi_2(y - x) + \phi_3(y - 2x) + ye^y$.

44. $z = e^{-x} \cos y - \frac{x}{2} e^{-x} \cos y = \frac{1}{2}(2 - x)e^{-x} \cos y$.

45. $(2x + y)z = 3x$.

▲▲

6

CHAPTER

Non-Homogeneous Linear Partial Differential Equations with Constant Coefficients

Definition: If the partial derivatives occurring in the given equation are not of same order then it is called a non-homogenous linear partial differential equation with constant coefficients.

For example,

$$\frac{\partial^2 z}{\partial x^2} - \frac{\partial z}{\partial y} = \sin(2x + 3y)$$

$$(D_x^2 - 3D_y + 5)(D_x - D_y)z = (x + 3y)e^{y - 3x}$$

$$3\frac{\partial^2 z}{\partial x \partial y} + \frac{\partial^2 z}{\partial y^2} - 5\frac{\partial z}{\partial y} = e^{x + 4y}$$

are non-homogeneous linear partial differential equations as the polynomial expression $F(D_x, D_y)$ in D_x, D_y is not homogeneous. As we have discussed in earlier chapter if $F(D_x, D_y)$ is reducible to linear factor then $F(D_x, D_y)z = f(x, y)$ is known as a reducible linear partial differential equation with constant coefficients otherwise as irreducible linear partial differential equation with constant coefficients.

For example, $(D_x^2 - D_y^2)z = e^{x + 2y}$ is a reducible linear partial differential equation with constant coefficients since $D_x^2 - D_y^2 = (D_x + D_y)(D_x - D_y)$ whereas $(D_x^2 - D_y^3)z = e^{x + 2y}$ is an irreducible linear partial differential equation with constant coefficient since $D_x^2 - D_y^3$ cannot be decomposed into linear factors. As in case of homogeneous case, the complete solution is the sum of complementary function and particular integral first, we will discuss methods of finding complementary function for non-homogeneous linear partial differential equations with constant coefficients.

Case I: *When $F(D_x, D_y)$ can be factorized in terms of distinct linear factors in D_x and D_y.*

Now, if $F(D_x, D_y)$ can be decomposed into distinct linear factors then $F(D_x, D_y)z = 0$ becomes

$$(\alpha_1 D_x + \beta_1 D_y + \gamma_1)(\alpha_2 D_x + \beta_2 D_y + \gamma_2) \ldots (\alpha_n D_x + \beta_n D_y + \gamma_n)z = 0$$

Clearly, any solution of $(\alpha_i D_x + \beta_j D_y + \gamma_j)z = 0$ where $1 \le i \le n$ is the solution of $F(D_x, D_y)z = 0$.

Hence, we find solution of

$$(\alpha_i D_x + \beta_j D_y + \gamma_j)z = 0$$

or
$$\alpha_i p + \beta_i q = -\gamma_i z$$

which is of Lagrange's form. So, its auxiliary equation is

$$\frac{dx}{\alpha_i} = \frac{dy}{\beta_i} = \frac{dz}{-\gamma_j z}$$

Taking the first two members, we have

$$\beta_i dx = \alpha_i dy.$$

Integrating, we obtain $\beta_i x - \alpha_i y = c_1$

Again, taking the first and third members, we get

$$\frac{-\gamma_i}{\alpha_j} dx = \frac{dz}{z} \text{ provided } a_i \neq 0.$$

Integrating, we have $\log z - \log c_2 = \dfrac{-\gamma_1}{\alpha_j} x$

$\Rightarrow$
$$\frac{z}{c_2} = e^{\frac{-\gamma_i}{\alpha_j} x}$$

$\Rightarrow$
$$z = c_2 e^{\frac{-\gamma_i}{\alpha_j} x}$$

As complete solution is $\phi(c_1, c_2) = 0$

$\Rightarrow$
$$c_2 = \psi_i(c_1)$$

$\therefore$
$$z = \phi_i(c_1) e^{\frac{-\gamma_i}{\alpha_j} x}$$

$\Rightarrow$
$$z = e^{\frac{-\gamma_i}{\alpha_j} x} \psi_i(\beta_i x - \alpha_i y)$$

If $\alpha_i = 0$ and $\beta_i \neq 0$ then taking the second and third members, we have

$$\frac{dy}{\beta_i} = \frac{dz}{-\gamma_i z} \Rightarrow \frac{dz}{z} = -\frac{\gamma_i}{\beta_i} dy$$

Integrating, we obtain $\log z - \log c_2 = -\dfrac{\gamma_i}{\beta_i} y$

$$\Rightarrow \qquad z = c_2 \, e^{-\frac{\gamma_i}{\beta_i} y}$$

$$\Rightarrow \qquad z = \psi_i(c_1) \, e^{-\frac{\gamma_i}{\beta_i} y}$$

$$\Rightarrow \qquad z = e^{-\frac{\gamma_i}{\beta_i} y} \psi_j(\beta_i x - \alpha_i y).$$

So, for each factor $(1 \leq i \leq n)$, we get a solution and hence complete solution of $F(D_x, D_y)z - 0$ is

$$z = e^{-\frac{\gamma_i}{\alpha_j} x} \psi_1(\beta_1 x - \alpha_1 y) + e^{-\frac{\gamma_2}{\alpha_2} x} \psi_2(\beta_2 x - \alpha_2 y) + \ldots + e^{-\frac{\gamma_n}{\alpha_n} x} \psi_n(\beta_n x - \alpha_n y)$$

provided $\alpha_1, \alpha_2 \ldots \alpha_n \neq 0$

or
$$z = e^{-\frac{\gamma_1}{\beta_1} y} \psi_1(\beta_1 x - \alpha_1 y) + e^{-\frac{\gamma_2}{\alpha_2} x} \psi_2(\beta_2 x - \alpha_2 y) + \ldots + e^{-\frac{\gamma_n}{\beta_n} x} \psi_n(\beta_n x - \alpha_n y)$$

provided $\beta_1, \beta_2 \ldots \beta_n \neq 0$.

Case II: *When $F(D_x, D_y)$ contains repeated factors.*

First, suppose the first two factors are same. Then $F(D_x, D_y)z = 0$ becomes

$$(\alpha_1 D_x + \beta_1 D_y + \gamma_1)^2 (\alpha_3 D_x + \beta_3 D_y + \gamma_3) \ldots (\alpha_n D_x + \beta_n D_y + \gamma_n)z = 0.$$

Clearly, for $3 \leq i \leq n$, the solution can be found by using case I.

Now consider

$$(\alpha_1 D_x + \beta_1 D_y + \gamma_1)(\alpha_1 D_x + \beta_1 D_y + \gamma_1)z = 0$$

Let $u = (\alpha_1 D_x + \beta_1 D_y + \gamma_1)z$

$$\therefore \qquad (\alpha_1 D_x + \beta_1 D_y + \gamma_1)u = 0$$

$$\Rightarrow \qquad u = e^{-\frac{\gamma_1}{\alpha_1} x} \psi_1(\beta_1 x - \alpha_1 y) \text{ if } \alpha_1 \neq 0.$$

or
$$u = e^{-\frac{\gamma_1}{\beta_1} y} \psi_1(\beta_1 x - \alpha_1 y) \text{ if } \beta_1 \neq 0.$$

Let $\alpha_1 \neq 0$

$$\therefore \quad (\alpha_1 D_x + \beta_1 D_y + \gamma_1)z = u = e^{-\frac{\gamma_1}{\alpha_1}x} \psi_1(\beta_1 x - \alpha_1 y)$$

$$\Rightarrow \quad \alpha_1 p + \beta_1 q = e^{-\frac{\gamma_1}{\alpha_1}x} \psi_1(\beta_1 x - \alpha_1 y) - \gamma_1 z$$

which is of Lagrange's form. So, its auxiliary equation is

$$\frac{dx}{\alpha_1} = \frac{dy}{\beta_1} = \frac{dz}{e^{-\frac{\gamma_1}{\alpha_1}x} \psi_1(\beta_1 x - \alpha_1 y) - \gamma_1 z}$$

Taking the first two members, we have

$$\beta_1 dx = \alpha_1 dy$$

Integrating, we obtain $\beta_1 x - \alpha_1 y = c_1$

Again, taking the first and third members, we get

$$\frac{dx}{\alpha_1} = \frac{dz}{e^{-\frac{\gamma_1}{\alpha_1}x} \psi_1(c_1) - \gamma_1 z}$$

$$\Rightarrow \quad \frac{dz}{dx} + \frac{\gamma_1}{\alpha_1} z = \frac{1}{\alpha_1} e^{-\frac{\gamma_1}{\alpha_1}x} \psi_1(c_1)$$

which is a linear differential equation of the first order with integrating factor

$$\text{IF} = e^{\int -\frac{\gamma_1}{\alpha_1} dx} = e^{-\frac{\gamma_1}{\alpha_1}x}$$

So, its solution is

$$z e^{\frac{\gamma_1}{\alpha_1}x} = \int \frac{1}{\alpha_1} \psi_1(c_1) dx + c_2$$

$$\Rightarrow \quad z = \psi_1'(c_1)\, x e^{-\frac{\gamma_1}{\alpha_1}x} + c_2 e^{-\frac{\gamma_1}{\alpha_1}x}$$

Again we can take $c_2 = \psi_2'(c_1)$

$$\therefore \quad z = e^{-\frac{\gamma_1}{\alpha_1}x} [x\psi_1'(c_1) + \psi_2'(c_1)]$$

$$\Rightarrow \quad z = e^{-\frac{\gamma_1}{\alpha_1}x} [x\psi_1'(\beta_1 x - \alpha_1 y) + \psi_2'(\beta_1 x - \alpha_1 y)]$$

Similarly, if $\beta_1 \neq 0$ then

$$(\alpha_1 D_x + \beta_1 D_y + \gamma_1)z = u = e^{-\frac{\gamma_1}{\beta_1}y}\psi_1(\beta_1 x - \alpha_1 y)$$

$$\Rightarrow \qquad \alpha_1 p + \beta_1 q = e^{-\frac{\gamma_1}{\beta_1}y}\psi_1(\beta_1 x - \alpha_1 y) - \gamma_1 z$$

which is of Lagrange's form. So, its auxiliary equation is

$$\frac{dx}{\alpha_1} = \frac{dy}{\beta_1} = \frac{dz}{e^{-\frac{\gamma_1}{\beta_1}y}\psi_1(\beta_1 x - \alpha_1 y) - \gamma_1 z}$$

Taking the first two members, we get $\beta_1 dx = \alpha_1 dy$ which on integration yields $\beta_1 x - \alpha_1 y = c_1$

Again, taking the second and third members, we have

$$\frac{dy}{\beta_1} = \frac{dz}{e^{-\frac{\gamma_1}{\beta_1}y}\psi_1(\beta_1 x - \alpha_1 y) - \gamma_1 z}$$

$$\Rightarrow \qquad \frac{dz}{dy} + \frac{\alpha_1}{\beta_1}z = \frac{1}{\beta_1}e^{-\frac{\gamma_1}{\alpha_1}y}\psi(c_1)$$

which is a linear differential equation of the first order with integrating factor,

$$\text{IF} = e^{\int\frac{\gamma_1}{\beta_1}dy} = e^{-\frac{\gamma_1}{\beta_1}y}$$

So, its solution is

$$ze^{-\frac{\gamma_1}{\alpha_1}y} = \int\frac{1}{\beta_1}\psi_1(c_1)dy + c_2$$

$$\Rightarrow \qquad ze^{-\frac{\gamma_1}{\beta_1}y} = \psi_1'(c_1)y + c_2$$

Again, let $\qquad c_2 = \psi_2'(c_1)$

$$\Rightarrow \qquad ze^{-\frac{\gamma_1}{\beta_1}y} = y\psi_1'(c_1) + \psi_2'(c_1)$$

$$\therefore \qquad z = e^{-\frac{\gamma_1}{\beta_1}y}[y\psi_1'(\beta_1 x - \alpha_1 y) + \psi_2'(\beta_1 x - \alpha_1 y)].$$

Note: If $\alpha_1 D_x + \beta_1 D_y + \gamma_1$ is repeated k-times, then we can write the general solution as

$$z = e^{-\frac{\gamma_1}{\alpha_1}x}[x^{k-1}\psi_1(\beta_1 x - \alpha_1 y) + x^{k-2}\psi_2(\beta_1 x - \alpha_1 y) + \dots + x\psi_{k-1}(\beta_1 x - \alpha_1 y) + \psi_k(\beta_1 x - \alpha_1 y)] \text{ if } \alpha_1 \neq 0$$

or

$$z = e^{-\frac{\gamma_1}{\beta_1}y}[y^{k-1}\psi_1(\beta_1 x - \alpha_1 y) + y^{k-2}\psi_2(\beta_1 x - \alpha_1 y) + \dots + y\psi_{k-1}(\beta_1 x - \alpha_1 y) + \psi_k(\beta_1 x - \alpha_1 y)] \text{ if } \beta_1 \neq 0.$$

Case III: *When $F(D_x, D_y)$ cannot be factorized into linear factors.*

When $F(D_x, D_y)$ is irreducible, it is not always possible to find a solution with the full number of arbitrary functions, but it is possible to obtain solution which contains as many arbitrary constants as we wish. We can easily prove that $F(D_x, D_y)e^{ax+by} = F(a, b)\, e^{ax+by}$.

From this relation, clearly e^{ax+by} is a solution of $F(D_x, D_y)z = 0$ if $F(a, b) = 0$.

So that

$$z = \Sigma A_i e^{aix + biy}$$

where, A_i, a_i, b_i are all constants and for any value of a_i we can find the corresponding b_i such that $F(a_i, b_i) = 0$ or vice versa.

Therefore, $z = \Sigma A_i e^{aix + biy}$ where $F(a_i, b_i) = 0$ is general solution.

EXAMPLE 6.1

Solve $(D_x^2 - D_y^2 + D_x - D_y)z = 0$.

Solution: Here, $(D_x - D_y)(D_x + D_y + 1)z = 0$.

Hence, the complete solution is

$$z = e^{-\frac{0}{1}x}\psi_1(-x - y) + e^{-\frac{1}{1}x}\psi_2(1\cdot x - 1\cdot y)$$

$$\Rightarrow \qquad z = \psi_1'(x + y) + e^{-x}\psi_2(x - y).$$

EXAMPLE 6.2

Solve $(D_x + 2D_y + 3)^2 z = 0$.

Solution: Here, we have factor $D_x + 2D_y + 3$ repeating two times.

The complete solution is

$$z = e^{-\frac{3}{1}x}[x\psi_1(2x - y) + \psi_2(2x - y)]$$

or

$$z = e^{-3x}(x\psi_1(2x - y) + \psi_2(2x - y)].$$

EXAMPLE 6.3

Solve $(D_x + 1)(D_x + D_y - 1)z = 0$.

Solution: The complete solution is

$$z = e^{-\frac{1}{1}x}\,\psi_1(0\cdot x - 1\cdot y) + e^{-\left(\frac{-1}{1}\right)x}\,\psi_2(1\cdot x - 1\cdot y)$$

or $$z = e^{-x}\,\psi_1'(y) + e^{x}\psi_2(x - y).$$

EXAMPLE 6.4

Solve $D_xD_y(D_x - 3D_y - 4)z = 0$.

Solution: The complete solution is

$$z = e^{-\frac{0}{1}x}\,\psi_1(0\cdot x - 1\cdot y) + e^{-\frac{0}{1}y}\,\psi_2(1\cdot x - 0\cdot y) + e^{-\left(\frac{-4}{1}\right)x}\,\psi_3(-3\cdot x - 1\cdot y)$$

or $$z = \psi_1'(y) + \psi_2(x) + e^{4x}\,\psi_3'(3x + y).$$

EXAMPLE 6.5

Solve $\dfrac{\partial^2 z}{\partial x^2} + 2\dfrac{\partial^2 z}{\partial x\partial y} + \dfrac{\partial^2 z}{\partial y^2} + 2\dfrac{\partial z}{\partial x} + 2\dfrac{\partial z}{\partial y} + z = 0.$

Solution: The given equation can be written as

$$(D_x^2 + 2D_xD_y + D_y^2 + 2D_x + 2D_y + 1)z = 0$$

$$\Rightarrow \quad ((D_x + D_y)^2 + 2(D_x + D_y) + 1)z = 0$$

$$\Rightarrow \quad (D_x + D_y + 1)^2 z = 0$$

The complete solution is

$$z = e^{-\frac{1}{1}x}\,(\psi_1(1\cdot x - 1\cdot y) + x\psi_2(1\cdot x - 1\cdot y))$$

or $$z = e^{-x}(\psi_1(x - y) + x\psi_2(x - y)).$$

EXAMPLE 6.6

Solve $3\dfrac{\partial^2 z}{\partial x^2} - \dfrac{\partial^2 z}{\partial y^2} + \dfrac{\partial z}{\partial x} = 0.$

Solution: The given equation can be written as

$$(3D_x^2 - D_y^2 + D_x)z = 0$$

Here $3D_x^2 - D_y^2 + D_x$ is irreducible

Let, $z = ZAe^{ax + by}$ be its solution with $3a^2 - b^2 + a = 0$

$$\Rightarrow \qquad b = \pm \sqrt{3a^2 + a}$$

$\therefore$ The complete solution is

$$z = \sum Ae^{ax \pm \sqrt{3a^2 + a}\, y}$$

where A, a are arbitrary constants.

EXAMPLE 6.7

Solve $\dfrac{\partial^2 z}{\partial y^2} + \dfrac{\partial z}{\partial x} + \dfrac{\partial z}{\partial y} = 0.$

Solution: The given equation can be written as

$$(D_y^2 + D_x + D_y)z = 0$$

Here, $D_y^2 + D_x + D_y$ is irreducible.

Let $z = \sum Ae^{ax + by}$ be its solution with $b^2 + a + b = 0 \Rightarrow a = -(b^2 + b)$

$\therefore$ The complete solution is $z = \sum Ae^{-(b^2 + b)x + by}$

where A, b are arbitrary constants.

EXAMPLE 6.8

Solve $(3D_x^4 + 4D_x^2 D_y + D_y^2)z = 0.$

Solution: Here $(3D_x^2 + 3D_x^2 D_y + D_x^2 D_y + D_y^2)z = 0$

$$\Rightarrow \qquad (3D_x^2(D_x^2 + D_y) + D_y(D_x^2 + D_y))z = 0$$

$$\Rightarrow \qquad (3D_x^2 + D_y)(D_x^2 + D_y)z = 0$$

Here, $3D_x^2 + D_y$ is irreducible.

Let, $z = \sum Ae^{ax + by}$ be solution of $(3D_x^2 + D_y)z = 0$

$$\therefore \qquad 3a^2 + b = 0 \Rightarrow b = -3a^2$$

$\therefore$ The corresponding complete solution is

$$z = \sum Ae^{ax - 3a^2 y}$$

Also, $D_x^2 + D_y$ is also irreducible.

Let, $z = \sum Be^{cx + dy}$ be solution of $(D_x^2 + D_y)z = 0$

$$\therefore \qquad c^2 + d = 0 \Rightarrow d = -c^2$$

∴ The corresponding complete solution is

$$z = \sum Be^{cx - c^2y}$$

Hence, the complete solution of the given equation is

$$z = \sum Ae^{ax - 3a^2y} + \sum Be^{cx - c^2y}.$$

where, A, a, B, c are arbitrary constants.

EXAMPLE 6.9

Solve $(D_x + 2D_y - 3)(D_x^2 + D_y)z = 0$.

Solution: Since $D_x + 2D_y - 3$ is a linear factor

$$\therefore \qquad z = e^{-\left(\frac{-3}{1}\right)x} \psi_1(2\cdot x - 1\cdot y) = e^{3x}\,\psi_1(2x - y)$$

is the complete solution corresponding to $(D_x + 2D_y - 3)z = 0$

Again $\qquad D_x^2 + D_y$ is irreducible.

Let, $\qquad z = \sum Ae^{ax + by}$ be solution corresponding to $(D_x^2 + D_y)z = 0$

$$\therefore \qquad a^2 + b = 0 \Rightarrow b = -a^2.$$

Thus, $\qquad z = \sum Ae^{ax - a^2y}$

Therefore, the complete solution of the given equation is $z = e^{3x}\,\psi_1(2x - y) + \sum Ae^{ax - a^2y}$.

where, ψ_1 is an arbitrary function and A, a are arbitrary constants.

EXAMPLE 6.10

Solve $(3D_x + D_y)^2 (D_x^2 + 5D_x + D_y)z = 0$.

Solution: Here $3D_x + D_y$ is a linear factor with multiplicity 2. So, corresponding to $(3D_x + D_y)^2 z = 0$, the solution is

$$z = e^{-\frac{0}{3}x}[\psi_1(1\cdot y - 3\cdot x) + x\psi_2(1\cdot y - 3\cdot x)]$$

or

$$z = \psi_1(y - 3x) + x\psi_2(y - 3x)$$

Also, $D_x^2 + 5D_x + D_y$ is irreducible. So, corresponding to $(D_x^2 + 5D_x + D_y)z = 0$, the solution is $z = \sum Ae^{ax + by}$ where, $a^2 + 5a + b = 0 \Rightarrow b = -(a^2 + 5a)$

Thus, $\qquad z = \sum Ae^{ax - (a^2 + 5a)y}.$

Therefore, the complete solution of the given equation is

$$z = \psi_1(y - 3x) + x\psi_2(y - 3x) + \Sigma A e^{ax - (a^2 + 5a)y}$$

where, ψ_1, ψ_2 are arbitrary functions and A, a are arbitrary constants.

EXAMPLE 6.11

Solve $(2D_x - 3D_y + 7)^2 (D_x^2 + 3D_y)z = 0$.

Solution: Here, $2D_x - 3D_y + 7$ is a linear factor with multiplicity 2. So, solution corresponding to $(2D_x - 3D_y + 7)^2 z = 0$ is given by

$$z = e^{\left(-\frac{7}{2}\right)x} [\psi_1(-3x - 2y) + x\psi_2(-3x - 2y)]$$

or
$$z = e^{-\frac{7x}{2}} [\psi_1'(3x + 2y) + x\psi_2(3x + 2y)]$$

Again, $D_x^2 + 3D_y$ is irreducible. So, solution corresponding to $(D_x^2 + 3D_y)z = 0$ is given by $z = \Sigma A e^{ax + by}$ where $a^2 + 3b = 0 \Rightarrow b = -\dfrac{a^2}{3}$

Thus,
$$z = \Sigma A e^{ax - \frac{a^2}{3}y}$$

Therefore, the complete solution of the given equation is

$$z = e^{-\frac{7x}{2}} [\psi_1'(3x + 2y) + x\psi_2'(3x + 2y)] + \Sigma A e^{ax - \frac{a^2}{3}y}$$

where ψ_1', ψ_2' are arbitrary functions and A, a are arbitrary constants.

▲▲

EXERCISE 6.1

Solve the following partial differential equations:

1. $(D_x^2 - D_y^2 + D_x - D_y)z = 0$.
2. $(D_x^2 - a^2D_y^2 + 2abD_x + 2a^2bD_y)z = 0$.
3. $r + 2s + t + 2p + 2q + z = 0$.
4. $D_xD_y (D_x - 2D_y - 3)z = 0$.
5. $(2D_x - 3) (3D_x - 5D_y - 7)^2 z = 0$.
6. $(3D_x - 5) (7D_y + 2)D_xD_y(2D_x + 3D_y + 5)z = 0$.
7. $r + s + q = 0$.
8. $(D_x^2 + D_x + D_y)z = 0$.
9. $(2D_x^2 - D_y^2 + D_x)z = 0$.
10. $(3D_x + 4D_y - 6)(D_x^2 + D_y)z = 0$.

11. $r + s + p + q + z = 0.$

12. $\dfrac{\partial z}{\partial x} - \dfrac{\partial^2 z}{\partial y^2} = 0.$

13. $\dfrac{\partial^2 z}{\partial x^2} + \dfrac{\partial^2 z}{\partial y^2} = n^2 z.$

14. $s + p - q - z = 0.$

15. $(D_x^2 + D_x D_y + Dy - 1)z = 0.$

ANSWERS

Exercise 6.1

1. $z = \phi_1(x + y) + e^{-x}\phi_2(x - y).$
2. $z = \phi_1(ax - y) + e^{-2abx}\phi_2(ax + y).$
3. $z = [\phi_1(x - y) + x\phi_2(x - y)]e^{-x}.$
4. $z = \phi_1(y) + \phi_2(x) + e^{3x}\phi_3(2x + y).$
5. $z = e^{3x/2}\phi_1(2y) + e^{7x/3}[\phi_2(5x + 3y) + x\phi_3(5x + 3y)].$
6. $z = e^{5x/3}\phi_1(3y) + e^{-2y/7}\phi_2(7x) + \phi_3(y) + \phi_4(x) + e^{-5x/2}\phi_5(3x - 2y).$
7. $z = \phi_1(x) + e^{-x}\phi_2(x - y).$
8. $z = \sum A e^{ax - (a^2 + a)y}.$
9. $z = \sum A\, e^{ax \pm \sqrt{2a^2 + a}\, y}.$
10. $z = e^{2x}\phi_1(4x - 3y) + \sum A e^{ax - a^2 y}.$
11. $z = \sum A\, e^{ax - \left(\frac{a^2 + a + 1}{a + 1}\right)y}.$
12. $z = \sum A e^{b^2 x + by}.$
13. $z = \sum A e^{n(x \cos a + y \sin a)}.$
14. $z = e^{x}\phi_1(y) + e^{-y}\phi_2(x).$
15. $z = e^{-x}\phi_1(y)e^{x}\phi_2\,(x - y).$

▲▲

Methods to find particular integral (z_p) of non-homogeneous linear partial differential equations with constant coefficients

The methods to evaluate particular integrals (z_p) of non-homogeneous linear partial differential equations with constant coefficients are quite similar to those used for solving homogeneous linear partial differential equations with constant coefficient. Here, we are discussing some rules to find particular integral (z_p) of $F(D_x, D_y)z = f(x, y)$.

Rule I

When $f(x, y) = e^{\alpha x + \beta y}$ then

$$z_p = \frac{1}{F(D_x, D_y)} e^{\alpha x+\beta y} = \frac{1}{F(\alpha, \beta)} e^{\alpha x + \beta y} \text{ provided } F(\alpha, \beta) \neq 0.$$

Rule II

When $f(x, y) = \sin(\alpha x + \beta y)$ or $\cos(\alpha x + \beta y)$ then

$$z_p = \frac{1}{F(D_x, D_y)} \sin(\alpha x + \beta y) \text{ or } \cos(\alpha x + \beta y)$$

$$= \frac{1}{\text{Change } D_x^2 \text{ by} -\alpha^2, D_y^2 \text{ by} -\beta^2 \text{ and } D_x D_y \text{ by} -\alpha\beta} \sin(\alpha x + \beta y) \text{ or } \cos(\alpha x + \beta y)$$

provided denominator $\neq 0$.

Rule III

When $f(x, y) =$ Sum of terms containing $x^l y^m$.

Then, $z_p = \dfrac{1}{F(D_x, D_y)}(x^l y^m) = [F(D_x, D_y)]^{-1}(x^l y^m)$.

Here, expand $[F(D_x, D_y)]^{-1}$ in ascending powers of $\dfrac{D_x}{D_y}$ or $\dfrac{D_y}{D_x}$.

Rule IV

When $f(x, y) = e^{\alpha x + \beta y} V(x, y)$

Then, $z_p = \dfrac{1}{F(D_x, D_y)}(e^{\alpha x+\beta y} V(x,y)) = e^{\alpha x+\beta y} \dfrac{1}{F(D_x+\alpha, D_y+\beta)} V(x,y)$

Note: $z_p = \dfrac{1}{F(D_x, D_y)} e^{\alpha x+\beta y} = e^{\alpha x+\beta y} \dfrac{1}{F(D_x+\alpha, D_y+\beta)}$ if $F(\alpha, \beta) = 0$.

EXAMPLE 6.12

Solve $(D_x - D_y - 1)(D_x - D_y - 2)z = e^{2x-2}$.

Solution: The complementary function is

$$z_c = e^x \psi_1(x + y) + e^{2x} \psi_2(x + y)$$

and IP is
$$z_p = \frac{1}{(D_x - D_y - 1)(D_x - D_y - 2)} e^{2x - y}$$

$$= \frac{1}{(2-(-1)-1)(2-(-1)-2)} e^{2x - y} \quad \text{\{using Rule I\}}$$

$$= \frac{1}{2} e^{2x - y}.$$

Hence, the complete solution is
$$z = z_c + z_p$$
$$z = e^x \psi_1(x + y) + e^{2x} \psi_2(x + y) + \frac{1}{2} e^{2x - y}.$$

EXAMPLE 6.13

Solve $(D_x^2 + D_x D_y + D_y - 1)z = \cos (x + 2y)$.

Solution: The given equation is
$$(D_x + 1)(D_x + D_y - 1)z = \cos (x + 2y)$$

$\therefore$ CF is
$$z_c = e^{-x} \psi_1(y) + e^x \psi_2(y - x)$$

and PI is
$$z_p = \frac{1}{D_x^2 + D_x D_y + D_y - 1} \cos (x + 2y)$$

$$= \frac{1}{-1^2 - 1 \cdot 2 + D_y - 1} \cos (x + 2y) \quad \text{\{using Rule II\}}$$

$$= \frac{1}{D_y - 4} \cos (x + 2y)$$

$$= \frac{D_y + 4}{D_y^2 - 16} \cos (x + 2y)$$

$$= \frac{D_y \cos(x + 2y) + 4 \cos(x + 2y)}{-4 - 16}$$

$$= -\frac{1}{20} (-2 \sin (x + 2y) + 4 \cos (x + 2y))$$

Hence, the complete solution is
$$z = z_c + z_p$$
$$z = e^{-x} \psi_1(y) + e^x \psi_2(y - x) + \frac{1}{10} (\sin (x + 2y) - 2 \cos (x + 2y)).$$

EXAMPLE 6.14

Solve $r - q = -z + x^2y$.

Solution: The given equation is $(D_x^2 - D_y + 1)z = x^2y$.

Here, $D_x^2 - D_y + 1$ is irreducible. So, CF is

$$z_c = \sum Ae^{ax + by} \text{ where } a^2 - b + 1 = 0$$

$$\Rightarrow \quad b = a^2 + 1$$

$$\therefore \quad z_c = \sum Ae^{ax + (a^2 + 1)y}$$

where, A, a are arbitrary constants

and PI is

$$z_p = \frac{1}{D_x^2 - D_y + 1}(x^2y) = (1 - (D_y - D_x^2))^{-1}(x^2y)$$

$$= [1 + (D_y - D_x^2) + (D_y - D_x^2)^2 + \dots]\,(x^2y)$$

$$= x^2y + x^2 - 2y - 4$$

Hence, the complete solution is

$$z = z_c + z_p$$

$$z = \sum Ae^{ax + (a^2 + 1)y} + x^2y + x^2 - 2y - 4.$$

where A, a are arbitrary constants.

EXAMPLE 6.15

Solve $(D_x^2 - D_y^2 - 3D_x + 3D_y)z = xy + e^{x + 2y}$.

Solution: The given equation is

$$(D_x - D_y)(D_x + D_y - 3)z = xy + e^{x + 2y}.$$

$\therefore$ CF is

$$z_c = \psi_1(y + x) + e^{3x}\psi_2(y - x)$$

and PI is

$$z_p = \frac{1}{(D_x - D_y)(D_x + D_y - 3)}(xy) + \frac{1}{(D_x - D_y)(D_x + D_y - 3)}(e^{x + 2y})$$

Now, $$\frac{1}{(D_x - D_y)(D_x + D_y - 3)}(xy) = -\frac{1}{3D_x}\left(1 - \frac{D_y}{D_x}\right)^{-1}\left(1 - \frac{D_x + D_y}{3}\right)(xy)$$

$$= -\frac{1}{3D_x}\left(1 + \frac{D_y}{D_x} + \frac{D_y^2}{D_x^2} + \dots\right)\left(1 + \frac{D_x + D_y}{3} + \left(\frac{D_x + D_y}{3}\right)^2 + \dots\right)(xy)$$

$$= \frac{-1}{3D_x}\left(1+\frac{D_x+D_y}{3}+\frac{D_y}{D_x}+2\frac{D_xD_y}{9}+\frac{D_y}{3}\right)(xy)$$

$$= \frac{-1}{3D_x}\left(xy+\frac{y+x}{3}+\frac{x^2}{2}+\frac{2}{9}+\frac{x}{3}\right)$$

$$= \frac{-1}{3D_x}\left(xy+\frac{y}{3}+\frac{2x}{3}+\frac{x^2}{2}+\frac{2}{9}\right)$$

$$= \frac{-1}{3}\left(\frac{x^2y}{2}+\frac{xy}{3}+\frac{x^2}{3}+\frac{x^3}{6}+\frac{2}{9}x\right)$$

and $\frac{1}{(D_x+D_y-3)(D_x-D_y)}e^{x+2y} = \frac{1}{D_x+D_y-3}\frac{1}{(1-2)}e^{x+2y}$

$$= -\frac{1}{D_x+D_y-3}e^{x+2y} = -e^x\frac{1}{1+D_y-3}e^{2y}$$

$$= -e^x\frac{1}{D_y-2}e^{2y} = -e^xe^{2y}\frac{1}{D_y+2-2} \tag{1}$$

$$= -e^xe^{2y}\frac{1}{D_y}(1) = -ye^{x+2y}$$

$\therefore$ $$z_p = -\frac{1}{3}\left(\frac{x^2y}{2}+\frac{xy}{3}+\frac{x^2}{3}+\frac{x^3}{6}+\frac{2}{9}x\right) - ye^{x+2y}$$

Hence, the complete solution is

$$z = z_c + z_p$$

$$z = \psi_1(y+x) + e^{3x}\psi_2(y-x) - \frac{1}{3}\left(\frac{x^2y}{2}+\frac{xy}{3}+\frac{x^2}{3}+\frac{x^3}{6}+\frac{2}{9}x\right) - ye^{x+2y}$$

EXAMPLE 6.16

Solve $\frac{\partial^2 z}{\partial x^2}-\frac{\partial^2 y}{\partial x\partial y}-2\frac{\partial^2 z}{\partial y^2}+2\frac{\partial z}{\partial x}+2\frac{\partial z}{\partial y} = e^{2x+3y}+\sin(2x+y)+xy.$

Solution: The given equation is

$$(D_x^2 - D_xD_y - 2D_y^2 + 2D_x + 2D_y)z = e^{2x+3y} + \sin(2x+y) + xy$$

$$\Rightarrow \qquad (D_x + D_y)(D_x - 2D_y + 2)z = e^{2x+3y} + \sin(2x+y) + xy.$$

$\therefore$ CF is

$$z_c = \phi_1(y-x) + e^{-2x}\phi_2(2x+y)$$

and PI is

$$z_p = \frac{1}{(D_x+D_y)(D_x-2D_y+2)}(e^{2x+3y} + \sin(2x+y) + xy)$$

Now, $$\frac{1}{(D_x+D_y)(D_x-2D_y+2)}(e^{2x+3y}) = \frac{1}{(2+3)(2-2(3)+2)}(e^{2x+3y})$$

$$= -\frac{1}{10}e^{2x+3y}$$

$$\frac{1}{(D_x+D_y)(D_x-2D_y+2)}(\sin(2x+y)) = \frac{1}{D_x^2 - D_xD_y - 2D_y^2 + 2D_x + 2D_y}(\sin(2x+y))$$

$$= \frac{1}{-2^2-(-2\cdot 1)-2(-1^2)+2D_x+2D_y}(\sin(2x+y))$$

$$= \frac{1}{2(D_x+D_y)}\sin(2x+y) = \frac{D_x-D_y}{2(D_x^2-D_y^2)}\sin(2x+y)$$

$$= \frac{D_x\sin(2x+y) - D_y\sin(2x+y)}{2(-2^2-(-1^2))}$$

$$= -\frac{1}{6}(2\cos(2x+y) - \cos(2x+y))$$

$$= -\frac{1}{6}\cos(2x+y)$$

and $$\frac{1}{(D_x+D_y)(D_x-2D_y+2}(xy) = \frac{\pm 1}{2D_x}\left(1+\frac{D_y}{D_x}\right)^{-1}\left(1+\frac{D_x-2D_y}{2}\right)^{-1}(xy)$$

$$= \frac{1}{2D_x}\left(1-\frac{D_y}{D_x}\right)\left(1-\frac{D_x-2D_y}{2}+\left(\frac{D_x-2D_y}{2}\right)^2\right)(xy)$$

$$= \frac{1}{D_x}\left(1-\frac{D_y}{D_x}\right)\left(xy-\frac{y-2x}{2}-1\right)$$

$$= \frac{1}{2D_x}\left(xy+\frac{2x-y}{2}-1-\frac{1}{D_x}\left(x-\frac{1}{2}\right)\right)$$

$$= \frac{1}{2D_x}\left(xy+x-\frac{y}{2}-1-\frac{x^2}{2}+\frac{x}{2}\right)$$

$$= \frac{1}{2D_x}\left(xy+\frac{3x}{2}-\frac{y}{2}-\frac{x^2}{2}-1\right)$$

$$= \frac{1}{2}\left(\frac{x^2}{2}y+\frac{3x^2}{4}-\frac{xy}{2}-\frac{x^3}{6}-x\right)$$

Hence, the complete solution is

$z = z_c + z_p$

$$z = \phi_1(y - x) + e^{-2x}\phi_2(2x + y) - \frac{1}{10}e^{2x + 3y} - \frac{1}{6}\cos(2x+y)+\frac{1}{2}\left(\frac{1}{2}x^2y+\frac{3}{4}x^2-\frac{1}{2}xy-\frac{x^3}{y}-x\right)$$

EXAMPLE 6.17

Solve $r - q = e^{x + y}$.

Solution: The given equation is

$$(D_x^2 - D_y)z = e^{x + y}$$

Here, $D_x^2 - D_y$ is irreducible.

So, CF is $\quad z_c = \Sigma A e^{ax + by}$

where, $\quad a^2 - b = 0 \Rightarrow b = a^2$

$\Rightarrow \quad z_c = \Sigma A e^{ax + a^2 y}$

where, A, a are arbitrary constants.

and PI is $\quad z_p = \frac{1}{D_x^2 - D_y}e^{x+y} = \frac{1}{1-1}e^{x + y}$

$$= e^{x + y}\frac{1}{(D_x+1)^2-(D_y+1)} \quad (1)$$

$$= e^{x + y}\frac{1}{D_x^2+2D_x-D_y} \quad (1)$$

$$= e^{x+y}\frac{1}{-D_y}\left(1+\left(\frac{2D_x+D_x^2}{D_y}\right)\right)^{-1}(1)$$

$$= e^{x+y}\frac{1}{-D_y}(1) = -ye^{x+y}.$$

Hence, the complete solution is

$$z = z_c + z_p$$

$$z = \Sigma Ae^{ax+a^2y} - ye^{x+y}.$$

where A, a are arbitrary constants.

EXAMPLE 6.18

Solve $r + s + p + q - z = e^{-2x}(x^2 + y^2)$.

Solution: The given equation is $(D_x^2 + D_xD_y + D_x + D_y - 1)z = e^{-2x}(x^2 + y^2)$

Here, $D_x^2 + D_xD_y + D_x + D_y - 1$ is irreducible.

So, CF is

$$z_c = \Sigma Ae^{ax+by} \text{ where, } a^2 + ab + a + b - 1 = 0$$

and PI is $$z_p = \frac{1}{D_x^2+D_xD_y+D_x+D_y-1}e^{-2x}(x^2+y^2)$$

$$= e^{-2x}\frac{1}{(D_x-2)^2+(D_x-2)D_y+(D_x-2)+D_y-1}(x^2+y^2)$$

$$= e^{-2x}\frac{1}{D_x^2-4D_x+4+D_xD_y-2D_y+D_x+D_y-3}(x^2+y^2)$$

$$= e^{-2x}\frac{1}{1-3D_x-D_y+D_xD_y+D_x^2}(x^2+y^2)$$

$$= e^{-2x}(1 - 3D_x + D_y - D_xD_y - D_x^2)^{-1}(x^2 + y^2)$$

$$= e^{-2x}(1 + (3D_x + D_y - D_xD_y - D_x^2) + (3D_x + D_y - D_xD_y - D_x^2)^2 + \ldots)(x^2 + y^2)$$

$$= e^{-2x}(x^2 + y^2 + (6x + 2y - 2) + (18 + 2))$$

$$= e^{-2x}(x^2 + y^2 + 6x + 2y + 18)$$

Hence, the complete solution is

$$z = z_c + z_p$$

$$z = \sum A e^{ax + by} + e^{-2x}(x^2 + y^2 + 6x + 2y + 18)$$

where $a^2 + ab + a + b - 1 = 0$.

EXAMPLE 6.19

Solve $(D_x + D_y - 1)(D_x + D_y - 3)(D_x + D_y)z = e^{x+y}\sin(x + 2y)$.

Solution: The complementary function is

$$z_c = e^x\phi_1(y - x) + e^{3x}\phi_2(y - x) + \phi_3(y - x)$$

where ϕ_1, ϕ_2, ϕ_3 are arbitrary functions

and PI is $z_p = \dfrac{1}{(D_x + D_y - 1)(D_x + D_y - 3)(D_x + D_y)} e^{x+y}\sin(x + 2y)$

$$= e^{x+y}\frac{1}{(\overline{D_x+1}+\overline{D_y+1}-1)(\overline{D_x+1}+\overline{D_y+1}-3)(\overline{D_x+1}+\overline{D_y+1})}\sin(x+2y)$$

$$= e^{x+y}\frac{1}{(D_x+D_y+1)(D_x+D_y-1)(D_x+D_y+2)}\sin(x+2y)$$

$$= e^{x+y}\frac{D_x+D_y-2}{((D_x+D_y)^2-1)((D_x+D_y)^2-4)}\sin(x+2y)$$

$$= e^{x+y}\frac{D_x\sin(x+2y)+D_y\sin(x+2y)-2\sin(x+2y)}{(D_x^2+2D_xD_y+D_y^2-1)(D_x^2+2D_xD_y+D_y^2-4)}$$

$$= e^{x+y}\frac{\cos(x+2y)+2\cos(x+2y)-2\sin(+2y)}{(-1-4-4-1)(-1-4-4-4)}$$

$$= e^{x+y}\frac{3\cos(x+2y)-2\sin(x+2y)}{130}$$

Hence, the complete solution is

$$z = z_c + z_p$$

$$z = e^x\phi_1(y - x) + e^{3x}\phi_2(y - x) + \phi_3(y - x) + \frac{e^{x+y}}{130}(3\cos(x + 2y) - 2\sin(x + 2y))$$

where ϕ_1, ϕ_2, ϕ_3 are arbitrary functions.

EXAMPLE 6.20

Solve the equation

$$(D_x^2 - 5D_y + 20)y = e^{5x + 9y}.$$

Solution: Here $D_x^2 - 5D_y + 20$ is irreducible.

So, CF is

$$z = \Sigma A e^{ax + by} \text{ where } a^2 - 5b + 20 = 0 \Rightarrow b = \frac{a^2 + 20}{5}$$

$$\therefore \quad z = \Sigma A\, e^{ax + \frac{a^2+20}{5}y}$$

where, A, a are arbitrary constants

and PI is
$$z_p = \frac{1}{D_x^2 - 5D_y + 20} e^{5x+9y} = \frac{1}{5^2 - 5(9) + 20} e^{5x + 9y}$$

$$= e^{5x + 9y} \frac{1}{(D_x + 5)^2 - 5(D_y + 9) + 20} (1)$$

$$= e^{5x + 9y} \frac{1}{D_x^2 + 10D_x - 5D_y} (1)$$

$$= e^{5x + 9y} \frac{1}{-5D_y} \left(1 - \frac{D_x^2 + 10D_x}{5D_y}\right)^{-1} (1)$$

$$= e^{5x + 9y} \frac{1}{-5D_y} \left(1 - \frac{D_x^2 + 10D_x}{5D_y} + \ldots\right) (1)$$

$$= e^{5x + 9y} \frac{1}{-5D_y} (1) = -\frac{1}{5} y e^{5x+9y}$$

Hence, the complete solution is

$$z = z_c + z_p$$

$$z = \Sigma A\, e^{ax + \frac{a^2 20}{5}y} - \frac{1}{5} y e^{5x + 9y}$$

where A, a are arbitrary constants.

EXAMPLE 6.21

Solve $(D_x^2 - D_y)z = xe^{ax + a^2y}$.

Solution: Here $D_x^2 - D_y$ is irreducible.

So, CF is

$$z_c = \Sigma A e^{\alpha x + \beta y} \text{ where } \alpha^2 - \beta = 0$$

$$\Rightarrow \quad \beta = \alpha^2$$

$$\therefore \quad z_c = \Sigma A e^{\alpha x + \alpha^2 y}$$

where, A, α are arbitrary constants

and PI is

$$z_p = \frac{1}{D_x^2 - D_y} x e^{ax + a^2 y}$$

$$= e^{ax + a^2 y} \frac{1}{(D_x + a)^2 - (D_y + a^2)} (x)$$

$$= e^{ax + a^2 y} \frac{1}{D_x^2 + 2aD_x - D_y} (x)$$

$$= e^{ax + a^2 y} \frac{1}{2aD_x} \left(1 + \frac{D_x^2 - D_y}{2aD_x}\right)^{-1} (x)$$

$$= e^{ax + a^2 y} \frac{1}{2aD_x} \left(1 + \left(\frac{D_x}{2a} - \frac{D_y}{2aD_x}\right)\right)^{-1} (x)$$

$$= e^{ax + a^2 y} \frac{1}{2aD_x} \left(1 - \left(\frac{D_x}{2a} - \frac{D_y}{2aD_x}\right) - \ldots\right) (x)$$

$$= e^{ax + a^2 y} \frac{1}{2aD_x} \left(x - \frac{1}{2a}\right)$$

$$= e^{ax + a^2 y} \left(\frac{x^2}{4a} - \frac{x}{4a^2}\right)$$

Hence, the complete solution is

$$z = z_c + z_p$$

$$z = \Sigma A e^{\alpha x + \alpha^2 y} + \left(\frac{x^2}{4a} - \frac{x}{4a^2}\right) e^{ax + a^2 y}$$

where A, α are arbitrary constants and a is given constant.

EXAMPLE 6.22

Find the complete solution of the partial differential equation

$$(D_x - D_y - D_y^2)(D_x^2 + D_y)z = \cos(2x + y).$$

Solution: Here $D_x - D_y - D_y^2$ is irreducible. So corresponding solution to this factor is

$$z = \sum Ae^{ax + by} \text{ where } a - b - b^2 = 0$$

$$\Rightarrow \quad a = b + b^2$$

$$\therefore \quad z = \sum Ae^{(b + b^2)x + by}$$

where A, b are arbitrary constants.

Also $D_x^2 + D_y$ is irreducible. So corresponding solution to this factor is

$$z = \sum Be^{cx + dy} \text{ where } c^2 + d = 0 \Rightarrow d = -c$$

$$\therefore \quad z = \sum Be^{cx - c^2 y}$$

where B, c are arbitrary constants.

Thus, CF is

$$z_c = \sum Ae^{(b + b^2)x + by} + \sum Be^{cx - c^2 y}$$

where A, b, B, c are arbitrary constants.

and PI is

$$z_p = \frac{1}{(D_x - D_y - D_y^2)(D_x^2 + D_y)} \cos(2x + y)$$

$$= \frac{1}{(D_x - D_y - (-1^2)(-2^2 + D_y)} \cos(2x + y)$$

$$= \frac{1}{(D_x - D_y + 1)(-4 + D_y)} \cos(2x + y)$$

$$= \frac{1}{-4D_x + 4D_y - 4 + D_x D_y - D_y^2 + D_y} \cos(2x + y)$$

$$= \frac{1}{-4D_x + 5D_y - 4 - 2 \cdot 1 - (-1^2)} \cos(2x + y)$$

$$= \frac{1}{-4D_x + 5D_y - 5} \cos(2x + y)$$

$$= \frac{-4D_x + 5D_y + 5}{(-4D_x + 5D_y)^2 - 25} \cos(2x + y)$$

$$= \frac{-4D_x \cos(2x+y) + 5D_y \cos(2x+y) + 5\cos(x+y)}{16D_x^2 + 25D_y^2 - 40D_xD_y - 25}$$

$$= \frac{8\sin(2x+y) - 5\sin(2x+y) + 5\cos(2x+y)}{16(-2^2) + 25(-1^2) - 40(-2-1) - 25}$$

$$= \frac{3\sin(2x+y) + 5\cos(2x+y)}{-34}$$

Hence, the complete solution is

$$z = z_c + z_p$$

$$z = \Sigma A e^{(b+b^2)x+by} + \Sigma B e^{cx-c^2y} - \frac{1}{34}(3\sin(2x+4) + 5\cos(2x+y))$$

where A, b, B, c are arbitrary constants.

EXAMPLE 6.23

Solve $(D_x - 1)(D_x - D_y + 1)z = e^y$.

Solution: The complementary function is

$$z_c = e^x\phi_1(y) + e^{-x}\phi_2(x+y)$$

where, ϕ_1, ϕ_2 are arbitrary functions.

Now, PI is

$$z_p = \frac{1}{(D_x-1)(D_x-D_y+1)}e^y = \frac{1}{D_x - D_y + 1}\frac{1}{0-1}e^y$$

$$= -e^y \frac{1}{(D_x+0)-(D_y+1)+1}(1)$$

$$= -e^y \frac{1}{D_x - D_y}(1) = -e^y \frac{1}{D_x}\left(1 - \frac{D_y}{D_x}\right)^{-1}(1)$$

$$= -e^y \frac{1}{D_x}\left(1 + \frac{D_y}{D_x} + \ldots\right)(1)$$

$$= -\frac{e^y}{D_x}(1)$$

$$= -xe^y$$

Hence, the complete solution is

$$z = z_c + z_p$$

$$z = e^x\phi_1(y) + e^{-x}\phi_2(x+y) - xe^y.$$

EXAMPLE 6.24

Solve $(D_x - 3D_y - 2)^2 z = 2e^{2n} \tan(y + 3x)$.

Solution: The complementary function is

$$z_c = e^{2x}[\phi_1(y + 3x) + x\phi_2(y + 3x)]$$

where ϕ_1, ϕ_2 are arbitrary functions.

Now, PI is

$$z_p = \frac{1}{(D_x - 3D_y - 2)^2} 2e^{2x} \tan(y + 3x)$$

$$= 2e^{2x} \frac{1}{(D_x + 2 - 3D_y - 2)^2} \tan(y + 3x)$$

$$= 2e^{2x} \frac{1}{(D_x - 3D_y)^2} \tan(y + 3x)$$

$$= 2e^{2x} \frac{x^2}{(1)^2(2!)} \tan(y + 3x)$$

Hence, the complete solution is

$$z = z_c + z_p$$

$$z = e^{2x}[\phi_1(y + 3x) + x\phi_2(y + 3x)] + x^2 e^{2x} \tan(y + 3x).$$

EXAMPLE 6.25

Solve $(D_x^2 - D_y)(D_x - 2D_y)z = e^{2x + y} + xy$.

Solution: Since $D_x^2 - D_y$ is irreducible. So, the corresponding solution of $(D_x^2 - D_y)z = 0$ is

$$z = \Sigma A e^{ax + by}$$

where $$a^2 - b = 0 \Rightarrow b = a^2$$

$\therefore$ $$z = \Sigma A e^{ax + a^2 y}$$

where A, a are arbitrary constants.

Now, $D_x - 2D_y$ is linear, so the corresponding solution of $(D_x - 2D_y)z = 0$ is

$$z = \phi_1(y + 2x)$$

Thus, CF is

$$z = \Sigma A e^{ax + a^2 y} + \phi_1(y + 2x).$$

Now, PI is

$$z_p = \frac{1}{(D_x - 2D_y)(D_x^2 - D_y)}(e^{2x+y} + xy)$$

$$\frac{1}{(D_x - 2D_y)(D_x^2 - D_y)} e^{2x+y} = \frac{1}{D_x - 2D_y} \frac{1}{(2)^2 - 1} e^{2x+y}$$

$$= \frac{1}{3} e^{2x+y} \frac{1}{D_x + 2 - 2D_y + 1} (1)$$

$$= \frac{1}{3}e^{2x+y}\frac{1}{D_x - 2D_y}(1)$$

$$= \frac{1}{3}e^{2x+y}\frac{1}{D_x}\left(1-2\frac{D_y}{D_x}\right)^{-1}(1)$$

$$= \frac{1}{3}e^{2x+y}\frac{1}{D_x}\left(1-2\frac{D_y}{D_x}+\ldots\right)(1)$$

$$= \frac{1}{3}e^{2x+y}\frac{1}{D_x}(1) = \frac{x}{3}e^{2x+y}$$

and $$\frac{1}{(D_x - 2D_y)(D_x^2 - D_y)}(xy) = \frac{1}{-2D_y}\left(1-\frac{D_x}{2D_y}\right)^{-1} \times -\frac{1}{D_y}\left(1-\frac{D_x^2}{D_y}\right)^{-1}(xy)$$

$$= \frac{1}{2D_y^2}\left(1+\frac{D_x}{2D_y}+\ldots\right)\left(1+\frac{D_x^2}{D_y}+\ldots\right)(xy)$$

$$= \frac{1}{2D_y^2}\left(1+\frac{D_x}{2D_y}+\frac{D_x^2}{D_y}+\ldots\right)(xy)$$

$$= \frac{1}{2D_y^2}\left(xy+\frac{y}{2D_y}\right)$$

$$= \frac{1}{2}\left(\frac{xy^2}{6}+\frac{y^4}{48}\right)$$

Hence, the complete solution is

$$z = z_c + z_p$$

$$z = \sum Ae^{ax + a^2y} + \phi_1(y + 2x) + \frac{x}{3}e^{2x+y} + \frac{xy^3}{12} + \frac{y^4}{96}$$

where ϕ is an arbitrary function and A, a are arbitrary constants.

EXAMPLE 6.26

Solve $(D_x^2 - D_xD_y + D_y - 1)z = \cos(x + 2y) + e^y + xy + 1$.

Solution: The given equation can be written as $(D_x - 1)(D_x - D_y + 1)z = \cos(x + 2y) + e^y + xy + 1$

$\therefore$ CF is

$$z_c = e^{-x}\phi_1(y) + e^{-x}\phi_2(y + x)$$

where ϕ_1, ϕ_2 are arbitrary functions

Now, PI is $$z_p = \frac{1}{D_x^2 - D_x D_y + D_y - 1}(\cos(x + 2y) + e^y + xy + 1)$$

$$\frac{1}{D_x^2 - D_x D_y + D_y - 1}\cos(x+2y) = \frac{1}{-1^2 - (-1 \cdot 2) + D_y - 1}\cos(x+2y)$$

$$= \frac{1}{D_y}\cos(x+2y) = \frac{\sin(x+2y)}{2}$$

$$\frac{1}{D_x^2 - D_x D_y + D_y - 1}e^y = e^y \frac{1}{D_x^2 - D_x(D_y+1)(D_y+1) - 1}(1)$$

$$= e^y \frac{1}{D_x^2 - D_x D_y - D_x + D_y}(1)$$

$$= \frac{e^y}{D_y}\left(1 + \frac{D_x^2 - D_x D_y - D_x}{D_y}\right)^{-1}(1)$$

$$= \frac{e^y}{D_y}\left(1 - \frac{d_x^2 - D_x D_y - D_x}{D_y} + \ldots\right)(1)$$

$$= \frac{e^y}{D_y}(1) = ye^y$$

and $$\frac{1}{(D_x - 1)(D_x - D_y + 1)}(xy+1) = -(1 - D_x)^{-1}(1 + \overline{D_x - D_y})^{-1}(xy+1)$$

$$= -(1 + D_x + D_x^2 + \ldots)(1 - (D_x - D_y) + (D_x - D_y)^2 - \ldots)(xy + 1)$$

$$= -(1 + D_x + \ldots)(1 - D_x + D_y - 2D_x Dy)(xy + 1)$$

$$= -(1 + D_x)(xy + 1 - y + x - 2)$$

$$= -(1 + D_x)(xy + x - y - 1)$$

$$= -(xy + x - y - 1 + y + 1)$$

$$= -(xy + x).$$

Hence, the complete solution is

$$z = z_c + z_p$$

$$z = e^x\phi_1(y) + e^{-x}\phi_2(y + x) + \frac{1}{2}\sin(x + 2y) + ye^y - (xy + x)$$

where ϕ_1, ϕ_2 are arbitrary functions. ▲▲

EXERCISE 6.2

Solve the following partial differential equations:

1. $(D_x - 1)(D_x - D_y + 1)z = \sin(x + 2y) + e^y$.
2. $(D_x - D_y - 1)(D_x - D_y - 2)z = e^{2x - y} + x$.
3. $\frac{\partial^2 z}{\partial x^2} - \frac{\partial^2 z}{\partial y^2} + \frac{\partial z}{\partial x} + 3\frac{\partial z}{\partial y} - 2z = e^{x - y} - x^2 y$.
4. $(D_x^2 - D_y^2 + D_x + D_y)z = e^{3x + 3y}$.
5. $(D_x D_y + \alpha D_x + \beta D_y + \alpha\beta)z = e^{ax + by}$.
6. $(D_x^2 - D_y^2 + D_x + 3D_y - 2)z = e^{3x - y}$.
7. $(D_x^2 - D_x D_y - 2D_y^2 + 2D_y + 2D_x)z = e^{3x + 4y}$.
8. $\frac{\partial^2 z}{\partial x^2} - \frac{\partial^2 z}{\partial y \partial y} + \frac{\partial z}{\partial x} = 2$.
9. $s + p - q = z + xy$.
10. $r - t - 3p + 3q = 54xy$.
11. $(D_x^2 - 2D_y - 1)z = x^2 y$.
12. $(D_x^2 - D_y^2 - 3D_x + 3D_y)z = xy$.
13. $p - t = \cos(x - 3y)$.
14. $(D_x + D_y)(D_x + D_y - 2)z = \cos(x + 2y)$.
15. $(2D_x D_y + D_y^2 - 3D_y)z = 50\sin(3x - 2y)$.
16. $r - s + q - z = 8\sin(x + 2y)$.
17. $r - s - 2p = \sin(3x + 4y)$.
18. $(D_x^2 - D_x D_y + D_y - 1)z = e^x$.
19. $r - s + q = z + e^y$.
20. $(D_x^2 - 4D_x D_y + 4D_y^2 - 2D_y)z = e^{2x + y}$.
21. $r - s + q - z = \cos(x + 2y) + e^y$.
22. $r - s + 2q - z = x^2 y^2 + e^{x + y}$.

23. $(D_x^2 - D_y)(D_x - 2D_y)z = 3e^{2x+y} + xy$.

24. $r - t - 3p + 3q = xy + e^{x+2y}$.

25. $(D_x^2 - 3D_xD_y + D_x + 1)z = e^{\frac{1}{2}x-y}$.

26. $D_x(D_x - 2D_y)(D_x + D_y)z = e^{x+2y}(x^2 + 4y^2)$.

ANSWERS

Exercise 6.2

1. $z = e^x\phi_1(y) + e^{-x}\phi_2(y + x) - \dfrac{1}{2}\cos(x + 2y) - xe^y$.

2. $z = e^x\phi_1(y + x) + e^{2x}\phi_2(y + x) + \dfrac{1}{2}e^{2x-y} \neq \dfrac{1}{2}x + \dfrac{1}{4}$.

3. $z = e^{-2x}\phi_1(y + x) + e^x\phi_2(y - x) - \dfrac{1}{4}e^{x-y} + \dfrac{1}{2}\left(x^2y + xy + \dfrac{3}{2}x^2 + \dfrac{3}{2}y + 3x + \dfrac{21}{4}\right)$.

4. $z = \phi_1(y - x) + e^{-x}\phi_2(y + x) + \dfrac{1}{6}e^{3x+3y}$.

5. $z = e^{-\beta x}\phi_1(y) + e^{-\alpha y}\phi_2(x) + \dfrac{1}{(\alpha+\beta)(b+a)}e^{ax+by}$.

6. $z = e^x\phi_1(y - x) + e^{-2x}\phi_2(y + x) + \dfrac{1}{6}e^{3x-y}$.

7. $z = \phi_1(y - x) + e^{-2x}\phi_2(y + 2x) - \dfrac{1}{21}e^{3x+4y}$.

8. $z = \phi_1(y) + e^{-x}\phi_2(y + x) + 2x$.

9. $z = e^x\phi_1(y) + e^{-y}\phi_2(x) - xy - y + x + 1$.

10. $z = e^{3x}\phi_1(y - x) + \phi_2(y + x) - (9x^2y + 3x^3 + 6xy + 6x^2 + 4x)$.

11. $z = \Sigma A\, e^{ax+\frac{a^2-1}{2}y} - xy^2 + 4xy - 8x$.

12. $z = \phi_1(y + x) + e^{3x}\phi_2(y - x) - \dfrac{1}{3}\left(\dfrac{x^2y}{2} + \dfrac{xy}{3} - \dfrac{2x}{9} + \dfrac{x^3}{6} + \dfrac{x^2}{3}\right)$.

13. $z = \Sigma Ae^{b^2x+by} + \dfrac{1}{82}(\sin(x - 3y) + 9\cos(x - 3y))$.

14. $z = \phi_1(y - x) + e^{2x}\phi_2(y - x) - \dfrac{1}{39}(2\sin(x + 2y) + 3\cos(x + 2y))$.

15. $z = \phi_1(x) + e^{3x/2}\phi_2(2y - x) + 4\sin(3x - 2y) - 3\cos(3x - 2y)$.

16. $z = e^{x}\phi_1(y) + e^{-x}\phi_2(y + x) - 4\cos(x + 2y)$.

17. $z = \phi_1(y) + e^{2x}\phi_2(y + x) + \dfrac{1}{15}[\sin(3x + 4y) + 2\cos(3x + 4y)]$.

18. $z = e^{x}\phi_1(y) + e^{-x}\phi_2(y + x) + \dfrac{1}{2}xe^{x}$.

19. $z = e^{x}\phi_1(y) + e^{-x}\phi_2(y + x) - xe^{y}$.

20. $z = \phi_1(y + 2x) + e^{-x}\phi_2(y + 2x) + xe^{2x + y}$.

21. $z = e^{x}\phi_1(y) + e^{-x}\phi_2(y + x) + \dfrac{1}{2}\sin(x + 2y) - xe^{y}$.

22. $z = \Sigma A\, e^{ax+\frac{a^2-1}{a-2}y} + e^{x + y} - x^2y^2 - 4x^2y + 4xy - 2y^2 - 8x^2 + 16x - 16y - 4$.

23. $z = \Sigma Ae^{ax + a^2y} + \phi_1(y + 2x) + \dfrac{1}{3}xe^{2x + y} + \dfrac{x^4y}{24} + \dfrac{x^5}{60} + \dfrac{x^6}{720}$.

24. $z = \phi_1(y + x) + e^{3x}\phi_2(y + x)\dfrac{x^2y}{6} + \dfrac{x^2}{9} - \dfrac{xy}{9} - \dfrac{x^3}{18} - \dfrac{2}{27}x - xe^{x + 2y}$.

25. $z = \Sigma A\, e^{ax+\frac{a^3+a+1}{3a}y}\, y + \dfrac{8}{25}e^{\frac{1}{2}x-y}$.

26. $z = \phi_1(y) + \phi_2(y + 2x) + \phi_3(y - x) - \dfrac{e^{x+2y}}{81}(9x^2 + 36y^2 - 18x - 72y + 84)$.

▲▲

7

CHAPTER

Partial Differential Equations Reducible to Equations with Constant Coefficients

Some partial differential equations having variable coefficients can be reduced to equations with constant coefficients by using suitable substitutions. Generally, we take

$$x = e^u \text{ and } y = e^v$$

so that $u = \log x$ and $v = \log y$

$$\therefore \qquad \frac{\partial z}{\partial x} = \frac{\partial z}{\partial u}\frac{\partial u}{\partial z} = \frac{1}{x}\frac{\partial z}{\partial u}$$

$$\Rightarrow \qquad x\frac{\partial z}{\partial x} = \frac{\partial z}{\partial u} = D_u z$$

and

$$\frac{\partial z}{\partial y} = \frac{\partial z}{\partial v}\frac{\partial v}{\partial y} = \frac{1}{y}\frac{\partial z}{\partial v}$$

$$\Rightarrow \qquad y\frac{\partial z}{\partial y} = \frac{\partial z}{\partial v} = D_v z$$

Now, $x\dfrac{\partial}{\partial x}\left(x^{n-1}\dfrac{\partial^{n-1} z}{\partial x^{n-1}}\right) = x^n \dfrac{\partial^n z}{\partial x^n} + (n-1)x^{n-1}\dfrac{\partial^{n-1} z}{\partial x^{n-1}}$

$$\Rightarrow \qquad x^n \frac{\partial^n z}{\partial x^n} = \left(x\frac{\partial}{\partial x} - n + 1\right)x^{n-1}\frac{\partial^{n-1} z}{\partial x^{n-1}}$$

$$\therefore \qquad x^2 \frac{\partial^2 z}{\partial x^2} = \left(x\frac{\partial}{\partial x} - 1\right)x\frac{\partial z}{\partial x}$$

$$= D_u(D_u - 1)z$$

$$x^3 \frac{\partial^3 z}{\partial x^3} = \left(x\frac{\partial}{\partial x} - 2\right)x^2 \frac{\partial^2 z}{\partial x^2}$$

$$= D_u(D_u - 1)\,(D_u - 2)z \text{ and so on.}$$

Similarly,
$$y\frac{\partial}{\partial y} = D_v$$

$$y^2 \frac{\partial^2}{\partial y^2} = D_v(D_v - 1)$$

and
$$xy\frac{\partial^2 z}{\partial x \partial y} = \left(x\frac{\partial}{\partial x}\right)\left(y\frac{\partial}{\partial y}\right)z = D_u D_v z$$

In general,

$$x^m y^n \frac{\partial^{m+n}}{\partial x^m \partial y^n} = \left(x^m \frac{\partial^m}{\partial x^m}\right)\left(y^n \frac{\partial^n}{\partial y^n}\right)$$

$$= D_u(D_u - 1) \ldots (D_u - m + 1)D_v\,(D_v - 1) \ldots (D_v - n + 1)$$

After making these substitutions, the equation is reduced to an equation having constant coefficients and can be easily solved by method discussed in earlier chapters.

▲▲

EXAMPLE 7.1

Solve the equation

$$x^2 \frac{\partial^2 z}{\partial x^2} + 2xy\frac{\partial^2 y}{\partial x \partial y} + y^2 \frac{\partial^2 z}{\partial y^2} = 0$$

Solution: Let, $x = e^u$ and $y = e^v$, so that

$u = \log x$ and $v = \log y$

If $D_u = \dfrac{\partial}{\partial u}$ and $D_v = \dfrac{\partial}{\partial v}$ then

$$x^2 \frac{\partial^2}{\partial x^2} = D_u(D_u - 1),\; y^2 \frac{\partial^2}{\partial y^2} = D_v(D_v - 1)$$

and
$$xy\frac{\partial^2}{\partial x \partial y} = D_u D_v$$

Making these substitutions, the given equation reduces to

$$[D_u(D_u - 1) + 2D_uD_v + D_v(D_v(D_v - 1)]z = 0$$

or

$$(D_u + D_v)\,(D_u + D_v - 1)z = 0$$

Hence, the complete solution is

$$\begin{aligned} z &= \phi_1(v - u) + e^u\phi_2(v - u) \\ &= \phi_2(\log y - \log x) + x\phi_2(\log y - \log x) \\ &= \phi_1\left(\log\frac{y}{x}\right) + x\phi_2\left(\log\frac{y}{x}\right) \\ &= \psi_1\left(\frac{y}{x}\right) + x\psi_2\left(\frac{y}{x}\right). \end{aligned}$$

EXAMPLE 7.2

Solve $x^2\dfrac{\partial^2 z}{\partial x^2} - y^2\dfrac{\partial^2 z}{\partial y^2} - y\dfrac{\partial z}{\partial y} + x\dfrac{\partial z}{\partial x} = 0.$

Solution: Let, $x = e^u$ and $v = e^v$ so that $u = \log x$ and $v = \log y$ and denoting $\dfrac{\partial}{\partial u}$ and $\dfrac{\partial}{\partial v}$ by D_u and D_v respectively, the given equation reduces to

$$[D_u(D_u - 1) - D_v(D_v - 1) - D_v + D_u]z = 0$$

or

$$(D_u^2 - D_v^2)z = 0$$

∴ The complete solution is

$$\begin{aligned} z &= \phi_1(v - u) + \phi_2(v + u) \\ &= \phi_1(\log y - \log x) + \phi_2(\log y + \log x) \\ &= \phi_1\left(\log\frac{y}{x}\right) + \phi_2(\log xy) \\ &= \psi_1\left(\frac{y}{x}\right) + \psi_2(xy). \end{aligned}$$

EXAMPLE 7.3

Solve $x^2\dfrac{\partial^2 z}{\partial x^2} - 3xy\dfrac{\partial^2 z}{\partial x\partial y} + 2y^2\dfrac{\partial^2 z}{\partial y^2} + 5y\dfrac{\partial z}{\partial y} - 2z = 0.$

Solution: Let $x = e^u$ and $y = e^v$ so that $u = \log x$ and $v = \log y$ and writing $\frac{\partial}{\partial u} = D_u$ and $\frac{\partial}{\partial v} = D_v$, the given equation reduces to

$$(D_u(D_u - 1) - 3D_uD_v + 2D_v(D_v - 1) + 5D_v - 2)z = 0$$

or
$$(D_u^2 - 3D_uD_v + 2D_v^2 - D_u + 3D_v - 2)z = 0$$

or
$$(D_u - D_v - 2)(D_u - 2D_v + 1)z = 0$$

∴ The complete solution is

$$\begin{aligned} z &= e^{2u}\phi_1(v + u) + e^{-u}\phi_2(2u + v) \\ &= x^2\phi_1(\log y + \log x) + x^{-1}\phi_2(2\log x + \log y) \\ &= x^2\phi_1(\log xy) + x^{-1}\phi_2(\log x^2y) \\ &= x^2\psi_1(xy) + x^{-1}\psi_2(x^2y). \end{aligned}$$

EXAMPLE 7.4

Solve $x^2\frac{\partial^2 z}{\partial x^2} + 2xy\frac{\partial^2 z}{\partial x\partial y} + y^2\frac{\partial^2 z}{\partial y^2} + x\frac{\partial z}{\partial x} + y\frac{\partial z}{\partial y} - z = 0.$

Solution: Let $x = e^u$ and $y = e^v$ so that $u = \log x$ and $v = \log y$ and denoting $\frac{\partial}{\partial u}, \frac{\partial}{\partial v}$ by D_u, D_v; the given equation reduces to

$$(D_u(D_u - 1) + 2D_uD_v + D_v(D_v - 1) + D_u + D_v - 1)z = 0$$

or
$$(D_u^2 + 2D_uD_v + D_v^2 - 1)z = 0$$

or
$$(D_u + D_v - 1)(D_u + D_v + 1)z = 0$$

∴ The complete solution is

$$\begin{aligned} z &= e^u\phi_1(u - v) + e^{-u}\phi_2(u - v) \\ &= x\phi_1(\log x - \log y) + x^{-1}\phi_2(\log x - \log y) \\ &= x\phi_1\left(\log\frac{x}{y}\right) + x^{-1}\phi_2\left(\log\frac{x}{y}\right) \\ &= x\psi_1\left(\frac{x}{y}\right) + x^{-1}\psi_2\left(\frac{x}{y}\right) \end{aligned}$$

where ψ_1 and ψ_2 are arbitrary functions.

EXAMPLE 7.5

Solve $\dfrac{1}{x^2}\dfrac{\partial^2 z}{\partial x^2}-\dfrac{1}{x^3}\dfrac{\partial z}{\partial x}=\dfrac{1}{y^2}\dfrac{\partial^2 z}{\partial y^2}-\dfrac{1}{y^3}\dfrac{\partial z}{\partial y}$.

Solution: Let, $\frac{1}{2}x^2 = u$ and $\frac{1}{2}y^2 = v$ so that

$$xdx = du \text{ and } ydy = dv$$

Hence,
$$\frac{\partial z}{\partial u}=\frac{\partial z}{\partial x}\frac{\partial x}{\partial u}=\frac{1}{x}\frac{\partial z}{\partial x}$$

and
$$\frac{\partial^2 z}{\partial u^2}=\frac{\partial}{\partial u}\left(\frac{1}{x}\frac{\partial z}{\partial x}\right)=\frac{\partial}{\partial x}\left(\frac{1}{x}\frac{\partial z}{\partial x}\right)\frac{\partial x}{\partial u}$$

$$=-\frac{1}{x^3}\frac{\partial z}{\partial x}+\frac{1}{x^2}\frac{\partial^2 z}{\partial x^2}$$

Similarly,
$$\frac{\partial z}{\partial v}=\frac{1}{y}\frac{\partial z}{\partial y}$$

and
$$\frac{\partial^2 z}{\partial v^2}=-\frac{1}{y^3}\frac{\partial z}{\partial y}+\frac{1}{y^2}\frac{\partial^2 z}{\partial y^2}$$

Thus, the given equation reduces to

$$\frac{\partial^2 z}{\partial u^2}=\frac{\partial^2 z}{\partial v^2} \text{ or } \frac{\partial^2 z}{\partial u^2}-\frac{\partial^2 z}{\partial v^2}=0$$

or
$$(D_u^2 - D_v^2)z = 0$$

∴ The complete solution is

$$z = \phi_1(v + u) + \phi_2(v - u)$$

$$= \phi_1\left(\frac{x^2+y^2}{2}\right)+\phi_2\left(\frac{y^2-x^2}{2}\right)$$

$$= \psi_1(x^2 + y^2) + \psi_2(y^2 - x^2).$$

EXAMPLE 7.6

Solve $x^2\dfrac{\partial^2 z}{\partial x^2}-y^2\dfrac{\partial^2 z}{\partial y^2}=xy$.

Solution: Let, $x = e^u$ and $y = e^v$ so that $u = \log x$ and $v = \log y$ and denoting $\frac{\partial}{\partial u}, \frac{\partial}{\partial v}$ by D_u, D_v; the given equation reduces to

$$[D_u(D_u - 1) - D_v(D_v - 1)]z = e^u e^v$$

or $$(D_u^2 - D_v^2 - D_u + D_v)z = e^{u+v}$$

or $$(D_u - D_v)(D_u + D_v - 1)z = e^{u+v}$$

$\therefore$ CF is

$$z_c = \phi_1(u + v) + e^u\phi_2(v - u)$$
$$= \phi_1(\log x + \log y) + x\phi_2(\log y - \log x)$$
$$= \phi_1(\log xy) + x\phi_2\left(\log \frac{y}{x}\right)$$
$$= \phi_1(xy) + x\phi_2\left(\frac{y}{x}\right)$$

Now, PI is

$$z_p = \frac{1}{(D_u - D_v)(D_u + D_v - 1)} e^{u+v}$$
$$= \frac{1}{D_u - D_v} \frac{1}{1+1-1} e^{u+v}$$
$$= \frac{1}{D_u - D_v} e^{u+v} = e^{u+v} \frac{1}{\overline{D_u + 1} - \overline{D_v + 1}} (1)$$
$$= e^{u+v} \frac{1}{D_u - D_v} (1)$$
$$= e^{u+v} \frac{1}{D_u}\left(1 - \frac{D_v}{D_u}\right)^{-1} (1)$$
$$= e^{u+v} \frac{1}{D_u}\left(1 + \frac{D_v}{D_u} + \ldots\right) (1)$$
$$= e^{u+v} \frac{1}{D_u} (1)$$
$$= ue^{u+v}$$
$$= (\log x)(xy) = xy \log x.$$

Hence, the complete solution is

$$z = z_c + z_p$$

$$z = \psi_1(xy) + x\psi_2\left(\frac{y}{x}\right) + xy \log x.$$

EXAMPLE 7.7

Solve $x^2 \frac{\partial^2 z}{\partial x^2} + 2xy \frac{\partial^2 y}{\partial x \partial y} - x \frac{\partial z}{\partial x} = \frac{x^3}{y^2}$.

Solution: Let, $x = e^u$ and $y = e^v$ so that $u = \log x$ and $v = \log y$ and denoting $\frac{\partial}{\partial u}, \frac{\partial}{\partial v}$ by D_u, D_v; the given equation reduces to

$$(D_u (D_u - 1) + 2D_u D_v - D_u)z = e^{3u - 2v}$$

or $$(D_u^2 + 2D_u D_v - 2D_u)z = e^{3u - 2v}$$

or $$D_u(D_u + 2D_v - 2)z = e^{3u - 2v}$$

$\therefore$ CF is

$$z_c = \phi_1(v) + e^{2u}\phi_2(v - 2u)$$

$$= \phi_1(\log y) + x^2\phi_2(\log y - 2 \log x)$$

$$= \psi_1(y) \mid x^2\psi_2\left(\frac{y}{x^2}\right)$$

Now, PI is

$$z_p = \frac{1}{3_u(D_u + 2D_v - 2)} e^{3u - 2v}$$

$$= \frac{1}{3(3 + 2(-2) - 2)} e^{3u-2v} = -\frac{1}{9} e^{3u-2v}$$

$$= -\frac{1}{9} \frac{x^3}{y^2}$$

Hence, the complete solution is

$$z = z_c + z_p$$

$$z = \psi_1(y) + x^2\psi_2\left(\frac{y}{x^2}\right) - \frac{1}{9} \frac{x^3}{y^2}.$$

EXAMPLE 7.8

Solve $x^2 \frac{\partial^2 z}{\partial x^2} + 2xy \frac{\partial^2 y}{\partial x \partial y} + y^2 \frac{\partial^2 z}{\partial y^2} = (x^2 + y^2)^{n/2}$.

Solution: Let, $x = e^u$ and $y = e^v$ so that $u = \log x$ and $v = \log y$ and denoting $\frac{\partial}{\partial u}, \frac{\partial}{\partial v}$ by D_u, D_v; the given equation reduces to

$$[D_u(D_u - 1) + 2D_uD_v + D_v(D_v - 1)]z = (e^{2u} + e^{2v})^{n/2}$$

or

$$(D_u + D_v)(D_u + D_v - 1)z = (e^{2u} + e^{2v})^{n/2}$$

$\therefore$ CF is

$$z_c = \phi_1(v - u) + \phi_2(v - u)e^u$$

$$= \psi_1\left(\frac{y}{x}\right) + x\psi_2\left(\frac{y}{x}\right)$$

Now, PI is $z_p = \dfrac{1}{(D_u + D_v)(D_u + D_v - 1)}(e^{2u} + e^{2v})^{n/2}$

$$= \frac{1}{(D_u + D_v)(D_u + D_v - 1)} e^{nu}(1 + e^{2(v-u)})^{n/2}$$

$$= \frac{1}{(D_u + D_v)(D_u + D_v - 1)} e^{nu}\left(1 + \frac{n}{2}e^{2(v-u)} + \frac{\left(\frac{n}{2}\right)\left(\frac{n}{2} - 1\right)}{2!}e^{4(v-u)} + \ldots\right)$$

$$= \frac{1}{(D_u + D_v)(D_u + D_v - 1)} e^{nu} + \frac{n}{2}\frac{1}{(D_u + D_v)(D_u + D_v - 1)} e^{(n-2)u+2v}$$

$$+ \frac{\left(\frac{n}{2}\right)\left(\frac{n}{2} - 1\right)}{2!}\frac{1}{(D_u + D_v)(D_u + D_v - 1)} e^{(n-4)u+4v} + \ldots$$

$$= \frac{1}{(n+0)(n+0-1)} e^{nu} + \frac{n}{2}\frac{1}{((n-2)+2)(n-2+2-1)} e^{(n-2)u+2v} + \frac{\left(\frac{n}{2}\right)\left(\frac{n}{2} - 1\right)}{2!}$$

$$\frac{1}{((n-4)+4)((n-4)+4-1)} e^{(n-4)u+4v} + \ldots$$

$$= \frac{e^{nu}}{n^2 - n}\left\{1 + \frac{n}{2}e^{2(v-u)} + \frac{\left(\frac{n}{2}\right)\left(\frac{n}{2} - 1\right)}{2!}e^{4(v-u)} + \ldots\right\}$$

$$= \frac{e^{nu}}{n^2 - n}(1 + e^{2(v-u)})^{n/2}$$

$$= \frac{1}{n^2 - n}(e^{2u} + e^{2v})^{n/2}$$

$$= \frac{1}{n^2 - n}(x^2 + y^2)^{n/2}.$$

Here, the required complete solution is

$$z = z_c + z_p$$

$$z = \psi_1\left(\frac{y}{x}\right) + x\psi_2\left(\frac{y}{x}\right) + \frac{1}{n^2 - n}(x^2 + y^2)^{n/2}.$$

EXAMPLE 7.9

Solve $x^2r - 3xys + 2y^2t + px + 2qy = x + 2y$.

Solution: The given equation can be written as

$$x^2\frac{\partial^2 z}{\partial x^2} - 3xy\frac{\partial^2 z}{\partial x\partial y} + 2y^2\frac{\partial^2 z}{\partial y^2} + x\frac{\partial z}{\partial x} + 2y\frac{\partial z}{\partial y} = x + 2y$$

Let $x = e^u$ and $y = e^v$ so that $u = \log x$ and $v = \log y$ and denoting $\frac{\partial}{\partial u}, \frac{\partial}{\partial v}$ by D_u, D_v, the given equation reduces to

$$(D_u(D_u - 1) - 3D_uD_v + 2D_v(D_v - 1) + D_u + 2D_v)z = e^u + 2e^v$$

or

$$(D_u - D_v)(D_u - 2D_v)z = e^u + 2e^v$$

∴ CF is

$$z_c = \phi_1(v + u) + \phi_2(2\cdot u + v)$$
$$= \phi_1(\log y + \log x) + \phi_2(2\log x + \log y)$$
$$= \phi_1(\log xy) + \phi_2(\log x^2y)$$
$$= \psi_1(xy) + \psi_2(x^2y)$$

Now PI is $$z_p = \frac{1}{(D_u - D_v)(D_u - 2D_v)}e^u + \frac{1}{(D_u - D_v)\cdot(D_u - 2D_v)}2e^v$$

$$= \frac{e^u}{(1-0)(1-0)} + \frac{2e^v}{(0-1)(0-2(1))}$$

$$= e^u + e^v = x + y$$

Hence, the complete solution is

$$z = \psi_1(xy) + \psi_2(x^2y) + x + y.$$

EXAMPLE 7.10

Solve $(x^2D_x^2 - xyD_xD_y - 2y^2D_y^2 + xD_x - 2yD_y)z = \log\left(\frac{y}{x}\right) - \frac{1}{2}$.

Solution: Let $x = e^u$ and $y = e^v$ so that $u = \log x$ and $v = \log y$ and denoting $\frac{\partial}{\partial u}, \frac{\partial}{\partial v}$ by D_u, D_v, respectively, the given equation reduces to

$$(D_u/D_u - 1) - D_uD_v - 2D_v(D_v - 1) + D_u - 2D_v)z = v - u - \frac{1}{2}$$

or
$$(D_u^2 - D_uD_v - 2D_v^2)z = v - u - \frac{1}{2}$$

or
$$(D_u - 2D_v)(D_u + D_v)z = v - u - \frac{1}{2}$$

$\therefore$ CF is

$$z = \phi_1(2u + v) + \phi_2(v - u)$$
$$= \psi_1(x^2y) + \psi_2(y/x)$$

Now PI is
$$z_p = \frac{1}{D_u^2 - D_uD_v - 2D_v^2}\left(v - u - \frac{1}{2}\right)$$

$$= \frac{1}{D_u^2}\left(1 - \left(\frac{D_v}{D_u} + \frac{2D_v^2}{D_u^2}\right)\right)^{-1}\left(v - u - \frac{1}{2}\right)$$

$$= \frac{1}{D_u^2}\left(1 + \left(\frac{D_v}{D_u} + \frac{2D_v^2}{D_u^2} + \ldots\right)\right)\left(v - u - \frac{1}{2}\right)$$

$$= \frac{1}{D_u^2}\left(v - u - \frac{1}{2} + \frac{1}{D_u}(1)\right)$$

$$= \frac{1}{D_u^2}\left(v - u - \frac{1}{2} + u\right) = \frac{1}{D_u^2}\left(v - \frac{1}{2}\right)$$

$$= \frac{u^2v}{2} - \frac{1}{4}u^2 = \frac{u^2}{4}(2v - 1) = \frac{(\log x)^2}{4}(2\log y - 1)$$

Hence, the required complete solution is

$$z = \psi_1(x^2y) + \psi_2(y/x) + \frac{(\log x)^2}{4}(2\log y - 1).$$

EXAMPLE 7.11

Solve $(x^2D_x^2 - 2xyD_xD_y - 3y^2D_y^2 + xD_x - 3yD_y)z = x2y \cos(\log x^2)$.

Solution: Let $x = e^u$ and $y = e^v$ so that $u = \log x$ and $v = \log y$ and denoting $\frac{\partial}{\partial u}, \frac{v}{\partial v}$ by D_u, D_v respectively; the given equation reduces to

$$[D_u (D_u - 1) - 2D_uD_v - 3D_v (D_v - 1) + D_u - 3D_v]z = e^{2u+v} \cos(2u)$$

or $$(D_u^2 - 2D_uD_v - 3D_v^2)z = e^{2u+v} \cos(2u)$$

or $$(D_u - 3D_v)(D_u + D_v)z = e^{2u+v} \cos(2u)$$

∴ CF is

$$z_c = \phi_1(3u + v) + \phi_2(v - u)$$
$$= \psi_1(x^3y) + \psi_2(y/x)$$

Now, PI is $$z_p = \frac{1}{(D_u - 3D_v)D_u + D_v)} e^{2u+v} \cos(2u)$$

$$= e^{2u+v} \frac{1}{((D_u + 2) - 3(D_v + 1))(\overline{D_u + 2 + D_v + 1})} \cos(2u)$$

$$= e^{2u+v} \frac{1}{((D_u - 3D_v - 1)(D_u + D_u + D_v + 3)} \cos(2u)$$

$$= e^{2u+v} \frac{1}{D_u^2 - 2D_uD_v - 3D_v^2 + 2D_u - 10D_v - 3} \cos(2u)$$

$$= e^{2u+v} \frac{1}{-2^2 - 2(-2 \cdot 0) - 3(-0^2) + 2D_u - 10D_v - 3} \cos(2u)$$

$$= e^{2u+v} \frac{1}{2D_u - 10D_v - 7} \cos(2u)$$

$$= e^{2u+v} \frac{2D_u - 10D_v + 7}{(2D_u - 10D_v)^2 - 49} \cos(2u)$$

$$= e^{2u+v} \frac{2D_u\cos(2u) - 10D_v\cos(2u) + 7\cos(2u)}{4D_u^2 - 100D_v^2 - 40D_uD_v - 49}$$

$$= e^{2u+v} \frac{-4\sin(2u)+7\cos(2u)}{4(-2^2)+100(-0^2)-40(-2\cdot 0)-49}$$

$$= e^{2u+v} \frac{-4\sin(2u)+7\cos(2u)}{-65}$$

$$= \frac{1}{65} x^2y[4 \sin (2 \log x) - 7 \cos (2 \log x)].$$

Hence, the required complete solution is

$$z = z_c + z_p$$

$$z = \psi_1(x^3y) + \psi_2(y/x) + \frac{1}{65} x^2y \, [4 \sin (2 \log x) - 7 \cos (2 \log x)].$$

▲▲

EXERCISE 7.1

Solve the following partial differential equations:

1. $x^2 \frac{\partial^2 z}{\partial x^2} - 4xy\frac{\partial^2 z}{\partial x \partial y} + 4y^2 \frac{\partial^2 z}{\partial y^2} + 6y\frac{\partial z}{\partial y} = x^2y^4.$

2. $x^2 \frac{\partial^2 z}{\partial x^2} - 3xy\frac{\partial^2 z}{\partial x \partial y} + 2y^2 \frac{\partial^2 z}{\partial y^2} + 3y\frac{\partial z}{\partial y} - 2z = 0.$

3. $x^2 \frac{\partial^2 z}{\partial x^2} + 2xy\frac{\partial^2 z}{\partial x \partial y} + y^2 \frac{\partial^2 z}{\partial y^2} - nx\frac{\partial z}{\partial y} - ny\frac{\partial z}{\partial y} + nz = x^2y^2.$

4. $(x^2D_x^2 + 2xyD_xD_y + y^2D_y^2)z = x^my^n$; where $m + n \neq 0$.

5. $x^2 \frac{\partial^2 z}{\partial x^2} + 2xy\frac{\partial^2 z}{\partial x \partial y} - x\frac{\partial z}{\partial x} = \frac{x^3}{y^2}.$

6. $x^2 \frac{\partial^2 z}{\partial x^2} + 2xy\frac{\partial^2 z}{\partial x \partial y} + y^2 \frac{\partial^2 z}{\partial y^2} + nz = n\left(x\frac{\partial z}{\partial x} + y\frac{\partial z}{\partial y}\right) + x^2 + y^2 + x^3.$

7. $x^2 \frac{\partial^2 z}{\partial x^2} - y^2 \frac{\partial^2 z}{\partial y^2} = xy.$

8. $yt - q = xy.$

9. $x^2r - y^2t + px - qy = \log x.$

10. $(x^2D_x^2 - 4y^2D_y^2 - 4yD_y - 1)z = x^2y^2 \log y.$

11. $(x^2D_x^2 + 2xyD_xD_y + y^2D_y^2)z = x^2y^2.$

12. $(x^2D_x^2 - 4xyD_xD_y + 4y^2D_y^2 + xD_x + 4yD_y)z = x^2y.$

13. $(x^2D_x^2 - 2xyD_xD_y + y^2D_y^2 - xD_x + 3yD_y)z = \dfrac{8y}{x}.$

14. $(x^2D_x^2 - y^2D_y^2)z = x^2y.$

15. $(x^2D_x^2 - xyD_xD_y - 2_y^2D_y^2 + xD_x - 2yD_y)z = \log(y/x).$

16. $x^2D_x^2 - 2xyD_xD_y - 3y^2D_y^2 + xD_x - 3yD_y)z = x^2y \sin(\log x^2).$

17. $x^2\dfrac{\partial^2 z}{\partial x^2} - y^2\dfrac{\partial^2 z}{\partial y^2} + x\dfrac{\partial z}{\partial x} - y\dfrac{\partial z}{\partial y} = x^2y^4.$

18. $x^2r - y^2t + xp - yp = \log x.$

19. $x^2\dfrac{\partial^2 z}{\partial x^2} - y^2\dfrac{\partial^2 z}{\partial y^2} = xy.$

20. $2x^2r - 5xys + 2y^2t + 2(px + qy) = 0.$

ANSWERS

Exercise 7.1

1. $z = x\phi_1(x^2y) + \phi_2(x^2y) + \dfrac{1}{30}x^3y^4.$

2. $z = x^2\phi_1(xy) + x\phi_2(x^2y).$

3. $z = x\phi_1(y/x) + x^n\phi_2(y/x) + \dfrac{x^2+y^2}{2-n}.$

4. $z = \phi_1(y/x) + x\phi_2(y/x) + \dfrac{1}{(m+n)(m+n-1)}(x^my^n).$

5. $z = \phi_1(y) + x^2\phi_2(y/x^2) - \dfrac{x^3}{9y^2}.$

6. $z = x\phi_1(y/x) + x^n\phi_2(y/x) + \dfrac{x^2+y^2}{2-n} + \dfrac{x^3}{2(3-n)}.$

7. $z = \phi_1(xy) + x\phi_2(y/x) + xy \log x.$

8. $z = \phi_1(x) + y^2\phi_2(x) + \dfrac{xy^2}{2}\log y.$

9. $z = \phi_1(xy) + \phi_2(y/x) + \dfrac{(\log x)^3}{6}$.

10. $z = \Sigma A x^a y^b + \dfrac{x^2y^2}{225}(16 - 15 \log y)$ where $a^2 - a - 4b^2 - 1 = 0$.

11. $z = \phi_1(y/x) + x\phi_2(y/x) + \dfrac{x^2y^2}{12}$.

12. $z = \phi_1(x^2y) + \log x\phi_2(x^2y) + \dfrac{1}{2}(\log x)^2 x^2y$.

13. $z = \phi_1(xy) + x^2\phi_2(xy) + \dfrac{y}{x}$.

14. $z = \phi_1(xy) + x\phi_2(y/x) + \dfrac{x^2y}{2}$.

15. $z = \phi_1(x^2y) + \phi_2(y/x) + \dfrac{1}{2}(\log x)^2 \log y$.

16. $z = \phi_1(x^3y) + \phi_2(y/x) - \dfrac{1}{65}[4 \cos(\log x^2) + 7 \sin(\log x^2)]$.

17. $z = \phi_1(xy) + \phi_2(y/x) - \dfrac{1}{12}x^2y^4$.

18. $z = \phi_1(xy) + \phi_2(y/x) + \dfrac{1}{6}(\log x)^3$.

19. $z = x\phi_1(y/x) + \phi_2(xy) + xy \log x$.

20. $z = \phi_1(xy^2) + \phi_2(x^2y)$. ▲▲

8

CHAPTER

Classification of Linear Partial Differential Equations of the Second Order

A partial differential equation is said to be of the second order when it involves at least one term containing second order partial derivatives, that is, it consists of any of the partial differential coefficient r, s and t but not of higher order. The second order partial differential equation is given by

$$f(x, y, z, p, q, r, s, t) = 0$$

where $$r = \frac{\partial^2 z}{\partial x^2} = \frac{\partial p}{\partial x},\ s = \frac{\partial^2 z}{\partial x \partial y} = \frac{\partial p}{\partial y} = \frac{\partial q}{\partial x}$$

and $$t = \frac{\partial^2 z}{\partial y^2} = \frac{\partial q}{\partial y}$$

The general linear partial differential equation of second order in two independent variables x and y with z as dependent variable is of the form

$$A\frac{\partial^2 z}{\partial x^2} + B\frac{\partial^2 z}{\partial x \partial y} + C\frac{\partial^2 z}{\partial y^2} + f\left(x, y, z, \frac{\partial u}{\partial y}, \frac{\partial u}{\partial y}\right) = 0$$

or $$Ar + Bs + Ct + f(x, y, z, p, q) = 0$$

where, A is positive.

Let, the operator $\phi = \sum_{i=1}^{2} \sum_{j=1}^{2} a_{ij}\, \partial_i\, \partial_j$

where $$\partial_i = \frac{\partial}{\partial x^i} \text{ and } \partial_j = \frac{\partial}{\partial y^j}$$

For equation involving two independent variables x and y; we have

$$\phi = A\partial_1^2 + B\partial_1\partial_2 + C\partial_2^2 \quad (1)$$

and $$-\Delta = B^2 - 4AC$$

At any point (x, y), the above differential Eq. (1) is said to be

(a) Elliptic if $B^2 - 4AC < 0$
(b) Hyperbolic if $B^2 - 4AC > 0$
(c) Parabolic if $B^2 - 4AC = 0$

Note: If all A, B, C are constants, the differential Eq. (1) will have the same nature throughout and if A, B, C are functions of x, y, the same Eq. (1) can be elliptic, hyperbolic or parabolic at different points of the region.

▲▲

EXAMPLE 8.1

Classify the equation $\dfrac{\partial^2 z}{\partial x^2} + \dfrac{\partial^2 u}{\partial x \partial y} + \dfrac{\partial^2 u}{\partial y^2} = 0.$

Solution: Here,
$$A = B = C = 1$$
$$\therefore \quad B^2 - 4AC = -3 < 0$$

⇒ Equation is elliptic.

EXAMPLE 8.2

Classify the equation $2\dfrac{\partial^2 u}{\partial x^2} + 4\dfrac{\partial^2 u}{\partial x \partial y} + 3\dfrac{\partial^2 u}{\partial y^2} = 0.$

Solution: Here,
$$A = 2, B = 4 \quad \text{and} \quad C = 3$$
$$\therefore \quad B^2 - 4AC = -820 < 0$$
⇒ Equation is elliptic.

EXAMPLE 8.3

Classify the partial differential equation $r = a^2 t \ (a \neq 0)$.

Solution: The given equation is $\dfrac{\partial^2 z}{\partial x^2} - a^2 \dfrac{\partial^2 z}{\partial y^2} = 0$

Here,
$$A = 1, B = 0 \quad \text{and} \quad C = -a^2$$
$$\therefore \quad B^2 - 4AC = 4a^2 > 0$$

⇒ Equation is hyperbolic.

EXAMPLE 8.4

Classify the partial differential equation $r + x^2 t = 0$.

Solution: The given equation is $\dfrac{\partial^2 z}{\partial x^2} + x^2 \dfrac{\partial^2 z}{\partial y^2} = 0$

Here, $A = 1, B = 0$ and $C = x^2$

$\therefore$ $B^2 - 4AC = -4x^2 < 0$

$\Rightarrow$ Equation is elliptic.

EXAMPLE 8.5

Classify the equation $t - 4s + 4r = 0$.

Solution: The given equation is

$$4\frac{\partial^2 z}{\partial x^2} - 4\frac{\partial^2 z}{\partial x \partial y} + \frac{\partial^2 z}{\partial y^2} = 0$$

Here, $A = 4, B = -4$ and $C = 1$

$\therefore$ $B^2 - 4AC = 0$

$\Rightarrow$ Equation is parabolic.

EXAMPLE 8.6

Classify the equation $xr + ys + t = 0$.

Solution: The given equation is $x\frac{\partial^2 z}{\partial x^2} + y\frac{\partial^2 z}{\partial x \partial y} + \frac{\partial^2 z}{\partial y^2} = 0$

Here, $A = x, B = y$ and $C = 1$

$\therefore$ $B^2 - 4AC = y^2 - 4x$

$\Rightarrow$ Equation is elliptic, if $y^2 < 4x$

Equation is parabolic if $y^2 = 4x$

and equation is hyperbolic if $y^2 > 4x$.

EXAMPLE 8.7

Classify the partial differential equation

$$xyr - (x^2 - y^2)s - xyt + py - qx - 2(x^2 - y^2) = 0.$$

Solution: The given equation is

$$xy\frac{\partial^2 z}{\partial x^2} - (x^2 - y^2)\frac{\partial^2 z}{\partial x \partial y} - xy\frac{\partial^2 z}{\partial y^2} + y\frac{\partial z}{\partial x} - x\frac{\partial z}{\partial y} - 2(x^2 - y^2) = 0$$

Here, $A = xy, B = -(x^2 - y^2)$ and $C = -xy$

$\therefore$ $B^2 - 4AC = (x^2 - y^2)^2 + 4x^2y^2 = (x^2 + y^2)^2 > 0$

$\Rightarrow$ Equation is hyperbolic at all points.

EXAMPLE 8.8

Classify the Laplace's equation

$$\frac{\partial^2 w}{\partial x^2} + \frac{\partial^2 w}{\partial y^2} + \frac{\partial^2 w}{\partial z^2} = 0$$

Solution: Here, the operator $\phi = \partial_1^2 + \partial_2^2 + \partial_3^2$

and $$a_{11} = 1 = a_{22} = a_{33}, a_{12} = 0 = a_{13} = a_{21} = a_{23} = a_{31} = a_{32}$$

Now, $$\Delta = \begin{vmatrix} a_{11} & a_{12} & a_{13} \\ a_{21} & a_{22} & a_{23} \\ a_{31} & a_{32} & a_{33} \end{vmatrix} = \begin{vmatrix} 1 & 0 & 0 \\ 0 & 1 & 0 \\ 0 & 0 & 1 \end{vmatrix} = 1 \neq 0$$

Since ϕ is positive for all non-zero real values of ∂_1, ∂_2, ∂_3 and reduces to zero only when $\partial_1 = 0 = \partial_2 = \partial_3$.

Hence, the given equation is elliptic.

EXAMPLE 8.9

Classify the wave equation

$$\frac{\partial^2 w}{\partial x^2} + \frac{\partial^2 w}{\partial y^2} + \frac{\partial^2 w}{\partial z^2} = \frac{1}{c^2}\frac{\partial^2 w}{\partial t^2}.$$

Solution: The given equation is of the second order with four independent variables x, y, z and t.

Here, the operator $\phi = \partial_1^2 + \partial_2^2 + \partial_3^2 - \frac{1}{c^2}\partial_4^2$

and $$a_{11} = 1 = a_{22} = a_{33}, a_{44} = -\frac{1}{c^2} \text{ and all } a_{ij} = 0 \text{ if } i \neq j, (i, j = 1, 2, 3, 4)$$

Now, $$\Delta = \begin{vmatrix} a_{11} & a_{12} & a_{13} & a_{14} \\ a_{21} & a_{22} & a_{23} & a_{24} \\ a_{31} & a_{32} & a_{33} & a_{34} \\ a_{41} & a_{42} & a_{43} & a_{44} \end{vmatrix} = \begin{vmatrix} 1 & 0 & 0 & 0 \\ 0 & 1 & 0 & 0 \\ 0 & 0 & 1 & 0 \\ 0 & 0 & 0 & -\frac{1}{c^2} \end{vmatrix}$$

Since, the operator ϕ can be both positive or negative, hence the given equation is hyperbolic.

EXAMPLE 8.10

Classify the heat equation

$$\frac{\partial^2 w}{\partial x^2} + \frac{\partial^2 w}{\partial y^2} + \frac{\partial^2 w}{\partial z^2} = \frac{1}{c^2}\frac{\partial w}{\partial t}.$$

Solution: The given equation is of the second order with four independent variables x, y, z and t.

Here, the operator $\phi = \partial_1^2 + \partial_2^2 + \partial_3^2$

and $a_{11} = 1 = a_{22} = a_{33}, a_{44} = 0$ and all $a_{ij} = 0$ if $i \neq j, (i, j = 1, 2, 3, 4)$

Now,
$$\Delta = \begin{vmatrix} a_{11} & a_{12} & a_{13} & a_{14} \\ a_{21} & a_{22} & a_{23} & a_{24} \\ a_{31} & a_{32} & a_{33} & a_{34} \\ a_{41} & a_{42} & a_{43} & a_{44} \end{vmatrix} = \begin{vmatrix} 1 & 0 & 0 & 0 \\ 0 & 1 & 0 & 0 \\ 0 & 0 & 1 & 0 \\ 0 & 0 & 0 & 0 \end{vmatrix} = 0$$

Since $\Delta = 0$, hence the given equation is parabolic.

EXAMPLE 8.11

Classify the equation

$$(1-x^2)\frac{\partial^2 z}{\partial x^2} - 2xy\frac{\partial^2 z}{\partial x \partial y} + (1-y^2)\frac{\partial^2 z}{\partial y^2} + 2x\frac{\partial z}{\partial x} + 6x^2 y\frac{\partial z}{\partial y} - 6z = 0.$$

Solution: Here, $A = 1 - x^2, B = -2xy$ and $C = 1 - y^2$

$\therefore$ $B^2 - 4AC = 4x^2y^2 - 4(1 - x^2)(1 - y^2) = 4(-1 + x^2 + y)$

As A, B and C are functions of x and y, the differential equation is hyperbolic in the region where $B^2 - 4AC > 0$

that is, where $x^2 + y^2 > 1$

parabolic in the region, where $B^2 - 4AC = 0$

that is, where $x^2 + y^2 = 1$

and elliptic in the region, where $B^2 - 4AC < 0$

that is $x^2 + y^2 < 1$.

Hence, the given differential equation is respectively hyperbolic, parabolic and elliptic in the region outside the circle, on the circle and inside the circle, respectively, whose equation is

$$x^2 + y^2 = 1.$$

▲▲

EXERCISE 8.1

Classify the following equations:

1. $u_{xx} + 2u_{xy} + 3u_{yy} = 0.$
2. $u_{xx} - 4u_{xy} + 2u_{yy} = 0.$
3. $u_{xx} + 4u_{ycy} + 4u_{yy} = 0.$
4. $\dfrac{\partial^2 u}{\partial x^2} + t\dfrac{\partial^2 u}{\partial x \partial t} + x\dfrac{\partial^2 u}{\partial t^2} = 0.$
5. $x^2 \dfrac{\partial^2 u}{\partial y^2} - \dfrac{\partial^2 u}{\partial x^2} + u = 0.$
6. $x\dfrac{\partial^2 u}{\partial x^2} + 2\dfrac{\partial^2 u}{\partial x \partial t} + y\dfrac{\partial^2 u}{\partial y^2} + \dfrac{\partial u}{\partial y} = 0.$
7. $8D_{xx}z - 2D_xD_yz - 3D_y^2z = 0.$
8. $\dfrac{\partial^2 z}{\partial x^2} + 2\dfrac{\partial^2 z}{\partial x \partial y} + 9\dfrac{\partial^2 z}{\partial y^2} = 0.$
9. $\dfrac{\partial^2 z}{\partial x^2} + 2\dfrac{\partial^2 z}{\partial x \partial y} + 5\dfrac{\partial^2 z}{\partial y^2} = 0.$
10. $u_{xx} + u_{yy} + u_{zz} + 2u_{yz} = 0.$
11. $\dfrac{\partial^2 u}{\partial x^2} + 2\dfrac{\partial^2 u}{\partial x \partial y} + y\dfrac{\partial^2 u}{\partial y^2} - xy\dfrac{\partial u}{\partial x} = 0.$
12. $(1 - x^2)\dfrac{\partial^2 u}{\partial x^2} + 2y\dfrac{\partial^2 u}{\partial x \partial y} + y\dfrac{\partial^2 u}{\partial y^2} = 0.$
13. $\dfrac{\partial^2 z}{\partial x^2} = \dfrac{\partial z}{\partial y}.$
14. $\dfrac{\partial^2 z}{\partial x^2} = \dfrac{\partial^2 z}{\partial y^2}.$
15. $\dfrac{\partial^2 z}{\partial x^2} + \dfrac{\partial^2 z}{\partial y^2} = 0.$

ANSWERS

Exercise 8.1

1. Elliptic.
2. Hyperbolic.
3. Parabolic.

4. Elliptic, if $t^2 > 4x$, parabolic, if $t^2 = 4x$ and hyperbolic if $t^2 < 4x$.
5. Hyperbolic, if $|x| > 0$, parabolic, if $x = 0$ and nowhere elliptic.
6. Hyperbolic if $xy < 1$, parabolic, if $xy = 1$ and elliptic if $xy > 1$.
7. Hyperbolic.
8. Parabolic.
9. Elliptic.
10. Parabolic.
11. Hyperbolic, if $y < x^2/4$, parabolic, if $y = x^2/4$ and elliptic, if $y > x^2/4$.
12. Hyperbolic, if $x^2 + y^2 > 1$, parabolic, if $x^2 + y^2 = 1$ and elliptic, if $x^2 + y^2 < 1$.
13. Parabolic.
14. Hyperbolic.
15. Elliptic.

▲▲

Laplace Transformation, Reduction to Canonical (or Normal) Forms

Consider the linear partial differential equation of the second order of the form

$$Rr + Ss + Tt + f(x, y, z, p, q) = 0 \tag{1}$$

where R, S, T are continuous functions of x and y possessing continuous partial derivatives of the order as high as necessary.

Equation (1) can be reduced to a more simple form by a suitable change of the independent variables and these simple forms are called the canonical (or normal) forms of Eq. (1).

Let, the independent variables x and y be changed to u and v by means of the transformations

$$u = u(x, y) \text{ and } v = v(x, y)$$

Now, we have

$$p = \frac{\partial z}{\partial x} = \frac{\partial z}{\partial u}\frac{\partial u}{\partial x} + \frac{\partial z}{\partial v}\frac{\partial v}{\partial x}$$

and $$q = \frac{\partial z}{\partial y} = \frac{\partial z}{\partial u}\frac{\partial u}{v y} + \frac{\partial z}{\partial v}\frac{\partial v}{\partial y}$$

$$\therefore \quad r = \frac{\partial^2 z}{\partial x^2} = \frac{\partial}{\partial x}\left(\frac{\partial z}{\partial x}\right) = \frac{\partial}{\partial x}\left(\frac{\partial z}{\partial u}\frac{\partial u}{\partial x} + \frac{\partial z}{\partial v}\frac{\partial v}{\partial x}\right)$$

$$= \frac{\partial}{\partial x}\left(\frac{\partial z}{\partial u}\right)\frac{\partial u}{\partial x} + \frac{\partial z}{\partial u}\frac{\partial^2 z}{\partial x^2} + \frac{\partial}{\partial x}\left(\frac{\partial z}{\partial v}\right)\frac{\partial v}{\partial x} + \frac{\partial z}{\partial v}\frac{\partial^2 v}{\partial x^2}$$

$$= \left[\frac{\partial}{\partial u}\left(\frac{\partial z}{\partial u}\right)\frac{\partial u}{\partial x} + \frac{\partial}{\partial v}\left(\frac{\partial z}{\partial u}\right)\frac{\partial v}{\partial x}\right]\frac{\partial u}{\partial x} + \frac{\partial z}{\partial u}\frac{\partial^2 u}{\partial x^2} + \left[\frac{\partial}{\partial u}\left(\frac{\partial z}{\partial v}\right)\frac{\partial u}{\partial x} + \frac{\partial}{\partial v}\left(\frac{\partial z}{\partial v}\right)\frac{\partial v}{\partial x}\right]\frac{\partial v}{\partial x} + \frac{\partial z}{\partial v}\frac{\partial^2 v}{\partial x^2}$$

$$= \left[\frac{\partial^2 z}{\partial u^2}\frac{\partial u}{\partial x} + \frac{\partial^2 z}{\partial v \partial u}\frac{\partial v}{\partial x}\right]\frac{\partial u}{\partial x} + \frac{\partial z}{\partial u}\frac{\partial^2 u}{\partial x^2} + \left[\frac{\partial^2 z}{\partial u \partial v}\frac{\partial u}{\partial x} + \frac{\partial^2 z}{\partial v^2}\frac{\partial v}{\partial x}\right]\frac{\partial v}{\partial x} + \frac{\partial z}{\partial v}\frac{\partial^2 v}{\partial x^2}$$

$$= \left(\frac{\partial u}{\partial x}\right)^2 \frac{\partial^2 z}{\partial u^2} + 2\frac{\partial^2 z}{\partial u \partial v}\frac{\partial u}{\partial x}\frac{\partial v}{\partial x} + \left(\frac{\partial v}{\partial x}\right)^2 \frac{\partial^2 z}{\partial v^2} + \frac{\partial z}{\partial u}\frac{\partial^2 u}{\partial x^2} + \frac{\partial z}{\partial v}\frac{\partial^2 v}{\partial x^2}$$

Similarly,

$$t = \frac{\partial^2 z}{\partial y^2} = \left(\frac{\partial u}{\partial y}\right)^2 \frac{\partial^2 z}{\partial u^2} + 2\frac{\partial^2 z}{\partial u \partial v}\frac{\partial u}{\partial y}\frac{\partial v}{\partial y} + \left(\frac{\partial v}{\partial y}\right)^2 \frac{\partial^2 z}{\partial v^2} + \frac{\partial z}{\partial u}\frac{\partial^2 u}{\partial y^2} + \frac{\partial z}{\partial v}\frac{\partial^2 v}{\partial y^2}$$

$$\text{and } s = \frac{\partial^2 z}{\partial x \partial y} = \frac{\partial}{\partial x}\left(\frac{\partial z}{\partial y}\right) = \frac{\partial}{\partial x}\left(\frac{\partial z}{\partial u}\frac{\partial u}{\partial y} + \frac{\partial z}{\partial v}\frac{\partial v}{\partial y}\right)$$

$$= \frac{\partial}{\partial x}\left(\frac{\partial z}{\partial u}\right)\frac{\partial u}{\partial y} + \frac{\partial z}{\partial u}\frac{\partial^2 y}{\partial x \partial y} + \frac{\partial}{\partial x}\left(\frac{\partial z}{\partial v}\right)\frac{\partial v}{\partial y} + \frac{\partial}{\partial v}\frac{\partial^2 v}{\partial x \partial y}$$

$$= \left[\frac{\partial}{\partial u}\left(\frac{\partial z}{\partial u}\right)\frac{\partial u}{\partial x} + \frac{\partial}{\partial v}\left(\frac{\partial z}{\partial u}\right)\frac{\partial v}{\partial x}\right]\frac{\partial u}{\partial y} + \frac{\partial z}{\partial u}\frac{\partial^2 u}{\partial x \partial y} + \left[\frac{\partial}{\partial u}\left(\frac{\partial z}{\partial v}\right)\frac{\partial u}{\partial x} + \frac{\partial}{\partial v}\left(\frac{\partial z}{\partial v}\right)\frac{\partial v}{\partial x}\right]\frac{\partial v}{\partial y} + \frac{\partial z}{\partial v}\frac{\partial^2 v}{\partial x \partial y}$$

$$= \frac{\partial u}{\partial x}\frac{\partial v}{\partial y}\frac{\partial^2 z}{\partial u^2} + \frac{\partial^2 z}{\partial u \partial v}\left(\frac{\partial u}{\partial y}\frac{\partial v}{\partial x} + \frac{\partial u}{\partial x}\frac{\partial v}{\partial y}\right) + \frac{\partial v}{\partial x}\frac{\partial v}{\partial y}\frac{\partial^2 z}{\partial v^2} + \frac{\partial z}{\partial u}\frac{\partial^2 u}{\partial x \partial y} + \frac{\partial z}{\partial v}\frac{\partial^2 v}{\partial x \partial y}$$

Putting these values of p, q, r, s and t in Eq. (1) and simplifying, we get

$$A\frac{\partial^2 z}{\partial u^2} + 2B\frac{\partial^2 z}{\partial u \partial v} + C\frac{\partial^2 z}{\partial v^2} + F\left(u, v, w, \frac{\partial z}{\partial u}, \frac{\partial z}{\partial v}\right) = 0 \tag{2}$$

Here,

$$A = R\left(\frac{\partial u}{\partial x}\right)^2 + S\frac{\partial u}{\partial x}\frac{\partial u}{\partial y} + T\left(\frac{\partial u}{\partial y}\right)^2$$

$$B = R\frac{\partial u}{\partial x}\frac{\partial v}{\partial x} + \frac{1}{2}S\left(\frac{\partial u}{\partial x}\frac{\partial v}{\partial y} + \frac{\partial u}{\partial y}\frac{\partial v}{\partial x}\right) + T\frac{\partial u}{\partial y}\frac{\partial v}{\partial y}$$

and

$$C = R\left(\frac{\partial v}{\partial x}\right)^2 + S\frac{\partial v}{\partial x}\frac{\partial v}{\partial y} + T\left(\frac{\partial v}{\partial y}\right)^2$$

and $F = \left(u, v, w, \frac{\partial z}{\partial u}, \frac{\partial z}{\partial v}\right)$ is the function of u, v, w, $\frac{\partial z}{\partial u}$ and $\frac{\partial z}{\partial v}$ where w is a function of u and v.

Now, we determine u and v so that Eq. (2) takes the simplest possible form. This equation reduces to the simplest integral form when the discriminant $S^2 - 4RT$ of the quadratic equation

$$R\lambda^2 + S\lambda + T = 0 \tag{3}$$

is either positive, negative or zero everywhere. The three cases are discussed as follows:

Case I: *If $S^2 - 4RT > 0$ that is (1) is hyperbolic.*
In this case, the roots λ_1, λ_2 (say) of the Eq. (3) are real and distinct. If we choose u and v such that

$$\frac{\partial u}{\partial x} = \lambda_1 \frac{\partial u}{\partial y} \tag{4}$$

and
$$\frac{\partial v}{\partial x} = \lambda_2 \frac{\partial v}{\partial y} \tag{5}$$

then
$$A = (R\lambda_1^2 + S\lambda_1 + T)\left(\frac{\partial u}{\partial y}\right)^2 = 0$$

$$\{\because R\lambda_1^2 + S\lambda_1 + T = 0 \text{ as } \lambda_1 \text{ is a root of the Eq. (3)}\}.$$

Similarly, $C = 0$.
Now Eq. (4) can be written as

$$\frac{\partial u}{\partial x} - \lambda_1 \frac{\partial u}{\partial y} = 0 \tag{6}$$

which is in Lagrange's form. So its auxiliary equations are

$$\frac{dx}{1} = \frac{dy}{-\lambda_1} = \frac{du}{0}$$

Now, $du = 0 \Rightarrow u =$ a constant c_1

and
$$\frac{dy}{dx} + \lambda_1 = 0$$

Let $f_1(x, y) = c_2$ be its solution
$\therefore$ The solution of Eq. (6) can be taken as

$$u = f_1(x, y) \tag{7}$$

which is a suitable choice for u.
Similarly, the solution of Eq. (5) that is

$$v = f_2(x, y) \tag{8}$$

is a suitable choice for v.

Now, it can be easily shown that

$$AC - B^2 = \frac{1}{4}(4RT - S^2)\left(\frac{\partial u}{\partial x}\frac{\partial v}{\partial y} - \frac{\partial u}{\partial y}\frac{\partial v}{\partial x}\right)^2 \tag{9}$$

Since $A = 0 = C$, therefore

$$B^2 = \frac{1}{4}(S^2 - 4RT)\left(\frac{\partial u}{\partial x}\frac{\partial v}{\partial y} - \frac{\partial u}{\partial y}\frac{\partial v}{\partial x}\right)^2$$

$\because$ $S^2 - 4RT > 0$ $\therefore$ $B^2 > 0$ $\Rightarrow$ $B \neq 0$

Thus, Eq. (2) reduces to

$$2B\frac{\partial^2 z}{\partial u \partial v} + F\left(u, v, w, \frac{\partial z}{\partial u}, \frac{\partial z}{\partial v}\right) = 0$$

On dividing by $2B$, it reduces to

$$\frac{\partial^2 z}{\partial u \partial v} \neq \phi\left(u, v, w, \frac{\partial z}{\partial u}, \frac{\partial z}{\partial v}\right)$$

Hence, if we make the substitutions of u defined by Eqs. (7) and (8), the given Eq. (1) reduces to the form

$$\frac{\partial^2 z}{\partial u \partial v} \neq \phi\left(u, v, w, \frac{\partial z}{\partial u}, \frac{\partial z}{\partial v}\right) \tag{10}$$

which is the canonical form of Eq. (1) and is simple to solve than the given Eq. (1).

Case II: *If $S^2 - 4RT = 0$ that is (1) is parabolic*

In this case the roots of Eq. (3), that is,

$$R\lambda^2 + S\lambda + T = 0$$

are real and equal, that is, $\lambda_1 = \lambda_2$

Here, we choose u as in case I such that

$$\frac{\partial u}{\partial x} = \lambda_1 \frac{\partial u}{\partial y}$$

which gives $$u = f_1(x, y) \tag{11}$$

Also, we take v to be any function of x and y which is independent of u.

As in case I $A = 0$.

Also from Eq. (9),

$$B^2 = 0 \quad \{\because S^2 - 4RT = 0\}$$

$$\Rightarrow \quad B = 0$$

▲▲

Hence, C cannot be zero otherwise v would be a function of u.
Putting $A = 0 = B$ in Eq. (2) and dividing by $C \neq 0$, the Eq. (2) becomes

$$\frac{\partial^2 z}{\partial v^2} = \phi\left(u, v, w, \frac{\partial z}{\partial u}, \frac{\partial z}{\partial v}\right)$$

Hence, if we make the substitution of u as defined in Eq. (11) and v any function of x and y, the given Eq. (1) in this case reduces to the form -

$$\frac{\partial^2 z}{\partial v^2} = \phi\left(u, v, w, \frac{\partial z}{\partial u}, \frac{\partial z}{\partial v}\right)$$

which is another canonical form of given Eq. (1) and is simpler to solve the given Eq. (1).

Case III: *If $S^2 - 4RT < 0$ that is Eq. (1) is elliptic.*

If this case, the roots of Eq. (3) that is $R\lambda^2 + S\lambda + T = 0$ are complex conjugates.

Proceeding as in case I, here Eq. (1) will reduce to the same canonical form as in case I but here the variables u and v are not real but are, in fact, the complex conjugates. To find the real canonical form let $u = \alpha + i\beta$ and $v = \alpha - j\beta$

$$\therefore \qquad \alpha = \frac{1}{2}(u + v) \quad \text{and} \quad \beta = \frac{1}{2}(u - v)$$

Now, we transform the independent variables u and v to α and β with the help of these relations

We have
$$\frac{\partial z}{\partial u} = \frac{\partial z}{\partial \alpha}\frac{\partial \alpha}{\partial u} + \frac{\partial z}{\partial \beta}\frac{\partial \beta}{\partial u} = \frac{1}{2}\left(\frac{\partial z}{\partial \alpha} + i\frac{\partial z}{\partial \beta}\right)$$

and
$$\frac{\partial z}{\partial v} = \frac{\partial z}{\partial \alpha}\frac{\partial \alpha}{\partial v} + \frac{\partial z}{\partial \beta}\frac{\partial \beta}{\partial v} = \frac{1}{2}\left(\frac{\partial z}{\partial \alpha} - j\frac{\partial z}{\partial \beta}\right)$$

$$\therefore \qquad \frac{\partial^2 z}{\partial u \partial v} = \frac{\partial}{\partial u}\left(\frac{\partial z}{\partial v}\right) = \frac{1}{4}\left(\frac{\partial}{\partial \alpha} + i\frac{\partial}{\partial \beta}\right)\left(\frac{\partial z}{\partial \alpha} - j\frac{\partial z}{\partial \beta}\right).$$

$$= \frac{1}{4}\left(\frac{\partial^2 z}{\partial \alpha^2} + \frac{\partial^2 z}{\partial \beta^2}\right)$$

Substituting in Eq. (10), the canonical form of Eq. (1) in this case is

$$\frac{\partial^2 z}{\partial \alpha^2} + \frac{\partial^2 z}{\partial \beta^2} = \phi\left(\alpha, \beta, w, \frac{\partial z}{\partial \alpha}, \frac{\partial z}{\partial \beta}\right)$$

EXAMPLE 8.12

Reduce the equation

$$y^2\frac{\partial^2 z}{\partial x^2}-2xy\frac{\partial^2 z}{\partial x\partial y}+x^2\frac{\partial^2 z}{\partial y^2}=\frac{y^2}{x}\frac{\partial z}{\partial x}+\frac{x^2}{y}\frac{\partial z}{\partial y}.$$

to canonical form and solve it.

Solution: The given equation can be written as

$$y^2r - 2xys + x^2t - \frac{y^2}{x}p-\frac{x^2}{y}q = 0$$

On comparing with $Rr + Ss + Tt + f(x, y, z, p, q) = 0$, we have

$$R = y^2,\ S = -2xy,\ T = x^2.$$

∴ The quadratic equation $R\lambda^2 + S\lambda + T = 0$ becomes

$$y^2\lambda^2 - 2xy\lambda + x^2 = 0$$

or $$(y\lambda - x)^2 = 0$$

⇒ $$\lambda=\frac{x}{y},\ \frac{x}{y}$$

∴ $$\lambda_1=\lambda_2=\frac{x}{y}\ \text{(equal roots)}$$

The equation $\frac{dy}{dx} + \lambda = 0$ becomes

$$\frac{dy}{dx}+\frac{x}{y}=0$$

⇒ $$ydy + xdx = 0$$

Integrating, we get

$$x^2 + y^2 = c_2$$

To change the independent variables x, y to u, v in the given equation, we take

$$u = x^2 + y^2$$

Also, we have to take v as some function of x and y independent of u,

Let, $v = x^2 - y^2$

∴ $$p=\frac{\partial z}{\partial x}=\frac{\partial z}{\partial u}\frac{\partial u}{\partial x}+\frac{\partial z}{\partial v}\frac{\partial v}{\partial x}=2x\left(\frac{\partial z}{\partial u}+\frac{\partial z}{\partial v}\right)$$

$$q=\frac{\partial z}{\partial y}=\frac{\partial z}{\partial u}\frac{\partial u}{\partial y}+\frac{\partial z}{\partial v}\frac{\partial v}{\partial y}=2y\left(\frac{\partial z}{\partial u}-\frac{\partial z}{\partial v}\right)$$

$$r = \frac{\partial}{\partial x}\left(\frac{\partial z}{\partial x}\right) = \frac{\partial}{\partial x}\left(2x\left(\frac{\partial z}{\partial u}+\frac{\partial z}{\partial v}\right)\right)$$

$$= 2\left(\frac{\partial z}{\partial u}+\frac{\partial z}{\partial v}\right) + 2x\frac{\partial}{\partial x}\left(\frac{\partial z}{\partial u}+\frac{\partial z}{\partial v}\right)$$

$$= 2\left(\frac{\partial z}{\partial u}+\frac{\partial z}{\partial v}\right) + 2x\left[\frac{\partial}{\partial u}\left(\frac{\partial z}{\partial u}+\frac{\partial z}{\partial v}\right)\frac{\partial u}{\partial x} + \frac{\partial}{\partial v}\left(\frac{\partial z}{\partial u}+\frac{\partial z}{\partial v}\right)\frac{\partial v}{\partial x}\right]$$

$$= 2\left(\frac{\partial z}{\partial u}+\frac{\partial z}{\partial v}\right) + 4x^2\left[\frac{\partial^2 z}{\partial u^2} + 2\frac{\partial^2 z}{\partial u \partial v} + \frac{\partial^2 z}{\partial v^2}\right]$$

$$s = \frac{\partial}{\partial x}\left(\frac{\partial z}{\partial y}\right) = \frac{\partial}{\partial x}\left(2y\left(\frac{\partial z}{\partial u}-\frac{\partial z}{\partial v}\right)\right)$$

$$= 2y\frac{\partial}{\partial x}\left(\frac{\partial z}{\partial u}-\frac{\partial z}{\partial v}\right)$$

$$= 2y\left[\frac{\partial}{\partial u}\left(\frac{\partial z}{\partial u}-\frac{\partial z}{\partial v}\right)\frac{\partial u}{\partial x} + \frac{\partial}{\partial v}\left(\frac{\partial z}{\partial u}-\frac{\partial z}{\partial v}\right)\frac{\partial v}{\partial x}\right]$$

$$= 4xy\left(\frac{\partial^2 z}{\partial u^2} - \frac{\partial^2 z}{\partial v^2}\right)$$

and
$$t = \frac{\partial}{\partial y}\left(\frac{\partial z}{\partial y}\right) = \frac{\partial}{\partial y}\left(2y\left(\frac{\partial z}{\partial u}-\frac{\partial z}{\partial v}\right)\right)$$

$$= 2\left(\frac{\partial z}{\partial u}-\frac{\partial z}{\partial v}\right) + 2y\frac{\partial}{\partial y}\left(\frac{\partial z}{\partial u}-\frac{\partial z}{\partial v}\right)$$

$$= 2\left(\frac{\partial z}{\partial u}-\frac{\partial z}{\partial v}\right) + 2y\left[\frac{\partial}{\partial u}\left(\frac{\partial z}{\partial u}-\frac{\partial z}{\partial v}\right)\frac{\partial u}{\partial y} + \frac{\partial}{\partial v}\left(\frac{\partial z}{\partial u}-\frac{\partial z}{\partial v}\right)\frac{\partial v}{\partial y}\right]$$

$$= 2\left(\frac{\partial z}{\partial u}-\frac{\partial z}{\partial v}\right) + 4y^2\left[\frac{\partial^2 z}{\partial u^2} - 2\frac{\partial^2 z}{\partial u \partial v} + \frac{\partial^2 z}{\partial v^2}\right]$$

Substituting in the given equation and simplifying, we have

$$\frac{\partial^2 z}{\partial v^2} = 0$$

which is the required canonical form.

Integrating, we obtain

$$z = v\phi(u) + \psi(u)$$

where $\phi(u)$, $\psi(u)$ are arbitrary functions of u.

$\therefore$ The required solution is

$$z = (x^2 - y^2)\, \phi(x^2 + y^2) + \psi(x^2 + y^2).$$

EXAMPLE 8.13

Reduce the following equation to canonical form

$$\frac{\partial^2 z}{\partial x^2} + x^2 \frac{\partial^2 z}{\partial y^2} = 0$$

Solution: The given equation can be written as

$$r + x^2 t = 0$$

On comparing with $Rr + Ss + Tt + f(x, y, z, p, q) = 0$, we have

$$R = 1,\ S = 0,\ T = x^2$$

$\therefore$ The quadratic equation $R\lambda^2 + S\lambda + T = 0$ becomes

$$\lambda^2 + x^2 = 0$$

$\Rightarrow$ $$\lambda = \pm\, jx$$

$\therefore$ $$\lambda_1 = ix,\ \lambda_2 = -ix \text{ (complex roots)}$$

Thus, $\dfrac{dy}{dx} + jx = 0$ and $\dfrac{dy}{dx} - ix = 0$

Integrating, we have

$$y + \frac{i}{2}x^2 = c_1 \quad \text{and} \quad y - \frac{j}{2}x^2 = c_2$$

$\therefore$ $$u = y + \frac{j}{2}x^2 = \alpha + j\beta$$

and $$v = y - \frac{i}{2}x^2 = \alpha - i\beta$$

so that $\alpha = y$ and $\beta = \dfrac{1}{2}x^2$

Now, $$p = \frac{\partial z}{\partial x} = \frac{\partial z}{\partial \alpha}\frac{\partial \alpha}{\partial x} + \frac{\partial z}{\partial \beta}\frac{\partial \beta}{\partial x} = x\frac{\partial z}{\partial \beta}$$

$$q = \frac{\partial z}{\partial y} = \frac{\partial z}{\partial \alpha}\frac{\partial \alpha}{\partial y} + \frac{\partial z}{\partial \beta}\frac{\partial \beta}{\partial y} = \frac{\partial z}{\partial \alpha}$$

$$r = \frac{\partial}{\partial x}\left(\frac{\partial z}{\partial x}\right) = \frac{\partial}{\partial x}\left(x\frac{\partial z}{\partial \beta}\right) = 1\cdot\frac{\partial z}{\partial \beta} + x\frac{\partial}{\partial x}\left(\frac{\partial z}{\partial \beta}\right)$$

$$= \frac{\partial z}{\partial \beta} + x\left[\frac{\partial}{\partial \alpha}\left(\frac{\partial z}{\partial \beta}\right)\frac{\partial \alpha}{\partial x} + \frac{\partial}{\partial \beta}\left(\frac{\partial z}{\partial \beta}\right)\frac{\partial \beta}{\partial x}\right]$$

$$= \frac{\partial z}{\partial \beta} + x^2\frac{\partial^2 z}{\partial \beta^2}$$

and

$$t = \frac{\partial}{\partial y}\left(\frac{\partial z}{\partial y}\right) = \frac{\partial}{\partial y}\left(\frac{\partial z}{\partial \alpha}\right) = \frac{\partial}{\partial \alpha}\left(\frac{\partial z}{\partial \alpha}\right)\frac{\partial \alpha}{\partial y} + \frac{\partial}{\partial \beta}\left(\frac{\partial z}{\partial \alpha}\right)\frac{\partial \beta}{\partial y}$$

$$= \frac{\partial^2 z}{\partial \alpha^2}$$

Substituting in the given equation, we have

$$\left(\frac{\partial z}{\partial \beta} + x^2\frac{\partial^2 z}{\partial \beta^2}\right) + x^2\frac{\partial^2 z}{\partial \alpha^2} = 0$$

or

$$\frac{\partial^2 z}{\partial \alpha^2} + \frac{\partial^2 z}{\partial \beta^2} = -\frac{1}{x^2}\frac{\partial z}{\partial \beta}$$

or

$$\frac{\partial^2 z}{\partial \alpha^2} + \frac{\partial^2 z}{\partial \beta^2} = -\frac{1}{2\beta}\frac{\partial z}{\partial \beta}$$

which is the required canonical form.

EXAMPLE 8.14

Reduce the following equation to canonical form in (a) hyperbolic region and (b) elliptic region

$$\frac{\partial^2 z}{\partial x^2} - x^2\frac{\partial^2 z}{\partial y^2} = 0.$$

Solution: The given equation can be written as

$$r - x^2 t = 0$$

On comparing with $Rr + Ss + Tt + f(x, y, z, p, q) = 0$, we have

$$R = 1,\ S = 0,\ T = -x^2$$

$$\therefore \quad S^2 - 4RT = 4x^2$$

Now, the given equation is

(a) Hyperbolic if $S^2 - 4RT > 0 \Rightarrow x^2 > 0$

(b) Elliptic if $S^2 - 4RT < 0 \Rightarrow x^2 < 0$

(c) Parabolic if $S^2 - 4RT = 0 \Rightarrow x^2 = 0$

The quadratic equation $R\lambda^2 + S\lambda + T = 0$ becomes

$$\lambda^2 - x^2 = 0$$

(a) Canonical form in hyperbolic region

$$\lambda^2 - x^2 = 0 \quad \Rightarrow \quad \lambda = \pm x \quad \{\because x^2 > 0\}$$

Let, $\lambda_1 = x$ and $\lambda_2 = -x$ [real and distinct roots]

$$\therefore \quad \frac{dy}{dx} + x = 0 \quad \text{and} \quad \frac{dy}{dx} - x = 0$$

Integrating, we have

$$y + \frac{1}{2}x^2 = c_1 \text{ and } y - \frac{1}{2}x^2 = c_2.$$

Thus, $u = y + \frac{1}{2}x^2$ and $v = y - \frac{1}{2}x^2$.

Now,
$$p = \frac{\partial z}{\partial x} = \frac{\partial z}{\partial u}\frac{\partial u}{\partial x} + \frac{\partial z}{\partial v}\frac{\partial v}{\partial x} = x\left(\frac{\partial z}{\partial u} - \frac{\partial z}{\partial v}\right)$$

$$q = \frac{\partial z}{\partial y} = \frac{\partial z}{\partial u}\frac{\partial u}{\partial y} + \frac{\partial z}{\partial v}\frac{\partial v}{\partial y} = \frac{\partial z}{\partial u} + \frac{\partial z}{\partial v}$$

$$r = \frac{\partial^2 z}{\partial x^2} = \frac{\partial}{\partial x}\left(\frac{\partial z}{\partial x}\right) = \frac{\partial}{\partial x}\left[x\left(\frac{\partial z}{\partial u} - \frac{\partial z}{\partial v}\right)\right]$$

$$= x\frac{\partial}{\partial x}\left(\frac{\partial z}{\partial u} - \frac{\partial z}{\partial v}\right) + 1\cdot\left(\frac{\partial z}{\partial u} - \frac{\partial z}{\partial v}\right)$$

$$= x\left[\frac{\partial}{\partial u}\left(\frac{\partial z}{\partial u} - \frac{\partial z}{\partial v}\right)\frac{\partial u}{\partial x} + \frac{\partial}{\partial v}\left(\frac{\partial z}{\partial u} - \frac{\partial z}{\partial v}\right)\frac{\partial v}{\partial x}\right] + \frac{\partial z}{\partial u} - \frac{\partial z}{\partial v}$$

$$= x^2\left(\frac{\partial^2 z}{\partial u^2} - 2\frac{\partial^2 z}{\partial u \partial v} + \frac{\partial^2 z}{\partial v^2}\right) + \frac{\partial z}{\partial u} - \frac{\partial z}{\partial v}$$

and $$t = \frac{\partial^2 z}{\partial y^2} = \frac{\partial}{\partial y}\left(\frac{\partial z}{\partial y}\right) = \frac{\partial}{\partial y}\left(\frac{\partial z}{\partial u} + \frac{\partial z}{\partial v}\right)$$

$$= \frac{\partial}{\partial u}\left(\frac{\partial z}{\partial u} + \frac{\partial z}{\partial v}\right)\frac{\partial u}{\partial y} + \frac{\partial}{\partial v}\left(\frac{\partial z}{\partial u} + \frac{\partial z}{\partial v}\right)\frac{\partial v}{\partial y}$$

$$= \frac{\partial^2 z}{\partial u^2} + 2\frac{\partial^2 z}{\partial u \partial v} + \frac{\partial^2 z}{\partial v^2}$$

Substituting these values, the given equation becomes

$$x^2\left(\frac{\partial^2 z}{\partial u^2} - 2\frac{\partial^2 z}{\partial u \partial v} + \frac{\partial^2 z}{\partial v^2}\right) + \frac{\partial z}{\partial u} - \frac{\partial z}{\partial v} - x^2\left(\frac{\partial^2 z}{\partial u^2} + 2\frac{\partial^2 z}{\partial u \partial v} + \frac{\partial^2 z}{\partial v^2}\right) = 0$$

$$\Rightarrow \qquad \frac{\partial^2 z}{\partial u \partial v} = \frac{1}{4x^2}\left(\frac{\partial z}{\partial u} - \frac{\partial z}{\partial v}\right)$$

$$\Rightarrow \qquad \frac{\partial^2 z}{\partial u \partial v} = \frac{1}{4(u-v)}\left(\frac{\partial z}{\partial u} - \frac{\partial z}{\partial v}\right)$$

which is the required canonical form of the given equation in the hyperbolic region.

(b) Canonical form in elliptic region

$$\lambda^2 + x^2 = 0 \Rightarrow \lambda = \pm\, ix \;\{\because x^2 < 0\}$$

By repeating the process as in Example 8.13, we obtain

$$\frac{\partial^2 z}{\partial \alpha^2} + \frac{\partial^2 z}{\partial \beta^2} = -\frac{1}{2\beta}\frac{\partial z}{\partial \beta}$$

where, $$\alpha = y \text{ and } \beta = \frac{1}{2}x^2$$

which is the required canonical form of the given equation in elliptic region.

EXAMPLE 8.15

Reduce the following equation to canonical form and solve it.

$$y(x + y)(r - s) - xp - yq - z = 0$$

Solution: Comparing the given equation with $Rr + Ss + Tt + f(x, y, z, p, q) = 0$, we have

$$R = y(x+y),\ S = -y(x+y);\ T = 0$$

$$\therefore \qquad S^2 - 4RT = y^2(x+y)^2 > 0$$

Hence, the quadratic equation $R\lambda^2 + S\lambda + T = 0$ becomes

$$y(x+y)\lambda^2 - y(x+y)\lambda + 0 = 0$$

$$\Rightarrow \qquad \lambda(\lambda - 1) = 0$$

$$\therefore \qquad \lambda = 0, 1$$

Thus, $\lambda_1 = 0$ and $\lambda_2 = 1$

$$\therefore \qquad \frac{dy}{dx} = 0 \quad \text{and} \quad \frac{dy}{dx} + 1 = 0$$

Integrating, we have

$$y = c_1 \text{ and } y + x = c_2$$

Hence, $u = y$ and $v = x + y$.

Now, $$p = \frac{\partial z}{\partial x} = \frac{\partial z}{\partial u}\frac{\partial u}{\partial x} + \frac{\partial z}{\partial v}\frac{\partial v}{\partial x} = \frac{\partial z}{\partial v}$$

and $$q = \frac{\partial z}{\partial y} = \frac{\partial z}{\partial u}\frac{\partial u}{\partial y} + \frac{\partial z}{\partial v}\frac{\partial v}{\partial y} = \frac{\partial z}{\partial u} + \frac{\partial z}{\partial v}$$

Also, $$r = \frac{\partial^2 z}{\partial x^2} = \frac{\partial}{\partial x}\left(\frac{\partial z}{\partial x}\right) = \frac{\partial}{\partial x}\left(\frac{\partial z}{\partial v}\right) = \frac{\partial}{\partial u}\left(\frac{\partial z}{\partial v}\right)\frac{\partial u}{\partial y} + \frac{\partial}{\partial v}\left(\frac{\partial z}{\partial v}\right)\frac{\partial v}{\partial y}$$

$$= \frac{\partial^2 z}{\partial v^2}$$

$$s = \frac{\partial^2 z}{\partial x \partial y} = \frac{\partial}{\partial x}\left(\frac{\partial z}{\partial y}\right) = \frac{\partial}{\partial x}\left(\frac{\partial z}{\partial u} + \frac{\partial z}{\partial v}\right)$$

$$= \frac{\partial}{\partial u}\left(\frac{\partial z}{\partial u} + \frac{\partial z}{\partial v}\right)\frac{\partial u}{\partial x} + \frac{\partial}{\partial v}\left(\frac{\partial z}{\partial u} + \frac{\partial z}{\partial v}\right)\frac{\partial z}{\partial x}$$

$$= \frac{\partial^2 z}{\partial v \partial u} + \frac{\partial^2 z}{\partial v^2}$$

and $$t = \frac{\partial^2 z}{\partial y^2} = \frac{\partial}{\partial y}\left(\frac{\partial z}{\partial y}\right) = \frac{\partial}{\partial y}\left(\frac{\partial z}{\partial u} + \frac{\partial z}{\partial v}\right)$$

$$= \frac{\partial}{\partial u}\left(\frac{\partial z}{\partial u}+\frac{\partial z}{\partial v}\right)\frac{\partial u}{\partial y}+\frac{\partial}{\partial v}\left(\frac{\partial z}{\partial u}+\frac{\partial z}{\partial v}\right)\frac{\partial v}{\partial y}$$

$$= \frac{\partial^2 z}{\partial u^2}+2\frac{\partial^2 z}{\partial u\partial v}+\frac{\partial^2 z}{\partial v^2}$$

Substituting these values in the given equation we have

$$-y(x+y)\frac{\partial^2 z}{\partial u\partial v}-x\frac{\partial z}{\partial v}-y\left(\frac{\partial z}{\partial u}+\frac{\partial z}{\partial v}\right)-z=0$$

$$\Rightarrow \quad -y(x+y)\frac{\partial^2 z}{\partial u\partial v}-y\frac{\partial z}{\partial u}-(x+y)\frac{\partial z}{\partial v}-z=0$$

$$\Rightarrow \quad uv\frac{\partial^2 z}{\partial u\partial v}+u\frac{\partial z}{\partial u}+v\frac{\partial z}{\partial v}+z=0$$

$$\rightarrow \quad \frac{\partial^2 z}{\partial u\partial v}+\frac{1}{v}\frac{\partial z}{\partial u}+\frac{1}{u}\frac{\partial z}{\partial v}+\frac{1}{uv}z=0$$

which is the required canonical form of the given equation.

We can write it as

$$\frac{\partial}{\partial v}\left(\frac{\partial z}{\partial u}+\frac{1}{u}z\right)+\frac{1}{v}\left(\frac{\partial z}{\partial u}+\frac{1}{u}z\right)=0$$

or $\frac{\partial}{\partial v}t+\frac{1}{v}t=0$ assuming $\frac{\partial z}{\partial u}+\frac{1}{u}z=t.$

which is a linear differential equation in t and v with v as independent variable.

Integrating factors (IF) $= e^{\int \frac{1}{v}dv} = e^{\log v} = v$

$\therefore$ Its solution is

$$vt = \int 0\,dv + f(u)$$

$$v\left(\frac{\partial z}{\partial u}+\frac{1}{u}z\right) = f(u)$$

$$\Rightarrow \quad \frac{\partial z}{\partial u}+\frac{1}{u}z=\frac{1}{v}f(u)$$

which is again a linear differential equation in z and u with u as independent variable.

Integrating factor (IF) $= e^{\int \frac{1}{u} du} = e^{\log u} = u$

So, its solution is

$$uz = \int u \frac{1}{v} f(u) du + \psi(v)$$

$$\Rightarrow \quad uz = \frac{1}{v}\phi(u) + \psi(v) \text{ where } \phi(u) = \int uf(u)du$$

$$\Rightarrow \quad z = \frac{1}{uv}\phi(u) + \frac{1}{u}\psi(v)$$

$$\Rightarrow \quad z = \frac{1}{y(x+y)}\phi(y) + \frac{1}{y}\psi(x+y)$$

which is the required solution with ϕ and ψ as arbitrary functions.

EXAMPLE 8.16

Reduce the following equation to canonical form and solve it.

$$x(xy-1)r - (x^2y^2-1)s + y(xy-1)t + (x-1)p + (y-1)q = 0.$$

Solution: Comparing the given equation with $Rr + Ss + Tt + f(x, y, z, p, q) = 0$, we have

$$R = x(xy-1),\ S = -(x^2y^2-1),\ T = y(xy-1)$$

$$\therefore \quad S^2 - 4RT = (xy-1)^4 > 0$$

Here, the quadratic equation $R\lambda^2 + S\lambda + T = 0$ becomes

$$x(xy-1)\lambda^2 - (x^2y^2-1)\lambda + y(xy-1) = 0$$

$$\Rightarrow \quad (xy-1)(x\lambda^2 - (xy+1)\lambda + y) = 0$$

$$\Rightarrow \quad (xy-1)(x\lambda - 1)(\lambda - y) = 0$$

$$\Rightarrow \quad \lambda = \frac{1}{x}, y$$

Let, $\lambda_1 = \frac{1}{x}$ and $\lambda_2 = y$

$$\therefore \quad \frac{dy}{dx} + \frac{1}{x} = 0 \quad \text{and} \quad \frac{dy}{dx} + y = 0$$

Integrating, we have

$$y + \log x = \log c_1 \text{ and } \log y + x = \log c_2$$

$$\therefore \quad xe^y = c_1 \text{ and } ye^x = c_2.$$

Thus, $u = xe^y$ and $v = ye^x$.

$$\therefore \quad p = \frac{\partial z}{\partial x} = \frac{\partial z}{\partial u}\frac{\partial u}{\partial x} + \frac{\partial z}{\partial v}\frac{\partial v}{\partial x} = e^y \frac{\partial z}{\partial u} + ye^x \frac{\partial z}{\partial v}$$

$$= e^y \frac{\partial z}{\partial u} + v\frac{\partial z}{\partial v}$$

$$q = \frac{\partial z}{\partial y} = \frac{\partial z}{\partial u}\frac{\partial u}{\partial y} + \frac{\partial z}{\partial v}\frac{\partial v}{\partial y} = xe^y \frac{\partial z}{\partial u} + e^x \frac{\partial z}{\partial v}$$

$$= u\frac{\partial z}{\partial u} + e^x \frac{\partial z}{\partial v}$$

$$r = \frac{\partial^2 z}{\partial x^2} = \frac{\partial}{\partial x}\left(\frac{\partial z}{\partial x}\right) = \frac{\partial}{\partial x}\left(e^y \frac{\partial z}{\partial u} + ye^x \frac{\partial z}{\partial v}\right)$$

$$= e^y\left[\frac{\partial}{\partial u}\left(\frac{\partial z}{\partial u}\right)\frac{\partial u}{\partial x} + \frac{\partial}{\partial v}\left(\frac{\partial z}{\partial u}\right)\frac{\partial v}{\partial x}\right] + y\left[e^x \frac{\partial z}{\partial v} + e^x\left(\frac{\partial}{\partial u}\left(\frac{\partial z}{\partial v}\right)\frac{\partial u}{\partial x} + \frac{\partial}{\partial v}\left(\frac{\partial z}{\partial v}\right)\frac{\partial v}{\partial x}\right)\right]$$

$$= e^{2y}\frac{\partial^2 z}{\partial u^2} + 2ye^{x+y}\frac{\partial^2 z}{\partial u \partial v} + y^2 e^{2x}\frac{\partial^2 z}{\partial v^2} + ye^x \frac{\partial z}{\partial v}$$

$$s = \frac{\partial^2 z}{\partial x \partial y} = \frac{\partial}{\partial x}\left(\frac{\partial z}{\partial y}\right) = \frac{\partial}{\partial x}\left(xe^y \frac{\partial z}{\partial u} + e^x \frac{\partial z}{\partial v}\right)$$

$$= e^y\left[1\cdot\frac{\partial z}{\partial u} + \left(\frac{\partial}{\partial u}\left(\frac{\partial z}{\partial u}\right)\frac{\partial u}{\partial x} + \frac{\partial}{\partial v}\left(\frac{\partial z}{\partial u}\right)\frac{\partial v}{\partial x}\right)\right] + e^x \frac{\partial z}{\partial v} + e^x\left[\frac{\partial}{\partial u}\left(\frac{\partial z}{\partial v}\right)\frac{\partial u}{\partial x} + \frac{\partial}{\partial v}\left(\frac{\partial z}{\partial v}\right)\frac{\partial v}{\partial x}\right]$$

$$= xe^{2y}\frac{\partial^2 z}{\partial u^2} + (1 + xy)e^{x+y}\frac{\partial^2 z}{\partial u \partial v} + ye^{2x}\frac{\partial^2 z}{\partial v^2} + e^y \frac{\partial z}{\partial u} + e^x \frac{\partial z}{\partial v}$$

and $$t = \frac{\partial^2 z}{\partial y^2} = \frac{\partial}{\partial y}\left(\frac{\partial z}{\partial y}\right) = \frac{\partial}{\partial y}\left(xe^y \frac{\partial z}{\partial u} + e^x \frac{\partial z}{\partial v}\right)$$

$$= x\left[e^y \frac{\partial z}{\partial u} + e^y\left(\frac{\partial}{\partial u}\left(\frac{\partial z}{\partial u}\right)\frac{\partial u}{\partial y} + \frac{\partial}{\partial v}\left(\frac{\partial z}{\partial u}\right)\frac{\partial v}{\partial y}\right)\right] + e^x\left[\frac{\partial}{\partial u}\left(\frac{\partial z}{\partial v}\right)\frac{\partial u}{\partial y} + \frac{\partial}{\partial v}\left(\frac{\partial z}{\partial v}\right)\frac{\partial v}{\partial y}\right]$$

$$= x^2 e^{2y}\frac{\partial^2 z}{\partial u^2} + 2xe^{x+y}\frac{\partial^2 z}{\partial u \partial v} + e^{2x}\frac{\partial^2 z}{\partial v^2} + xe^y \frac{\partial z}{\partial u}$$

Substituting these values in the given equation we have

$$x(xy-1)\left[e^{2y}\frac{\partial^2 z}{\partial u^2}+2ye^{x+y}\frac{\partial^2 z}{\partial u\partial v}+y^2e^{2x}\frac{\partial^2 z}{\partial v^2}+ye^x\frac{\partial z}{\partial v}\right]$$

$$-(x^2y^2-1)\left(xe^{2y}\frac{\partial^2 z}{\partial u^2}+(xy+1)e^{x+y}\frac{\partial^2 z}{\partial u\partial v}+ye^{2x}\frac{\partial^2 z}{\partial v^2}+e^y\frac{\partial z}{\partial u}+e^x\frac{\partial z}{\partial v}\right)$$

$$+\ y(xy-1)\left(x^2y^{2y}\frac{\partial^2 z}{\partial u^2}+2xe^{x+y}\frac{\partial^2 z}{\partial u\partial v}+e^{2x}\frac{\partial^2 z}{\partial v^2}+xe^y\frac{\partial z}{\partial u}\right)+(x-1)$$

$$\left(e^y\frac{\partial z}{\partial u}+ye^x\frac{\partial z}{\partial v}\right)+(y-1)\left(xe^y\frac{\partial z}{\partial u}+e^x\frac{\partial z}{\partial v}\right)=0$$

$$\Rightarrow \frac{\partial^2 z}{\partial u\partial v}=0$$

which is the required canonical form of the given equation.

Integrating, we obtain

$$z = \int f(v)dv + \psi(u)$$

$$\Rightarrow \quad z = \phi(v) + \psi(u)$$

$$\Rightarrow \quad z = \phi(ye^x) + \psi(xe^y)$$

where ϕ and ψ are arbitrary functions.

EXAMPLE 8.17

Reduce the equation

$$\frac{\partial^2 z}{\partial x^2}+2\frac{\partial^2 z}{\partial x\partial y}+\frac{\partial^2 z}{\partial y^2}=0$$

to canonical form and solve it.

Solution: The given equation can be written as

$$r + 2s + t = 0$$

On comparing with $Rr + Ss + Tt + f(x, y, z, p, q) = 0$, we have

$$R = 1,\ S = 2,\ T = 1$$

$$\therefore \quad R^2 - 4RT = 0$$

Here, the quadratic equation $R\lambda^2 + S\lambda + T = 0$ becomes

$$\lambda^2 + 2\lambda + 1 = 0 \quad \Rightarrow \quad (\lambda + 1)^2 = 0$$

$$\Rightarrow \qquad \lambda = -1, -1 \text{ (Equal roots).}$$

Let, $\lambda_1 = \lambda_2 = -1$.

$$\therefore \qquad \frac{dy}{dx} - 1 = 0$$

Integrating, we have

$$x - y = c_1$$

Thus, $\qquad u = x - y$

Also, we have to take v as some function of x and y independent of u.

Let, $\qquad v = x + y$

Now, $\qquad p = \dfrac{\partial z}{\partial x} = \dfrac{\partial z}{\partial u}\dfrac{\partial u}{\partial x} + \dfrac{\partial z}{\partial v}\dfrac{\partial v}{\partial x} = \dfrac{\partial z}{\partial u} + \dfrac{\partial z}{\partial v}$

$$q = \frac{\partial z}{\partial y} = \frac{\partial z}{\partial u}\frac{\partial u}{\partial y} + \frac{\partial z}{\partial v}\frac{\partial v}{\partial y} = -\frac{\partial z}{\partial u} + \frac{\partial z}{\partial v}$$

$$r = \frac{\partial}{\partial x}\left(\frac{\partial z}{\partial x}\right) = \frac{\partial}{\partial u}\left(\frac{\partial z}{\partial u} + \frac{\partial z}{\partial v}\right)\frac{\partial u}{\partial x} + \frac{\partial}{\partial v}\left(\frac{\partial z}{\partial u} + \frac{\partial z}{\partial v}\right)\frac{\partial v}{\partial x}$$

$$= \frac{\partial^2 z}{\partial u^2} + 2\frac{\partial^2 z}{\partial u \partial v} + \frac{\partial^2 z}{\partial v^2}$$

$$s = \frac{\partial}{\partial x}\left(\frac{\partial z}{\partial y}\right) = \frac{\partial}{\partial u}\left(\frac{-\partial z}{\partial u} + \frac{\partial z}{\partial v}\right)\frac{\partial u}{\partial x} + \frac{\partial}{\partial v}\left(\frac{-\partial z}{\partial u} + \frac{\partial z}{\partial v}\right)\frac{\partial v}{\partial x} = -\frac{\partial^2 z}{\partial u^2} + \frac{\partial^2 z}{\partial v^2}$$

and $\qquad t = \dfrac{\partial}{\partial y}\left(\dfrac{\partial z}{\partial y}\right) = \dfrac{\partial}{\partial u}\left(\dfrac{-\partial z}{\partial u} + \dfrac{\partial z}{\partial v}\right)\dfrac{\partial u}{\partial y} + \dfrac{\partial}{\partial v}\left(\dfrac{-\partial z}{\partial u} + \dfrac{\partial z}{\partial v}\right)\dfrac{\partial v}{\partial y}$

$$= \frac{\partial^2 z}{\partial u^2} - 2\frac{\partial^2 z}{\partial u \partial v} + \frac{\partial^2 z}{\partial v^2}.$$

Substituting these values in the given equation; we have $\dfrac{\partial^2 z}{\partial v^2} = 0$

which is the required canonical form of the given equation. Integrating, we obtain

$$z = v\phi(u) + \psi(u)$$

$$\Rightarrow \qquad z = (x + y)\,\phi(x - y) + \psi(x - y)$$

which is the required solution.

EXAMPLE 8.18

Reduce the equation

$$\frac{\partial^2 z}{\partial x^2}+2\frac{\partial^2 z}{\partial x\partial y}+5\frac{\partial^2 z}{\partial y^2}+\frac{\partial z}{\partial x}-2\frac{\partial z}{\partial y}-3z=0$$

to the canonical form.

Solution: The given equation can be written as

$$r + 2s + 5t + p - 24 - 3z = 0$$

On comparing with $Rr + Ss + Tt + f(x, y, z, p, q) = 0$, we have

$$R = 1, S = 2, T = 5$$

$\therefore$ $$S^2 - 4RT = -1 < 0$$

Here, the quadratic equation $R\lambda^2 + S\lambda + T = 0$ becomes

$$\lambda^2 + 2\lambda + 5 = 0 \Rightarrow \lambda = -1 \pm 2i$$

Let $\lambda_1 = -1 + 2i$ and $\lambda_2 = -1 - 2i$ (Complex roots)

$\therefore$ $$\frac{dy}{dx} - 1 + 2i = 0 \quad \text{and} \quad \frac{dy}{dx} - 1 - 2i = 0$$

Integrating, we have

$$y - x + 2xi = c_1 \text{ and } y - x - 2xi = c_2.$$

Thus, $u = y - x + 2xi$

and $v = y - x - 2xi = \alpha + i\beta = \alpha - i\beta$

So that $\alpha = y - x$ and $\beta = 2x$

Now, $$p = \frac{\partial z}{\partial x} = \frac{\partial z}{\partial \alpha}\frac{\partial \alpha}{\partial x} + \frac{\partial z}{\partial \beta}\frac{\partial \beta}{\partial x} = -\frac{\partial z}{\partial \alpha} + 2\frac{\partial z}{\partial \beta}$$

$$q = \frac{\partial z}{\partial y} = \frac{\partial z}{\partial \alpha}\frac{\partial \alpha}{\partial y} + \frac{\partial z}{\partial \beta}\frac{\partial \beta}{\partial y} = \frac{\partial z}{\partial \alpha}$$

$$r = \frac{\partial^2 z}{\partial x^2} = \frac{\partial}{\partial x}\left(\frac{\partial z}{\partial x}\right) = \frac{\partial}{\partial x}\left(-\frac{\partial z}{\partial \alpha} + 2\frac{\partial z}{\partial \beta}\right)$$

$$= \frac{\partial}{\partial \alpha}\left(-\frac{\partial z}{\partial \alpha} + 2\frac{\partial z}{\partial \beta}\right)\frac{\partial \alpha}{\partial x} + \frac{\partial}{\partial \beta}\left(\frac{\partial z}{\partial \alpha} + 2\frac{\partial z}{\partial \beta}\right)\frac{\partial \beta}{\partial x}$$

$$= \frac{\partial^2 z}{\partial \alpha^2} - 4\frac{\partial^2 z}{\partial \alpha\partial \beta} + 4\frac{\partial^2 z}{\partial \beta^2}$$

$$s = \frac{\partial^2 z}{\partial x \partial y} = \frac{\partial}{\partial x}\left(\frac{\partial z}{\partial y}\right) = \frac{\partial}{\partial x}\left(\frac{\partial z}{\partial \alpha}\right)$$

$$= \frac{\partial}{\partial \alpha}\left(\frac{\partial z}{\partial \alpha}\right)\frac{\partial \alpha}{\partial x} + \frac{\partial}{\partial \beta}\left(\frac{\partial z}{\partial \alpha}\right)\frac{\partial \beta}{\partial x}$$

$$= -\frac{\partial^2 z}{\partial \alpha^2} + 2\frac{\partial^2 z}{\partial \beta \partial \alpha}$$

and $$t = \frac{\partial^2 z}{\partial y^2} = \frac{\partial}{\partial y}\left(\frac{\partial z}{\partial y}\right) = \frac{\partial}{\partial y}\left(\frac{\partial z}{\partial \alpha}\right) = \frac{\partial}{\partial \alpha}\left(\frac{\partial z}{\partial \alpha}\right)\frac{\partial \alpha}{\partial y} + \frac{\partial}{\partial \beta}\left(\frac{\partial z}{\partial \alpha}\right)\frac{\partial \beta}{\partial y} = \frac{\partial^2 z}{\partial \alpha^2}.$$

Substituting these values in the given equation; we have

$$\left(\frac{\partial^2 z}{\partial \alpha^2} - 4\frac{\partial^2 z}{\partial \alpha \partial \beta} + 4\frac{\partial^2 z}{\partial \beta^2}\right) + 2\left(-\frac{\partial^2 z}{\partial \alpha^2} + 2\frac{\partial^2 z}{\partial \alpha \partial \beta}\right) + 5\frac{\partial^2 z}{\partial \alpha^2} + \left(-\frac{\partial z}{\partial \alpha} + 2\frac{\partial z}{\partial \beta}\right) - 2\frac{\partial z}{\partial \alpha} - 3z = 0$$

$$\Rightarrow \quad 4\frac{\partial^2 z}{\partial \alpha^2} + 4\frac{\partial^2 z}{\partial \beta^2} - 3\frac{\partial z}{\partial \alpha} + 2\frac{\partial z}{\partial \beta} - 3z = 0$$

$$\Rightarrow \quad \frac{\partial^2 z}{\partial \alpha^2} + \frac{\partial^2 z}{\partial \beta^2} = \frac{3}{4}\frac{\partial z}{\partial \alpha} - \frac{1}{2}\frac{\partial z}{\partial \beta} + \frac{3}{4}z$$

which is the required canonical form.

EXAMPLE 8.19

Reduce the equation

$$(y-1)r - (y^2-1)s + y(y-1)t + p - q = 2ye^{2x}(1-y)^3$$

to the canonical form and solve it.

Solution: On comparing the given equation with $Rr + Ss + Tt + f(x, y, z, p, q) = 0$, we have

$$R = y - 1,\ S = -(y^2 - 1);\ T = y(y-1)$$

$$\therefore \quad S^2 - 4RT = (y^2-1)^2 - 4y(y-1)^2$$

$$= y^4 - 2y^2 + 1 - 4y^3 - 4y + 8y^2$$

$$= y^4 - 4y^3 + 6y^2 - 4y + 1$$

Here, the quadratic equation $R\lambda^2 + S\lambda + T = 0$ becomes $(y-1)\lambda^2 - (y^2-1)\lambda + y(y-1) = 0$

$$\Rightarrow \quad \lambda^2 - (y+1)\lambda + y = 0$$

$$\Rightarrow \quad (\lambda - 1)(\lambda - y) = 0 \quad \Rightarrow \quad \lambda = 1, y$$

Let, $\lambda_1 = 1$ and $\lambda_2 = y$ (Real and distinct roots)

$$\therefore \qquad \frac{dy}{dx} + 1 = 0 \quad \text{and} \quad \frac{dy}{dx} + y = 0$$

Integrating, we have

$$x + y = c_1 \text{ and } ye^x = c_2$$

Thus, $u = x + y$ and $v = ye^x$.

Now,
$$p = \frac{\partial z}{\partial x} = \frac{\partial z}{\partial u}\frac{\partial u}{\partial x} + \frac{\partial z}{\partial v}\frac{\partial v}{\partial x} = \frac{\partial z}{\partial u} + ye^x\frac{\partial z}{\partial v} = \frac{\partial z}{\partial u} + v\frac{\partial z}{\partial v}$$

$$q = \frac{\partial z}{\partial y} = \frac{\partial z}{\partial u}\frac{\partial u}{\partial y} + \frac{\partial z}{\partial v}\frac{\partial v}{\partial y} = \frac{\partial z}{\partial u} + e^x\frac{\partial z}{\partial v}$$

$$r = \frac{\partial}{\partial x}\left(\frac{\partial z}{\partial x}\right) = \frac{\partial}{\partial u}\left(\frac{\partial z}{\partial u} + v\frac{\partial z}{\partial v}\right)\frac{\partial u}{\partial x} + \frac{\partial}{\partial v}\left(\frac{\partial z}{\partial u} + v\frac{\partial z}{\partial v}\right)\frac{\partial v}{\partial x}$$

$$= \frac{\partial^2 z}{\partial u^2} + 2v\frac{\partial^2 z}{\partial u \partial v} + v^2\frac{\partial^2 z}{\partial v^2} + v\frac{\partial z}{\partial v}$$

$$s = \frac{\partial}{\partial x}\left(\frac{\partial z}{\partial y}\right) = \frac{\partial}{\partial x}\left(\frac{\partial z}{\partial u} + e^x\frac{\partial z}{\partial v}\right)$$

$$= \frac{\partial}{\partial u}\left(\frac{\partial z}{\partial u}\right)\frac{\partial u}{\partial x} + \frac{\partial}{\partial v}\left(\frac{\partial z}{\partial u}\right)\frac{\partial v}{\partial x} + e^x\frac{\partial z}{\partial v}$$

$$= e^x\left[\frac{\partial}{\partial u}\left(\frac{\partial z}{\partial v}\right)\frac{\partial u}{\partial x} + \frac{\partial}{\partial v}\left(\frac{\partial z}{\partial v}\right)\frac{\partial v}{\partial x}\right]$$

$$= \frac{\partial^2 z}{\partial u^2} + (v + e^x)\frac{\partial^2 z}{\partial u \partial v} + ve^x\frac{\partial^2 z}{\partial v^2} + e^x\frac{\partial z}{\partial v}$$

and
$$t = \frac{\partial}{\partial y}\left(\frac{\partial z}{\partial y}\right) = \frac{\partial}{\partial y}\left(\frac{\partial z}{\partial u} + e^x\frac{\partial z}{\partial v}\right)$$

$$= \frac{\partial}{\partial u}\left(\frac{\partial z}{\partial u} + e^x\frac{\partial z}{\partial v}\right)\frac{\partial u}{\partial y} + \frac{\partial}{\partial v}\left(\frac{\partial z}{\partial u} + e^x\frac{\partial z}{\partial v}\right)\frac{\partial v}{\partial y}$$

$$= \frac{\partial^2 z}{\partial u^2} + 2e^x\frac{\partial^2 z}{\partial u \partial v} + e^{2x}\frac{\partial^2 z}{\partial v^2}$$

Substituting these values in the equation, we have

$$(y-1)\left(\frac{\partial^2 z}{\partial u^2}+2v\frac{\partial^2 z}{\partial u\partial v}+v^2\frac{\partial^2 z}{\partial v^2}+v\frac{\partial z}{\partial v}\right)-(y^2-1)\left(\frac{\partial^2 z}{\partial u^2}+(v+e^x)\frac{\partial^2 z}{\partial u\partial v}+ve^x\frac{\partial^2 z}{\partial v^2}+e^x\frac{\partial z}{\partial v}\right)$$

$$+y(y-1)\left(\frac{\partial^2 z}{\partial u^2}+2e^x\frac{\partial^2 z}{\partial u\partial v}+e^{2x}\frac{\partial^2 z}{\partial v^2}\right)+(v-e^x)$$

$$\frac{\partial z}{\partial v}=2ye^{2x}(1-y)^3$$

$$\Rightarrow \qquad (1-y)^3 e^x\frac{\partial^2 z}{\partial u\partial v}=2ye^{2x}(1-y)^3$$

$$\Rightarrow \qquad \frac{\partial^2 z}{\partial u\partial v}=2ye^x \Rightarrow \frac{\partial^2 z}{\partial u\partial v}=2v$$

which is the required canonical form of the given equation. Integrating, we obtain

$$z = uv^2 + \int f(u)du + \psi(v)$$

$$\Rightarrow \qquad z = uv^2 + \phi(u) + \psi(v)$$

$$\Rightarrow \qquad z = (x + y)y^2e^{2x} + \phi(x + y) + \psi(ye^x)$$

which is the required solution of the given equation.

EXAMPLE 8.20

Reduce $\frac{\partial^2 z}{\partial x^2}=(1+y)^2\frac{\partial^2 z}{\partial y^2}$ to the canonical form.

Solution: The given equation can be written as

$$r - (1 + y)^2 t = 0$$

On comparing with $Rr + Ss + Tt + f(x, y, z, p, q) = 0$, we have

$$R = 1,\ S = 0,\ T = -(1 + y)^2$$

$$\therefore \qquad S^2 - 4RT = 4(1 + y)^2 > 0$$

Here, the quadratic equation $R\lambda^2 + S\lambda + T = 0$ becomes

$$\lambda^2 - (1 + y)^2 = 0$$

$$\Rightarrow \qquad \lambda = 1 + y,\ -(1 + y)$$

$\therefore$ $$\frac{dy}{dx} + (1 + y) = 0 \quad \text{and} \quad \frac{dy}{dx} - (1 + y) = 0$$

Integrating, we have

$$\log (1 + y) + x = c_1 \quad \text{and} \quad \log (1 + y) - x = c_2.$$

Thus, $u = \log (1 + y) + x$ and $v = \log (1 + y) - x$.

Now, $$p = \frac{\partial z}{\partial x} = \frac{\partial z}{\partial u}\frac{\partial u}{\upsilon x} + \frac{\partial z}{\partial v}\frac{\partial v}{\partial x} = \frac{\partial z}{\partial u} - \frac{\partial z}{\partial v}$$

$$q = \frac{\partial z}{\partial y} = \frac{\partial z}{\partial u}\frac{\partial u}{\upsilon y} + \frac{\partial z}{\partial v}\frac{\partial v}{\partial y} = \frac{1}{1+y}\left(\frac{\partial z}{\partial u} + \frac{\partial z}{\partial v}\right)$$

$$r = \frac{\partial^2 z}{\partial x^2} = \frac{\partial}{\partial x}\left(\frac{\partial z}{\partial x}\right) = \left(\frac{v}{\partial u} - \frac{\partial}{\partial v}\right)\left(\frac{\partial z}{\partial u} - \frac{\partial z}{\partial v}\right) = \frac{\partial^2 z}{\partial u^2} - 2\frac{\partial^2 z}{\partial u \partial v} + \frac{\partial^2 z}{\partial v^2}$$

and $$t = \frac{\partial^2 z}{\partial y^2} = \frac{\partial}{\partial y}\left(\frac{\partial z}{\partial y}\right) = \frac{\partial}{\partial y}\left[\frac{1}{1+y}\left(\frac{\partial z}{\partial u} + \frac{\partial z}{\partial v}\right)\right]$$

$$= -\frac{1}{(1+y)^2}\left(\frac{\partial z}{\partial u} + \frac{\partial z}{\partial v}\right) + \frac{1}{1+y}\left\{\frac{\partial}{\partial u}\left(\frac{\partial z}{\partial u} + \frac{\partial z}{\partial v}\right)\frac{\partial u}{\partial y} + \frac{\partial}{\partial v}\left(\frac{\partial z}{\partial u} + \frac{\partial z}{\partial v}\right)\frac{\partial v}{\partial y}\right\}$$

$$= -\frac{1}{(1+y)^2}\left(\frac{\partial z}{\partial u} + \frac{\partial z}{\partial v}\right) + \frac{1}{(1+y)^2}\left(\frac{\partial^2 z}{\partial u^2} + 2\frac{\partial^2 z}{\partial u \partial v} + \frac{\partial^2 z}{\partial v^2}\right)$$

Substituting these values in the given equation, we have

$$\frac{\partial^2 z}{\partial u^2} - 2\frac{\partial^2 z}{\partial u \partial v} + \frac{\partial^2 z}{\partial v^2} + \left(\frac{\partial z}{\partial u} + \frac{\partial z}{\partial v}\right) - \left(\frac{\partial^2 z}{\partial u^2} + \frac{\partial^2 z}{\partial u \partial v} + \frac{\partial^2 z}{\partial v^2}\right) = 0$$

$\Rightarrow$ $$4\frac{\partial^2 z}{\partial u \partial v} = \frac{\partial z}{\partial u} + \frac{\partial z}{\partial v}$$

which is the required canonical form.

EXAMPLE 8.21

Reduce the equation

$$r - (2 \sin x)s - (\cos^2 x)t - (\cos x)q = 0$$

to the canonical form and solve it.

Solution: On comparing the given equation with $Rr + Ss + Tt + f(x, y, z, p, q) = 0$, we have

$$R = 1,\ S = -2 \sin x,\ T = -\cos^2 x$$

$\therefore \qquad S^2 - 4RT = 4\sin^2 x + 4\cos^2 x = 4 > 0.$

Here, the quadratic equation $R\lambda^2 + S\lambda + T = 0$ becomes

$$\lambda^2 - 2\sin x\lambda - \cos^2 x = 0$$

$\Rightarrow \qquad \lambda = \sin x + 1, \sin x - 1.$

$\therefore \qquad \frac{dy}{dx} + \sin x + 1 = 0$ and $\frac{dy}{dx} + \sin x - 1 = 0.$

Integrating, we have

$$y - \cos x + x = c_1 \text{ and } y - \cos x - x = c_2.$$

Thus, $u = y - \cos x + x$ and $v = y - \cos x - x$.

Now, $p = \frac{\partial z}{\partial x} = \frac{\partial z}{\partial u}\frac{\partial u}{\partial x} + \frac{\partial z}{\partial v}\frac{\partial v}{\partial x}(\sin x + 1)\frac{\partial z}{\partial u} + (\sin x - 1)\frac{\partial z}{\partial v}$

$$q = \frac{\partial z}{\partial y} = \frac{\partial z}{\partial u}\frac{\partial u}{\partial y} + \frac{\partial z}{\partial v}\frac{\partial v}{\partial y} = \frac{\partial z}{\partial u} + \frac{\partial z}{\partial v}$$

$$r = \frac{\partial^2 z}{\partial x^2} = \frac{\partial}{\partial x}\left(\frac{\partial z}{\partial x}\right) = \frac{\partial}{\partial x}\left[(1+\sin x)\frac{\partial z}{\partial u} + (\sin x - 1)\frac{\partial z}{\partial v}\right]$$

$$= \cos x\frac{\partial z}{\partial u} + (1+\sin x)\left[\frac{\partial}{\partial u}\left(\frac{\partial z}{\partial u}\right)\frac{\partial u}{\partial x} + \frac{\partial}{\partial v}\left(\frac{\partial z}{\partial u}\right)\frac{\partial}{\partial x}\right] + \cos x\frac{\partial z}{\partial v} + (\sin x - 1)$$

$$\left[\frac{\partial}{\partial u}\left(\frac{\partial z}{\partial v}\right)\frac{\partial u}{\partial x} + \frac{\partial}{\partial v}\left(\frac{\partial z}{\partial v}\right)\frac{\partial v}{\partial x}\right]$$

$$= \cos x\left(\frac{\partial z}{\partial u} + \frac{\partial z}{\partial v}\right) + (1+\sin x)^2\frac{\partial^2 z}{\partial u^2} - 2\cos^2 x\frac{\partial^2 z}{\partial u\partial v} + (\sin x - 1)^2\frac{\partial^2 z}{\partial v^2}$$

$$s = \frac{\partial^2 z}{\partial x\partial y} = \frac{\partial}{\partial x}\left(\frac{\partial z}{\partial y}\right) = \frac{\partial}{\partial x}\left(\frac{\partial z}{\partial u} + \frac{\partial z}{\partial v}\right)$$

$$= \frac{\partial}{\partial u}\left(\frac{\partial z}{\partial u} + \frac{\partial z}{\partial v}\right)\frac{\partial u}{\partial x} + \frac{\partial}{\partial v}\left(\frac{\partial z}{\partial u} + \frac{\partial z}{\partial v}\right)\frac{\partial v}{\partial x}$$

$$= (\sin x + 1)\left(\frac{\partial^2 z}{\partial u^2} + \frac{\partial^2 z}{\partial u\partial v}\right) + (\sin x - 1)\left(\frac{\partial^2 z}{\partial v\partial u} + \frac{\partial^2 z}{\partial v^2}\right)$$

$$= \sin x\left(\frac{\partial^2 z}{\partial u^2} + 2\frac{\partial^2 z}{vu\partial v} + \frac{\partial^2 z}{\partial v^2}\right) + \frac{\partial^2 z}{\partial u^2} - \frac{\partial^2 z}{\partial v^2}$$

and $$t = \frac{\partial^2 z}{\partial y^2} = \frac{\partial}{\partial y}\left(\frac{\partial z}{\partial y}\right) = \left(\frac{\partial}{\partial u} + \frac{\partial}{\partial v}\right)\left(\frac{\partial z}{\partial u} + \frac{\partial z}{\partial v}\right) = \frac{\partial^2 z}{\partial u^2} + 2\frac{\partial^2 z}{\partial u \partial v} + \frac{\partial^2 z}{\partial v^2}$$

Substituting these values in the given equation, we have

$$\cos x\left(\frac{\partial z}{\partial u} + \frac{\partial z}{\partial v}\right) + (1 + 2\sin x + \sin^2 x)\frac{\partial^2 z}{\partial u^2} + (\sin^2 x + 1 - 2\sin x)\frac{\partial^2 z}{\partial v^2} - 2\cos^2 x\frac{\partial^2 z}{\partial u \partial v} - 2\sin x$$

$$\left\{\sin x\left(\frac{\partial^2 z}{\partial u^2} + 2\frac{\partial^2 z}{\partial u \partial v} + \frac{\partial^2 z}{\partial v^2}\right) + \frac{\partial^2 z}{\partial u^2} - \frac{\partial^2 z}{\partial v^2}\right\} - \cos^2 x\left(\frac{\partial^2 z}{\partial u^2} + 2\frac{\partial^2 z}{\partial u \partial v} + \frac{\partial^2 z}{\partial v^2}\right) - \cos x$$

$$\left(\frac{\partial z}{\partial u} + \frac{\partial z}{\partial v}\right) = 0$$

$$\Rightarrow \qquad \frac{\partial^2 z}{\partial u \partial v} = 0$$

which is the required canonical form.

Integrating, we obtain

$$z = \int f(v)dv + \psi(u)$$

$$\Rightarrow \qquad z = \phi(v) + \psi(u)$$

$$\therefore \qquad z = \phi(y - \cos x - x) + \psi(y - \cos x + 1)$$

which is the required solution of the given equation.

EXAMPLE 8.22

Reduce the equation

$$yr + (x + y)s + xt = 0$$

to the canonical form and find its solution.

Solution: On comparing the given equation with $Rr + Ss + Tt + f(x, y, z, p, q) = 0$, we have

$$R = y,\ S = x + y,\ T = x$$

$$\therefore \qquad S^2 - 4RT = (x + y)^2 - 4xy = (x - y)^2 > 0.$$

Here, the quadratic equation $R\lambda^2 + S\lambda + T = 0$ becomes

$$y\lambda^2 + (x + y)\lambda + x = 0$$

$$\Rightarrow \qquad (y\lambda + x)(\lambda + 1) = 0 \quad \Rightarrow \quad \lambda = -1, -x/y.$$

$$\therefore \qquad \frac{dy}{dx} - 1 = 0 \quad \text{and} \quad \frac{dy}{dx} - \frac{x}{y} = 0$$

Integrating, we have $y - x = c_1$ and $\frac{y^2}{2} - \frac{x^2}{2} = c_2$.

Thus, $u = y - x$ and $v = \frac{y^2}{2} - \frac{x^2}{2}$.

Now,
$$p = \frac{\partial z}{\partial x} = \frac{\partial z}{\partial u}\frac{\partial u}{\partial x} + \frac{\partial z}{\partial v}\frac{\partial v}{\partial x} = -\left(\frac{\partial z}{\partial u} + \frac{\partial z}{\partial v}\right)$$

$$q = \frac{\partial z}{\partial y} = \frac{\partial z}{\partial u}\frac{\partial u}{\partial y} + \frac{\partial z}{\partial v}\frac{\partial v}{\partial y} = \frac{\partial z}{\partial u} + y\frac{\partial z}{\partial v}$$

$$r = \frac{\partial^2 z}{\partial x^2} = \frac{\partial}{\partial x}\left(\frac{\partial z}{\partial x}\right) = -\frac{\partial}{\partial x}\left(\frac{\partial z}{\partial u} + x\frac{\partial z}{\partial v}\right)$$

$$= -\left[\frac{\partial}{\partial u}\left(\frac{\partial z}{\partial u}\right)\frac{\partial u}{\partial x} + \frac{\partial}{\partial v}\left(\frac{\partial z}{\partial u}\right)\frac{\partial v}{\partial x}\right] - \frac{\partial z}{\partial v}$$

$$= -x\left[\frac{\partial}{\partial u}\left(\frac{\partial z}{\partial v}\right)\frac{\partial u}{\partial x} + \frac{\partial}{\partial v}\left(\frac{\partial z}{\partial v}\right)\frac{\partial v}{\partial x}\right]$$

$$= -\left(-\frac{\partial^2 z}{\partial u^2} - x\frac{\partial^2 z}{\partial u \partial v}\right) - \frac{\partial z}{\partial v} - x\left(-\frac{\partial^2 z}{\partial u \partial v} - x\frac{\partial^2 z}{\partial v^2}\right)$$

$$= \frac{\partial^2 z}{\partial u^2} + 2x\frac{\partial^2 z}{\partial u \partial v} + x^2\frac{\partial^2 z}{\partial v^2} - \frac{\partial z}{\partial v}$$

$$s = \frac{\partial^2 z}{\partial x \partial y} = \frac{\partial}{\partial x}\left(\frac{\partial z}{\partial y}\right) = \frac{\partial}{\partial x}\left(\frac{\partial z}{\partial u} + y\frac{\partial z}{\partial v}\right)$$

$$= \left[\frac{\partial}{\partial u}\left(\frac{\partial z}{\partial u}\right)\frac{\partial u}{\partial x} + \frac{\partial}{\partial v}\left(\frac{\partial z}{\partial u}\right)\frac{\partial v}{\partial x}\right] + y\left[\frac{\partial}{\partial u}\left(\frac{\partial z}{\partial v}\right)\frac{\partial u}{\partial x} + \frac{\partial}{\partial v}\left(\frac{\partial z}{\partial v}\right)\frac{\partial v}{\partial x}\right]$$

$$= -\frac{\partial^2 z}{\partial u^2} - (x + y)\frac{\partial^2 z}{\partial u \partial v} - xy\frac{\partial^2 z}{\partial v^2}$$

and
$$t = \frac{\partial^2 z}{\partial y^2} = \frac{\partial}{\partial y}\left(\frac{\partial z}{\partial y}\right) = \frac{\partial}{\partial y}\left(\frac{\partial z}{\partial u} + y\frac{\partial z}{\partial v}\right)$$

$$= \left[\frac{\partial}{\partial u}\left(\frac{\partial z}{\partial u}\right)\frac{\partial u}{\partial y} + \frac{\partial}{\partial v}\left(\frac{\partial z}{\partial u}\right)\frac{\partial v}{\partial y}\right] + 1 - \frac{\partial z}{\partial v} + y\left[\frac{\partial}{\partial u}\left(\frac{\partial z}{\partial v}\right)\frac{\partial u}{\partial y} + \frac{\partial}{\partial v}\left(\frac{\partial z}{\partial v}\right)\frac{\partial v}{\partial y}\right]$$

$$= \frac{\partial^2 z}{\partial u^2} + 2y\frac{\partial^2 z}{\partial u \partial v} + y^2\frac{\partial^2 z}{\partial v^2} + \frac{\partial z}{\partial v}$$

Substituting these values in the given equation, we have

$$y\left(\frac{\partial^2 z}{\partial u^2}+2x\frac{\partial^2 z}{\partial u\partial v}+x^2\frac{\partial^2 z}{\partial v^2}-\frac{\partial z}{\partial v}\right)+(x+y)\left(-\frac{\partial^2 z}{\partial u^2}-(x+y)\frac{\partial^2 z}{\partial u\partial v}-xy\frac{\partial^2 z}{\partial v^2}\right)$$

$$+x\left(\frac{\partial^2 z}{\partial u^2}+2y\frac{\partial^2 z}{\partial u\partial v}+y^2\frac{\partial^2 z}{\partial v^2}+\frac{\partial z}{\partial v}\right)=0$$

$$\Rightarrow \quad [4xy-(x+y)^2]\frac{\partial^2 z}{\partial u\partial v}-y\frac{\partial z}{\partial v}+x\frac{\partial z}{\partial v}=0$$

$$\Rightarrow \quad (y-x)^2\frac{\partial^2 z}{\partial u\partial v}+(y-x)\frac{\partial z}{\partial v}=0$$

$$\Rightarrow \quad u\frac{\partial^2 z}{\partial u\partial v}+u\frac{\partial z}{\partial v}=0$$

$$\Rightarrow \quad u\frac{\partial^2 z}{\partial u\partial v}+\frac{\partial z}{\partial v}=0 \text{ as } u\neq 0$$

which is the required canonical form.

Multiplying this by v, we have

$$(uvD_uD_v+vD_v)z=0$$

Let, $\quad u=e^X$ and $v=e^Y$

so that $\quad X=\log u$ and $Y=\log v$

and denoting $\frac{\partial}{\partial X},\frac{\partial}{\partial Y}$ by D_X, D_Y respectively, we have

$$(D_XD_Y+D_Y)z=0$$

$$\Rightarrow \quad D_Y(D_X+1)z=0$$

$\therefore$ Its complete solution is

$$\begin{aligned} z &= \phi_1(X)+e^{-X}\phi_2(Y) \\ &= \phi_1(\log u)+u^{-1}\phi_2(\log v) \\ &= \psi_1(u)+u^{-1}\psi_2(v) \\ &= \psi_1(y-x)+(y-x)^{-1}\,\psi_2(y^2-x^2) \end{aligned}$$

where, ψ_1 and ψ_2 are arbitrary functions.

EXAMPLE 8.23

Reduce $r-2s+t+p-q=e^x(2y-3)-e^y$, to the canonical form and hence solve it.

Solution: On comparing the given equation with $Rr+Ss+Tt+f(x,y,z,p,q)=0$, we have

$$R = 1,\ S = -2,\ T = 1 \quad \therefore\ S^2 - 4RT = 0$$

Here, the quadratic equation $R\lambda^2 + S\lambda + T = 0$ becomes

$$\lambda^2 - 2\lambda + 1 = 0$$

$$\Rightarrow \quad (\lambda - 1)^2 = 0$$

$$\Rightarrow \quad \lambda = 1, 1 \text{ (Equal roots)}$$

$$\therefore \quad \frac{dy}{dx} + 1 = 0 \Rightarrow dx + dy = 0$$

Integrating, we have $x + y = c_1$

Thus, $u = x + y$ and let $v = y$ so that u and v are independent functions as

$$w(u, v) = \begin{vmatrix} \dfrac{\partial u}{\partial x} & \dfrac{\partial u}{\partial y} \\ \dfrac{\partial v}{\partial x} & \dfrac{\partial v}{\partial y} \end{vmatrix} = \begin{vmatrix} 1 & 1 \\ 0 & 1 \end{vmatrix} = 1 \neq 0$$

Now,
$$p = \frac{\partial z}{\partial x} = \frac{\partial z}{\partial u}\frac{\partial u}{\partial x} + \frac{\partial z}{\partial v}\frac{\partial v}{\partial x} = \frac{\partial z}{\partial u}$$

$$q = \frac{\partial z}{\partial y} = \frac{\partial z}{\partial u}\frac{\partial u}{\partial y} + \frac{\partial z}{\partial v}\frac{\partial v}{\partial y} = \frac{\partial z}{\partial u} + \frac{\partial z}{\partial v}$$

$$r = \frac{\partial^2 z}{\partial x^2} = \frac{\partial}{\partial x}\left(\frac{\partial z}{\partial x}\right) = \frac{\partial}{\partial u}\left(\frac{\partial z}{\partial u}\right) = \frac{\partial^2 z}{\partial u^2}$$

$$s = \frac{\partial^2 z}{\partial x \partial y} = \frac{\partial}{\partial x}\left(\frac{\partial z}{\partial y}\right) = \frac{\partial}{\partial u}\left(\frac{\partial z}{\partial u} + \frac{\partial z}{\partial v}\right) = \frac{\partial^2 z}{\partial u^2} + \frac{\partial^2 z}{\partial u \partial v}$$

and
$$t = \frac{\partial^2 z}{\partial y^2} = \frac{\partial}{\partial y}\left(\frac{\partial z}{\partial y}\right) = \left(\frac{\partial}{\partial u} + \frac{\partial}{\partial v}\right)\left(\frac{\partial z}{\partial u} + \frac{\partial z}{\partial v}\right) = \frac{\partial^2 z}{\partial u^2} + 2\frac{\partial^2 z}{\partial u \partial v} + \frac{\partial^2 z}{\partial v^2}$$

Substituting these values in the given equation, we have

$$\frac{\partial^2 z}{\partial u^2} - 2\left(\frac{\partial^2 z}{\partial u^2} + \frac{\partial^2 z}{\partial u \partial v}\right) + \frac{\partial^2 z}{\partial u^2} + 2\frac{\partial^2 z}{\partial u \partial v} + \frac{\partial^2 z}{\partial v^2} + \frac{\partial z}{\partial u} - \left(\frac{\partial z}{\partial u} + \frac{\partial z}{\partial v}\right) = e^{u-v}(2v - 3) - e^v$$

$$\Rightarrow \quad \frac{\partial^2 z}{\partial v^2} - \frac{\partial z}{\partial v} = e^{u-v}(2v - 3) - e^v$$

which is the required canonical form and can be written as

$$(D_v^2 - D_v)z = e^{u-v}(2v - 3) - e^v$$

∴ CF is

$$z_c = \phi(u) + e^{v}\psi(u)$$
$$= \phi(x + y) + e^{y}\psi(x + y)$$

Now, PI is $$z_p = \frac{1}{D_v(D_v-1)}[e^{u-v}(2v-3) - e^{v}]$$

$$\frac{1}{D_v(D_v-1)}e^{u-v}(2v-3) = e^{u-v}\frac{1}{(D_v-1)(D_v-1-1)}(2v-3)$$

$$= e^{u-v}\frac{1}{(D_v-1)(D_v-2)}(2v-3)$$

$$= \frac{1}{2}e^{u-v}(1-D_v)\left(1-\frac{D_v}{2}\right)^{-1}(2v-3)$$

$$= \frac{1}{2}e^{u-v}(1+D_v+\ldots)\left(1+\frac{D_v}{2}+\ldots\right)(2v-3)$$

$$= \frac{1}{2}e^{u-v}\left(1+3\frac{D_v}{2}+\ldots\right)(2v-3)$$

$$= \frac{1}{2}e^{u-v}(2v-3+3) = ve^{u-v} = ye^{x}$$

and $$\frac{1}{D_v(D_v-1)}(-e^{v}) = -\frac{1}{D_v-1}\frac{1}{D_v}e^{v} = -\frac{1}{D_v-1}e^{v}$$

$$= -e^{v}\frac{1}{D_v+1-1}(1) = -e^{v}\frac{1}{D_v}(1) = -ve^{v} = -ye^{y}$$

Hence, the required complete solution is $z = z_c + z_p$

$$z = \phi(x + y) + e^{y}\psi(x + y) + ye^{x} - ye^{y}.$$

EXAMPLE 8.24

Reduce the equation $x^2r - 2xys + y^2t - xp + 3yq = \frac{8y}{x}$ to the canonical form.

Solution: On comparing the given equation with $Rr + Ss + Tt + f(x, y, z, p, q) = 0$, we have

$$R = x^2,\ S = -2xy,\ T = y^2$$

∴ $$S^2 - 4RT = 4x^2y^2 - 4x^2y^2 = 0$$

Here, the quadratic equation $R\lambda^2 + S\lambda + T = 0$ becomes

$$x^2\lambda^2 - 2xy\lambda + y^2 = 0$$

$$\Rightarrow \qquad (x\lambda - y)^2 = 0 \quad \Rightarrow \quad \lambda = y/x, y/x \text{ (equal roots).}$$

$$\therefore \qquad \frac{dy}{dx} + \frac{y}{x} = 0 \Rightarrow \frac{dy}{y} + \frac{dx}{x} = 0$$

Integrating, we have

$$xy = c_1$$

Thus, $u = xy$ and $v = x$ so that u and v are independent functions as

$$w(u,v) = \begin{vmatrix} \frac{\partial u}{\partial x} & \frac{\partial u}{\partial y} \\ \frac{\partial v}{\partial x} & \frac{\partial v}{\partial y} \end{vmatrix} = \begin{vmatrix} y & x \\ 1 & 0 \end{vmatrix} = -x \neq 0$$

Now,
$$p = \frac{\partial z}{\partial x} = \frac{\partial z}{\partial u}\frac{\partial u}{\partial x} + \frac{\partial z}{\partial v}\frac{\partial v}{\partial x} = y\frac{\partial z}{\partial u} + \frac{\partial z}{\partial v}$$

$$q = \frac{\partial z}{\partial y} = \frac{\partial z}{\partial u}\frac{\partial u}{\partial y} + \frac{\partial z}{\partial v}\frac{\partial v}{\partial y} = x\frac{\partial z}{\partial u}$$

$$r = \frac{\partial^2 z}{\partial x^2} = \frac{\partial}{\partial x}\left(\frac{\partial z}{\partial x}\right) = \frac{\partial}{\partial x}\left(y\frac{\partial z}{\partial u} + \frac{\partial z}{\partial v}\right)$$

$$= y\left[\frac{\partial}{\partial u}\left(\frac{\partial z}{\partial u}\right)\frac{\partial u}{\partial x} + \frac{\partial}{\partial v}\left(\frac{\partial z}{\partial u}\right)\frac{\partial v}{\partial x}\right] + \frac{\partial}{\partial u}\left(\frac{\partial z}{\partial v}\right)\frac{\partial u}{\partial x} + \frac{\partial}{\partial v}\left(\frac{\partial z}{\partial v}\right)\frac{\partial v}{\partial x}$$

$$= y^2\frac{\partial^2 z}{\partial u^2} + 2y\frac{\partial^2 z}{\partial u \partial v} + \frac{\partial^2 z}{\partial v^2}$$

$$s = \frac{\partial^2 z}{\partial x \partial y} = \frac{\partial}{\partial x}\left(\frac{\partial z}{\partial y}\right) = \frac{\partial}{\partial x}\left(x\frac{\partial z}{\partial u}\right)$$

$$= 1\cdot\frac{\partial z}{\partial u} + x\left\{\frac{\partial}{\partial u}\left(\frac{\partial z}{\partial u}\right)\frac{\partial u}{\partial x} + \frac{\partial}{\partial v}\left(\frac{\partial z}{\partial u}\right)\frac{\partial v}{\partial x}\right\}$$

$$= \frac{\partial z}{\partial u} + xy\frac{\partial^2 z}{\partial u^2} + x\frac{\partial^2 z}{\partial v^2}$$

and $$t = \frac{\partial^2 z}{\partial y^2} = \frac{\partial}{\partial y}\left(\frac{\partial z}{\partial y}\right) = \frac{\partial}{\partial y}\left(x\frac{\partial z}{\partial u}\right)$$

$$= x\left\{\frac{\partial}{\partial u}\left(\frac{\partial z}{\partial u}\right)\frac{\partial u}{\partial y} + \frac{\partial}{\partial v}\left(\frac{\partial z}{\partial u}\right)\frac{\partial v}{\partial y}\right\}$$

$$= x^2\frac{\partial^2 z}{\partial u^2}$$

Substituting these values in the given equation, we have

$$x^2\left[y^2\frac{\partial^2 z}{\partial u^2} + 2y\frac{\partial^2 z}{\partial u \partial v} + \frac{\partial^2 z}{\partial v^2}\right] - 2xy\left(\frac{\partial z}{\partial u} + x\frac{\partial^2 z}{\partial u \partial v} + xy\frac{\partial^2 z}{\partial u^2}\right) + y^2x^2\frac{\partial^2 z}{\partial u^2} - x\left(y\frac{\partial z}{\partial u} + \frac{\partial z}{\partial v}\right)$$

$$+ 3xy\frac{\partial z}{\partial u} - \frac{8y}{x} = 0$$

$$\Rightarrow \qquad x^2\frac{\partial^2 z}{\partial v^2} - x\frac{\partial z}{\partial v} = \frac{8y}{x}$$

$$\Rightarrow \qquad v^2\frac{\partial^2 z}{\partial v^2} - v\frac{\partial z}{\partial v} = \frac{8u}{v^2}$$

$$\Rightarrow \qquad (v^2D_v^2 - vD_v)z = \frac{8u}{v^2}$$

Let $u = e^X$ and $v = e^Y$

so that $X = \log u$ and $Y = \log v$

and denoting $\frac{\partial}{\partial X}, \frac{\partial}{\partial Y}$ by D_X, D_Y respectively, we have

$$(D_Y(D_Y - 1) - D_Y)z = 8e^{X-2Y}$$

$$\Rightarrow \qquad D_Y(D_Y - 2)z = 8e^{X-2Y}$$

$\therefore$ CF is

$$z_c = \phi(X) + e^{2Y}\psi(X)$$
$$= \phi(\log u) + v^2\psi(\log u)$$
$$= F(u) + v^2G(u)$$
$$= F(xy) + x^2G(xy)$$

Now, PI is

$$z_p = \frac{1}{D_Y(D_Y - 2)}(8e^{X-2Y})$$

$$= 8e^{X-2Y}\frac{1}{(D_Y - 2)(\overline{D_Y - 2} - 2)} \qquad (1)$$

$$= 8e^{X-2Y}\frac{1}{8}\left(1-\frac{D_Y}{2}\right)^{-1}\left(1-\frac{D_Y}{4}\right)^{-1}(1)$$

$$= e^{X-2Y}\left(1+\frac{D_Y}{2}+\ldots\right)\left(1+\frac{D_Y}{4}+\ldots\right)(1)$$

$$= e^{X-2Y}\left(1+\frac{3D_Y}{4}+\ldots\right)(1) = \frac{e^X}{e^{2Y}} = \frac{u}{v^2} = \frac{y}{x}$$

Hence, the required complete solution is

$$z = z_c + z_p$$

$$z = F(xy) + x^2G(xy) + \frac{y}{x}.$$

where F and G are arbitrary functions.

EXAMPLE 8.25

Reduce $\frac{\partial^2 z}{\partial x^2} + y^2 \frac{\partial^2 z}{\partial y^2} = y$ to the canonical form.

Solution: The given equation can be written as

$$r + y^2 t = y$$

On comparing with $Rr + Ss + Tt + f(x, y, z, p, q) = 0$, we have

$$R = 1,\ S = 0,\ T = +y^2 \quad \Rightarrow \quad S^2 - 4RT = -4y^2 < 0.$$

Here, the quadratic equation $R\lambda^2 + S\lambda + T = 0$ becomes

$$\lambda^2 + y^2 = 0 \quad \Rightarrow \quad \lambda = iy, -iy \text{ (Complex roots)}$$

$\therefore$
$$\frac{dy}{dx} + iy = 0 \quad \text{and} \quad \frac{dy}{dx} - iy = 0$$

Integrating, we have

$$\log y + ix = c_1 \text{ and } \log y - ix = c_2.$$

Thus, $u = \log y + ix = \alpha + i\beta$

and $v = \log y - ix = \alpha - i\beta$

$\therefore$ $\alpha = \log y$ and $\beta = x$

Now,
$$p = \frac{\partial z}{\partial x} = \frac{\partial z}{\partial \alpha}\frac{\partial \alpha}{\partial x} + \frac{\partial z}{\partial \beta}\frac{\partial \beta}{\partial x} = \frac{\partial z}{\partial \beta}$$

$$q = \frac{\partial z}{\partial y} = \frac{\partial z}{\partial \alpha}\frac{\partial \alpha}{\partial y} + \frac{\partial z}{\partial \beta}\frac{\partial \beta}{\partial y} = \frac{1}{y}\frac{\partial z}{\partial \alpha}$$

$$r = \frac{\partial^2 z}{\partial x^2} = \frac{\partial}{\partial x}\left(\frac{\partial z}{\partial x}\right) = \frac{\partial}{\partial \beta}\left(\frac{\partial z}{\partial \beta}\right) = \frac{\partial^2 z}{\partial \beta^2}$$

and
$$t = \frac{\partial^2 z}{\partial y^2} = \frac{\partial}{\partial y}\left(\frac{\partial z}{\partial y}\right) = \frac{\partial}{\partial y}\left(\frac{1}{y}\frac{\partial z}{\partial \alpha}\right)$$

$$= -\frac{1}{y^2}\frac{\partial z}{\partial \alpha} + \frac{1}{y}\left[\frac{\partial}{\partial \alpha}\left(\frac{\partial z}{\partial \alpha}\right)\frac{\partial \alpha}{\partial y} + \frac{\partial}{\partial \beta}\left(\frac{\partial z}{\partial \alpha}\right)\frac{\partial \beta}{\partial y}\right]$$

$$= -\frac{1}{y^2}\frac{\partial z}{\partial \alpha} + \frac{1}{y^2}\frac{\partial^2 z}{\partial \alpha^2}.$$

Substituting, these values in the given equation, we have

$$\frac{\partial^2 z}{\partial \beta^2} + \frac{\partial^2 z}{\partial \alpha^2} - \frac{\partial z}{\partial \alpha} - y = 0$$

$$\Rightarrow \qquad \frac{\partial^2 z}{\partial \alpha^2} + \frac{\partial^2 z}{\partial \beta^2} = \frac{\partial z}{\partial \alpha} + e^{\alpha}$$

which is the required canonical form.

EXAMPLE 8.26

Reduce $y^2 \frac{\partial^2 z}{\partial x^2} + x^2 \frac{\partial^2 z}{\partial y^2} = 0$ to the canonical form.

Solution: The given equation can be written as

$$y^2 r + x^2 t = 0$$

On comparing with $Rr + Ss + Tt + f(x, y, z, p, q) = 0$, we have

$$R = y^2,\ 4 = 0,\ T = x^2$$

$$\therefore \qquad S^2 - 4RT = -4x^2y^2 < 0$$

Here, the quadratic equation $R\lambda^2 + S\lambda + T = 0$ becomes

$$y^2\lambda^2 + x^2 = 0$$

$$\Rightarrow \qquad \lambda = \frac{ix}{y}, -\frac{ix}{y} \text{ (Complex roots)}$$

The corresponding characteristic equations are

$$\frac{dy}{dx}+\frac{ix}{y}=0 \quad \text{and} \quad \frac{dy}{dx}-\frac{ix}{y}=0$$

Integrating, we have

$$y^2 + ix^2 = c_1 \text{ and } y^2 - ix^2 = c_2.$$

Thus, $\quad u = y^2 + ix^2 = \alpha + i\beta \quad \text{and} \quad v = y^2 - ix^2 = \alpha - i\beta$

$\therefore \quad \alpha - y^2$ and $\beta - x^2$.

Now, $$p = \frac{\partial z}{\partial x} = \frac{\partial z}{\partial \alpha}\frac{\partial \alpha}{\partial x} + \frac{\partial z}{\partial \beta}\frac{\partial \beta}{\partial x} = 2x\frac{\partial z}{\partial \beta}$$

$$q = \frac{\partial z}{\partial y} = \frac{\partial z}{\partial \alpha}\frac{\partial \alpha}{\partial y} + \frac{\partial z}{\partial \beta}\frac{\partial \beta}{\partial y} = 2y\frac{\partial z}{\partial \alpha}$$

$$= \frac{\partial^2 z}{\partial x^2} = \frac{\partial}{\partial x}\left(\frac{\partial z}{\partial x}\right) = \frac{\partial}{\partial x}\left(2x\frac{\partial z}{\partial \beta}\right)$$

$$= 2\frac{\partial z}{\partial \beta} + 2x\left\{\frac{\partial}{\partial \alpha}\left(\frac{\partial z}{\partial \beta}\right)\frac{\partial \alpha}{\partial x} + \frac{\partial}{\partial \beta}\left(\frac{\partial z}{\partial \beta}\right)\frac{\partial \beta}{\partial x}\right\} = 2\frac{\partial z}{\partial \beta} + 4x^2\frac{\partial^2 z}{\partial \beta^2}$$

and $$t = \frac{\partial^2 z}{\partial y^2} = \frac{\partial}{\partial y}\left(\frac{\partial z}{\partial y}\right) = \frac{\partial}{\partial y}\left(2y\frac{\partial z}{\partial \alpha}\right)$$

$$= 2\frac{\partial z}{\partial \alpha} + 2y\left\{\frac{\partial}{\partial \alpha}\left(\frac{\partial z}{\partial \alpha}\right)\frac{\partial \alpha}{\partial y} + \frac{\partial}{\partial \beta}\left(\frac{\partial z}{\partial \alpha}\right)\frac{\partial \beta}{\partial y}\right\}$$

$$= 2\frac{\partial z}{\partial \alpha} + 4y^2\frac{\partial^2 z}{\partial \alpha^2}$$

Substituting these values in the given equation, we have

$$2y^2\frac{\partial z}{\partial \beta} + 4x^2y^2\frac{\partial^2 z}{\partial \beta^2} + 2x^2\frac{\partial^2 z}{\partial \alpha} + 4x^2y^2\frac{\partial^2 z}{\partial \alpha^2} = 0$$

or $$2\alpha\beta\left(\frac{\partial^2 z}{\partial \alpha^2} + \frac{\partial^2 z}{\partial \beta^2}\right) + \alpha\frac{\partial z}{\partial \beta} + \beta\frac{\partial z}{\partial \alpha} = 0$$

$\Rightarrow$ $$\frac{\partial^2 z}{\partial \alpha^2} + \frac{\partial^2 z}{\partial \beta^2} + \frac{1}{2}\left(\frac{1}{\alpha}\frac{\partial z}{\partial \alpha} + \frac{1}{\beta}\frac{\partial z}{\partial \beta}\right) = 0$$

which is the required canonical form.

▲▲

EXERCISE 8.2

Reduce the following equations to the canonical form:

1. $\dfrac{\partial^2 z}{\partial x^2}+4\dfrac{\partial^2 z}{\partial x\partial y}-5\dfrac{\partial^2 z}{\partial y^2}+6\dfrac{\partial z}{\partial x}+3\dfrac{\partial z}{\partial y}-9z=0.$

2. $6\dfrac{\partial^2 z}{\partial x\partial y}+3\dfrac{\partial^2 z}{\partial y^2}+2\dfrac{\partial z}{\partial x}=0.$

3. $\dfrac{\partial^2 z}{\partial x^2}-2\dfrac{\partial^2 z}{\partial x\partial y}-8\dfrac{\partial^2 z}{\partial y^2}+9\dfrac{\partial z}{\partial x}=0.$

4. $\dfrac{\partial^2 z}{\partial x^2}+2\dfrac{\partial^2 z}{\partial x\partial y}+5\dfrac{\partial^2 z}{\partial y^2}+\dfrac{\partial z}{\partial x}-2\dfrac{\partial z}{\partial y}-3z=0.$

5. $\dfrac{\partial^2 z}{\partial x^2}-5\dfrac{\partial^2 z}{\partial x\partial y}+6\dfrac{\partial^2 z}{\partial y^2}=0.$

6. $\dfrac{\partial^2 z}{\partial x^2}-4\dfrac{\partial^2 z}{\partial x\partial y}+13\dfrac{\partial^2 z}{\partial y^2}-9\dfrac{\partial z}{\partial y}=0.$

7. $\dfrac{\partial^2 z}{\partial x^2}-6\dfrac{\partial^2 z}{\partial x\partial y}+9\dfrac{\partial^2 z}{\partial y^2}+2\dfrac{\partial z}{\partial x}+3\dfrac{\partial z}{\partial y}-z=0.$

8. $t-s+p-q\left(1+\dfrac{1}{x}\right)+\dfrac{z}{x}=0.$

9. $x\dfrac{\partial^2 z}{\partial x^2}+\dfrac{\partial^2 z}{\partial y^2}=x^2\ (x>0).$

10. $\dfrac{\partial^2 z}{\partial x^2}+2x\dfrac{\partial^2 z}{\partial x\partial y}+x^2\dfrac{\partial^2 z}{\partial y^2}=0.$

11. $r-2s+t+p-q=0.$

Reduce the following equations to the canonical form, and solve them.

12. $r+2s+t=0.$
13. $xyr-(x^2-y^2)s-xyt+py-qx=2(x^2-y^2).$
14. $(a-1)^2 z_{xx}-y^{2a} z_{xy}=ay^{2a-1} z_y.$
15. $x^2(y-1)r-x(y^2-1)s+y(y-1)t+xyp-q=0.$

16. $x^2r - y^2t + px - qy = x^2$.
17. $x^2r - xys + px + qy - z = 2x^2y$.
18. $xys - x^2r - px - qy + z = -2x^2y$.
19. $x^2r + 2xys + y^2t = 0$.
20. $r - 4s + 4t = 0$.
21. $x^2r + 2xys + y^2t + xyp + y^2q = 0$.
22. $2r - 4s + 2t + 3z = 0$.
23. $\frac{\partial^2 z}{\partial x^2} + 2\frac{\partial^2 z}{\partial x \partial y} + \frac{\partial^2 z}{\partial y^2} = 0$.
24. $y^2 \frac{\partial^2 z}{\partial x^2} - 2xy\frac{\partial^2 z}{\partial x \partial y} + x^2\frac{\partial^2 z}{\partial y^2} = \frac{y^2}{x}\frac{\partial z}{\partial x} + \frac{x^2}{y}\frac{\partial z}{\partial y}$.
25. $3\frac{\partial^2 z}{\partial x^2} + 10\frac{\partial^2 z}{\partial x \partial y} + 3\frac{\partial^2 z}{\partial y^2} = 0$.
26. $x(y - x)r - (y^2 - x^2)s + y(y - x)t + (y + x)(p - x) = 2x + 2y + 2$.

ANSWERS

Exercise 8.2

1. $\frac{\partial^2 z}{\partial u \partial v} = \frac{1}{4}\frac{\partial z}{\partial u} - \frac{3}{4}\frac{\partial z}{\partial v} - \frac{1}{4}z$.

2. $\frac{\partial^2 z}{\partial u \partial v} = \frac{1}{6}\left(\frac{\partial z}{\partial u} + \frac{\partial z}{\partial v}\right)$.

3. $\frac{\partial^2 z}{\partial u \partial v} = \frac{\partial z}{\partial u} - \frac{1}{2}\frac{\partial z}{\partial v}$.

4. $\frac{\partial^2 z}{\partial \alpha^2} + \frac{\partial^2 z}{\partial \beta^2} = \frac{3}{4}\frac{\partial z}{\partial \alpha} - \frac{1}{2}\frac{\partial z}{\partial \beta} + \frac{3}{4}z$.

5. $\frac{\partial^2 z}{\partial u \partial v} = 0$.

6. $\frac{\partial^2 z}{\partial \alpha^2} + \frac{\partial^2 z}{\partial \beta^2} = \frac{\partial z}{\partial \alpha}$.

7. $\frac{\partial^2 z}{\partial v^2} = -\frac{\partial z}{\partial u} - \frac{1}{3}\frac{\partial z}{\partial v} + \frac{1}{9}u$.

8. $\dfrac{\partial^2 z}{\partial u \partial v} - \dfrac{\partial z}{\partial v} + \dfrac{1}{v}\dfrac{\partial z}{\partial u} - \dfrac{z}{v} = 0$.

9. $\dfrac{\partial^2 z}{\partial \alpha^2} + \dfrac{\partial^2 z}{\partial \beta^2} = \dfrac{\beta^2}{4} + \dfrac{1}{\beta}\dfrac{\partial z}{\partial \beta}$.

10. $\dfrac{\partial^2 z}{\partial v^2} = \dfrac{\partial z}{\partial u}$.

11. $\dfrac{\partial^2 z}{\partial v^2} = \dfrac{\partial z}{\partial v}$.

12. $\dfrac{\partial^2 z}{\partial v^2}$ 0; $z = (x + y)\, \phi_1(x - y) + \phi_2(x - y)$.

13. $\dfrac{\partial^2 z}{\partial u \partial v} = \dfrac{v^2 - 1}{(v^2 + 1)^2}$; $z = -xy + \phi_1(x^2 + y^2) + \phi_2\left(\dfrac{y}{x}\right)$.

14. $\dfrac{\partial^2 z}{\partial u \partial v} = 0$; $z = \phi_1(x - y^{-a+1}) + \phi_2(x + y^{-a+1})$.

15. $\dfrac{\partial^2 z}{\partial u \partial v} = 0$; $z = \phi_1(xy) + \phi_2(xe^y)$.

16. $\dfrac{\partial^2 z}{\partial u \partial v} = \dfrac{1}{4}$; $z = \dfrac{1}{4}x^2 + \phi_1(xy) + \phi_2\left(\dfrac{x}{y}\right)$.

17. $\dfrac{\partial^2 z}{\partial u \partial v} = \dfrac{1}{v}\dfrac{\partial z}{\partial u} + \dfrac{1}{u}\dfrac{\partial z}{\partial v} - \dfrac{2u}{v^2} - \dfrac{z}{uv}$; $z = x^2 y + xy\, \phi_1(y) + y\phi_2(xy)$.

18. $\dfrac{\partial^2 z}{\partial u \partial v} = \dfrac{1}{v}\dfrac{vz}{\partial u} + \dfrac{1}{u}\dfrac{\partial z}{\partial v} - \dfrac{z}{uv} - \dfrac{2u}{v^2}$, $z = x^2 y + xy\phi_1(y) + y\phi_2(xy)$.

19. $\dfrac{\partial^2 z}{\partial v^2} = 0; z = y\phi_1\left(\dfrac{y}{x}\right) + \phi_2\left(\dfrac{y}{x}\right)$.

20. $\dfrac{\partial^2 z}{\partial v^2} = 0$; $z = y\phi_1(y + 2x) + \phi_2(y + 2x)$.

21. $\dfrac{\partial^2 z}{\partial v^2} = -\dfrac{\partial z}{\partial v}; z = \phi_1\left(\dfrac{y}{x}\right) + e^{-y}\phi_2\left(\dfrac{y}{x}\right)$.

22. $\dfrac{\partial^2 z}{\partial v^2} = -\dfrac{3z}{2}; z = e^{(\sqrt{3}/2)y}\phi_1(y + x) + e^{(-\sqrt{3}/2)y}\phi_2(y + x)$.

23. $\dfrac{\partial^2 z}{\partial v^2} = 0$; $z = (x + y)\, \phi_1(x - y) + \phi_2(x - y)$.

24. $\dfrac{\partial^2 z}{\partial v^2} = 0$; $z = (x^2 - y^2)\, \phi_1(x^2 + y^2) + \phi_2(x^2 + y^2)$.

25. $\dfrac{\partial^2 z}{\partial u \partial v} = 0$; $z = \phi_1(y - 3x) + \phi_2(3x - x)$.

26. $\dfrac{\partial^2 z}{\partial u \partial v} = \dfrac{2(v+1)}{(v^2 - 4u)^{3/2}}$; $z = x + y + \log(2x) + \phi_1(x + 4) + \phi_2(xy)$.

▲▲

9

CHAPTER

Monge's Methods

Most of the nonlinear second order equations of the form $f(x, y, z, p, q, r, s, t) = 0$ cannot be integrated exactly. In this chapter, we will study Monge's methods to solve nonlinear partial differential equation of the second order of the form

$$Rr + Ss + Tt = V$$

and
$$Rr + Ss + Tt + U(rt - s^2) = V$$

where R, S, T, U and V are functions of x, y, z, p and q.

Monge's method for solving $\boldsymbol{Rr + Ss + Tt = V}$

Consider the equation

$$Rr + Ss + Tt = V \tag{1}$$

Since z is to be a function of x and y, we have

$$dz = \frac{\partial z}{\partial x}dx + \frac{\partial z}{\partial y}dy = pdx + qdy \tag{2}$$

$$dp = \frac{\partial p}{\partial x}dx + \frac{\partial p}{\partial y}dy = rdx + sdy \tag{3}$$

and
$$dq = \frac{\partial q}{\partial x}dx + \frac{\partial q}{\partial y}dy = sdx + tdt \tag{4}$$

From Eqs. (3) and (4), we get

$$r = \frac{dp - sdy}{dx} \text{ and } t = \frac{dq - sdx}{dy}$$

Putting these values in Eq. (1), we get

$$R\left(\frac{dp - sdy}{dx}\right) + Ss + T\left(\frac{dq - sdx}{dy}\right) = V$$

or
$$Rdpdy + Tdqdx - Vdxdy + s(R(dy)^2 - Sdxdy + T(dx)^2) = 0 \tag{5}$$

The equations

$$Rdpdy + Tdqdx - Vdxdy = 0 \tag{6}$$

and

$$R(dy)^2 - Sdxdy + T(dx)^2 = 0 \tag{7}$$

are known as Monge's auxiliary equations.

Any relation between x, y, z, p and q satisfying Eqs. (6) and (7) will satisfy Eq. (5) and hence Eq. (1). Equation (7) being quadratic in $\frac{dy}{dx}$ can be decomposed into two equations.

Case I: *When the left hand member of Eq. (7) is a perfect square.*

Let $$R(dy)^2 - Sdxdy + T(dx)^2 = (Ady + Bdx)^2.$$

Then, the system of Eqs. (6) and (7) is equivalent to the system

$$\left.\begin{aligned} Rdpdy + Tdqdx - Vdxdy &= 0 \\ Ady + Bdx &= 0 \end{aligned}\right\} \tag{8}$$

If $u(x, y, z, p, q) = c_1$ and $v(x, y, z, p, q) = c_2$ satisfy Eq. (8) then we can write

$$u = \phi(v) \tag{9}$$

where ϕ is an arbitrary function. Using Eq. (9) complete integral of Eq. (1) is obtained by applying the method of solution of the first order equations.

Case II: *When the left hand member of Eq. (7) is not a complete square.*

Let, $R(dy)^2 - Sdxdy + T(dx)^2 = (A_1dy + B_1dx)\,(A_2dy + B_2dx)$

where $A_1B_2 - A_2B_1 \neq 0$ identically.

Then the system of Eqs. (6) and (7) is equivalent to the systems

$$\left.\begin{aligned} Rdydy + Tdqdx - Vdxdy &= 0 \\ A_1dy + B_1dx &= 0 \end{aligned}\right\} \tag{10}$$

and

$$\left.\begin{aligned} Rdpdy + Tdqdx - Vdxdy &= 0 \\ A^2dy + B^2dx &= 0 \end{aligned}\right\} \tag{11}$$

In case, only one system is integrable, then the procedure for finding complete integral is same as in case I. If both the systems are integrable then proceed as to solve these integrals for p and q and then substitute these values in Eq. (2); which when integrated will give the complete integral.

EXAMPLE 9.1

Solve the equation $r = a^2t$.

Solution: The given equation is $r - a^2t = 0$.

On comparing with $Rr + Ss + Tt = V$, we have

$$R = 1, S = 0, T = -a^2, V = 0.$$

Hence, Monge's auxiliary equations

$$Rdpdy + Tdqdx - Vdxdy = 0$$

and $$R(dy)^2 - Sdxdy + T(dx)^2 = 0$$

becomes $$dpdy - a^2dqdx = 0 \quad (1)$$

and $$(dy)^2 - a^2(dx)^2 = 0 \quad (2)$$

$\Rightarrow$ $$(dy + adx)(dy - adx) = 0$$

Hence, the system of equations to be considered is

$$dpdy - a^2dqdx = 0, dy - adx = 0 \quad (3)$$

and $$dpdy - a^2dqdx = 0, dy + adx = 0 \quad (4)$$

Second equation of Eq. (3) on integration gives

$$y - ax = c_1$$

Eliminating $\frac{dy}{dx}$ between the Eq. (3), we get

$$dp - adq = 0$$

which on integration gives $$p - aq = c_1$$

Hence, the intermediate integral corresponding to Eq. (3) is

$$p - aq = \phi_1(y - ax) \quad (5)$$

Similarly, another intermediate integral corresponding to Eq. (4) is

$$p + aq = \phi_2(y + ax) \quad (6)$$

Solving Eqs. (5) and (6) for p and q, we have

$$p = \frac{1}{2}[\phi_2(y + ax) + \phi_1(y - ax)]$$

and $$q = \frac{1}{2a}[\phi_2(y + ax) - \phi_1(y - ax)]$$

Substituting in $dz = pdx + qdy$, we get

$$dz = \frac{1}{2}[\phi_2(y + ax) + \phi_1(y - ax)]dx + \frac{1}{2a}[\phi_2(y + ax) - \phi_1(y - ax)]dy$$

$$= \frac{1}{2a}\phi_2(y + ax)\,(dy + adx) - \frac{1}{2a}\phi_1(y - ax)\,(dy - adx)$$

Integrating, we have

$$z = \psi_2(y + ax) + \psi_1(y - ax)$$

where ψ_1 and ψ_2 are arbitrary functions.

EXAMPLE 9.2

Solve $r - t\cos^2 x + p + \tan x = 0$.

Solution: On comparing the given equation with $Rr + Ss + Tt = V$, we have

$$R = 1,\ S = 0,\ T = -\cos^2 x,\ V = -p\tan x$$

Hence, the Monge's auxiliary equations become

$$dpdy - \cos^2 x\, dqdx + p\tan x\, dxdy = 0 \quad (1)$$

and $$(dy)^2 - \cos^2 x\,(dx)^2 = 0 \quad (2)$$

$\Rightarrow$ $$(dy - \cos x dx)(dy + \cos x dx) = 0$$

$\therefore$ $$dy - \cos x dx = 0 \quad (3)$$

or $$dy + \cos x dx = 0 \quad (4)$$

Substituting $\frac{dy}{dx}$ from Eq. (3) in Eq. (1), we have

$$dp - \cos x dq + p\tan x dx = 0$$

$\Rightarrow$ $$\sec x dp - dq + p\sec x\tan x dx = 0$$

$\Rightarrow$ $$d(p\sec x) - dq = 0$$

Integrating, we have

$$p\sec x - q = c_1$$

Integrating Eq. (3), we have $y - \sin x = c_2$

Hence, the intermediate solution is

$$p\sec x - q = \phi_1(y - \sin x) \quad (5)$$

Similarly, Eqs. (2) and (4) gives

$$p\sec x + q = \phi_2(y + \sin x) \quad (6)$$

Solving Eqs. (5) and (6) for p and q, we get

$$p = \frac{1}{2}\cos x[\phi_1(y - \sin x) + \phi_2(y + \sin x)]$$

and $$q = \frac{1}{2}[\phi_2(y + \sin x) - \phi_1(y - \sin x)]$$

Putting these values in $dz = pdx + qdy$, we get

$$dz = \frac{1}{2}\cos x[\phi_1(y - \sin x) + \phi_2(y + \sin x)]dx + \frac{1}{2}[\phi_2(y + \sin x) - \phi_1(y - \sin x)]dy$$

$$= \frac{1}{2}\phi_2(y + \sin x)\,(dy + \cos x dx) - \frac{1}{2}\phi_1(y - \sin x)\,(dy - \cos x dx)$$

Integrating, we have the complete solution

$$z = \psi_2(y + \sin x) + \psi_1(y - \sin x).$$

where, ψ_1 and ψ_2 are arbitrary functions.

EXAMPLE 9.3

Solve $pt - qs = q^3$.

Solution: On comparing the given equation with $Rr + Ss + Tt = V$, we have

$$R = 0,\ S = -q,\ T = p,\ V = q^3.$$

Hence, the Monge's auxiliary equations become

$$pdqdx - q^3dxdy = 0$$

and

$$qdxdy + p(dx)^2 = 0$$

$\Rightarrow$

$$pdq - q^3dy = 0 \tag{1}$$

and

$$pdx + qdy = 0 \tag{2}$$

Now, $\quad dz = pdx + qdy = 0 \Rightarrow z = c_1$ (where c_1 is a constant)

Again, combining Eq. (2) with Eq. (1), we get

$$pdq - q^2(-pdx) = 0$$

$$\Rightarrow \quad dq + q^2dx = 0$$

$$\Rightarrow \quad \frac{dq}{q^2} + dx = 0$$

Integrating, we have $\quad -\frac{1}{q} + x = c_2$

Thus, $\quad -\frac{1}{q} + x = \phi(z)$

$$\Rightarrow \qquad -\frac{\partial y}{\partial z} + x = \phi(z)$$

$$\Rightarrow \qquad \frac{\partial y}{\partial z} = x - \phi(z)$$

Integrating, we have

$$y = xz - \int \phi(z)dz + c_3$$

$$\Rightarrow \qquad y = xz - \phi_1(z) + \phi_2(x)$$

which is the required solution.

EXAMPLE 9.4

Solve $r + (a + b)s + abt = xy$.

Solution: On comparing with $Rr + Ss + Tt = V$, we have

$$R = 1,\ S = a + b,\ T = ab,\ V = xy$$

Here, the Monge's auxiliary equations become

$$dpdy + abdqdx - xydxdy = 0 \qquad (1)$$

and

$$(dy)^2 - (a + b)dxdy + ab(dx)^2 = 0 \qquad (2)$$

$$\Rightarrow \qquad (dy - adx)\ (dy - bdx) = 0$$

$$\therefore \qquad dy - adx = 0$$

or

$$dy - bdx = 0$$

Integrating, we get

$$y - ax = c_1 \qquad (3)$$

or

$$y - bx = c_2 \qquad (4)$$

Substituting $\frac{dy}{dx}$ from Eq. (3) in Eq. (1), we have

$$adp + abdq - ax(c_1 + ax)dx = 0$$

$$\Rightarrow \qquad dp + bdq - x(c_1 + ax)dx = 0$$

Integrating, we have

$$p + bq - \left(c_1\frac{x^2}{2} + a\frac{x^3}{3}\right) = A$$

$$\Rightarrow \qquad p + bq - \frac{1}{2}x^2(y - ax) - \frac{a}{3}x^3 = \phi_1(y - ax)$$

$$\Rightarrow \qquad p + bq + \frac{1}{2}ax^3 - \frac{1}{2}x^2y = \phi_1(y - ax) \tag{5}$$

Similarly, from Eqs. (4) and (1), we obtain

$$p + aq + \frac{1}{6}bx^3 - \frac{1}{2}x^2y = \phi_2(y - bx) \tag{6}$$

Solving Eqs. (5) and (6), we get

$$p = \frac{1}{2}\, x^2y - \frac{1}{6}\,(a + b)x^3 - \frac{1}{b-a}[a\phi_1(y - ax) - b\phi_2(y - bx)]$$

and

$$q = \frac{1}{6}x^3 + \frac{1}{b-a}[\phi_1(y - ax) - \phi_2(y - bx)]$$

Thus, $dz = pdx + qdx$ becomes

$$dz = \frac{1}{2}\,x^2ydx + \frac{1}{6}\,x^3dy - (a + b)\,\frac{1}{6}\,x^3dx + \frac{1}{b-a}[\phi_1(y - ax)\,(dy - adx) - \phi_2(y - bx)\,(dy - bdx)]$$

Integrating, we get

$$z = \frac{1}{6}\,x^3y - (a + b)\,\frac{x^4}{24} + \psi_1(y - ax) + \psi_2(y - bx)$$

which is the complete solution with ψ_1 and ψ_2 as arbitrary functions.

EXAMPLE 9.5

Solve $t - r\sec^4 y = 2q\tan y$.

Solution: Comparing with $Rr + Ss + Tt = V$, we have

$$R = \sec^4 y,\ S = 0,\ T = 1,\ V = 2q\tan y.$$

Here, the Monge's auxiliary equations are

$$-\,dpdy\sec^4 y + dqdx - 2q\tan dydx = 0 \tag{1}$$

and

$$-\sec^4 y\,(dy)^2 + (dx)^2 = 0 \tag{2}$$

$$\Rightarrow \qquad (dx - \sec^2 ydy)\,(dx + \sec^2 ydy) = 0$$

$$\therefore \qquad dx - \sec^2 ydy = 0 \tag{3}$$

or

$$dx + \sec^2 ydy = 0 \tag{4}$$

Integrating, we get

$$x - \tan y = c_1 \text{ or } x + \tan y = c_2$$

Substituting value of $\frac{dy}{dx}$ from Eq. (3) in Eq. (1), we have

$$-dp\sec^2 y + dq - 2q\tan y dy = 0$$

$$\Rightarrow \quad dp - dq\cos^2 y + 2q\sin y\cos y dy = 0$$

Integrating, we have

$$p - q\cos^2 y = A$$

Hence,

$$p - q\cos^2 y = \phi_1(x - \tan y) \tag{5}$$

Similarly, from Eqs. (4) and (1), we obtain

$$p + q\cos^2 y = \phi_2(x + \tan y) \tag{6}$$

Solving Eqs. (5) and (6), we get

$$p = \frac{1}{2}[\phi_1(x - \tan y) + \phi_2(x + \tan y)]$$

and

$$q = \frac{1}{2}\sec^2 y\,[\phi_2(x + \tan y) - \phi_1(x - \tan y)].$$

Hence, $dz = pdx + qdy$ becomes

$$dz = \frac{1}{2}[\phi_1(x - \tan y) + \phi_2(x + \tan y)]dx + \frac{1}{2}\sec^2 y\,[\phi_2(x + \tan y) - \phi_1(x - \tan y)]dy$$

$$= \frac{1}{2}\phi_1(x - \tan y)\,(dx - \sec^2 y dy) + \frac{1}{2}\phi_2(x + \tan y)\,(dx + \sec^2 y dy)$$

Integrating, we have the complete solution of the given differential equation

$$z = \psi_1(x - \tan y) + \psi_2(x + \tan y)$$

where ψ_1 and ψ_2 are arbitrary functions.

EXAMPLE 9.6

Solve $(1 + q)^2 r - 2(1 + p + pq)s + (1 + p)^2 t = 0$.

Solution: On comparing with $Rr + Ss + Tt = V$, we have

$$R = (1 + q)^2,$$

$$S = -2(1 + p + q + pq),$$

$$T = (1 + p)^2,\ V = 0$$

Hence, the Monge's auxiliary equations are

$$(1 + q)^2\,dpdy + (1 + p)^2\,dqdx = 0 \tag{1}$$

and

$$(1 + q)^2\,(dy)^2 + 2(1 + p + q + pq)dxdy + (1 + p)^2\,(dx)^2 = 0 \tag{2}$$

$$\Rightarrow \qquad [(1+q)dy + (1+p)dx]^2 = 0$$

$$\therefore \qquad (1+q)dy + (1+p)dx = 0 \qquad (3)$$

From Eqs. (3) and (1), we get

$$(1+q)dp - (1+p)dq = 0$$

$$\Rightarrow \qquad \frac{dp}{1+p} - \frac{dq}{1+q} = 0$$

Integrating, we have $\dfrac{1+p}{1+q} = A$

From Eq. (3), we have

$$dx + dy + (pdx + qdy) = 0$$

$$\Rightarrow \qquad dx + dy + dz = 0$$

Integrating, we get $\qquad x + y + z = B$

Hence, the intermediate integral is

$$(1+p) = (1+q)\,\phi(x+y+z)$$

$$\Rightarrow \qquad p - q\,\phi(x+y+z) = \phi(x+y+z) - 1$$

which is of Lagrange's form. So, its auxiliary equations are

$$\frac{dx}{1} = \frac{dy}{-\phi(x+y+z)} = \frac{dz}{\phi(x+y+z)-1} = \frac{dx+dy+dz}{0}$$

Hence, $\qquad dx + dy + dz = 0$

$$\Rightarrow \qquad x + y + z = C$$

Again, from the first two members, we have

$$\frac{dx}{1} = \frac{dy}{-\phi(c)} \Rightarrow y + x\phi(c) = D$$

Therefore, the complete integral is

$$y + x\phi(x+y+z) = \psi(x+y+z)$$

where ϕ and ψ are arbitrary functions.

EXAMPLE 9.7

Solve $(b + cq)^2 r - 2(b + cq)(a + cp)s + (a + cp)^2 t = 0$.

Solution: On comparing with $Rr + Ss + Tt = V$, we have

$$R = (b + cq)^2,$$

$$S = -2(b + cq)(a + cp),$$
$$T = (a + cp)^2\ V = 0.$$

Thus, the Monge's auxiliary equations are

$$(b + cq)^2 dpdy + (a + cp)^2 dqdx = 0 \quad (1)$$

and $(b + cq)^2 (dy)^2 + 2(b + cq)(a + cp)dxdy + (a + cp)^2 (dx) = 0$ (2)

$$\Rightarrow \quad [(b + cq)dy + (a + cp)dx]^2 = 0$$

$$\therefore \quad (b + cq)dy + (a + cp)dx = 0 \quad (3)$$

From Eqs. (3) and (1), we have

$$(b + cq)dp - (a + cp)dq = 0$$

$$\Rightarrow \quad \frac{dp}{a+cp} - \frac{dq}{b+cq} = 0$$

Integrating, we have $\dfrac{a+cp}{b+cq} = A$

Also,

$$\Rightarrow \quad a + cp = A(b + cq)$$

From Eq. (3), we have

$$adx + bdy + c(pdx + qdy) = 0$$

$$\Rightarrow \quad adx + bdy + cdz = 0$$

Integrating, we get

$$ax + by + cz = B$$

Thus, the intermediate integral is

$$a + cp = (b + cq)\,\phi(ax + by + cz)$$

$$\Rightarrow \quad cp - c\phi(ax + by + cz)q = -a + b\phi(ax + by + cz)$$

which is of Lagrange's form. So, its auxiliary equations are

$$\frac{dx}{c} = \frac{dy}{-c\phi(ax+by+cz)} = \frac{dz}{-a+b\phi(ax+by+cz)} = \frac{adx+bdy+cdz}{0}$$

Therefore, $adx + bdy + cdz = 0$

$$\Rightarrow \quad ax + by + cz = c_1$$

Again, from the first two members, we have

$$\frac{dx}{c} = \frac{dy}{-c\phi(c_1)}$$

$\Rightarrow$ $$dy + \phi(c_1)dx = 0$$

Integrating, we get $y + \phi(c_1)x = c_2.$

Hence, the complete solution is

$$y + x\phi(ax + by + cz) = \psi(ax + by + cz)$$

where ϕ and ψ are arbitrary functions.

EXAMPLE 9.8

Solve $q(1 + q)r - (p + q + 2pq)s + p(1 + p)t = 0.$

Solution: On comparing with $Rr + Ss + Tt = V$, we have

$$R = q(1 + q),$$

$$S = -(p + q + 2pq),$$

$$T = p(1 + p),\ v = 0.$$

Hence, the Monge's auxiliary equations are

$$q(1 + q)dpdy + p(1 + p)dqdx = 0 \quad (1)$$

and $q(1 + q)\ (dy)^2 + (p + q + 2pq)dxdy + p(1 + p)\ (dx)^2 = 0$ (2)

$\Rightarrow$ $$(pdx + qdy)\ ((1 + p)dx + (1 + q)dy) = 0$$

$\therefore$ $$pdx + qdy = 0 \quad (3)$$

and $$(1 + p)dx + (1 + q)dy = 0 \quad (4)$$

From Eqs. (3) and (1), we have

$$-(1 + q)dp + (1 + p)dq = 0$$

$\Rightarrow$ $$\frac{dp}{1+p} - \frac{dq}{1+q} = 0$$

Integrating, we have $$\frac{1+p}{1+q} = A$$

$\Rightarrow$ $$1 + p = A(1 + q)$$

Also, from Eq. (3), we have

$$pdx + qdy = 0 = dz$$

$\therefore$ $$z = B$$

Thus, one intermediate integral is

$$(1 + p) = (1 + q)\,\phi_1(z) \tag{5}$$

Again, from Eqs. (4) and (1), we get

$$-qdp + pdq = 0 \Rightarrow \frac{dp}{p} - \frac{dq}{q} = 0$$

Integrating, we get $\quad \dfrac{p}{q} = c \quad \Rightarrow \quad p = cq$

Again, from Eq. (4), we have

$$dx + dy + (pdx + qdy) = 0$$

$$\Rightarrow \qquad dx + dy + dz = 0$$

Integrating, we have $\quad x + y + z = D$

Thus, another intermediate integral is

$$p = q\,\phi_2(x + y + z) \tag{6}$$

Now, solving Eqs. (5) and (6), we get

$$p = \frac{(\phi_1 - 1)\phi_2}{\phi_2 - \phi_1} \quad \text{and} \quad q = \frac{\phi_1 - 1}{\phi_2 - \phi_1}$$

Thus, $\quad dz = pdx + qdy$

becomes $\quad (\phi_2 - \phi_1)dz = (\phi_1 - 1)\phi_2 dx + (\phi_1 - 1)dy$

$$\Rightarrow \qquad (\phi_2 - 1)dz = (\phi_1 - 1)(dx + dy + dz) + (\phi_1 - 1)\,[\phi_2 - 1]dx$$

$$\Rightarrow \qquad \frac{dz}{\phi_1 - 1} = \frac{1}{\phi_2 - 1} d(x + y + z) + dx$$

Integrating, we have

$$\psi_1 = \psi_2 + x$$

$$\Rightarrow \qquad \psi_1(z) = \psi_2(x + y + z) + x$$

where ψ_1 and ψ_2 are arbitrary functions.

EXAMPLE 9.9

Solve $xy(t - r) + (x^2 - y^2)\,(s - 2) = py - qx$.

Solution: The given equations can be written as

$$xyr - (x^2 - y^2)s - xyt + 2(x^2 - y^2) + py - qx = 0$$

On comparing with $Rr + Ss + Tt = V$; we have

$$R = xy,\ S = -(x^2 - y^2),\ T = -xy,\ V = -2(x^2 - y^2) - py + qx.$$

Thus, the Monge's auxiliary equations are

$$xydpdy - xydqdx + [2(x^2 - y^2) + py - qx]dxdy = 0 \tag{1}$$

and
$$xy(dy)^2 + (x^2 - y^2)dxdy - xy(dx)^2 = 0 \tag{2}$$

$\Rightarrow$
$$(xdx + ydy)\,(xdy + ydx) = 0$$

$\therefore$
$$xdx + ydy = 0 \tag{3}$$

or
$$xdy + ydx = 0 \tag{4}$$

Integrating, we get

$$x^2 + y^2 = c_1$$

From Eqs. (3) and (1), we have

$$xdp + ydq - 2xdy - 2ydx + pdx + qdy = 0$$

Integrating, we have

$$xp + yq - 2xy = c_2$$

Hence, the intermediate integral is

$$xp + yq - 2xy = \phi(x^2 + y^2)$$

which is of Lagrange's form. So, its auxiliary equations are

$$\frac{dx}{x} = \frac{dy}{y} = \frac{dz}{2xy + \phi(x^2 + y^2)}$$

From the first two members, we get

$$-\frac{dx}{x} + \frac{dy}{y} = 0$$

$\Rightarrow$
$$\frac{y}{x} = c_2$$

Also,
$$\frac{dx}{x} = \frac{dz}{2c_2x^2 + \phi(x^2 + c_2^2x^2)}$$

$\Rightarrow$
$$dz = 2c_2xdx + \frac{1}{x^2}\phi(x^2 + c_2^2x^2)xdx$$

$\Rightarrow$
$$dz = 2c_2xdx + \phi_1(x^2 + c_2^2x^2)xdx$$

Integrating, we have

$$z = c_2x^2 + \psi_1(x^2 + c_2^2x^2) + c^3$$

$$\Rightarrow \qquad z = xy + \psi_1(x^2 + y^2) + \psi_2\left(\frac{y}{x}\right)$$

where, ψ_1 and ψ_2 are arbitrary functions.

EXAMPLE 9.10

Solve $y^2r + 2xys + x^2t + px + qy = c$.

Solution: On comparing with $Rr + Ss + Tt = V$, we have

$$R = y^2, S = 2xy, T = x^2, V = -px - qy$$

Hence, the Monge's auxiliary equations are

$$y^2dpdy + x^2dqdx + (px + qy)dxdy = 0 \quad (1)$$

and
$$y^2(dy)^2 - 2xydxdy + x^2(dx)^2 = 0 \quad (2)$$

$$\Rightarrow \qquad (ydy - xdx)^2 = 0$$

$$\therefore \qquad ydy - xdx = 0 \quad (3)$$

Integrating, we have $x^2 - y^2 = c_1$

From Eqs. (3) and (1), we get

$$ydp + xdq + pdy + qdx = 0$$

Integrating, we get $py + qx = c_2$.

Hence, the intermediate integral is

$$py + qx = \phi(x^2 - y^2)$$

which is of Lagrange's form. So, its auxiliary equations are

$$\frac{dx}{y} = \frac{dy}{x} = \frac{dz}{\phi(x^2 - y^2)}$$

From the first two members, we have

$$xdx - ydy = 0 \Rightarrow x^2 - y^2 = c_1$$

Also,
$$\frac{dy}{x} = \frac{dz}{\phi(c_1)}$$

$$\Rightarrow \qquad \frac{dy}{\sqrt{c_1 + y^2}} = \frac{dz}{\phi(c_1)}$$

$$\Rightarrow \qquad dz = \phi(c_1)\frac{dy}{\sqrt{c_1 + y^2}}$$

Integrating, we get

$$z = \phi(c_1)\ \frac{1}{c_1}\log(y + \sqrt{c_1 + y^2}\,) + c_3$$

$$\Rightarrow \qquad z = \psi_1(x^2 - y^2) \log (y + x) + \psi_2(x^2 - y^2)$$

which is the required complete integral with ψ_1 and ψ_2 as arbitrary functions.

EXAMPLE 9.11

Solve $(r - s)y + (s - t)x + q - p = 0$.

Solution: The given equation can be written as

$$ry + s(x - y) - tx + q - p = 0$$

On comparing with $Rr + Ss + Tt = V$, we have

$$R = y,\ S = x - y,\ T = -x,\ V = -(q - p).$$

Hence, the Monge's auxiliary equations are

$$ydpdy - xdqdx + (q - p)dxdy = 0 \qquad (1)$$

and

$$y(dy)^2 - (x - y)dxdy - x(dx)^2 = 0 \qquad (2)$$

$$(dy + dx)\ (ydy - x(dx) = 0$$

$$\therefore \qquad dy + dx = 0 \qquad (3)$$

or

$$ydy - xdx = 0 \qquad (4)$$

Integrating, we get $\quad y + x = c_1$

From Eqs. (3) and (1), we get

$$ydp + xdq + qdx + pdy = 0$$

Integrating, we get $\quad py + qx = c_2.$

Hence, $\quad py + qx = \phi(y + x)$

which is of Lagrange's form. So, its auxiliary equations are

$$\frac{dx}{y} = \frac{dy}{x} = \frac{dz}{\phi(y + x)}$$

From the first two members, we have

$$xdx - ydy = 0$$

Integrating, we get $\quad x^2 - y^2 = c_3$

Also, $$\frac{dx}{y} = \frac{dz}{\phi(x+y)}$$

$$\Rightarrow \qquad dz = \frac{\phi(x+\sqrt{c^2 - c_3})}{\sqrt{x^2 - c_3}}dx$$

Integrating, we get

$$\Rightarrow \qquad z = \psi_1 (x+\sqrt{x^2 - c_3}) + c_4$$

$$\Rightarrow \qquad z = \psi_1(x+ y) + \psi_2(x^2 - y^2)$$

where ψ_1 and ψ_2 are arbitrary functions.

EXAMPLE 9.12

Solve $(x - 2y)(2xr - (x + 2y)s + yt) = (x + 2y)(2p - q)$.

Solution: On comparing with $Rr + Ss + Tt = V$, we have

$$R = 2x(x - 2y),$$
$$S = -(x^2 - 4y^2),$$
$$T = y(x - 2y),$$
$$V = (x + 2y)(2p - q).$$

Hence, the Monge's auxiliary equations are

$$2x(x - 2y)dpdy + y(x - 2y)dqdx - (x + 2y)(2p - q)dxdy = 0$$

$$\Rightarrow \qquad 2xdpdy + ydqdx - \frac{x+2y}{x-2y}(2p - 4)dxdy = 0 \qquad (1)$$

and $$2x(x - 2y)(dy)^2 + (x + 4y^2)dxdy + y(x - 2y)(dx)^2 = 0$$

$$\Rightarrow \qquad 2x(dy)^2 + (x + 2y)dxdy + y(dx)^2 = 0 \qquad (2)$$

$$\Rightarrow \qquad (xdy + ydx)(2dy + dx) = 0$$

$$\therefore \qquad xdy + ydx = 0 \qquad (3)$$

or $$2dy + dx = 0 \qquad (4)$$

Integrating, we get $xy = c_1$

From Eqs. (3) and (1),we get

$$2dp - dq - \frac{x+2y}{x-2y}(2p - q)dydx = 0$$

$$\Rightarrow \qquad \frac{2dp - dq}{2p-q} = \frac{dx - 2dy}{x-2y}$$

Integrating, we get

$$2p - q = c_2(x - 2y)$$

Hence, the intermediate integral is

$$2p - q = (x - 2y)\ \phi(xy)$$

which is of Lagrange's form. So, its auxiliary equations are

$$\frac{dx}{2} = \frac{dy}{-1} = \frac{dz}{(x-2y)\phi(xy)}$$

$$= \frac{y\phi(xy)dx + x\phi(xy)dy + dz}{0}$$

Hence, $\quad y\phi(xy)dx + x\phi(xy)dy + dz = 0$

$\Rightarrow \quad dz + \phi(xy)d(xy) = 0$

Integrating, we get $\quad z = \psi_1(xy) + c_3$

Also, from the first two members, we have

$$\frac{dx}{2} = \frac{dy}{-1} \Rightarrow dx + 2dy = 0$$

Integrating, we get $\quad x + 2y = c_4$.

Hence, the complete solution is

$$z = \psi_1(xy) + \psi_2(x + 2y)$$

where ψ_1 and ψ_2 are arbitrary functions.

EXAMPLE 9.13

Solve $(x - y)\ (xr - (x + y)s + yt) = (x + y)\ (p - q)$.

Solution: On comparing with $Rr + Ss + Tt = V$, we have

$$R = x(x - y),$$
$$S = -(x^2 - y^2),$$
$$T = y(x - y),$$
$$V = (x + y)\ (p - q)$$

Hence, the Monge's auxiliary equations are

$$x(x - y)dpdy + y(x - y)dqdx - (x + y)\ (p - q)dxdy = 0 \qquad (1)$$

and $$x(x-y)(dy)^2 + (x^2 - y^2)dxdy + y(x-y)(dx)^2 = 0$$

$\Rightarrow$ $$x(dy)^2 + (x+y)dxdy + y(dy)^2 = 0 \quad (2)$$

$\Rightarrow$ $$(dx+dy)(xdy+ydx) = 0$$

$\therefore$ $$xdy + ydx = 0 \quad (3)$$

or $$dx + dy = 0 \quad (4)$$

Integrating, we get $$xy = c_1$$

From Eqs. (3) and (1), we get

$$-y(x-y)dpdx + y(x-y)dqdx - (p-q)(-y(dx)^2 + ydxdy) = 0$$

$\Rightarrow$ $$(x-y)(dp-dq) - (p-q)(dx-dy) = 0$$

$\Rightarrow$ $$\frac{dp-dq}{p-q} = \frac{dx-dy}{x-y}$$

Integrating, we get $p - q = c_2(x-y)$

Hence, the intermediate integral is

$$p - q = (x-y)\,\phi(xy)$$

which is of the Lagrange's form. Hence, the Lagrange's auxiliary equations are

$$\frac{dx}{1} = \frac{dy}{-1} = \frac{dz}{(x-y)\phi(xy)} = \frac{\phi(xy)\cdot(ydx+xdy)+dz}{0}$$

From the first two members, we get

$$dx + dy = 0 \qquad \Rightarrow x + y = c_3$$

and $$dz + \phi(xy)d(xy) = 0$$

$\Rightarrow$ $$z = \psi_1(xy) + c_4$$

Hence, the complete solution is

$$z = \psi_1(xy) + \psi_2(x+y)$$

where ψ_1 and ψ_2 are arbitrary functions.

EXAMPLE 9.14

Solve $x^2r - y^2t - 2xp + 2z = 0$.

Solution: On comparing with $Rr + Ss + Tt = V$, we have

$$R = x^2,\ S = 0,\ T = -y,\ V = 2xp - 2z$$

Hence, the Monge's auxiliary equations are

$$x^2dpdy - y^2dqdx - (2xp - 2z)dxdy = 0 \quad (1)$$

and

$$x^2(dy)^2 - y^2(dx)^2 = 0 \quad (2)$$

$$\Rightarrow \quad (xdy - ydx)(xdy + ydx) = 0$$

$$\therefore \quad xdy - ydx = 0 \quad (3)$$

or

$$xdy + ydx = 0 \quad (4)$$

Integrating, we get $\frac{y}{x} = c_1$

From Eqs. (3) and (1), we get

$$xdp(xdy) - ydq(ydx) - 2(xp - z)(xdx)\left(\frac{1}{x}dy\right) = 0$$

$$\Rightarrow \quad xdp - ydq - 2(xp - z)\frac{1}{x}dx = 0$$

$$\Rightarrow \quad d(xp - z) + dz - pdx - ydq - 2(xp - z)\frac{dx}{x} = 0$$

$$\Rightarrow \quad d(xp - z) + qdy - ydq - 2(xp - z)\frac{dx}{x} = 0$$

$$\Rightarrow \quad d(xp - z) - d(yq) + 2qdy - 2(xp - z)\frac{dx}{x} = 0$$

$$\Rightarrow \quad d(xp - z) - d(yq) + 2qy\frac{dx}{x} - 2(xp - z)\frac{dx}{x} = 0$$

$$\Rightarrow \quad d(xp - z - yq) - 2(xp - yq - z)\frac{dx}{x} = 0$$

$$\Rightarrow \quad \frac{d(xp - z - yq)}{xp - z - yq} - 2\frac{dx}{x} = 0$$

Integrating, we get $xp - yq - z = x^2c_1$

Thus, the intermediate integral is

$$xp - yq = z + x^2\phi_1\left(\frac{y}{x}\right)$$

which is of Lagrange's form. So, its auxiliary equations are

$$\frac{dx}{x} = \frac{dy}{-y} = \frac{dz}{z + x^2\phi_{1x}\left(\frac{y}{x}\right)}$$

From the first two members, we get

$$\frac{dx}{x}+\frac{dy}{y}=0 \Rightarrow xy = c_2$$

From the second and third members, we have

$$\frac{dz}{dy}+\frac{z}{y}=-\frac{x^2}{y}\phi_1\left(\frac{y}{x}\right)=-\frac{c_2^2}{y^3}\phi_1\left(\frac{y^2}{c_2}\right)$$

which is a linear differential equation with integrating factor (IF) $= e^{\int \frac{1}{y}dy} = e^{\log y} = y.$ So its solution is

$$z \cdot y=-\int \frac{c_2^2}{y^2}\phi_1\left(\frac{y^2}{c_2}\right)dy + c_3$$

$$= -\frac{{c_2}^{3/2}}{2}\int\left(\frac{c_2}{y^2}\right)\phi_1\left(\frac{y^2}{c_2}\right)\left(\frac{\sqrt{c_2}}{y}\right)d\left(\frac{y^2}{c_2}\right)+c_3$$

$$= c_2^{3/2}\phi_2\left(\frac{y^2}{c_2}\right)+c_3$$

Hence, the complete solution is

$$zy = x^{3/2}y^{3/2}\,\phi_2\left(\frac{y}{x}\right) + \phi_3(xy)$$

$$\Rightarrow \qquad z = x^{3/2}\sqrt{y}\,\phi_2\left(\frac{y}{x}\right)+\frac{1}{y}\phi_3(xy)$$

$$\Rightarrow \qquad z = x^2\sqrt{\frac{y}{x}}\,\phi_2\left(\frac{y}{x}\right)+x\frac{1}{xy}\,\phi_3(xy)$$

$$\Rightarrow \qquad z = x^2\psi_2\left(\frac{y}{x}\right) + x\phi_3(xy)$$

where ϕ_1, ϕ_2, ϕ_3, ψ_2 and ψ_3 are arbitrary functions.

EXAMPLE 9.15

Solve $q^2r - 2pqs + p^2t = pq^2$.

Solution: On comparing with $Rr + Ss + Tt = V$, we have

$$R = q^2,$$
$$S = -2pq,$$
$$T = p^2,$$
$$V = pq^2.$$

Here, the Monge's auxiliary equations are

$$q^2dpdy + p^2dqdx - pq^2dxdy = 0 \quad (1)$$

and $$q^2(dy)^2 + 2pqdxdy + p^2(dx)^2 = 0 \quad (2)$$

$\Rightarrow$ $$(qdy + pdx)^2 = 0$$

$\therefore$ $$qdy + pdx = 0 \quad (3)$$

$\Rightarrow$ $$dz = 0$$

so that $$z = c_1$$

From Eq. (1), we have

$$(qdp)(qdy) + (pdx)(pdq) - (qdy)(pqdx) = 0$$

$\Rightarrow$ $$qdp - pdq - pqdx = 0 \quad \{\text{Using Eq. (3)}\}$$

$\Rightarrow$ $$\frac{1}{p}dp - \frac{1}{q}\, dq - dx = 0$$

so that $$\log p - \log q - x = \log c_2$$

$\Rightarrow$ $$\frac{p}{q}e^{-x} = c_2$$

Thus, $$\frac{p}{q}e^{-x} = \phi(z)$$

$\Rightarrow$ $$pe^{-x} - \phi(z)q = 0$$

which is Lagrange's equation. So, its auxiliary equations are

$$\frac{dx}{e^{-x}} = \frac{dy}{-\phi(z)} = \frac{dz}{0}$$

$\therefore$ $$dz = 0 \Rightarrow z = c_1$$

Also, $$\frac{dx}{e^{-x}} = \frac{dy}{-\phi(c_1)}$$

$\Rightarrow$ $$e^x\phi(c_1)dx + dy = 0$$

Thus, $$e^x\phi(c_1) + y = c_3$$

Hence, the required complete solution is

$$e^x\phi(z) + y = \psi(z)$$

where, ϕ and ψ are arbitrary functions.

Monge's method for solving $Rr + Ss + Tt + U(rt - s^2) = V$

Consider the equation

$$Rr + Ss + Tt + U(rt - s^2) = V \tag{1}$$

where U is non-zero identically.

We have $$dp = rdx + sdy$$

and $$dq = sdx + tdy.$$

So that $$r = \frac{dp - sdy}{dx}$$

and $$t = \frac{dq - sdx}{dy}$$

Substituting these values of r and t in Eq. (1), we get

$$R\left(\frac{dp - sdy}{dx}\right) + Ss + T\left(\frac{dq - sdx}{dy}\right) + U\left(\frac{dp - sdy}{dx} - dq - \frac{sdx}{dy} - s^2\right) - V$$

or $dpdy + Tdqdx + Udpdq - Vdxdy = s\,(R(dy)^2 - Sdxdy + T(dx)^2 + Udpdx + Udqdy$ (2)

Hence, Monge's auxiliary equations are

$$Rdpdy + Tdqdx + Udpdq - Vdxdy = 0 \tag{3}$$

and $$R(dy)^2 - Sdxdy + T(dx)^2 + Udpdx + Udqdy = 0 \tag{4}$$

But in this case Eq. (4) cannot be factorized on account of the term $Udpdx + udqdy$ present in it. So, we try to factorize the equation obtained on multiplying Eq. (4) by λ and adding it to Eq. (3) where λ is some multiplier to be determined later.

Now, $$Rdpdy + Tdqdx + Udpdq - Vdxdy + \lambda[R(dy)^2 - Sdxdy + T(dx)^2 + Udpdx + Udqdy] = 0 \tag{5}$$

Let Eq. (5) be equivalent to

$$(ady + bdx + cdp)\,(\alpha dy + \beta dx + \gamma dq) = 0$$

or $a\alpha(dy)^2 + (a\beta + b\alpha)dxdy + b\beta(dx)^2 + c\alpha dpdy + c\beta dpdx + c\gamma dpdq + dydq + b\gamma dxdq = 0$ (6)

Comparing the coefficients in Eqs. (5) and (6), we have

$$a\alpha = \lambda R,\quad a\beta + b\alpha = -V - S\lambda,\quad b\beta = \lambda T,\quad c\alpha = R,$$

$$c\beta = \lambda U,\quad c\gamma = U,\quad \alpha\gamma = \lambda U \quad \text{and} \quad b\gamma = T.$$

The first relation will be satisfied by taking $a = \lambda$ and $\alpha = R$

with this choice $c = 1$, $\gamma = U$, $b = \dfrac{T}{U}$ and $\beta = \lambda U$

Substituting these values in the second relation, we have

$$U\lambda^2 + \frac{T}{U}R = -V - S\lambda$$

or
$$U^2\lambda^2 + SU\lambda + TR + UV = 0 \tag{7}$$

Let λ_1, λ_2 be the roots of Eq. (7)

In general λ_1 and λ_2 will be distinct.

Thus, Eq. (5) can be factorized as

$$(\lambda_1 Udy + Tdx + Udp)\,(Rdy + \lambda_1 Udx + Udq) = 0$$

and $(\lambda_2 Udy + Tdx + Udp)\,(Rdy + \lambda_2 Udx + Udq) = 0$

There are four systems to be considered. The system

$$\lambda_1 Udy + Tdx + Udp = 0$$

$$\lambda_2 Udy + Tdx + Udp = 0$$

implies
$$(\lambda_1 - \lambda_2)Udy = 0$$

and hence
$$Udy = 0$$

if
$$\lambda_1 \neq \lambda_2.$$

which will not furnish any solution.

Similarly, the system

$$Rdy + \lambda_1 Udx + Udq = 0$$

$$Rdy + \lambda_2 Udx + Udq = 0$$

implies
$$Udx = 0$$

Thus, the only two systems to be considered are

$$\left.\begin{aligned} \lambda_1 U dy + T dx + U dp &= 0 \\ R dy + \lambda_2 U dx + U dq &= 0 \end{aligned}\right\}$$

and

$$\left.\begin{aligned} \lambda_2 U dy + T dx + U dp &= 0 \\ R dy + \lambda_1 U dx + U dq &= 0 \end{aligned}\right\}$$

Each system if integrable yields an intermediate integral of Eq. (1) and then the procedure is same as discussed in the previous method.

EXAMPLE 9.16

Solve $qr + (p + x)s + yt + y(rt - s^2) + q = 0$.

Solution: On comparing the given equation with $Rr + Ss + Tt + U(rt - s^2) = V$, we have

$$R = q,\ S = p + x,\ T = y,\ U = y,\ V = -q.$$

Here, λ-equation $U^2\lambda^2 + SU\lambda + TR + UV = 0$

becomes $$y^2\lambda^2 + (p + x)y\lambda + (-yq + yq) = 0$$

$\Rightarrow$ $$\lambda = 0,\ -\frac{(x+p)}{y}.$$

$\therefore$ Two systems to be considered are

$$\left.\begin{aligned} (o)(y)dy + ydx + ydp &= 0 \\ qdy - \frac{(x+p)}{y} ydx + ydq &= 0 \end{aligned}\right\} \qquad (1)$$

and

$$\left.\begin{aligned} -\frac{(x+p)}{y}(y)dy + ydx + ydp &= 0 \\ qdy - (0)ydx + ydq &= 0 \end{aligned}\right\} \qquad (2)$$

From Eq. (1), we have $$dx + dp = 0$$

$\Rightarrow$ $$x + p = c_1$$

Thus, $$qdy - xdx + pdp + ydq = 0$$

$\Rightarrow$ $$d(qy) + d\left(\frac{p^2 - x^2}{2}\right) = 0$$

$\therefore$ $$2qy + p^2 - x^2 = c_2$$

Thus, the intermediate integral corresponding to Eq. (1) is

$$2qy + p^2 - x^2 = \phi_1(x + p) \tag{3}$$

From Eq. (2), we have

$$qdy + ydq = 0$$

$$\Rightarrow \quad d(qy) = 0 \quad \therefore qy = c_3.$$

Thus, $\quad -(x + p)dy + ydx + ydp = 0$

$$\Rightarrow \quad -\frac{dy}{y} + \frac{d(x+p)}{x+p} = 0$$

$$\therefore \quad \frac{x+p}{y} = c_4$$

Therefore, the intermediate integral corresponding to Eq. (2) is

$$yq = \phi_2\left(\frac{x+p}{y}\right) \tag{4}$$

Using Charpit's method to solve Eq. (4), the auxiliary equations are

$$\frac{dx}{-\frac{1}{y}\phi_2\left(\frac{x+p}{y}\right)} = \frac{dp}{\frac{1}{y}\phi_2\left(\frac{x+p}{y}\right)}$$

$$\therefore \quad dx + dp = 0$$

Integrating, we get

$$x + p = c_1$$

$$\therefore \quad p = c_1 - x$$

and Eq. (4) $\Rightarrow$

$$q = \frac{1}{y}\phi_2\left(\frac{c_1}{y}\right)$$

Thus,

$$dz = pdx + qdy$$

gives

$$dz = (c_1 - x)dx + \frac{1}{y}\phi_2\left(\frac{c_1}{y}\right)dy$$

$$= (c_1 - x)dx + \left(-\frac{y}{c_1}\right)\phi_2\left(\frac{c_1}{y}\right)d\left(\frac{c_1}{y}\right)$$

Integrating, we have

$$z = c_1x - \frac{x^2}{2} + \psi_2\left(\frac{c_1}{y}\right) + c_5$$

$$\Rightarrow \qquad z = c_1 x - \frac{x^2}{2} + \psi_2\left(\frac{c_1}{y}\right) + \psi_1(c_1)$$

where ψ_1 and ψ_2 are arbitrary functions.

EXAMPLE 9.17

Solve $s^2 - rt = a^2$.

Solution: On comparing with $Rr + Ss + Tt + U(rt - s^2) = V$,

we have $\quad R = O = S = T, U = -1, V = a^2$.

Here, λ equation $\quad U^2\lambda^2 + SU\lambda + TR + UV = 0$

becomes $\quad (y)^2\lambda^2 - a^2 = 0 \Rightarrow \lambda^2 = a^2$

$$\therefore \qquad \lambda = \pm a.$$

$\therefore$ Two systems to be considered are

$$\left.\begin{aligned} -ady - dp &= 0 \\ adx - dq &= 0 \end{aligned}\right\} \tag{1}$$

and

$$\left.\begin{aligned} ady - dp &= 0 \\ -adx - dq &= 0 \end{aligned}\right\} \tag{2}$$

From Eq. (1), we have $\quad dp + ady = 0$

so that $\quad p + ay = c_1$

and $\quad dq - adx = 0$

$$\Rightarrow \qquad q - ax = c_2.$$

$\therefore$ The intermediate integral corresponding to Eq. (1) is

$$p + ay = \phi_1(q - ax) \tag{3}$$

From Eq. (2), we have

$$dp - ady = 0 \quad \Rightarrow \quad p - ay = c_3$$

and $\quad dq + adx = 0 \quad \Rightarrow \quad q + ax = c_4$.

$\therefore$ The intermediate integral corresponding to Eq. (2) is

$$p - ay = \phi_2(q + ax) \tag{4}$$

From Eqs. (3) and (4), we have

$$p = \frac{1}{2}[\phi_1(q - ax) + \phi_2(q + ax)]$$

and

$$y = \frac{1}{2a}[\phi_1(q - ax) - \phi_2(q + ax)]$$

Now, $dz = pdx + qdy$ gives

$$z = \int pdx + qy - \int ydq$$

$$\Rightarrow \qquad z = -yq = \frac{1}{2}\int[\phi_1(q - ax) + \phi_2(q + ax)]dx - \frac{1}{2a}\int[\phi_1(q - ax) - \phi_2(q + ax)]dq$$

$$= \frac{1}{2a}\int\phi_1(q-ax)\frac{d(ax-q)}{a} + \frac{1}{2}\int\phi_2(q+ax)\frac{d(ax+q)}{a}$$

$$= \psi_1(q - ax) + \psi_2(q + ax)$$

Thus, the complete solution is

$$z - qy = \psi_1(q - ax) + \psi_2(q + ax)$$

where

$$q = -\psi_1'(q-ax) - \psi_2'(q+ax)$$

$$\psi_1(q-ax) = \frac{1}{2}\int\phi_1(q-ax)\frac{d(ax-q)}{a}$$

$$\psi_2(q+ax) = \frac{1}{2}\int\phi_2(q+ax)\frac{d(ax+q)}{a}$$

$$\psi_1'(q-ax) = \frac{-1}{2a}\phi_1(q-ax)$$

$$\psi_2'(q+ax) = \frac{1}{2a}\phi_2(q+ax)$$

EXAMPLE 9.18

Solve $z(1 + q^2)r - 2pqzs + z(1 + p^2)t - z^2(s^2 - rt) + 1 + p^2 + q^2 = 0.$

Solution: On comparing with $Rr + Ss + Tt + U(rt - s^2) = V$;

we have $R = z(1 + q^2)$, $S = -2pqz$, $T = z(1 + p^2)$, $U = yz^2$, $V = -(1 + p^2 + q^2)$.

Here, λ-equation $U^2\lambda^2 + SU\lambda + TR + UV = 0$

becomes $(z^2)^2\lambda^2 - 2pqz^3\lambda + z^2(1 + p^2)(1 + q^2) - z^2(1 + p^2 + q^2) = 0$

$$\Rightarrow \qquad z^2\lambda^2 - 2pqz\lambda + p^2q^2 = 0$$

$$\Rightarrow \qquad (z\lambda - pq)^2 = 0$$

$$\Rightarrow \qquad \lambda = \frac{pq}{z}, \frac{pq}{z}$$

$\therefore$ The system to be considered is

$$\left.\begin{aligned} pqzdy + z(1 + p^2)dx + z^2dp &= 0 \\ z(1 + q^2)dy + zpqdx + z^2dq &= 0 \end{aligned}\right\} \qquad (1)$$

$$\therefore \qquad pqdy + (1 + p^2)dx + zdp = 0$$

$$\Rightarrow \quad p(qdy + pdx) + dx + zdp = 0$$

$$\Rightarrow \quad pdz + dx + zdp = 0$$

$$\Rightarrow \quad d(pz + x) = 0$$

Integrating, we get

$$pz + x = c_1$$

Similarly, $$(1 + q^2)dy + pqdx + zdq = 0$$

$$\Rightarrow \quad qz + y = c_2$$

∴ The intermediate integral corresponding to Eq. (1) is

$$qz + y = \phi(pz + x)$$

Let $pz + x = a$ and $qz + y = b$

$$\therefore \quad p = \frac{a-x}{z} \quad \text{and} \quad q = \frac{b-y}{z}$$

Thus, $dz = pdx + qdy$ gives

$$zdz = (a - x)dx + (b - y)dy$$

Integrating, we get

$$z^2 = -(a - x)^2 - (b - y)^2 + c^3$$

which is the general solution of the given equation.

EXAMPLE 9.19

Solve $2s + (rt - s^2) = 1$.

Solution: On comparing with $Rr + Ss + Tt + U(rt - s^2) = V$;

we have $R = 0, S = 2, T = 0, U = 1, V = 1$

Here, the λ-equation $U^2\lambda^2 + SU\lambda + TR + UV = 0$

becomes $\lambda^2 + 2\lambda + 1 = 0 \Rightarrow (\lambda + 1)^2 = 0$

$$\Rightarrow \quad \lambda = -1, -1$$

The system to be considered is

$$\left.\begin{aligned} -dy + (0)dx + dp &= 0 \\ (0)dy - dx + dq &= 0 \end{aligned}\right\} \quad (1)$$

which gives $y - p = a$ and $x - q = b$.

Hence, the only intermediate integral is

$$y - p = \phi(x - q)$$

Also, $p = y - a$ and $q = x - b$

Thus, $$dz = pdx + qdy \text{ gives}$$

$$dz = (y - a)dx + (x - b)dy$$

Integrating, we have

$$z = xy - ax - by + c$$

which is the general solution.

EXAMPLE 9.20

Solve $5r + 6s + 3t + 2(rt - s^2) + 3 = 0$.

Solution: On comparing with $Rr + Ss + Tt + U(rt - s^2) = V$;

we have $R = 5, S = 6, T = 3, U = 2, Vv = -3$

$\therefore$ The λ-equation $U^2\lambda^2 + SU\lambda + TR + UV = 0$ becomes

$$4\lambda^2 + 12\lambda + 15 - 6 = 0$$

$$\Rightarrow \quad 4\lambda^2 + 12\lambda + 9 = 0$$

$$\Rightarrow \quad (2\lambda + 3)^2 = 0$$

$$\therefore \quad \lambda = -3/2, -3/2.$$

$\therefore$ The system to be considered is

$$\left.\begin{aligned} -\frac{3}{2}(2)dy + 3dx + 2dp &= 0 \\ 5dy - \frac{3}{2}(2)dx + 2dq &= 0 \end{aligned}\right\} \quad (1)$$

which gives $$-3y + 3x + 2p = a \text{ and } 5y - 3x + 2q = b.$$

Thus, $$p = \frac{1}{2}(a - 3x + 3y) \quad \text{and} \quad q = \frac{1}{2}(b + 3x + 5y)$$

Hence, $$dz = pdx + qdy \text{ gives}$$

$$dz = \frac{1}{2}(a - 3x + 3y)dx + \frac{1}{2}(b + 3x - 5y)dy$$

Integrating, we have

$$2z = ax - \frac{3}{2}x^2 + 3xy + by - \frac{5y^2}{2} + c$$

$$\Rightarrow \quad 4z = 2ax - 3x^2 + 6xy + 2by - 5y^2 + c'.$$

which is the required general solution.

EXAMPLE 9.21

Solve $ar + bs + ct + e(rt - s^2) = h$ where, a, b, c, e, h are constants.

Solution: On comparing with $Rr + Ss + Tt + U(rt - s^2) = V$, we have $R = a, S = b, T = c, U = e, V = h$.

Here, the λ-equation $U^2\lambda^2 + SU\lambda + TR + UV = 0$ becomes

$$e^2\lambda^2 + be\lambda + ac + eh = 0$$

Let λ_1, λ_2 be its roots, which are constants.

$\therefore$ The systems to be considered are

$$\left.\begin{aligned} \lambda_1 edy + cdx + edp &= 0 \\ ady + \lambda^2 edx + edq &= 0 \end{aligned}\right\} \quad (1)$$

and

$$\left.\begin{aligned} \lambda^2 edy + cdx + edp &= 0 \\ ady + \lambda_1 edx + edq &= 0 \end{aligned}\right\} \quad (2)$$

From Eq. (1), we have

$$\lambda_1 ey + cx + ep = c_1$$

and

$$ay + \lambda_2 ex + eq = c_2$$

Hence, the intermediate integral corresponding to Eq. (1) is

$$\lambda_1 ey + cx + ep = \phi_1(ay + \lambda_2 ex + eq) \quad (3)$$

From Eq. (2), we have

$$\lambda_2 ey + cx + ep = 0 \quad \text{and} \quad ay + \lambda_1 ex + eq = 0$$

Hence, the intermediate integral corresponding to Eq. (2) is

$$\lambda^2 ey + cx + ep = \phi_2(ay + \lambda_1 ex + eq) \quad (4)$$

Here, p and q cannot be directly found out, so we combine any particular integral of the second with the general integral of the first system. Thus, we take

$$\lambda_2 eg + cx + ep = k_1 \quad (5)$$

$\therefore$ Eq. (3) gives

$$\lambda_1 ey + cx + ep = \phi_1(ay + \lambda_2 ex + eq)$$

$$\Rightarrow \quad \lambda_1 ey + (k_1 - \lambda_2 ey) = \phi_1(ay + \lambda_2 ex + eq)$$

$$\Rightarrow \quad (\lambda_1 - \lambda_2)ey + k_1 = \phi_1(ay + \lambda_2 ex + eq)$$

$$\Rightarrow \quad \phi^{-1}[(\lambda_1 - \lambda_2)ey + k_1] = ay + \lambda_2 ex + eq$$

$$\Rightarrow \quad \psi((\lambda_1 - \lambda_2)ey + k_1) = ay + \lambda_2 ex + eq \quad (6)$$

From Eqs. (5) and (6), we have

$$dz = pdx + qdy$$

$$edz = (k_1 - cx - \lambda_2 ey)dx + (\psi - ay - \lambda_2 ex)dy$$

Integrating, we get

$$ez = k_1x - \frac{cx^2}{2} - \lambda_2 exy + \int \psi dy - \frac{ay^2}{2} + k_2$$

$$\Rightarrow \quad ez = k_1x - \frac{cx^2}{2} - \lambda_2 exy + \psi_1((\lambda_1 - \lambda_2)ey + k_1) - \frac{ay^2}{2} + k_2$$

where ψ_1 is an arbitrary functions and k_2 is an arbitrary constant.

EXAMPLE 9.22

Solve $r + 3s + t + (rt - s^2) = 1$.

Solution: On comparing with $Rr + Ss + Tt + U(rt - s^2) = V$;
we have $R = 1, S = 3, T = 1, U = 1, V = 1$

Here, the λ-equation $U^2\lambda^2 + SU\lambda + TR + UV = 0$ becomes

$$\lambda^2 + 3\lambda + 1 + 1 = 0$$

$$\Rightarrow \quad \lambda^2 + 3\lambda + 2 = 0$$

$$\Rightarrow \quad \lambda = -1, -2$$

$\therefore$ The systems to be considered are

$$\left.\begin{aligned} -dy + dx + dp &= 0 \\ dy - 2dx + dq &= 0 \end{aligned}\right\} \quad (1)$$

and

$$\left.\begin{aligned} -2dy + dx + dp &= 0 \\ dy - dx + dq &= 0 \end{aligned}\right\} \quad (2)$$

From Eq. (1), we have

$$y - x - p = c_1$$

and

$$y - 2x + q = c_2.$$

Hence, the intermediate integral corresponding to Eq. (1) is

$$y - x - p = \phi_1(y - 2x + q) = \phi_1(a)$$

From Eq. (2), we have

$$2y - x - p = c_3$$

and

$$y - x + q = c_4.$$

Hence, the intermediate integral corresponding to Eq. (2) is

$$2y - x - p = \phi_2(y - x + q) = \phi_2(b)$$

Thus,

$$a = y - 2x + q$$

$$b = y - x + q$$

$$\therefore \qquad x = b - a \quad \text{and} \quad y = \phi_2(b) - \phi_1(a)$$

$$p = y - x - \phi_1(a) \quad \text{and} \quad q = b - y + x$$

Thus, $dz = pdx + qdy$ gives

$$dz = (y - x - \phi_1(a))dx + (b - y + x)dy$$

$$= -(x - y)\ (dx - dy) - \phi_1(a)\ (db - da) + b(\phi_2'(b)db - \phi_1'(a)da)$$

Integrating, we get

$$z = -\frac{1}{2}(x - y)^2 + \int \phi_1(a)da - \int (\phi_1(a)db + b\phi_1'(a)da + \int b\phi_2'(b)db$$

$$\Rightarrow \qquad z = -\frac{1}{2}(x - y)^2 + \psi_1(a) - b\phi_1(a) + \psi_2(b)$$

which is the required general solution.

EXAMPLE 9.23

Solve $rt - s^2 - s(\sin x + \sin y) = \sin x \sin y$.

Solution: On comparing with $Rr + Ss + Tt + U(rt - s^2) = V$;
we have

$$R = O = T,\ S = -(\sin x + \sin y),\ U = 1;\ V = \sin x \sin y$$

$\therefore$ λ-equation $U^2\lambda^2 + SU\lambda + TR + UV = 0$

becomes $\lambda^2 - (\sin x + \sin y)\lambda + \sin x \sin y = 0$

$$= \lambda = \sin x, \sin y$$

$\therefore$ The systems to be considered are

$$\left.\begin{aligned} \sin x dy + dp &= 0 \\ \sin y dx + dq &= 0 \end{aligned}\right\} \qquad (1)$$

and

$$\left.\begin{aligned} \sin y dy + dp &= 0 \\ \sin x dx + dq &= 0 \end{aligned}\right\} \qquad (2)$$

From Eq. (2), we have $\qquad p - \cos y = c_1$

and $\qquad q - \cos x = c_2$

$\therefore$ The intermediate integral corresponding to Eq. (2) is

$$p - \cos y = \phi(q - \cos x) \qquad (3)$$

Using Charpit's method to solve Eq. (3), the auxiliary equations are

$$\frac{dx}{-1} = \frac{dy}{\phi(q - \cos x)} = \frac{dp}{-\sin x\phi'(q - \cos x)} = \frac{dq}{\sin y} \tag{4}$$

Evidently, Eq. (4) cannot be integrated, so to obtain an integral of most general form, assume the arbitrary function ϕ to be linear

Let

$$p - \cos y = m(q - \cos x) + n \tag{5}$$

where m and n are constants.

Now, Lagrange's auxiliary equations are

$$\frac{dx}{1} = \frac{dy}{-m} = \frac{dz}{\cos y - m\cos x + n}$$

From the first two members, we get

$$y + mx = c_1$$

From the first and third members, we have

$$dz = (\cos(c_1 - mx) - m(\cos x + n)dx.$$

Integrating, we get

$$z = \frac{\sin(c_1 - mx)}{-m} - m \sin x + nx + c_2$$

$$\Rightarrow \quad mz + \sin y + m^2 \sin x - mnx \quad c_2 m = 0$$

$\therefore$ The most general integral possible is

$$mz + \sin y + m^2 \sin x - mnx = m\phi(y + mx).$$

EXAMPLE 9.24

Solve $2pr + 2qt - 4pq(rt - s^2)$.

Solution: On comparing with $Rr + Ss + Tt + U(Rt - s^2) = V$;

we have $R = 2p$, $S = 0$, $T = 2q$, $U = -4pq$, $V = 1$

$\therefore$ λ-equation $U^2\lambda^2 + SU\lambda + TR + UV = 0$ becomes

$$16p^2q^2\lambda^2 + 4pq - 4pq = 0$$

$$\Rightarrow \quad \lambda = 0, 0$$

$\therefore$ the system to be considered is

$$\left.\begin{aligned} 2qdx - 4pqdp &= 0 \\ \text{and} \quad 2pdy - 4pqdq &= 0 \end{aligned}\right\} \tag{1}$$

$$\Rightarrow \quad dx - 2pdp = 0$$

and $$dy - 2qdq = 0$$

Integrating, we have $$p^2 - x = a$$

and $$q^2 - y = b$$

Thus, $$p = \pm(a + x)^{1/2}$$

and $$q = \pm(b + y)^{1/2}$$

Hence, $$dz = pdx + qdy \text{ gives}$$

$$dz = \pm(a + x)^{1/2}\, dx \pm (b + y)^{1/2} dy$$

$\Rightarrow$ $$3z = \pm 2(a + x)^{3/2} \pm 2(b + y)^{3/2} + c$$

which is the complete integral with a, b, c as arbitrary constants.

EXAMPLE 9.25

Solve $2r + e^x t - (rt - s^2) = 2e^x$.

Solution: On comparing with $Rr + Ss + Tt + U(rt - s^2) = V$,

we have $R = 2$, $S = 0$, $T = e^x$, $U = -1$, $V = 2e^x$.

$\therefore$ λ-equation $U^2\lambda^2 + SU\lambda + TR + UV = 0$ gives $\lambda^2 + 2e^x - 2e^x = 0$

$\Rightarrow$ $$\lambda = 0, 0$$

Thus, the system to be considered is

$$\left.\begin{aligned} e^x dx - dp &= 0 \\ 2dy - dq &= 0 \end{aligned}\right\} \quad (1)$$

Integrating, we have

$$e^x - p = a \quad \text{and} \quad 2y - q = b$$

Thus $$p = e^x - a \quad \text{and} \quad q = 2y - b$$

Hence, $$dz = pdx + qdy \text{ gives}$$

$$dz = (e^x - a)dx + (2y - b)dy$$

Integrating, we get

$$z = e^x - ax + y^2 - by + c$$

which is the required complete integral with a, b, c as arbitrary constants.

EXAMPLES 9.26

Solve $rt - s^2 + 1 = 0$.

Solution: On comparing with $Rr + Ss + Tt + U(rt - s^2) = V$,

we have $R = S = T = 0$, $U = 1$, $V = -1$

$\therefore$ λ-equation $U^2\lambda^2 + SU\lambda + TR + UV = 0$ becomes $\lambda^2 - 1 = 0 \Rightarrow \lambda = 1, -1$

Since the two values of λ are distinct, we shall get two intermediate integrals which are given by the following set of system of equations

$$\left.\begin{aligned} dy - dp &= 0 \\ -dx - dq &= 0 \end{aligned}\right\} \qquad (1)$$

and

$$\left.\begin{aligned} -dy - dp &= 0 \\ dx - dq &= 0 \end{aligned}\right\} \qquad (2)$$

Integrating, we get

$$p - y = c_1; \quad q + x = c_2$$

and

$$p + y = c_3; \quad q - x = c_4$$

Hence, the two intermediate integrals are given by

$$p - y = \phi_1(q + x) \quad \text{and} \quad p + y = \phi_2(q - x)$$

Let

$$q + x = k_1 \quad \text{and} \quad q - x = k_2$$

$\therefore$

$$p - y = \phi_1(k_1) \quad \text{and} \quad p + y = \phi_2(k)$$

Thus,

$$x = \frac{k_1 - k_2}{2}, \; y = \frac{\phi_2(k_2) - \phi_1(k_1)}{2}$$

$$p - y + \phi_1(k_1), \quad q = x + k_2$$

Hence, $dz = pdx + qdy$ gives

$$dz = (y + \phi_1(k_1)dx + (x + k_2)dy$$

$$\Rightarrow \qquad z = d(xy) + \phi_1(k_1)\left(\frac{dk_1 - dk_2}{2}\right) + k_2\left(\frac{\phi_2{}'(k_2)dk_2 - \phi'{}_1(k_1)dk_1}{2}\right)$$

$$\Rightarrow \qquad 2z = d(2xy) + \phi(k_1)dk_1 - (\phi_1(k_1)dk_2 + k_2\phi_1'(k_1)dk_1) + \phi_2'(k_2)(k_2^2)dk_2$$

Integrating, we have

$$2z = 2xy + \psi_1(k_1) - k_2\phi_1(k_1) + \psi_2(k_2)$$

which is the required solution in parametric form with k_1 and k_2 being parameter and ϕ_1 and ψ_1, ψ_2 being arbitrary functions.

▲▲

EXERCISE 9.1

Solve the following differential equations using Monge's methods:

1. $r = t$.
2. $(r - s)x = (t - s)y$.
3. $r + ka^2t = 2as$.

4. $y^2r - 2ys + t = p + by$.
5. $3r + 4s + t + (rt - s^2) = 1$.
6. $(b + cq^2)r - 2(b + cq)(a + cp)s + (a + cp)^2t = 0$.
7. $x^2r + 2xys + y^2t = 0$.
8. $q(yq + z)r - p(2yq + z)s + yp^2t + p^2q = 0$.
9. $x^{-2}r - y^{-2}t = x^{-3}p - y^{-3q}$.
10. $rx^2 - 3sxy + 2ty^2 + px + 2qy = x + 2y$.
11. $(r - s)y + (s - t)x + q - p = 0$.
12. $(x - y)(xr - xs - ys + yt) = (x + y)(p - q)$.
13. $(r - t)xy - s(x^2 - y^2) = qx - py$.
14. $xr + (x + y)s + yt + p + q = 0$.
15. $x^2r - y^2t = xp - yq$.
16. $y^2r + 2xys + x^2t + px + qy = 0$.
17. $q^2r - 2pqs + p^2t = 0$.
18. $x^2r - 2xs + t + q = 0$.
19. $x^2r + 2xys + y^2t = 0$.
20. $(e^x - 1)(pr - qs) = pqe^x$.
21. $2s + (rt - s^2) = 1$.
22. $3s + (rt - s^2) = 2$.
23. $r + 4s + t + rt - s^2 = 2$.
24. $7r - 8s - 3t + (rt - s^2) = 36$.
25. $(1 + q^2)r - 2pqs + (1 + p^2)t + (1 + p^2 + q^2)^{-1/2}(rt - s^2) = -(1 + p^2 + q^2)^{3/2}$.
26. $qxr + (x + y)s + pyt + xy(rt - s^2) = 1 - pq$.
27. $r + t - (rt - s^2) = 1$.
28. $rt - s^2 + 1 = 0$.

ANSWERS

Exercise 9.1

1. $z = \phi_1(y + x) + \phi_2(y - x)$.
2. $z = \phi_1(x + y) + (x + y)\phi_2(y/x)$.
3. $z = \phi_1(y + a(1 + k)x) + \phi_2(y + a(1 - k)x)$.
4. $z = y^3 - y\phi_1(y^2 + 2x) + \phi_2(y^2 + 2x)$.

5. $z = 2xy - \frac{1}{2}(x^2 + 3y^2) + nx + \phi(y + mx)$.
6. $y + x\phi_1(ax + by + cz) = \phi_2(ax + by + cz)$.
7. $z + \phi_1(y/x) = x\phi_2(y/x)$.
8. $\phi_1(z) = x + \phi_2(yz)$.
9. $z = \phi_1(x^2 + y^2) + \phi_2(x^2 - y^2)$.
10. $z = x + y + \phi_1(x^2y) + \phi_2(xy)$.
11. $z - \phi_1(x + y) = \phi_2(x^2 - y^2)$.
12. $\phi_1(xy) + z = \phi_2(x + y)$.
13. $z = \phi_1(x^2 + y^2) + \phi_2(y/x)$.
14. $z = \phi_1(x/y) + \phi_2(x - y)$.
15. $z = x^2\phi_1(y/x) + \phi_2(xy)$.
16. $z - \phi_1(x^2 - y^2) \log (x + y) = \phi_2(x^2 - y^2)$.
17. $y + x\phi_1(z) = \phi_2(z)$.
18. $z - x\phi_1(y + \log x) = \phi_2(y + \log x)$.
19. $z = y\phi_1(y/x) + \phi_2(y/x)$.
20. $x = e^x + \phi_1(y) + \phi_2(z)$.
21. $z = xy - ax - by + c$.
22. $z = xy + \beta[\phi'(\alpha) - \phi'(\beta)] - \psi(\alpha) - \psi(\beta)$.
23. $2z = 2xy - x^2 - y^2 + \beta[\phi'(\alpha) - \phi'(\beta)] - \psi(\alpha) - \psi(\beta)$.
24. $2z = 3x^2 - 7y^2 - \log y - \beta[\phi'(\alpha) - \phi'(\beta)] - \psi(\alpha) - \psi(\beta)$.
25. $(x - a)^2 + (y - b)^2 + (z - c)^2 = 1$.
26. $z + mx + \frac{y}{m} - n \log x = \phi(x^m y)$.
27. $z = \frac{1}{2}(x^2 + y^2) + ax + by + c$.
28. $z = xy + \frac{1}{2}[\phi(\alpha) - \psi(\beta)] + \beta y$.

where $x = \frac{1}{2}(\alpha - \beta)$, $y = \frac{1}{2}[\psi'(\beta) - \psi'(\alpha)]$.

▲▲

10

CHAPTER

Applications of Partial Differential Equations

In theoretical physics and engineering, partial differential equations generally arise from the mathematical formulation of real physical problems.

In these problems, we always seek a solution which satisfies some specified conditions known as the boundary conditions. The differential equation together with these boundary conditions constitutes a boundary value problem. In problems involving ordinary differential equations, we may first find the general solution and then determine the arbitrary constants from the initial values. But the same process is not applicable to problems involving partial differential equations for the general solution of a partial differential equation contains arbitrary functions which are difficult to adjust so as to satisfy the given boundary conditions.

Method of Separation of Variables

Separation of variables is a powerful technique to solve partial differential equations. For a partial differential equation let the required solution $u(x, y)$ be separable as

$$u(x, y) = X(x)Y(y) \tag{1}$$

where $X(x)$ is a function of x only and $Y(y)$ is a function of y only. Then substitution of $u(x, y)$ from Eq. (1) and its derivatives, reduces the partial differential equation to the form

$$f(X, X', X'', \ldots) = g(Y, Y', Y'', \ldots) \tag{2}$$

which is separation in X and Y. Since the left hand side of Eq. (2) is a function of x only and the right hand side of Eq. (2) is a function of y only, Eq. (2) must be equal to a common constant say k. Thus, Eq. (2) reduces to

$$f(X, X', X'', \ldots) = k \tag{3}$$

$$g(Y, Y', Y'', \ldots) = k \tag{4}$$

Thus, the determination of solution to partial differential equation reduces to the determination of solution to two ordinary differential equations with appropriate conditions. ▲▲

EXAMPLE 10.1

Solve $u_{xx} - u_y = 0$ by separation of variables.

Solution: Let $u(x, y) = X(x)\, Y(y)$ be the solution.

$$\therefore \qquad u_{xx} = X''Y \text{ and } u_y = XY'$$

So, the given partial differential equation reduces to

$$2X''y - Xy' = 0$$

$$\Rightarrow \qquad \frac{2X''}{X} = \frac{y'}{Y} = 2k = \text{constant.}$$

Solving $X'' - kX = 0$, we have

$$X(x) = c_1 e^{\sqrt{k}x} + c_2 e^{-\sqrt{k}x}$$

and solving $Y' - 2kY = 0$, we have

$$Y(y) = c_3 e^{2ky}$$

Hence, the required solution is

$$u(x, y) = X(x)Y(y) = (c_1 e^{\sqrt{k}x} + c_2 e^{-\sqrt{k}x})\,(c_3 e^{2ky})$$

$$\Rightarrow \qquad u(x, y) = (Ae^{\sqrt{k}x} + Be^{-\sqrt{k}x})e^{2ky}$$

EXAMPLE 10.2

Solve $\frac{\partial^2 z}{\partial x^2} - 2\frac{\partial z}{\partial x} + \frac{\partial z}{\partial y} = 0$ by the method of separation of variables.

Solution: Let $u(x, y) = X(x)\, Y(y)$ be the solution.

Thus, the given equation reduces to

$$X''Y - 2X'Y + XY' = 0$$

$$\Rightarrow \qquad (X'' - 2X')Y + XY' = 0$$

$$\Rightarrow \qquad \frac{X'' - 2X'}{X} = -\frac{Y'}{Y} = k = \text{constant.}$$

Solving $X'' - 2X' - kX = 0$, we have

$X(x) = c_1 e^{(-1+\sqrt{1+k})x} + c_2 e^{(-1-\sqrt{1+k})x}$ and solving $Y' + kY = 0$, we have

$Y(y) = c_3 e^{-ky}$.

Hence, the required solution is

$$u(x, y) = X(x)\,Y(y) = c_1 e^{(-1+\sqrt{1+k})x} + c_2 e^{(-1-\sqrt{1+k})x} \times c_3 e^{-ky}$$

$$\Rightarrow \quad u(x, y) = (Ae^{(-1+\sqrt{1+k})x} + Be^{(-1-\sqrt{1+k})x})e^{-ky}$$

EXAMPLE 10.3

Use the separation of variables technique to solve $3u_x + 2u_y = 0$ with $u(x, 0) = ye^{-x}$.

Solution: Let $u(x, y) = X(x) \cdot Y(y)$ be the solution.

Then, the given partial differential equation becomes

$$3X'Y + 2XY' = 0$$

$$\Rightarrow \quad \frac{X'}{X} = -\frac{2}{3}\frac{Y'}{Y} = k = \text{constant.}$$

Solving $X' - kX = 0$, we have $X(x) = c_1 e^{kx}$

and solving $Y' \mid \frac{3}{2} Y = 0$, wc havc $Y(y) = c_2 c^{-\frac{3}{2}ky}$.

So $u(x, y) = X(x)\,Y(y) = Ae^{kx - \frac{3}{2}ky}$

$$= Ae^{\frac{k}{2}(2x-3y)}$$

Given that $u(x, 0) = Ae^{kx} = 4e^{-x}$

Thus, $\quad A - 4, K - -1.$

Hence, the required solution is

$$u(x, y) = Ye^{-\frac{1}{2}(2x-3y)}.$$

▲▲

ONE-DIMENSIONAL WAVE EQUATION

The classical one-dimensional wave equation arises in the study of transverse vibrations of an elastic string or torsional oscillations or longitudinal vibrations of a rod. We will derive one-dimensional wave equation by studying the problem of a vibrating string. Consider an elastic string, stretched to its length l and aligned along the x-axis with its two ends $x = 0$ and $x = l$. Let ρ be the constant density of the string, that is, the string is homogeneous. Let the function $u(x, t)$ denote the displacement of string at any point x and at any time $t > 0$ from the equilibrium position (x-axis). When the string is distorted, then it vibrates. The small transverse vibrations of such a vibrating string are mathematically modelled by one-dimensional wave equation.

Assume that the string is perfectly flexible and offers no resistance to bending. Applying Newton's second law of motion for a small portion of the string between x and $x + \Delta x$, the equation is derived. Let T_1 and T_2 be the tensions at the end points P and Q of this portion of the string.

Assuming that points on the string move only in the vertical direction and, these is no motion in the horizontal direction. Thus, the sum of the forces in the horizontal direction must be zero.

$$\therefore \qquad -T_1 \cos \alpha + T_2 \cos \beta = 0$$

$$\Rightarrow \qquad T_1 \cos \alpha = T_2 \cos \beta = T = \text{a constant} \tag{5}$$

Neglecting the gravitational force on the string the only two forces acting on the string are the vertical components of tension $-T_1 \sin \alpha$ at P and $T_2 \sin \beta$ at Q with upward direction taken as positive. By Newton's second law

Resultant of forces = Mass × Acceleration.

$$\therefore \qquad T_2 \sin \beta - T_1 \sin \alpha = (\rho \Delta x)\left(\frac{\partial^2 x}{\partial t^2}\right) \tag{6}$$

From Eqs. (5) and (6), we obtain

$$\frac{T_2 \sin\beta}{T_2 \cos\beta} - \frac{T_1 \sin\alpha}{T_1 \cos\alpha} = \frac{\rho\Delta x}{T}\frac{\partial^2 x}{\partial t^2}$$

$$\Rightarrow \qquad \tan\beta - \tan\alpha = \frac{\rho\Delta x}{T}\frac{\partial^2 x}{\partial t^2} \tag{7}$$

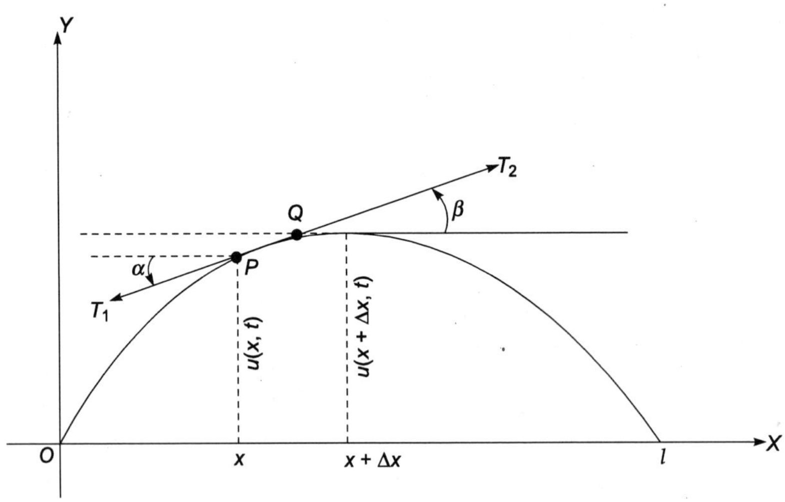

Fig. 10.1

Now $\tan \alpha = \dfrac{\partial u(x,t)}{\partial x}$ and $\tan \beta = \dfrac{\partial u(x+\Delta x,t)}{\partial x}$

$$\therefore \qquad \frac{\partial u(x+\Delta x,t)}{\partial x}-\frac{\partial u(x,t)}{\partial x}=\frac{\rho\Delta x}{T}\frac{\partial^2 u}{\partial t^2}$$

Taking the limit as $\Delta x \to 0$, we obtain

$$\frac{\partial^2 u}{\partial x^2}=\frac{\rho}{T}\frac{\partial^2 u}{\partial t^2}=\frac{1}{c^2}\frac{\partial^2 u}{\partial t^2} \tag{8}$$

where, $c^2=\dfrac{T}{\rho}$ Eq. (8) is known as one-dimensional wave equation.

We have the boundary conditions

$$u(0, t) = 0, u(l, t) = 0.$$

The force of motion of the string will depend on the initial displacement

$$\therefore \qquad u(x, 0) = f(x)$$

and the initial velocity, $\left.\dfrac{\partial u}{\partial l}\right|_{t=0}=g(x)$

This initial boundary value problem can be solved by the method of separation of variables.

Let $\qquad u(x, t) = X(x)T(t)$

Thus, Eq. (8) becomes

$$X''T = c^2XT''$$

$$\Rightarrow \qquad \frac{X''}{X}=\frac{T''}{c^2T}$$

Since left hand side is a function of x only and right hand side is a function of t only, so both sides must be equal to common constant k. This results in two ordinary differential equations as

$$X'' - kX = 0 \quad \text{and} \quad T'' - kc^2T = 0$$

with boundary conditions

$$0 = u(0, t) = X(0)T(t) \;\Rightarrow\; X(0) = 0$$

$$0 = u(l, t) = X(l)T(t) \;\Rightarrow\; X(l) = 0.$$

It can be shown that for $k \geq 0$, only trivial solution $u(x, t) = 0$ exists. So for non-trivial solution, consider $k = -\lambda^2 < 0$.

Then, the solution of $X'' + \lambda^2X = 0$ is

$$X(x) = A\cos\lambda x + B\sin\lambda x$$

Using $X(0) = 0$, we have $\quad A = 0$

$$\therefore \qquad X(x) = B\sin\lambda x$$

Using $X(l) = 0$, we have $B\sin\lambda l = 0$

For $u(x, t) \neq 0$, B should no be zero.

$$\therefore \qquad \sin \lambda l = 0$$

$$\Rightarrow \qquad \lambda l = n\pi$$

$$\therefore \qquad \lambda_n = \frac{n\pi}{l} \text{ for } n = 1, 2, 3, \ldots$$

Thus, $X_n(x) = \sin\left(\frac{n\pi}{l}x\right)$ are the infinitely many solutions of $X'' + \lambda^2 X = 0$. With $k = -\lambda^2 = -\frac{n^2\pi^2}{l^2}$, the solution of $T'' - kc^2T = 0$; is

$$T_n(t) = A_n \cos \lambda_n ct + B \sin \lambda_n ct$$

$$\therefore \qquad u_n(x, t) = X_n(x)T_n(t) \qquad \{n = 1, 2, 3, \ldots\}$$

By the principle of superposition, the general solution of (8) is

$$u(x, t) = \sum_{n=0}^{\infty} u_n(x, t) = \sum_{n=0}^{\infty} \{A_n \cos (\lambda_n ct) + B_n \sin (\lambda_n ct)\}\, x \sin (\lambda_n x)$$

The unknown constant A_n and B_n are determined by using the initial conditions.

Case I: *When initial displacement is given*

$$u(x, 0) = f(x)$$

Then,

$$f(x) = u(x, 0) = \sum_{n=0}^{\infty} A_n \sin\frac{n\pi x}{l}$$

Thus, A_n's are the Fourier coefficients (which will be discussed later) in the half-range Fourier sine series expansion of $f(x)$ in the interval $(0, l)$.

Hence,

$$A_n = \frac{2}{l}\int_0^l f(x)\sin\left(\frac{n\pi x}{l}\right)dx \qquad (9)$$

with $n = 1, 2, 3, \ldots$

Case II: *When initial velocity is given*

$$\frac{\partial u}{\partial t}(x,0) = u_t(x, 0) = g(x)$$

Now,

$$\frac{\partial u}{\partial t} = \sum_{n=1}^{\infty}\{-A_n\lambda_n c\sin(\lambda_n ct) + B_n\lambda_n c \times \cos(\lambda_n ct)\}\sin\left(\frac{n\pi x}{l}\right)$$

For $t = 0$, we have

$$g(x) = \sum_{n=1}^{\infty} \mathrm{B}_n\, \lambda_n c \sin\left(\frac{n\pi x}{l}\right)$$

Then, $$B_n \lambda_n c = \frac{2}{l}\int_0^l g(x)\sin\left(\frac{n\pi x}{l}\right)dx.$$

$$\Rightarrow \quad B_n = \frac{2}{c \cdot n\pi}\int_0^l g(x)\sin\left(\frac{n\pi x}{l}\right)dx \tag{10}$$

Hence, the general solution of the one-dimensional wave Eq. (8) is

$$u(x,t) = \sum_{n=1}^{\infty}\left\{A_n \cos\left(\frac{n\pi ct}{l}\right) + B_n \sin\left(\frac{n\pi ct}{l}\right)\right\}\sin\left(\frac{n\pi x}{l}\right)$$

where, A_n and B_n are given by Eq. (10).

Note 1: When only the displacement is given $f(x) \neq 0$ and $g(x) = 0$, in which case all B_n's are zero. Thus, the general solution is

$$u(x,t) = \sum_{n=1}^{\infty} A_n \cos\left(\frac{n\pi ct}{l}\right)\sin\left(\frac{n\pi x}{l}\right)$$

with A_n's given by Eq. (9).

Note 2: When only initial velocity is prescribed $g(x) \neq 0$ and $f(x) = 0$; then the solution is

$$u(x,t) = \sum_{n=1}^{\infty} B_n \sin\left(\frac{n\pi ct}{l}\right)\sin\left(\frac{n\pi x}{l}\right)$$

with B_n's given by Eq. (10).

Note 3: One-dimensional wave equation is usually written in the form

$$\frac{\partial^2 u}{\partial t^2} = c^2 \frac{\partial^2 u}{\partial x^2} \quad \text{where } c^2 = \frac{T}{\rho}.$$

D′ Alembert's Solution of the Wave Equation

As we have derived, the one dimensional wave equation is

$$\frac{\partial^2 u}{\partial t^2} = c^2 \frac{\partial^2 u}{\partial x^2} \tag{11}$$

Let $v = x + ct$ and $w = x - ct$ so that u becomes a function of v and w.

Then, $$\frac{\partial u}{\partial x} = \frac{\partial u}{\partial v}\frac{\partial v}{\partial x} + \frac{\partial u}{\partial w}\frac{\partial w}{\partial x} = \frac{\partial u}{\partial v} + \frac{\partial u}{\partial w}$$

and $$\frac{\partial^2 u}{\partial x^2} = \frac{\partial}{\partial x}\left(\frac{\partial u}{\partial v} + \frac{\partial u}{\partial w}\right) = \frac{\partial}{\partial v}\left(\frac{\partial u}{\partial v} + \frac{\partial u}{\partial w}\right) + \frac{\partial}{\partial w}\left(\frac{\partial u}{\partial v} + \frac{\partial u}{\partial w}\right)$$

$$= \frac{\partial^2 u}{\partial v^2} + 2\frac{\partial^2 u}{\partial v \partial w} + \frac{\partial^2 u}{\partial w^2}$$

Similarly, $$\frac{\partial^2 u}{\partial t^2} = c^2\left(\frac{\partial^2 u}{\partial v^2} - 2\frac{\partial^2 u}{\partial v \partial w} + \frac{\partial^2 u}{\partial w^2}\right)$$

By using these Eq. (11) reduces to

$$\frac{\partial^2 u}{\partial v \partial w} = 0$$

Integrating, we get

$$u = \int \phi_1(v)dv + \psi(w)$$

$$\Rightarrow \quad u = \phi(v) + \psi(w)$$

$$\therefore \quad u(x, t) = \phi(x + ct) + \psi(x - ct)$$

This is the general solution of the wave Eq. (11).

Now, to determine ϕ and ψ, suppose initially

$$v(x, 0) = f(x) \quad \text{and} \quad \frac{\partial u}{\partial t}(x,0) = 0.$$

Then, $$\frac{\partial u}{\partial t} = c\phi'(x + ct) - c\psi'(x - ct)$$

$\therefore$ At $t = 0$ $\phi'(x) = \psi'(x) \Rightarrow \phi(x) = \psi(x) + k$

Also, $$u(x, 0) = \phi(x) + \psi(x) = f(x)$$

$$\Rightarrow \quad 2\psi(x) + k = f(x)$$

$$\therefore \quad \psi(x) = \frac{1}{2}[f(x) - k] \quad \text{and} \quad \phi(x) = \frac{1}{2}[f(x) + k].$$

Therefore, $u(x, t) = \frac{1}{2}[f(x + ct) + k] + \frac{1}{2}[f(x - ct) - k] = \frac{1}{2}[f(x - ct) + f(x - ct)]$

which is the D'Alembert's solution of the one-dimensional wave Eq. (11).

Note: The above solution gives a very useful method of solving partial differential equations by change of variables.

Two-Dimensional Wave Equation

The small transverse vibrations of a membrane tightly stretched over a flat frame (in the xy-plane) are governed by the two-dimensional wave equation. Let $u(x, y, t)$ denote the transverse displacement of the membrane from its equilibrium position at time t. Assume that (a) surface tension T is constant and is independent of position, (b) membrane is homogeneous with surface density ρ (mass per unit area) constant and (c) deflections are small so that all angles of inclination are small. The wave equation is derived by applying Newton's law of motion to a small rectangle of dimensions Δx and

Δy of the membrane. On each edge of the rectangle a concentrated force of magnitude $T\Delta x$ or $T\Delta y$ is exerted.

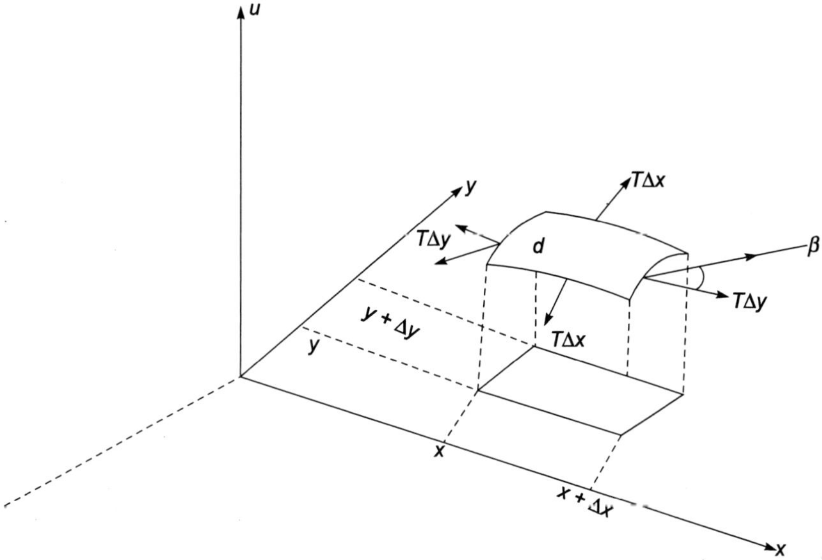

Fig. 10.2 Vibrating membrane

Looking at the projections on the xu- and yu-planes, the resultant of horizontal forces in x-direction is $T\Delta y(\cos \beta - \cos \alpha)$ and in the y-direction is $T\Delta x(\cos \delta - \cos \gamma)$. For small $\alpha, \beta, \delta, \gamma$ their cosine are close to 1 and hence, the horizontal components at opposite sides are equal.

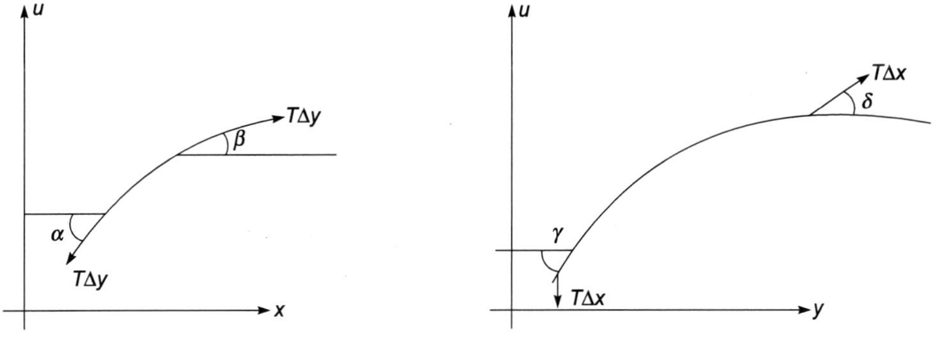

Fig. 10.3

Equating the sum of the forces in the vertical direction to mass × acceleration, we get

$$T\Delta y(\sin \beta - \sin \alpha) + T\Delta x(\sin \delta - \sin \gamma) = \rho \Delta x \Delta y \frac{\partial^2 u}{\partial t^2}$$

Assuming $\alpha, \beta, \delta, \gamma$ to be small, we have

$$T\Delta y(\tan\beta - \tan\alpha) + T\Delta x(\tan\delta - \tan\gamma) = \rho\Delta x\Delta y\ \frac{\partial^2 u}{\partial t^2}$$

Thus, Eq. (11) becomes

$$g''f = c^2(gf_{xx} + gf_{yy})$$

$$\Rightarrow \qquad \frac{g''}{c^2 g} = \frac{f_{xx} + f_{yy}}{f} = -\lambda^2$$

where we have introduced a separation constant

$$k = -\lambda^2 \text{ for } k < 0$$

Therefore, $$g'' + c^2\lambda^2 g = 0 \tag{12}$$

and $$f_{xx} + f_{yy} + \lambda^2 f = 0 \text{ (two-dimensional Helmholtz equation)} \tag{13}$$

Also, boundary condition reduces to

$$f(0, y) = 0 = f(x, 0) \tag{14}$$

$$f(a, y) = 0 = f(x, b) \tag{15}$$

Equating Eq. (13) with boundary conditions (14), (15) constitute a two-dimensional eigen-value problem. So, we assume that

$$f(x, y) = X(x)Y(y).$$

$$\therefore \qquad X''Y + XY'' = -\lambda^2 XY$$

$$\Rightarrow \qquad \frac{X''}{X} = -\lambda^2 - \frac{Y''}{Y} = -\mu^2$$

where we have introduced a second separation constant $-\mu^2$

Thus, we get

$$X'' + \mu^2 X = 0 \quad \text{and} \quad Y'' + (\lambda^2 - \mu^2)Y = 0$$

with boundary conditions.

$$X(0) = 0 = X(a) \quad \text{and} \quad Y(0) = 0 = Y(b)$$

$$\Rightarrow \quad T\Delta y\left[\frac{\partial u}{\partial x}(x+\Delta x, y, t) - \frac{\partial u}{\partial x}(x, y, t)\right] + T\Delta x\left[\frac{\partial u}{\partial y}(x, y+\Delta y, t) - \frac{\partial u}{\partial y}(x, y, t)\right] = \rho\Delta x\Delta y\frac{\partial^2 u}{\partial t^2}$$

Taking the limits $\Delta x \to 0$ and $\Delta y \to 0$, we obtain

$$T\left(\frac{\partial^2 u}{\partial x^2} + \frac{\partial^2 u}{\partial y^2}\right) = \rho\frac{\partial^2 u}{\partial t^2}$$

$$\Rightarrow \qquad \frac{\partial^2 u}{\partial t^2} = c^2\left(\frac{\partial^2 u}{\partial x^2} + \frac{\partial^2 u}{\partial y^2}\right) \tag{16}$$

where, $c^2 = \frac{T}{\rho}$ is a positive constant.

This equation is known as two-dimensional wave equation.

Since membrane is fixed to a flat frame so that

$$u(0, y, t) = 0 = u(x, 0, t)$$

If the dimensions of the rectangular membrane are a and b, then

$$u(a, y, t) = 0 = u(x, b, t)$$

Let the given initial displacement be

$$u(x, y, 0) = \alpha(x, y)$$

and the given initial velocity be

$$\frac{\partial u}{\partial t}(x, y, 0) = \beta(x, y)$$

We apply the method of separation of variables and first assume

$$u(x, y, t) = f(x, y)\, g(t)$$

For $X'' + \mu^2 X = 0$, the solution is

$$X(x) = A \cos \mu x + B \sin \mu x$$

using $X(0) = 0$, we have $A = 0$

$$\therefore \qquad X(x) = B \sin \mu x$$

using $X(a) = 0$, we have

$$0 = B \sin \mu a$$

For $X \neq 0$, B must be non-zero. Thus,

$$\sin \mu a = 0$$

$$\Rightarrow \qquad \mu a = n\pi$$

$$\therefore \qquad \mu_n = \frac{n\pi}{a} \text{ with } n = 1, 2, 3, \ldots$$

So, $\mu_n^2 = \left(\frac{n\pi}{a}\right)^2$ $(n = 1, 2, 3, \ldots)$ are the eigenvalues with corresponding eigenfunctions

$$X_n(x) = \sin\left(\frac{n\pi x}{a}\right)$$

Similarly, for the eigenvalue problem

$$Y'' + (\lambda^2 - \mu^2)Y = 0 \text{ with } Y(0) = 0 = Y(b);$$

We get the eigenvalues as

$$\lambda_m^2 - \mu_n^2 = \left(\frac{m\pi}{b}\right)^2; \; m = 1, 2, 3, \ldots$$

with corresponding eigenfunctions as

$$Y_m(y) = \sin\left(\frac{m\pi y}{b}\right)$$

then $\lambda_m^2 = \mu_n^2 + \left(\frac{m\pi}{b}\right)^2 = \left(\frac{n\pi}{a}\right)^2 + \left(\frac{m\pi}{b}\right)^2$ where $m, n = 1, 2, 3, \ldots$

Thus,
$$f(x, y) = \sin\left(\frac{n\pi x}{a}\right)\sin\left(\frac{m\pi y}{b}\right)$$

The solution of Eq. (2) is

$$g(t) = c\cos(c\lambda_{mn}t) + D\sin(c\lambda_{mn}t)$$

So $u_{mn}(x, y, t)$ consists of two doubly infinite families

$$\sin\left(\frac{n\pi x}{a}\right)\sin\left(\frac{m\pi y}{b}\right)\sin(c\lambda_{mn}t)$$

and
$$\sin\left(\frac{n\pi x}{a}\right)\sin\left(\frac{m\pi y}{b}\right)\cos(c\lambda_{mn}t)$$

which are known as modes of vibrations. As time varies, the shape remains the same but only the amplitude varies periodically. Applying the principle of superposition, we get the displacement function as

$$u(x, y, t) = \sum_{m=1}^{\infty}\sum_{n=1}^{\infty} u_{mn}(x, y, t)$$

$$= \sum_{m=1}^{\infty}\sum_{n=1}^{\infty} \sin\left(\frac{n\pi x}{a}\right)\sin\left(\frac{m\pi y}{b}\right) \times (A_{mn}\cos(c\lambda_{mm}t) + B_{mn}\sin(c\lambda_{mn}t) \qquad (17)$$

The functions $u_{mn}(x, y, t)$ is known as the (n, m)th harmonic of the rectangular drum. *For* $t = 0$

$$u(x, y, 0) = \alpha(x, y) = \sum_{m=1}^{\infty}\left[\sum_{n=1}^{\infty} A_{mn}\sin\left(\frac{n\pi x}{a}\right)\right]\sin\left(\frac{m\pi y}{b}\right)$$

Put
$$A'_m = \sum_{m=1}^{\infty} A_{mn}\sin\left(\frac{n\pi x}{a}\right)$$

Thus,
$$A'_m = \frac{2}{b}\int_0^b \alpha(x, y)\sin\left(\frac{m\pi y}{b}\right)dy$$

and
$$A_{mn} = \frac{2}{a}\int_0^a A_m'(x)\sin\left(\frac{n\pi x}{a}\right)dx$$

$\Rightarrow$
$$A_{mn} = \frac{4}{ab}\int_0^a\int_0^b \alpha(x,y)\sin\left(\frac{m\pi y}{b}\right)dy \times \sin\left(\frac{n\pi x}{a}\right)dx \tag{18}$$

Also, for $t = 0$

$$\left.\frac{\partial u}{\partial t}\right|_{(x,y,0)} = \beta(x,y) = \sum_{m=1}^{\infty}\sum_{n=1}^{\infty} c\lambda_{mn}B_{mn}\sin\left(\frac{n\pi x}{a}\right)\times\sin\left(\frac{m\pi y}{b}\right)$$

$$c\lambda_{mn}B_{mn} = \frac{4}{ab}\int_0^a\int_0^b \beta(x,y)\sin\left(\frac{n\pi x}{a}\right)dx \times \sin\left(\frac{m\pi y}{b}\right)dy \tag{19}$$

Thus, the solution of Eq. (16) is the doubly infinite series given by Eq. (17) with the coefficient A_{mn} and B_{mn} determined by Eqs. (18) and (19), respectively.

Note 1: If the membrane starts from rest then $\beta(x, y) = 0$ so all B_{mn} are zero.
The solution (17) reduces to

$$u(x, y, t) = \sum_{m=1}^{\infty}\sum_{n=1}^{\infty} A_{mn}\sin\left(\frac{n\pi x}{a}\right)\sin\left(\frac{m\pi y}{b}\right)\times\cos(c\lambda_{mn}t)$$

with A_{mn}'s determined by Eq. (18).

Note 2: If the initial displacement is zero, then $\alpha(x, y) = 0$ so all A_{mn}'s are zero.
Then, Eq. (17) reduced to

$$u(x, y, t) \sum_{m=1}^{\infty}\sum_{n=1}^{\infty} B_{mn}\sin\left(\frac{n\pi x}{a}\right)\sin\left(\frac{m\pi y}{b}\right)\times\sin(c\lambda_{mn}t)$$

with B_{mn}'s determined by Eq. (19). ▲▲

EXAMPLE 10.4

A string of length l is fastened at both ends A and C. At a distance b from the end A, the string is transversely displaced to a distance d and is released from rest; when it is in this position. Find the equation of the subsequent motion.

Solution: Let $u(x, t)$ denote the displacement of the string. The initial displacement is given by ABC. Equation of AB is $y = \dfrac{dx}{b}$ and equation of BC is $y = \dfrac{d}{b-l}(x-l)$.

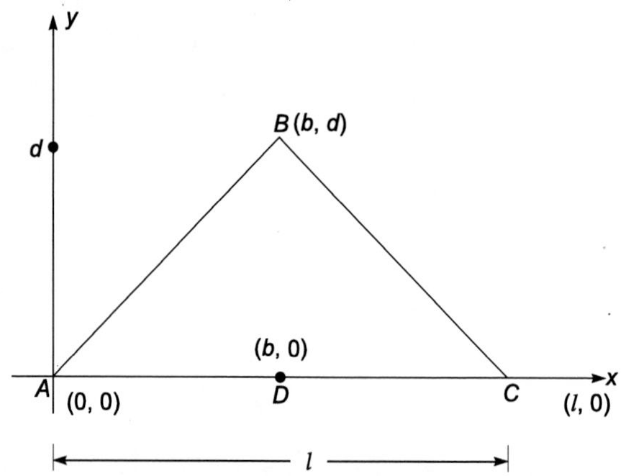

Fig. 10.4

Thus, the problem is to solve the one-dimensional wave equation

$$\frac{\partial^2 u}{\partial t^2} = a^2 \frac{\partial^2 u}{\partial x^2}$$

with boundary conditions

$$u(0, t) = 0, u(d, t) = 0$$

and with initial displacement

$$u(x, 0) = f(x) = \begin{cases} \dfrac{dx}{b} & \text{if } 0 \le x \le b \\ \dfrac{d}{b-l}(x-l) & \text{if } b \le x \le l \end{cases}$$

The solution is given by

$$u(x, t) = \sum_{n=1}^{\infty} A_n \sin\left(\frac{n\pi x}{l}\right) \cos\left(\frac{n\pi at}{l}\right)$$

where $A_n \frac{2}{l}\int_0^l f(x)\sin\left(\frac{n\pi x}{l}\right)dx$

$$= \frac{2}{l}\int_0^b \frac{d}{b} x \cdot \sin\left(\frac{n\pi x}{l}\right)dx + \frac{2}{l}\int_b^l \frac{d}{b-l}(x-l)\sin\left(\frac{n\pi x}{l}\right) \times dx$$

$$= \frac{2d}{bl}\left\{x\left(\frac{-l}{n\pi}\right)\cos\left(\frac{n\pi x}{l}\right) + \left(\frac{l}{n\pi}\right)^2 \sin\left(\frac{n\pi x}{l}\right)\right\}_0^b + \frac{2d}{l(b-l)}$$

$$\left\{(x-l)\left(\frac{-l}{n\pi}\right)\cos\left(\frac{n\pi x}{l}\right) + \left(\frac{-l}{n\pi}\right)^2 \sin\left(\frac{n\pi x}{l}\right)\right\}_b^l$$

$$= -\frac{2d}{n\pi}\cos\left(\frac{n\pi b}{l}\right)+\frac{2dl}{bn^2\pi^2}\sin\left(\frac{n\pi b}{l}\right)+\frac{2d}{n\pi}\cos\left(\frac{n\pi b}{l}\right)-\frac{2dl}{(b-l)n^2\pi^2}\sin\left(\frac{n\pi b}{l}\right)$$

Then,
$$A_n = -\frac{2dl^2}{b(b-l)n^2\pi^2}\sin\left(\frac{n\pi b}{l}\right)$$

Hence, the subsequent displacement of the string is given by the displacement function

$$u(x, l) = -\frac{2dl^2}{b(b-l)\pi^2}\sum_{n=1}^{\infty}\frac{1}{n^2}\sin\left(\frac{n\pi b}{l}\right)\sin\left(\frac{n\pi x}{l}\right)\times\cos\left(\frac{n\pi at}{l}\right).$$

EXAMPLE 10.5

A string is stretched between two fixed points at a distance l apart. Motion is started by displacing the string in the form $y = a\sin\left(\frac{\pi x}{l}\right)$ from which it is released at time $t = 0$. Show that the displacement of any point at a fixed distance x from one end at time t is given by $y(x, t) = a\sin\left(\frac{\pi x}{l}\right)\cos\left(\frac{\pi ct}{l}\right)$.

Solution: Let an elastic string be tightly stretched between two points O and A. Let O be the origin and OA along x-axis. Giving the string a small displacement perpendicular to its length (parallel to y-axis), let $y(x, t)$ be its displacement at the point $P(x, y)$ at any time t. The vibrations of the string will be governed by the wave equation.

$$\frac{\partial^2 y}{\partial t^2} = c^2\frac{\partial^2 y}{\partial x^2} \tag{1}$$

where, c is a constant.

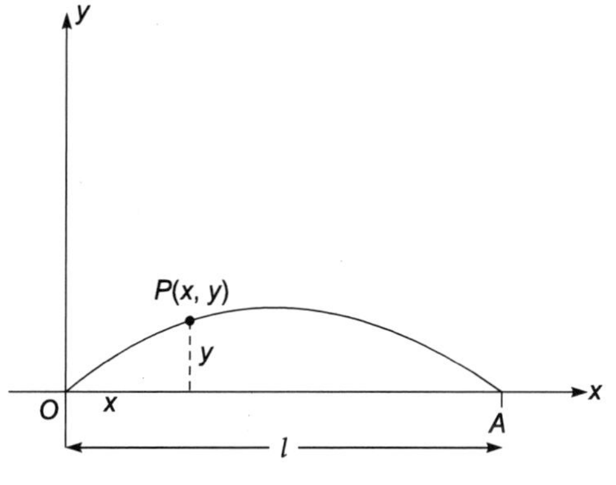

Fig. 10.5

Since the ends O and A of the string are fixed for all time.

$\therefore$ The boundary conditions are $y(0, t) = 0 = y(l, t)$ and the initial conditions are

$$y(x, 0) = a \sin\left(\frac{\pi x}{l}\right), \left(\frac{\partial y}{\partial t}\right)_{t=0} = 0.$$

Since the vibrations of the string are periodic, so the solution will be of the form

$$y(x, t) = (c_1 \cos kx + c_2 \sin kx)(c_3 \cos ckt + c_4 \sin ckt)$$

Now, $y(0, t) = 0$

$\Rightarrow$ $c_1(c_3 \cos ckt + c_4 \sin ckt) = 0$

$\Rightarrow$ $c_1 = 0$

$\{\because c_3 \cos ckt + c_4 \sin ckt \neq 0$ otherwise we get $y(x, t) = 0$, which is trivial solution$\}$

$\therefore$ $y(x, t) = c_2 \sin kx(c_3 \cos ckt + c_4 \sin ckt)$

Also, $y(l, t) = 0$

$\Rightarrow$ $c_2 \sin kl(c_3 \cos ckt + c_4 \sin ckt) = 0$

$\Rightarrow$ $\sin kl = 0$ $\{\because$ if $c_2 = 0 \Rightarrow y(x, t) = 0\}$

$\therefore$ $kl = n\pi \Rightarrow k = \dfrac{n\pi}{l}$ $\{n = 1, 2, 3, \ldots\}$

Thus, $y_n(x, t) = \left\{c_n \cos\left(\frac{cn\pi t}{l}\right) + d_n \sin\left(\frac{cn\pi t}{l}\right)\right\} \times \sin\left(\frac{n\pi x}{l}\right)$ with $n = 1, 2, 3, \ldots$

$\therefore$ the general solution of Eq. (1) is

$$y(x, t) = \sum_{n=1}^{\infty} y_n(x, t)$$

$$= \sum_{n=1}^{\infty}\left\{c_n \cos\left(\frac{cn\pi t}{l}\right) + d_n \sin\left(\frac{cn\pi t}{l}\right)\right\}\sin\left(\frac{n\pi x}{l}\right)$$

Now, $$\frac{\partial y}{\partial t} = \sum_{n=1}^{\infty}\left\{-c_n \cos\left(\frac{cn\pi}{l}\right)\sin\left(\frac{cn\pi t}{l}\right) + d_n\left(\frac{cn\pi}{l}\right)\cos\left(\frac{cn\pi t}{l}\right)\right\}\sin\left(\frac{n\pi x}{l}\right).$$

Now, $\left(\frac{\partial y}{\partial t}\right)_{t=0} = 0$

$$\Rightarrow \sum_{n=1}^{\infty} d_n\left(\frac{cn\pi t}{l}\right)\sin\left(\frac{cn\pi}{l}\right) = 0$$

$\therefore$ $d_n = 0$

Thus, $$y(x, t) = \sum_{n=1}^{\infty} c_n \cos\left(\frac{cn\pi t}{l}\right)\sin\left(\frac{cn\pi}{l}\right)$$

Now, $$y(x, 0) = \sum_{n=1}^{\infty} c_n \sin\left(\frac{n\pi x}{l}\right) = a \sin\left(\frac{\pi x}{l}\right)$$

$\Rightarrow$ $$c_1 = a,\ c_2 = c_3 = \ldots = 0$$

$\therefore$ $$y(x, t) = a \cos\left(\frac{c\pi t}{l}\right)\sin\left(\frac{\pi x}{l}\right)$$

which is the required solution.

EXAMPLE 10.6

Find the deflection of a vibrating string of unit length having fixed ends with initial velocity zero and initial deflection

$$f(x) = k(\sin x - \sin 2x).$$

Solution: By D'Alembert's method, the solution is

$$y(x, t) = \frac{1}{2}[f(x + ct) + f(x - ct)]$$

$$= \frac{1}{2}[k(\sin(x + ct) - \sin 2(x + ct)) + k(\sin(x - ct) - \sin 2(x - ct))]$$

$$= k[\sin x \cos ct - \sin 2x \cos 2ct]$$

Also, $$y(x, 0) = k(\sin x - \sin 2x) = f(x)$$

and $\frac{\partial y}{\partial t}(x,0) = \{k(-c\sin x \sin ct + 2c\sin 2x \sin 2ct)\}_{t=0} = 0$, that is, the given boundary conditions are satisfied.

EXAMPLE 10.7

The points of trisection of a string are parallel aside through the same distance on the opposite sides of the position of equilibrium and the string is released from rest. Derive an expression for the displacement of the string at subsequent time and show that the mid-point of the string always remains at rest.

Solution: Let B and C be the points of the trisection of the string $OA = l$. Initially, the string is held in the form $OB'C'A$ where, $BB' = CC' = a$(say). The displacement $y(x, t)$ of any point of the string is given by

$$\frac{\partial^2 y}{\partial t^2} = c^2 \frac{\partial^2 y}{\partial x^2}$$

and the boundary conditions are $y(o, t) = 0 = y(l, t)$

and $$\left(\frac{\partial y}{\partial t}\right)_{t=0} = 0$$

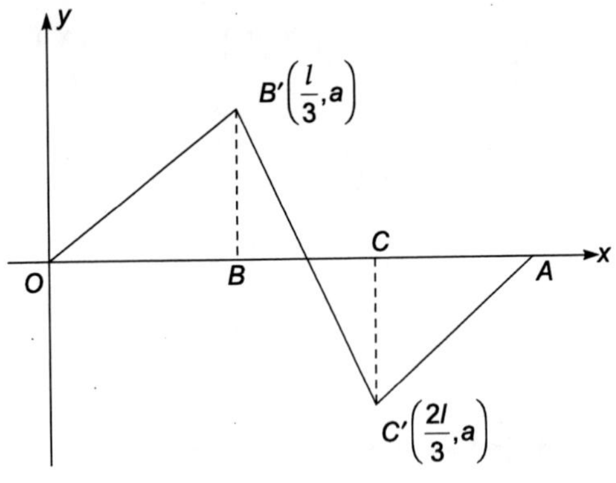

Fig. 10.6

Also, at $t = 0$, the string rests in the form $OB'C'A$. The equation of OB' is $y = \frac{3a}{l}x$ the equation of $B'C'$ is $y - a = -\frac{2a}{(l/3)}\left(l - \frac{l}{3}\right)$.

$$\Rightarrow \qquad y = \frac{3a}{l}(l - 2x)$$

and the equation of $C'A$ is $y = \frac{3a}{l}(x - l)$.

Hence,
$$y(x, 0) = \begin{cases} \frac{3a}{l} & 0 \le x \le \frac{l}{3} \\ \frac{3a}{l}(l - 2x) & \frac{l}{3} \le x \le \frac{2l}{3} \\ \frac{3a}{l}(x - l) & \frac{2l}{3} \le x \le l \end{cases}$$

Now,
$$y(x,t) = \sum_{n=1}^{\infty} b_n \sin\left(\frac{n\pi x}{l}\right)\cos\left(\frac{cn\pi t}{l}\right)$$

Let, $t = 0$, we have

$$y(x, 0) = \sum_{n=1}^{\infty} b_n \sin\left(\frac{n\pi x}{l}\right)$$

and
$$b_n = \frac{2}{l}\left\{\int_0^{l/3} \frac{3ax}{l}\sin\left(\frac{n\pi x}{l}\right)dx + \int_{l/3}^{2l/3} \frac{3a}{l}(l - 2x)\sin\left(\frac{n\pi x}{l}\right)dx + \int_{2l/3}^{l} \frac{3a}{l}(x - l) \times \sin\left(\frac{n\pi x}{l}\right)dx\right\}$$

$$= \frac{6a}{l^2}\left[\left[+x\left(\frac{-l}{n\pi}\right)\cos\left(\frac{n\pi x}{l}\right)+\left(\frac{l}{n\pi}\right)^2 \sin\left(\frac{n\pi x}{l}\right)\right]_0^{l/3}\right.$$

$$+\left[(l-2x)\left(\frac{-l}{n\pi}\right)\cos\left(\frac{n\pi x}{l}\right)-2\left(\frac{l}{n\pi}\right)^2 \sin\left(\frac{n\pi x}{l}\right)\right]_{l/3}^{2l/3}$$

$$\left.+\left[(x-l)\left(\frac{-l}{n\pi}\right)\cos\left(\frac{n\pi x}{l}\right)+\left(\frac{l}{n\pi}\right)^2 \sin\left(\frac{n\pi x}{l}\right)\right]_{2l/3}^{l}\right]$$

$$= \frac{6a}{l^2}\left\{\frac{-l^2}{3n\pi}\cos\left(\frac{n\pi}{3}\right)+\frac{l^2}{n^2\pi^2}\sin\left(\frac{n\pi}{3}\right)+\frac{l^2}{3n\pi}\cos\left(\frac{2n\pi}{3}\right)\right.$$

$$-\frac{2l^2}{n^2\pi^2}\sin\left(\frac{2n\pi}{3}\right)+\frac{l^2}{3n\pi}\cos\left(\frac{n\pi}{3}\right)+2\left(\frac{l^2}{n^2\pi^2}\right)\sin\left(\frac{n\pi}{3}\right)$$

$$\left.-\frac{l^2}{3n\pi}\cos\left(\frac{2n\pi}{3}\right)-\frac{l^2}{n^2\pi^2}\sin\left(\frac{2n\pi}{3}\right)\right\}$$

$$= \frac{6a}{l^2}\cdot\frac{3l^2}{n^2\pi^2}\left[\sin\left(\frac{n\pi}{3}\right)-\sin\left(\frac{2n\pi}{3}\right)\right]$$

$$= \frac{18a}{n^2\pi^2}\sin\left(\frac{n\pi}{3}\right)[1+(-1)^n]$$

$$\left\{\because \sin\left(\frac{2n\pi}{3}\right)=\sin\left(n\pi=\frac{n\pi}{3}\right)=(-1)^n\sin\left(\frac{n\pi}{3}\right)\right\}$$

Thus, $b_n = \begin{cases} 0 & \text{when } n \text{ is odd} \\ \dfrac{36a}{n^2\pi^2}\sin\left(\dfrac{n\pi}{3}\right) & \text{when } n \text{ is even} \end{cases}$

Hence, $y(x,t) = \sum\limits_{n=2,4,\ldots}^{\infty} \dfrac{36a}{n^2\pi^2}\sin\left(\dfrac{n\pi}{3}\right)\sin\left(\dfrac{n\pi x}{l}\right)\cos\left(\dfrac{cn\pi t}{l}\right)$

$$= \frac{9a}{\pi^2}\sum_{m=1}^{\infty}\frac{1}{m^2}\sin\left(\frac{2m\pi}{3}\right)\sin\left(\frac{2m\pi x}{l}\right)\times\cos\left(\frac{c2m\pi t}{l}\right)$$

Let $x = l/2$, we have

$$y(l/2, t) = 0 \qquad \{\because \sin n\pi = 0 \text{ for every integer } m\}$$

This shows that the mid-point of the string is always at rest.

EXAMPLE 10.8

A tightly stretched string with fixed end points $x = 0$ and $x = l$ is initially at rest in its equilibrium position. If the string is set in motion by giving to each of its points an initial velocity $\lambda x(l - x)$, find the displacement of the string at any distance x from one end point at any time t.

Solution: The motion of the string is governed by the partial differential equation

$$\frac{\partial^2 y}{\partial t^2} = c^2 \frac{\partial^2 y}{\partial x^2} \tag{1}$$

Given that the end points of the string are fixed for all time t, we have

$$y(0, t) = 0 = y(l, t)$$

Also,
$$y(x, 0) = 0$$

$$\left(\frac{\partial y}{\partial t}\right)_{t=0} = \lambda x(l - x)$$

We know that the solution of Eq. (1) is given by

$$y(x, t) = (A \cos kx + B \sin kx)\,(C \cos ckt + D \sin ckt)$$

Now,
$$y(0, t) = A\,(C \cos ckt + D \sin ckt) = 0$$

$\Rightarrow$
$$A = 0$$

$\therefore$
$$y(x, t) = B \sin kx(C \cos ckt + D \sin ckt)$$

Also,
$$y(l, t) = B \sin kl(C \cos ckt + D \sin ckt) = 0$$

As for $y(x, t) \neq 0$, we must have $B \neq 0$

$\therefore$
$$\sin kl = 0$$

$\Rightarrow$
$$kl = n\pi \quad \therefore k = \frac{n\pi}{l}$$

Thus,
$$y(x, t) = B \sin\left(\frac{n\pi x}{l}\right)\left\{c \cos\left(\frac{cn\pi t}{l}\right) + D \sin\left(\frac{cn\pi t}{l}\right)\right\}$$

Also
$$y(x, 0) = B \sin\left(\frac{n\pi x}{l}\right) C = 0$$

$\Rightarrow$
$$C = 0$$

$\therefore$
$$y(x, t) = BD \sin\left(\frac{n\pi x}{l}\right)\sin\left(\frac{cn\pi t}{l}\right)$$

and the general solution is

$$y(x, t) = \sum_{n=1}^{\infty} b_n \sin\left(\frac{n\pi x}{l}\right)\sin\left(\frac{cn\pi t}{l}\right)$$

Now,
$$\frac{\partial y}{\partial t} = \sum_{n=1}^{\infty} b_n \sin\left(\frac{n\pi x}{l}\right)\left(\frac{cn\pi}{l}\right)\cos\left(\frac{cn\pi t}{l}\right)$$

For $t = 0$, we have
$$\left(\frac{\partial y}{\partial t}\right)_{t=0} = \sum_{n=1}^{\infty} b_n\left(\frac{cn\pi}{l}\right)\sin\left(\frac{n\pi x}{l}\right) = \lambda x(l - x)$$

So,
$$b_n = \frac{2}{l}\int_0^l \lambda x(l-x)\sin\left(\frac{n\pi x}{l}\right)dx$$

which after simplification, can be expressed as
$$b_n = \begin{cases} \dfrac{8\lambda l^3}{cn^4\pi^4} & \text{when } n \text{ is odd} \\ 0 & \text{when } n \text{ is even} \end{cases}$$

Therefore, the required solution is given by
$$y(x, t) = \frac{8\lambda l^3}{c\pi^4}\sum_{n=\text{odd}} \frac{1}{n^4}\sin\left(\frac{n\pi x}{l}\right)\sin\left(\frac{cn\pi t}{l}\right)$$
$$= \frac{8\lambda l^3}{c\pi^4}\left\{\frac{1}{1^4}\sin\left(\frac{\pi x}{l}\right)\sin\left(\frac{c\pi t}{l}\right) + \frac{1}{3^4}\sin\left(\frac{3\pi x}{l}\right)\times\sin\left(\frac{c3\pi t}{l}\right) + \ldots\right\}.$$

EXAMPLE 10.9

Find the deflection $u(x, y, t)$ of the square membrane with $a = b = 1$ and $c = 1$. If the initial velocity is zero and the initial deflection is $f(x, y) = A \sin \pi x \sin 2\pi y$.

Solution: Here,
$$A_{mn} = \frac{4}{ab}\int_0^a\int_0^b f(x,y)\sin\left(\frac{m\pi x}{a}\right)\sin\left(\frac{n\pi y}{b}\right)dxdy$$
$$= 4A\int_0^1\int_0^1 \sin \pi x \sin 2\pi y \sin(m\pi x) \sin(n\pi y)dxdy.$$
$$= 4A\int_0^1 \sin \pi x \sin(m\pi x)dx \int_0^1 \sin 2\pi y \sin(n\pi y)dy$$

For $m \neq 1$, $\int_0^1 \sin \pi x \sin (m\pi x)dx = 0$

For $m = 1$, $\int_0^1 \sin \pi x \sin \pi x dx = \frac{1}{2}$

Similarly, for $n \neq 2$, $\int_0^1 \sin 2\pi y \sin(n\pi y)dy = 0$

for $n = 2$, $\int_0^1 \sin 2\pi y \sin(2\pi y)dy = \frac{1}{2}$

$$\therefore \quad A_{12} = 4A\left(\frac{1}{2}\right)\left(\frac{1}{2}\right) = A$$

and $$A_{mn} = 0 \text{ when } m \neq 1, n \neq 2$$

Also, $$\lambda_{mn}^2 = \pi^2\left(\frac{m^2}{a^2} + \frac{n^2}{b^2}\right) = \pi^2(1^2 + 2^2) = 5\pi^2$$

$$\Rightarrow \quad \lambda_{mn} = \sqrt{5}\pi$$

Hence, the required solution is

$$u(x, y, t) = \sum_{m=1}^{\infty}\sum_{n=1}^{\infty} A_{mn} \sin\left(\frac{m\pi x}{a}\right)\sin\left(\frac{n\pi y}{b}\right)\cos(c\lambda_{mn}t)$$

$$= A_{12} \sin(\pi x)\sin(2\pi y)\sin(\sqrt{5}\pi t)$$

$$= A \sin(\pi x)\sin(2\pi y)\cos(\sqrt{5}\pi t)$$

EXAMPLE 10.10

Find the vibrations $u(x, y, t)$ of a rectangular membrane $(0 < x < a, 0 < y < b)$ whose boundary is fixed. Given that it starts from rest and $u(x, y, 0) = hxy(a - x)(b - y)$.

Solution: Since the membrane starts from rest, so

$$u(x,y,t) = \sum_{m=1}^{\infty}\sum_{n=1}^{\infty} A_{mn} \sin\left(\frac{m\pi x}{a}\right)\sin\left(\frac{n\pi y}{b}\right) \times \cos(c\lambda_{mn}t)$$

Then, $$u(x, y, 0) = \sum_{m=1}^{\infty}\sum_{n=1}^{\infty} A_{mn} \sin\left(\frac{m\pi x}{a}\right)\sin\left(\frac{n\pi y}{b}\right)$$

$$= hxy(a - x)(b - y)$$

where $$A_{mn} = \frac{4}{ab}\int_0^a\int_0^b hxy(a-x)(b-y)\sin\left(\frac{m\pi x}{a}\right) \times \sin\left(\frac{n\pi y}{b}\right)dxdy$$

$$= \frac{4b}{ab}\left\{\left[x(a-x)\times\frac{\cos\left(\frac{m\pi x}{a}\right)}{\left(-\frac{m\pi}{a}\right)}-(a-2x)\times\frac{\sin\left(\frac{m\pi x}{a}\right)}{-\left(\frac{m\pi}{a}\right)^2}+(-2)\frac{\cos\left(\frac{m\pi x}{a}\right)}{\left(\frac{m\pi}{a}\right)^3}\right]_0^a\right.$$

$$\left.\left[y(b-y)\times\frac{\cos\left(\frac{n\pi y}{b}\right)}{\left(-\frac{n\pi}{b}\right)}-(b-2y)\times\frac{\sin\left(\frac{n\pi y}{b}\right)}{-\left(\frac{n\pi}{b}\right)^2}+(-2)\frac{\cos\left(\frac{n\pi y}{b}\right)}{\left(\frac{n\pi}{b}\right)^3}\right]_0^b\right\}$$

$$= \frac{4b}{ab} - \frac{2a^3}{m^3\pi^3}(\cos m\pi - 1) - \frac{2b^3}{n^3\pi^3}(\cos n\pi - 1)$$

Hence, $$u(x, y, t) = \sum_{m=1}^{\infty}\sum_{n=1}^{\infty} A_{mn} \sin\left(\frac{m\pi x}{a}\right)\sin\left(\frac{n\pi y}{b}\right)\cos(c\lambda_{mn} t)$$

where $$A_{mn} = \frac{16ha^2b^2}{m^3n^3\pi^6}(1 \cos m\pi)(1 - \cos n\pi)$$

and $$\lambda_{mn} = \pi\sqrt{(m/a)^2 + (n/b)^2}.$$

EXAMPLE 10.11

Discuss the solution of transverse vibrations of a thin membrane boundary by a circle of radius a in xy-plane described by the function $u(x, y, t)$ satisfying the wave equation

$$\frac{\partial^2 u}{\partial r^2} + \frac{1}{r}\frac{\partial u}{\partial r} + \frac{1}{r^2}\frac{\partial^2 u}{\partial \theta^2} = \frac{1}{c^2}\frac{\partial^2 u}{\partial t^2} \quad (1)$$

with conditions $u = 0$ on $r = a$ and $u = f(r)$, $\frac{\partial u}{\partial t}$ at $t = 0$.

Here due to the radial symmetry $\frac{\partial^2 u}{\partial \theta^2} = 0$

Solution: Let the solution of the given wave Eq. (1) be

$$u(r, t) = R(r)T(t)$$

$\therefore$ Equation (1) becomes

$$TR'' + \frac{T}{r}R' = \frac{R}{c^2}T'' \Rightarrow \frac{R''}{R} + \frac{R'}{Rr} = \frac{T''}{c^2T} = -k^2 = \text{constant}$$

Thus, we have

$$R'' + \frac{R'}{r} + k^2R = 0 \quad \text{and} \quad T'' + c^2k^2T = 0$$

Let $s = kr$ so that $\frac{1}{r} = \frac{k}{s}$

So, $R'' + \frac{R'}{r} + k^2R = 0$ reduces to

$$\frac{d^2R}{ds^2} + \frac{1}{s}\frac{dR}{ds} + R = 0 \quad \text{or} \quad s^2\frac{d^2R}{ds^2} + s\frac{dR}{ds} + s^2R = 0$$

which is Bessel's equation of zero order with general solution as

$$R = c_1J_0(s) + c_2Y_0(s)$$

where J_0 and Y_0 are Bessel's functions of first and second kind of order zero, respectively.

$$\therefore \qquad R = c_1J_0(kr) + c_2Y_0(kr)$$

Since the deflection of the membrane is always finite while $Y_0(kr)$ becomes infinite as r approaches to zero, so we cannot use Y_0 and must choose $c_2 = 0$

$$\therefore \qquad R = c_1J_0(kr)$$

The solution of $T'' + c^2k^2T = 0$ is

$$T = c_3 \cos ckt + c_4 \sin ckt$$

Thus, $u(r, t) = J_0(kr)(c_1c_3 \cos ckt + c_1c_4 \sin ckt)$

Now, $u(a, t) = J_0(ka)\,(c_1c_3 \cos ckt + c_1c_4 \sin ckt) = 0$

$\Rightarrow \qquad J_0(ka) = 0$

Let $k_n(n = 1, 2, 3, \ldots)$ be the positive roots of $J_0(ka) = 0$

Hence, $u_n(r, t) = (c_n \cos ck_nt + d_n \sin ck_nt)\, J_0(k_nr)$ with $n = 1, 2, 3, \ldots$ {choose $c_n = c_1c_3$ and $d_n = c_1c_4$}

By the principle of superposition, we have

$$u(r, t) = \sum_{n=0}^{\infty} u_n(r,t) = \sum_{n=0}^{\infty}(c_n \cos ck_nt + d_n \sin ck_nt) \times J_0(k_nr) \qquad (2)$$

Now, $$\frac{\partial u}{\partial t} = \sum_{n=0}^{\infty} (ck_n)\,(-c_n \sin ck_nt + d_n \cos ck_nt) \times J_0(k_nr)$$

$$\therefore \qquad u(r, 0) = f(r) = \sum_{n=1}^{\infty} c_nJ_0(k_nr)$$

and $$\left(\frac{\partial u}{\partial t}\right)_{t=0} = g(r) = \sum_{n=1}^{\infty} ck_nd_nJ_0(k_nr)$$

By using Fourier–Bessel series functions, we have

$$c_n = \frac{2}{a^2 J_1^2(k_n a)} \int_0^a r f(r) J_0(k_n r) dr \tag{3}$$

and

$$c k_n d_n = \frac{2}{a^2 J_1^2(k_n a)} \int_0^a r g(r) J_0(k_n r) dr$$

$\Rightarrow$

$$d_n = \frac{2}{c k_n a^2 J_1^2(k_n a)} \int_0^a r g(r) J_0(k_n r) dr \tag{4}$$

Hence, the required solution of the wave Eq. (1) is given by Eq. (2) where c_n and d_n are given by Eqs. (3) and (4), respectively.

Note: If the initial velocity $g(r)$ of the membrane is zero

$\therefore$

$$\left(\frac{\partial u}{\partial t}\right)_{t=0} = g(r) = 0$$

then

$$d_n = 0$$

In this case, the solution of wave Eq. (1) is given by

$$u(r, t) = \sum_{n=1}^{\infty} c_n \cos(c k_n t) J_0(k_n r)$$

where

$$c_n = \frac{2}{a^2 J_1^2(k_n a)} \int_0^a r f(r) J_0(k_n r) dr$$

▲▲

One-Dimensional Heat Equation

Consider a heat conducting homogenous rod of length l, placed along the x-axis with one end of the rod at $x = 0$ (origin) and the other end of the rod at $x = l$. Assume that the rod has constant density ρ and uniform area of cross section A. Also, assume that the lateral surface of the rod is impenetrable to heat transfer, that is, the rod is insulated laterally and therefore heat flows only in the x-direction. The rod is sufficiently thin so that the temperature is same at all points of any cross-sectional area of the rod. Let $u(x, t)$ be the temperature of the cross section at the point x at any time t.

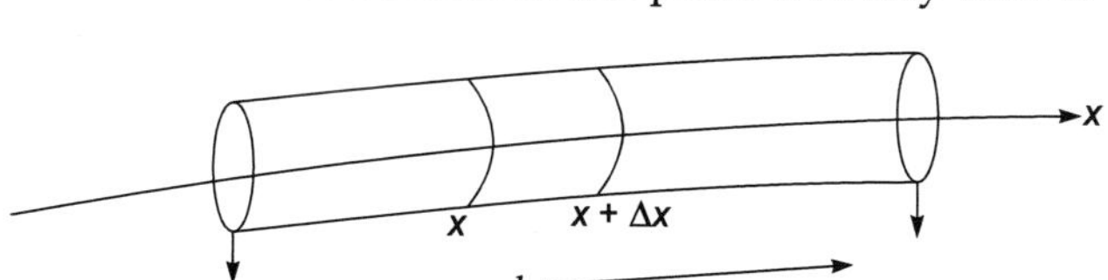

Fig. 10.7

The amount of heat $Q(t)$ in a small segment of the rod between the cross sections at x and $x + \Delta x$ is

$$Q(t) = \int_x^{x+\Delta x} c\rho A u(x, t)dx \tag{1}$$

where c is the specific heat of the rod. By Fourier law of heat conduction, rate of propagation of heat, that is, the quantity of heat passing through a cross section at x in unit time is

$$q = -k\frac{\partial u}{\partial x}A \tag{2}$$

where, k is the coefficient of thermal conductivity. Since heat flows in the direction of decreasing temperature, a negative sign appears in Eq. (2). The rate of heat flow at cross section $x + \Delta x$ is equal to $-KA\dfrac{\partial u}{\partial x}(x + \Delta x, t)$.

The rate of change of heat content in the segment of the rod between x and $x + \Delta x$ must be equal to the net heat flow into this segment of the rod. Thus,

$$\frac{\partial Q}{\partial t} = kA\left[\left.\frac{\partial u}{\partial x}\right|_{(x+\Delta x,t)} - \left.\frac{\partial u}{\partial x}\right|_{(x,t)}\right]$$

By using mean value theorem, we obtain

$$\frac{\partial Q}{\partial t} = \frac{\partial}{\partial t}\int_x^{x+\Delta x} c\rho A u(x,t)dx = \int_x^{x+\Delta x} c\rho A\frac{\partial u}{\partial t}dx$$

$$= c\rho A u_t(\varepsilon, t)\Delta x \text{ where } x < \varepsilon < x + \Delta x$$

Therefore,

$$c\rho A\frac{\partial u}{\partial t}(\varepsilon,t)\Delta x = KA\left[\left.\frac{\partial u}{\partial x}\right|_{(x+\Delta x,t)} - \left.\frac{\partial u}{\partial x}\right|_{(x,t)}\right]$$

Taking the limit as $\Delta x \to 0$, we have

$$\frac{\partial u(x,t)}{\partial t} = a^2\frac{\partial^2 u}{\partial x^2} \tag{3}$$

where $a^2 = \dfrac{k}{c\rho}$ known as diffusivity constant. Equation (3) is the one-dimensional heat equation. Assume, the boundary conditions be

$$u(0, t) = 0 \; u(l, t) = 0$$

that is, the end points of the rod are at zero temperature and the initial temperature distribution in the rod is given by

$$u(x, 0) = f(x)$$

where $f(x)$ is a given function of x. We will obtain solution of Eq. (3) by using variable separable method. So, let $u(x, t) = X(x)T(t)$

$$\therefore \quad \frac{\partial u}{\partial t} = XT', \frac{\partial u}{\partial x} = X'T, \frac{\partial^2 u}{\partial x^2} = x''T$$

So, Eq. (3) reduces to

$$XT' = a^2X''T$$

$$\Rightarrow \quad \frac{X''}{X} = \frac{T'}{a^2T} = k = \text{constant}$$

$$\therefore \quad X'' - KX = 0 \quad \text{and} \quad T' - a^2KT = 0$$

The boundary conditions reduce to

$$0 = u(0, t) = X(0)T(t) \Rightarrow X(0)$$

and $$0 = u(l, t) = X(l)T(t) \Rightarrow X(l) = 0.$$

Case 1: $k - 0$.

In this case $$X'' = 0$$

$$\therefore \quad X(x) = a_1x + a_2$$

Now, $$X(0) = 0 = a_1 \cdot 0 + a_2 \quad \Rightarrow \quad a_2 = 0$$

$$\therefore \quad X(x) = a_1x$$

Also, $$X(l) = 0 = a_1l$$

as $$l \neq 0 \quad \therefore a_1 = 0$$

Thus, $X(x) = 0$ for all x

$\Rightarrow u(x, t) = 0$ for all x and t, which is a trivial solution.

Case 2: $k = \lambda^2 > 0$

In this case $$X'' - \lambda^2X = 0$$

$$\therefore \quad X(x) = Ae^{\lambda x} + Be^{-\lambda x}$$

Now, $$X(0) = 0 = A + B \quad \Rightarrow \quad A = -B$$

Also, $$X(l) = 0 = A(e^{kl} - e^{-kl})$$

$$\Rightarrow \quad A = 0 = B$$

Thus, $X(x) = 0$ for all x

$\Rightarrow u(x, t) = 0$ for all x and t, which is a trivial solution.

Case 3: $k = -\lambda^2 < 0$

In this case $$X'' + \lambda^2X = 0$$

$$\therefore \quad X(x) = A\cos\lambda x + B\sin\lambda x$$

Now, $X(0) = 0 = A + B\cdot 0 \Rightarrow A = 0$

$$\therefore \quad X(x) = B\sin\lambda x$$

Also, $X(l) = 0 = B\sin\lambda l$

For $u \neq 0$ we must have $B \neq 0$

$$\therefore \quad \sin\lambda l = 0$$

$$\Rightarrow \quad \lambda l = n\pi \quad \therefore \lambda = \frac{n\pi}{l} \text{ with } n = 1, 2, 3, \ldots$$

$$\therefore \quad X_n(x) = B_n \sin\frac{n\pi x}{l} \quad (n = 1, 2, 3 \ldots)$$

which gives infinitely many solutions.

Also, $T' + a^2\lambda^2 T = 0$ gives

$$T_n(t) = c_n e^{-a^2\lambda^2 t} = c_n e^{-\frac{a^2n^2\pi^2}{l^2}t}$$

Hence, the eigenfunctions are

$$u_n(x, t) = X_n(x)T_n(t) = B_n{}'\sin\frac{n\pi x}{l}e^{-\frac{a^2n^2\pi^2}{l^2}t} \qquad \{n = 1, 2, 3, \ldots\}$$

where $B_n{}' = B_n c_n$

By using the principle of superposition, we have

$$u(x, t) = \sum_{n=1}^{\infty} u_n(x, t)$$

$$= \sum_{n=1}^{\infty} B_n{}'\sin\frac{n\pi x}{l}e^{-\frac{a^2n^2\pi^2}{l^2}t} \qquad (4)$$

Also, $$u(x, 0) = f(x) = \sum_{n=1}^{\infty} B_n{}'\sin\frac{n\pi x}{l}$$

Since the right hand members forms a half range Fourier sine series expansion of $f(x)$ in the interval (o, l), therefore the Fourier coefficients B_n's are given by

$$B_n' = \frac{2}{l}\int_0^l f(x)\sin\frac{n\pi x}{l}dx \qquad \{n = 1, 2, 3, \ldots\} \qquad (5)$$

Thus, the temperature distribution in the rod is given by the solution (4) with B_n's calculated from Eq. (5).

Note 1: Since solution (4) contains negative exponential terms so u deceases as t increases which is consistent with physical nature of the heat conduction problem.

Note 2: If we want to find the general solution without any specific boundary conditions, then the process is similar to the above.

For $k = 0$, $u = (a_1x + a_2)a_3$

For $k > 0$, $u = (Ae^{\lambda x} + Be^{-\lambda x})e^{a^2\lambda^2 t}$

For $k < 0$, $u = (A\cos\lambda x + B\sin\lambda x)e^{-a^2\lambda^2 t}$

Out of these three solutions, we have to choose that condition which is consistent with the physical nature of the problem, that is, u decreases with increase in time t. So, accordingly, the solution will be of the form

$$u(x, t) = (A\cos\lambda x + B\sin\lambda x)e^{-a^2\lambda^2 t}$$

which is the only suitable solution of the one-dimensional heat Eq. (3).

Two-Dimensional Heat Flow

Consider the flow of heat in a metal plate of uniform thickness d, density ρ, specific heat s and thermal conductivity h. Let xy-plane be taken in one face of the plate. If the temperature at any point is independent of the z-coordinate and depends only on x, y and time t, then the flow is said to be two dimensional. In this case, the heat flow is in the xy-plane only and is zero along the normal to the xy-plane. Consider a rectangular element of dimensions Δx and Δy.

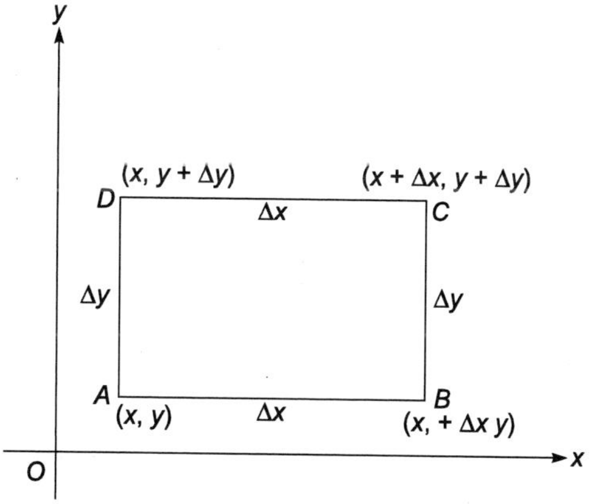

Fig. 10.8

Now, the amount of heat entering the element in unit time from the face $AB =$ $-hd\Delta x\left(\frac{\partial u}{\partial y}\right)_y$ and the amount of heat entering the element in unit time from the face $AD = -hd\Delta y\left(\frac{\partial u}{\partial x}\right)_x$. The quantity of heat flowing out through the face CD in unit time $= -hd\Delta x\left(\frac{\partial u}{\partial y}\right)_{y+\Delta y}$ and the quantity of heat flowing out through the face BC in unit time $= -hd\Delta y\left(\frac{\partial u}{\partial x}\right)_{x+\Delta x}$.

Hence, the total gain of heat by the element in unit time

$$= -hd\Delta x\left(\frac{\partial u}{\partial y}\right)_y - hd\Delta y\left(\frac{\partial u}{\partial x}\right)_x + hd\Delta x\left(\frac{\partial u}{\partial y}\right)_{y+\Delta y} + hd\Delta y\left(\frac{\partial u}{\partial x}\right)_{x+\Delta x}$$

$$= hd\Delta x\left\{\left(\frac{\partial u}{\partial y}\right)_{y+\Delta y} - \left(\frac{\partial u}{\partial y}\right)_y\right\} + hd\Delta y\left\{\left(\frac{\partial u}{\partial x}\right)_{x+\Delta x} - \left(\frac{\partial u}{\partial x}\right)_x\right\}$$

$$= hd\Delta x\Delta y\left\{\frac{\left(\frac{\partial u}{\partial y}\right)_{y+\Delta y} - \left(\frac{\partial u}{\partial y}\right)_y}{\Delta y} + \frac{\left(\frac{\partial u}{\partial x}\right)_{x+\Delta x} - \left(\frac{\partial u}{\partial x}\right)_x}{\Delta x}\right\}$$

Also, the rate of gain of heat by the element $= \rho\Delta x\Delta y ds\dfrac{\partial u}{\partial t}$.

Thus, we obtain on taking limits as $\Delta x, \Delta y \to 0$

$$h\left(\frac{\partial^2 u}{\partial y^2} + \frac{\partial^2 u}{\partial x^2}\right) = \rho s\frac{\partial u}{\partial t}$$

$$\Rightarrow \qquad \frac{\partial u}{\partial t} = a^2\left(\frac{\partial^2 u}{\partial x^2} + \frac{\partial^2 u}{\partial y^2}\right) \qquad (1)$$

where $a^2 = \dfrac{h}{\rho s}$ is the diffusivity.

Hence, the Eq. (1) gives the temperature distribution of the plate in the transient state and is known as the two-dimensional heat equation.

Now, consider the boundary conditions

$$u(x, 0, t) = 0 = u(x, b, t) \; \{0 < x < l, 0 < t]$$

and $$u(0, y, t) = 0 = u(l, y, t) \; \{0 < y < b, 0 < t\}$$

where, l and b are dimensions of the plate with thickness d.

Also, $$u(x, y, 0) = f(x, y) \; \{0 < x < l, 0 < y < b\}$$

Now, we will be obtaining solution of Eq. (1) by using the method of separation of variables. So, let $u(x, y, t) = \phi(x, y)T(t)$.

Thus, Eq. (1) reduces to

$$\left(\frac{\partial^2\phi}{\partial x^2}+\frac{\partial^2\phi}{\partial y^2}\right)T=\frac{1}{a^2}\times\phi T'$$

$$\Rightarrow \qquad \frac{1}{\phi}\left(\frac{\partial^2\phi}{\partial x^2}+\frac{\partial^2\phi}{\partial y^2}\right)=\frac{T'}{a^2T}=-\lambda^2=\text{constant}$$

where negative sign is used since the time dependent part of the product solution exponentially decays.

Thus, we have

$$T'+\lambda^2a^2T=0 \qquad \{t>0\}$$

and
$$\frac{\partial^2\phi}{\partial x^2}+\frac{\partial^2\phi}{\partial y^2}=-\lambda^2\phi \qquad \{0<x<l,\ 0<y<b\}$$

Also, we obtain

$$u(x,0,t)-0-\phi(x,0)T(l)\Rightarrow \psi(x,0)=0$$

$$u(x,b,t)=0=\phi(x,b)T(t)\Rightarrow \phi(x,b)=0$$

$$u(0,y,t)=0=\phi(0,y)T(t)\Rightarrow \phi(0,y)=0$$

$$u(l,y,t)=0=\phi(l,y)T(t)\Rightarrow \phi(l,y)=0$$

Let
$$\phi(x,y)=X(x)Y(y)$$

So,
$$\frac{\partial^2\phi}{\partial x^2}+\frac{\partial^2\phi}{\partial y^2}=-\lambda^2\phi \text{ becomes}$$

$$\frac{X''}{X}+\frac{Y''}{Y}=-\lambda^2$$

Thus,
$$\frac{X''}{X}=-\mu^2 \quad\text{and}\quad \frac{Y''}{Y}=-\gamma^2$$

$$\Rightarrow \qquad X''+\mu^2X=0 \tag{2}$$

and
$$Y''+\nu^2Y=0 \tag{3}$$

where
$$\mu^2+\nu^2=\lambda^2$$

Also, the boundary conditions becomes

$$X(x)Y(0)=0 \quad\Rightarrow\quad Y(0)=0$$

$$X(x)Y(b)=0 \quad\Rightarrow\quad Y(b)=0$$

$$X(0)Y(y)=0 \quad\Rightarrow\quad X(0)=0$$

$$X(l)Y(y)=0 \quad\Rightarrow\quad X(l)=0$$

Equations (2) and (3) form two independent eigenvalue problems, with the following solutions

$$X_m(x) = \sin\left(\frac{m\pi x}{l}\right),\ \mu_m^2 = \left(\frac{m\pi}{l}\right)^2$$

and
$$Y_n(x) = \sin\left(\frac{n\pi y}{b}\right),\ \nu_n^2 = \left(\frac{n\pi}{b}\right)^2$$

Hence, the indices m and n are independent.

Thus,
$$\phi_{mn}(x, y) = X_m(x)Y_n(y)$$

with
$$\lambda_{mn}^2 = \mu_m^2 + \nu_n^2$$

and the corresponding solution of $T' + \lambda^2 a^2 T = 0$ is

$$T_{mn} = e^{-\lambda_{mn}^2 a^2 t}$$

Hence,
$$u_{mn}(x, y, t) = \phi_{mn}(x, y)T_{mn}(t)$$

$$= \sin\left(\frac{m\pi x}{l}\right)\sin\left(\frac{n\pi x}{b}\right)e^{-\lambda_{mn}^2 a^2 t}$$

for each pair of indices m and n with $m, n = 1, 2, 3, \ldots$

Using the superposition principle, we obtain the double series

$$u(x, y, t) = \sum_{m=1}^{\infty}\sum_{n=1}^{\infty} A_{mn}\phi_{mn}(x,y)T_{mn}(t)$$

Now,
$$u(x, y, 0) = f(x, y) = \sum_{m=1}^{\infty}\sum_{n=1}^{\infty} A_{mn}\phi_{mn}(x, y)$$

Thus, we have

$$u(x, y, t) = \sum_{m=1}^{\infty}\sum_{n=1}^{\infty} A_{mn}\sin\left(\frac{m\pi x}{l}\right)\sin\left(\frac{n\pi y}{b}\right)\times e^{-\lambda_{mn}^2 a^2 t} \tag{4}$$

and
$$f(x, y) = \sum_{m=1}^{\infty}\sum_{n=1}^{\infty} A_{mn}\sin\left(\frac{m\pi x}{l}\right)\sin\left(\frac{n\pi y}{b}\right)$$

which is a double Fourier sine series.

Here, A_{mn}'s are determined by

$$A_{mn} = \frac{4}{lb}\int_0^l\int_0^b f(x,y)\sin\left(\frac{m\pi x}{l}\right)\sin\left(\frac{n\pi y}{b}\right)dxdy \text{ with } m, n = 1, 2, 3 \ldots \tag{5}$$

Thus, the complete solution is given by Eq. (4) with coefficients A_{mn} determined by Eq. (5).

Note: When u is independent of time t, temperature distribution is said to be in the steady-state and in this case the Eq. (1) reduces to

$$\frac{\partial^2 u}{\partial x^2} + \frac{\partial^2 u}{\partial y^2} = 0 \tag{6}$$

Equation (6) is known as two-dimensional Laplace's equation or steady-state heat equation.

Let $u = X(x)Y(y)$ be a solution of Eq. (6)

So, Eq. (6) becomes $\frac{X''}{X} = -\frac{Y''}{Y} = k = \text{constant}$

$$\Rightarrow \quad X'' - kX = 0 \quad \text{and} \quad Y'' + kY = 0$$

Solving these equations, we get

(1) When $k = \lambda^2 > 0$

$$X(x) = c_1 e^{\lambda x} + c_2 e^{-\lambda x},\ Y(x) = c_3 \cos \lambda y + c_4 \sin \lambda y$$

(2) When $k = -\lambda^2 < 0$

$$X(x) = c_1 \cos \lambda x + c_2 \sin \lambda x,\ Y(y) = c_3 e^{\lambda y} + c_4 e^{-\lambda y}$$

(3) When $k = 0$

$$X(x) = c_1 x + c_2,\ Y(y) = c_3 y + c_4$$

Thus, the various possible solution of Eq. (6) are

$$u = (c_1 e^{\lambda x} + c e^{-\lambda x})(c_3 \cos \lambda y + c_4 \sin \lambda y)$$

$$u = (c_1 \cos \lambda x + c_2 \sin \lambda x)(c_3 e^{\lambda y} + c_4 e^{-\lambda y})$$

$$u = (c_1 x + c_2)(c_3 y + c_4)$$

Of these, we take that solution which is consistent with the given boundary conditions. By using the substitutions $x = r \cos \theta$ and $y = r \sin \theta$, Eq. (6) reduces in polar form as

$$r^2 \frac{\partial^2 u}{\partial r^2} + r \frac{\partial u}{\partial r} + \frac{\partial^2 u}{\partial \theta^2} = 0 \tag{7}$$

Let $u = R(r)\,\phi(\theta)$ be a solution of Eq. (7).
So, Eq. (7) becomes

$$r^2 R''\phi + rR'\phi + R\phi'' = 0$$

$$\Rightarrow \quad \frac{r^2 R'' + rR'}{R} = -\frac{\phi''}{\phi} = k = \text{constant}$$

Thus,

$$r^2 R'' + rR' - kR = 0 \quad \text{and} \quad \phi'' + k\phi = 0$$

Taking $r = e^z$, we have

$$\frac{d^2R}{dz^2} - KR = 0 \quad \text{and} \quad \phi'' + k\phi = 0$$

Solving these, we get

(1) When $k = \lambda^2 > 0$

$$R = c_1e^{\lambda z} + c_2e^{-\lambda z} = c_1r^{\lambda} + c_2r^{-\lambda}$$

$$\phi = c_3 \cos \lambda\theta + c_4 \sin \lambda\theta$$

(2) When $k = -\lambda^2 < 0$

$$R = c_1 \cos \lambda z + c_2 \sin \lambda z = c_1 \cos (\lambda \log r) + c_2 \sin (\lambda \log r),\ \phi = c_3e^{+\lambda\theta} + c_4e^{-\lambda\theta}$$

(3) When $k = 0$

$$R = c_1z + c_2 = c_1 \log r + c_2,\ \phi = c_3\theta + c_4$$

Thus, the three possible solutions of Eq. (7) are

$$u = (c_1r^{\lambda} + c_2r^{-\lambda})(c_3 \cos \lambda\theta + c_4 \sin \lambda\theta) \tag{8}$$

$$u = (c_1 \cos (\lambda \log r) + c_2 \sin (\lambda \log r))(c_3e^{\lambda\theta} + c_4e^{-\lambda\theta}) \tag{9}$$

$$u = (c_1 \log r + c_2)(c_3\theta + c_4) \tag{10}$$

of these solutions, we have to take that solution which is consistent with the physical nature of the problem. The general solution will consist of a sum of terms of type Eqs. (8), (9) or (10). ▲▲

EXAMPLE 10.12

Show that the solution of the equation $\frac{\partial^2 u}{\partial x^2} = \frac{\partial u}{\partial t}$ satisfying the conditions

(a) $u \to 0$ as $t \to \infty$

(b) $u = 0$ when $x = \pm a$ for all values of $t > 0$.

(c) $u = x$ when $t = 0$ and $-a < x < a$ is

$$u = \frac{2a}{\pi}\sum_{n=1}^{\infty}\frac{(-1)^{n-1}}{n}\sin\left(\frac{n\pi x}{a}\right)e^{-\frac{n^2\pi^2 t}{a^2}}.$$

Solution: Given equation is $\frac{\partial^2 u}{\partial x^2} = \frac{\partial u}{\partial t}$ (1)

By using method of separation of variables,

Let $$u(x, t) = X(x)T(t)$$

So, Eq. (1) becomes

$$X''T = XT'$$

$$\Rightarrow \qquad \frac{X''}{X} = \frac{T'}{T} = -k^2 \text{ (a constant)} \qquad \{\because u \to 0 \text{ as } t \to \infty\}$$

As we have found earlier the solution of $X'' + K^2X = 0$ and $T' + K^2T = 0$ will be $X(x) = c_1 \cos kx + c_2 \sin kx$ and $T(t) = c_3e^{-k^2t}$

So, $$u(x, t) = (A \cos kx + (B \sin kx)e^{-k^2t}$$

where, $A = c_1c_3$ and $B = c_2c_3$

Now, we are given $u(-a, t) = 0 = u(a, t)$

$$\therefore \qquad u(-a, t) = (A \cos ka - B \sin ka)e^{-k^2t} = 0$$

and $$u(a, t) = (A \cos ka + B \sin ka)e^{-k^2t} = 0$$

On adding, we get $2A \cos ka\, e^{-k^2t} = 0$

$$\therefore \qquad A = 0$$

$$\Rightarrow \qquad u(x, t) = B \sin kx\, e^{-k^2t}$$

Also, on subtracting, we get $2B \sin kae^{-k^2t} = 0$

Now, for $u \neq 0$, we must have $B \neq 0$

$$\therefore \qquad \sin ka = 0$$

$$\Rightarrow \qquad ka = n\pi$$

$$\therefore \qquad k = \frac{n\pi}{a} \text{ with } n = 1, 2, 3, \ldots$$

Thus, for integral values of n, we have

$$u_n(x, t) = B_n \sin\left(\frac{n\pi x}{a}\right)e^{-\frac{n^2\pi^2t}{a^2}}$$

By the principle of superpositive, we obtain

$$u(x,t) = \sum_{n=1}^{\infty} u_n(x,t) = \sum_{n=1}^{\infty} B_n \sin\left(\frac{n\pi x}{a}\right)e^{-\frac{n^2\pi^2t}{a^2}}$$

Also, we have $u(x, 0) = x$ for $-a < x < a$

$$\Rightarrow \qquad \sum_{n=1}^{\infty} B_n \sin\left(\frac{n\pi x}{a}\right) = x$$

So, by the Fourier sine series formula, we have

$$B_n = \frac{2}{a}\int_0^a x \sin\left(\frac{n\pi x}{a}\right)dx$$

$$= \frac{2}{a}\left\{x\frac{\cos\left(\frac{n\pi x}{a}\right)}{\left(-\frac{n\pi}{a}\right)} - 1\cdot\frac{\sin\left(\frac{n\pi x}{a}\right)}{-\left(\frac{n\pi}{a}\right)^2}\right\}_0^a$$

$$= \frac{2}{a}\frac{a^2}{-n\pi}\cos n\pi = \frac{2a}{n\pi}(-1)^{n-1}$$

Hence, the required solution is

$$u(x, t) = \sum_{n=1}^{\infty} \frac{2a}{n\pi}(-1)^{n-1}\sin\left(\frac{n\pi x}{a}\right)e^{-n^2\pi^2 t/a^2}$$

EXAMPLE 10.13

Find the temperature in a thin metal rod of length l, with both the ends insulated so that there is no passage of heat through the ends and with initial temperature in the rod $\sin\left(\frac{\pi x}{l}\right)$.

Solution: Let $u(x, t)$ be the temperature in the rod.

Then, the initial boundary value problem is

$$\frac{\partial^2 u}{\partial t} = c^2\frac{\partial^2 u}{\partial x^2} \tag{1}$$

with
$$\left(\frac{\partial u}{\partial x}\right)_{x=0} = 0 = \left(\frac{\partial u}{\partial x}\right)_{x=l}$$

and
$$u(x, 0) = \sin\left(\frac{\pi x}{l}\right)$$

Its solution will be found by using method of separation of variables.

Let
$$u(x, t) = X(x)T(t)$$

So, Eq. (1) becomes
$$c^2X''T = XT'$$

with
$$u_x(0, t) = X'(0)T(t) = 0 \Rightarrow X'(0) = 0$$

and
$$u_x(l, t) = X'(l)T(t) = 0 \Rightarrow X'(l) = 0$$

$$u(x, 0) = X(x)T(0) = \sin\left(\frac{\pi x}{l}\right) = f(x) \text{ (say)}$$

Now, $\dfrac{X''}{X} = \dfrac{T}{c^2T} = -k^2$ (a constant)

$\therefore$ Solution of $X'' + k^2X = 0$ and $T' + c^2k^2T = c$ is

$$X(x) = c_1 \cos kx + c_2 \sin kx \text{ and } T(t) = c_3e^{-c^2k^2t}$$

$\therefore$ $$X'(x) = k(-c_1 \sin kx + c_2 \cos kx)$$

Thus, $$X'(0) = kc_2 = 0 \quad \therefore c_2 = 0$$

$\therefore$ $$X(x) = c_1 \cos kx$$

and $$X'(x) = -k\, c_1 \sin kx$$

Again, $$X'(l) = 0 = -kc_1 \sin kl$$

So, for $u \neq 0$, we must have $c_1 \neq 0$

$\therefore$ $$\sin kl = 0 \Rightarrow kl = n\pi \quad \therefore k = \frac{n\pi}{l}$$

with $$n - 0, 1, 2, 3, \ldots$$

$\therefore$ For integral values of n, we have

$$X_n(x) = \cos\left(\frac{n\pi x}{l}\right)$$

$\Rightarrow$ $$u_n(x, t) = A_n \cos\left(\frac{n\pi x}{l}\right) e^{-n^2\pi^2c^2t/l^2}$$

By the linearity principal, we have

$$u(x, t) = \sum_{n=0}^{\infty} u_n(x,t) = \sum_{n=0}^{\infty} A_n \cos\left(\frac{n\pi x}{l}\right) e^{-\frac{n^2\pi^2c^2t}{l^2}}$$

$\therefore$ $$u(x, t) = \frac{A_0}{2} + \sum_{n=1}^{\infty} A_n \cos\left(\frac{n\pi x}{l}\right) e^{-n^2\pi^2c^2t/l^2}$$

where $$A_0 = \frac{2}{l}\int_0^l f(x)dx = \frac{2}{l}\int_0^l \sin\left(\frac{\pi x}{l}\right) dx$$

{by using Fourier series formulae}

$$= \frac{2}{l}\left[\frac{\cos\left(\frac{\pi x}{l}\right)}{\left(-\frac{\pi}{l}\right)}\right]_0^l - \frac{2}{\pi}(\cos\pi - \cos 0) = \frac{4}{\pi}$$

and
$$A_n = \frac{2}{l}\int_0^l f(x)\cos\left(\frac{n\pi x}{l}\right)dx$$

$$= \frac{2}{l}\int_0^l \sin\left(\frac{\pi x}{l}\right)\cos\left(\frac{n\pi x}{l}\right)dx$$

For $n = 1$
$$A_1 = \frac{2}{l}\int_0^l \sin\left(\frac{\pi x}{l}\right)\cos\left(\frac{\pi x}{l}\right)dx = 0$$

For $n \neq 1$
$$A_n = \frac{2}{l}\int_0^l \sin\left(\frac{\pi x}{l}\right)\cos\left(\frac{n\pi x}{l}\right)dx$$

$$= \frac{1}{l}\int_0^l \left\{\sin(1+n)\frac{\pi x}{l} + \sin(1-n)\frac{\pi x}{l}\right\}dx$$

$$= \frac{1}{l}\left\{\frac{\cos(n+1)\frac{\pi x}{l}}{-(n+1)\frac{\pi}{l}} + \frac{\cos(1-n)\frac{\pi x}{l}}{-(1-n)\frac{\pi}{l}}\right\}_0^l$$

$$= \frac{1}{\pi}\left\{\frac{\cos(n+1)\pi - 1}{-(n+1)} + \frac{\cos(1-n)\pi - 1}{-(1-n)}\right\}$$

$$= \frac{1}{\pi}\left\{\frac{(-1)^{n+1} - 1}{-(n+1)} + \frac{(-1)^{n-1} - 1}{n-1}\right\}$$

$$= \frac{1}{\pi}\left\{\frac{(-1)^{n+1} - 1}{-(n+1)} + \frac{(-1)^{n+1} - 1}{n-1}\right\}$$

$$= \frac{1}{\pi}\{(-1)^{n+1} - 1\}\left(\frac{-1}{n+1} + \frac{1}{n-1}\right)$$

$$= \frac{2}{\pi(n^2-1)}[(-1)^{n+1} - 1]$$

Thus, the temperature $u(x, t)$ in the rod is

$$u(x,t) = \frac{2}{\pi} + \frac{2}{\pi}\sum_{n=1}^{\infty}\frac{1}{(n^2-1)}[(-1)^{n+1} - 1]\cos\left(\frac{n\pi x}{l}\right)\times e^{-c^2n^2\pi^2t/l^2}.$$

EXAMPLE 10.14

A homogeneous rod of conducting material of length l has its ends kept at zero temperature and the initial temperature is

$$u(x, 0) = \begin{cases} x & \text{if } 0 \le x \le l/2 \\ l - x & \text{if } \frac{l}{2} \le x \le l \end{cases}$$

Find the temperature $u(x, t)$ at any time t.

Solution: Here, we have to solve the one-dimensional heat equation $\frac{\partial u}{\partial t} = c^2 \frac{\partial^2 u}{\partial x^2}$ with the conditions.

$$u(0, t) = 0 = u(l, t)$$

and

$$u(x, 0) = \begin{cases} x & \text{if } 0 \le x \le \frac{l}{2} \\ l - x & \text{if } \frac{l}{2} \le x \le l \end{cases}$$

As we have done in the previous Example 10.13, the solution by using method of separation of variables will be

$$u(x, t) = X(x)T(t) = (c_1 \cos Kx + c_2 \sin Kx)e^{-c^2k^2t}$$

where K is a constant.

Now, $\quad u(0, t) = X(0)T(t) = 0$

$\Rightarrow \quad c_1 e^{-c^2k^2t} = 0$

$\therefore \quad c_1 = 0$

Thus, $\quad u(x, t) = X(x)T(t) = c_2 \sin Kx\, e^{-c^2k^2t}$

Again, $\quad u(l, t) = X(l)T(t) = 0$

$\Rightarrow \quad c_2 \sin Kl\, e^{-c^2k^2t} = 0$

Now, for $u(x, t) \neq 0$, we must have $c_2 \neq 0$

$\therefore \quad \sin Kl = 0$

$\Rightarrow \quad Kl = n\pi \qquad \therefore K = \frac{n\pi}{l}$ with $n = 1, 2, 3, \ldots$

So, for integral values of n, we have

$$u_n(x, t) = B_n \sin\left(\frac{n\pi x}{l}\right) e^{-c^2n^2\pi^2t/l^2}$$

By the linearity principle, the solution will be

$$u(x,t)=\sum_{n=1}^{\infty}u_n(x,t)=\sum_{n=1}^{\infty}B_n\sin\left(\frac{n\pi x}{l}\right)e^{-\frac{c^2n^2\pi^2t}{l^2}}$$

Now,
$$u(x,0)=\sum_{n=1}^{\infty}B_n\sin\left(\frac{n\pi x}{l}\right)=\begin{cases}x\quad, & \text{if } 0\le x\le \dfrac{l}{2}\\ l-x, & \text{if } \dfrac{l}{2}\le x\le l\end{cases}$$

By using Fourier series formulae, we have

$$B_n=\frac{2}{l}\int_0^l u(x,0)\sin\left(\frac{n\pi x}{l}\right)dx$$

$$=\frac{2}{l}\left\{\int_0^{l/2}x\sin\left(\frac{n\pi x}{l}\right)dx+\int_{l/2}^{l}(l-x)\sin\left(\frac{n\pi x}{l}\right)dx\right\}$$

$$=\frac{4l}{n^2\pi^2}\sin\left(\frac{n\pi}{2}\right) \qquad \{\text{After simplification}\}$$

Therefore, the required temperature is

$$u(x,t)=\frac{4l}{\pi^2}\sum_{n=1}^{\infty}\frac{1}{n^2}\sin\left(\frac{n\pi}{2}\right)\sin\left(\frac{n\pi x}{l}\right)e^{-\frac{c^2n^2\pi^2t}{l^2}}.$$

EXAMPLE 10.15

The temperature at one end of a 50 cm long bar with insulated sides is kept at 0°C and the other end is kept at 100°C until steady state condition prevails.

The two ends are the suddenly insulated, so that the temperature gradient is zero at each end thereafter. Find the temperature distribution.

Solution: Let the rod be along the axis of x with one end which is kept at 0°C be at the origin. The temperature distribution $u(x, t)$ in the bar at any time t at any distance x is governed by one-dimensional heat equation

$$\frac{\partial u}{\partial t}=c^2\frac{\partial^2 u}{\partial x^2} \tag{1}$$

Initially, $u = 0$ at $x = 0$ and $u = 100$°C at $x = 50$. In the steady state, u is independent of t,

$$\therefore \qquad \frac{\partial u}{\partial t}=0$$

So, Eq. (1) becomes $\dfrac{\partial^2 u}{\partial x^2} = 0 \qquad \Rightarrow u(x) = Ax + B$

Since at $x = 0$, $u = 0 \Rightarrow 0 = B$ and at $x = 50$, $u = 100 \Rightarrow 100 = A(50) \Rightarrow A = 2$.

Therefore, the initial temperature distribution in the bar is given by

$$u = 2x \quad \text{or} \quad u(x, 0) = 2x$$

Now, let $u(x, t)$ represent the temperature distribution in the rod at time t measured from the instant when the two ends of the bar are suddenly insulated after the rod reaches the steady state.

Thus, we are required to solve

$$\frac{\partial u}{\partial t} = c^2 \frac{\partial^2 u}{\partial x^2} \text{ under the conditions:}$$

(a) $u_x(0, t) = 0 = u_x(50, t)$

(b) $u(x, 0) = 2x$.

By using method of separation of variables,

Let $\qquad u(x, t) = X(x)T(t)$

Now, proceeding as in Example 10.13, we have

$$u(x, t) = \frac{A_0}{2} + \sum_{n=1}^{\infty} A_n \cos\left(\frac{n\pi x}{50}\right) e^{-\frac{c^2 n^2 \pi^2 t}{50}}$$

where $$A_0 = \frac{2}{50}\int_0^{50} u(x,0) = \frac{2}{50}\int_0^{50} 2x\,dx = 100$$

and $$A_n = \frac{2}{50}\int_0^{50} u(x,0)\cos\left(\frac{n\pi x}{50}\right)dx$$

$$= \frac{2}{50}\int_0^{50} 2x\cos\left(\frac{n\pi x}{50}\right)dx$$

$$= \frac{1}{25}\left\{2x\frac{\sin\left(\frac{n\pi x}{50}\right)}{\left(\frac{n\pi}{50}\right)} - 2\frac{\cos\left(\frac{n\pi x}{50}\right)}{-\left(\frac{n\pi}{50}\right)^2}\right\}_0^{50}$$

$$= \frac{2}{25}\left(\frac{50}{n\pi}\right)^2 (\cos n\pi - 1) = \frac{200}{n^2\pi^2}[(-1)^n - 1]$$

$$= \begin{cases} 0 & , \text{ when } n \text{ is odd} \\ -\dfrac{400}{n^2\pi^2}, & \text{when } n \text{ is odd} \end{cases}$$

Hence, the required solution is

$$u(x, t) = \frac{100}{2} + \sum_{n=1}^{\infty} -\frac{400}{(2n-1)^2\pi^2} \cos\frac{(2n-1)\pi x}{50} \times e^{-(2n-1)^2 \pi^2 c^2 t/(50)^2}$$

$$= 50 - \frac{400}{\pi^2} \sum_{n=1}^{\infty} \frac{1}{(2n-1)^2} \cos\frac{(2n-1)\pi x}{50} \times e^{-(2n-1)^2 \pi^2 c^2 t/2500}.$$

EXAMPLE 10.16

A rod of length 10 cm has its ends P and Q kept at 50°C and 100°C, respectively and steady state condition prevails. The temperature at P is suddenly raised to 90°C and that at Q is lowered to 60°C and these temperatures are maintained. Find the temperature at a distance x from one end at time t.

Solution: We know that the one-dimensional heat flow is governed by the partial differential equation

$$\frac{\partial u}{\partial t} = c^2 \frac{\partial^2 u}{\partial x^2} \qquad (1)$$

Given that $u(0, t) = 50°\text{C}$, $u(10, t) = 100°\text{C}$

For steady state condition $\dfrac{\partial u}{\partial t} = 0$ and Eq. (1) reduces to

$\dfrac{\partial^2 u}{\partial x^2} = 0$ whose solution is

$$u(x) = Ax + B$$

$\because$ At $x = 0$ u 50 $\Rightarrow$ $50 = B$

and at $x = 10$, $u = 100$ $\Rightarrow$ $100 = 10A + 50$

$\therefore$ $A = 5$

$\therefore$ $u(x) = 5x + 50 = u(x, 0)$

Also, given that there is a sudden change in temperature at the end points P and Q, so we again have the transient state. If $u_1(x, t)$ denote the subsequent temperature, then the boundary conditions are

$$u_1(0, t) = 90°\text{C},\ u_1(10, t) = 60°\text{C}$$

and the initial condition is

$$u_1(x, 0) = 5x + 50$$

If $u_s(x)$ denoted the subsequent steady state function, then $\dfrac{\partial^2 u_s}{\partial x^2} = 0$ which has solution as $u_s(x) = Ax + B$.

As $u_1 \to u_s$, at the end points, we have

$$\text{at } x = 0,\ u_s = 90 \quad \Rightarrow \quad B = 90$$

and $$\text{at } x = 10,\ u_s = 160 \quad \Rightarrow \quad 160 = A(10) + 90$$

$\Rightarrow$ $$-3 = A$$

Thus, $$u_s(x) = -3x + 90$$

Therefore, the total temperature distribution in the rod at time t is

$$u(x, t) = u_s(x) + u_T(x, t)$$

where $u_T(x, t)$ is the temperature in the transient state.

$\therefore$ $$u(x, t) = -3x + 90 + u_T(x, t)$$

Moreover, we have $$\frac{\partial u_T}{\partial t} = c^2 \frac{\partial^2 u_T}{\partial x^2}$$

whose solution is

$$u_T(x, t) = (e_1 \cos kx + e_2 \sin kx)e^{-c^2k^2t}$$

where $u_T(x, t)$ satisfies the following boundary conditions:

$$u_T(0, t) = u_1(0, t) - u_s(0) = 0$$
$$u_T(10, t) = u_1(10, t) - u_s(10) = 0$$
$$u_T(x, 0) = u_1(x, 0) - u_s(x) = 8x - 40$$

Now, $$u_T(0, t) = 0 = e_1 e^{-c^2k^2t} \quad \Rightarrow \quad e_1 = 0$$

$\therefore$ $$u_T(x, t) = e_2 \sin kx\, e^{-c^2k^2t}$$

Again, $$u_T(10, t) = 0 = e_2 \sin 10\, k e^{-c^2k^2t}$$

For $u_T \neq 0$, we must have $e_2 \neq 0$

$\therefore$ $$\sin 10\, k = \theta$$

$\Rightarrow$ $$10\, k = n\pi \Rightarrow k = \frac{n\pi}{10} \quad \text{with } n = 1, 2, 3, \ldots$$

$\therefore$ $$u_T(x, t) = \sum_{n=1}^{\infty} b_n \sin\left(\frac{n\pi x}{10}\right) e^{-c^2n^2\pi^2t/100}$$

Also, $$b_n = \frac{2}{10}\int_0^{10} u_T(x, 0) \sin\left(\frac{n\pi x}{10}\right) dx$$

$$= \frac{2}{10}\int_0^{10} (8x - 40) \sin\left(\frac{n\pi x}{10}\right) dx$$

$$= \frac{1}{5}\left[(8x-40)\frac{\cos\left(\frac{n\pi x}{10}\right)}{\left(-\frac{n\pi}{10}\right)} - 8\frac{\sin\left(\frac{n\pi x}{10}\right)}{-\left(\frac{n\pi}{10}\right)^2}\right]_0^{10}$$

$$= \frac{1}{5}\left(-\frac{10}{n\pi}\right)[40\cos n\pi + 40]$$

$$= -\frac{80}{n\pi}[(-1)^n + 1] = \begin{cases} 0 & , \text{ when } n \text{ is odd} \\ -\frac{160}{n\pi}, & \text{when } n \text{ is even} \end{cases}$$

Thus,
$$u_T(x, y) = -\frac{80}{\pi}\sum_{n=1}^{\infty}\frac{1}{n}\sin\left(\frac{n\pi x}{5}\right) e^{-c^2n^2\pi^2t/25}$$

(changing index of summation from n to $2n$)

Hence, the required temperature distribution is

$$u(x, t) = -3x + 90 - \frac{80}{\pi}\sum_{n=1}^{\infty}\frac{1}{n}\sin\left(\frac{n\pi x}{5}\right) e^{-c^2n^2\pi^2t/25}$$

EXAMPLE 10.17

A thin rectangular plane whose surface is impervious to heat flow has at $t = 0$ an arbitrary distribution of temperature $f(x, y)$. Its four edges $x = 0$, $x = a$, $y = 0$, $y = b$ are kept at zero temperature. Determine the temperature at a point of the plate as t increases.

Solution: The temperature in the plane is governed by two-dimensional heat equation

$$\frac{\partial^2 u}{\partial x^2} + \frac{\partial^2 u}{\partial y^2} = \frac{1}{c^2}\frac{\partial u}{\partial t} \quad (1)$$

where, $c^2 = \frac{h}{\rho s}$ and $u(x, y, t)$ is the temperature at the point (x, y) at time t.

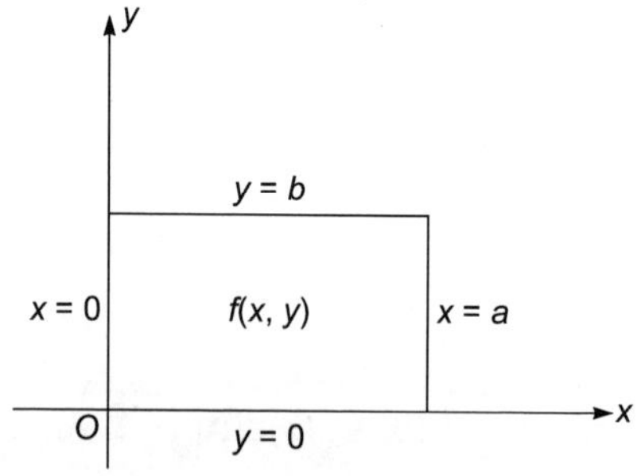

Fig. 10.9

Here, the four faces of the plate are kept at zero temperature and the initial temperature of the plate is $f(x, y)$. Therefore, the function $u(x, y, t)$ is required to satisfy Eq. (1) and the following boundary and initial conditions.

Boundary conditions are

$$u(0, y, t) = 0 = u(a, y, t)$$
$$u(x, 0, t) = 0 = u(x, b, t)$$

and initial condition is $u(x, y, 0) = f(x, y)$.

Let the solution of Eq. (1) be of the form, by using method of separation of variables

$$u(x, y, t) = X(x)Y(y)T(t)$$

So, Eq. (1) becomes

$$X''yT + Xy''T = \frac{1}{c^2} XyT'$$

$$\Rightarrow \quad \frac{X''}{x} + \frac{Y''}{y} = \frac{T'}{c^2T} = K \text{ (a constant)}$$

$$\therefore \quad \frac{X''}{x} + \frac{Y''}{y} = K \quad \text{and} \quad \frac{T'}{c^2T} = K$$

Again, $\frac{X''}{x} = K_1$, $\frac{Y''}{y} = K_2$ where $K = K_1 + K_2$

Now, there are three possibilities

$$K = 0 \quad \text{or} \quad K = \lambda^2 > 0 \quad \text{or} \quad K = -\lambda^2 < 0$$

Thus, we get the following three sets of three ordinary differential equations

(a) $X'' = 0$ $Y'' = 0$ and $T' = 0$

(b) $X'' - \lambda_1^2 X = 0$ $Y'' - \lambda_1^2 Y = 0$ $T' = X^2c^2T$
where $\lambda^2 = \lambda_1^2 + \lambda_2^2$

(c) $X'' + \lambda_1^2 X = 0$ $Y'' + \lambda_1^2 Y = 0$ $T' = -\lambda^2 c^2 T$
where $\lambda^2 = \lambda_1^2 + \lambda_2^2$

The general solution in these cases are

(a) $X = Ax + B$, $Y = Cx + D$ and $T = E$

(b) $X = Ae^{\lambda_1 x} + Be^{-\lambda_1 x}$, $Y = Ce^{\lambda_2 x} + De^{-\lambda_2 y}$
and $T = Ee^{c^2\lambda^2 t}$, where $\lambda^2 = \lambda_1^2 + \lambda_2^2$

(c) $X = A \cos \lambda_1 x + B \sin \lambda_1 x$, $Y = C \cos \lambda_2 y + D \sin \lambda_2 y$ and $T = Ee^{-c^2\lambda^2 t}$, where $\lambda^2 = \lambda_1^2 + \lambda_2^2$

Since, we must have as $t \to \infty$, temperature should remain constant, so only (c) gives us the required solution.

Now $u(0, y, t) = 0 = A(c \cos \lambda_2 y + D \sin \lambda_2 y) \times Ee^{-c^2\lambda^2 t}$

$$\Rightarrow \quad A = 0$$

Again, $u(a, y, t) = 0 = B \sin \lambda_1 a(c \cos \lambda_2 y + D \sin \lambda_2 y) Ee^{-c^2\lambda^2 t}$

Now, for $u \neq 0$, $X \neq 0$ so we must have $B \neq 0$

$$\therefore \qquad \sin \lambda_1 a = 0$$

$$\Rightarrow \qquad \lambda_1 a = m\pi \qquad \therefore \lambda_1 = \frac{m\pi}{a}$$

where m is any integer.

Similarly, $$u(x, 0, t) = 0 \Rightarrow C = 0$$

and $u(x, b, y) = 0 \Rightarrow D \neq 0$ and $\lambda_2 = \dfrac{n\pi}{b}$

where n is any integer.

Thus, for different values of m and n, the solution is given by

$$u_{mn}(x, y, t) = A_{mn} \sin\left(\frac{m\pi x}{a}\right)\sin\left(\frac{n\pi y}{b}\right)e^{-c^2\lambda_{mn}^2 t}$$

where $$\lambda_{mn}^2 = \pi^2\left(\frac{m^2}{a^2} + \frac{n^2}{b^2}\right)$$

By the principle of superposition, the more general solution of Eq. (1) is given by

$$u(x, y, t) = \sum_{m=1}^{\infty}\sum_{n=1}^{\infty} u_{mn}(x, y, t)$$

$$= \sum_{m=1}^{\infty}\sum_{n=1}^{\infty} A_{mn} \sin\left(\frac{m\pi x}{a}\right)\sin\left(\frac{n\pi y}{b}\right)e^{-c^2\lambda_{mn}^2 t} \tag{2}$$

Now, $$u(x, y, 0) = f(x, y) = \sum_{m=1}^{\infty}\sum_{n=1}^{\infty} A_{mn} \sin\left(\frac{m\pi x}{a}\right)\times \sin\left(\frac{n\pi y}{b}\right)$$

By using the double Fourier sine series of $f(x, y)$, we have

$$A_{mn} = \frac{2}{a}\frac{2}{b}\int_0^a\int_0^b f(x,y)\sin\left(\frac{m\pi x}{a}\right)\sin\left(\frac{n\pi y}{b}\right)\times dxdy \tag{3}$$

Hence, the solution of Eq. (1) is given by Eq. (2) with the coefficient given by Eq. (3) where

$$\lambda_{mn}^2 = \pi^2\left(\frac{m^2}{a^2} + \frac{n^2}{b^2}\right)$$

EXAMPLE 10.18

A rectangular plate bounded by the lines $x = 0, y = 0, x = a, y = b$ has an initial temperature distribution given by

$$V = A\sin\left(\frac{\pi x}{a}\right)\sin\left(\frac{\pi y}{b}\right)$$

The edges are kept at zero temperature and the plate faces are impervious to heat. Find the temperature at any point and time.

Solution: The temperature $u(x, y, t)$ at any point and time are governed by the two-dimensional heat equation

$$\frac{\partial u}{\partial t} = c^2\left(\frac{\partial^2 u}{\partial x^2} + \frac{\partial^2 u}{\partial y^2}\right)$$

Here, the boundary and initial conditions are as follows:

Boundary conditions

$$u(0, y, t) = 0 = u(a, y, t)$$
$$u(x, 0, t) = 0 = u(x, b, t)$$

Initial condition $\quad u(x, y, 0) = V = A \sin\left(\frac{\pi x}{a}\right)\sin\left(\frac{\pi y}{b}\right).$

By using the method of separation of variables, let $u(x, y, t) = X(x)Y(y)T(t)$

Proceeding similarly as in Example 10.17, we have

$$u(x, y, t) = \sum_{m=1}^{\infty}\sum_{n=1}^{\infty} u_{mn}(x, y, t)$$

$$= \sum_{m=1}^{\infty}\sum_{n=1}^{\infty} A_{mn} \sin\left(\frac{m\pi x}{a}\right)\sin\left(\frac{n\pi y}{b}\right)\times e^{-c^2\lambda_{mn}^2 t}$$

where $\quad \lambda_{mn}^2 = \pi^2\left(\frac{m^2}{a^2} + \frac{n^2}{b^2}\right)$

and $\quad A_{mn} = \frac{2}{b}\frac{2}{b}\int_0^a\int_0^b f(x,y)\sin\left(\frac{m\pi x}{a}\right)\sin\left(\frac{n\pi y}{b}\right)\times dxdy$

$$= \frac{4A}{ab}\int_0^a\int_0^b \sin\left(\frac{\pi x}{a}\right)\sin\left(\frac{\pi y}{b}\right)\sin\left(\frac{m\pi x}{a}\right)\sin\left(\frac{n\pi y}{b}\right)\times dxdy$$

$$= \frac{4A}{ab}\left(\int_0^a \sin\left(\frac{\pi x}{a}\right)\sin\left(\frac{m\pi x}{a}\right)dx\right)\left(\int_0^b \sin\left(\frac{\pi y}{b}\right)\times\sin\left(\frac{n\pi y}{b}\right)dy\right)$$

Now, $\int_0^b \sin\left(\frac{\pi y}{b}\right)\sin\left(\frac{m\pi y}{b}\right)dy = 0$ for $m \neq 1$

For $m = 1$

$$\int_0^b \sin\left(\frac{\pi y}{b}\right)\sin\left(\frac{\pi y}{b}\right)dy = \frac{b}{2}$$

Similarly, $$\int_0^b \sin\left(\frac{\pi x}{a}\right)\sin\left(\frac{n\pi x}{a}\right)dx = \begin{cases} 0 & n \neq 1 \\ \frac{a}{2} & n = 1 \end{cases}$$

$$\therefore \qquad A_{11} = \frac{4A}{ab}\left(\frac{a}{2}\right)\left(\frac{b}{2}\right) = A$$

and
$$\lambda_{11}^2 = \pi^2\left(\frac{1}{a^2}+\frac{1}{b^2}\right)$$

Thus,
$$u(x, y, t) = \sum_{m=1}^{\infty}\sum_{n=1}^{\infty} A_{mn} \sin\left(\frac{m\pi x}{a}\right)\sin\left(\frac{n\pi y}{b}\right)e^{-c^2\lambda_{mn}^2 t}$$
$$= A_{11}\sin\left(\frac{\pi x}{a}\right)\sin\left(\frac{\pi y}{b}\right)e^{-c^2k^2t}$$
$$= A\sin\left(\frac{\pi x}{a}\right)\sin\left(\frac{\pi y}{b}\right)e^{-c^2\pi^2\left(\frac{1}{a^2}+\frac{1}{b^2}\right)t}$$

which gives the temperature of the plate at any point and time.

EXAMPLE 10.19

Find the temperature distribution in a rectangular plate $0 < x < 1$, $0 < y < 2$ with both its faces insulated, top and bottom edges insulated and right and left edges kept at zero temperature. The initial temperature distribution is given by

$$u(x, y, 0) = x(1 - x)\, y^2(3 - y) \; [0 < x < 1, 0 < y < 2].$$

Solution: The temperature distribution in the plate is governed by two-dimensional heat equation

$$\frac{\partial u}{\partial t} = c^2\left(\frac{\partial^2 u}{\partial x^2}+\frac{\partial^2 u}{\partial y^2}\right) \tag{1}$$

with the following conditions

$$u(0, y, t) = 0 \text{ for all } t \text{ and } 0 < y < 2 = u(l, y, t)$$

$$\frac{\partial u}{\partial y}(x, 0, t) = 0 = \frac{\partial u}{\partial y}(x, 2, t)$$

for all t and $0 < x < 1$ and $u(x, y, 0) = x(1 - x)\, y^2(3 - y)$ $[0 < x < 1, 0 < y < 2]$.

By using the method of separation of variables, let $u(x, y, t) = X(x)Y(y)T(t)$.

Now, proceeding as in the previous examples, we will obtain the solution as

$$u(x, y, t) = (A\cos\lambda_1 x + B\sin\lambda_1 x)(C\cos\lambda_2 y + D\sin\lambda_2 y)Ee^{-c^2\lambda^2 t}$$

where, $\lambda^2 = \lambda_1^2 + \lambda_2^2$.

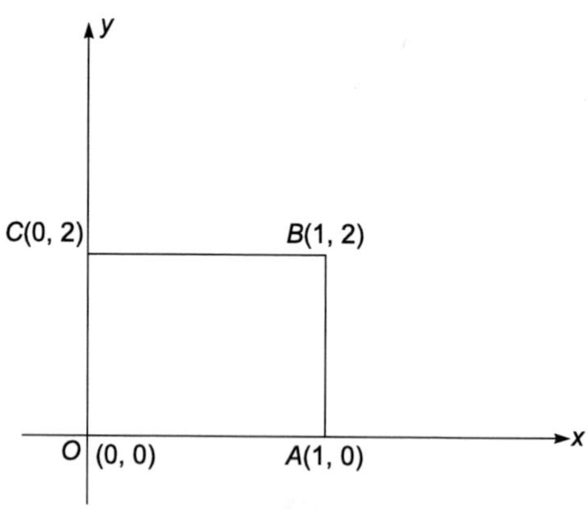

Fig. 10.10

Now, $u(0, y, t) = 0 = A(C \cos \lambda_2 y + D \sin \lambda_2 y) \times Ee^{-c^2\lambda^2 t}$

$\Rightarrow \qquad A = 0$

Again, $u(l, y, t) = 0 = B \sin \lambda_1(C \cos \lambda_2 y + D \sin \lambda_2 y)Ee^{-c^2\lambda^2 t}$

Now, for $u \neq 0$, $X \neq 0$, we must have $B \neq 0$

$$\therefore \qquad \sin \lambda_1 = 0 \Rightarrow \lambda_1 = \frac{m\pi}{1} \text{ with } m \text{ an integer.}$$

$$\text{Now, } \frac{\partial u}{\partial y} = B \sin \lambda_1 x(\lambda_2)\,(-C \sin \lambda_2 y + D \cos \lambda_2 y)Ee^{-c^2\lambda^2 t}$$

$$\therefore \qquad \frac{\partial u}{\partial y}(x, 0, t) = 0 = \lambda_2\, B \sin \lambda_1 x(D)Ee^{-c^2\lambda^2 t}$$

$$\Rightarrow \qquad D = 0$$

$$\text{Again, } \frac{\partial u}{\partial y}(x, 2, t) = 0 = \lambda_2 B \sin \lambda_1 x\,(-C \sin 2\lambda_y) \times Ee^{-c^2\lambda^2 t}$$

Now, for $u \neq 0$, $y \neq 0$, we must have $c \neq 0$

$$\therefore \qquad \sin 2\lambda_2 = 0$$

$$\Rightarrow 2\lambda_2 = n\pi \Rightarrow \lambda_2 = \frac{n\pi}{2} \text{ with } n \text{ an integer}$$

$$\text{Thus, } u_{mn}(x, y, t) = A_{mn} \sin\left(\frac{m\pi x}{1}\right)\cos\left(\frac{n\pi y}{2}\right)e^{-c^2\lambda_{mn}^2 t}.$$

Hence, the most general solution is given by

$$u(x, y, t) = \sum_{m=1}^{\infty}\sum_{n=1}^{\infty} A_{mn} \sin\left(\frac{m\pi x}{1}\right)\cos\left(\frac{n\pi y}{2}\right) \times e^{-c^2\lambda_{mn}^2 t}$$

Now, $u(x, y, 0) = x(1 - x)\, y^2(3 - y)$

$$= \sum_{m=1}^{\infty}\sum_{n=1}^{\infty} A_{mn} \sin\left(\frac{m\pi x}{1}\right)\cos\left(\frac{n\pi y}{2}\right)$$

$$\text{Let } A_m(y) = \sum_{n=10}^{\infty} A_{mn} \cos\left(\frac{n\pi y}{2}\right)$$

$$\text{Then, } \qquad x(1 - x)\, y^2(3 - y) = \sum_{m=1}^{\infty} A_m(y) \sin\left(\frac{m\pi x}{1}\right)$$

with y treated as constant, we have

$$A_m(y) = \frac{2}{(1)}\int_0^1 x(1-x)y^2(3-y)\sin\left(\frac{m\pi x}{1}\right)dx$$

$$= 2y^2(3-y)\left\{x(1-x)\frac{\cos(m\pi x)}{(-m\pi)} - (1-2x)\frac{\sin(m\pi x)}{-(m\pi)^2} - 2\frac{\cos(m\pi x)}{(m\pi)^3}\right\}_0^1$$

$$= -\frac{2}{m^3\pi^3}(\cos m\pi - 1)2y^2(3-y)$$

$$\therefore \quad A_m(y) = \begin{cases} \dfrac{8y^2(3-y)}{m^3\pi^3} & \text{when } m \text{ is odd} \\ 0 & \text{when } m \text{ is even} \end{cases}$$

Thus, from $A_m(y) = \sum\limits_{n=0}^{\infty} A_{mn} \cos\left(\dfrac{n\pi y}{2}\right)$;

we have

$A_m(y) = A_{mn} = 0$ for m even

For odd m

$$A_{mn} = \frac{2}{2(2)}\int_0^2 A_m(y)\cos\left(\frac{n\pi y}{2}\right)dy$$

For $n = 0$

$$A_{m0} = \frac{2}{2(2)}\int_0^2 \frac{8y^2(3-y)}{m^3\pi^3}dy$$

$$= \frac{2}{2(2)}\cdot\frac{8}{m^3\pi^3}\left[y^3 - \frac{y^4}{4}\right]_0^2$$

$$= \frac{2}{2(2)}\frac{8}{m^3\pi^3}\left[8 - \frac{16}{4}\right] = \frac{16}{m^3\pi^3}$$

For $n > 0$

$$A_{mn} = \frac{2}{2(2)}\int_0^2 \frac{8y^2(3-y)}{m^3\pi^3}\cos\left(\frac{n\pi y}{2}\right)dy$$

$$= \frac{4}{m^3\pi^3}\left[y^2(3-y)\frac{\sin\left(\frac{n\pi y}{2}\right)}{\left(\frac{n\pi}{2}\right)} - (6y-3y^2)\frac{\cos\left(\frac{n\pi y}{2}\right)}{-\left(\frac{m\pi}{2}\right)^2} + (6-6y)\frac{\sin\left(\frac{n\pi y}{2}\right)}{-\left(\frac{n\pi}{2}\right)^3} - (-6)\frac{\cos\left(\frac{n\pi y}{2}\right)}{\left(\frac{n\pi}{2}\right)^4}\right]_0^2$$

$$= \frac{4}{m^3\pi^3}\left[\left(\frac{2}{n\pi}\right)^2(0)+6\left(\frac{2}{n\pi}\right)^4(\cos n\pi - 1)\right] = \frac{758}{m^3 n^4 \pi^7}(\cos n\pi - 1)$$

$$= \begin{cases} \dfrac{1566}{m^3 n^4 \pi^7} & \text{if } n \text{ is odd} \\ 0 & \text{if } n \text{ is even} \end{cases}$$

Thus, the required solution of Eq. (1) is given by

$$u(x, y, t) = \sum_{m=1}^{\infty}\sum_{n=1}^{\infty} A_{mn} \operatorname{Sin}\left(\frac{m\pi x}{1}\right)\sin\left(\frac{n\pi y}{2}\right)\times e^{-c^2\lambda_{mn}^2 t}$$

where
$$A_{mn} = \begin{cases} 0 & \text{if } m \text{ is even} \\ \dfrac{16}{m^3\pi^3} & \text{if } m \text{ is odd and } n = 0 \\ \dfrac{1566}{m^3 n^4 \pi^7} & \text{if } m \text{ and } n \text{ both are odd} \end{cases}$$

and
$$\lambda_{mn}^2 = \pi^2\left(\frac{m^2}{1^2}+\frac{n^2}{2^2}\right).$$

▲▲

EXERCISE 10.1

1. Find the deflection $u(x, t)$ of the vibrating string with length π and fixed ends. Corresponding to zero initial velocity and initial deflection $f(x) = k(\sin x - \sin 2x)$ {take $c^2 = 1$}.
2. If a string is released from rest in the position $y = 4ax(l - x)/l^2$, show that its motion is described by the equation

$$y = \frac{32a}{\pi^3}\sum_{n=0}^{\infty}\frac{1}{(2n+1)^3}\sin\frac{(2n+1)\pi x}{l}\cos\frac{(2n+1)\pi ct}{l}$$

3. A tightly stretched string with fixed ends at $x = 0$ and $x = l$ is initially in equilibrium position. It is set vibrating by giving to each of its points an initial velocity $b\sin^3\left(\frac{n\pi}{l}\right)$. Find the displacement of the string at any time t.
4. A string is stretched between the fixed points (0, 0) and (1, 0) and released from rest from the position $u = A \sin \pi x$. Find the formula for the subsequent displacement, $u(x, t)$.

5. The vibrations of an elastic string is governed by the partial differential equation $\frac{\partial^2 u}{\partial t^2} = \frac{\partial^2 u}{\partial x^2}$. The length of the string is π and the ends are fixed. The initial velocity is zero and the initial deflection is $f(x) = 2(\sin x + \sin 3x)$. Find the deflection $u(x, t)$ of the vibrating string for $t > 0$.
6. A string is stretched between the fixed points (0, 0) and $(l, 0)$ and released from rest and the initial defection is given by

$$f(x) = \begin{cases} \left(\frac{2k}{l}\right)x & \text{when } 0 < x < l/2 \\ \left(\frac{2k}{l}\right)(l - x) & \text{when } \frac{1}{2} < x < l \end{cases}$$

Find the deflection of the string at any time t.
7. A string of length l is tightly stretched between two end points. Obtain the solution of this equation if the string is originally placed at the middle point by giving it an initial displacement h from the mean position.
8. By separating the variables, show that one-dimensional wave equation $\frac{\partial^2 u}{\partial x^2} = \frac{1}{c^2}\frac{\partial^2 u}{\partial t^2}$ has solution of the form $A \exp(\pm cnx \pm \pi ct)$ where, A and C are constants.
9. Solve the differential equation

$$\frac{\partial^2 u}{\partial x^2} + \frac{\partial^2 u}{\partial y^2} = \frac{1}{c^2}\frac{\partial^2 u}{\partial t^2}$$

with the condition that $u = 0$ along the lines $x = 0 = y$, $x = a = y$ for $t \geq 0$.
10. Solve the two-dimensional wave equation $\frac{\partial^2 u}{\partial t^2} = c^2\left(\frac{\partial^2 u}{\partial x^2} + \frac{\partial^2 u}{\partial y^2}\right)$ subject to the following conditions $u = 0$ along the lines $x = 0 = y$, $x = a = y$ for $t \geq 0$ and when $t = 0$ $u = f(x, y)$ and $\frac{\partial u}{\partial t} = 0$ for $0 \leq x, y \leq a$.
11. A circular membrane of unit radius fixed along its boundary starts vibrating from rest and has initial deflation $u(r, 0) = f(r)$. Show that the deflection $u(r, t)$ of the membrane at any instant is given by

$$u(r, t) = \sum_{n=1}^{\infty} [A_n \cos(p_n t) \times J_0(a_n r)]$$

where $$A_n = \frac{2}{J_1^2(a_n)}\int_0^1 rf(r)J_1(a_n r)dr$$

and $\quad p_n = ca_n (n = 1, 2, \ldots)$ and $a_n (n = 1, 2, \ldots)$

are the positive roots of the Bessel's function $J_0(k) = 0$.

12. A tightly stretched unit square membrane starts vibrating from rest and its initial displacement is $k \sin 2\pi x \sin \pi y$. Show that the deflection at any instant is

$$k \sin 2\pi x \sin \pi y \cos(\sqrt{5}\pi ct).$$

13. Find the deflection $u(r, t)$ of the circular membrane of unit radius if $c = 1$, the initial velocity is zero and the initial deflection is $0 - 25(1 - r^2)$.

14. Solve the equation $\dfrac{\partial^2 u}{\partial x^2} = \dfrac{\partial u}{\partial t}$ given that $u = 0$ when $t \to \infty$ and when $x = 0, l$ $u = 0$ for any $t \geq 0$.

15. Solve one-dimensional diffusion equation $\dfrac{\partial^2 u}{\partial x^2} = \dfrac{1}{K}\dfrac{\partial u}{\partial t}$ in the region $0 \leq x \leq \pi$, $t \geq 0$, when
    (a) u remains finite as $t \to \infty$.
    (b) $u = 0$ if $x = 0$ or π, for all values of t.
    (c) at $t = 0$

$$u = \begin{cases} x\ ; & 0 \leq x \leq \pi/2 \\ \pi - x; & \pi/2 \leq x \leq \pi \end{cases}$$

16. Find the solution of $\dfrac{\partial^2 u}{\partial x^2} = k^2 \dfrac{\partial u}{\partial t}$ for which $u(0, t) = u(l, t) = 0$ and $u(x, 0) = \sin\left(\dfrac{\pi x}{l}\right)$ by the method of variable separable.

17. Solve the one-dimensional diffusion equation in the region $0 \leq x \leq 2\pi$, $t > 0$, subject to the conditions $u(x, 0) = \sin^3 x$ for $0 \leq x \leq 2\pi$ and $u(0, t) = u(2\pi, t) = 0$ for $t \geq 0$.

18. An insulated rod of length l has its ends A and B maintained at 0°C and 100°C respectively untill steady state prevails. If B is suddenly reduced to 0°C and maintained at 0°C, find the temperature at a distance x from A at time t.

19. A bar of length unity, has its end $x = 0$ insulated and its end at $x = 1$, is kept at temperature zero. Find an expression for the temperature $u(x, t)$, if

$$u(x,0) = \begin{cases} 1 & \text{when } 0 \leq x \leq 1/2 \\ 2(1-x) & \text{when } 1/2 \leq x \leq 1 \end{cases}$$

20. Find the temperature $u(x, t)$ in a slab of length π which is insulated laterally and whose ends $x = 0$ and $x = \pi$ are also insulated. The initial temperature in the bar is $\sin x$.

21. The ends A and B of a rod 20 cm long have the temperature 30°C and 80°C, respectively untill steady state prevails. The temperature of the ends are changed to 40°C and 60°C, respectively. Find the temperature distribution in the rod at time t.

22. Solve the Laplace's equation $\frac{\partial^2 u}{\partial x^2} + \frac{\partial^2 u}{\partial y^2} = 0$ subject to the conditions $u(0, y) = u(l, y) = u(x, 0) = 0$ and $u(x, a) = \sin\left(\frac{n\pi x}{l}\right)$.

23. Solve the Laplace's equation $\frac{\partial^2 u}{\partial x^2} + \frac{\partial^2 u}{\partial y^2} = 0$, subject to the following boundary conditions $u = 0$ when y is very large for all x, $u = 0$ when $x = 0$ and 1, or all y and $u = x(1 - x)$ when $y = 0$, for $0 < x < 1$.

24. A rectangular plate with insulated surface is 8 cm wide and so long compared to its width that it may be considered infinite in length. If the temperature along one short edge $y = 0$ is given by $u(x, 0) = 100 \sin(\pi x/8)$, $(0 < x < 8)$, while the two long edges $x = 0$ and $x = 8$, as well as the other short edge are kept at 0°C, find the steady state temperature $u(x, y)$.

25. Find the steady state temperature distribution $(u(x, y))$ in the uniform square $0 \le x \le 1, 0 \le y \le 1$ when edge $y = 1$ is kept at temperature u_0 and the remaining three sides are kept at temperature zero.

26. The boundary diameter of a semicircular plate of radius a cm is kept at 0°C and the temperature along the semicircular boundary is given by

$$u(a,\theta) = \begin{cases} 50\theta & \text{when } 0 \le \theta \le \pi/2 \\ 50(\pi - \theta) & \text{when } \pi/2 \le \theta \le \pi \end{cases}.$$

Find the steady-state temperature function $u(r, \theta)$.

ANSWERS

Exercise 10.1

1. $u(x, t) = k(\cos t \sin x - \cos 2t \sin 2x)$.

3. $u(x, t) = \frac{bl}{12\pi x}\left[9\sin\left(\frac{\pi ct}{l}\right)\sin\left(\frac{\pi x}{l}\right) - \sin\left(\frac{3\pi ct}{l}\right)\sin\left(\frac{3\pi x}{l}\right)\right]$.

4. $u(x, t) = A \cos c\pi t \sin \pi x$.

5. $u(x, t) = 4 \cos x \cos 2t \sin 2x$.

6. $u(x, t)\ \frac{8k}{\pi^2}\sum_{n=1}^{\infty}\frac{(-1)^{n+1}}{(2n-1)^2}\cos\frac{(2n-1)\pi ct}{l} \times \sin\frac{(2n-1)\pi x}{l}$.

7. $u(x, t)\ \frac{8k}{\pi^2}\sum_{n=1}^{\infty}\frac{(-1)^{n+1}}{(2n-1)^2}\cos\frac{(2n-1)\pi ct}{l} \times \sin\frac{(2n-1)\pi x}{l}$.

9. $u(x, y, t) \sum_{m=1}^{\infty} \sum_{n=1}^{\infty} [c_{mn} \cos(ck_{mn}t) + D_{mn} \sin(ck_{mn}t)] \sin\left(\frac{m\pi x}{a}\right)\sin\left(\frac{n\pi y}{a}\right)$ where $k_{mn}^2 =$
$\pi^2(m^2 + n^2)/a^2$.

10. $u(x, y, t) \sum_{m=1}^{\infty}\sum_{n=1}^{\infty} c_{mn} \cos(ck_{mn}t)\sin\left(\frac{m\pi x}{a}\right)\times\sin\left(\frac{n\pi y}{a}\right)$ where $c_{mn} = \frac{4}{a^2}$

$$\int_{x=0}^{a} \int_{y=0}^{a} f(x,y)\sin\left(\frac{m\pi x}{a}\right)\sin\left(\frac{n\pi y}{a}\right)\times dxdy$$

and $k_{mn}^2 = \pi^2(m^2 + n^2)/a^2$.

13. $u(r, t) = \sum_{n=1}^{\infty}\frac{J_2(a_n)}{a_n^2 J_1^2(a_n)} \cos(a_n t)\ J(a_n r)$.

14. $u(x, t) = \sum_{n=1}^{\infty} D_n \sin\left(\frac{n\pi x}{l}\right)e^{\ n^2\pi^2 t/l^2}$.

15. $u(x, t) = \frac{4}{\pi}\sum_{n=1}^{\infty}\frac{(-1)^{n-1}}{(2n-1)^2} \sin\ (2n - 1)x \times e^{-k(2n-1)^2 t}$.

16. $u(x, t) = \sin\left(\frac{\pi x}{l}\right)e^{-\pi^2 t/l^2 k^2}$.

17. $u(x, t) = \frac{1}{4}(3 \sin xe^{-c^2 t} - \sin 3x\ e^{-9c^2 t})$.

18. $u(x, t) = \frac{200}{\pi}\sum_{n=1}^{\infty}\frac{(-1)^{n+1}}{n}\sin\left(\frac{n\pi x}{l}\right)\times e^{-(n\pi/l)^2 t}$

19. $u(x, t) = \frac{16}{2}\sum_{n=1}^{\infty}\frac{1}{(2n-1)^2}\cos\left(\frac{(2n-1)\pi}{4}\right)\times\cos\left(\frac{(2n-1)\pi x}{2}\right)e^{-(2n-1)^2\pi^2 c^2 t/4}$.

20. $u(x, t) = \frac{2}{\pi}-\frac{4}{\pi}\sum_{n=1}^{\infty}\frac{1}{(4n^2-1)}\cos 2nx \times e^{-4n^2c^2t}$.

21. $u(x, t) = x + 40 - \frac{20}{\pi}\sum_{n=1}^{\infty}[2(-1)^n + 1]\sin\left(\frac{n\pi x}{20}\right)\times e^{-(n\pi c/20)^2 t}$.

22. $u(x, y) = \frac{\sinh(n\pi y/l)}{\sinh(n\pi a/l)}\sin\left(\frac{n\pi x}{l}\right)$.

23. $u(x, y) = \dfrac{8}{\pi^3} \sum_{n=0}^{\infty} \dfrac{1}{(2n+1)^3} e^{-(2n+1)\pi y} \times \sin(2n+1)\pi x$.

24. $u(x, y) = 100 e^{-\pi y/8} \sin\left(\dfrac{\pi x}{8}\right)$.

25. $u(x, y) = \dfrac{440}{\pi} \sum_{n=1}^{\infty} \dfrac{1}{(2n-1)} \dfrac{\sin(2n-1)\pi y \times \sin(2n-1)\pi x}{\sin(2n-1)\pi}$.

26. $u(r, \theta) = \dfrac{200}{\pi} \sum_{n=1}^{\infty} \dfrac{(-1)^{n-1}}{(2n-1)^2} \left(\dfrac{r}{a}\right)^{2n-1} \sin(2n-1)\theta$.

▲▲

SECTION-C

Laplace Transforms, Fourier Transforms and Fourier Series

1

CHAPTER

Introduction to Laplace Transforms

Laplace transformation has had a profound influence over the development of mathematical analysis and solving linear constant coefficient ordinary or partial differential equations under suitable initial and boundary conditions. The concept of a transform is that it turns a given function into another function.

Definition: Let $f(t)$ be a function defined for $t \geq 0$.

Then, the integral

$$\int_0^\infty e^{-st} f(t)dt = F(s)$$

is called the Laplace transform of $f(t)$, provided the integral exists. We can also write

$$F(s) = L(f(t))$$

where, L is called the Laplace transform operator. The original function $f(t)$ is called the inverse Laplace transform of $F(s)$.

Thus, $$L^{-1}(F(s)) = f(t).$$

Linearity of Laplace and Inverse Laplace Transformations

Let $f(t)$ and $g(t)$ be any two function whose Laplace transforms $F(s)$ and $G(s)$ exist. Then for any two constants a and b, we have

$$L(af(t) + bg(t)) = aL(f(t)) + bL(g(t))$$

and $$L^{-1}(aF(s) + bG(s)) = aL^{-1}(F(s)) + bL^{-1}(G(s))$$

Proof: Using the definition, we have

$$L[af(t) + bg(t)] = \int_0^\infty e^{-st}(af(t) + bg(t))dt$$

$$= a\int_0^\infty e^{-st}f(t)dt + b\int_0^\infty e^{-st}g(t)dt$$

$$= aL(f(t)) + bL(g(t)).$$

We can also write it as

$$L(af(t) + bg(t)) = aF(s) + bG(s)$$

$$\Rightarrow \quad af(t) + bg(t) = L^{-1}(aF(s) + bG(s))$$

$$\Rightarrow \quad aL^{-1}(F/s) + bL^{-1}(G(s)) = L^{-1}(aF(s) + bG(s))$$

Definition: A function $f(t)$ is said to be of exponential order a if there exist constants a and $M > 0$ such that

$$|f(t)| \leq Me^{at} \text{ for } t \geq 0$$

Geometrically, this condition implies that the graph of $f(t)$, $t > 0$ does not grow faster than the graph of the exponential function Me^{at}, $a > 0$.

e.g., $f(t) = t^2$ is of exponential order 3 as $|t^2| < e^{3t}$ for all $t > 0$.

$f(t) = t$ is of exponential order 1 as $|t| < e^t$ also for all $t > 0$.

$f(t) = e^{-2t}$ is also of exponential order 1 as $|e^{-2t}| \leq e^t$ for all $t > 0$.

Similarly, $f(t) = \cos t$ is of exponential order 1.

But, $f(t) = e^{t^2}$ is not of exponential order as it is not possible to determine constants a and M such that $e^{t^2} \leq Me^{at}$ for $t \geq 0$.

Sufficient Condition for Existence of Laplace Transform

If $f(t)$ is a piecewise continuous function on the interval $[0, \infty)$ and is of exponential order a for $t \geq 0$, then $L[f(t)]$ exists for $s > a$.

Proof: If $f(t)$ is piecewise continuous on $[0, l]$ for $t > 0$ then $e^{-st}f(t)$ is also piecewise continuous on $[0, l]$. Hence, $\int_0^\infty e^{-st}f(t)dt$ exists.

Therefore, the existence of Laplace transform of $f(t)$ depends on whether this integral converges or has a finite limit as $t \to \infty$.

As $f(t)$ is of exponential order a so there exist some numbers M such that

$$|f(t)| \leq Me^{at}, \text{ for } t > 0$$

Then,

$$|e^{-st}f(t) \leq Me^{-(s-a)t}$$

Now,

$$\left|\int_0^\infty e^{st} f(t)dt\right| \leq \int_0^\infty e^{-st}|f(t)|\, dt \leq \int_0^\infty Me^{-(s-a)t}dt$$

$$= M\left\{\frac{e^{-(s-a)t}}{-(s-a)}\right\}_0^\infty$$

$$= \frac{M(e^{-\infty} - e^0)}{-(s-a)}$$

$$= \frac{M(0-1)}{-(s-a)} = \frac{M}{s-a} \text{ with } s > a \qquad (\because e^{-\infty} = 0)$$

$$\Rightarrow \qquad |L(f(t))| \leq \frac{M}{s-a}.$$

which proves the existence of Laplace transform of $f(t)$.

Note: It is important to note that the above condition is only sufficient condition for the existence of Lapalace transform. That is, a function may have Laplace transform even if it does not satisfy the existence condition.

For example, consider the function $f(t) = t^{-1/2}$. This function is not continuos on any interval $[0, l]$ because $t^{-1/2} \to \infty$ as $t \to 0$, from the right.

However, $\int_0^l e^{-st}\, t^{-1/2}\, dt$ exists for all $t > 0$ as $\left|\int_0^\infty e^{-st} t^{-1/2} dt\right| \leq \int_0^\infty t^{-1/2} dt$ which is convergent.

Beta and Gamma Functions

Before proceeding further, we would like to recall some results on Beta and Gamma functions.

Gamma function $\Gamma(a) = \int_0^\infty x^{a-1}\, e^{-x} dx.$

Beta function $\beta(m, n) = \int_0^1 x^{m-1}(1 - x)^{n-1} dx$

Some identities of Gamma function

(a) $\Gamma(1) = 1$

(b) $\Gamma(a + 1) = a\Gamma(a)$

(c) $\Gamma(n + 1) = n!$ for any positive integer n.

(d) $\Gamma(1/2) = \sqrt{\pi}$

(e) $\Gamma(-1/2) = -2\sqrt{\pi}$

Some identities of Beta function

(a) $\beta(m, n) = \beta(n, m)$

(b) $\beta(m, n) = 2\int_0^{\pi/2} \sin^{2m-1}\theta \cos^{2n-1}\theta d\theta$

$$= 2\int_0^{\pi/2} \cos^{2m-1}\theta \sin^{2n-1}\theta d\theta.$$

(c) $\beta(m, n) = \int_0^{\infty} \frac{x^{m-1}}{(1+x)^{m+n}} dx = \int_0^{\infty} \frac{x^{m-1}}{(1+x)^{m+n}} dx$

(d) $\beta(m, n) = \frac{\Gamma(m)\Gamma(n)}{\Gamma(m+n)}$

(e) $\beta(m, n) = \beta(m+1, n) + \beta(m, n+1)$

Note: Also, we have

$$\Gamma(2n) = \frac{2^{2n-1}}{\sqrt{\pi}} \Gamma(n)\,\Gamma\left(n+\frac{1}{2}\right)$$

and

$$\Gamma\left(\frac{1}{4}\right)\Gamma\left(\frac{3}{4}\right) = \pi\sqrt{2}.$$

Laplace Transforms of Some Elementary Functions

(a) $L(1) = \frac{1}{s}$

By using definition for $f(t) = 1$, we have

$$L(1) = \int_0^{\infty} 1\cdot e^{-st} dt = \left[\frac{e^{-st}}{-s}\right]_0^{\infty} = \frac{e^{-\infty} - e^0}{-s} = \frac{1}{s} \qquad \{\because e^{-\infty} = 0\}$$

(b) $L(t^a) = \frac{\Gamma(a+1)}{s^{a+1}}$

Taking $f(t) = t^a$, we obtain

$$L(t^a) = \int_0^{\infty} t^a e^{-st}\, dt$$

Let $st = u \Rightarrow dt = du/s$

$$\therefore \qquad L(t^a) = \int_0^\infty \left(\frac{u}{s}\right)^a e^{-u} \frac{du}{s}$$

$$= \frac{1}{s^{a+1}} \int_0^\infty u^a e^{-u} du$$

$$= \frac{1}{s^{a+1}} \Gamma(a+1)$$

(c) $L(t^n) = \dfrac{n!}{s^{n+1}}$ for positive integer n.

From (b), we have

$$L(t^n) = \frac{\Gamma(n+1)}{s^{n+1}} = \frac{n!}{s^{n+1}}$$

(d) $L(e^{at}) = \dfrac{1}{s-a}$

Taking $f(t) = e^{at}$, we have

$$L(e^{at}) = \int_0^\infty e^{at} e^{-s} dt = \int_0^\infty e^{-(s-a)t} dt$$

$$= \frac{1}{s-a}$$

(e) $L(\sin at) = \dfrac{a}{s^2+a^2}$ and $L(\cos at) = \dfrac{s}{s^2+a^2}$.

We know $\cos at + i \sin at = e^{jat}$

$\therefore \; L(\cos at + i \sin at) = L(e^{jat})$

$\Rightarrow \; L(\cos at) + iL(\sin at) = \dfrac{1}{s-ia} = \dfrac{s+ia}{s^2+a^2}$

Therefore, equating the real and imaginary parts on both sides, we obtain the required result.

(f) $L(\sinh at) = \dfrac{a}{s^2-a^2}$ and $L(\cosh at) = \dfrac{s}{s^2-a^2}$

We know $\sinh at = \dfrac{e^{at}-e^{-at}}{2}$ and $\cosh at = \dfrac{e^{at}+e^{-at}}{2}$

$\therefore \; L(\sinh at) = \dfrac{1}{2} L(e^{at} - e^{-at}) = \dfrac{1}{2} [L(e^{at}) - L(e^{-at})]$

$$= \frac{1}{2}\left\{\frac{1}{s-a} - \frac{1}{s+a}\right\} = \frac{a}{s^2-a^2}$$

Similarly, we can get $L(\cosh at) = \dfrac{s}{s^2 - a^2}$

Using the definition, we can write

(a) $L^{-1}\left(\dfrac{1}{s}\right) = 1$

(b) $L^{-1}\left(\dfrac{1}{s^{a+1}}\right) = \dfrac{t^a}{\Gamma(a+1)}$

(c) $L^{-1}\left(\dfrac{1}{s^{n+1}}\right) = \dfrac{t^n}{n!}$ for positive integer n.

(d) $L^{-1}\left(\dfrac{1}{s-a}\right) = e^{at}$

(e) $L^{-1}\left(\dfrac{a}{s^2+a^2}\right) = \sin at$

(f) $L^{-1}\left(\dfrac{s}{s^2+a^2}\right) = \cos at$

(g) $L^{-1}\left(\dfrac{a}{s^2-a^2}\right) = \sinh at$

(h) $L^{-1}\left(\dfrac{s}{s^2-a^2}\right) = \cosh at$ ▲▲

EXAMPLE 1.1

Find the Laplace transform of $f(t) = t;\ t \geq 0$.

Solution: We have

$$L(t) = \int_0^{\infty} te^{-st}dt$$

$$= \left[t\frac{e^{-st}}{-s} - 1 \times \frac{e^{-st}}{s^2}\right]_0^{\infty} = \frac{1}{s^2}$$

EXAMPLE 1.2

Evaluate: $L(f(t))$, if

$$f(t) = \begin{cases} (t-1)^2 & t > 1 \\ 0 & 0 < t < 1 \end{cases}.$$

Solution: We have

$$L(f(t)) = \int_0^\infty f(t)e^{-st}dt$$

$$= \int_0^1 0 \cdot e^{-st}dt + \int_1^\infty (t-1)^2 e^{-st}dt$$

$$= 0 + \left[(t-1)^2 \frac{e^{-st}}{-s} - 2(t-1)\frac{e^{-st}}{s^2} + 2\frac{e^{-st}}{-s^3}\right]_1^\infty$$

$$= -\frac{2}{s^3}(e^{-\infty} - e^{-s}) = \frac{2e^{-s}}{s^3} \qquad (\because e^{-\infty} = 0)$$

EXAMPLE 1.3

Find the Laplace transform of $\frac{1}{\sqrt{t}}$.

Solution: From the definition, we have

$$L\left(\frac{1}{\sqrt{t}}\right) = \int_0^\infty t^{-1/2} e^{-st}dt$$

Let,

$$st = u \Rightarrow dt = du/s$$

$\therefore$

$$L\left(\frac{1}{\sqrt{t}}\right) = \int_0^\infty \left(\frac{u}{s}\right)^{-1/2} e^{-u}\frac{du}{s}$$

$$= \frac{1}{\sqrt{s}}\int_0^\infty u^{-1/2} e^{-u}du$$

$$= \frac{1}{\sqrt{s}}\Gamma\left(-\frac{1}{2}+1\right) = \frac{\Gamma(1/2)}{\sqrt{s}} = \frac{\sqrt{\pi}}{\sqrt{s}} = \sqrt{\frac{\pi}{s}}.$$

EXAMPLE 1.4

Find the Laplace transform of $\sin\sqrt{t}$.

Solution: We know

$$\sin x = x - \frac{x^3}{3!} + \frac{x^5}{5!} - \ldots$$

$\therefore$

$$\sin\sqrt{t} = t^{1/2} - \frac{t^{3/2}}{3!} + \frac{t^{5/2}}{5!} - \ldots$$

Thus,

$$L(\sin\sqrt{t}) = L(t^{1/2}) - \frac{1}{3!}L(t^{3/2}) + \frac{1}{5!}L(t^{5/2}) - \ldots$$

$$= \frac{\Gamma(3/2)}{s^{3/2}} - \frac{1}{3!}\frac{\Gamma(5/2)}{s^{5/2}} + \frac{1}{5!}\frac{\Gamma(7/2)}{s^{7/2}} - \ldots$$

$$= \frac{\Gamma(3/2)}{s^{3/2}}\left\{1 - \frac{3/2}{3!}\frac{1}{s} + \frac{\frac{5}{2}\cdot\frac{3}{2}}{5!}\frac{1}{s^2} - \dots\right\} \qquad \text{[Using } \Gamma(a+1) = a\Gamma(a)]$$

$$= \frac{\frac{1}{2}\Gamma\left(\frac{1}{2}\right)}{s^{3/2}}\left\{1 - \left(\frac{1}{2s}\right) + \frac{1}{2!}\left(\frac{1}{2s}\right)^2 - \dots\right\}$$

$$= \frac{\sqrt{\pi}}{2s^{3/2}} e^{-1/2s}$$

EXAMPLE 1.5

Find the Laplace transform of $\cos^2(at)$.

Solution: We have $\cos^2(at) = \frac{(1+\cos(2at))}{2}$. Therefore,

$$L(\cos^2(at)) = \frac{1}{2}\{L(1) + L(\cos(2at))\}$$

$$= \frac{1}{2}\left\{\frac{1}{s} + \frac{s}{s^2 + (2a)^2}\right\}$$

$$= \frac{s^2 + 2a^2}{s(s^2 + 4a^2)}.$$

EXAMPLE 1.6

Find the Laplace transform of

$$4e^{5t} + 6t^3 - 3\sin 4t + 2\cos 2t.$$

Solution: We have

$$L(4e^{5t} + 6t^3 - 3\sin 4t + 2\cos 2t)$$

$$= 4L(e^{5t}) + 6L(t^3) - 3L(\sin 4t) + 2L(\cos 2t)$$

$$= \frac{4}{s-5} + 6\frac{3!}{s^4} - 3\frac{4}{s^2+4^2} + 2\frac{s}{s^2+2^2}$$

$$= \frac{4}{s-5} + \frac{36}{s^4} - \frac{12}{s^2+16} + \frac{2s}{s^2+4}.$$

EXAMPLE 1.7

Find the Laplace transform of sin 2t sin 3t.

Solution: We have

$$\sin 2t \sin 3t = \frac{1}{2}\{\cos t - \cos 5t)\}$$

$$\therefore \quad L(\sin 2t \sin 3t) = \frac{1}{2}\{L(\cos t) - L(\cos 5t)\}$$

$$= \frac{1}{2}\left\{\frac{s}{s^2+1^2} - \frac{s}{s^2+5^2}\right\} - \frac{12s}{(s^2+1)(s^2+25)}.$$

EXAMPLE 1.8

Prove that

$$L(2^t) = \frac{1}{s - \log 2}$$

Solution: We have

$$2^t = e^{\log 2^t} = e^{t \log 2}$$

Therefore,

$$L(2^t) = L(e^{t \log 2}) = \frac{1}{s - \log 2}.$$

EXAMPLE 1.9

Find $L(f(t))$ where $f(t) = \begin{cases} t & 0 \le t \le 1/2 \\ t-1 & 1/2 \le t \le 1 \\ 0 & t > 1 \end{cases}$

Solution: We have

$$L(f(t)) = \int_0^\infty f(t)e^{-st}dt$$

$$= \int_0^{1/2} te^{-st}dt + \int_{1/2}^{1} (t-1)e^{-st}dt + \int_1^\infty 0 \cdot e^{-st}dt$$

$$= \left\{t\frac{e^{-st}}{-s} - \frac{e^{-st}}{s^2}\right\}_0^{1/2} + \left\{(t-1)\frac{e^{-st}}{-s} - \frac{e^{-st}}{s^2}\right\}_{1/2}^{1}$$

$$= \left\{\frac{1}{2}\frac{e^{-s/2}}{-s} - \frac{e^{-s/2}}{s^2} + \frac{1}{s^2}\right\} + \left\{\frac{e^{-s}}{s^2} + \left(-\frac{1}{2}\right)\frac{e^{-s/2}}{s} + \frac{e^{-s/2}}{s^2}\right\}$$

$$= -\frac{e^{-s/2}}{s} + \frac{1}{s^2}(1 - e^{-s}).$$

EXAMPLE 1.10

Find the Laplace transform of $f(t) = \begin{cases} 0 & 0 \le t \le \pi \\ \sin t & t \ge \pi \end{cases}$.

Solution: We have

$$L(f(t)) = \int_0^\infty f(t)\, e^{-st} dt = \int_0^\pi 0 \cdot e^{-st} dt + \int_\pi^\infty \sin te^{-st} dt$$

$$= \left[\frac{e^{st}}{s^2+1}(-s\sin t - \cos t) \right]_\pi^\infty$$

$$= 0 - \frac{e^{-\pi s}}{s^2+1}(-s\sin\pi - \cos\pi) = -\frac{e^{-\pi s}}{s^2+1}.$$

EXAMPLE 1.11

Find $L^{-1}\left(\frac{2s+5}{s^2+25}\right)$.

Solution: We have by the linearity principle

$$L^{-1}\left(\frac{2s+5}{s^2+25}\right) = 2L^{-1}\left(\frac{s}{s^2+5^2}\right) + L^{-1}\left(\frac{5}{s^2+5^2}\right)$$

$$= 2 \cos (5t) + \sin (5t).$$

EXAMPLE 1.12

Find $L^{-1}\left(\frac{5s^2+3s-16}{(s-1)(s-2)(s+3)}\right)$.

Solution: Since, $\frac{5s^2+3s-16}{(s-1)(s-2)(s+3)} = \frac{2}{s-1} + \frac{2}{s-2} + \frac{1}{s+3}$

$$\therefore \quad L^{-1}\left(\frac{5s^2+3s-16}{(s-1)(s-2)(s+3)}\right) = 2L^{-1}\left(\frac{1}{s-1}\right) + 2L^{-1}\left(\frac{1}{s-2}\right) + L^{-1}\left(\frac{1}{s+3}\right)$$

$$= 2e^t + 2e^{2t} + e^{-3t}.$$

EXAMPLE 1.13

Prove that

$$L^{-1}\left(\frac{1}{s} \sin \frac{1}{s}\right) = t - \frac{t^3}{(3!)^2} + \frac{t^5}{(5!)^2} - \frac{t^7}{(7!)^2} + \ldots.$$

Solution: We have

$$L^{-1}\left(\frac{1}{s}\sin\frac{1}{s}\right) = L^{-1}\left(\frac{1}{s}\left(\frac{1}{s} - \frac{1}{3!}\frac{1}{s^3} + \frac{1}{5!}\frac{1}{s^5} - \ldots\right)\right)$$

$$= L^{-1}\left(\frac{1}{s^2}\right) - \frac{1}{3!}L^{-1}\left(\frac{1}{s^4}\right) + \frac{1}{5!}L^{-1}\left(\frac{1}{s^6}\right) - \ldots$$

$$= \frac{t}{1!} - \frac{1}{3!}\frac{t^3}{3!} + \frac{1}{5!}\frac{t^5}{5!} - \ldots$$

$$= t - \frac{t^3}{(3!)^2} + \frac{t^5}{(5!)^2} - \ldots.$$

EXAMPLE 1.14

Evaluate: $L^{-1}\left(\frac{3s-2}{s^{3/2}} + \left(\frac{\sqrt{s}-1}{s}\right)^2 + \frac{4s-18}{9-s^2}\right)$.

Solution: We have

$$L^{-1}\left(\frac{3}{s^{1/2}} - \frac{2}{s^{3/2}} + \frac{s-2s^{1/2}+1}{s^2} - \frac{4s}{s^2-9} + \frac{18}{s^2-9}\right)$$

$$= L^{-1}\left(\frac{3}{s^{1/2}} - \frac{2}{s^{3/2}} + \frac{1}{s} - \frac{2}{s^{3/2}} + \frac{1}{s^2} - 4\frac{s}{s^2-9} + 6\frac{3}{s^2-9}\right)$$

$$= 3L^{-1}\left(\frac{1}{s^{1/2}}\right) - 4L^{-1}\left(\frac{1}{s^{3/2}}\right) + L^{-1}\left(\frac{1}{s}\right) + L^{-1}\left(\frac{1}{s^2}\right) - 4L^{-1}\left(\frac{s}{s^2-9}\right) + 6L^{-1}\left(\frac{3}{s^2-9}\right)$$

$$= 3\frac{t^{-1/2}}{\Gamma(1/2)} - 4\frac{t^{1/2}}{\Gamma(3/2)} + 1 + t - 4\cosh(3t) + 6\sinh(3t).$$

$$= \frac{3}{\sqrt{\pi}}t^{-1/2} - \frac{8}{\sqrt{\pi}}t^{1/2} + 1 + t - 4\cosh(3t) + 6\sinh(3t).$$

EXAMPLE 1.15

Evaluate: $L^{-1}\left(\frac{1}{(s+1)(s^2+1)}\right)$.

Solution: We have

$$\frac{1}{(s+1)(s^2+1)} = \frac{A}{s+1} + \frac{Bs+c}{s^2+1}$$

$$\Rightarrow \qquad 1 = A(s^2+1) + (Bs+c)(s+1)$$

Coefficient of s^2} $\quad 0 = A + B$

Coefficient of s} $\quad 0 = B + C$

Constant term} $\quad 1 = A + C$

Solving, we get $\quad A = 1/2,\ B = -\frac{1}{2}$ and $C = \frac{1}{2}$.

$$\therefore \qquad \frac{1}{(s+1)(s^2+1)} = \frac{\frac{1}{2}}{s+1} + \frac{-\frac{1}{2}s+\frac{1}{2}}{s^2+1}$$

Thus,
$$L^{-1}\left(\frac{1}{(s+1)(s^2+1)}\right) = \frac{1}{2}L^{-1}\left(\frac{1}{s+1}\right) - \frac{1}{2}L^{-1}\left(\frac{s}{s^2+1}\right) + \frac{1}{2}L^{-1}\left(\frac{1}{s^2+1}\right)$$

$$= \frac{1}{2}e^{-t} - \frac{1}{2}\cos t + \frac{1}{2}\sin t$$

$$= \frac{1}{2}(e^{-t} - \cos t + \sin t).$$ ▲▲

EXERCISE 1.1

Find the Laplace transform of the following functions:

1. $(t + 3)^2$.
2. $\sinh 2t + \cosh 2t$.
3. $e^t \sin t$.
4. $e^t \cosh t$.
5. $\sinh^2 t$.
6. te^{2t}.
7. $t \sin t$.
8. $(\cos t + \sin t)^2$.
9. $f(t) = \begin{cases} 1 & 0 \le t < 1 \\ -1 & t \ge 1 \end{cases}$.
10. $f(t) = \begin{cases} \cos t & 0 \le t < \pi \\ 0 & t \ge \pi \end{cases}$.
11. $\left(\sqrt{t} - \frac{1}{\sqrt{t}}\right)^2$.

12. $\dfrac{\cos\sqrt{t}}{\sqrt{t}}$.

13. $-[t]$.

14. $\sin(at + b)$ $(t \geq 0)$.

15. $\sin at \sin bt$.

Find the inverse Laplace transform of the following functions:

16. $\dfrac{3}{s^2+2s}$.

17. $\dfrac{s^2+2s+5}{(s-1)(s-2)(s-3)}$.

18. $\dfrac{s}{(s+1)(s+2)(s-3)}$.

19. $\dfrac{s^2-3s+4}{s^3}$.

20. $\dfrac{5s+3}{(s-1)(s^2+1)}$.

21. $\dfrac{3}{s^2-3}+\dfrac{3s+2}{s^3}-\dfrac{3s-27}{s^2+9}+\dfrac{6-30\sqrt{s}}{s^4}$.

22. $\dfrac{1}{s}\cos\dfrac{1}{s}$.

23. $\dfrac{1}{s^3+1}$.

24. Find $L(e^{(a+ib)t})$. Hence, show that

$$L(e^{at}\sin bt) = \frac{b}{(s-a)^2+b^2} \quad \text{and} \quad L(e^{at}\cos bt) = \frac{s-a}{(s-a)^2+b^2}.$$

25. Prove that if $L(f(t)) = F(s)$ then $L(f(at)) = \dfrac{1}{a}\,F\left(\dfrac{s}{a}\right)$.

26. Find $L\left(\dfrac{\sin at}{t}\right)$; given that $L\left(\dfrac{\sin t}{t}\right) = \tan^{-1}\left(\dfrac{1}{s}\right)$.

ANSWERS

Exercise 1.1

1. $\frac{2}{s^3}+\frac{6}{s^2}+\frac{9}{s}$.

2. $\frac{1}{s-2}$.

3. $\frac{1}{(s-1)^2+1}$.

4. $\frac{s-1}{s(s-2)}$.

5. $\frac{2}{s(s^2-4)}$.

6. $\frac{1}{(s-2)^2}$.

7. $\frac{2s}{(s^2+1)^2}$.

8. $\frac{1}{s}+\frac{2}{s^2+4}$.

9. $\frac{(1-2e^{-s})}{s}$.

10. $\frac{s(1+e^{-\pi s})}{s^2+1}$.

11. $\frac{\sqrt{\pi}}{4}\left(\frac{3}{s^{5/2}}-\frac{6}{s^{3/2}}+\frac{12}{s^{1/2}}+\frac{8}{s^{-1/2}}\right)$.

12. $\frac{\sqrt{\pi}}{s}\left(1-\left(\frac{1}{ys}\right)+\frac{1}{2!}\left(\frac{1}{ys}\right)^2-\frac{1}{3!}\left(\frac{1}{ys}\right)^3+\ \ldots\right)=\sqrt{\frac{\pi}{s}}e^{-1/4s}$.

13. $\frac{1}{s(e^s-1)}$.

14. $\frac{a\cos b+s\sin b}{s^2+a^2}$.

15. $\dfrac{2abs}{[s^2+(a-b)^2][s^2+(a+b)^2]}$.

16. $\dfrac{3(1-e^{-2t})}{2}$.

17. $4e^t - 13e^{2t} + 10e^{3t}$.

18. $\dfrac{1}{4}e^{-t} - \dfrac{2}{5}e^{-2t} + \dfrac{3}{20}e^{3t}$.

19. $1 - 3t + 2t^2$.

20. $4e^t - 4\cos t + \sin t$.

21. $3t + t^2 + t^3 - 16t^2(t/\pi)^{1/2} + \sqrt{3}\sin h\sqrt{3}\,t - 3\cos 3t + 9\sin 3t$.

22. $1-\dfrac{t^2}{(2!)^2}+\dfrac{t^4}{(4!)^2}-\dfrac{t^6}{(6!)^2}+\ldots$.

23. $\dfrac{t^2}{2!}-\dfrac{t^5}{5!}+\dfrac{t^8}{8!}-\dfrac{t^{11}}{11!}+\ldots$.

24. $\dfrac{(s-a)+ib}{(s-a)^2+b^2}$.

26. $\tan^{-1}\left(\dfrac{a}{s}\right)$.

2

CHAPTER

Laplace Transforms of Derivatives and Integrals

As we have discussed earilier, Laplace transformation is one of the important tools for solving the initial value problems. Before, we use this to solve differential equations, we will find Laplace transform of derivatives and integrals.

Laplace Transform of Derivatives

Let $f(t)$, $t \geq 0$ be a continuous function and be of exponential order on $[0, \infty)$. Let $f'(t)$ be also of exponential order and at least piecewise continuous on $[0, \infty)$. Then,

$$L(f'(t)) = sL(f(t)) - f(0).$$

Proof: First, assume that $f'(t)$ is continuous on $[0, \infty)$. Then,

$$L(f'(t)) = \int_0^\infty f'(t)\, e^{-st} dt \text{ (Integrating by parts)}$$

$$= [f(t)\, e^{-st}]_0^\infty - \int_0^\infty - se^{-st} f(t) dt$$

$$= 0 - f(0) + s\int_0^\infty e^{-st} f(t) dt \qquad \{\because e^{-\infty} = 0\}$$

$$= - f(0) + sL[f(t)]$$

Hence, we have obtained the required result when $f'(t)$ is continuous on $[0, \infty)$.

Suppose now that $f'(t)$ is piecewise continuous on $[0, \infty)$, let at $t = T$, $f'(t)$ has a finite jump.

Then,

$$L(f'(t)) = \int_0^\infty f'(t)\, e^{-st} dt$$

$$= \lim_{\varepsilon \to 0} \left[\int_0^{T-\varepsilon} f'(t) e^{-st} dt + \int_{T+\varepsilon}^\infty f'(t)\, e^{-st} dt \right]$$

$$= \lim_{\varepsilon \to 0} \left\{ [f(t)e^{-st}]_0^{T-\varepsilon} - \int_0^{T-\varepsilon} -se^{-st} f(t)dt + [f(t)e^{-st}]_{T+\varepsilon}^{\infty} - \int_{T+\varepsilon}^{\infty} -se^{-st} f(t)dt \right\}$$

$$= \lim_{\varepsilon \to 0} \left\{ f(T^-)e^{-sT} - f(0) + s\int_0^{T-\varepsilon} e^{-st} f(t)dt + 0 - f(T^+)e^{-sT} + s\int_{T+\varepsilon}^{\infty} e^{-st} f(t)dt \right\}$$

{$\because f(t)$ is continuous on $[0, \infty)$

$\therefore \quad f(T^-) = f(T^+)$

where $f(T^-) = f(T - \varepsilon)$ and

$f(T^+) = f(T + \varepsilon)$}

$$= \lim_{\varepsilon \to \infty} \left\{ f(T)e^{-sT} - f(0) + s\int_0^{\infty} e^{-st} f(t)dt - f(T)e^{-sT} \right\}$$

$= -f(0) + sL(f(t))$.

So, we have proved the result when $f'(t)$ is piecewise continuous on $[0, \infty)$.

Note 1. We can also write

$$L(f''(t)) = L\left(\frac{d}{dt}(f'(t))\right) = sL(f'(t)) - f'(0)$$

$$= s[sL(f(t)) - f(0)] - f'(0)$$

$$= s^2L(f(t)) - sf(0) - f'(0)$$

Similarly, $$L(f'''(t)) = s^3L(f(t)) - s^2f(0) - sf'(0) - f''(0)$$

By induction, we have

$$L(f^{(n)}(t)) = s^nL(f(t)) - s^{n-1}f(0) - s^{n-2}f'(0) - \ldots - f^{(n-1)}(0)$$

Note 2. Let $f(t)$ satisfy the condition

$$f(0) = 0 \text{ and } f^{(n)}(0) = 0 \text{ for any } n.$$

Then, $$L(f^{(n)}(t)) = s^nL(f(t))$$

$$\Rightarrow \quad L(f^{(n)}(t)) = s^nF(s)$$

$$\therefore \quad L^{-1}(s^nF(s)) = f^{(n)}(t) = \frac{d^n}{dt^n}(f(t))$$

$$= \frac{d^n}{dt^n}(L^{-1}(F(s))).$$

▲▲

EXAMPLE 2.1

Using the formula of Laplace transform of derivatives, find the Laplace transform of $f(t) = t^3$.

Solution: We have $f(t) = t^3\ f'(t) = 3t^2\ \ f''(t) = 6t\ \ f'''(t) = 6$

Also, $f(0) = f'(0) = f''(0) = 0$

Thus, $L(f'''(t)) = s^3\ L(f(t)) - s^2 f(0) - sf'(0) - f''(0)$

$$\Rightarrow \quad L(6) = s^3 L(t^3) - 0 - 0 - 0$$

$$\Rightarrow \quad 6L(1) = s^3 L(t^3)$$

$$\Rightarrow \quad 6\frac{1}{s} = s^3 L(t^3)$$

$$\therefore \quad \frac{6}{s^4} = \frac{3!}{s^4} = L(t^3).$$

EXAMPLE 2.2

Find $L^{-1}\left(\dfrac{s}{s^2+9}\right)$, using the formula for Laplace transform of derivatives.

Solution: Let $F(s) = \dfrac{1}{s^2+9}$

$$\therefore \quad L^{-1}(F(s)) = L^{-1}\left(\frac{1}{s^2+9}\right) = \frac{1}{3}\sin 3t = f(t)$$

Also, $f(0) = 0$

Thus, $L^{-1}(sF(s)) = \dfrac{d}{dt}L^{-1}(F(s))$

$$\Rightarrow \quad L^{-1}\left(s\frac{1}{s^2+9}\right) = \frac{d}{dt}\left(\frac{1}{3}\sin 3t\right)$$

$$\Rightarrow \quad L^{-1}\left(\frac{s}{s^2+9}\right) = \frac{1}{3} \times \cos 3t \times 3 = \cos 3t$$

which is same as we had got by using the formulas discussed earlier under Laplace transform of elementary functions.

EXAMPLE 2.3

Evaluate: $L(\cos^2 2t)$.

Solution: We have $f(t) = \cos^2 2t\ f(0) = 1$

$$f'(t) = 2\cos 2t\,(-\sin 2t)2 = -2\sin 4t$$

Now, $L(f'(t)) = sL(f(t)) - f(0)$

$$\Rightarrow \quad L(-2\sin 4t) = sL(\cos^2 2t) - 1.$$

$$\Rightarrow \qquad -2\cdot\frac{4}{s^2+16}+1 = sL(\cos^2 2t)$$

$$\therefore \qquad \frac{s^2+8}{s(s^2+16)} = L(\cos^2 2t)$$

EXAMPLE 2.4

If $L(\sin\sqrt{t})=\dfrac{\sqrt{\pi}}{2s^{3/2}}e^{-1/4s}$, show that $L\left(\dfrac{\cos\sqrt{t}}{\sqrt{t}}\right)=\dfrac{\sqrt{\pi}}{s}e^{-1/4s}$.

Solution: Here, $\qquad f(t) = \sin\sqrt{t} \;\; f(0) = 0$

$$\therefore \qquad f'(t) = \cos\sqrt{t}\cdot\frac{1}{2}t^{-1/2} = \frac{1}{2}\,\frac{\cos\sqrt{t}}{\sqrt{t}}$$

Now, we know that $L(f'(t)) = sL(f(t)) - f(0)$

$$\Rightarrow \qquad L\left(\frac{1}{2}\,\frac{\cos\sqrt{t}}{\sqrt{t}}\right) = sL(\sin\sqrt{t})-0$$

$$\Rightarrow \qquad L\left(\frac{\cos\sqrt{t}}{\sqrt{t}}\right) = 2s-\frac{\sqrt{\pi}}{2s^{3/2}}e^{-1/4s} = \sqrt{\frac{\pi}{s}}\,e^{-1/4s}.$$

EXAMPLE 2.5

Find $L(f(t))$, if $f''(t) + 3f'(t) + 2f(t) = 0$, $f(0) = 1$, $f'(0) = 2$.

Solution: Here, $\qquad L(f''(t) + 3f'(t) + 2f(t)) = 0$

$$\Rightarrow \; s^2L(f(t)) - sf(0) - f'(0) + 3(sL(f(t)) - f(0)) + 2L(f(t)) = 0$$

$$\Rightarrow \qquad (s^2 + 3s + 2)\,L(f(t)) - s - 2 - 3 = 0$$

$$\Rightarrow \qquad (s^2 + 3s + 2)\,L(f(t)) = s + 5 \;\;\Rightarrow\;\; L(f(t)) = \frac{s+5}{s^2+3s+2}.$$

EXAMPLE 2.6

Solve the initial value problem

$$y'' + 4y = 0,\; y(0) = 1,\; y'(0) = 0.$$

Solution: Taking Laplace transform of the given differential equation, we obtain.

$$L(y'') + 4L(y) = 0$$

$$\Rightarrow \qquad s^2L(y) - sy(0) - y'(0) + 4L(y) = 0$$

$$\Rightarrow \qquad (s^2 + 4)\,L(y) - s(1) - 0 = 0$$

$$\Rightarrow \qquad Y(s) = L(y) = \frac{s}{s^2+4}$$

$$\therefore \qquad y = L^{-1}(Y(s)) = L^{-1}\left(\frac{s}{s^2+4}\right) = \cos 2t.$$

EXAMPLE 2.7

Solve the initial value problem

$$y'' - 5y' + 4y = e^{2t},\ y(0) = 19/12,\ y'(0) = 8/3.$$

Solution: Taking Laplace transform of the given equation, we obtain

$$s^2L(y) - sy(0) - y'(0) - 5[sL(y) - y(0)] + 4L(y) = L(e^{2t})$$

$$\Rightarrow \qquad (s^2 - 5s + 4)\,Y(s) - s\left(\frac{19}{12}\right) - \frac{8}{3} + \frac{95}{12} = \frac{1}{s-2}$$

$$\therefore \qquad Y(s) = \frac{1}{(s-2)(s-1)(s-4)} + \frac{19s-63}{12(s-1)(s-4)}$$

$$\Rightarrow \qquad Y(s) = \frac{-1/2}{s-2} + \frac{1/3}{s-1} + \frac{1/6}{s-4} + \frac{11/9}{s-1} + \frac{13/36}{s-4}$$

$$= -\frac{1}{2(s-2)} + \frac{14}{9(s-1)} + \frac{19}{36(s-4)}$$

Therefore, $$y(t) = L^{-1}(Y(s)) = -\frac{1}{2}e^{2t} + \frac{14}{9}e^{t} + \frac{19}{36}e^{4t}$$

EXAMPLE 2.8

Solve the initial value problem

$$y'' + 2y = \sin t,\ y(0) = 0,\ y'(0) = 0.$$

Solution: Taking the Laplace transform of the given differential equation, we obtain

$$s^2L(y) - sy(0) - y'(0) + 2L(y) = L(\sin t)$$

$$\Rightarrow \qquad (s^2 + 2)\,Y(s) - s(0) - 0 = \frac{1}{s^2+1}$$

$$\Rightarrow \qquad Y(s) = \frac{1}{(s^2+1)(s^2+2)} = \frac{1}{s^2+1} - \frac{1}{s^2+2}$$

$$\therefore \quad y(t) = L^{-1}(Y(s)) = L^{-1}\left(\frac{1}{s^2+1}\right) - L^{-1}\left(\frac{1}{s^2+2}\right)$$

$$= \sin t - \frac{1}{\sqrt{2}} \sin \sqrt{2}\, t.$$

EXAMPLE 2.9

Solve by the method of Laplace transforms, the initial value problem

$$y_1' - y_2 = 1, \quad y_1(0) = -1$$

$$y_2' + 4y_1 = t, \quad y_2(0) = 1$$

Solution: Taking Laplace transforms, we obtain

$$sL(y_1) - y_1(0) - L(y_2) = L(1)$$

$$\Rightarrow \quad sY_1 - (-1) - Y_2 = \frac{1}{s}$$

$$\therefore \quad sY_1 - Y_2 = \frac{1}{s} - 1 \tag{1}$$

and

$$sL(y_2) - y_2(0) + 4L(y_1) = L(t)$$

$$\Rightarrow \quad sY_2 - 1 + 4Y_1 = \frac{1}{s^2}$$

$$\therefore \quad sY_2 + 4Y_1 = \frac{1}{s^2} + 1 \tag{2}$$

Now, multiplying Eq. (1) by s and adding Eq. (1) and Eq. (2), we get

$$(s^2 + 4)Y_1 = \left(\frac{1}{s} - 1\right)s + \frac{1}{s^2} + 1$$

$$\Rightarrow \quad (s^2 + 4)Y_1 = 1 - s + \frac{1}{s^2} + 1 = 2 - s + \frac{1}{s^2}$$

$$\therefore \quad Y_1 = \frac{2}{s^2+4} - \frac{s}{s^2+4} + \frac{1}{s^2(s^2+4)}$$

$$= \frac{2}{s^2+4} - \frac{s}{s^2+4} + \frac{1}{4}\left\{\frac{1}{s^2} - \frac{1}{s^2+4}\right\}$$

$$= \frac{1}{4s^2} - \frac{s}{s^2+4} + \frac{7}{4(s^2+4)}$$

Thus, $$y_1(t) = L^{-1}(Y_1) = \frac{t}{4} - \cos 2t + \frac{7}{8}\sin 2t$$

So, $$y_1'(t) = \frac{1}{4} + 2\sin 2t + \frac{7}{4}\cos 2t$$

Substituting in $y_1' - y_2 = 1$, we get

$$\frac{1}{4} + 2\sin 2t + \frac{7}{4}\cos 2t - y_2 = 1$$

$$\Rightarrow \quad y_2 = -\frac{3}{4} + \frac{7}{4}\cos 2t + 2\sin 2t.$$

EXAMPLE 2.10

Solve the initial value problem

$$y_1' - 3y_2 = 3\sin 3t,\ y_1(0) = -3$$
$$3y_1 + y_2' = 3(1 - \cos 3t),\ y_2(0) = 1$$

Solution: Taking the Laplace transforms, we get

$$sL(y_1) - y_1(0) - 3L(y_2) = 3L(\sin 3t)$$

$$\Rightarrow \quad sY_1 - (-3) - 3Y_2 = \frac{9}{s^2+9}$$

$$\Rightarrow \quad sY_1 - 3Y_2 = \frac{9}{s^2+9} - 3 \qquad (1)$$

and $$3L(y_1) + sL(y_2) - y_2(0) = 3L(1 - \cos 3t)$$

$$\Rightarrow \quad 3Y_1 + sY_2 - 1 = 3\left(\frac{1}{s} - \frac{s}{s^2+9}\right)$$

$$\Rightarrow \quad 3Y_1 + sY_2 = \frac{3}{s} - \frac{3s}{s^2+9} + 1 \qquad (2)$$

Multiplying Eq. (1) by 3 and Eq. (2) by s and subtracting, we get

$$(-9 - s^2)Y_2 = \frac{27}{s^2+9} - 9 - \left(3 - \frac{3s^2}{s^2+9} + s\right)$$

$$\Rightarrow \quad -(s^2+9)Y_2 = \frac{3(s^2+9)}{s^2+9} - 12 - s$$

$$\Rightarrow \quad -(s^2+9)Y_2 = -9 - s$$

$$\therefore \quad Y_2 = \frac{s}{s^2+9} + \frac{9}{s^2+9}$$

Thus, $$y_2(t) = L^{-1}(Y_2(s)) = L^{-1}\left(\frac{s}{s^2+9}\right) + 9L^{-1}\left(\frac{1}{s^2+9}\right)$$

$$= \cos 3t + 3 \sin 3t$$

So, $$y_2'(t) = -3 \sin 3t + 9 \cos 3t$$

Substituting in $3y_1 + y_2' = 3(1 - \cos 3t)$, we get

$$3y_1 - 3 \sin 3t + 9 \cos 3t = 3(1 - \cos 3t)$$

$$\Rightarrow \quad y_1(t) = 1 + \sin 3t - 4 \cos 3t.$$

EXAMPLE 2.11

The coordinates (x, y) of a particle moving along a plane curve at any time t, are given by $\frac{dy}{dt} + 2x = \sin 2t$, and $\frac{dx}{dt} - 2y = \cos 2t$ $(t > 0)$. If at $t = 0$, $x = 1$ and $y = 0$, show by using the Laplace transforms that the particle moves along the curve $4x^2 + 4xy + 5y^2 = 4$.

Solution: Taking the Laplace transforms of the given equations, we get

$$sL(y) - y(0) + 2L(x) = \frac{2}{s^2+4}$$

$$\Rightarrow \quad sY + 2X = \frac{2}{s^2+4} \tag{1}$$

and $$sL(x) - x(0) - 2L(y) = \frac{2}{s^2+4}$$

$$\Rightarrow \quad sX - 2Y = \frac{s}{s^2+4} + 1 \tag{2}$$

Multiplying Eq. (1) by s and Eq. (2) by 2 and subtracting we get,

$$(s^2 + 4)Y = -2$$

$$\Rightarrow \quad Y = -\frac{2}{s^2+4}$$

Thus, $$y(t) = L^{-1}(Y(s)) = L^{-1}\left(\frac{2}{s^2+4}\right) = -\sin 2t$$

So, $$\frac{dy}{dt} = -2 \cos 2t,$$

Substituting in $\frac{dy}{dt} + 2x = \sin 2t$, we get

$$-2\cos 2t + 2x = \sin 2t$$

$$\Rightarrow \quad 2x = \sin 2t + 2\cos 2t$$

Now, $4x^2 + 4xy + 5y^2 = (\sin 2t + 2\cos 2t)^2 + 2(\sin 2t + 2\cos 2t)(-\sin 2t) + 5(-\sin 2t)$

$$= 4\cos^2 2t + 4\sin 2t\cos 2t - 4\sin 2t\cos 2t - \sin^2 2t + 5\sin^2 2t$$

$$= 4(\sin^2 2t + \cos^2 2t) = 4.$$

Laplace Transform of Integrals

Let $f(t)$; $t \geq 0$ be a piecewise continuous function and be of exponential order on $[0, \infty)$. Then,

$$L\left(\int_0^t f(\tau)d\tau\right) = \frac{1}{s}L[f(t)]$$

Proof: Since $f(t)$ is piecewise continuous on $[0, \infty)$ and is of exponential order. Therefore, $\phi(t) = \int_0^t f(\tau)d\tau$ is at least piecewise continuous on $[0, \infty)$ and is of exponential order.

Further, $\phi'(t) = f(t)$ except for points at which $f(t)$ is discontinuous. That is, $\phi'(t)$ is piecewise continuous on each finite sub-interval. Hence, we can write

$$L(\phi'(t)) = sL(\phi(t)) - \phi(0)$$

$$\therefore \quad L(\phi(t)) = sL(\phi(t)) - 0$$

$$\Rightarrow \quad L(f(t)) = sL\left(\int_0^t f(\tau)d\tau\right)$$

$$\Rightarrow \quad \frac{1}{s}L(f(t)) = L\left(\int_0^t f(\tau)d\tau\right)$$

Further, we can have

$$\frac{1}{s}F(s) = L\left(\int_0^t f(\tau)d\tau\right)$$

$$\Rightarrow \quad L^{-1}\left\{\frac{1}{s}F(s)\right\} = \int_0^t f(\tau)d\tau$$

$$= \int_0^t L^{-1}(F(s))d\tau.$$

EXAMPLE 2.12

Find $L^{-1}\left(\frac{1}{s(s^2+9)}\right)$.

Solution: We have

$$L^{-1}\left(\frac{1}{s}\frac{1}{(s^2+9)}\right)=\int_0^t L^{-1}\left(\frac{1}{s^2+9}\right)dt$$

$$=\int_0^t \frac{\sin 3t}{3}dt$$

(We can use τ in the integrand, for sale of differentiation with variable t in limits)

Also, $$f(t) = \frac{\sin 3t}{3} = 0 \text{ at } t = 0$$

$\therefore$ $$L^{-1}\left(\frac{1}{s(s^2+9)}\right)=\left[-\frac{\cos 3t}{3\times 3}\right]_0^t$$

$$= -\frac{1}{9}(\cos 3t-1)=\frac{1}{9}(1-\cos 3t).$$

EXAMPLE 2.13

Find $L^{-1}\left(\frac{1}{s^2(s^2+4)}\right)$ by using formula of Laplace transform of integrals

Solution: We have

$$L^{-1}\left(\frac{1}{s^2(s^2+4)}\right)=L^{-1}\left(\frac{1}{s}\frac{1}{s(s^2+4)}\right)=\int_0^t L^{-1}\left(\frac{1}{s(s^2+4)}\right)dt$$

Now, $$L^{-1}\left(\frac{1}{s}\frac{1}{s^2+4}\right)=\int_0^t L^{-1}\left(\frac{1}{s^2+4}\right)dt$$

$$=\int_0^t \frac{1}{2}\sin 2t dt$$

$$=\left[-\frac{\cos 2t}{4}\right]_0^t=\frac{1-\cos 2t}{4}$$

Also, $$f(t) = L^{-1}\left(\frac{1}{s^2+4}\right)=\frac{\sin 2t}{2} = 0 \text{ at } t = 0$$

Thus,
$$L^{-1}\left(\frac{1}{s^2(s^2+4)}\right)=\int_0^t \frac{1-\cos 2t}{4}dt$$
$$=\frac{1}{4}\left\{t-\frac{\sin 2t}{2}\right\}_0^t=\frac{1}{4}\left(t-\frac{\sin 2t}{2}\right)$$
$$=\frac{1}{8}(2t-\sin 2t)$$

Here, also $L^{-1}\left(\frac{1}{s(s^2+4)}\right)=\frac{1-\cos 2t}{4}=0$ at $t=0$.

EXAMPLE 2.14

Evaluate: $L^{-1}\left(\frac{1}{s^4+3s^3}\right)$.

Solution: Here,
$$L^{-1}\left(\frac{1}{s^3(s+3)}\right)=L^{-1}\left(\frac{1}{s}\,\frac{1}{s^2(s+3)}\right)=\int_0^t L^{-1}\left(\frac{1}{s^2(s+3)}\right)dt$$

Now,
$$L^{-1}\left(\frac{1}{s^2(s+3)}\right)=L^{-1}\left(\frac{1}{s}\,\frac{1}{s(s+3)}\right)$$
$$=\int_0^t \frac{1}{3}\,L^{-1}\left(\frac{1}{s}-\frac{1}{s+3}\right)dt=\frac{1}{3}\int_0^t(1-e^{-3t})dt$$
$$=\frac{1}{3}\left[t-\frac{e^{-3t}}{-3}\right]_0^t$$
$$=\frac{1}{3}\left[t-\frac{e^{-3t}-1}{-3}\right]=\frac{(e^{-3t}+3t-1)}{9}$$

Here, we have $L^{-1}\left(\frac{1}{s(s+3)}\right)=\frac{1-e^{-3t}}{3}=0$ at $t=0$

Thus,
$$L^{-1}\left(\frac{1}{s^3(s+3)}\right)=\int_0^t \frac{e^{-3t}+3t-1}{9}dt$$
$$=\frac{1}{9}\left[\frac{e^{-3t}}{-3}+\frac{3t^2}{2}-t\right]_0^t$$

$$= \frac{1}{9}\left[\frac{e^{-3t}-1}{-3}+\frac{3t^2}{2}-t\right]_0^t$$

$$= \frac{1}{27}(1-e^{-3t}) + \frac{1}{18}(3t^2-2t)$$

Here, also $L^{-1}\left(\frac{1}{s^2(s+3)}\right) = \frac{e^{-3t}+3t-1}{9} = 0$ at $t = 0$.

EXAMPLE 2.15

Find the solution of the initial value problem

$$y' + 3y + 2\int_0^t y(\tau)d\tau = t,\ y(0) = 0.$$

Solution: Taking the Laplace transform on both sides of the differential equation, we obtain

$$sL(y) + 3L(y) + 2\frac{L(y)}{s} = L(t)$$

$$\Rightarrow \quad \left(s+3+\frac{2}{s}\right)Y = \frac{1}{s^2}$$

$$\Rightarrow \quad (s^2 + 3s + 2)Y = \frac{1}{s}$$

$$\therefore \quad Y = \frac{1}{s(s^2+3s+2)}$$

Thus,

$$y(t) = L^{-1}(Y) = L^{-1}\left(\frac{1}{s(s+1)(s+2)}\right)$$

$$= \int_0^t L^{-1}\left(\frac{1}{s+1}-\frac{1}{s+2}\right)dt$$

$$= \int_0^t (e^{-t} - e^{-2t})dt$$

$$= \left[\frac{e^{-t}}{-1}-\frac{e^{-2t}}{-2}\right]_0^t$$

$$= \frac{e^{-t}-1}{-1}+\frac{e^{-2t}-1}{2} = \frac{1}{2}e^{-2t} - e^{-t} + \frac{1}{2}$$

Here, we have $L^{-1}\left(\frac{1}{s^2+3s+2}\right) = e^{-t} - e^{-2t} = 0$ at $t = 0$ ▲▲

EXERCISE 2.1

Using the formula of Laplace transform of derivatives, find the Laplace transforms of the following functions:

1. $\sin^2 t$.
2. t^5.
3. te^{at}.
4. $t \cos at$.
5. $t \sinh at$.
6. $t \sin at$.
7. $(t + 1)^3$.
8. $(t - 1)^2$.

Find the inverse Laplace transform of the following functions:

9. $\dfrac{1}{s^2 + 5s}$.
10. $\dfrac{3}{s^2(s^2 + 4)(s^2 + 1)}$.
11. $\dfrac{16}{s^3 + 9s}$.
12. $\dfrac{s - a}{s(s^2 + a^2)}$.
13. $\dfrac{s - 2}{s(s + 3)}$.
14. $\dfrac{1}{s^3 + s^2}$.
15. $\dfrac{w}{s^2(s^2 + w^2)}$.

Solve the following initial value problems:

16. $y'' + 4y = 0; y(0) = 1, y'(0) = 6$.
17. $y'' + 2y' - 3y = 3, y(0) = 4, y'(0) = -7$.
18. $y''' + 2y'' - y' - 2y = 0; y(0) = y'(0) = 0, y''(0) = 6$.
19. $y'' - 2y' + y = e^t; y(0) = 2, y'(0) = -1$.
20. $(D^2 + a^2)y = a \sin (nt + b); y(0) = y'(0) = 0$.

21. $y' - 4y + 3\int_0^t y(\tau)d\tau = t;\ y(0) = 1.$

22. $y' + 6y + 5\int_0^t y(\tau)d\tau = 1 + t;\ y(0) = 1.$

23. $y' + y - 6\int_0^t y(\tau)d\tau = \sin t;\ y(0) = 2.$

24. $y_1' + y_2 = 2 - \sin t,\ y_1(0) = 3.$

    $y_1 - y_2' = t + \cos t,\ y_2\ (0) = 0.$

25. $y_1' + y_2 = 3t,\ y_1(0) = 0.$

    $y_1 - y_2' = t^2 - 1,\ y_2(0) = 1.$

26. The small oscillations of a certain system with two degrees of freedom are given by the equations

$$\frac{d^2x}{dt^2} + 3x - 2y = 0,\ \frac{d^2x}{dt^2} + \frac{d^2y}{dt^2} - 3x + 5y = 0$$

If $x = 0, y = 0; x = 3, y = 2$ when $t = 0$; find x and y when $t = 1/2$.

ANSWERS

Exercise 2.1

1. $\dfrac{2}{s(s^2+4)}$.

2. $\dfrac{5!}{s^6}$.

3. $\dfrac{1}{(s-a)^2}$.

4. $\dfrac{(s^2-a^2)}{(s^2+a^2)^2}$.

5. $\dfrac{2as}{(s^2-a^2)^2}$.

6. $\dfrac{2as}{(s^2+a^2)^2}$.

7. $\dfrac{3!}{s^4}+\dfrac{3!}{s^3}+\dfrac{3}{s^2}+\dfrac{1}{s}$.

8. $\frac{2!}{s^3} - \frac{2}{s^2} + \frac{1}{s}$.

9. $\frac{1-e^{-5t}}{5}$.

10. $\frac{3}{4}t - \sin t + \frac{1}{8}\sin 2t$.

11. $\frac{16}{9}(1 - \cos 3t)$.

12. $\frac{1}{a^2}(\sin at - \cos at + 1) - \frac{t}{a}$.

13. $-\frac{1}{3}(2 - 5e^{-3t})$.

14. $t + e^{-t} - 1$.

15. $\frac{1}{\omega^2}(\omega t - \sin \omega t)$.

16. $y(t) = \cos 2t + 3 \sin 2t$.

17. $y(t) = -1 + 2e^t + 3e^{-3t}$.

18. $y(t) = e^t - 3e^{-t} + 2e^{-2t}$.

19. $y(t) = 2e^t - 3te^t + \frac{1}{2}t^2e^t$.

20. $y(t) = a[\sin nt \cos b - nt \cos (nt + b)]/2n^2$.

21. $y(t) = \frac{1}{3} - e^t + \frac{5}{3}e^{3t}$.

22. $y(t) = \frac{1}{5} - \frac{1}{4}e^{-t} + \frac{21}{20}e^{-5t}$.

23. $y(t) = -\frac{7}{50}\cos t - \frac{1}{50}\sin t + \frac{63}{50}e^{3t} \frac{22}{25}e^{-2t}$.

24. $y_1(t) = \cos 3t + 3 \sin 3t$; $y_2(t) = 2 \sin t - \cos t + 1$.

25. $y_1(t) = t^2 - \sin t$; $y_2(t) = t + \cos t$.

26. $x = \frac{11}{4}\sin t + \frac{1}{12}\sin 3t$; $y = \frac{11}{4}\sin t + \frac{1}{4}\sin 3t$. ▲▲

3

CHAPTER

First and Second Shifting Theorems

In the previous chapter, we have solved some initial value problems. In the present chapter, to find the inverse Laplace transform, we will discuss some more formulas and results, which simplify the computation.

First Shifting Theorem

Let $L(f(t)) = F(s)$ $\{s > a \geq 0\}$ and c be a real number. Then,

$$L(e^{ct}f(t)) = F(s - c)\ \{s > c + a\}.$$

Proof: By definition, we have

$$\begin{aligned} L(e^{ct}f(t)) &= \int_0^\infty e^{-st}e^{ct}f(t)dt \\ &= \int_0^\infty e^{-(s-c)t} f(t)dt \\ &= F(s - c),\ \{s - c > a \Rightarrow s > c + a\} \end{aligned}$$

$$\left\{\because L(f(t)) = \int_0^\infty e^{-st} f(t)dt = F(s)\ \{s > a\}\right\}$$

Note:

$$L(e^{ct} f(t)) = F(s - c)$$

$$\Rightarrow \quad f(t) = e^{-ct}L^{-1}(F(s - c))$$

$$\Rightarrow \quad L^{-1}(F(s)) = e^{-ct}L^{-1}(F(s - c)).$$

▲▲

EXAMPLE 3.1

Find the Laplace transforms of $e^{at} \sin bt$ and $e^{at}t^3$.

Solution: By using First Shifting Theorem, we have

$$L(e^{at} \sin bt) = F(s - a) \quad \text{where} \quad F(s) = L(\sin bt) = \frac{b}{s^2 + b^2}$$

$$= \frac{b}{(s - a)^2 + b^2}.$$

Similarly, $$L(e^{at}t^3) = \frac{3!}{(s-a)^4}.$$

EXAMPLE 3.2

Find the inverse Laplace transform of $\dfrac{3s-1}{(s-2)^2}$ and $\dfrac{6+s}{s^2+6s+13}$.

Solution: We have
$$L^{-1}\left(\frac{3s-1}{(s-2)^2}\right) = e^{2t}L^{-1}\left(\frac{3(s+2)-1}{s^2}\right)$$
$$= e^{2t}L^{-1}\left(\frac{3s+5}{s^2}\right)$$
$$= e^{2t}L^{-1}\left(\frac{3}{s}+\frac{5}{s^2}\right) = e^{2t}(3+5t)$$

Also,
$$L^{-1}\left(\frac{6+s}{s^2+6s+13}\right) = L^{-1}\left(\frac{s+3+3}{(s+3)^2+4}\right)$$
$$= e^{-3t}L^{-1}\left(\frac{s+3}{s^2+4}\right)$$
$$= e^{-3t}L^{-1}\left(\frac{s}{s^2+4}+\frac{3}{s^2+4}\right)$$
$$= e^{-3t}\left(\cos 2t + \frac{3}{2}\sin 2t\right).$$

EXAMPLE 3.3

Show that
$$L(t\sin at) = \frac{2as}{(s^2+a^2)^2} \quad \text{and} \quad L(t\cos at) = \frac{s^2-a^2}{(s^2+a^2)^2}.$$

Solution: Since $L(t) = 1/s^2$

$\therefore$ $$L(te^{iat}) = \frac{1}{(s-ia)^2} = \frac{(s+ia)^2}{(s-ia)^2(s+ia)^2}$$

$\Rightarrow$ $$L(t(\cos at + i\sin at)) = \frac{s^2-a^2+i2as}{(s^2+a^2)^2}$$

Equating the real and imaginary parts, we get the required result.

EXAMPLE 3.4

Find the inverse transformation of $\dfrac{s}{s^4+4a^4}$.

Solution: Since
$$s^4 + 4a^4 = (s^2 + 2a^2)^2 - (2as)^2$$
$$= (s^2 + 2a^2 + 2as)(s^2 + 2a^2 - 2as)$$

$\therefore$
$$\frac{s}{s^4+4a^4} = \frac{As+B}{s^2+2as+2a^2} + \frac{Cs+D}{s^2-2as+2a^2}$$

$\Rightarrow$
$$s = (As + B)(s^2 - 2as + 2a^2) + (Cs + D)(s^2 + 2as + 2a^2)$$

Coefficient of s^3} $0 = A + C$

Coefficient of s^2} $0 = -2aA + B + 2aC + D$

Coefficient of s} $1 = 2a^2A \quad 2aB + 2a^2C + 2aD$

Constant term} $0 = 2a^2B + 2a^2D$

$\Rightarrow$
$$B + D = 0$$

So, we have $0 = -A + C$

On solving, we get $A = 0 = C$

Thus, $B + D = 0$ and $B - D = -1/2a$

$\therefore$
$$B = -\frac{1}{4a},\ D = \frac{1}{4a}$$

$\therefore$
$$L^{-1}\left(\frac{s}{s^4+4a^4}\right) = -\frac{1}{4a}L^{-1}\left(\frac{1}{s^2+2as+2a^2}\right) + \frac{1}{4a}L^{-1}\left(\frac{1}{s^2-2as+2a^2}\right)$$

$$= -\frac{1}{4a}L^{-1}\left(\frac{1}{(s+a)^2+a^2}\right) + \frac{1}{4a}L^{-1}\left(\frac{1}{(s-a)^2+a^2}\right)$$

$$= -\frac{1}{4a}\left\{e^{-at}L^{-1}\left(\frac{1}{s^2+a^2}\right)\right\} + \frac{1}{4a}e^{at}L^{-1}\left(\frac{1}{s^2+a^2}\right)$$

$$= \frac{1}{4a}(e^{at} - e^{-at})\frac{\sin at}{a} = \frac{1}{2a^2}\sinh at \sin at$$

EXAMPLE 3.5

Find the solution of the initial value problem
$$y'' + 4y' + 13y = e^{-t},\ y(0) = 0,\ y'(0) = 2.$$

Solution: Taking the Laplace transform of the differential equation, we obtain
$$(s^2L/y) - sy(0) - y'(0)) + 4(sL(y) - y(0)) + 13L(y) = \frac{1}{s+1}$$

$\Rightarrow$
$$(s^2 + 4s + 13)\,Y(s) = \frac{2s+3}{s+1}$$

$\Rightarrow$
$$Y(s) = \frac{2s+3}{(s+1)(s^2+4s+13)} = \frac{1}{10}\left[\frac{1}{s+1} - \frac{s-17}{s^2+4s+13}\right]$$

Thus,
$$y(t) = \frac{1}{10} L^{-1}\left[\frac{1}{s+1} - \frac{s-17}{(s+2)^2+9}\right]$$
$$= \frac{1}{10}\left[e^{-t} - e^{-2t}L^{-1}\left(\frac{s-19}{s^2+9}\right)\right]$$
$$= \frac{1}{10}\left[e^{-t} - e^{-2t}\left(\cos 3t - \frac{19}{3}\sin 3t\right)\right]$$

Unit Step Function

$$u(t) = \begin{cases} 0, & \text{if } t < 0 \\ 1, & \text{if } t \geq 0 \end{cases}$$

is called a unit step function.

If the discontinuity is at the point $x = a$ then we have

$$u(t-a) = \begin{cases} 0, & \text{if } t < a \\ 1, & \text{if } t \geq a \end{cases} = u_a(t)$$

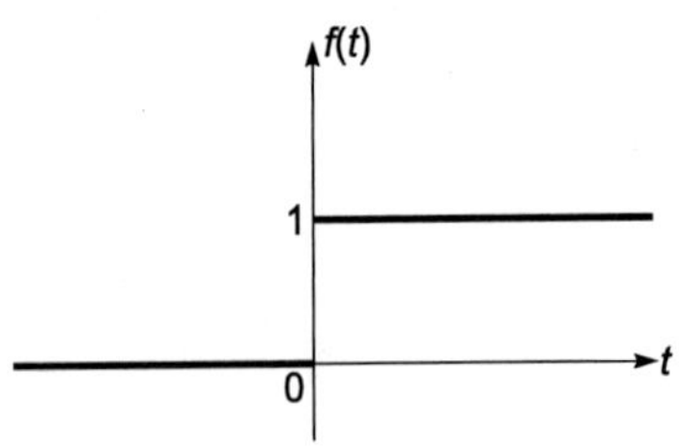

$u(t)$

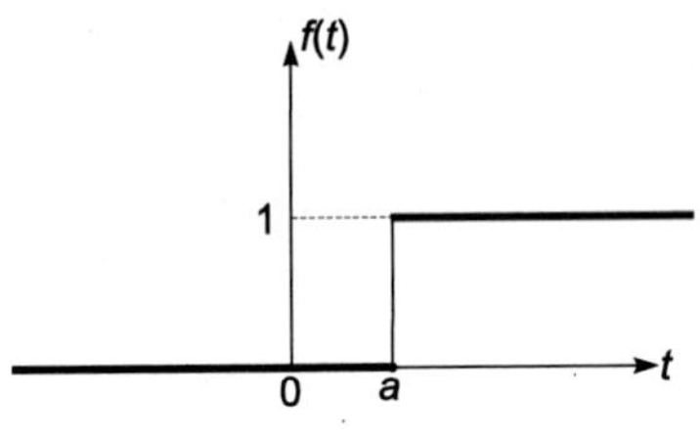

$u_a(t)$

The following are some examples of functions represented in terms of unit step functions.

(a) $f(t) = \begin{cases} 0, & \text{if } t < \pi \\ \sin t, & \text{if } t \geq \pi \end{cases} = \sin t \cdot u_\pi(t)$

(b) $f(t) = \begin{cases} 0, & \text{if } t < 2 \\ t^2, & \text{if } t \geq 2 \end{cases} = t^2 u_2(t)$

(c) Let $a < b$. We define a pulse function as

$$f(t) = \begin{cases} 0, & \text{if } t < a \\ 1, & \text{if } a \leq t < b \\ 0, & \text{if } t \geq b \end{cases} = u_a(t) - u_b(t)$$

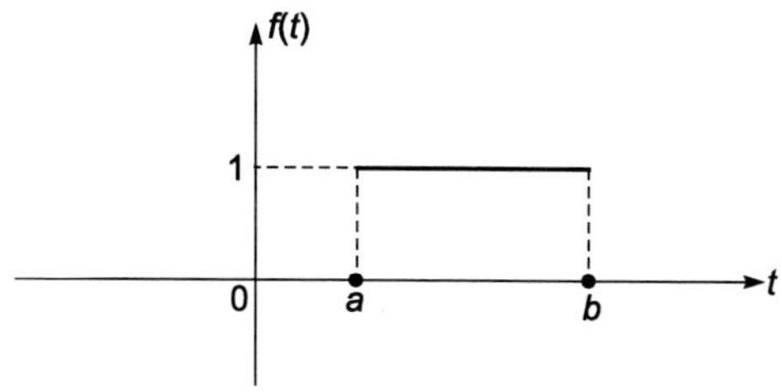

(d) $f(t) = \begin{cases} 5t, & 0 \le t < 2 \\ 0, & t \ge 2 \end{cases} = 5t(1 - u_2(t)) = 5t[u_0(t) - u_2(t)]$

(e) $f(t) = \begin{cases} 5, & 0 \le t < 2 \\ -5, & t \ge 2 \end{cases} = 5[1 - 2u_2(t)] = 5[u_0(t) - 2u_2(t)]$.

Second Shifting Theorem

Let $L(f(t)) = F(s)$ $\{s > \alpha\}$ and $a \ge 0$ be a real number. Then, $L(f(t-a)u_a(t)) = e^{-as}F(s)$

Proof: Using the definition of Laplace transforms, we get

$$L(f(t-a)u_a(t)) = \int_0^\infty e^{-st}f(t-a)u_a(t)dt$$

$$= \int_a^\infty e^{-st}f(t-a)dt$$

Let $\tau = t - a \Rightarrow d\tau = dt$

$$\therefore \quad L(f(t-a)u_a(t)) = \int_0^\infty e^{-s(\tau + a)} f(\tau)d\tau$$

$$= e^{-s\tau}\int_0^\infty e^{-s\tau} f(\tau)d\tau = e^{-as}L[f(t)]$$

$$= e^{-as}F(s)$$

Note 1. $L^{-1}(e^{-as}F(s)) = u_a(t)\, f(t-a)$

Also, $L(f(t-a)u_a(t)) = e^{-as}L(f(t))$

$\Rightarrow \quad L(f(t)u_a(t)) = e^{-as}L(f(t+a))$

Note 2.
$$L(u_a(t)) = L(u_a(t)\cdot 1) = e^{-as}L(1) = \frac{e^{-as}}{s}$$

EXAMPLE 3.6

Find the Laplace transform of the function

$$f(t) = \begin{cases} 0 & 0 \le t < 3 \\ (t-3)^2 & t \ge 3. \end{cases}$$

Solution: Here, $f(t) = (t-3)^2\, u_3(t)$

Therefore,

$$\begin{aligned} L(f(t)) &= L((t-3)^2\, u_3(t)) \\ &= e^{-3\lambda}\, L(\overline{t+3-3})^2 \\ &= e^{-3s}\, L(t^2) = e^{-3s}\frac{2!}{s^3}. \end{aligned}$$

EXAMPLE 3.7

Find $L(f(t))$ where

$$f(t) = \begin{cases} 0, & 0 \le t < \pi/2 \\ \sin t, & t \ge \pi/2. \end{cases}$$

Solution: Here, $f(t) = \sin u_{\pi/2}(t)$

Therefore,

$$\begin{aligned} L(f(t)) &= L(\sin t u_{\pi/2}(t)) \\ &= e^{-\pi/2s}\, L(\sin (t + \pi/2)) \\ &= e^{-\pi s/2}\, L(\cos t) \\ &= e^{-\pi s/2}\frac{s}{s^2+1}. \end{aligned}$$

EXAMPLE 3.8

Find $L(f(t))$ for

$$f(t) = \begin{cases} 0, & 0 \le t < 1 \\ t-1, & 1 \le t < 2 \\ 0, & t \ge 2 \end{cases}$$

Solution: Here, $f(t) = (t-1)\,(u_1(t) - u_2(t))$

Therefore,

$$\begin{aligned} L(f(t)) &= L((t-1)\,(u_1(t) - u_2(t))) \\ &= e^{-s}\, L(\overline{t+1}-1) - e^{-2s}\, L(\overline{t+2}-1) \\ &= e^{-s}\, L(t) - e^{-2s}\, L(t+1) \\ &= e^{-s}\left(\frac{1}{s^2}\right) - e^{-2s}\left(\frac{1}{s^2}+\frac{1}{s}\right) = \frac{e^{-s} - e^{-2s}(1+s)}{s^2}. \end{aligned}$$

EXAMPLE 3.9

Find the inverse Laplace transform of $\dfrac{4e^{-s\pi/2}}{s^2+16}$.

Solution: We have $L^{-1}\left(\frac{4e^{-s\pi/2}}{s^2+16}\right)=u_{\pi/2}(t)\sin 4\left(t-\frac{\pi}{2}\right)$ $\left[\because f(t)=L^{-1}\left(\frac{4}{s^2+16}\right)=\sin 4t\right]$

$= \sin 4t \cdot u_{\pi/2}(t).$

EXAMPLE 3.10

Evaluate: $L^{-1}\left(\frac{3(1+e^{-s\pi})}{s^2+9}\right)$.

Solution: We have $L^{-1}\left(\frac{3}{s^2+9}\right)+L^{-1}\left(\frac{3e^{-\pi s}}{s^2+9}\right)$

$$= \sin 3t + u_\pi(t)\sin 3(t-\pi)$$

$$= \sin 3t - u_\pi(t)\sin 3t$$

$$= \sin 3t(1-u_\pi(t))$$

$$= \sin 3t\,(u_0(t)-u_\pi(t)).$$

EXAMPLE 3.11

Evaluate: $L^{-1}\left(\frac{(s+1)e^{-\pi s}}{s^2+s+1}\right)$.

Solution: We have $L^{-1}\left(e^{-\pi s}\frac{s+1}{s^2+s+1}\right)=u_\pi(t)f(t-\pi)$

$\because$ $$f(t)=L^{-1}\left(\frac{s+1}{s^2+s+1)}\right)=L^{-1}\left(\frac{s+\frac{1}{2}+\frac{1}{2}}{\left(s+\frac{1}{2}\right)^2+\frac{3}{4}}\right)$$

$$= e^{-1/2t}\,L^{-1}\left(\frac{s+1/2}{s^2+(\sqrt{3}/2)^2}\right)$$

$$= e^{-1/2t}\left(\cos\frac{\sqrt{3}}{2}t+\frac{1}{2}\,\frac{2}{\sqrt{3}}\sin\frac{\sqrt{3}}{2}t\right)$$

$\therefore$ $$L^{-1}\left(e^{-\pi s}\frac{s+1}{s^2+s+1}\right)=\frac{1}{\sqrt{3}}\,u_\pi(t)e^{-\frac{(t-\pi)}{2}}\times\left\{\sqrt{3}\cos\frac{\sqrt{3}(t-\pi)}{2}+\sin\frac{\sqrt{3}(t-\pi)}{2}\right\}.$$

EXAMPLE 3.12

Find the solution of the initial value problem,

$$y'+y=f(t),\ y(0)=2 \text{ where } f(t)=\begin{cases}0, & 0\le t<\pi/2\\ \cos t, & t\ge\pi/2\end{cases}$$

Solution: We have $$f(t) = \cos t\, u_{\pi/2}(t)$$

Taking Laplace transform of the differential equation, we obtain

$$sL(y) - y(0) + L(y) = L(f(t)) = L(\cos t u_{\pi/2}(t))$$

$$\Rightarrow \quad sY(s) - 2 + Y(s) = e^{-\pi s/2} L(\cos (t + \pi/2))$$

$$\Rightarrow \quad (s + 1)\, Y(s) - 2 = e^{-\pi s/2} L(-\sin t)$$

$$\Rightarrow \quad (s + 1)\, Y(s) = -e^{-\pi s/2} \frac{1}{s^2+1} + 2$$

$$\Rightarrow \quad Y(s) = \frac{2}{s+1} - \frac{e^{-\pi s/2}}{(s+1)(s^2+1)}$$

$$= \frac{2}{s+1} - \frac{e^{-\pi s/2}}{2}\left(\frac{1}{s+1} - \frac{s-1}{s^2+1}\right)$$

Therefore, $$y(t) = L^{-1}(Y(s)) = 2e^{-t} - \frac{1}{2} u_{\pi/2}(t)\, f\left(t - \frac{\pi}{2}\right)$$

$$= 2e^{-t} - \frac{1}{2} u_{\pi/2}(t)\, (e^{-(t-\pi/2)} - \sin t - \cos t).$$

where $$f(t) = L^{-1}\left(\frac{1}{s+1} - \frac{s}{s^2+1} + \frac{1}{s^2+1}\right)$$

$$= e^{-t} - \cos t + \sin t$$

EXAMPLE 3.13

Find the solution of the integral differential equation

$$y' + 5y + 4\int_0^t y(\tau)d\tau = f(t)$$

with $y(0) = 2$ and $f(t) = k(u_a(t) - u_b(t))$.

Solution: Taking Laplace transform of the given equation, we obtain

$$sy(s) - y(0) + 5Y(s) + u\,\frac{y(s)}{s} = kL(u_a(t) - u_b(t))$$

$$\Rightarrow \quad \left(s+5+\frac{4}{s}\right)Y(s) - 2 = k\left(\frac{e^{-as}}{s} - \frac{e^{-bs}}{s}\right)$$

$$\Rightarrow \quad (s^2 + 5s + 4)\, Y(s) = k(e^{-as} - e^{-bs}) + 2s$$

$$\Rightarrow \quad Y(s) = \frac{k(e^{-as} - e^{-bs})}{(s+1)(s+4)} + \frac{2s}{(s+1)(s+4)}$$

$$= k\frac{(e^{-as} - e^{-bs})}{3}\left(\frac{1}{s+1} - \frac{1}{s+4}\right) + \frac{2}{-3}\left(\frac{1}{s+1} - \frac{4}{s+4}\right)$$

$$\text{Therefore, } y(t) = L^{-1}(Y(s)) = \frac{k}{3}\, L^{-1}(e^{-as} - e^{-bs})\left(\frac{1}{s+1} - \frac{1}{s+4}\right) - \frac{2}{3}\, L^{-1}\left(\frac{1}{s+1} - \frac{4}{s+4}\right)$$

$$= \frac{k}{3}(u_a(t)\, f(t-a) - u_b(t)\, f(t-b)) - \frac{2}{3}\,(e^{-t} - 4e^{-Yt})$$

where $$f(t) = L^{-1}\left(\frac{1}{s+1} - \frac{1}{s+4}\right) = e^{-t} - e^{-Yt}.$$

EXAMPLE 3.14

Solve the initial value problem

$$y' + 4\int_0^t y(\tau)d\tau = 1 + u_{\pi/2}(t) \text{ with } y(0) = 0.$$

Solution: Taking Laplace transform of the given differential equation, we obtain

$$sL(y) - y(0) + Y\frac{L(y)}{s} = L(1 + u_{\pi/2}(t))$$

$$\Rightarrow \qquad \left(s + \frac{Y}{s}\right)Y(s) = \frac{1}{s} + \frac{e^{-\pi/2s}}{s}$$

$$\Rightarrow \qquad Y(s) = \frac{1 + e^{-\pi s/2}}{s^2 + 4}$$

Therefore,

$$y(t) = L^{-1}(Y(s)) = L^{-1}\left(\frac{1}{s^2+4} + e^{-\pi s/2}\frac{1}{s^2+4}\right)$$

$$= \frac{1}{2}\sin 2t + u_{\pi/2}(t)\,\frac{1}{2}\sin 2\left(t - \frac{\pi}{2}\right)$$

$$= \frac{1}{2}\,[\sin 2t + u_{\pi/2}(t)\,(-\sin 2t)]$$

$$= \frac{1}{2}\sin 2t\,(1 - u_{\pi/2}(t))$$

$$= \frac{1}{2}\sin 2t\,(u_0(t) - u_{\pi/2}(t))$$

EXAMPLE 3.15

Solve the initial value problem

$$y' + y = f(t),\ y(0) = 3;\ f(t) = \begin{cases} 1, & 0 \le t < 1 \\ -1, & t \ge 1 \end{cases}$$

Solution: We have $$f(t) = 1 - 2\,u_1(t) = u_0(t) - 2u_1(t)$$

Taking the Laplace transform of the given differential equation, we obtain

$$sL(y) - y(0) + L(y) = L(u_0(t) - 2u_1(t))$$

$$\Rightarrow \quad sY(s) - 3 + Y(s) = \frac{e^{-os}}{s} - 2\frac{e^{-s}}{s}$$

$$\Rightarrow \quad (s + 1)\, Y(s) - 3 = \frac{1-2e^{-s}}{s}$$

$$\Rightarrow \quad Y(s) = \frac{1-2e^{-s}+3s}{s(s+1)}$$

$$\Rightarrow \quad Y(s) = \frac{1}{s(s+1)} - \frac{2e^{-s}}{s(s+1)} + \frac{3}{s+1}$$

$$\Rightarrow \quad Y(s) = \frac{1}{s} - \frac{1}{s+1} - 2e^{-s}\left(\frac{1}{s} - \frac{1}{s+1}\right) + \frac{3}{s+1}$$

Therefore,

$$y(t) = L^{-1}(Y(s)) = L^{-1}\left(\frac{1}{s} + \frac{2}{s+1} - 2e^{-s}\left(\frac{1}{s} - \frac{1}{s+1}\right)\right)$$

$$= 1 + 2e^{-t} - 2u_1(t)\,(1 - e^{-(t-1)})$$

$$= (1 + 2e^{-t})u_0(t) - 2(1 - e^{-(t-1)})u_1(t).$$

EXAMPLE 3.16

Solve the initial value problem

$$y' + 3y = f(t),\ y(0) = 2,\ f(t) = \begin{cases} t, & 0 \le t < 1 \\ 0, & t \ge 1 \end{cases}.$$

Solution: We have

$$f(t) = t - \begin{cases} 0, & 0 \le t < 1 \\ t, & t \ge 1 \end{cases}$$

$$= t - tu_1(t) = t(u_0(t) - u_1(t)).$$

Taking Laplace transform of the given differential equation, we obtain

$$sL(y) - y(0) + 3L(y) = L(t) - e^{-s}L(t + 1)$$

$$\Rightarrow \quad (s + 3)\, Y(s) - 2 = \frac{1}{s^2} - e^{-s}\left(\frac{1}{s^2} + \frac{1}{s}\right)$$

$$\Rightarrow \quad (s + 3)\, Y(s) = \frac{1-e^{-s}(1+s)+2s}{s^2}$$

$$\Rightarrow \qquad Y(s) = \frac{1+2s-e^{-s}(1+s)}{s^2(s+3)}$$

$$= \frac{1}{s^2(s+3)} + \frac{2}{s(s+3)} - e^{-s}\left(\frac{1}{s^2(s+3)} + \frac{1}{s(s+3)}\right)$$

$$= \frac{(1-e^{-s})}{9}\left(\frac{1}{s+3} - \frac{1}{s} + \frac{3}{s^2}\right) + \frac{(2-e^{-s})}{2}\left(\frac{1}{s} - \frac{1}{s+3}\right).$$

Therefore,

$$y(t) = L^{-1}(Y(s)) = \frac{1}{9}(e^{-3t} - 1 + 3t - u_1(t)\,(e^{-3(t-1)} - 1 + 3(t-1))$$

$$+ \frac{1}{2}(2(1 - e^{-3t}) - u_1(t)\,(1 - e^{-3(t-1)})$$

$$= \left(3t + \frac{8}{9} - \frac{8}{9}e^{-3t}\right)u_0(t) - u_1(t)\left(-\frac{8}{9}e^{-3(t-1)} + \frac{8}{9} + 3(t-1)\right)$$

$$= \frac{1}{9}((27t + 8 - 8e^{-3t})\,u_0(t) - (27(t-1)) + 8 - 8e^{-3(t-1)}) \times u_1(t)).$$

Dirac Delta Function

The idea of a very large force acting for a very short interval of time is of frequent occurrence in mechanics. For these cases, we use Dirac-Delta function (also called unit impulse function) which is considered the limiting form of the function

$$\delta_k(t-a) = \begin{cases} 0 & t < a \\ 1/k & a \le t \le a+k \\ 0 & t < a+k \end{cases} = \frac{1}{k}[u_a(t) - u_{a+k}(t)] \text{ as } k \to 0$$

So,

$$\delta(t-a) = \lim_{k\to 0} \delta_k(t-a)$$

$$= \lim_{k\to 0} \frac{1}{k}(u_a(t) - u_{a+k}(t))$$

Thus,

$$L[\delta(t-a)] = \lim_{k\to 0} \frac{1}{k}(L(u_a(t)) - L(u_{a+k}(t)))$$

$$= \lim_{k\to 0} \frac{1}{k}\left\{\frac{e^{-as}}{s} - \frac{e^{-(a+k)s}}{s}\right\}$$

$$= \lim_{k \to 0} \frac{e^{-as}}{s}\left(\frac{1-e^{-ks}}{k}\right)$$

$$= \frac{e^{-as}}{s}(s) = e^{-as}$$

$$\therefore \qquad L[\delta(t-a)] = e^{-as}.$$

EXAMPLE 3.17

Evaluate the integral $\int_0^\infty \cos t\,\delta\left(t-\frac{\pi}{4}\right)dt$.

Solution: We have $\int_0^\infty \cos t\,\delta\left(t-\frac{\pi}{4}\right)dt = \int_0^\infty \left(\frac{e^{it}-e^{-it}}{2}\right)\delta\left(t-\frac{\pi}{4}\right)dt$

$$= \frac{1}{2}\left(\int_0^\infty e^{it}\delta\left(t-\frac{\pi}{4}\right)dt - \int_0^\infty e^{-it}\delta\left(t-\frac{\pi}{4}\right)dt\right).$$

Now, by definition, we have

$$\int_0^\infty e^{-st}\delta\left(t-\frac{\pi}{4}\right)dt = L\left(\delta\left(t-\frac{\pi}{4}\right)\right) = e^{-\pi/4s}$$

Thus,

$$\int_0^\infty \cos t\,\delta\left(t-\frac{\pi}{4}\right)dt = \frac{1}{2}\{e^{-\pi/4(i)} - e^{-\pi/4(-i)}\} = \cos\left(-\frac{\pi}{4}\right) = \cos\frac{\pi}{4} = \frac{1}{\sqrt{2}}.$$

EXAMPLE 3.18

Find the solution of the initial value problem.

$$y'' + 2y' + 5y = \delta(t-2);\ y(0) = 0,\ y'(0) = 0.$$

Solution: Taking the Laplace transform of the differential equation, we obtain

$$s^2L(y) - sy(0) - y'(0) + 2(sL(y) - y(0)) + 5L(y) = L(\delta(t-2))$$

$$\Rightarrow \qquad (s^2+2s+5)Y(s) = e^{-2s}$$

$$\Rightarrow \qquad Y(s) = e^{-2s}\frac{1}{s^2+2s+5}.$$

Therefore,

$$y(t) = L^{-1}(Y(s)) = L^{-1}\left(e^{-2s}\frac{1}{s^2+2s+5}\right)$$

$$= u_2(t)\,f(t-2) = \frac{1}{2}u_2(t)\,e^{-(t-2)}\sin 2(t-2)$$

where
$$f(t) = L^{-1}\left(\frac{1}{s^2+2s+5}\right) = L^{-1}\left(\frac{1}{(s+1)^2+4}\right)$$
$$= e^{-t}L^{-1}\left(\frac{1}{s^2+4}\right) = \frac{e^{-t}\sin 2t}{2}.$$

EXAMPLE 3.19

Solve $y' + y\int_0^t y(\tau)d\tau = \delta(t) + u_{\pi/2}(t)$ with $y(0) = 0$.

Solution: Taking the Laplace transform of the given differential equation, we obtain

$$sL(y) - y(0) + y\frac{L(y)}{s} = L(\delta(t)) + L(u_{\pi/2}(t)).$$

$$\Rightarrow \qquad \left(s+\frac{y}{s}\right)Y(s) = e^{-os} + e^{-\pi s/2}$$

$$\Rightarrow \qquad Y(s) = \frac{s(1+e^{-\pi s/2})}{s^2+4}$$

Therefore,
$$y(t) = L^{-1}(Y(s)) = L^{-1}\left(\frac{s}{s^2+4} + e^{-\pi s/2}\frac{s}{s^2+4}\right)$$
$$= \cos 2t + u_{\pi/2}(t)\cos 2\left(t-\frac{\pi}{2}\right)$$
$$= \cos 2t + u_{\pi/2}(t)(-\cos 2t)$$
$$= \cos 2t\,(1 - u_{\pi/2}(t))$$
$$= \cos 2t\,(u_0(t) - u_{\pi/2}(t))$$

▲▲

EXERCISE 3.1

Find the Laplace transform of the following functions:

1. $e^{at}\cos bt$.
2. $e^{-at}t^2$.
3. $t^2u_3(t)$.
4. $\sinh t \sin t$.
5. $(t^2 - 1)u_2(t)$.
6. $(\sin t)u_\pi(t)$.
7. $f(t) = \begin{cases} 5; & 0 \le t < 2 \\ -5; & t \ge 2 \end{cases}$.

8. $f(t) = \begin{cases} t, & 0 \le t < 3 \\ 0, & t \ge 3 \end{cases}.$

9. $f(t) = \begin{cases} k, & 0 \le t < 2 \\ 0, & 2 \le t < 4 \\ k, & t \ge 4 \end{cases}.$

10. $f(t) = \begin{cases} t^2, & 0 \le t < 1 \\ 0, & t \ge 1 \end{cases}.$

Find the inverse Laplace transform of the following functions:

11. $\dfrac{1}{s^2 - 4s + 8}.$

12. $\dfrac{3s - 1}{(s-2)^2}.$

13. $\dfrac{e^{-3s}}{s+2}.$

14. $\dfrac{e^{-2s}}{s^2}.$

15. $\dfrac{(s+1)^2}{(s-2)^4}.$

16. $\dfrac{3s+5}{s^2 + 6s + 12}.$

17. $\dfrac{(1+e^{-\pi s})^2}{s+5}.$

18. $\dfrac{e^{-\pi s} - e^{-2\pi s}}{s^2 - 8s + 25}.$

19. Using unit-step function, find the Laplace transform of

$$f(t) = \begin{cases} \sin t, & 0 \le t < \pi \\ \sin 2t, & \pi \le t < 2\pi \\ \sin 3t, & t \ge 2\pi \end{cases}.$$

20. Using Laplace transform, evaluate:

$$\int_0^\infty e^{-t}(1 + 2t - t^2 + t^3)u_1(t)dt.$$

Solve the following initial value problems:

21. $y'' + 4y' + 4y = 12t^2 e^{-2t}$; $y(0) = 2$, $y'(0) = 1$.
22. $y'' + y = e^t \sin t$; $y(0) = 0$, $y'(0) = 0$.
23. $y' + \int_0^t y(\tau)d\tau = tu_1(t)$; $y(0) = 2$.
24. $y'' + 4y' = 8y = 16\delta(t-1); + 8\delta(t-2)$; $y(0) = 2$, $y'(0) = 0$.
25. $y'' + 4y = 8\delta(t - \pi/6)$; $y(0) = 2$, $y'(0) = 0$.
26. $y'' + y - f(t)$; $y(0) - 1$; $y'(0) - 0$,

    and where $f(t) = \begin{cases} 3, & 0 \le t \le 4 \\ 2t-5, & t > 4 \end{cases}$.

ANSWERS

Exercise 3.1

1. $\dfrac{s-a}{(s-a)^2 + b^2}$.
2. $\dfrac{2}{(s+a)^3}$.
3. $e^{-3s}\dfrac{(2+6s+9s^2)}{s^3}$.
4. $\dfrac{2s}{s^2+4}$.
5. $e^{-2s}(2 + 4s + 3s^2)/s^3$.
6. $-e^{-\pi s}/(s^2 + 1)$.
7. $5(1 - 2e^{-2s})/s$.
8. $\dfrac{1-e^{-3s}(1+3s)}{s^2}$.
9. $k(1 - e^{-2s} + e^{-ys})/s$.
10. $\dfrac{e^{-s} - e^{-2s}(1+s)}{s^2}$.
11. $\dfrac{1}{2}e^{2t} \sin 2t$.
12. $3e^{2t} + 5te^{2t}$.
13. $u_3(t)\, e^{-2(t-3)}$.
14. $(t-2)u_2(t)$.

15. $(2t + 6t^2 + 3t^3)e^{2t/2}$.

16. $(3\sqrt{3}\cos\sqrt{3}\,t - 4\sin\sqrt{3}\,t)e^{-3t/\sqrt{3}}$.

17. $e^{-5t}(u_0(t) + 2e^{5\pi}\, u_\pi(t) + e^{10\pi}\, u_{2\pi}(t))$.

18. $-\, e^{4t}\sin 3t\,(e^{-4\pi}\, u_\pi(t) + e^{-8\pi}\, u_{2\pi}(t))/3$.

19. $\dfrac{1}{s^2+1} + e^{-\pi s}\left(\dfrac{2}{s^2+4} - \dfrac{1}{s^2+1}\right) + e^{-2\pi s}\left(\dfrac{3}{s^2+9} - \dfrac{2}{s^2+4}\right)$.

20. $16/e$.

21. $y(t) = (2 + 5t + t^4)e^{-2t}$.

22. $y(t) = \dfrac{1}{5}((2\cos t + \sin t) + e^t(\sin t - 2\cos t))$.

23. $y(t) = 2\cos 2t\; u_0(t) + (1 + \cos 2(t-1) + 2\sin 2(t-1))\; u_1(t)/4$.

24. $y(t) = 8e^{-2(t-1)}\sin 2(t-1)\; u_1(t) + ye^{-2(t-2)}\sin 2(t-2)\; u_2(t)$.

25. $y(t) = 2\cos 2t + 4\sin 2(t - \pi/6)\; u_{\pi/6}(t)$.

26. $y(t) = 3 - 2\cos t + 2\,((t-4) - \sin(t-4))\; u_y(t)$. ▲▲

4

CHAPTER

Differentiation and Integration of Laplace Transforms

In the earlier chapters, we have discussed the solution of constant coefficient differential equations along with its initial conditions. The formulas of Laplace transforms discussed so for are not sufficient to solve the differential equation with variable coefficients in terms of the independent variable. Here, the concept of differentiation and integration of Laplace transforms helps us in the solution of such type of differential equations.

Differentiation of Laplace Transforms

Let $f(t)$ be piecewise continuous on $[0, \infty)$ and be of exponential order. Then,

$$L(tf(t)) = -F'(s) = -\frac{d}{ds}(L(f(t)))$$

and
$$L(t^n f(t)) = (-1)^n \frac{d^n}{ds^n} F(s) = (-1)^n \frac{d^n}{ds^n}(L(f(t)).$$

Proof: We have by using Leibnitz's formula with constant limits

$$\frac{dF}{ds} = \frac{d}{ds}(L(f(t))) = \frac{d}{ds}\left(\int_0^\infty e^{-st} f(t)dt\right)$$

$$= \int_0^\infty \frac{\partial}{\partial s}(e^{-st}) f(t)dt = \int_0^\infty e^{-st}(-tf(t))dt$$

$$= -L(tf(t))$$

Now,
$$L(t^2 f(t)) = L(t \cdot tf(t)) = -\frac{d}{ds}(L(tf(t)))$$

$$= -\frac{d}{ds}\left\{-\frac{d}{ds}L(f(t))\right\} = (-1)^2 \frac{d^2}{ds^2}(L(f(t)))$$

So, by induction, we can prove that

$$L(t^n f(t)) = (-1)^n \frac{d^n}{ds^n}(L(f(t))).$$

Note. We can have

$$L^{-1}(F'(s)) = -tf(t) = -tL^{-1}(F(s))$$

and
$$L^{-1}\left(\frac{d^n}{ds^n}F(s)\right) = (-1)^n t^n f(t)$$
$$= (-1)^n t^n L^{-1}(F(s)).$$

Integration of Laplace Transforms

Let $f(t)$ be piecewise continuous on $[0, \infty)$ and be of exponential order. Further, let $\lim_{t\to 0^+}\left(\frac{f(t)}{t}\right)$ exist.

Then,
$$L\left\{\frac{f(t)}{t}\right\} = \int_s^\infty L^{-1}(f(t))ds'$$
$$= \int_s^\infty F(s')ds'.$$

Proof: From the definition, we have

$$F(s) = L(f(t)) = \int_0^\infty e^{-st}f(t)dt$$

Integrating from s to ∞, we obtain

$$\int_s^\infty F(s')ds' = \int_s^\infty\left(\int_0^\infty e^{-s't}f(t)dt\right)ds'$$
$$= \int_0^\infty\left(\int_s^\infty e^{-s't}ds'\right)f(t)dt$$
$$= \int_0^\infty\left(\frac{e^{-s't}}{-t}\right)_s^\infty f(t)dt$$
$$= \int_0^\infty e^{-st}\left(\frac{f(t)}{t}\right)dt = L\left(\frac{f(t)}{t}\right)$$

Note. $L^{-1}\left(\int_s^\infty F(s')ds'\right) = \frac{f(t)}{t}$

$$\Rightarrow \quad L^{-1}\left|\int_s^\infty F(s')ds'\right| = \frac{1}{t}L^{-1}(F(s)).$$

▲▲

EXAMPLE 4.1

Find the Laplace transform of $t^2 \cos 3t$.

Solution: Since
$$L(\cos 3t) = \frac{s}{s^2+9}$$

$\therefore$
$$L(t^2 \cos 3t) = (-1)^2 \frac{d^2}{ds^2} L(\cos 3t)$$
$$= \frac{d^2}{ds^2}\left(\frac{s}{s^2+9}\right) = \frac{2s(s^2-27)}{(s^2+9)^4}.$$

EXAMPLE 4.2

Evaluate: $L(t^2e^{-2t})$.

Solution: Since
$$L(e^{-2t}) = \frac{1}{s+2}$$

$\therefore$
$$L(t^2e^{-2t}) - (-1)^2 \frac{d^2}{ds^2} L(e^{-2t})$$
$$= \frac{d^2}{ds^2}\left(\frac{1}{s+2}\right) = \frac{2}{(s+2)^3}.$$

EXAMPLE 4.3

Find the Laplace transform of $(\sin at)/t$.

Solution: We have
$$L\left(\frac{\sin at}{t}\right) = \int_s^\infty L(\sin at)ds$$
$$= \int_s^\infty \frac{a}{s^2+a^2} ds$$
$$= \left[\tan^{-1}\frac{a}{s}\right]_s^\infty = \frac{\pi}{2} - \tan^{-1}\frac{s}{a} = \cot^{-1}\frac{s}{a}.$$

EXAMPLE 4.4

Evaluate: $L(te^{-t} \sin 3t)$.

Solution: Since
$$L(\sin 3t) = \frac{3}{s^2+9},$$

therefore,
$$L(t \sin 3t) = -\frac{d}{ds}\left(\frac{3}{s^2+9}\right) = \frac{6s}{(s^2+9)^2}$$

Now using the First shifting theorem, we obtain

$$L(e^{-t}\, t \sin 3t) = \frac{6(s+1)}{((s+1)^2+9)^2}.$$

EXAMPLE 4.5

Find the Laplace transform of $\dfrac{\cos at - \cos bt}{t} + t \sin at$

Solution: Since $\quad L(\cos at - \cos bt) = \dfrac{s}{s^2+a^2} - \dfrac{s}{s^2+b^2}$

and $\quad L(\sin at) = \dfrac{a}{s^2+a^2}$

$$\therefore \quad L\left(\frac{\cos at - \cos bt}{t}\right) + L(t \sin at) = \int_s^{\infty}\left(\frac{s}{s^2+a^2} - \frac{s}{s^2+b^2}\right)ds - \frac{d}{ds}\left(\frac{a}{s^2+a^2}\right)$$

$$= \frac{1}{2}[\log(s^2+a^2) - \log(s^2+b^2)]_s^{\infty} - a - \frac{2s}{(s^2+a^2)^2}$$

$$= \frac{1}{2}\operatorname*{Lt}_{s\to\infty} \log\frac{s^2+a^2}{s^2+b^2} - \frac{1}{2}\log\frac{s^2+a^2}{s^2+b^2} + \frac{2as}{(s^2+a^2)^2}$$

$$= \frac{1}{2}\log\left(\frac{s^2+b^2}{s^2+a^2}\right) + \frac{2as}{(s^2+a^2)^2}.$$

EXAMPLE 4.6

Evaluate: $L\left(e^{-t}\int_0^t \dfrac{\sin t}{t}dt\right)$.

Solution: We have $\quad L\left(\dfrac{\sin t}{t}\right) = \displaystyle\int_s^{\infty} L(\sin t)dt$

$$= \int_s^{\infty}\frac{1}{s^2+1}ds = \left[\tan^{-1}s\right]_s^{\infty} = \frac{\pi}{2} - \tan^{-1}s = \cot^{-1}s$$

Thus, $\quad L\left(\displaystyle\int_0^t \frac{\sin t}{t}dt\right) = \dfrac{1}{s}L\left(\dfrac{\sin t}{t}\right) = \dfrac{1}{s}\cot^{-1}s$

So, $\quad L\left(e^{-t}\displaystyle\int_0^t \frac{\sin t}{t}dt\right) = \dfrac{1}{s+1}\cot^{-1}(s+1)$

(By using First shifting theorem).

EXAMPLE 4.7

Evaluate: $L\left(t\int_0^t \frac{e^{-t}\sin t}{t}dt\right)$.

Solution: Since
$$L\left(\frac{\sin t}{t}\right) = \cot^{-1}s$$

∴
$$L\left(e^{-t}\frac{\sin t}{t}\right) = \cot^{-1}(s+1)$$

(By using First shifting theorem)

Thus,
$$L\left(\int_0^t e^{-t}\frac{\sin t}{t}dt\right) = \frac{1}{s}\cot^{-1}(s+1)$$

Hence,
$$L\left(t\int_0^t e^{-t}\frac{\sin t}{t}dt\right) = -\frac{d}{ds}\left(\frac{1}{s}\cot^{-1}(s+1)\right)$$

$$= -\frac{s\cdot\dfrac{-1}{1+(s+1)^2} - \cot^{-1}(s+1)\cdot 1}{s^2}$$

$$= \frac{s+(s^2+2s+2)\cot^{-1}(s+1)}{s^2(s^2+2s+2)}.$$

EXAMPLE 4.8

Evaluate: $\int_0^\infty te^{-2t}\sin t\,dt$.

Solution: Now,
$$L(t\sin t) = -\frac{d}{ds}L(\sin t) = -\frac{d}{ds}\left(\frac{1}{s^2+1}\right) = \frac{2s}{(s^2+1)^2}$$

∴
$$\int_0^\infty e^{-st}t\sin t\,dt = \frac{2s}{(s^2+1)^2}$$

Let
$$s = 2$$

$$\int_0^\infty e^{-2t}t\sin t\,dt = \frac{2(2)}{((2)^2+1)^2} = \frac{4}{25}.$$

EXAMPLE 4.9

Evaluate: $\int_0^\infty e^{-t}\left(\frac{\cos at - \cos bt}{t}\right)dt.$

Solution: Now,

$$L(\cos at - \cos bt) = \frac{s}{s^2+a^2} - \frac{s}{s^2+b^2}$$

$$\therefore \quad L\left(\frac{\cos at - \cos bt}{t}\right) = \int_s^\infty \left(\frac{s}{s^2+a^2} - \frac{s}{s^2+b^2}\right)ds$$

$$= \frac{1}{2}\log\left(\frac{s^2+b^2}{s^2+a^2}\right)$$

Thus,

$$\int_0^\infty e^{-st}\left(\frac{\cos at - \cos bt}{t}\right)dt = \frac{1}{2}\log\left(\frac{s^2+b^2}{s^2+a^2}\right)$$

Let $s = 1$

$$\therefore \quad \int_0^\infty e^{-t}\left(\frac{\cos at - \cos bt}{t}\right)dt = \frac{1}{2}\log\frac{1+b^2}{1+a^2}.$$

EXAMPLE 4.10

Find the inverse Laplace transform of $\frac{2(s+1)}{(s^2+2s+2)^2}$.

Solution: Let $L^{-1}\left(\frac{2(s+1)}{(s^2+2s+2)^2}\right) = f(t)$

$$\therefore \quad \frac{f(t)}{t} = L^{-1}\left(\int_s^\infty \frac{2(s+1)}{(s^2+2s+2)^2}ds\right)$$

$$= L^{-1}\left(\left[-\frac{1}{s^2+2s+2}\right]_s^\infty\right)$$

$$L\left(\frac{f(t)}{t}\right) = \lim_{s\to\infty} -\frac{1}{s^2+2s+2} - \frac{-1}{s^2+2s+2} = \frac{1}{s^2+2s+2}$$

$$\therefore \quad \frac{f(t)}{t} = L^{-1}\left(\frac{1}{(s+1)^2+1}\right) = e^{-t}L^{-1}\left(\frac{1}{s^2+1}\right)$$

$$= e^{-t}\sin t$$

Hence,

$$f(t) = e^{-t}\, t\sin t.$$

EXAMPLE 4.11

Evaluate: $L^{-1}\left(\dfrac{1}{(s+5)^4}\right)$.

Solution: We have

$$\frac{d^3}{ds^3}\left(\frac{1}{s+5}\right) = -\frac{\frac{-1}{2}\cdot\frac{-1}{3}}{(s+5)^4}$$

$$= -\frac{1}{6(s+5)^4}$$

Thus,

$$L^{-1}\left(\frac{1}{(s+5)^4}\right) = L^{-1}\left(-6\frac{d^3}{ds^3}\left(\frac{1}{s+5}\right)\right)$$

$$= -6 \times (-1)^3\, t^3\, L^{-1}\left(\frac{1}{s+5}\right)$$

$$= 6t^3\, e^{-5t}.$$

EXAMPLE 4.12

Find the inverse Laplace transform of $\log \dfrac{s+a}{s+b}$; a, b constants.

Solution: Let $L^{-1}(\log(s + a) - \log(s + b) = f(t)$

$$\therefore \qquad -tf(t) = L^{-1}\left(\frac{d}{ds}(\log(s+a) - \log(s+b)\right)$$

$$= L^{-1}\left(\frac{1}{s+a} - \frac{1}{s+b}\right) = e^{-at} - e^{-bt}$$

Hence,

$$f(t) = \frac{e^{-bt} - e^{-at}}{t}$$

EXAMPLE 4.13

Evaluate: $L^{-1}\left(\cot^{-1}\left(\dfrac{s}{2}\right)\right)$.

Solution: Let $L^{-1}\left(\cot^{-1}\left(\dfrac{s}{2}\right)\right) = f(t)$

$$\therefore \qquad -tf(t) = L^{-1}\left(\frac{d}{ds}\left(\cot^{-1}\left(\frac{s}{2}\right)\right)\right)$$

$$= L^{-1}\left(\frac{-\frac{1}{2}}{1+\left(\frac{s}{2}\right)^2}\right) = L^{-1}\left(-\frac{2}{s^2+4}\right) = -\sin 2t$$

Thus,

$$f(t) = \frac{\sin 2t}{t}.$$

EXAMPLE 4.14

Find the inverse Laplace transform of $\tan^{-1}\left(\frac{2}{s^2}\right)$.

Solution: Let $L^{-1}\left(\tan^{-1}\left(\frac{2}{s^2}\right)\right) = f(t)$

$$\therefore \qquad -tf(t) = L^{-1}\left(\frac{d}{ds}\left(\tan^{-1}\left(\frac{2}{s^2}\right)\right)\right)$$

$$= L^{-1}\left(\frac{1}{1+\frac{4}{s^4}} \times -\frac{4}{s^3}\right) = L^{-1}\left(-\frac{4s}{s^4+4}\right)$$

$$= -L^{-1}\left(\frac{4s}{(s^2+2)^2-(2s)^2}\right) = L^{-1}\left(\frac{4s}{(s^2+2+2s)(s^2+2-2s)}\right)$$

$$= -L^{-1}\left(-\frac{1}{s^2+2s+2} + \frac{1}{s^2-2s+2}\right)$$

$$= L^{-1}\left(\frac{1}{(s+1)^2+1} - \frac{1}{(s-1)^2+1}\right)$$

$$= e^{-t} L^{-1}\left(\frac{1}{s^2+1}\right) - e^{t} L^{-1}\left(\frac{1}{s^2+1}\right)$$

$$= (e^{-t} - e^{t}) \sin t$$

Hence,

$$f(t) = (e^{t} - e^{-t}) \frac{\sin t}{t} = \frac{2 \sinh t \cdot \sin t}{t}$$

EXAMPLE 4.15

Find the solution of the initial value problem $y'' + ty' - 2y = 6 - t$; $y(0) = 0$, $y'(0) = 1$ given that $L(y(t))$ exists.

Solution: Talking the Laplace transform of the given differential equation, we get

$$L(y'') + L(ty') - 2L(y) = L(6 - t)$$

$$\Rightarrow \qquad L(y'') - \frac{d}{ds}L(y') - 2L(y) = L(6) - L(t)$$

$$\Rightarrow \qquad (s^2L(y) - sy(0) - y'(0)) - \frac{d}{ds}(sL(y) - y(0)) - 2L(y) = \frac{6}{s} - \frac{1}{s^2}$$

$$\Rightarrow \qquad (s^2L(y) - 1) - \left(1 \cdot L(y) + s\frac{d}{ds}L(y)\right) - 2L(y) = \frac{6}{s} - \frac{1}{s^2}$$

$$\Rightarrow \qquad -s\frac{dY}{ds} + (-3 + s^2)Y(s) = \frac{6}{s} - \frac{1}{s^2} + 1$$

$$\Rightarrow \qquad \frac{dY}{ds} + \left(-s + \frac{3}{s}\right)Y = -\left(\frac{1}{s} + \frac{6}{s^2} - \frac{1}{s^3}\right)$$

which is a first order linear ordinary differential equation with IF

$$= e^{\int\left(\frac{3}{s} - s\right)ds}$$

$$= e^{3\log s - \frac{s^2}{2}} = s^3e^{-s^2/2}$$

Hence, we obtain

$$s^3e^{-s^2/2}Y(s) = \int e^{-s^2/2}ds - 6\int se^{-s^2/2}ds - \int s^2e^{-s^2/2}ds$$

$$= \int e^{-s^2/2}ds + 6e^{-s^2/2} - \left[-se^{-s^2/2} - \int -1 \cdot e^{-s^2/2}ds\right]$$

$$= (6 + s)e^{-s^2/2} + c$$

Therefore, $\qquad Y(s) = \dfrac{6}{s^3} + \dfrac{1}{s^2} + \dfrac{c}{s^3}e^{s^2/2}$

Since $L(y(t))$ exists, therefore, $y(t)$ satisfies the conditions of existence theorem that is $\lim\limits_{s\to\infty} Y(s) = 0$

$\Rightarrow \quad C = 0$

Thus, $\qquad Y(s) = \dfrac{6}{s^3} + \dfrac{1}{s^2}$

$$\therefore \qquad y(t) = L^{-1}(Y(s)) = L^{-1}\left(\frac{6}{s^3} + \frac{1}{s^2}\right)$$

$$= 3t^2 + t$$

which is the solution of the given initial value problem.

EXAMPLE 4.16

Solve $ty'' + (6t - 2)y' - 6y = 0$; $y(0) = 1$ and $y'(0)$ is an arbitrary constant.

Solution: Taking Laplace transform of the given equation, we obtain

$$L(ty'') + 6L(ty') - 2L(y') - 6L(y) = 0$$

$$\Rightarrow \quad -\frac{d}{ds}L(y'') - 6\frac{d}{ds}L(y') - 2L(y') - 6L(y) = 0$$

$$\Rightarrow \quad -\frac{d}{ds}[s^2L(y) - sy(0) - y'(0)] - 6\frac{d}{ds}[sL(y) - y(0)] - 2[sL(y) - y(0)] - 6L(y) = 0$$

$$\Rightarrow \quad -\frac{d}{ds}[s^2Y(s) - s - y'(0)] - 6\frac{d}{ds}[sY(s) - 1] - 2[sY(s) - 1] - 6Y(s) = 0$$

$$\Rightarrow \quad -\left[2sY + s^2\frac{dY}{ds} - 1\right] - 6\left[1 \cdot Y + s\frac{dY}{ds} - 0\right] - 2[sY - 1] - 6Y = 0$$

{$\because$ $y'(0)$ is a constant]

$$\Rightarrow \quad -(s^2 + 6s)\frac{dY}{ds} - (Ys + 12)Y = 3$$

$$\Rightarrow \quad \frac{dY}{ds} + \frac{s+3}{s(s+6)}Y = \frac{3}{s(s+6)}$$

which is a first order linear ordinary differential equation with IF

$$= e^{4\int \frac{s+6-3}{s(s+6)}ds} = e^{4\int \left(\frac{1}{s} - \frac{3}{s(s+6)}\right)ds}$$

$$= e^{4\int \left[\frac{1}{s} - \frac{3}{6}\left(\frac{1}{s} - \frac{1}{s+6}\right)\right]ds}$$

$$= e^{4\int \left[\frac{1}{2s} + \frac{1}{2(s+6)}\right]ds}$$

$$= e^{2[\ln s + \ln(s+6)]} = s^2(s+6)^2$$

Therefore,

$$s^2(s+6)^2Y = \int 3s(s+6)\,ds + c$$

$$= 3\left[\frac{s^3}{3} + 6\frac{s^2}{2}\right] + c$$

$$= s^3 + 9s^2 + c$$

$$= s^2(s+9) + c$$

Thus, $$Y = \frac{s+9}{(s+6)^2} + \frac{c}{s^2(s+6)^2}$$

Hence, $$y(t) = L^{-1}(Y(s)) = L^{-1}\left(\frac{\overline{s+6}+3}{(s+6)^2} + c\left(\frac{1}{s(s+6)}\right)^2\right)$$

$$= L^{-1}\left(\frac{1}{s+6} + \frac{3}{(s+6)^2} + \frac{c}{36}\left(\frac{1}{s} - \frac{1}{s+6}\right)^2\right)$$

$$= e^{-6t} + 3L^{-1}\left(-\frac{d}{ds}\left(\frac{1}{s+6}\right)\right) + \frac{c}{36}L^{-1}\left(\frac{1}{s^2} + \frac{1}{(s+6)^2} - \frac{2}{s(s+6)}\right)$$

$$= e^{-6t} + 3t\,L^{-1}\left(\frac{1}{s+6}\right) + \frac{c}{36}\left[t + L^{-1}\left(-\frac{d}{ds}\left(\frac{1}{s+6}\right)\right) - \frac{2}{6}L^{-1}\left(\frac{1}{s} - \frac{1}{s+6}\right)\right]$$

$$= e^{-6t} + 3t\,e^{-6t} + \frac{c}{36}\left[t + te^{-6t} - \frac{1}{3}(1 - e^{-6t})\right]$$

$$= (1+3t)e^{-6t} + \frac{c}{108}(-1 + 3t + (1+3t)e^{-6t})$$

where c is an arbitrary constant.

EXAMPLE 4.17

Solve $ty'' + (8t - 2)y' - 8y = 0$ with $y(0) = 2$ and $y'(0)$ an arbitrary constant.

Solution: Taking the Laplace transform of the given differential equation, we obtain

$$L(ty'') + 8L(ty') - 2L(y') - 8L(y) = 0$$

$$\Rightarrow \quad -\frac{d}{ds}L(y'') - 8\frac{d}{ds}L(y') - 2L(y') - 8L(y) = 0$$

$$\Rightarrow \quad -\frac{d}{ds}(s^2L(y) - sy(0) - y'(0)) - 8\frac{d}{ds}(sL(y) - y(0)) - 2(sL(y) - y(0)) - 8L(y) = 0$$

$$\Rightarrow \quad -\frac{d}{ds}(s^2Y - 2s - y'(0)) - 8\frac{d}{ds}(sY - 2) - 2(sY - 2) - 8Y = 0$$

$$\Rightarrow \quad -\left(2sY+s^2\frac{dY}{ds}-2\right)-8\left(1\cdot Y+s\frac{dY}{ds}-0\right)-2(sY-2)-8Y = 0$$

$\{\because y'(0)$ is a constant$\}$

$$\Rightarrow \quad -(s^2+8s)\frac{dY}{ds}-(4s+16)Y = -6$$

$$\Rightarrow \quad \frac{dY}{ds}+\frac{Y(s+4)}{s(s+8)}Y = \frac{6}{s(s+8)}$$

which is a first order ordinary linear differential equation with IF

$$= e^{4\int\frac{\overline{s+8}-4}{s(s+8)}ds} = e^{4\int\left(\frac{1}{s}-\frac{4}{8}\left(\frac{1}{s}-\frac{1}{s+8}\right)\right)ds}$$

$$= e^{4\int\left(\frac{1}{2s}+\frac{1}{2(s+8)}\right)ds}$$

$$= e^{2(\ln s+\ln(s+8))} = s^2(s+8)^2$$

Therefore $\quad s^2(s+8)^2Y = 16\int s(s+8)\,ds+c$

$$= 16\left[\frac{s^3}{3}+\frac{8s^2}{2}\right]+c$$

$$= \frac{16}{3}s^2(s+12)+c$$

$$\Rightarrow \quad Y = \frac{16}{3}\frac{s+12}{(s+8)^2}+\frac{c}{s^2(s+8)^2}$$

Hence, $y(t) = L^{-1}(Y(s)) = \frac{16}{3}L^{-1}\left(\frac{\overline{s+8}+4}{(s+8)^2}\right)+\frac{c}{64}L^{-1}\left(\left(\frac{1}{s}-\frac{1}{s+8}\right)^2\right)$

$$= \frac{16}{3}L^{-1}\left(\frac{1}{s+8}+4\frac{d}{ds}\left(\frac{-1}{s+8}\right)\right)+\frac{c}{64}L^{-1}\left(\frac{1}{s^2}+\frac{1}{(s+8)^2}-\frac{2}{s(s+8)}\right)$$

$$= \frac{16}{3}\left(e^{-8t}+4tL^{-1}\left(\frac{1}{s+8}\right)\right)+\frac{c}{64}\left(t+c^{-1}\left(\frac{d}{ds}\left(\frac{-1}{s+8}\right)\right)-\frac{2}{8}L^{-1}\left(\frac{1}{s}-\frac{1}{s+8}\right)\right)$$

$$= \frac{16}{3}\left(e^{-8t}+4te^{-8t}\right)+\frac{c}{64}\left(t+te^{-8t}-\frac{1}{4}\left(1-e^{-8t}\right)\right)$$

$$= \frac{16}{3}(1+4t)e^{-8t}+\frac{c}{256}\left(-1+4t+(1+4t)e^{-8t}\right)$$

where c is an arbitrary constant.

EXAMPLE 4.18

Find the solution of the initial value problem $ty'' + 2ty' + 2y = 2$; $y(0) = 1$ and $y'(0)$ is an arbitrary constant.

Solution: Taking Laplace transform of the given differential equation, we obtain

$$L(ty'') + 2L(ty') + 2L(y) = L(2)$$

$$\Rightarrow \quad -\frac{d}{ds}L(y'') - 2\frac{d}{ds}L(y') + 2L(y) = \frac{2}{s}$$

$$\Rightarrow \quad -\frac{d}{ds}(s^2L(y) - sy(0) - y'(0)) - 2\frac{d}{ds}(sL(y) - y(0)) + 2L(y) = \frac{2}{s}$$

$$\Rightarrow \quad -\frac{d}{ds}(s^2Y - s - y'(0)) - 2\frac{d}{ds}(sY - 1) + 2Y = \frac{2}{s}$$

$$\Rightarrow \quad -\left(2sY + s^2\frac{dY}{ds} - 1\right) - 2\left(1 \cdot Y + s\frac{dY}{ds}\right) + 2Y = \frac{2}{s}$$

{$\because$ $y'(0)$ is an arbitrary constant}

$$\Rightarrow \quad -(s^2 + 2s)\frac{dY}{ds} - (2s)Y = \frac{2}{s} - 1$$

$$\Rightarrow \quad \frac{dY}{ds} + \left(\frac{2}{s+2}\right)Y = \frac{1}{s(s+2)} - \frac{2}{s^2(s+2)} = \frac{s-2}{s^2(s+2)}$$

which is a first order ordinary linear differential equation with

$$\text{IF} = e^{\int \frac{2}{s+2}ds}$$

$$= e^{2\ln(s+2)} = (s+2)^2$$

Therefore,

$$(s+2)^2Y = \int \frac{(s+2)(s-2)}{s^2}ds + c$$

$$= \int \left(\frac{s^2-4}{s^2}\right)ds + c$$

$$= \int \left(1 - \frac{4}{s^2}\right)ds + c$$

$$= s + \frac{4}{s} + c$$

$$\Rightarrow \qquad Y(s) = \frac{s}{(s+2)^2} + \frac{4}{s(s+2)^2} + \frac{c}{(s+2)^2}$$

$$= \frac{\overline{s+2}-2}{(s+2)^2} + \frac{4}{s(s+2)^2} + \frac{c}{(s+2)^2}$$

$$= \frac{1}{s+2} - \frac{2}{(s+2)^2} + \frac{4}{s(s+2)^2} + \frac{c}{(s+2)^2}$$

$$= \frac{1}{s+2} + \frac{4}{s(s+2)^2} + \frac{c-2}{(s+2)^2}$$

Hence, $y(t) = L^{-1}(Y(s)) = L^{-1}\left(\frac{1}{s+2} + \frac{4}{s(s+2)^2} + \frac{c-2}{(s+2)^2}\right)$

$$= e^{-2t} + 4\int_0^t L^{-1}\left(\frac{1}{(s+2)^2}\right)dt + (c-2)L^{-1}\left(-\frac{d}{ds}\left(\frac{1}{s+2}\right)\right)$$

$$= e^{-2t} + 4\int_0^t L^{-1}\left(-\frac{d}{ds}\left(\frac{1}{s+2}\right)\right)dt + (c-2) + tL^{-1}\left(\frac{1}{s+2}\right)$$

$$= e^{-2t} + 4\int_0^t t\,L^{-1}\left(\frac{1}{s+2}\right)dt + (c-2)te^{-2t}$$

$$= e^{-2t} + 4\int_0^t te^{-2t}dt + (c-2)te^{-2t}$$

$$= e^{-2t} + 4\left[t\cdot\frac{e^{-2t}}{-2} - \frac{e^{-2t}}{4}\right]_0^t + (c-2)te^{-2t}$$

$$= e^{-2t} + 4\left[-\frac{te^{-2t}}{2} - \frac{e^{-2t}}{4} + \frac{1}{4}\right] + (c-2)te^{-2t}$$

$$= 1 + e^{-2t}(c-4)t$$

where c is an arbitrary constant.

EXAMPLE 4.19

Solve $ty'' + 2y' + ty = 0$ with $y(0) = 1$, $y(\infty) = 0$ and $y'(0)$ an arbitrary constant.

Solution: Taking the Laplace transform of the given differential equation, we obtain

$$L(ty'') + 2L(y') + L(ty) = 0$$

$$\Rightarrow \quad -\frac{d}{ds}L(y'') + 2L(y') - \frac{d}{ds}L(y) = 0$$

$$\Rightarrow \quad -\frac{d}{ds}(s^2L(y) - sy(0) - y'(0)) + 2(sL(y) - y(0)) - \frac{d}{ds}L(y) = 0$$

$$\Rightarrow \quad -\frac{d}{ds}(s^2Y - s - y'(0)) + 2(sY - 1) - \frac{dY}{ds} = 0$$

$$\Rightarrow \quad -\left(2sY + s^2\frac{dY}{ds} - 1\right) + 2(sY - 1) - \frac{dY}{ds} = 0$$

$\{\because y'(0)$ is a constant$\}$

$$\Rightarrow \quad -(s^2 + 1)\frac{dY}{ds} - 1 = 0$$

$$\Rightarrow \quad \frac{dY}{ds} = \frac{-1}{1+s^2}$$

Integrating, we get

$$Y = -\tan^{-1}s + c$$

Since $y(\infty) = 0 \Rightarrow L(y(\infty)) = Y(\infty) = 0$

$$\Rightarrow c = \frac{\pi}{2}$$

$$\therefore Y = -\tan^{-1}s + \frac{\pi}{2} = \cot^{-1}s$$

Thus, $\quad y(t) = L^{-1}(Y) = L^{-1}(\cot^{-1}s)$

let $\quad L^{-1}(\cot^{-1}s) = f(t)$

$$\Rightarrow \quad -tf(t) = L^{-1}\left(\frac{d}{ds}(\cot^{-1}s)\right)$$

$$= L^{-1}\left(\frac{-1}{1+s^2}\right) = -\sin t$$

$$\Rightarrow \quad f(t) = \frac{\sin t}{t}$$

Hence, $$y(t) = \frac{\sin t}{t}$$

which is the required solution of the given boundary value problem.

EXERCISE 4.1

Find the Laplace transform of the following functions:

1. $t \sin 4t$.
2. te^{5t}.
3. $t^2 \sinh 3t$.
4. $\sinh t/t$.
5. $\frac{1-\cos bt}{t}$.
6. $te^{-4t} \sin 3t$.
7. $t^2 \int_0^t e^{-5\tau} d\tau$.
8. $\frac{e^{-2t} \sin 3t}{t}$.
9. $t \int_0^t e^{-2\tau} \cos 3\tau d\tau$.
10. $t \int_0^t e^{-\tau} \sin 2\tau d\tau$.

Find the inverse Laplace transform of the following functions:

11. $\frac{1}{(s^2+9)^2}$.
12. $\frac{s}{(s^2-9)^2}$.
13. $\frac{8(s+2)}{(s^2+4s+8)^2}$.
14. $\frac{6s+5}{(s^2+2s+2)^2}$.
15. $\log\left(\frac{s^2+1}{s^2}\right)$.

16. $\log\left(\dfrac{s^2+1}{s(s+1)}\right)$.

17. $\dfrac{1}{s}\tan^{-1}\left(\dfrac{1}{s}\right)$.

18. $\tanh^{-1}\left(\dfrac{1}{s}\right)$.

19. $\dfrac{s}{(s+2)(s^2+9)}$.

20. $\dfrac{1}{(s^2+4s+13)^2}$.

Find the solution of the following initial value problems:

21. $ty' - 2y = 6$.
22. $ly' - 3y = 2t$.
23. $y'' + 6ty' - 30y = 0$, $y(0) = 0$, $y'(0) = 2$.
24. $ty'' + 2y' + ty = \cos t$; $y(0) = 1$.
25. $ty'' + (1 - 2t)y' - 2y = 0$; $y(0) = 1$, $y'(0) = 2$.

ANSWERS

Exercise 4.1

1. $\dfrac{8s}{(s^2+16)^2}$.
2. $\dfrac{1}{(s-5)^2}$.
3. $\dfrac{18(s^2+3)}{(s^2-9)^3}$.
4. $\dfrac{1}{2}\log\left(\dfrac{s+1}{s-1}\right)$.
5. $\log\left(\dfrac{\sqrt{s^2+b}}{s}\right)$.
6. $\dfrac{6(s+4)}{(s^2+8s+25)^2}$.
7. $\dfrac{2(3s^2+15s+25)}{s^3(s+5)^3}$.

8. $\cot^{-1}\left(\frac{s+2}{3}\right)$.

9. $\frac{2s^3+5s^2+8s+13}{s^2(s^2+4s+13)^2}$.

10. $\frac{2(3s^2+4s+5)}{s^2(s^2+2s+5)^2}$.

11. $\frac{1}{18}\left(\frac{1}{3}\sin 3t - t\cos 3t\right)$.

12. $\frac{1}{6}t \sinh 3t$.

13. $2e^{-2t} \sin 2t$.

14. $3te^{-t} \sin t - (e^{-t} \sin t - te^{-t} \cos t)/2$.

15. $2(1 - \cos t)/t$.

16. $\frac{1+e^{-t}-2\cos t}{t}$.

17. $\int_0^t \frac{\sin\tau}{\tau} d\tau$.

18. $\frac{\sinh t}{t}$.

19. $\frac{1}{13}(3 \sin 3t + 2 \cos 2t - 2e^{-2t})$.

20. $\frac{e^{-2t}}{54}(\sin 3t - 3t \cos 3t)$.

21. $y(t) = -3 + ct^{2/2}$.

22. $y(t) = -t + \frac{ct^3}{6}$.

23. $y(t) = 2(5t + 20t^3 + 12t^5)/5$.

24. $y(t) = \frac{1}{2}\left(1+\frac{2}{t}\right)\sin t$.

25. $y = e^{2t}$.

▲▲

5

CHAPTER

Convolution Theorem and Laplace Transform of Periodic Functions

In this chapter, we will learn about convolution theorem which is used for finding inverse Laplace transforms of product of two functions. First, we define convolution and then prove convolution theorem and use it to find inverse Laplace transform of some functions.

Definition: Let $f(t)$ and $g(t)$ be defined on $[0, \infty)$. Then, the convolution of $f(t)$ and $g(t)$ is defined by

$$(f * g)(t) = \int_0^t f(\tau)\, g(t - \tau)d\tau, \quad (t \geq 0)$$

Convolution satisfies the following relations.

(a) $f * g = g * f$
(b) $f * (g_1 + g_2) = f * g_1 + f * g_2$
(c) $(f * g) * h = f * (g * h)$
(d) $f * 0 = 0 = 0 * f$

Note: $f * 1 \neq f$

For $f(t) = t$, we have

$$(1 * f)(t) = \int_0^t 1(t - \tau)d\tau = \frac{t^2}{2} \neq f(t).$$

▲▲

EXAMPLE 5.1

Find the convolution of sin ωt and cos ωt.

Solution: We have

$$\sin \omega t * \cos \omega t = \int_0^t \sin (\omega\tau) \cos \omega(t - \tau)d\tau$$

$$= \frac{1}{2}\int_0^t [\sin \omega t + \sin \omega(2\tau - t)]d\tau$$

$$= \frac{1}{2}\left[\tau \sin \omega t - \frac{1}{2w}\cos \omega(2\tau - t)\right]_0^t$$

$$= \frac{1}{2}\left\{t \sin \omega t - \frac{1}{2w}\cos \omega t + \frac{1}{2\omega}\cos \omega t\right\}$$

$$= \frac{1}{2} t \sin \omega t.$$

EXAMPLE 5.2

Evaluate: $1 * e^{-2t}$.

Solution: We have

$$1 * e^{-2t} = \int_0^t 1 \cdot e^{-2(t-\tau)} dz$$

$$= \left[\frac{e^{-2(t-\tau)}}{2}\right]_0^t$$

$$= \frac{1}{2}(1 - e^{-2t}).$$

EXAMPLE 5.3

Verify $t * e^{at} = e^{at} * t$.

Solution: We have

$$t * e^{at} = \int_0^t \tau e^{a(t-\tau)} d\tau$$

$$= \left[\tau \frac{e^{a(t-\tau)}}{-a} - 1 \cdot \frac{e^{a(t-\tau)}}{(-a)^2}\right]_0^t$$

$$= -\frac{t}{a} - \frac{1-e^{at}}{a^2} = \frac{e^{at} - 1 - at}{a^2}$$

Now, $$e^{at} * t = \int_0^t e^{a\tau}(t - \tau) d\tau$$

$$= \left[(t-\tau)\frac{e^{a\tau}}{a} - (-1)\frac{e^{a\tau}}{a^2}\right]_0^t$$

$$= 0 - \frac{t}{a} + \frac{e^{at} - 1}{a^2} = \frac{e^{at} - 1 - at}{a^2}$$

Thus, $$t * e^{at} = e^{at} * t.$$

We now prove the convolution theorem.

Convolution theorem: Let $f(t)$, $g(t)$ be piecewise continuous functions on $[0, \infty)$ and be of exponential order. Then

$$L((f * g)(t)) = L(f(t))\, L(g(t)) = F(s) \cdot G(s)$$

Proof: We have

$$F(s)\, G(s) = L(f(t))\, L(g(t))$$

$$= \left(\int_0^\infty e^{-su} f(u)du\right)\left(\int_0^\infty e^{-sv} g(v)dv\right)$$

$$= \int_0^\infty \int_0^\infty e^{-s(u+v)} f(u)\, g(v)dudv$$

$$= \int_0^\infty f(u)\left(\int_0^\infty e^{-s(u+v)} g(v)dv\right)du$$

Let $u + v = t$ where u is fixed.

Therefore, we obtain

$$F(s)G(s) = \int_0^\infty f(u)\left(\int_u^\infty e^{-st} g(t-u)dt\right)du$$

The region of integration is $0 \le u < \infty;\ u \le t < \infty$.

Changing the order of integration, we obtain the region of integration as

$$0 \le u \le t;\ 0 \le t < \infty$$

Therefore,

$$F(s)G(s) = \int_0^\infty e^{-st}\left(\int_0^t f(u)g(t-u)du\right)dt$$

$$= \int_0^\infty e^{-st}\, ((f * g)(u))dt = L((f * g)(t))$$

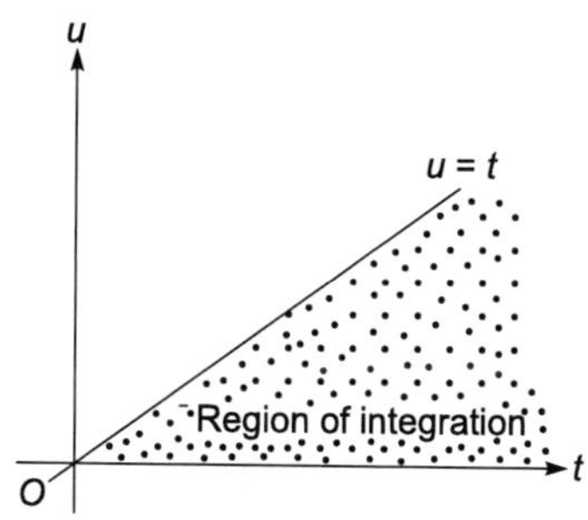

Note. The inverse Laplace transform of product of two functions $F(s)$ and $G(s)$ is given by

$$\begin{aligned} L^{-1}(F(s)\,G(s)) &= (f * g)\,(t) \\ &= L^{-1}(F(s)) * L^{-1}(G(s)). \end{aligned}$$

EXAMPLE 5.4

Find the inverse Laplace transform of $\dfrac{1}{(s^2+a^2)^2}$.

Solution: We have
$$\frac{1}{(s^2+a^2)^2} = \frac{1}{s^2+a^2}\cdot\frac{1}{s^2+a^2} = F(s)\cdot G(s)$$

Now,
$$f(t) = g(t) = L^{-1}\left(\frac{1}{s^2+a^2}\right) = \frac{1}{a}\sin at$$

Therefore,
$$\begin{aligned} L^{-1}(F(s)\,G(s)) &= (f * g)\,(t) \\ &= \int_0^t \frac{1}{a}\sin a\tau\cdot\frac{1}{a}\sin a(t-\tau)\,d\tau \\ &= \frac{1}{2a^2}\int_0^t [\cos a(2\tau - t) - \cos at]\,d\tau \\ &= \frac{1}{2a^2}\left[\frac{\sin a(2\tau - t)}{2a} - \tau\cos at\right]_0^t \\ &= \frac{1}{2a^2}\left(\frac{\sin at - \sin(-at)}{2a} - t\cos at\right) \\ &= \frac{1}{2a^3}(\sin at - at\cos at) \end{aligned}$$

EXAMPLE 5.5

Apply convolution theorem to evaluate
$$L^{-1}\left(\frac{s^2}{(s^2+a^2)(s^2+b^2)}\right).$$

Solution: Since
$$f(t) = L^{-1}\left(\frac{s}{s^2+a^2}\right) = \cos at$$

and
$$g(t) = L^{-1}\left(\frac{s}{s^2+b^2}\right) = \cos bt$$

By convolution theorem, we have

$$L^{-1}\left(\frac{s}{s^2+a^2}\cdot\frac{s}{s^2+b^2}\right)=\int_0^t \cos a\tau \cos b(t-\tau)d\tau$$

$$=\frac{1}{2}\int_0^t [\cos((a-b)\tau+bt)+\cos((a+b)\tau-bt)]d\tau$$

$$=\frac{1}{2}\left[\frac{\sin((a-b)\tau+bt)}{a-b}+\frac{\sin((a+b)\tau-bt)}{a+b}\right]_0^t$$

$$=\frac{1}{2}\left(\frac{\sin at-\sin bt}{a-b}+\frac{\sin at-\sin(-bt)}{a+b}\right)$$

$$=\frac{a\sin at-b\sin bt}{a^2-b^2}.$$

EXAMPLE 5.6

Evaluate: $L^{-1}\left(\dfrac{s}{(s^2+1)(s^2+4)(s^2+9)}\right)$.

Solution: Here, $L^{-1}\left(\dfrac{1}{s^2+4}\right)=\cos 2t,\ g(t)=L^{-1}\left(\dfrac{1}{s^2+1}\right)=\sin t$

and $h(t)=L^{-1}\left(\dfrac{1}{s^2+9}\right)=\dfrac{1}{3}\sin 3t$

Now,

$$L^{-1}\left(\frac{1}{(s^2+1)(s^2+9)}\right)=\int_0^t \sin\tau-\frac{1}{3}\sin 3(t-\tau)d\tau$$

$$=\frac{1}{6}\int_0^t [\cos(4\tau-3t)-\cos(3t-2\tau)d\tau$$

$$=\frac{1}{6}\left[\frac{\sin(4\tau-3t)}{4}-\frac{\sin(3t-2\tau)}{-2}\right]_0^t$$

$$=\frac{1}{6}\left(\frac{1}{4}(\sin t+\sin 3t)+\frac{1}{2}(\sin t-\sin 3t)\right)$$

$$=\frac{1}{8}\left(\sin t-\frac{1}{3}\sin 3t\right)$$

Therefore, $L^{-1}\left(\dfrac{s}{s^2+4}\cdot\dfrac{1}{(s^2+1)(s^2+9)}\right)=\cos 2t * \dfrac{1}{8}\left(\sin t-\dfrac{1}{3}\sin 3t\right).$

$$= \int_0^t \frac{1}{8}\left(\sin\tau - \frac{1}{3}\sin 3\tau\right)\cos 2(t-\tau)d\tau$$

$$= \frac{1}{8}\int_0^t \left(\sin\tau\cos 2(t-\tau) - \frac{1}{3}\sin 3\tau\cos 2(t-\tau)\right)dz$$

$$= \frac{1}{8}\int_0^t \left[\left(\frac{1}{2}(\sin(2t-\tau)+\sin(3t-2t)\right) - \frac{1}{6}\{\sin(2t+\tau)+\sin(5\tau-2t)\}\right]d\tau$$

$$= \frac{1}{16}\left[-\frac{\cos(2t+\tau)}{-1} - \frac{\cos(5\tau-2t)}{3}\right]_0^t - \frac{1}{48}\left[-\frac{\cos(2t+\tau)}{1} - \frac{\cos(5\tau-2t)}{5}\right]_0^t$$

$$= \frac{1}{16}\left(\cos 2t - \cos 2t - \frac{1}{3}(\cos t - \cos 2t)\right) - \frac{1}{48}\left(-\cos 3t + \cos 2t - \frac{1}{3}(\cos 3t - \cos 2t)\right)$$

$$= \frac{1}{24}\cos t - \frac{1}{5}\cos 2t + \frac{1}{40}\cos 3t.$$

EXAMPLE 5.7

Find $L^{-1}\left(\frac{1}{(s^2+4s+13)^2}\right)$.

Solution: Here, $f(t) = g(t) = L^{-1}\left(\frac{1}{s^2+4s+13}\right) = L^{-1}\left(\frac{1}{(s+2)^2+9}\right)$

$$= e^{-2t}L^{-1}\left(\frac{1}{s^2+9}\right) = e^{-2t}\frac{\sin 3t}{3}$$

Thus, $L^{-1}\left(\frac{1}{(s^2+4s+9)^2}\right) = e^{-2t}\frac{\sin 3t}{3} * e^{-2t}\frac{\sin 3t}{3}$

$$= \int_0^t e^{-2\tau}\frac{\sin 3\tau}{3}\cdot e^{-2(t-\tau)}\frac{\sin 3(t-\tau)}{3}d\tau$$

$$= \int_0^t \frac{e^{-2t}}{9}\sin 3\tau\cdot\sin 3(t-\tau)d\tau$$

$$= \int_0^t \frac{e^{-2t}}{18}[\cos(6\tau-3t) - \cos 3t]d\tau$$

$$= \frac{e^{-2t}}{18}\left[\frac{\sin(6\tau-3t)}{6} - \tau\cos 3t\right]_0^t$$

$$= \frac{e^{-2t}}{18}\left[\frac{\sin 3t + \sin 3t}{6} - t\cos 3t\right]$$

$$= \frac{e^{-2t}}{54}(\sin 3t - 3t \cos 3t).$$

EXAMPLE 5.8

Find $f(t)$ as the solution of the integral equation

$$f(t) = t + e^{-2t} + \int_0^t f(\tau)e^{2(t-\tau)}d\tau.$$

Solution: Taking Laplace transform of the given equation, we obtain

$$F(s) = \frac{1}{s^2} + \frac{1}{s+2} + F(s)\left(\frac{1}{s-2}\right)$$

$$\Rightarrow \qquad \left[1 - \frac{1}{s-2}\right]F(s) = \frac{1}{s^2} + \frac{1}{s+2}$$

$$\Rightarrow \qquad F(s) = \frac{s-2}{s-3}\left\{\frac{s^2+s+2}{s^2(s+2)}\right\}$$

Using the partial fractions, we get

$$F(s) = \frac{1}{45}\left[\frac{14}{s-3} - \frac{5}{s} + \frac{30}{s^2} + \frac{36}{s+2}\right]$$

Taking the inverse Laplace transform; we obtain

$$f(t) = \frac{1}{45}(14\, e^{3t} - 5 + 30t + 36\, e^{-2t}).$$

EXAMPLE 5.9

Using convolution theorem, solve the initial value problem

$$y'' + 9y = \sin 3t;\ y(0) = 0,\ y'(0) = 0$$

Solution: Taking Laplace transform of the given differential equation, we obtain

$$s^2L(y) - sy(0) - y'(0) + 9L(y) = \frac{3}{s^2+9}$$

$$\Rightarrow \qquad Y(s) = \frac{3}{(s^2+9)^2}$$

$$\therefore \qquad y(t) = L^{-1}(Y(s)) = 3L^{-1}\left(\frac{1}{(s^2+3^2)^2}\right)$$

{Using example - 5.4 with $a = 3$}

$$= 3\left\{\frac{1}{2(3)^3}(\sin 3t - 3t\cos 3t)\right\}$$

$$= \frac{1}{18}(\sin 3t - 3t \cos 3t).$$

EXAMPLE 5.10

Using convolution theorem, solve the initial value problem

$$y' - y = te^t \cos t;\ y(0) = 0$$

Solution: Taking Laplace transform of the given differential equation, we obtain

$$sL(y) - y(0) - L(y) = L(te^t \cos t)$$

$$sY(s) - 0 - Y(s) = -\frac{d}{ds}L(e^t \cos t)$$

$$(s - 1)\, Y(s) = -\frac{d}{ds}\left(\frac{s-1}{(s-1)^2+1}\right)$$

$$= -\frac{d}{ds}\left(\frac{s-1}{(s^2-2s+2}\right) = -\frac{(s^2-2s+2)\cdot 1-(s-1)(2s-2)}{(s^2-2s+2)^2}$$

$$= -\frac{s^2-2s+2-2(s-1)^2}{(s^2-2s+2)^2}$$

$$= -\frac{-s^2-2s}{(s^2-2s+2)^2} = \frac{s(s+2)}{(s^2-2s+2)^2}$$

$$\therefore \qquad Y(s) = \frac{s(s+2)}{(s-1)((s-1)^2+1)^2}$$

Thus,

$$y(t) = L^{-1}(Y(s)) = L^{-1}\left(\frac{s(s+2)}{(s-1)((s-1)^2+1)^2}\right)$$

$$= e^t L^{-1}\left(\frac{(s+1)(s+3)}{s(s^2+1)^2}\right)$$

$$= e^t L^{-1}\left(-\frac{1}{s}+\frac{s}{s^2+1}+\frac{1}{(s^2+1)^2}\right)$$

$$= e^t\left(-1+\cos t - L^{-1}\left(\frac{d}{ds}\left(\frac{1}{s^2+1}\right)\right)\right)$$

$$= e^t\left(-1+\cos t + tL^{-1}\left(\frac{1}{s^2+1}\right)\right)$$

$$= e^t(-1 + \cos t + t \sin t).$$

EXAMPLE 5.11

Show that

$$1 * 1 * 1 * \ldots n \text{ times} = \frac{t^n}{n!}$$

Solution: We have

$$1 * 1 = \int_0^t 1 \cdot 1 \cdot d\tau = t$$

Now,

$$1 * 1 * 1 = 1 * (1 * t) = 1 * t$$

$$= \int_0^t \tau \cdot 1 \cdot d\tau = \frac{t^2}{2}$$

and

$$1 * 1 * 1 * 1 = 1 * (1 * 1 * 1) = 1 * \frac{t^2}{2}$$

$$= \int_0^t \frac{\tau^2}{2} \cdot 1 \cdot d\tau = \frac{t^3}{3 - 2} = \frac{t^3}{3!}$$

So, by induction, we obtain

$$1 * 1 * 1 * \ldots n \text{ times} = \frac{t^n}{n!}$$

Laplace Transform of Periodic Functions

Periodic functions occur in many applications in science and engineering. A function $f(t)$ is said to be periodic with period T if

$$f(t + T) = f(t)$$

where, T is the smallest positive value of t which satisfies the above condition. The following are some examples of periodic functions.

(a) $f(t) = \begin{cases} 1, & 0 \le t < \pi \\ -1, & \pi \le t < 2\pi \end{cases}$. Here $T = 2\pi$.

(b) $f(t) = \begin{cases} t/a, & 0 \le t < a \\ (2a - t)/a, & a \le t \le 2a \end{cases}$. Here $T = 2a$.

(c) $f(t) = t(0 \le t < a)$. Here $T = a$.

The Laplace transform of a periodic function can be obtained by integrating over one period as shown in the following theorem.

Theorem: Let $f(t)$ be piecewise continuous on $[0, \infty)$, be of exponential order and periodic with period T. Then,

$$L(f(t)) = \frac{1}{1-e^{-sT}} \int_0^T e^{-st} f(t)dt \ (t > 0)$$

Proof: We have by definition

$$L(f(t)) = \int_0^\infty e^{-st} f(t)dt$$

$$= \int_0^T e^{-st} f(t)dt + \int_T^{2T} e^{-st} f(t)dt + \int_{2T}^{3T} e^{-st} + f(t)dt + \ldots.$$

For having same limit of integration in each integral, substitute

$t = u$ in the first integral,

$t = u + T$ in the second integral,

$t = u + 2T$ in the third integral ...,

$t = u + nT$ in the $(n + 1)$th integral and so on.

Thus, we obtain

$$L[f(t)] = \int_0^T e^{-su} f(u)du + \int_0^T e^{-s(u+T)} f(u + T)du + \int_0^T e^{-s(u+2T)} f(u + 2T)du + \ldots$$

Since $f(t)$ is periodic with period T, so we have

$$f(u + T) = f(u + 2T) = \ldots = f(u)$$

Therefore, we obtain

$$L(f(t)) = \int_0^T e^{-su} f(u)du + e^{-sT} \int_0^T e^{-su} f(u)du + e^{-2sT} \int_0^T e^{-su} f(u)du + \ldots$$

$$= (1 + e^{-sT} + e^{-2sT} + \ldots) \int_0^T e^{-su} f(u)du$$

$$= \frac{1}{1-e^{-sT}} \int_0^T e^{-st} f(t)dt.$$

EXAMPLE 5.12

Find the Laplace transform of the periodic function

$$f(t) = t(0 \le t \le a)$$

and $$f(t + a) = f(t)$$

Solution: Here, $T = a$. Therefore,

$$L[f(t)] = \frac{1}{1-e^{-sa}} \int_0^a e^{-st} t dt$$

$$= \frac{1}{1-e^{-as}}\left[t\frac{e^{-st}}{-s} - 1 - \frac{e^{-st}}{(-s)^2}\right]_0^a$$

$$= \frac{1}{1-e^{-as}}\left[\frac{ae^{-sa}-0}{-s} - \frac{1}{s^2}(e^{-as}-1)\right]$$

$$= \frac{1}{s^2} - \frac{ae^{-as}}{s(1-e^{-as})}.$$

EXAMPLE 5.13

Find the Laplace transform of the periodic function

$$f(t) = \begin{cases} t, & 0 < t < \pi \\ \pi - t, & \pi < t < 2\pi \end{cases} \quad \text{and} \quad f(t + 2\pi) = f(t)$$

Solution: Here, $T = 2\pi$. Therefore,

$$L[f(t)] = \frac{1}{1-e^{-2\pi s}}\int_0^{2\pi} e^{-st}f(t)dt$$

$$= \frac{1}{1-e^{-2\pi s}}\left[\int_0^{\pi} e^{-st}tdt + \int_{\pi}^{2\pi} e^{-st}(\pi - t)dt\right]$$

$$= \frac{1}{1-e^{-2\pi s}}\left[\left(t\frac{e^{-st}}{-s} - \frac{e^{-st}}{s^2}\right)_0^{\pi} + \left(\frac{e^{-st}}{-s}(\pi - t) - (-1)\frac{e^{-st}}{s^2}\right)_{\pi}^{2\pi}\right]$$

$$= \frac{1}{1-e^{-2\pi s}}\left[-\frac{\pi e^{-\pi s}}{s} - \frac{e^{-\pi s}-1}{s^2} + e^{-2\pi s}\frac{\pi}{s} + \frac{e^{-2\pi s}-e^{-\pi s}}{s^2}\right]$$

$$= \frac{1}{1-e^{-2\pi s}}\left[\frac{\pi}{s}(e^{-2\pi s} - e^{-\pi s}) + \frac{1}{s^2}(1 + e^{-2\pi s} - 2e^{-\pi s})\right]$$

$$= \frac{1}{1+e^{-\pi s}}\left[-\frac{\pi e^{-\pi s}}{s} + \frac{1-e^{-\pi s}}{s^2}\right]$$

EXAMPLE 5.14

Find the Laplace transform of the periodic function

$$f(t) = \begin{cases} t/a, & 0 \le t \le a \\ (2a - t)/a, & a \le t \le 2a \end{cases}$$

and $$f(t + 2a) = f(t).$$

Solution: Here, $T = 2a$. Therefore,

$$L[f(t)] = \frac{1}{1-e^{-2as}}\int_0^{2a} e^{-st}f(t)dt$$

$$= \frac{1}{1-e^{-2as}}\left\{\int_0^a e^{-st}\frac{t}{a}dt + \int_a^{2a} e^{-st}\frac{(2a-t)}{a}dt\right\}$$

$$= \frac{1}{1-e^{-2as}}\left[\frac{1}{a}\left(\frac{e^{-st}}{-s}t - \frac{e^{-st}}{s^2}\right)_0^a + \frac{1}{a}\left(\frac{e^{-st}}{-s}(2a-t) - (-1)\frac{e^{-st}}{s^2}\right)_a^{2a}\right]$$

$$= \frac{1}{1-e^{-2as}}\left[\frac{1}{a}\left(\frac{ae^{-as}-0}{-s} - \frac{e^{-as}-1}{s^2}\right) + \frac{1}{a}\left(\frac{0-e^{-st}a}{-s} + \frac{e^{-2as}-e^{-as}}{s^2}\right)\right]$$

$$= \frac{1}{as^2}\frac{1-2e^{-as}+e^{-2as}}{1-e^{-2as}}$$

$$= \frac{1}{as^2}\frac{(1-e^{-as})^2}{1-e^{-2as}} = \frac{1-e^{-as}}{as^2(1+e^{-as})}$$

$$= \frac{1}{as^2}\frac{e^{as/2}-e^{-as/2}}{e^{as/2}+e^{-as/2}} = \frac{1}{as^2}\tanh\left(\frac{as}{2}\right).$$

EXAMPLE 5.15

Find the Laplace transform of the periodic function

$$f(t) = \begin{cases} t, & 0 < t < a \\ 0, & a < t < 2a \end{cases}$$

Solution: Here, $f(t + 2a) = f(t)$. Therefore, $T = 2a$.

Now,
$$L[f(t)] = \frac{1}{1-e^{-2as}}\int_0^{2a} e^{-st}f(t)dt$$

$$= \frac{1}{1-e^{-2as}}\left[\int_0^a e^{-st}t\,dt\right]$$

$$= \frac{1}{1-e^{-2as}}\left[\frac{e^{-st}}{-s}t - \frac{e^{-st}}{s^2}\right]_0^a$$

$$= \frac{1}{1-e^{-2as}}\left[\frac{ae^{-st}-0}{-s} - \frac{e^{-as}-1}{s^2}\right]$$

$$= \frac{(1-e^{-as}) - ase^{-as}}{s^2(1-e^{-2as})}$$

EXAMPLE 5.16

Find the Laplace transform of the periodic function

$$f(t) = \begin{cases} t, & 0 < t \le \pi \\ 2\pi - t, & \pi < t < 2\pi \end{cases}$$

Solution: Here, $\quad f(t + 2a) = f(t)$

Therefore, $\quad T = 2\pi$

Now, $\quad L(f(t)) = \dfrac{1}{1-e^{-2\pi s}} \displaystyle\int_0^{2\pi} e^{-st} f(t)dt$

$$= \frac{1}{1-e^{-2\pi s}} \left[\int_0^{\pi} e^{-st} t\,dt + \int_{\pi}^{2\pi} e^{-st}(2\pi - t)dt \right]$$

$$= \frac{1}{1-e^{-2\pi s}} \left[\left(\frac{e^{-st}}{-s} t - \frac{e^{-st}}{s^2} \right)_0^{\pi} + \left(\frac{e^{-st}}{-s}(2\pi - t) - \frac{e^{-st}}{s^2}(-1) \right)_{\pi}^{2\pi} \right]$$

$$= \frac{1}{1-e^{-2\pi s}} \left[\frac{e^{-\pi s}\pi - 0}{-s} - \frac{e^{-\pi s} - 1}{s^2} + \frac{0 - e^{-st}\pi}{-s} + \frac{e^{-2\pi s} - e^{-\pi s}}{s^2} \right]$$

$$= \frac{1}{1-e^{-2\pi s}} \left(\frac{1 - 2e^{-\pi s} + e^{-2\pi s}}{s^2} \right)$$

$$= \frac{1}{1-e^{-2\pi s}} \frac{(1-e^{-\pi s})^2}{s^2} = \frac{1-e^{-\pi s}}{s^2(1+e^{-\pi s})}.$$

EXAMPLE 5.17

Solve the initial value problem with $f(t)$ as a periodic function.

$$y'' + 4y' + 5y = f(t);\ y(0) = 0,\ y'(0) = 0 \quad \text{and} \quad f(t) = \begin{cases} 1, & 0 \le t < \pi \\ -1, & \pi \le t < 2\pi \end{cases}$$

Solution: Taking Laplace transform of the given differential equation, we obtain

$$s^2L(y) - sy(0) - y'(0) + 4y(sL(y) - y(0)) + 5L(y) = L(f(t))$$

$$\Rightarrow \quad (s^2 + ys + 5)\,Y(s) = 2L(f(t)$$

Now,
$$L(f(t)) = \frac{1}{1-e^{-2\pi s}} \int_0^{2\pi} e^{-st} f(t)dt$$
$$= \frac{1}{1-e^{-2\pi s}} \left[\int_0^{\pi} e^{-st} \cdot 1dt + \int_{\pi}^{2\pi} e^{-st} (-1)dt \right]$$
$$= \frac{1}{1-e^{-2\pi s}} \left[\left(\frac{e^{-st}}{-s} \right)_0^{\pi} - \left(\frac{e^{-st}}{-s} \right)_{\pi}^{2\pi} \right]$$
$$= \frac{1}{1-e^{-2\pi s}} \frac{-e^{-\pi s} + 1 + e^{-2\pi s} - e^{-\pi s}}{s}$$
$$= \frac{1}{1-e^{-2\pi s}} \frac{1 - 2e^{-\pi s} + e^{-2\pi s}}{s}$$
$$= \frac{1}{1-e^{-2\pi s}} \frac{(1-e^{-\pi s})^2}{s} = \frac{1-e^{-\pi s}}{s(1+e^{-\pi s})}$$
$$= \frac{2-(1+e^{-\pi s})}{s(1+e^{-\pi s})} = \frac{1}{s} \left[\frac{2}{1+e^{-\pi s}} - 1 \right]$$
$$= \frac{1}{s} [2(1 + e^{-\pi s})^{-1} - 1]$$
$$= \frac{1}{s} [2(2(1 - e^{-\pi s} + e^{-2\pi s} - e^{-3\pi s} + \ldots) - 1]$$
$$= \frac{1}{s} (1 - 2e^{-\pi s} + 2e^{-2\pi s} - 2e^{-3\pi s} + \ldots)$$

Therefore,
$$Y(s) = \frac{1}{s[(s+2)^2+1]} (1 - 2e^{-\pi s} + 2e^{-2\pi s} - 2e^{-3\pi s} + \ldots)$$
$$\Rightarrow \quad y(t) = L^{-1}(Y(s)) = L^{-1}\left[\frac{1}{s[(s+2)^2+1]} (1 - 2e^{-\pi s} + 2e^{-2\pi s} - 2e^{-3\pi s} + \ldots) \right]$$

Now,
$$L^{-1}\left\{ \frac{1}{s[(s+2)^2+1]} \right\} = \int_0^t L^{-1}\left(\frac{1}{(s+2)^2+1} \right) dt$$

$$= \int_0^t e^{-2t} L^{-1}\left(\frac{1}{s^2+1}\right) dt$$

$$= \int_0^t e^{-2t} \sin t dt$$

$$= \left\{\frac{e^{-2t}}{4+1}(-2\sin t - \cos t)\right\}_0^t$$

$$\left[\because \int e^{at} \sin bt\, dt = \frac{e^{at}}{a^2+b^2}(a \sin bt - b \cos bt)\right]$$

$$= \frac{e^{-2t}}{5}(-2 \sin t - \cos t) - \frac{1}{5}(0 - 1)$$

$$= \frac{1}{5}[1 - e^{-2t}(2 \sin t + \cos t)] = g(t)$$

So, by using the second shifting theorem, we get

$$y(t) = u_0(t)\, g(t - 0) - 2u_\pi(t)g(t - \pi) + 2u_{2\pi}(t)g(t - 2\pi) - 2u_{3\pi}(t)g(t - 3\pi) + \ldots$$

EXAMPLE 5.18

Find the solution of the initial value problem

$$y'' + 8y' + 17y = f(t);\ y(0) = 0;\ y'(0) = 0 \text{ where } f(t) \text{ is the periodic function}$$

$$f(t) = \begin{cases} 1 & 0 < t < \pi \\ 0 & \pi < t < 2\pi \end{cases} \quad \text{and} \quad f(t + 2\pi) = f(t)$$

Solution: Taking the Laplace transform of the given differential equation, we obtain

$$s^2L(y) - sy(0) - y'(0) + 8(sL(y) - y(0)) + 17L(y) = L(f(t))$$

$$\Rightarrow \qquad (s^2 + 8s + 17)\, Y(s) = L(f(t))$$

Now,

$$L(f(t)) = \frac{1}{1-e^{-2\pi s}} \int_0^{2\pi} e^{-st} f(t) dt$$

$$= \frac{1}{1-e^{-2\pi s}} \left\{\int_0^{\pi} e^{-st} \cdot 1\, dt\right\}$$

$$= \frac{1}{1-e^{-2\pi s}} \left\{\frac{e^{-st}}{-s}\right\}_0^{\pi} = \frac{1}{1-e^{-2\pi s}} \frac{e^{-\pi s} - 1}{-s}$$

$$= \frac{1}{s(1+e^{-2\pi s})} = \frac{1}{s}(1+e^{-\pi s})^{-1} = \frac{1}{s}(1-e^{-\pi s}+e^{-2\pi} - \ldots)$$

Thus,
$$Y(s) = \frac{1}{s(s^2+8s+17)} \times (1 - e^{-\pi s} + e^{-2\pi s} - \ldots)$$

$$\Rightarrow \quad y(t) = L^{-1}(Y(s)) = L^{-1}\left(\frac{1}{s((s+4)^2+1)}(1-e^{-\pi s}+e^{-2\pi s} - \ldots)\right)$$

Now, $L^{-1}\left(\frac{1}{s((s+4)^2+1)}\right) = \int_0^t L^{-1}\left(\frac{1}{(s+4)^2+1}\right)dt = \int_0^t e^{-4t}L^{-1}\left(\frac{1}{s^2+1}\right)dt = \int_0^t e^{-4t}\sin t\,dt$

$$= \left(\frac{e^{-4t}}{16+1}(-4\sin t - \cos t)\right)_0^t$$

$$= \frac{e^{-4t}}{17}(-4\sin t - \cos t) - \frac{1}{17}(0-1)$$

$$= \frac{1}{17}(1 - e^{-4t}(4\sin t + \cos t)) = g(t)$$

Using the second shifting theorem, we get

$$y(t) = u_0(t)\, g(t-0) - u_\pi(t)\, g(t-\pi) + u_{2\pi}(t)\, g(t-2\pi) - \ldots.$$

EXAMPLE 5.19

Solve the initial value problem where $f(t)$ is a periodic function

$$y'' + y = f(t);\; y(0) = 0,\; y'(0) = 0 \text{ and}$$

$$f(t) = \begin{cases} 0, & 0 \le t < \pi \\ \sin t, & \pi \le t < 2\pi \end{cases}$$

Solution: Taking Laplace transform of the given differential equation, we obtain

$$s^2L(y) - sy(0) - y'(0) + L(y) = L(f(t))$$

$$\Rightarrow \quad (s^2+1)\,Y(s) = L(f(t))$$

Now,
$$L(f(t)) = \frac{1}{1-e^{-2\pi s}}\int_0^{2\pi} e^{-st}f(t)dt$$

$$= \frac{1}{1-e^{-2\pi s}}\int_\pi^{2\pi} e^{-st}\sin dt$$

$$= \frac{1}{1-e^{-2\pi s}}\left[\frac{e^{-st}}{s^2+1}(-\sin t-\cos t)\right]_{\pi}^{2\pi}$$

$$= \frac{1}{1-e^{-2\pi s}}\,1\left(\frac{e^{-2\pi s}(-1)-e^{-\pi s}}{s^2+1}\right)$$

$$= -\frac{e^{-\pi s}(1+e^{-\pi s})}{(s^2+1)(1-e^{-2\pi s})}$$

$$= -\frac{e^{-\pi s}}{(s^2+1)(1-e^{-\pi s})}$$

$$= -\frac{e^{-\pi s}}{(s^2+1)}(1-e^{-\pi s})^{-1}$$

$$= -\frac{e^{-\pi s}}{(s^2+1)}(1+e^{-\pi s}+e^{-2\pi s}+\ldots)$$

$$= -\frac{1}{s^2+1}(e^{-\pi s}+e^{-2\pi s}+e^{-3\pi s}+\ldots)$$

Thus, $$Y(s) = -\frac{1}{(s^2+1)^2}(e^{-\pi s}+e^{-2\pi s}+e^{-3\pi s}+\ldots)$$

$\Rightarrow$ $$y(s) - L^{-1}(Y(s)) - L^{-1}\left(-\frac{1}{(s^2+1)^2}(e^{-\pi s}+e^{-2\pi s}+e^{-3\pi s}+\ldots)\right)$$

Now, $$L^{-1}\left(-\frac{1}{(s^2+1)^2}\right) = -\frac{1}{2}(\sin t - t\cos t) = g(t)$$

(Using example 5.4 with $a = 1$)

So, by using the second shifting theorem, we get

$$y(t) = u_{\pi}(t)\, g(t-\pi) + u_{2\pi}(t)\, g(t-2\pi) + u_{3\pi}(t)\, g(t-3\pi) + \ldots$$

EXERCISE 5.1

In the following problems, use the convolution theorem to find the inverse Laplace transforms:

1. $\dfrac{1}{(s-2)(s+3)}$.

2. $\dfrac{1}{s^2(s^2+16)}$.

3. $\dfrac{1}{(s^2+9)^2}$.

4. $\dfrac{s}{(s^2+4)^2}$.

5. $\dfrac{1}{(s+a)\ 2(s+b)^2}\ (a \neq b)$.

6. $\dfrac{s}{(s^2+4)(s^2+9)}$.

7. $\dfrac{8s}{(s^2+16)(s^2+1)^2}$.

8. $\dfrac{1}{(s-2)(s+2)^2}$.

Using convolution theorem, solve the following initial value problems:

9. $y' + y = t \cos t;\ y(0) = 0.$

10. $y'' + 16y = \cos yt;\ y(0) = 0;\ y'(0) = 0.$

11. $y'' - 5y' - 6y = e^{-t};\ y(0) = 1,\ y'(0) = 1.$

Solve the following integral equations using convolution theorem:

Find the Laplace transform of the periodic functions defined as

12. $f(t) = \cos t + e^{-t}\int_0^t f(\tau)e^{\tau}d\tau.$

13. $f(t) = e^t + \int_0^t \tau f(t-\tau)d\tau.$

14. $f(t) = 1 + t + 2\int_0^t \sin \tau f(t-\tau)d\tau.$

15. Find the Laplace transform of the sawtoothed wave of period T, given

$$f(t) = t/T \text{ for } 0 < t < T.$$

16. Find the Laplace transform of the square-wave function of period a defined as

$$f(t) = \begin{cases} k & 0 < t < a \\ -k & a < t < 2a \end{cases}.$$

17. Using the Laplace transform of periodic functions, verify that $L(\sin at) = \dfrac{a}{s^2+a^2}$.

Solve the following initial value problems with the periodic function $f(t)$.

18. $f(t) = kt\ (0 < t < 2\pi)$ and k is a constant.

19. $f(t) = t^2\ (0 < t < 2\pi)$.

20. $f(t) = \begin{cases} 0 & 0 < t < a \\ t-a & a < t < 2a \end{cases}$.

21. $f(t) = \begin{cases} \cos t & 0 < t < \pi/2 \\ -\cos t & \pi/2 < t < 3\pi/2 \\ \cos t & 3\pi/2 < t < 2\pi \end{cases}$

22. $f(t) = \begin{cases} 0 & 0 < t < a \\ 1 & a < t < 2a \\ 2 & 2a < t < 3a \end{cases}$

Solve the following initial value problems with the periodic function $f(t)$.

23. $y'' + 4y = f(t);\ y(0) = 0,\ y'(0) = 0$ and $f(t) = \begin{cases} \sin t & 0 \le t < \pi \\ 0 & \pi \le t < 2\pi \end{cases}$

24. $y'' + 3y' + 2y = f(t);\ y(0) = 0,\ y'(0) = 1$ and $f(t) = t(0 \le t \le a)$.

25. $y'' + 4y' + 3y = f(t);\ y(0) = 0;\ y'(0) = 0$ and $f(t) = \begin{cases} t & 0 < t \le \pi \\ 2\pi - t & \pi < t < 2\pi \end{cases}$

26. Apply the convolution theorem, to prove that

$$\beta(m, n) = \int_0^1 x^{m-1}(1-x)^{n-1}dx = \frac{\Gamma(m)\Gamma(n)}{\Gamma(m+n)} \quad (m, n > 0)$$

where $\beta(m, n)$ is the Beta function.

$\left[\textbf{Hint: } L\left(\int_0^t \tau^{m-1}(t-\tau)^{n-1}d\tau\right) = L(t^{m-1}) \times L(t^{n-1}) \text{ and take } t = 1 \text{ in the result}\right]$

ANSWERS

Exercise 5.1

1. $\frac{1}{5}(e^{2t} - e^{-3t})$.

2. $\frac{(4t - \sin 4t)}{64}$.

3. $\frac{(\sin 3t - 3t\cos 3t)}{54}$.

4. $\dfrac{t\sin 2t}{4}$.

5. $\dfrac{[2+(a-b)t]e^{-as}-[2-(a-b)t]e^{-bs}}{(a-b)^3}$.

6. $\dfrac{\cos 2t-\cos 3t}{5}$.

7. $\dfrac{60t\sin t-8\cos t+8\cos 4t}{225}$.

8. $\dfrac{1}{16}(e^{2t}-e^{-2t}-4te^{-2t})$.

9. $(t\cos t-(1-t)\sin t)/2$.

10. $(t\sin yt)/8$.

11. $((34-7t)e^{-t}+15e^{6t})/49$.

12. $\cos t+\sin t$.

13. $(3e^t+2te^t+e^{-t})/4$.

14. $(2e^t-t-1)$.

15. $\dfrac{1}{s^2T}-\dfrac{e^{-sT}}{s(1-e^{-sT})}$.

16. $\dfrac{a}{s}\tanh\left(\dfrac{as}{2}\right)$.

18. $\dfrac{k}{s^2}-\dfrac{2k\pi e^{-2\pi s}}{s(1-e^{-2\pi s})}$.

19. $\dfrac{2}{s^2}-\dfrac{4\pi(\pi s+1)e^{-2\pi s}}{s^2(1-e^{-2\pi s})}$.

20. $\dfrac{e^{-as}(as+1)e^{-2as}}{s^2(1-e^{-2as})}$.

21. $\left(\dfrac{1}{s}+\dfrac{2e^{-\pi s/2}}{s^2(1-e^{-\pi s})}\right)\left(\dfrac{s^2}{1+s^2}\right)$.

22. $\dfrac{e^{-as}+e^{-2as}-2e^{-3as}}{s(1-e^{-3as})}=\dfrac{e^{-as}(1-2e^{-as})}{s(1+e^{-as}+e^{-2as})}$.

23. $y(t) = g(t-0)\, u_0(t) + g(t-\pi)\, u_\pi(t) + g(t-2\pi)\, u_{2\pi}(t) + \ldots$

where $g(t) = \dfrac{1}{6}(2 \sin t - \sin 2t)$.

24. $y(t) = g_1(t-0)\, u_0(t) - a\left\{\left(\dfrac{1}{2} + g_2(t-a)\right)u_a(t) + \left(\dfrac{1}{2} + g_2(t-2a)\right)u_{2a}(t) + \ldots\right\}$

where $y_1(t) = -\dfrac{3}{4} + \dfrac{t}{2} + 2e^{-t} - \dfrac{5}{4}e^{-2t}$ and $g_g(t) = \dfrac{1}{2}e^{-2t} - e^{-t}$.

25. $y(t) = \left\{-\dfrac{4}{9} + g(t)\right\}u_0(t) - 2\left\{\left(-\dfrac{4}{9} + g(t-\pi)\right) \times u_\pi(t)\left(-\dfrac{4}{9} + g(t-2\pi)\right)u_{2\pi}(t) + \ldots\right\}$ where

$g(t) = \dfrac{1}{3} - t + \dfrac{1}{2}e^{-t} - \dfrac{1}{18}e^{-3t}$.

▲▲

6

CHAPTER

Solution of Partial Differential Equations using Laplace Transforms

The Laplace transform is also useful in solving various partial differential equations subject to boundary conditions. Such problems are often referred to as boundary-value problems. The method depends on the following results of Laplace transform and is illustrated with the examples.

THEOREM: Let $u(x, t)$ be a function defined for $a \le x \le b$, $t > 0$. Show that

(a) $$L\left(\frac{\partial u}{\partial t}\right) = sL(u(x, t) - u(x, 0)$$
$$= sU(x, s) - u(x, 0)$$

(b) $$L\left(\frac{\partial u}{\partial x}\right) = \frac{dU}{dx}$$

(c) $$L\left(\frac{\partial^2 u}{\partial t^2}\right) = s^2\, U(x, s) - su(x, 0) - u_t(x, 0)$$

(d) $$L\left(\frac{\partial^2 u}{\partial x^2}\right) = \frac{d^2U}{dx^2}$$

Proof:

(a) $$L\left(\frac{\partial u}{\partial t}\right) = \int_0^\infty e^{-st}\frac{\partial u}{\partial t}dt$$
$$= [e^{-st}u]_0^\infty - \int_0^\infty -se^{-st}udt$$
$$= 0 - u(x, 0) + sL(u)$$
$$= sU(x, s) - u(x, 0).$$

(b) $L\left(\frac{\partial u}{\partial x}\right) = \int_0^\infty e^{-st} \frac{\partial u}{\partial x} dt$

$$= \frac{d}{dx}\left\{\int_0^\infty e^{-st} u dt\right\} = \frac{d}{dx} L(u) = \frac{dU}{dx}$$

(c) Put $v = \frac{\partial u}{\partial t}$

$$\therefore \quad L\left(\frac{\partial^2 u}{\partial t^2}\right) = L\left(\frac{\partial v}{\partial t}\right) = sL(v) - v(x, 0)$$

$$= sL\left(\frac{\partial u}{\partial t}\right) - v(x, 0)$$

$$= s[sL(u) - u(x, 0)] - u_t(x, 0)$$

$$= s^2U(x, s) - su(x, 0) - u_t(x, 0)$$

(d) Let $v = \frac{\partial u}{\partial x}$

$$\therefore \quad L\left(\frac{\partial^2 u}{\partial x^2}\right) = L\left(\frac{\partial v}{\partial x}\right) = \frac{d}{dx} L(v)$$

$$= \frac{d}{dx} L\left(\frac{\partial u}{\partial x}\right) = \frac{d}{dx}\left(\frac{dU}{dx}\right) = \frac{d^2U}{dx^2}.$$

▲▲

EXAMPLE 6.1

Using Laplace transforms, find the solution of the initial value problem

$$x\frac{\partial u}{\partial t} + \frac{\partial u}{\partial x} = xt; \; u(x, 0) = 0, \; u(0, t) = t$$

Solution: Taking Laplace transform of the differential equation, we obtain

$$L\left(\frac{\partial u}{\partial t}\right) + L\left(\frac{\partial u}{\partial x}\right) = xL(t)$$

$$\Rightarrow \quad s[sL(u) - u(x, 0)] + \frac{dU}{dx} = x\left(\frac{1}{s^2}\right)$$

$$\Rightarrow \quad s[sU - 0] + \frac{dU}{dx} = \frac{x}{s^2}$$

$$\Rightarrow \quad \frac{dU}{dx} + sxU = \frac{x}{s^2}$$

which is an ordinary linear differential equation of the first order with IF = $e^{\int sxdx} = e^{sx^2/2}$ and so its solution is given by

$$e^{sx^2/2}\,U = \int \frac{x}{s^2} e^{sx^2/2} dx + a(s)$$

$$= \frac{1}{s^3} e^{sx^2/2} + a(s)$$

$$\Rightarrow \qquad U = \frac{1}{s^3} + a(s)e^{-sx^2/2}$$

where $a(s)$ is an arbitrary function of s.

Now, we are given $u(0, t) = t$. Therefore,

$$L(u(0, t)) = U(0, s) = \frac{1}{s^2}$$

Thus, $$U(0, s) = \frac{1}{s^2} = \frac{1}{s^3} + a(s)$$

$$\Rightarrow \qquad a(s) = \frac{1}{s^2} - \frac{1}{s^3}$$

Hence, $$U(x, s) = \frac{1}{s^3} + \left(\frac{1}{s^2} - \frac{1}{s^3}\right)e^{-sx^2/2}$$

Therefore, $$u(x, t) = L^{-1}(U(x, s)) = L^{-1}\left(\frac{1}{s^3} + \left(\frac{1}{s^2} - \frac{1}{s^3}\right) \times e^{-sx^2/2}\right)$$

$$= \frac{t^2}{2!} + u_{x^2/2}(t)\left[\left(t - \frac{x^2}{2}\right) - \frac{1}{2!}\left(t - \frac{x^2}{2}\right)^2\right]$$

$$\left\{\because \; F = \frac{1}{s^2} - \frac{1}{s^3} \Rightarrow f = L^{-1}(F) = t - \frac{t^2}{2!}\right\}$$

EXAMPLE 6.2

Solve $\dfrac{\partial u}{\partial t} = 2\dfrac{\partial^2 u}{\partial x^2}$ given that $u(0, t) = 0$, $u(5, t) = 0$, $u(x, 0) = \sin \pi x$.

Solution: Taking Laplace transform of the given partial differential equation

$$L\left(\frac{\partial u}{\partial t}\right) = 2L\left(\frac{\partial^2 u}{\partial x^2}\right)$$

$$\Rightarrow \qquad sU - u(x, 0) = 2\frac{d^2U}{dx^2}$$

$$\Rightarrow \qquad \frac{d^2U}{dx^2} - \frac{s}{2}U = -\frac{1}{2}\sin \pi x$$

Thus, $$U = Ae^{\sqrt{\frac{s}{2}}x} + Be^{-\sqrt{\frac{s}{2}}x} - \frac{1}{2}\frac{\sin \pi x}{-\pi^2 - \frac{s}{2}}$$

$$= Ae^{\sqrt{\frac{s}{2}}x} + Be^{-\sqrt{\frac{s}{2}}x} + \frac{\sin \pi x}{s + 2\pi^2}$$

Now, $$u(0, t) = 0 = u(5, t)$$

$\Rightarrow$ $$L(u(0, t)) = 0 = L(u(5, t))$$

$\Rightarrow$ $$U(0, s) = 0 = U(5, s)$$

Therefore, $$A + B = 0$$

and $$Ae^{5\sqrt{s/2}} + Be^{-5\sqrt{s/2}} = 0$$

$\Rightarrow$ $$A = 0 = B$$

So, $$U = \frac{\sin \pi x}{s + 2\pi^2}$$

Hence, $$u(x, t) = L^{-1}(U(x, s)) = \sin \pi x \, L^{-1}\left(\frac{1}{s + 2\pi^2}\right) = e^{-2\pi^2 t} \sin \pi x.$$

EXAMPLE 6.3

Solve $\frac{\partial^2 u}{\partial t^2} = 9\frac{\partial^2 u}{\partial x^2}$ given that $u(0, t) = u(2, t) = 0$, $u_t(x, 0) = 0$ and

$$u(x, 0) = 10 \sin 2\pi x - 20 \sin 5\pi x$$

Solution: Taking the Laplace transform of the given partial differential equation, we get

$$L\left(\frac{\partial^2 u}{\partial t^2}\right) = 9L\left(\frac{\partial^2 u}{\partial x^2}\right)$$

$\Rightarrow$ $$s^2 U(x, s) - su(x, 0) - u_t(x, 0) = 9\frac{d^2 U}{dx^2}$$

$\Rightarrow$ $$\frac{d^2 U}{dx^2} - \frac{s^2}{9}U = -\frac{s}{9}(10 \sin 2\pi x - 20 \sin 5\pi x)$$

Thus, $$U = Ae^{sx/3} + Be^{-sx/3} - \frac{s}{9}\left(10 \cdot \frac{\sin 2\pi x}{-4\pi^2 - \frac{s^2}{9}} - 20\frac{\sin 5\pi x}{-25\pi^2 - \frac{s^2}{9}}\right)$$

$$= Ae^{sx/3} + Be^{-sx/3} + \frac{10s}{s^2+36\pi^2}\sin 2\,\pi x - \frac{20s}{s^2+225\pi^2}\sin 5\,\pi x$$

We have $\qquad u(0, t) = 0 = u(2, t)$

$\therefore \qquad L(u(0, t)) = 0 = L(u(2, t))$

$\Rightarrow \qquad U(0, s) = 0 = U(2, s)$

Therefore, $\qquad A + B = 0$

and $\qquad Ae^{2s/3} + Be^{-2s/3} = 0$

$\Rightarrow \qquad A = 0 = B$

Thus, $$U = \frac{10s}{s^2+36\pi^2}\sin 2\pi x - \frac{20s}{s^2+225\pi^2}\sin 5\pi.$$

$\therefore\ u(x, t) = L^{-1}(U(x, s)) = 10 \cos 6\,\pi t \sin 2\,\pi x - 20 \cos 15\,\pi t \sin 5\,\pi x.$

EXAMPLE 6.4

Solve: $\dfrac{\partial^2 u}{\partial t^2} = \dfrac{\partial^2 u}{\partial x^2}$; $x > 0$, $t > 0$, given that $u(0, t) = 10 \sin 2t$, $u(x, 0) = 0$, $u_t(x, 0) = 0$ and $\lim\limits_{x\to\infty} u(x, t) = 0$.

Solution: Taking Laplace transform of the given partial differential equation, we get

$$L\left(\frac{\partial^2 u}{\partial t^2}\right) = L\left(\frac{\partial^2 u}{\partial x^2}\right)$$

$$\Rightarrow \qquad s^2U(x, s) - su(x, 0) - u_t(x, 0) = \frac{d^2U}{dx^2}$$

$$\Rightarrow \qquad \frac{d^2U}{dx^2} - s^2U = 0$$

$$\therefore \qquad U = Ae^{+sx} + Be^{-sx}$$

Also, $u(0, t) = 10 \sin 2t$

$$\therefore \qquad L(u(0, t)) = U(0, s) = \frac{10\times 2}{s^2+4}$$

$$\therefore \qquad A + B = \frac{20}{s^2+4}$$

Again, $\qquad \lim\limits_{x\to\infty} u(x, t) = 0$

$$\Rightarrow \qquad \lim_{x\to\infty} L(u(x, t)) = \lim_{x\to\infty} U(x, s) = 0$$

which gives $\qquad A = 0,\ B = \dfrac{20}{s^2+4}$

Therefore, $$U = \frac{20}{s^2+4}e^{-sx}.$$

Thus, $$u(x, t) = L^{-1}(U(x, t)) = 20\, u_x(t) \times \frac{1}{2} \sin 2(t-x)$$

$$\left\{\because F = \frac{1}{s^2+4} \Rightarrow f = L^{-1}(F) = \frac{\sin 2t}{2}\right\}$$

$$= 10\, u_x(t) \sin 2(t-x).$$

EXAMPLE 6.5

Solve $\frac{\partial^2 u}{\partial x^2} - \frac{\partial^2 u}{\partial t^2} = xt$ when $u(x, t)$ is bounded for all x and t with $u(x, 0) = 0 = u_t(x, 0)$ and $u(0, t) = 0$.

Solution: Taking the Laplace transform of the give partial differential equation, we get

$$L\left(\frac{\partial^2 u}{\partial x^2}\right) - L\left(\frac{\partial^2 u}{\partial t^2}\right) = xL(t)$$

$$\Rightarrow \quad \frac{d^2U}{dx^2} - s^2U(x, s) + su(x, 0) + u_t(x, 0) = x\left(\frac{1}{s^2}\right)$$

$$\Rightarrow \quad \frac{d^2U}{dx^2} - s^2U = \frac{x}{s^2}$$

Thus, $$U = Ae^{sx} + Be^{-sx} + \frac{1}{s^2}\frac{x}{D^2 - s^2}$$

$$= Ae^{sx} + Be^{-sx} - \frac{1}{s^2}\frac{1}{-s^2}\left(1 - \frac{D^2}{s^2}\right)^{-1} x$$

$$= Ae^{sx} + Be^{-sx} + \frac{1}{s^4}\left(1 + \frac{D^2}{s^2} + \dots\right)x$$

$$= Ae^{sx} + Be^{-sx} + \frac{x}{s^4}$$

Now, $$u(0, t) = 0$$

$$\therefore \quad L(u(0, t)) = U(0, s) = 0$$

$$\Rightarrow \quad A + B = 0$$

Since $u(x, t)$ is bounded for all x and t.

$\therefore$ $L(u(x, t)) = U(x, s)$ is also bounded

$\Rightarrow \lim_{x \to \infty} U(x, s)$ is finite.

which gives $\qquad A = 0; B = 0$

Therefore, $\qquad U(x, s) = x/s^4$

Hence, $\qquad u(x, t) = L^{-1}(U(x, s)) = \dfrac{xt^3}{6}$

which is the required solution.

EXAMPLE 6.6

Solve the boundary value problem

$\dfrac{\partial u}{\partial x} + 4\dfrac{\partial u}{\partial t} = 8t$ $(t, x > 0)$ with the condition $u(x, 0) = 0$ and $u(0, t) = 2t^2$.

Solution: Taking the Laplace transform of the given differential equation, we obtain

$$\frac{dU}{dx} + 4sU - 4u(x, 0) = -\frac{8}{s^2}$$

$$\Rightarrow \qquad \frac{dU}{dx} + 4sU = -\frac{8}{s^2}$$

which is a linear differential equation with IF $= e^{\int 4sdx} = e^{4sx}$ and so its solution is given by

$$Ue^{4sx} = \int -\frac{8}{s^2}e^{4sx}dx$$

$$= -\frac{8}{s^2}\frac{e^{4sx}}{Ys} + C$$

$$\therefore \qquad U = -\frac{2}{s^3} + ce^{-4sx}$$

Now, $u(0, t) = 2t^2$.

$$\therefore \qquad L[u(0, t)] = U(0, s) = 4/s^3$$

Thus, $\qquad U(0, s) = \dfrac{4}{s^3} = -\dfrac{2}{s^3} + C$

$$\therefore \qquad C = 6/s^3$$

Therefore, $\qquad U = -\dfrac{2}{s^3} + \dfrac{6}{s^3}e^{-4sx}$

Hence, $\qquad u(x, t) = L^{-1}[U(x, s)] = -t^2 + u_{4x}(t)\, 3(t - 4x)$

$$\left\{\because F = \frac{6}{s^3} \Rightarrow f = L^{-1}(F) = 3\cdot t^2\right\}$$

EXAMPLE 6.7

A semi-infinite solid $x > 0$ is initially at temperature zero. At time $t > 0$, a constant temperature $V_0 > 0$ is applied and maintained at the face $x = 0$. Find the temperature at any point of the solid at any time $t > 0$.

Solution: Here, the temperature $u(x, t)$ at any point of the solid at any time $t > 0$ is governed by one-dimensional heat equation

$$\frac{\partial u}{\partial t} = c^2 \frac{\partial^2 u}{\partial x^2} \qquad (x, t > 0)$$

with the boundary and initial conditions

$$u(0, t) = V_0, u(x, 0) = 0$$

Taking the Laplace transform, we obtain

$$sU - u(x, 0) = c^2 \frac{d^2 U}{dx^2}$$

$$\Rightarrow \quad \frac{d^2 U}{dx^2} - \frac{s}{c^2} U = 0$$

$$\therefore \quad U = Ae^{\frac{\sqrt{s}}{c}x} + Be^{-\frac{\sqrt{s}}{c}x}$$

Since u is finite when $x \to \infty$

$\therefore \lim_{x \to \infty} L(u) = \lim_{x \to \infty} U$ is also finite

which gives $A = 0$.

Also, $$u(0, t) = V_0$$

$$\therefore \quad L(u(0, t)) = U(0, s) = \frac{V_0}{s}$$

Thus, $$U(0, s) = \frac{V_0}{s} = B$$

Hence, $$U(x, s) = \frac{V_0}{s} e^{-\frac{\sqrt{s}}{c}x}$$

$$\therefore \quad u(x, t) = L^{-1}(U(x, s)) = V_0 L^{-1}\left(\frac{e^{-\frac{\sqrt{s}}{c}x}}{s}\right)$$

$$= V_0 \, erfc\left(\frac{x}{2c\sqrt{t}}\right) \qquad \{\text{with } \lambda = x/c\}$$

which is the required solution.

Note. Since $L^{-1}\left(\frac{e^{-\lambda\sqrt{s}}}{s}\right) = erfc\left(\frac{\lambda}{2\sqrt{t}}\right)$ as given below.

$$L^{-1}(e^{-\sqrt{s}}) = L^{-1}\left(1 - \sqrt{s} + \frac{s}{2!} - \frac{s^{3/2}}{3!} + \frac{s^2}{4!} - \frac{s^{5/2}}{5!} + \ldots\right)$$

$$= L^{-1}(1) - L^{-1}(\sqrt{s}) + \frac{1}{2!}L^{-1}(s) - \frac{1}{3!}L^{-1}(s^{3/2}) + \frac{1}{4!}L^{-1}(s^2) - \frac{1}{5!}L^{-1}(s^{5/2}) + \ldots$$

Now, $$L^{-1}\left(s^{n+\frac{1}{2}}\right) = L^{-1}\left(\frac{1}{s^{-n-\frac{1}{2}}}\right) = \frac{t^{-n-\frac{3}{2}}}{\Gamma\left(-n-\frac{1}{2}\right)}$$

$$= \frac{(-1)^{n+1}}{\sqrt{\pi}}\left(\frac{1}{2}\right)\left(\frac{3}{2}\right)\left(\frac{5}{2}\right)\cdots\left(\frac{2n+1}{2}\right)t^{n-3/2}$$

Also, $L^{-1}(s^n) = 0$ if n is a positive integer

$$\therefore\ L^{-1}(e^{-\sqrt{s}}) = -\frac{(-1)t^{-3/2}}{\sqrt{\pi}}\frac{1}{2} - \frac{1}{3!}\frac{(-1)^2}{\sqrt{\pi}}\left(\frac{1}{2}\right)\left(\frac{3}{2}\right)t^{-5/2} - \frac{1}{5!}\frac{(-1)^3}{\sqrt{\pi}}\left(\frac{1}{2}\right)\left(\frac{3}{2}\right)\left(\frac{5}{2}\right)t^{-7/2} + \ldots$$

$$= \frac{1}{2\sqrt{\pi}\,t^{3/2}}\left\{1 - \left(\frac{1}{4t}\right) + \frac{(1/4t)^2}{2!} - \frac{(1/4t)^3}{3!} + \ldots\right\}$$

$$= \frac{1}{2\sqrt{\pi}\,t^{3/2}}\,e^{-1/4t}$$

Since
$$L^{-1}\left(\frac{F(s)}{s}\right) = \int_0^t F(s)dt$$

$$\therefore\quad L^{-1}\left(\frac{e^{-\sqrt{s}}}{s}\right) = \int_0^t \frac{1}{2\sqrt{\pi}\,t^{3/2}}e^{-1/4}dt$$

Let
$$t = \frac{1}{4y^2} \text{ so that } dt = -\frac{dy}{2y^3}$$

Thus,
$$L^{-1}\left(\frac{e^{-\sqrt{s}}}{s}\right) = \int_{\infty}^{1/2\sqrt{t}} -\frac{2}{\sqrt{\pi}}e^{-y^2}dy = \frac{2}{\sqrt{\pi}}\int_{1/2\sqrt{t}}^{\infty} e^{-y^2}dy$$

$$= erfc\left(\frac{1}{2\sqrt{t}}\right).$$

where the complementary error function is given by

$$erf\,c(t) = 1 - erf(t)$$

$$= 1 - \frac{2}{\sqrt{\pi}}\int_0^t e^{-y^2}dy$$

$$= \frac{2}{\sqrt{\pi}}\left\{\int_0^{\infty} e^{-y^2}dy - \int_0^t e^{-y^2}dy\right\}$$

$$= \frac{2}{\sqrt{\pi}}\int_t^{\infty} e^{y^2}dy$$

So, we have

$$L^{-1}\left(\frac{e^{-\sqrt{s}}}{s}\right) = erfc\left(\frac{1}{2\sqrt{t}}\right).$$

By change of scale property, we obtain

$$L^{-1}\left(\frac{e^{-\sqrt{\lambda^2 s}}}{\lambda^2 s}\right)=\frac{1}{\lambda^2}\,erfc\left(\frac{1}{2\sqrt{t/\lambda^2}}\right)$$

$\Rightarrow$
$$L^{-1}\left(\frac{e^{-\lambda\sqrt{s}}}{s}\right)=erfc\left(\frac{\lambda}{2\sqrt{t}}\right).$$

EXAMPLE 6.8

A string is stretched between two fixed points (0, 0) and (c, 0). If it is displaced into the curve $y = b\sin\left(\frac{\pi x}{c}\right)$ and released from rest in that position at time $t = 0$, find its displacement at any time $t > 0$ and any point $0 < x < c$.

Solution: Here the displacement $u(x, t)$ of the string is governed by the differential equation

$$\frac{\partial^2 u}{\partial t^2}=a^2\frac{\partial^2 u}{\partial x^2}$$

with boundary conditions $u(0, t) = 0 = u(c, t)$ and initial condition $u(x, 0) = b\sin\frac{\pi x}{c}$

and $u_t(x, 0) = 0$

Taking the Laplace transform, we obtain

$$s^2U - su(x, 0) - u_t(x, 0) = a^2\frac{d^2U}{dx^2}$$

$\Rightarrow$
$$\frac{d^2U}{dx^2}-\frac{s^2}{a^2}U=-\frac{bs}{a^2}\sin\frac{\pi x}{c}$$

$\therefore$
$$U = Ae^{sx/a} + Be^{sx/a} - \frac{bs}{a^2}\,\frac{\sin\frac{\pi x}{c}}{-\frac{\pi^2}{c^2}-\frac{s^2}{a^2}}$$

$$= Ae^{sx/a} + Be^{-sx/a} + bs\,\frac{\sin\frac{\pi x}{c}}{s^2+\frac{\pi^2 a^2}{c^2}}$$

Now, $u(0, t) = 0 = u(c, t)$

$\therefore$ $L(u(0, t)) = 0 = L(u(c, t))$

$\Rightarrow$ $U(0, s) = 0 = U(c, s)$

Thus, $A + B = 0$

and $Ae^{sc/a} + Be^{-sc/a} = 0$

Solving, we have $$A = 0 = B$$

Hence, $$U = bs\,\frac{\sin\dfrac{\pi x}{c}}{s^2 + \dfrac{\pi^2 a^2}{c^2}}$$

$\therefore$ $$u(x, t) = L^{-1}(U(x, s)) = b\cos\frac{\pi a t}{c}\sin\frac{\pi x}{c}$$

which is the required solution.

EXAMPLE 6.9

Solve the boundary value problem $\partial^2 u/\partial t^2 = a^2\,\partial^2 u/\partial x^2 - g$, $(x > 0, t > 0)$ with the boundary condition $u(x, 0) = 0 = u_t(x, 0)$ $(x > 0)$ and $u(0, t) = 0$, $\lim_{x \to \infty} u_x(x, t) = 0$ $(t \geq 0)$.

Solution: Taking Laplace transform of the given partial differential equation, we obtain

$$s^2 U - s u(x, 0) - u_t(x, 0) = a^2 \frac{d^2 U}{dx^2} - \frac{g}{s}$$

$\Rightarrow$ $$\frac{d^2 U}{dx^2} - \frac{s^2}{a^2} U = \frac{g}{a^2 s}$$

$\therefore$ $$U = Ae^{sx/a} + Be^{-sx/a} + \frac{1}{a^2 s}\,\frac{g}{D^2 - \dfrac{s^2}{a^2}}$$

$$= Ae^{sx/a} + Be^{-sx/a} + \frac{1}{a^2 s}\,\frac{1}{-\dfrac{s^2}{a^2}}\,\frac{g}{\left(1 - \dfrac{a^2}{s^2} D^2\right)}$$

$$= Ae^{sx/a} + Be^{-sx/a} - \frac{g}{s^3}\left(1 - \frac{a^2}{s^2} D^2\right)^{-1} (1)$$

$$= Ae^{sx/a} + Be^{-sx/a} - \frac{g}{s^3}\left(1 - \frac{a^2}{s^2} D^2 + \ldots\right)(1)$$

$$= Ae^{sx/a} + Be^{-sx/a} - \frac{g}{s^3}$$

Now, $$U_x = Ae^{sx/a}\,\frac{s}{a} + Be^{-sx/a}\left(-\frac{s}{a}\right)$$

Since $$U_x(x, t) = 0 \text{ as } x \to \infty$$

$\therefore$ $$L(u_x(x, t)) = U_x(x, s) = 0 \text{ as } x \to \infty$$

which gives $$A = 0$$

Thus, $$U = Be^{-sx/a} - \frac{g}{s^3}$$

Also, $$u(0, t) = 0 \Rightarrow L(u(0, t)) = U(0, s) = 0$$

$$0 = B - \frac{g}{s^3} \Rightarrow B = \frac{g}{s^3}$$

$\therefore$ $$U = \frac{g}{s^3}e^{-sx/a} - \frac{g}{s^3}$$

Therefore, $$u(x, t) = L^{-1}(U(x, s)) = u_{x/a}(t)\,\frac{g}{2}\left(t - \frac{x}{a}\right)^2 - \frac{1}{2}gt^2$$

which is the required solution.

EXAMPLE 6.10

Solve $\frac{\partial u}{\partial t} = \frac{\partial^2 u}{\partial x^2} - 4u;\, u(0, t) = 0 = u(\pi, t)$ and $u(x, 0) = 3 \sin x - 4 \sin 2x$.

Solution: Taking Laplace transform of the given partial differential equation, we obtain

$$sU - u(x, 0) = \frac{d^2U}{dx^2} - 4U$$

$\Rightarrow$ $$\frac{d^2U}{dx^2} - (s + 4)U = -3 \sin x + 4 \sin 2x$$

$\therefore$ $$U = Ae^{\sqrt{s+4}\,x} + Be^{-\sqrt{s+4}\,x} - \frac{3\sin x}{-1-(s+4)} + \frac{4\sin 2x}{-4-(s+4)}$$

$$= Ae^{\sqrt{s+4}\,x} + Be^{-\sqrt{s+4}\,x} + \frac{3\sin x}{s+5} - \frac{4\sin 2x}{s+8}$$

Now, $$u(0, t) = 0 = u(\pi, t)$$

$\Rightarrow$ $$L(u(0, t)) = 0 = L(u(\pi, t))$$

$\Rightarrow$ $$U(0, s) = 0 = U(\pi, s)$$

So, $$A + B = 0 \quad \text{and} \quad Ae^{\pi\sqrt{s+y}} + Be^{-\pi\sqrt{s+y}} = 0$$

which gives $$A = 0 = B$$

Thus, $$U = \frac{3\sin x}{s+5} - 4\frac{\sin 2x}{s+8}$$

$\therefore$ $$u(x, t) = L^{-1}(U(x, s)) = 3e^{-5t} \sin x - 4e^{-8t} \sin 2x$$

which is the required solution.

EXAMPLE 6.11

Solve: $\frac{\partial u}{\partial x} + 6x\frac{\partial u}{\partial t} = 12x$; $u(x, 0) = 2$, $u(0, t) = 2$.

Solution: Taking Laplace transform of the given partial differential equation, we obtain

$$\frac{dU}{dx} + 6x[sU - u(x, 0)] = \frac{12x}{s}$$

$$\Rightarrow \qquad \frac{dU}{dx} + 6sxU = \frac{12x}{s} + 12x$$

which is a linear differential equation of the first order with IF $= e^{\int 6sxdx} = e^{3sx^2}$

So, its solution is

$$Ue^{3sx^2} = \int\left(\frac{1}{s}+1\right)12xe^{3sx^2}dx$$

$$= \left(\frac{1}{s}+1\right)\frac{2e^{3sx^2}}{s} + a$$

$$\therefore \qquad U = 2\left(\frac{1}{s^2}+\frac{1}{s}\right) + ae^{-3sx^2}$$

Now, $\qquad u(0, t) = 2 \Rightarrow L[u(0, t)] = U(0, s) = \frac{2}{s}$

Thus, $\qquad U(0, s) = \frac{2}{s} = 2\left(\frac{1}{s^2}+\frac{1}{s}\right) + a \Rightarrow a = -\frac{2}{s^2}$

Hence, $\qquad U = \frac{2}{s^2} + \frac{2}{s} - \frac{2}{s^2}e^{-3sx^2}$

$$\therefore \qquad u(x, t) = L^{-1}(U(x, s)) = 2(t + 1) - U_{3x^2}(t) \times 2(t - 3x^2)$$

which is the required solution. ▲▲

EXERCISE 6.1

Solve the following partial differential equations using Laplace transform method:

1. $\frac{\partial u}{\partial t} = \frac{\partial^2 u}{\partial x^2}$; $u(x, 0) = \sin 2\pi x$, $u(1, t) = 0$ where $0 < x < 1$.

2. $\frac{\partial u}{\partial t} = 3\frac{\partial^2 u}{\partial x^2}$; $u_x(0, t) = 0$, $u\left(\frac{\pi}{2}, t\right) = 0$, $u(x, 0) = \cos 3x - 2\cos 9x$.

3. $\frac{\partial u}{\partial x} + x\frac{\partial u}{\partial t} = 0$, $u(x, 0) = 1$, $u(0, t) = t$.

4. $\frac{\partial u}{\partial x}+cx\frac{\partial u}{\partial t}$ = 0, $u(x, 0) = 0$, $u(0, t) = pt^2$ where c and p are constants.

5. $\frac{\partial u}{\partial x}+x\frac{\partial u}{\partial t}$ = xt^2, $u(x, 0) = 0$, $u(0, t) = t$.

6. $\frac{\partial u}{\partial x}+x\frac{\partial u}{\partial t}$ = $4xt$, $u(x, 0) = 0$, $u(0, t) = 0$.

7. $\frac{\partial^2 u}{\partial t^2}=\frac{\partial^2 u}{\partial x^2}$; $x > 0, t > 0$

$u(x, 0) = 0$, $u_t(x, 0) = e^{-x}$, $u(0, t) = \sin t$, $u(x, t)$ is bounded as $x \to \infty$.

8. $\frac{\partial^2 u}{\partial t^2}=\frac{\partial^2 u}{\partial x^2}$; $0 < x < 1, t > 0$, $u(x, 0) = 0$, $u_t(x, 0) = \sin 3\pi x(0 < x < 1)$; $u(0, t) = 0$, $u(1, t) = 0$.

9. $u_{tt} = u_{xx}$; $x > 0$, $t > 0$. $u(x, 0) = 0$, $u_t(x, 0) = 0$, $u(0, t) = \begin{cases} \sin t & 0 \le t \le \pi \\ 0 & t > \pi \end{cases}$

$u(x, t)$ is bounded as $x \to \infty$.

10. $u_{tt} = u_{xx}$ $(x > 0, t > 0)$ $u(x, 0) = e^{-x}$, $u_t(x, 0) = 0$; $u(0, t) = 0$ and $u(x, t)$ is bounded as $x \to \infty$.

11. $\partial u/\partial x - \partial u/\partial t = 1 - e^{-t}$, $u(x, 0) = x$ and $u(x, t)$ is bounded on $0 < x < 1, t > 0$.

12. $\frac{\partial^2 u}{\partial x^2}=\frac{1}{k}\frac{\partial u}{\partial t}$ with the conditions $u = 0$ when $x = \infty$, $u = 0$ when $t = 0$.

13. $\partial u/\partial t = \partial^2 u/\partial x^2$ where $y(\pi/2, t) = 0$ $(\partial u/\partial x)_{x=0}$ and $u(x, 0) = \cos 5x$.

14. $\partial u/\partial t = \partial^2 u/\partial x^2$; $y(x, 0) = 3 \sin 2\pi x$, $y(0, t) = 0 = y(1, t)$ $(t > 0, 0 < x < 1)$.

15. Find the bounded solution $u(x, t)$, $0 < x < 5$, $t > 0$ of the boundary-value problem

$$\frac{\partial u}{\partial t}=2\frac{\partial^2 u}{\partial x^2};\ u(0, t) = 0,\ u(5, t) = 0;\ u(x, 0) = 10 \sin 4\pi x - 5 \sin 6\pi x.$$

ANSWERS

Exercise 6.1

1. $u(x, t) = e^{-4\pi^2 t} \sin 2\pi x$.
2. $u(x, t) = e^{-27t} \cos 3x - 2e^{-243t} \cos 9x$.
3. $u(x, t) = 1+\left[\left(t-\frac{x^2}{2}\right)-1\right]u_{\frac{x^2}{2}}(t)$.

4. $u(x, t) = p\left(t - \frac{cx^2}{2}\right)^2 u_{\frac{cx^2}{2}}(t).$

5. $u(x, t) = \frac{t^3}{3} + \left[\left(t - \frac{x^2}{2}\right) - \frac{1}{3}\left(t - \frac{x^2}{2}\right)^3\right] u_{\frac{x^2}{2}}(t).$

6. $u(x, t) = 2t^2 - 2\left(t - \frac{x^2}{2}\right)^2 u_{x^2/2}(t).$

7. $u(x, t) = e^{-x} \sinh t + (\sin (t - x) - \sinh (t - x))u_x(t).$

8. $u(x, t) = \frac{1}{3\pi} \sin 3\pi x \ \sin 3\pi t.$

9. $u(x, t) = \sin (t - x)u_x(t) \sin(t - x)u_{\pi + x}(t).$

10. $u(x, t) = e^{-x} \cos ht - \cos h(t - x)u_x(t).$

11. $u(x, t) = x + 1 - e^{-t}.$

12. $u(x, t) = \frac{c_2(t)x}{\sqrt{k}} \times \frac{1}{2\sqrt{\pi x^3}} e^{-x^2/4kt}.$

13. $u(x, t) = e^{-25} \cos 5x.$

14. $u(x, t) - 3e^{-4\pi^2 t} \sin 4x.$

15. $u(x, t) = 20e^{-32\pi^2 t} \cos 3x - 5e^{-72\pi^2 t} \cos 9x.$

▲▲

7

CHAPTER

Introduction to Fourier Series

A series expansion in terms of the trigonometric sine and cosine functions is called a Fourier series. Many functions including some discontinuous periodic functions can be expanded in a Fourier series. It is useful in the study of heat conduction, mechanics, concentration of chemicals and pollutants, electrostatics, acoustics, etc.

Fourier series is powerful method to solve ordinary and partial differential equations particularly with periodic functions appearing as non-homogeneous terms. While, Taylor's series expansion is valid only for functions which are continuous and differentiable, Fourier series is possible not only for continuous functions but for periodic functions, functions—discontinuous in their values and derivatives. Further, because of the periodic nature, Fourier series constructed for one period is valid for all values. Harmonic analysis is the theory of expanding functions in Fourier series.

Definition. The trigonometric series of the form

$$f(x) = \frac{a_0}{2} + \sum_{n=1}^{\infty}\left(a_n \cos\frac{n\pi x}{l} + b_n \sin\frac{n\pi x}{l}\right) \tag{1}$$

is called the Fourier series of $f(x)$ in the interval $-l \le x \le l$.

The coefficients a_0, a_n, b_n are given by

$$a_0 = \frac{1}{l}\int_{-l}^{l} f(x)dx \tag{2}$$

$$a_n = \frac{1}{l}\int_{-l}^{l} f(x)\cos\frac{n\pi x}{l}dx \tag{3}$$

$$b_n = \frac{1}{l}\int_{-l}^{l} f(x)\sin\frac{n\pi x}{l}dx \tag{4}$$

These formulae are called the Euler's formulae.

Note 1. The function $f(x)$ satisfies the Dirichlet condition in the interval $[-l, l]$ viz:

(a) $f(x)$ is defined and is single-valued except possibly at finite number of points in the interval $[-l, l]$ and

(b) $f(x)$ and $f'(x)$ are piecewise continuous in the interval $[-l, l]$.

Note 2. At a point of discontinuity $x = a$ (say), $f(a) = \frac{1}{2}[f(a+0) + f(a-0)]$. ▲▲

EXAMPLE 7.1

Find the Fourier series expansion of the periodic function $f(x) = x - x^2$, $-\pi \le x \le \pi$ and $f(x + 2\pi) = f(x)$.

Solution: The Fourier coefficients are obtained as follows

$$a_0 = \frac{1}{\pi}\int_{-\pi}^{\pi}(x - x^2)dx = \frac{1}{\pi}\left[\frac{x^2}{2} - \frac{x^3}{3}\right]_{-\pi}^{\pi} = -\frac{2\pi^2}{3}$$

$$a_n = \frac{1}{\pi}\int_{-\pi}^{\pi}(x - x^2)\cos\frac{n\pi x}{\pi}dx$$

$$= \frac{1}{\pi}\left[(x - x^2)\frac{\sin nx}{n} - (1 - 2x)\left(-\frac{\cos nx}{n^2}\right) + (-2)\left(-\frac{\sin nx}{n^3}\right)\right]_{-\pi}^{\pi}$$

$$= -\frac{4(-1)^n}{n^2} \quad \{\because \cos n\pi = (-1)^n \text{ and } \sin n\pi = 0\}$$

and $$b_n = \frac{1}{\pi}\int_{-\pi}^{\pi}(x - x^2)\sin\frac{n\pi x}{\pi}dx$$

$$= \frac{1}{\pi}\left[(x - x^2)\left(-\frac{\cos nx}{n}\right) - (1 - 2x)\left(-\frac{\sin nx}{n^2}\right) + (-2)\left(\frac{\cos nx}{n^3}\right)\right]_{-\pi}^{\pi}$$

$$= -\frac{2(-1)^n}{n}.$$

Therefore, the Fourier expansion of the given function on $[-\pi, \pi]$ is given by

$$x - x^2 = -\frac{2\pi^{2/3}}{2} + \sum_{n=1}^{\infty}\left(-\frac{4(-1)^n}{n^2}\cos\frac{n\pi x}{\pi} - \frac{2(-1)^n}{n}\sin\frac{n\pi x}{n}\right)$$

$$= -\frac{\pi^2}{3} + 4\left[\frac{\cos x}{1^2} - \frac{\cos 2x}{2^2} + \frac{\cos 3x}{3^2} - \frac{\cos 4x}{4^2} + \ldots\right]$$

$$+2\left[\frac{\sin x}{1} - \frac{\sin 2x}{2} + \frac{\sin 3x}{3} - \frac{\sin 4x}{4} + \ldots\right]$$

Note. Putting $x = 0$, we find another interesting series

$$0 = -\frac{\pi^2}{3} + 4\left(\frac{1}{1^2} - \frac{1}{2^2} + \frac{1}{3^2} - \frac{1}{4^2} + \ldots\right)$$

$$\Rightarrow \qquad \frac{1}{1^2} - \frac{1}{2^2} + \frac{1}{3^2} - \frac{1}{4^2} + \ldots = \frac{\pi^2}{12}.$$

EXAMPLE 7.2

Find the Fourier series expansion of the following periodic function of period 4.

$$f(x) = \begin{cases} 2+x; & -2 \le x \le 0 \\ 2-x; & 0 < x \le 2 \end{cases} \quad \text{and } f(x+4) = f(x).$$

Solution: The Fourier coefficients are obtained as follows

$$a_0 = \frac{1}{2}\int_{-2}^{2} f(x)dx = \frac{1}{2}\left[\int_{-2}^{0}(2+x)dx + \int_{0}^{2}(2-x)dx\right]$$

$$= \frac{1}{2}\left[\left(2x + \frac{x^2}{2}\right)_{-2}^{0} + \left(2x - \frac{x^2}{2}\right)_{0}^{2}\right]$$

$$= \frac{1}{2}[0 - (-4+2) + (4-2) - 0] = 2$$

$$a_n = \frac{1}{2}\int_{-2}^{2} f(x)\cos\frac{n\pi x}{2}dx$$

$$= \frac{1}{2}\left[\int_{-2}^{0}(2+x)\cos\frac{n\pi x}{2}dx + \int_{0}^{2}(2-x)\cos\frac{n\pi x}{2}dx\right]$$

$$= \frac{1}{2}\left[\left((2+x)\frac{\sin(n\pi x/2)}{n\pi/2} + \frac{\cos(n\pi x/2)}{(n\pi/2)^2}\right)_{-2}^{0} + \left((2-x)\frac{\sin(n\pi x/2)}{x\pi/2} - \frac{\cos(n\pi x/2)}{(n\pi/2)^2}\right)_{0}^{2}\right]$$

$$= \frac{1}{2}\left[\left(\frac{1}{(n\pi/2)^2} - \frac{\cos(n\pi)}{(n\pi/2)^2}\right) + \left(\frac{1}{(n\pi/2)^2} - \frac{\cos(n\pi)}{(n\pi/2)^2}\right)\right]$$

$$= \frac{4}{n^2\pi^2}(1-(-1)^n) = \begin{cases} 0, & \text{for } n \text{ even} \\ \dfrac{8}{n^2\pi^2}, & \text{for } n \text{ odd} \end{cases}$$

$$b_n = \frac{1}{2}\int_{-2}^{2} f(x)\sin\left(\frac{n\pi x}{2}\right) dx$$

$$= \frac{1}{2}\left[\int_{-2}^{0} (2+x)\sin\left(\frac{n\pi x}{2}\right) dx + \int_{0}^{2} (2-x)\sin\left(\frac{n\pi x}{2}\right) \times dx\right]$$

$$= \frac{1}{2}\left[\left(-(2+x)\frac{\cos(n\pi x/2)}{n\pi/2} + \frac{\sin(n\pi x/2)}{(n\pi/2)^2}\right)_{-2}^{0} + \left(-(2-x)\frac{\cos(n\pi x/2)}{n\pi/2} - \frac{\sin(n\pi x/2)}{(n\pi/2)^2}\right)_{0}^{2}\right]$$

$$= \frac{1}{2}\left[-\frac{2}{n\pi/2} + \frac{2}{n\pi/2}\right] = 0.$$

Therefore, the Fourier series expansion is given by

$$f(x) = \frac{1}{2} + \sum_{n=1}^{\infty} \frac{4}{n^2\pi^2}(1-(-1)^n)\cos\left(\frac{n\pi x}{2}\right)$$

$$= 1 + \frac{8}{\pi^2}\left[\frac{\cos \pi x/2}{1^2} + \frac{\cos 3\pi x/2}{3^2} + \ldots\right]$$

$$= 1 + \frac{8}{\pi^2}\sum_{m=1}^{\infty} \frac{1}{(2m-1)^2}\cos\left[(2m-1)\frac{\pi x}{2}\right]$$

Note. At the point $x = 0$, the given function is continuous. Therefore, the series converges to $f(0) = 2$. Hence, taking $x = 0$, we obtain

$$2 = 1 + \frac{8}{\pi^2}\left\{\frac{1}{1^2} + \frac{1}{3^2} + \frac{1}{5^2} + \ldots\right\}$$

$$\Rightarrow \qquad \frac{1}{1^2} + \frac{1}{3^2} + \frac{1}{5^2} + \ldots = \frac{\pi^2}{8}.$$

EXAMPLE 7.3

Find the Fourier series for the function defined by

$$f(x) = \begin{cases} 0, & -\pi \le x < 0 \\ \sin x, & 0 \le x \le \pi \end{cases}.$$

Solution: The Fourier coefficients are obtained as follows

$$a_0 = \frac{1}{\pi}\left\{\int_{-\pi}^{0} 0\,dx + \int_{0}^{\pi} \sin x dx\right\}$$

$$= \frac{1}{\pi}[-\cos x]_0^{\pi} = \frac{1}{\pi}[-\cos\pi + \cos o] = \frac{2}{\pi}$$

$$a_n = \frac{1}{\pi}\left[\int_{-\pi}^{0} 0\cdot\cos\frac{n\pi x}{\pi}dx + \int_{0}^{\pi}\sin x\cos\frac{n\pi x}{n}dx\right]$$

$$= \frac{1}{\pi}\int_{0}^{\pi} \sin x \cos nx dx$$

$$= \frac{1}{2\pi}\int_{0}^{\pi}[\sin (n+1)x - \sin (n-1)x]dx$$

$$= \frac{1}{2\pi}\left[-\frac{\cos(n+1)x}{n+1} + \frac{\cos(n-1)x}{n-1}\right]_0^{\pi}$$

$$= \frac{1}{2\pi}\left\{-\frac{(-1)^{n+1}+1}{n+1} + \frac{(-1)^{n-1}-1}{n-1}\right\} \quad \text{(for } n > 1\text{)}$$

$$= \begin{cases} 0, & \text{if } n \text{ is odd} \\ -\dfrac{-2}{\pi(n^2-1)}, & \text{if } n \text{ is even} \end{cases}.$$

$$b_n = \frac{1}{\pi}\left[\int_{-\pi}^{0} 0\cdot\sin\frac{n\pi x}{\pi}dx + \int_{0}^{\pi}\sin\frac{n\pi x}{\pi}dx\right]$$

$$= \frac{1}{2\pi}\int_{0}^{\pi}[\cos (n-1)x - \cos (n+1)x]dx$$

$$= \frac{1}{2\pi}\left[\frac{\sin(n-1)x}{n-1} - \frac{\sin(n+1)x}{n+1}\right]_0^{\pi}$$

$= 0$ for $n > 1$.

For $n = 1$

$$a_1 = \frac{1}{2\pi}\int_0^{\pi}\sin 2x dx = \frac{1}{2\pi}\left\{-\frac{\cos 2x}{-2}\right\}_0^{\pi} = 0$$

$$b_1 = \frac{1}{2\pi}\int_0^{\pi}2\sin^2 x dx = \frac{1}{2\pi}\int_0^{\pi}(1-\cos 2x)dx = \frac{1}{2}$$

Therefore, the Fourier series is given by

$$f(x) = \frac{2/\pi}{2} + \left(-\frac{2}{\pi}\right)\left[\frac{\cos 2x}{2^2-1} + \frac{\cos 4x}{4^2-1} + \ldots\right] + \frac{1}{2}\sin x$$

$$= \frac{1}{\pi} + \frac{1}{2}\sin x - \frac{2}{\pi}\sum_{m=1}^{\infty}\frac{\cos 2mx}{4m^2-1}.$$

EXAMPLE 7.4

Find the Fourier series expansion of $f(x) = e^{-x}$ in $(0, 2\pi)$.

Solution: The Fourier coefficients are obtained as follows

$$a_0 = \frac{1}{\pi}\int_0^{2\pi}e^{x}dx = \frac{1}{\pi}\left[\frac{e^{-x}}{-1}\right]_0^{2\pi} = \frac{1-e^{-2\pi}}{\pi}$$

$$a_n = \frac{1}{\pi}\int_0^{2\pi}e^{-x}\cos\left(\frac{n\pi x}{\pi}\right)dx$$

$$= \frac{1}{\pi}\frac{1}{n^2+1}[e^{-x}(-\cos nx + n\sin nx)]_0^{2\pi}$$

$$= \left(\frac{1-e^{-2\pi}}{\pi}\right)\frac{1}{n^2+1}$$

$$b_n = \frac{1}{\pi}\int_0^{2\pi}e^{-x}\sin\left(\frac{n\pi x}{\pi}\right)dx$$

$$= \frac{1}{\pi}\frac{1}{n^2+1}[e^{-x}(-\sin nx - n\cos nx)]_0^{2\pi}$$

$$= \left(\frac{1-e^{-2\pi}}{\pi}\right)\frac{n}{n^2+1}$$

The required Fourier series is

$$e^{-x} = \frac{1-e^{-2\pi}}{\pi}\left\{\frac{1}{2} + \sum_{n=1}^{\infty}\frac{1}{n^2+1}\cos\left(\frac{n\pi x}{\pi}\right) + \frac{n}{n^2+1}\sin\left(\frac{n\pi x}{\pi}\right)\right\}$$

$$= \frac{1-e^{-2\pi}}{\pi}\left[\frac{1}{2}+\left(\frac{1}{2}\cos x+\frac{1}{5}\cos 2x+\frac{1}{10}\cos 3x+\ \ldots\right)\right.$$
$$\left.+\left(\frac{1}{2}\sin x+\frac{2}{5}\sin 2x+\frac{3}{10}\sin 3x+\ \ldots\right)\right]$$

EXAMPLE 7.5

Expand $f(x) = \sqrt{1-\cos x}$; $0 < x < 2\pi$ in a Fourier series. Hence, deduce that

$$\frac{1}{1\cdot 3}+\frac{1}{3\cdot 5}+\frac{1}{5\cdot 7}+\ \ldots=\frac{1}{2}.$$

Solution: We have $f(x) = \sqrt{1-\cos x} = \sqrt{2}\sin(x/2)$

The Fourier coefficients are obtained as follows

$$a_0 = \frac{1}{\pi}\int_0^{2\pi}\sqrt{2}\sin\frac{x}{2}dx = \frac{\sqrt{2}}{\pi}\left[-\frac{\cos x/2}{1/2}\right]_0^{2\pi} = \frac{4\sqrt{2}}{\pi}$$

$$a_n = \frac{1}{\pi}\int_0^{2\pi}\sqrt{2}\sin\frac{x}{2}\cos\left(\frac{n\pi x}{\pi}\right)dx$$

$$= \frac{\sqrt{2}}{2\pi}\int_0^{2\pi}\left[\sin\left(n+\frac{1}{2}\right)x-\sin\left(n-\frac{1}{2}\right)x\right]dx$$

$$= \frac{1}{\sqrt{2}\pi}\left[-\frac{\cos\left(\frac{2n+1}{2}\right)x}{\frac{2n+1}{2}}+\frac{\cos\left(\frac{2n-1}{2}\right)x}{\frac{2n-1}{2}}\right]_0^{2\pi}$$

$$= \frac{2}{\sqrt{2}\pi}\left[-\frac{1}{2n+1}(\cos(2n+1)\pi-1)+\frac{1}{2n-1}\{\cos(2n-1)\pi-1\}\right]$$

$$= \frac{\sqrt{2}}{\pi}\left\{\frac{2}{2n+1}-\frac{2}{2n-1}\right\} = -\frac{4\sqrt{2}}{\pi(4n^2-1)}$$

$$\{\because \cos(2n+1)\pi = \cos(2n-1)\pi = -1\}$$

$$b_n = \frac{1}{\pi}\int_0^{2\pi}\sqrt{2}\sin\frac{x}{2}\sin\left(\frac{n\pi x}{\pi}\right)dx.$$

$$= \frac{\sqrt{2}}{\sqrt{2}\pi}\int_0^{2\pi}\left[\cos\left(n-\frac{1}{2}\right)x - \cos\left(n+\frac{1}{2}\right)x\right]dx$$

$$= \frac{1}{\sqrt{2}\pi}\left[\frac{\sin\left(\frac{2n-1}{2}\right)x}{\frac{2n-1}{2}} - \frac{\sin\left(\frac{2n+1}{2}\right)x}{\frac{2n+1}{2}}\right]_0^{2\pi}$$

$$= \frac{2}{\sqrt{2}\pi}\left\{\frac{1}{2n-1}\{\sin(2n-1)\pi - 0\} - \frac{1}{2n+1}\{\sin(2n+1)\pi - 0\}\right\} = 0.$$

The, required Fourier series expansion is

$$\sqrt{1-\cos x} = \frac{2\sqrt{2}}{\pi} - \sum_{n=1}^{\infty}\frac{4\sqrt{2}}{(4n^2-1)\pi}\cos nx$$

When $x = 0$; we have

$$0 = \frac{2\sqrt{2}}{\pi} - \frac{4\sqrt{2}}{\pi}\sum_{n=1}^{\infty}\frac{1}{4n^2-1}$$

$$\Rightarrow \quad \frac{1}{1\cdot 3} + \frac{1}{3\cdot 5} + \frac{1}{5\cdot 7} + \ldots \text{ to } \infty = \frac{1}{2}.$$

▲▲

Fourier Series for Even and Odd Functions

We know that a function $f(x)$ is said to be even if $f(-x) = f(x)$ for all x.

For example, x^2 and $\cos x$ are even functions. The graph of an even function is symmetric about the y-axis.

A function $f(x)$ is said to be odd if $f(-x) = -f(x)$ for all x.

For example, x^3 and $\sin x$ are odd functions. The graph of an odd function is skew symmetric about the origin and $f(0) = 0$.

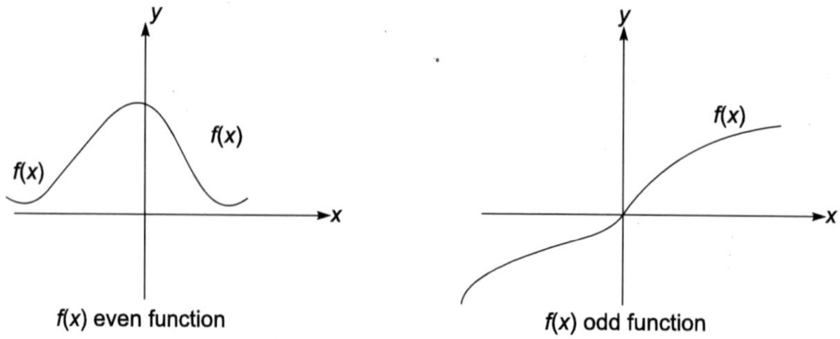

Fig. 7.1

It is clear from the figure that

$$\int_{-a}^{a} f(x)dx = \begin{cases} 2\int_{0}^{a} f(x)dx, & \text{if } f(x) \text{ is even} \\ 0, & \text{if } f(x) \text{ is odd} \end{cases}.$$

because the integrand represents the algebraic (signed) area under the curves.

If $f(x)$ is an even function in $[-l, l]$, then all b_n's will be zero since the integrand Eq. (4) is an odd function. Thus, the Fourier series of an even function contains only cosine terms and is known as Fourier cosine series given by

$$f(x) = \frac{a_0}{2} + \sum_{n=1}^{\infty} a_n \cos\frac{n\pi x}{l} \tag{5}$$

where,

$$a_0 = \frac{2}{l}\int_{0}^{l} f(x)dx \tag{6}$$

and

$$a_n = \frac{2}{l}\int_{0}^{l} f(x)\cos\frac{n\pi x}{l}dx \tag{7}$$

If $f(x)$ is an odd function in $[-l, l]$, then all a_n's are zero because the integrand in Eq. (3) is an odd function. Also, a_0 is zero since $f(x)$ is odd. Thus, the Fourier series of an odd function contains only sine term and is known as Fourier sine series given by

$$f(x) = \sum_{n=1}^{\infty} b_n \sin\left(\frac{n\pi x}{l}\right) \tag{8}$$

where,

$$b_n = \frac{2}{l}\int_{-l}^{l} f(x)\sin\left(\frac{n\pi x}{l}\right)dx \tag{9}$$

Procedure

(a) Identify whether the given function $f(x)$ is an even or odd function in the given interval.

(b) If $f(x)$ is even, calculate only a_0 and a_n's (from Eqs. (6) and (7)) (no need to calculate b_n's). The Fourier cosine series is given by Eq. (5).

(c) If $f(x)$ is odd, calculate only b_n's (from Eq. (9)). The Fourier sine series is given by Eq. (8).

▲▲

EXAMPLE 7.6

Find the Fourier series expansion of the periodic function

$$f(x) = x; -\pi \le x \le \pi, f(x + 2\pi) = f(x).$$

Solution: Since $f(x) = x$ is an odd function, therefore, $a_0 = 0 = a_n$

$$b_n = \frac{2}{\pi}\int_0^{\pi} x \sin\frac{n\pi x}{\pi}\,dx$$

$$= \frac{2}{\pi}\left[-x\frac{\cos nx}{n} + 1\cdot\frac{\sin nx}{n^2}\right]_0^{\pi} = \frac{2}{n}(-1)^{n+1}$$

Therefore, the Fourier expansion is given by

$$x = 2\sum_{n=1}^{\infty}\frac{(-1)^{n+1}}{n}\sin\frac{n\pi x}{\pi}$$

$$= 2\left\{\sin x - \frac{\sin 2x}{2} + \frac{\sin 3x}{3} - \frac{\sin 4x}{4} + \ldots\right\}.$$

EXAMPLE 7.7

Find the Fourier series expansion of the function $f(x) = x^2; -2 \le x \le 2$.

Solution: The given function $f(x) = x^2$ is an even function, therefore, $b_n = 0$

$$a_0 = \frac{2}{2}\int_0^2 x^2 dx = \left[\frac{x^3}{3}\right]_0^2 = \frac{8}{3}$$

$$a_n = \frac{2}{2}\int_0^2 x^2 \cos\frac{n\pi x}{2}\,dx$$

$$= \left[x^2\frac{\sin\frac{n\pi x}{2}}{n\pi/2} - 2x\frac{-\cos\frac{n\pi x}{2}}{(n\pi/2)^2} + 2\frac{-\sin\frac{n\pi x}{2}}{(n\pi/2)^3}\right]_0^2$$

$$= \frac{16}{n^2\pi^2}\cos n\pi = \frac{16(-1)^n}{n^2\pi^2}$$

Therefore, the Fourier series is given by

$$f(x) = \frac{4}{3} + \frac{16}{\pi^2}\sum_{n=1}^{\infty}\frac{(-1)^n}{n^2}\cos\left(\frac{n\pi x}{2}\right).$$

EXAMPLE 7.8

Find the Fourier series expansion of $f(x) = |\cos x|$ in the interval $(-\pi, \pi)$.

Solution: Since $f(x) = |\cos x|$ is an even function, therefore $b_n = 0$

$$a_0 = \frac{2}{\pi}\int_0^{\pi}|\cos x|\,dx$$

$$= \frac{2}{\pi}\left[\int_0^{\pi/2} \cos x dx + \int_{\pi/2}^{\pi} (-\cos x) dx\right]$$

$$= \frac{2}{\pi}\left[(\sin x)_0^{\pi/2} - (\sin x)_{\pi/2}^{\pi}\right] = \frac{4}{\pi}$$

and $$a_n = \frac{2}{\pi}\int_0^{\pi} (\cos x)\cos\frac{n\pi x}{\pi} dx$$

$$= \frac{2}{\pi}\left[\int_0^{\pi/2} \cos x \cos nx dx - \int_{\pi/2}^{\pi} \cos x \cos nx dx\right]$$

$$= \frac{1}{\pi}\left[\int_0^{\pi/2} [\cos(n+1)x + \cos(n-1)x] dx - \int_{\pi/2}^{\pi} [\cos(n+1)x + \cos(n-1)x] dx\right]$$

$$= \frac{1}{\pi}\left[\left(\frac{\sin(n+1)x}{n+1} + \frac{\sin(n-1)x}{n-1}\right)_0^{\pi/2} - \left(\frac{\sin(n+1)x}{n+1} + \frac{\sin(n-1)x}{n-1}\right)_{\pi/2}^{\pi}\right]$$

$$= \frac{2}{\pi}\left[\frac{\sin(n+1)\pi/2}{n+1} + \frac{\sin(n-1)\pi/2}{n-1}\right]$$

$$= \frac{2}{\pi}\left[\frac{\cos n\pi/2}{n+1} - \frac{\cos n\pi/2}{n-1}\right] = -\frac{4\cos n\pi/2}{\pi(n^2-1)} \quad (n > 1)$$

For $n = 1$

$$a_1 = \frac{2}{\pi}\int_0^{\pi} (\cos x)\cos x dx$$

$$= \frac{2}{\pi}\left[\int_0^{\pi/2} \cos^2 x dx - \int_{\pi/2}^{\pi} \cos^2 x dx\right] = 0$$

Hence, the required Fourier series expansion is

$$|\cos x| = \frac{2}{\pi} + \frac{4}{\pi}\left\{\frac{1}{3}\cos 2x - \frac{1}{15}\cos 4x + \dots\right\}$$

EXAMPLE 7.9

Find the Fourier series expansion of $f(x) = |x|$ in the interval $[-\pi, \pi]$

Solution: Since $f(x) = |x|$ is an even function, therefore, $b_n = 0$

$$a_0 = \frac{2}{\pi}\int_0^{\pi}(x)dx = \frac{2}{\pi}\left[\frac{x^2}{2}\right]_0^{\pi} = \pi$$

$$a_n = \frac{2}{\pi}\int_0^{\pi}(x)\cos\frac{n\pi x}{\pi}dx = \frac{2}{\pi}\int_0^{\pi}x\cos nxdx$$

$$= \frac{2}{\pi}\left[x\frac{\sin nx}{n} - 1 - \frac{\cos nx}{n^2}\right]_0^{\pi}$$

$$= \frac{2}{\pi n^2}[(-1)^n - 1] = \begin{cases} 0, & \text{if } n \text{ is even} \\ -\frac{4}{\pi n^2}, & \text{if } n \text{ is odd} \end{cases}.$$

Thus, the required Fourier expansion is

$$(x) = \frac{\pi}{2} - \frac{4}{\pi}\left[\frac{\cos x}{1^2} + \frac{\cos 3x}{3^2} + \frac{\cos 5x}{5^2} + \ldots\right].$$

EXAMPLE 7.10

Obtain Fourier series for the function $f(x)$ given by

$$f(x) = \begin{cases} 1 + 2x/\pi & -\pi \le x \le 0 \\ 1 - 2x/\pi & 0 \le x \le \pi \end{cases}.$$

Deduce that $\frac{1}{1^2} + \frac{1}{3^2} + \frac{1}{5^2} + \ldots = \frac{\pi^2}{8}$

Solution: Here,
$$f(-x) = \begin{cases} 1 - 2x/\pi, & -\pi \le x \le 0 \\ 1 + 2x/\pi, & 0 \le x \le \pi \end{cases}$$

$\therefore f(x)$ is an even function in $[\pi, \pi]$

So,
$$b_n = 0$$

$$a_0 = \frac{2}{\pi}\int_0^{\pi}\left(1 - \frac{2x}{\pi}\right)dx = \frac{2}{\pi}\left\{x - \frac{x^2}{\pi}\right\}_0^{\pi} = 0$$

and
$$a_n = \frac{2}{\pi}\int_0^{\pi}\left(1 - \frac{2x}{\pi}\right)\cos\frac{n\pi x}{\pi}dx$$

$$= \frac{2}{\pi}\left[\left(1-\frac{2x}{\pi}\right)\frac{\sin nx}{n}-\left(-\frac{2}{\pi}\right)-\frac{\cos nx}{n^2}\right]_0^{\pi}$$

$$= \frac{4}{n^2\pi^2}(1-(-1)^n) = \begin{cases} 0, & \text{if } n \text{ is even} \\ 8/n^2\pi^2, & \text{if } n \text{ is odd} \end{cases}.$$

Thus, the Fourier expansion is

$$f(x) = \frac{8}{\pi^2}\left\{\frac{\cos x}{1^2}+\frac{\cos 3x}{3^2}+\frac{\cos 5x}{5^2}+\ldots\right\}.$$

▲▲

Half Range Fourier Cosine and Sine Series

So far, we have considered the Fourier series expansion of a function which is periodic, defined in an interval $-l$ to l of length $2l$. Now, we consider the procedure to expand a non-periodic function $f(x)$ defined in half of the above interval, say $(0, l)$ of length l. Such expansions are known as half range expansions or half range Fourier series. In particular, a half range expansion containing only cosine terms is known as half range Fourier cosine series of $f(x)$ in the interval $(0, l)$. In a similar way, half range Fourier sine series contains only sine terms. Note that the given function is neither periodic nor even nor odd. In order to obtain a Fourier cosine series for $f(x)$ in the interval $(0, l)$, we define a new function $g(x)$ such that

(a) $g(x) \equiv f(x)$ in the interval $(0, l)$

(b) $g(x)$ is even function in $(-l, l)$ and is periodic with period $2l$.

Such a function $g(x)$ is known as the even periodic continuation (or extension) of $f(x)$. The Fourier cosine series for $g(x)$ valid in $(-l, l)$ (or in fact for all x) is readily obtained as

$$g(x) = \frac{a_0}{2}+\sum_{n=1}^{\infty} a_n \cos\frac{n\pi x}{l} \tag{1}$$

where,

$$a_0 = \frac{2}{l}\int_0^l g(x)dx = \frac{2}{l}\int_0^l f(x)dx \tag{2}$$

$$a_n = \frac{2}{l}\int_0^l g(x)\cos\frac{n\pi x}{l}dx = \frac{2}{l}\int_0^l f(x)\cos\frac{n\pi x}{l}dx \tag{3}$$

Since by our supposition, $f(x)$ and $g(x)$ are equal in $(0, l)$, the required half-range Fourier cosine series of $f(x)$ is given by Eq. (1) with a_0, a_n given by Eqs. (2) and (3).

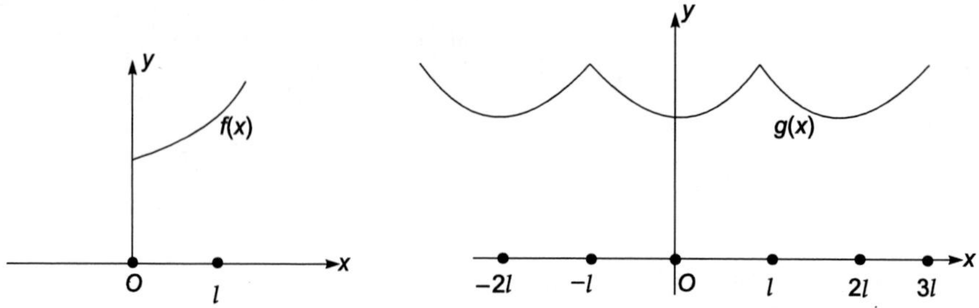

Even periodic extension
g(x) of f(x)

Fig. 7.2

Here,
$$g(x) = \begin{cases} f(x), & \text{in } (0,l) \\ f(-x), & \text{in } (-l,0) \end{cases}$$

Note. The series expansion of $f(x)$ given by Eq. (1) is valid for $f(x)$ only in the interval $(0, l)$, but not outside this interval.

In a similar way, to obtain the half-range Fourier sine series for $f(x)$ in $(0, l)$ define $g(x)$ such that

(a) $g(x) = f(x)$ in $(0, l)$ and

(b) $g(x)$ is an odd function in $(-l, l)$ periodic with period $2l$. The new function $g(x)$ is called odd periodic continuation of $f(x)$.

Now the Fourier sine series of $g(x)$ is given by

$$g(x) = \sum_{n=1}^{\infty} b_n \sin\frac{n\pi x}{l} \tag{4}$$

with
$$b_n = \frac{2}{l}\int_0^l g(x)\sin\frac{n\pi x}{l}dx = \frac{2}{l}\int_0^l f(x)\sin\frac{n\pi x}{l}dx \tag{5}$$

Since $f(x)$ and $g(x)$ are equal in $(0, l)$, the required half-range Fourier sine series expansion of $f(x)$ in the interval $(0, l)$ is given by Eq. (4) with b_n's given by Eq. (5). This expansion (4) is valid for $f(x)$ only in the interval $(0, l)$. Thus, a given non-periodic function $f(x)$ can be expanded in cosine series or sine series.

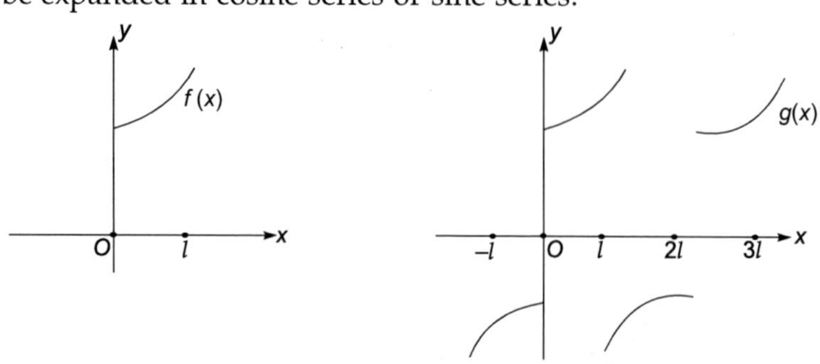

Odd periodic continuation
(extension) g(x) of f(x)

Fig. 7.3

EXAMPLE 7.11

Find the Fourier cosine and sine series of the function $f(x) = 1, 0 \le x \le 2$.
Solution: The Fourier coefficients for the cosine series are obtained as follows

$$a_0 = \frac{2}{2}\int_0^2 dx = 2$$

$$a_n = \frac{2}{2}\int_0^2 1\cdot\cos\left(\frac{n\pi x}{2}\right)dx = \left[\frac{\sin\left(\frac{n\pi x}{2}\right)}{n\pi/2}\right]_0^2 = 0$$

Therefore, the Fourier cosine series of $f(x) = 1$; itself

For the Fourier sine series, we have

$$b_n = \frac{2}{2}\int_0^2 1\cdot\sin\left(\frac{n\pi x}{2}\right)dx = \left[-\frac{\cos(n\pi x/2)}{n\pi/2}\right]_0^2$$

$$= \frac{2}{n\pi}[1 - \cos n\pi]$$

$$= \frac{2}{n\pi}[1 - (-1)^n] = \begin{cases} 0, & \text{if } n \text{ is even} \\ \frac{4}{n\pi}, & \text{if } n \text{ is odd} \end{cases}.$$

Therefore, the Fourier sine series is

$$f(x) = \frac{4}{\pi}\left\{\sin\left(\frac{\pi x}{2}\right) + \frac{1}{3}\sin\left(\frac{3\pi x}{2}\right) + \frac{1}{5}\sin\left(\frac{5\pi x}{2}\right) + \dots\right\}$$

EXAMPLE 7.12

Obtain the Fourier expansion of $f(x) = x \sin x$ as a cosine series in $(0, \pi)$.

Hence, show that $\frac{1}{1\cdot 3} + \frac{1}{3\cdot 5} + \frac{1}{5\cdot 7} + \dots\infty = \frac{\pi - 2}{4}$

Solution: Here, $a_0 = \frac{2}{\pi}\int_0^2 x \sin x dx$

$$= \frac{2}{\pi}[-x\cos x + 1\cdot\sin x]_0^\pi = 2$$

and $a_n = \frac{2}{\pi}\int_0^\pi x\sin x\cos\left(\frac{n\pi x}{\pi}\right)dx = \frac{2}{\pi}\int_0^\pi x\sin x\cos nx dx$

$$= \frac{1}{\pi}\int_0^\pi x(\sin(n+1)x - \sin(n-1)x)dx$$

$$= \frac{1}{\pi}\left[x\left(-\frac{\cos(n+1)x}{n+1}+\frac{\cos(n-1)x}{n-1}\right)-\left(-\frac{\sin(n+1)x}{n+1}+\frac{\sin(n-1)x}{n-1}\right)\right]_0^{\pi}$$

$$= \frac{1}{\pi}\left[\pi\left(-\frac{\cos(n+1)\pi}{n+1}+\frac{\cos(n-1)\pi}{n-1}\right)\right] \quad (n \neq 1)$$

For $n = 1$

$$a_1 = \frac{2}{\pi}\int_0^{\pi} x\sin x\cos x dx = \frac{1}{\pi}\int_0^{\pi} x\sin 2x dx$$

$$= \frac{1}{\pi}\left[-x\frac{\cos 2x}{2}+1\cdot\frac{\sin 2x}{4}\right]_0^{\pi} = -\frac{1}{2}$$

Hence, $x \sin x = 1 - \frac{1}{2}\cos x - 2\left(\frac{\cos 2x}{1\cdot 3}-\frac{\cos 3x}{3\cdot 5}+\frac{\cos 5x}{5\cdot 7}- \dots \infty\right)$

Let $x = \pi/2$, we get

$$\frac{\pi}{2} = 1-\frac{1}{2}(0)-2\left(-\frac{1}{1\cdot 3}-\frac{-1}{3\cdot 5}+\frac{1}{5\cdot 7}- \dots \infty\right)$$

Hence, $\frac{1}{1\cdot 3}-\frac{1}{3\cdot 5}+\frac{1}{5\cdot 7}- \dots \infty = \frac{\frac{\pi}{2}-1}{2} = \frac{\pi-2}{4}.$

EXAMPLE 7.13

Find the Fourier cosine series of the function

$$f(x) = \begin{cases} x^2, & 0 \le x \le 2 \\ 4, & 2 \le x \le 4 \end{cases}.$$

Solution: Note that $f(x)$ is to be extended as an even function.

We have

$$a_0 = \frac{2}{4}\int_0^4 f(x)dx = \frac{1}{2}\left\{\int_0^2 x^2 dx + \int_2^4 4dx\right\}$$

$$= \frac{1}{2}\left\{\left[\frac{x^3}{3}\right]_0^2 + [4x]_2^4\right\} = \frac{1}{2}\left\{\frac{8}{3}+8\right\} = \frac{16}{3}.$$

and

$$a_n = \frac{2}{4}\left[\int_0^4 f(x)\cos\left(\frac{n\pi x}{4}\right)dx\right]$$

$$= \frac{1}{2}\left[\int_0^2 x^2\cos\left(\frac{n\pi x}{4}\right)dx + \int_2^4 4\cdot\cos\left(\frac{n\pi x}{4}\right)dx\right]$$

$$= \frac{1}{2}\left[\left(x^2\frac{\sin(n\pi x/4)}{n\pi/4} - 2x - \frac{\cos(n\pi x/4)}{(n\pi/4)^2} + 2 - \frac{\sin(n\pi x/4)}{(n\pi/4)^2}\right)_0^2 + 4\left(\frac{\sin(n\pi x/4)}{n\pi/4}\right)_2^4\right]$$

$$= \frac{1}{2}\left[\frac{4\sin(n\pi/2)}{n\pi/4} + \frac{4\cos(n\pi/2)}{(n\pi/4)^2} - 2\frac{\sin(n\pi/2)}{(n\pi/4)^3} - 4\frac{\sin(n\pi/2)}{n\pi/4}\right]$$

$$= \frac{32}{n^2\pi^2}\left[\cos\left(\frac{n\pi}{2}\right) - \left(\frac{2}{n\pi}\right)\sin\left(\frac{n\pi}{2}\right)\right]$$

Therefore, $f(x) = \frac{8}{3} + \frac{32}{\pi^2}\sum_{n=1}^{\infty}\frac{1}{n^2}\left[\cos\left(\frac{n\pi}{2}\right) - \left(\frac{2}{n\pi}\right)\sin\left(\frac{n\pi}{2}\right)\right] \times \cos\left(\frac{n\pi x}{4}\right).$

EXAMPLE 7.14

Find the Fourier cosine and sine series of the function $f(x) = x$ in $0 < x < 2$.

Solution: For Fourier cosine series, we obtain

$$a_0 = \frac{2}{2}\int_0^2 x\,dx = \int_0^2 x\,dx = 2$$

and $$a_n = \frac{2}{2}\int_0^2 x\cos\left(\frac{n\pi x}{2}\right)dx$$

$$= \left\{x\frac{\sin(n\pi x/2)}{n\pi/2} - 1\cdot\frac{\cos(n\pi x/2)}{(n\pi/2)}\right\}_0^2$$

$$= \frac{4}{n^2\pi^2}[\cos n\pi - 1] = \frac{4}{n^2\pi^2}[(-1)^n - 1] = \begin{cases} 0, & \text{if } n \text{ is even} \\ -\frac{8}{n^2\pi^2}, & \text{if } n \text{ is odd} \end{cases}.$$

Hence, the Fourier cosine series is

$$f(x) = 1 - \frac{8}{\pi^2}\left[\frac{\cos(\pi x/2)}{1^2} + \frac{\cos(3\pi x/2)}{3^2} + \frac{\cos(5\pi x/2)}{5^2} + \dots\right]$$

For, the Fourier sine series, we have

$$b_n = \frac{2}{2}\int_0^2 x\sin\left(\frac{n\pi x}{2}\right)dx$$

$$= \left\{x\cdot\frac{\cos\left(\frac{n\pi x}{2}\right)}{n\pi/2} - 1\cdot\frac{\sin\left(\frac{n\pi x}{2}\right)}{(n\pi/2)^2}\right\}_0^2$$

$$= -\frac{4(-1)^n}{n\pi}$$

Therefore,

$$f(x) = \frac{4}{\pi}\left[\frac{\sin\left(\frac{n\pi}{2}\right)}{1} - \frac{1}{2}\sin\left(\frac{2\pi x}{2}\right) + \frac{1}{3}\sin\left(\frac{3\pi x}{2}\right) - \frac{1}{4}\sin\left(\frac{4\pi x}{2}\right) + \dots\right].$$

EXAMPLE 7.15

Expand $f(x) = \begin{cases} \frac{1}{4} - x & \text{if } 0 < x < \frac{1}{2} \\ x - \frac{3}{4} & \text{if } \frac{1}{2} < x < 1 \end{cases}$

as the Fourier series of sine terms.

Solution: Here, $b_n = \frac{2}{1}\int_0^1 f(x)\sin\left(\frac{n\pi x}{1}\right)dx$

$$= 2\left[\int_0^{1/2}\left(\frac{1}{4} - x\right)\sin n\pi x dx + \int_{1/2}^1\left(x - \frac{3}{4}\right)\sin n\pi x dx\right]$$

$$= 2\left[\left(-\left(\frac{1}{4} - x\right)\frac{\cos n\pi x}{n\pi} - \frac{\sin n\pi x}{n^2\pi^2}\right)_0^{1/2} + \left(-\left(x - \frac{3}{4}\right)\frac{\cos n\pi x}{n\pi} + \frac{\sin n\pi x}{n^2\pi^2}\right)_{1/2}^1\right]$$

$$= 2\left[\frac{1}{4n\pi}\cos\left(\frac{n\pi}{2}\right)+\frac{1}{4n\pi}-\frac{\sin\left(\frac{n\pi}{2}\right)}{n^2\pi^2}-\frac{1}{4n\pi}\cos n\pi-\frac{1}{4n\pi}\cos\left(\frac{n\pi}{2}\right)-\frac{\sin(n\pi/2)}{n^2\pi^2}\right]$$

$$= \frac{1}{2n\pi}[1-(-1)^n]-4\frac{\sin(n\pi/2)}{n^2\pi^2}$$

Hence, $f(x) = \sum_{n=1}^{\infty}\left(\frac{1}{2n\pi}[1-(-1)^n]-4\frac{\sin(n\pi/2)}{n^2\pi^2}\right)\times\sin\left(\frac{n\pi x}{1}\right)$

$$= \left(\frac{1}{\pi}-\frac{4}{\pi^2}\right)\sin(\pi x)+\left(\frac{1}{3\pi}+\frac{4}{3^2\pi^2}\right)\sin(3\pi x)+\left(\frac{1}{5\pi}-\frac{4}{5^2\pi^2}\right)\sin(5\pi x)+\ldots$$

which is the required Fourier sine series expansion.

EXERCISE 7.1

Find the Fourier series of the given function on the given interval:

1. $f(x) = \begin{cases}\pi + x, & -\pi < x < 0\\ 0, & 0 \le x < \pi\end{cases}; f(x + 2\pi) = f(x).$

2. $f(x) = \begin{cases}\pi, & -\pi < x < 0\\ \pi - x, & 0 \le x < \pi\end{cases}.$

3. $f(x) = \begin{cases}-k, & -\pi < x < 0\\ k, & 0 \le x < \pi\end{cases}.$

4. $f(x) = \begin{cases}2, & -\pi < x < 0\\ 4, & 0 \le x < 4\end{cases}.$

5. $f(x) = \begin{cases}-(\pi + x), & -\pi < x < 0\\ -(\pi - x), & 0 \le x < \pi\end{cases}.$

6. $f(x) = 1 - |x|, -\pi < x < \pi.$

7. $f(x) = x^2, -\pi < x < \pi.$

8. $f(x) = x^3, -\pi < x < \pi.$

9. $f(x) = \begin{cases}x^2, & -\pi < x < 0\\ -x^2, & 0 \le x \le \pi\end{cases}.$

10. $f(x)=\begin{cases}0, & -\pi \le x \le 0\\ \sin x, & 0 \le x \le \pi\end{cases}$

11. $f(x) = x, -2 < x < 2.$

12. $f(x) = x^2, -3 < x < 3.$

13. $f(x) = |x|, -1 < x < 1.$

14. $f(x) = 9 - x^2, -3 \le x \le 3.$

15. $f(x) = x \sin x; 0 < x < 2\pi.$

16. $f(x) = |\sin x|, -\pi < x < \pi.$

17. $f(x) = x^2, 0 \le x \le \pi$ (cosine series only).

18. $f(x) = x \cos x$ in $(0, \pi)$ (sine series only).

19. $f(x) = e^x$ in $0 < x < 1$ (sine series only).

20. Find the Fourier series expansion of the function $f(x) = \pi + x, -\pi < x < \pi.$

    Also show that $\frac{\pi}{4}=1-\frac{1}{3}+\frac{1}{5}-\frac{1}{7}+\ldots.$

21. A periodic function of period 2π is defined as $f(x)=\begin{cases}1, & -\pi/2<x<\pi/2\\ 0, & \pi/2<x<3\pi/2\end{cases}.$

    Obtain the Fourier series expansion of $f(x)$ and show that $\frac{\pi}{4}=1-\frac{1}{3}+\frac{1}{5}-\frac{1}{7}+\ldots.$

22. A periodic function of period 2 is defined as $f(x) = 1 + x, -1 < x < 1.$

    Obtain the Fourier series expansion of $f(x)$ and show that $\frac{\pi}{4}=1-\frac{1}{3}+\frac{1}{5}-\frac{1}{7}+\ldots.$

23. Find the Fourier series expansion of the following periodic function of period 2π

$$f(x)=\begin{cases}0, & -\pi<x<0\\ x^2, & 0<x<\pi\end{cases}.$$

    Also show that

$$\frac{\pi^2}{6}=1+\frac{1}{2^2}+\frac{1}{3^2}+\frac{1}{4^2}+\ldots \quad \text{and} \quad \frac{\pi^2}{12}=1-\frac{1}{2^2}+\frac{1}{3^2}-\frac{1}{4^2}+\ldots.$$

24. Obtain the Fourier series expansion of the periodic function $f(x) = e^x; -\pi < x < \pi,$ $f(x + 2\pi) = f(x).$

    Hence, find the sum of the series

$$\frac{1}{1+2^2}-\frac{1}{1+3^2}+\frac{1}{1+4^2}-\ldots+\frac{(-1)^n}{1+n^2}+\ldots.$$

25. Obtain the Fourier series expansion of the following periodic function of period 4.

$f(x) = 4 - x^2, -2 \le x \le 2$. Hence, show that $\frac{\pi^2}{12} = 1-\frac{1}{2^2}+\frac{1}{3^2}-\frac{1}{4^2}+\ldots$

ANSWERS

Exercise 7.1

1. $f(x) = \frac{\pi}{4}+\frac{2}{\pi}\left\{\frac{\cos x}{1^2}+\frac{\cos 3x}{3^2}+\ldots\right\}-\left\{\frac{\sin x}{1}+\frac{\sin 2x}{2}+\ldots\right\}.$

2. $f(x) = \frac{3\pi}{2}+\sum_{n=1}^{\infty}\left\{-\frac{1}{\pi n^2}\{\cos n\pi\}\cos(nx)+\frac{1}{n}\cos(n\pi)\sin(nx)\right\}.$

3. $f(x) = \frac{2k}{\pi}\sum_{n=1}^{\infty}\frac{1}{n}(1 - \cos(n\pi))\sin(nx).$

4. $f(x) = 1 + \frac{2}{\pi}\sum_{n=1}^{\infty}\frac{1}{n}(1 - \cos(n\pi))\sin(nx).$

5. $f(x) = -\frac{\pi}{2}-\frac{2}{\pi}\sum_{n=1}^{\infty}\frac{1}{n^2}(1 - \cos(n\pi))\cos(nx).$

6. $f(x) = \frac{1}{2}(2-\pi)+\frac{2}{\pi}\sum_{n=1}^{\infty}\frac{1}{n^2}(1 - \cos(n\pi))\cos(nx).$

7. $f(x) = \frac{\pi^2}{3}+4\sum_{n=1}^{\infty}\frac{1}{n^2}\cos(n\pi)\cos(nx).$

8. $f(x) = \frac{2}{\pi}\sum_{n=1}^{\infty}\left[\left(\frac{6\pi}{n^3}-\frac{\pi^2}{n}\right)\cos(n\pi)\sin(nx)\right].$

9. $f(x) = \frac{2}{\pi}\sum_{n=1}^{\infty}\left[\frac{\pi^2}{n}\cos(n\pi)+\frac{2}{n^3}[1-\cos(n\pi)]\sin(nx)\right].$

10. $f(x) = \frac{1}{\pi}+\frac{1}{2}\sin x+\frac{1}{\pi}\sum_{n=1}^{\infty}\left[\frac{(-1)^{n-1}-1}{(n^2-1)}\cos(nx)\right].$

11. $f(x) = \frac{4}{\pi}\sum_{n=1}^{\infty}\left[\frac{(-1)^{n+1}}{n}\sin\left(\frac{n\pi x}{2}\right)\right].$

12. $f(x) = 3 + \frac{36}{\pi^2}\sum_{n=1}^{\infty}\left[\frac{1}{n^2}\cos(n\pi)\cos\left(\frac{n\pi x}{3}\right)\right].$

13. $f(x) = \frac{1}{2} - \frac{2}{\pi^2} \sum_{n=1}^{\infty} \left[\frac{1}{n^2} [1 - \cos(n\pi)\cos(n\pi x)) \right].$

14. $f(x) = 6 - \frac{36}{\pi^2} \sum_{n=1}^{\infty} \left[\frac{1}{n^2} \cos(n\pi) \cos\left(\frac{n\pi x}{3}\right) \right].$

15. $f(x) = -1 + \pi \sin x - \frac{1}{2} \cos x + \frac{2}{2^2 - 1} \cos 2x + \frac{2}{3^2 - 1} \cos 3x.$

16. $f(x) = \frac{2}{\pi} - \frac{4}{\pi} \left(\frac{\cos 2x}{3} + \frac{\cos 4x}{15} + \ldots + \frac{\cos 2nx}{4n^2 - 1} + \ldots \right).$

17. $f(x) = \frac{\pi^2}{3} - 4\left(\cos x - \frac{1}{2^2} \cos 2x + \frac{1}{3^2} \cos 3x - \frac{1}{4^2} \cos 4x + \ldots \infty \right).$

18. $f(x) = -\frac{1}{2} \sin x + \sum_{n=2}^{\infty} \frac{(-1)^n \cdot 2n}{n^2 - 1} \sin nx.$

19. $f(x) = \sum_{n=1}^{\infty} \frac{2n\pi}{1 + n^2\pi^2} (1 - e \cos n\pi) \sin n\pi x.$

20. $f(x) = \pi + 2\sum_{n=1}^{\infty} \left[\frac{(-1)^{n+1}}{n} \sin(nx) \right]$ and let $x = \frac{\pi}{2}$.

21. $f(x) = \frac{1}{2} + \frac{2}{\pi} \sum_{n=1}^{\infty} \left[\frac{1}{n} \sin\left(\frac{n\pi}{2}\right) \cos(nx) \right]$ and let $x = 0$.

22. $f(x) = 1 - \frac{2}{\pi} \sum_{n=1}^{\infty} \left[\frac{1}{n} \cos(n\pi) \sin(n\pi x) \right]$ and let $x = 1/2$.

23. $f(x) = \frac{\pi^2}{6} + \sum_{n=1}^{\infty} \left[\frac{2}{n^2} (-1)^n \cos(nx) + \left(\frac{\pi}{n} (-1)^{n+1} - \frac{2}{\pi n^3} (1 - \cos n\pi) \right) \sin(nx) \right]$ and let $x =$ 0, π.

24. $f(x) = \frac{1}{\pi} \sinh \pi + \frac{2 \sinh \pi}{\pi} \sum_{n=1}^{\infty} \left[\frac{(-1)^n}{n^2 + 1} (\cos(nx) - n \sin(nx) \right]$ and let $x = 0$; $\pi/2 \sinh \pi$.

25. $f(x) = \frac{8}{3} - \frac{16}{\pi^2} \sum_{n=1}^{\infty} \left[\frac{(-1)^n}{n^2} \cos\left(\frac{n\pi x}{2}\right) \right]$ and let $x = 0$.

8

CHAPTER

Fourier Transforms

In this chapter, we define the Fourier transform which is an integral transform similar to Laplace transform. Fourier transforms are used in many areas of science, engineering, medicine, etc.

Definition: Let $f(t)$ be piecewise continuous on $(-\infty, \infty)$. Assume that $f(t)$ is absolutely convergent, that is, $\int_{-\infty}^{\infty} |f(t)|\, dt$ converges. Then, the Fourier transform of $f(t)$ denoted by $F[f(t)]$ is defined as

$$F(\omega) = \frac{1}{\sqrt{2\pi}} \int_{-\infty}^{\infty} f(t)e^{i\omega t}dt = F[f(t)]$$

Assume, now that $\int_{-\infty}^{\infty} |F(\omega)|\, d\omega$ converges. Then, we define the inverse Fourier transform of $F(\omega)$ as

$$F^{-1}(F(\omega)) = \frac{1}{\sqrt{2\pi}} \int_{-\infty}^{\infty} F(\omega)e^{-i\omega t}dt = f(t)$$

The Fourier transform and its inverse are called a transform pair.

Now, suppose
$$\sqrt{\frac{2}{\pi}} \int_{0}^{\infty} f(t)\sin \omega t dt = F_s(\omega)$$

and
$$\sqrt{\frac{2}{\pi}} \int_{0}^{\infty} F(\omega)\sin \omega t d\omega = f_s(t)$$

Here, $F(\omega)$ is called the Fourier sine transform of $f(t)$ and $f_s(t)$ is called the inverse Fourier sine transform of $F(\omega)$

Similarly,
$$\sqrt{\frac{2}{\pi}} \int_{0}^{\infty} f(t) \cos \omega t\, dt = F_c(\omega)$$

and
$$\sqrt{\frac{2}{\pi}} \int_{0}^{\infty} F(\omega)\cos \omega t d\omega = f_c(t)$$

Here, $F_c(\omega)$ is called the Fourier cosine transform of $f(t)$ and $f_c(t)$ is called the inverse Fourier cosine transform of $F_c(\omega)$. ▲▲

EXAMPLE 8.1

Find the Fourier transform of

$$f(t) = \begin{cases} a, & -1<t<0 \\ 0, & \text{otherwise} \end{cases} \qquad (a > 0).$$

Solution: We have

$$F(f(t)) = \frac{1}{\sqrt{2\pi}} \int_{-\infty}^{\infty} f(t)e^{i\omega t}\,dt$$

$$= \frac{1}{\sqrt{2\pi}} \int_{-1}^{0} ae^{i\omega t}\,dt$$

$$= \frac{1}{\sqrt{2\pi}} \left[\frac{ae^{i\omega t}}{i\omega}\right]_{-1}^{0} = \frac{-ia}{\sqrt{2\pi}\,\omega}(1-e^{-i\omega})$$

$$= \frac{ia}{\omega\sqrt{2\pi}}(e^{-i\omega}-1).$$

EXAMPLE 8.2

Find the Fourier transform of the function $f(t) = \begin{cases} 0, & t<0 \\ e^{-\alpha t}, & t\geq 0 \end{cases} \quad (\alpha > 0)$

Solution: Since

$$\int_{-\infty}^{\infty} |f(t)|\,dt = \int_{0}^{\infty} e^{-\alpha t}\,dt = \frac{1}{\alpha}$$

Therefore, Fourier transform of $f(t)$ exists, we have

$$F(f(t)) = \frac{1}{\sqrt{2\pi}} \int_{-\infty}^{\infty} f(t)e^{i\omega t}\,dt$$

$$= \frac{1}{\sqrt{2\pi}} \int_{0}^{\infty} e^{-\alpha t}e^{-i\omega t}\,dt$$

$$= \frac{1}{\sqrt{2\pi}} \left[\frac{e^{-(a+i\omega)t}}{-(a+i\omega)}\right]_{0}^{\infty}$$

$$= \frac{1}{\sqrt{2\pi}}\frac{1}{a+i\omega}$$

EXAMPLE 8.3

Find the Fourier transform of $e^{-\alpha t^2}$ $(\alpha > 0)$.

Solution: From the definition, we obtain

$$F(e^{-\alpha t^2}) = \frac{1}{\sqrt{2\pi}} \int_{-\infty}^{\infty} e^{-\alpha t^2} e^{i\omega t}\, dt$$

$$= \frac{1}{\sqrt{2\pi}} \int_{-\infty}^{\infty} e^{-\alpha\left(t^2 - \frac{i\omega t}{\alpha}\right)} dt$$

$$= \frac{1}{\sqrt{2\pi}} \int_{-\infty}^{\infty} e^{-\alpha\left\{\left(t - \frac{i\omega}{2\alpha}\right)^2 + \frac{\omega^2}{4\alpha^2}\right\}} dt$$

$$= \frac{1}{\sqrt{2\pi}} \int_{-\infty}^{\infty} e^{-\frac{\omega^2}{4\alpha}} e^{-\alpha\left(t - \frac{i\omega}{2\alpha}\right)^2} dt$$

Let $$\sqrt{\alpha}\left(t - \frac{i\omega t}{2\alpha}\right) = \tau \Rightarrow dt = \frac{d\tau}{\sqrt{\alpha}}$$

$$\therefore \quad F(e^{-\alpha t^2}) = \frac{1}{\sqrt{2\pi}} e^{-\omega^2/4\alpha} \int_{-\infty}^{\infty} e^{-\tau} \frac{d\tau}{\sqrt{\alpha}}$$

$$= \frac{1}{\sqrt{2\pi}} e^{-\omega^2/4\alpha} \frac{2}{\sqrt{\alpha}} \int_{-\infty}^{\infty} e^{-\tau^2} d\tau$$

Let, $$\tau^2 = z \Rightarrow \tau = \sqrt{z} \Rightarrow d\tau = \frac{dz}{2\sqrt{z}}$$

$$= \sqrt{\frac{2}{\pi\alpha}} e^{-\omega^2/4\alpha} \int_{0}^{\infty} e^{-z} \frac{dz}{2\sqrt{z}}$$

$$= \sqrt{\frac{2}{\pi\alpha}} \frac{e^{-\omega^2/4\alpha}}{2} \int_{-\infty}^{\infty} z^{-1/2} e^{-z} dz$$

$$= \frac{e^{-\omega^2/4\alpha}}{\sqrt{2\pi\alpha}} \Gamma\left(\frac{1}{2}\right) = \frac{e^{-\omega^2/4\alpha}}{\sqrt{2\alpha}}$$

EXAMPLE 8.4

Find the inverse Fourier transform of $e^{-|\omega| y}$.

Solution: We know that $|\omega| = \begin{cases} -\omega & \omega < 0 \\ \omega & \omega \geq 0 \end{cases}$

Therefore,

$$f(t) = F^{-1}[F(\omega)]$$

$$= \frac{1}{\sqrt{2\pi}} \int_{-\infty}^{\infty} F(\omega) e^{-i\omega t} d\omega$$

$$= \frac{1}{\sqrt{2\pi}} \int_{-\infty}^{\infty} e^{-|\omega| y} e^{-i\omega t} d\omega$$

$$= \frac{1}{\sqrt{2\pi}} \left[\int_{-\infty}^{0} e^{(\omega y - i\omega t)} d\omega + \int_{0}^{\infty} e^{-(\omega y + j\omega t)} d\omega \right]$$

$$= \frac{1}{\sqrt{2\pi}} \left\{ \left[\frac{e^{(y-it)\omega}}{y-it} \right]_{-\infty}^{0} + \left[\frac{e^{-y+jt)\omega}}{-(y+jt)} \right]_{0}^{\infty} \right\}$$

$$= \frac{1}{\sqrt{2\pi}} \left\{ \frac{1}{y-it} + \frac{1}{y+jt} \right\} = \sqrt{\frac{2}{\pi}} \frac{1}{(y^2+t^2)}.$$

EXAMPLE 8.5

Find the Fourier transform of $f(t)$ defined by

$$f(t) = \begin{cases} 1-t^2, & |t| < 1 \\ 0, & |t| > 1 \end{cases}.$$

Hence, evaluate

(a) $\int_{0}^{\infty} \frac{t\cos t - \sin t}{t^3} \cos\frac{t}{2} dt$

(b) $\int_{0}^{\infty} \frac{t\cos t - \sin t}{t^3} dt.$

Solution: We have $F[f(t) = \frac{1}{\sqrt{2\pi}} \int_{-\infty}^{\infty} f(t) e^{-j\omega t} dt$

$$= \frac{1}{\sqrt{2\pi}} \left\{ \int_{-\infty}^{-1} 0 e^{-i\omega t} dt + \int_{-1}^{1} (1-t^2) e^{-\omega t} dt + \int_{1}^{\infty} 0 e^{j\omega t} dt \right\}$$

$$= \frac{1}{\sqrt{2\pi}}\left\{\frac{(t-t^2)e^{i\omega t}}{i\omega} - \frac{-2te^{j\omega t}}{-\omega^2} - 2\frac{e^{i\omega t}}{-\omega^3}\right\}_{-1}^{1}$$

$$= \frac{1}{\sqrt{2\pi}}\left\{\frac{2e^{i\omega} + 2e^{-j\omega}}{-\omega^2} + \frac{2}{\omega^3}(e^{j\omega} - e^{-i\omega})\right\}$$

$$= \frac{1}{\sqrt{2\pi}}\left(\frac{4\cos\omega}{-\omega^2} + \frac{4\sin\omega}{\omega^3}\right) = 2\sqrt{\frac{2}{\pi}}\left(\frac{-\omega\cos\omega + \sin\omega}{\omega^3}\right)$$

Now, by inversion formula, we have

$$f(t) = \frac{1}{\sqrt{2\pi}}\int_{-\infty}^{\infty} F(\omega)e^{-i\omega t}d\omega$$

$$= \frac{1}{\sqrt{2\pi}}2\sqrt{\frac{2}{\pi}}\int_{-\infty}^{\infty} - \frac{\omega\cos\omega + \sin\omega}{\omega^3}\cdot e^{i\omega t}d\omega$$

$$= \frac{2}{\pi}\int_{\infty}^{\infty}\left(\frac{\sin\omega - \omega\cos\omega}{\omega^3}\right)(\cos\,\omega t - i\,\sin\,\omega t)d\omega$$

$$= \frac{4}{\pi}\int_{0}^{\infty}\left(\frac{\sin\omega - \omega\cos\omega}{\omega^3}\right)\cos\,\omega t\,d\omega$$

(Using property of definite integrals since $\sin\omega t$ is an odd function)

$$= \begin{cases} 1-t^2, & |t|<1 \\ 0, & |t|>1 \end{cases} \tag{1}$$

Let, $t = \frac{1}{2}$ in Eq. (1)

$$\therefore \quad \frac{4}{\pi}\int_{0}^{\infty}\frac{\sin\omega - \omega\cos\omega}{\omega^3}\cos\frac{\omega}{2}d\omega = \frac{3}{4}$$

$$\Rightarrow \quad \int_{0}^{\infty}\frac{\sin\omega - \omega\cos\omega}{\omega^3}\cos\frac{\omega}{2}d\omega = \frac{3\pi}{16}$$

Thus,

$$\int_{0}^{\infty}\frac{t\cos t - \sin t}{t^3}\cos\frac{t}{2}dt = -\frac{3\pi}{16}$$

Let, $t = 0$ in Eq. (1)

$$\frac{4}{\pi}\int_{0}^{\infty}\frac{\sin\omega - \omega\cos\omega}{\omega^3}d\omega = 1$$

$$\Rightarrow \qquad \int_0^{\infty} \frac{t\cos t - \sin t}{t^3}\,dt = -\frac{\pi}{4}.$$

EXAMPLE 8.6

Find the Fourier transform of $\dfrac{1}{\sqrt{|x|}}$.

Solution: By definition

$$F[f(x)] = \frac{1}{\sqrt{2\pi}} \int_{-\infty}^{\infty} f(x) e^{i\omega x}\,dx$$

$$= \frac{1}{\sqrt{2\pi}} \left[\int_{-\infty}^{0} \frac{1}{\sqrt{-x}} e^{i\omega x}\,dx + \int_0^{\infty} \frac{1}{\sqrt{x}} e^{j\omega x}\,dx \right]$$

In the first integral, put $-x = p \Rightarrow -dx = dp$
So, we have

$$\int_{-\infty}^{0} \frac{1}{\sqrt{-x}} e^{i\omega x}\,dx = -\int_{-\infty}^{0} \frac{1}{\sqrt{p}} e^{-j\omega p}\,dp$$

$$= \int_0^{\infty} \frac{1}{\sqrt{x}} e^{i\omega x}\,dx \quad \text{(Changing } p \text{ to } x\text{)}$$

$$\therefore \qquad F(f(x)) = \frac{1}{\sqrt{2\pi}} \int_0^{\infty} \frac{e^{-i\omega x} + e^{j\omega x}}{\sqrt{x}}\,dx$$

$$= \frac{1}{\sqrt{2\pi}} \int_0^{\infty} \frac{2\cos \omega x}{\sqrt{x}}\,dx$$

$$= 2 \times \text{R.P. of } \frac{1}{\sqrt{2\pi}} \int_0^{\infty} \frac{e^{i\omega x}}{\sqrt{x}}\,dx$$

Let $\qquad i\omega x = t \Rightarrow dx = \dfrac{dt}{i\omega}$

Thus, $\qquad F[f(x)] = 2 \times \text{R.P. of } \dfrac{1}{\sqrt{2\pi}} \displaystyle\int_0^{\infty} \frac{e^{-t}}{\sqrt{\frac{t}{i\omega}}} \frac{dt}{i\omega}$

$$= \text{R.P. of } \frac{1}{\sqrt{2\pi}} \frac{2}{\sqrt{i\omega}} \int_0^{\infty} e^{-t} t^{-1/2}\,dt$$

$$= \text{R.P. of } \frac{\sqrt{2}}{\sqrt{\pi\omega}}(i)^{-1/2}\ \Gamma(1/2) \qquad \left(\because \frac{1}{\sqrt{i}} = \frac{\sqrt{i}}{\sqrt{-1}}\right)$$

$$= \text{R.P. of } \frac{\sqrt{2}}{\sqrt{\pi\omega}}\left(\cos\frac{\pi}{2} + i\sin\frac{\pi}{2}\right)^{-1/2} \sqrt{\pi}$$

$$= \text{R.P. of } \frac{\sqrt{2}}{\sqrt{\omega}}\left(\cos\frac{\pi}{4} - i\sin\frac{\pi}{4}\right)$$

$$= \frac{\sqrt{2}}{\sqrt{\omega}}\cos\frac{\pi}{4} = \frac{\sqrt{2}}{\sqrt{\omega}}\frac{1}{\sqrt{2}} = \frac{1}{\sqrt{\omega}}.$$

EXAMPLE 8.7

Find the Fourier sine and cosine transforms of $f(x) = e^{-ax}$.

Solution: The Fourier sine transform of $f(x)$ is given by

$$F_s(\omega) = \sqrt{\frac{2}{\pi}} \int_0^\infty f(x) \sin \omega x \, dx$$

$$= \sqrt{\frac{2}{\pi}} \int_0^\infty e^{-ax} \sin \omega x \, dx$$

$$- \sqrt{\frac{2}{\pi}} \left[\frac{e^{-ax}}{a^2 + \omega^2} (\ a\sin\omega x - \omega\cos\omega x)\right]_0^\infty$$

$$= \sqrt{\frac{2}{\pi}} \left[0 - \frac{1}{a^2 + \omega^2}(0 - \omega)\right] = \frac{\sqrt{2}}{\pi}\frac{\omega}{a^2 + \omega^2}$$

The Fourier cosine transform is given by

$$F_c(\omega) = \sqrt{\frac{2}{\pi}} \int_0^\infty f(x) \cos \omega x \, dx$$

$$= \sqrt{\frac{2}{\pi}} \int_0^\infty e^{-ax} \cos \omega x \, dx$$

$$= \sqrt{\frac{2}{\pi}} \left[\frac{e^{-ax}}{a^2 + w^2} (-a\cos\omega x + \omega\sin\omega x)\right]_0^\infty$$

$$= \sqrt{\frac{2}{\pi}} \left\{0 - \frac{1}{a^2 + \omega^2}(-a + 0)\right\} = \sqrt{\frac{2}{\pi}}\frac{a}{a^2 + \omega^2}.$$

EXAMPLE 8.8

Find the Fourier sine and cosine transform of

$$f(x) = \begin{cases} 1, & 0 < x < a \\ 0, & x > a \end{cases}.$$

Solution: Fourier sine transform is

$$F_s(\omega) = \sqrt{\frac{2}{\pi}} \int_0^\infty f(x) \sin \omega x \, dx$$

$$= \sqrt{\frac{2}{\pi}} \int_0^a 1 \cdot \sin \omega x \, dx$$

$$= \sqrt{\frac{2}{\pi}} \left[-\frac{\cos \omega x}{\omega} \right]_0^a = \sqrt{\frac{2}{\pi}} \left\{ -\frac{\cos \omega a}{\omega} + \frac{1}{\omega} \right\}$$

Fourier cosine transform is

$$F_c(\omega) = \sqrt{\frac{2}{\pi}} \int_0^\infty f(x) \cos \omega x \, dx$$

$$= \sqrt{\frac{2}{\pi}} \int_0^a 1 \cdot \cos \omega x dx = \sqrt{\frac{2}{\pi}} \left[\frac{\sin \omega x}{\omega} \right]_0^a = \sqrt{\frac{2}{\pi}} \frac{\sin \omega a}{\omega}.$$

Finite Fourier Sine and Cosine Transform

The finite Fourier sine transform of $f(x)$ in $0 < x < a$ is defined as

$$F_s(n) = \int_0^a f(x) \sin\left(\frac{n\pi x}{a}\right) dx$$

where, n is integer. The function $f(x)$ is called the inverse finite Fourier cosine transform of $F_s(n)$ and is given by

$$f(x) = \frac{2}{a} \sum_{n=1}^{\infty} F_s(n) \sin\left(\frac{n\pi x}{a}\right)$$

The finite Fourier cosine transform of $f(x)$ in $0 < x < a$ is defined as

$$F_c(n) = \int_0^a f(x) \cos\left(\frac{n\pi x}{a}\right) dx$$

where, n is an integer. The function $f(x)$ is called the inverse finite Fourier cosine transform of $F_c(n)$ and is given by

$$f(x) = \frac{1}{a}F_c(0) + \frac{2}{a}\sum_{n=1}^{\infty} F_c(n)\cos\left(\frac{n\pi x}{a}\right).$$

EXAMPLE 8.9

Find the Finite Fourier sine and cosine transforms of $f(x) = x$ where $0 < x < 4$.

Solution: Finite Fourier sine transform is

$$F_s(n) = \int_0^4 x\sin\left(\frac{n\pi x}{4}\right)dx$$

$$= \left[x \cdot \frac{\cos\left(\frac{n\pi x}{4}\right)}{\frac{n\pi}{4}} + \frac{\sin\left(\frac{n\pi x}{4}\right)}{\left(\frac{n\pi}{4}\right)^2}\right]_0^4$$

$$- -\frac{16}{n\pi}\cos n\pi + \frac{16}{n^2\pi^2}\sin n\pi = -\frac{16}{n\pi}(-1)^n$$

Also, finite Fourier cosine transform is

$$F_c(n) = \int_0^4 x\cos\left(\frac{n\pi x}{4}\right)dx$$

$$= \left[x\frac{\sin\left(\frac{n\pi x}{4}\right)}{n\pi/4} + \frac{\cos\left(\frac{n\pi x}{4}\right)}{(n\pi/4)^2}\right]_0^4$$

$$= \frac{16}{n^2\pi^2}(\cos n\pi - 1) = \frac{16}{n^2\pi^2}\{(-1)^n - 1\} \quad (n = 0, 1, 2, 3, \ldots).$$

EXAMPLE 8.10

Find the Finite Fourier sine and cosine transform of $f(x) = e^{ax}$ in $(0, l)$.

Solution: Finite Fourier sine transform is given by

$$F_s(n) = \int_0^l e^{ax}\sin\left(\frac{n\pi x}{l}\right)dx$$

$$= \left[\frac{e^{ax}}{a^2 + \frac{n^2\pi^2}{l^2}}\left(a\sin\frac{n\pi x}{l} - \frac{n\pi}{l}\cos\frac{n\pi x}{l}\right)\right]_0^l$$

$$= \frac{1}{a^2 + \frac{n^2\pi^2}{l^2}}\left(-e^{al}\frac{n\pi}{l}\cos n\pi + \frac{n\pi}{l}\right)$$

$$= \frac{n\pi l}{a^2 l^2 + n^2\pi^2}[(-1)^{n+1}e^{al} + 1]$$

Finite Fourier cosine transform is given by

$$F_c(n) = \int_0^l e^{ax}\cos\left(\frac{n\pi x}{l}\right)dx$$

$$= \left[\frac{e^{ax}}{a^2 + \frac{n^2\pi^2}{l^2}}\left(a\cos\frac{n\pi x}{l} + \frac{n\pi}{l}\sin\frac{n\pi x}{l}\right)\right]_0^l$$

$$= \frac{1}{a^2 + \frac{n^2\pi^2}{l^2}}(e^{al}a\cos n\pi - a) = \frac{al^2}{a^2 l^2 + n^2\pi^2}[(-1)^n e^{al} - 1]$$

EXAMPLE 8.11

Find the inverse finite Fourier cosine transform $f(x)$ if

$$F_c(n) = \frac{\cos\left(\frac{2n\pi}{3}\right)}{(2n+1)^2}; \text{ where } 0 < x < 4.$$

Solution: From the inverse Finite Fourier cosine transform formula, we have

$$f(x) = \frac{1}{4}F_c(0) + \frac{2}{4}\sum_{n=1}^{\infty}F_c(n)\cos\left(\frac{n\pi x}{4}\right)$$

$$= \frac{1}{4}\cdot 1 + \frac{1}{2}\sum_{n=1}^{\infty}\frac{\cos\left(\frac{2n\pi}{3}\right)}{(2n+1)^2}\cos\left(\frac{n\pi x}{4}\right)$$

$$= \frac{1}{4} + \frac{1}{2}\sum_{n=1}^{\infty}\frac{\cos\left(\frac{2n\pi}{3}\right)}{(2n+1)^2}\cos\left(\frac{n\pi x}{4}\right).$$

Properties of Fourier Transforms

(a) ***Linear property:*** If $F_1(\omega)$ and $F_2(\omega)$, are Fourier transform of $f_1(x)$ and $f_2(x)$, respectively then

$$F(af_1(x) + bf_2(x)) = aF_1(\omega) + bF_2(\omega)$$

where a and b are constants.

Proof: Now, $F(af_1(x)) + bf_2(x)) = \frac{1}{\sqrt{2\pi}} \int_{-\infty}^{\infty} e^{i\omega x}(af_1(x) + bf_2(x))dx$

$$= \frac{1}{\sqrt{2\pi}} \int_{-\infty}^{\infty} af_1(x)e^{j\omega x}dx + \frac{1}{\sqrt{2\pi}} \int_{-\infty}^{\infty} bf_2(x)e^{i\omega x}dx$$

$$= aF(f_1(x)) + bF(f_2(x))$$

$$= aF_1(\omega) + bF_2(\omega)$$

(b) ***Change of scale property:*** If $F(\omega)$ is the complex Fourier transform of $f(x)$ then

$$F[f(ax)] = \frac{1}{a}F\left(\frac{\omega}{a}\right) \quad (a \neq 0)$$

Proof: We have $F(f(ax)) = \frac{1}{\sqrt{2\pi}} \int_{-\infty}^{\infty} e^{i\omega x} f(ax)dx \quad \left\{\text{Put } ax = t \text{ so that } dx = \frac{dt}{a}\right\}$

$$= \frac{1}{\sqrt{2\pi}} \int_{-\infty}^{\infty} e^{i\omega t/a} f(t)\frac{dt}{a} = \frac{1}{a}\frac{1}{\sqrt{2\pi}} \int_{-\infty}^{\infty} e^{i(\omega/a)t} f(t)dt$$

$$= \frac{1}{a}F\left(\frac{\omega}{a}\right)$$

(c) ***Shifting property:*** If $F(\omega)$ is the complex Fourier transform of $f(x)$ then

$$F[f(x-a) = e^{ia\omega}F(\omega)$$

Proof: We have

$$F[f(x-a)] = \frac{1}{\sqrt{2\pi}} \int_{-\infty}^{\infty} e^{i\omega x} f(x-a)dx \quad [\text{Put } x - a = t \text{ so that } dx = dt].$$

$$= \frac{1}{\sqrt{2\pi}} \int_{-\infty}^{\infty} e^{i\omega(t+a)} f(t)dt = \frac{1}{\sqrt{2\pi}} e^{ia\omega} \int_{-\infty}^{\infty} e^{i\omega t} f(t)dt = e^{ia\omega}F(\omega)$$

Note. $F(e^{iax}f(x)] = \frac{1}{\sqrt{2\pi}} \int_{-\infty}^{\infty} e^{i\omega x} e^{jax} f(x)dx$

$$= \frac{1}{\sqrt{2\pi}} \int_{-\infty}^{\infty} e^{i(\omega+a)x} f(x)dx = F(\omega + a)$$

(d) ***Modulation theorem:*** If $F(\omega)$ is the complex Fourier transform of $f(x)$ then

$$F[f(x)\cos ax] = \frac{1}{2}[F(\omega+a)+F(\omega-a)]$$

Proof: We have

$$F[f(x)\cos ax] = \frac{1}{\sqrt{2\pi}}\int_{-\infty}^{\infty} e^{i\omega x} f(x)\cos ax\, dx$$

$$= \frac{1}{\sqrt{2\pi}}\int_{-\infty}^{\infty} e^{i\omega x} f(x)\frac{e^{iax}+e^{-iax}}{2}dx$$

$$= \frac{1}{2}\left\{\frac{1}{\sqrt{2\pi}}\int_{-\infty}^{\infty} e^{i(\omega+a)x} f(x)dx + \frac{1}{\sqrt{2\pi}}\int_{-\infty}^{\infty} e^{i(\omega-a)x} f(x)dx\right\}$$

$$= \frac{1}{2}[F(\omega+a)+F(\omega-a)].$$

Note. It can be easily deduced that if $F_s(\omega)$ and $F_c(\omega)$ are Fourier sine and cosine transforms of $f(x)$ respectively then

(i) $F_s[f(x)\cos ax) = \frac{1}{2}[F_s(\omega+a)+F_s(\omega-a)]$

(ii) $F_c(f(x)\sin ax] = \frac{1}{2}[F_s(\omega+a)-F_s(\omega-a)]$

(iii) If $F_s(f(x)\sin ax) = \frac{1}{2}[F_c(\omega-a)-F_c(\omega+a)]$

(e) If $F(f(x)) = F(\omega)$ then $F(x^n f(x)) = (-1)^n \frac{d^n}{d\omega^n}F(\omega)$

Proof: We know that

$$F(\omega) = \frac{1}{\sqrt{2\pi}}\int_{-\infty}^{\infty} e^{i\omega x} f(x)dx$$

Differentiating with respect to ω both sides n-times, we get

$$\frac{d^n F(\omega)}{d\omega^n} = \frac{1}{\sqrt{2\pi}}\int_{-\infty}^{\infty} (ix)^n e^{j\omega x} f(x)dx$$

$$= (i)^n F[x^n f(x)]$$

$$\therefore \qquad F(x^n f(x)) = (-i)^n \frac{d^n F(\omega)}{d\omega^n}$$

(f) $F\left(\frac{d^n}{dx^n}f(x)\right) = (-i\omega)^n F(\omega)$ if $f(x)\to 0$ as $x\to\infty$

Proof: We have by using definition

$$F[f'(x)] = F\left(\frac{d}{dx}f(x)\right) = \frac{1}{\sqrt{2\pi}}\int_{-\infty}^{\infty} f'(x)e^{i\omega x}dx$$

$$= \frac{1}{\sqrt{2\pi}}\left[\left[f(x)e^{i\omega x}\right]_{-\infty}^{\infty} - \int_{-\infty}^{\infty} e^{i\omega x} i\omega\, f(x)dx\right]$$

$$= -i\omega\, F(\omega)$$

Now,

$$F(f''(x)) = F\left(\frac{d^2}{dx^2}f(x)\right) = \frac{1}{\sqrt{2\pi}}\int_{-\infty}^{\infty} f''(x)e^{iwx}dx$$

$$= \frac{1}{\sqrt{2\pi}}\left\{\left[f'(x)e^{i\omega x}\right]_{-\infty}^{\infty} - \int_{-\infty}^{\infty} e^{i\omega x} j\omega\, f'(x)dx\right\}$$

$$= -i\omega\, F(f'(x))$$

$$= -i\omega\,(-i\omega\ \ F(\omega)) = (-i\omega)^2\, F(\omega)$$

Similarly, we obtain

$$F\left(\frac{d^n}{dx^n}f(x)\right) = (-i\omega)^n F(\omega)$$

(g) $F\left(\int_a^x f(t)dt\right) = \dfrac{F(\omega)}{(-j\omega)}$

Proof: Let

$$f_1(x) = \int_a^x f(t)dt \Rightarrow f'_1(x) = f(x)$$

$\therefore$

$$F(F'_1(x)) = -i\omega\, F_1(\omega) = -i\omega F\left(\int_a^x f(t)dt\right)$$

Thus,

$$F\left(\int_a^x f(t)dt\right) = \frac{1}{(-i\omega)}\, F(f_1'(x))$$

$$= \frac{1}{(-i\omega)} F(f(x)) = \frac{F(\omega)}{(-i\omega)}$$

(h) (i) $F_s(f'(x)) = -\omega\, F_c(\omega)$

(ii) $F_c(f'(x)) = -\sqrt{\dfrac{2}{\pi}}\, f(0) + \omega F_s(\omega)$

Proof: By definition, we have

$$F_s(f'(x)) = \sqrt{\frac{2}{\pi}} \int_0^\infty f'(x) \sin \omega x \, dx$$

$$= \sqrt{\frac{2}{\pi}} \left(\left[f(x) \sin \omega x \right]_0^\infty - \int_0^\infty \cos \omega x \; \omega f(x) dx \right)$$

$$= -\sqrt{\frac{2}{\pi}} \, \omega \int_0^\infty f(x) \cos \omega x \, dx \qquad (\because f(x) \to 0 \text{ as } x \to \infty)$$

$$= -\,\omega F_c(\omega) \text{ which is (i)}$$

For proving (ii), we have

$$F_c(f'(x)) = \sqrt{\frac{2}{\pi}} \int_0^\infty f'(x) \cos \omega x \, dx$$

$$= \sqrt{\frac{2}{\pi}} \left[\left[f(x) \cos \omega x \right]_0^\infty - \int_0^\infty - \sin \omega x \, \omega f(x) dx \right]$$

$$= \sqrt{\frac{2}{\pi}} \left[0 - f(0) + \omega \int_0^\infty \sin \omega x \, f(x) dx \right] \qquad (\because f(x) \to 0 \text{ as } x \to \infty)$$

$$= -\sqrt{\frac{2}{\pi}} \, f(0) + \omega F_s(\omega)$$

EXAMPLE 8.12

Find the Fourier cosine transform of e^{-x^2}.

Solution: We have

$$F_c(e^{-x^2}) = \sqrt{\frac{2}{\pi}} \int_0^\infty e^{-x^2} \cos \omega x \, dx = F(\omega)$$

Differentiating with respect to ω, we get

$$\frac{dF}{dw} = \sqrt{\frac{2}{\pi}} \int_0^\infty e^{-x^2} (-x \sin \omega x) dx$$

$$= \sqrt{\frac{2}{\pi}} \, \frac{1}{2} \int_0^\infty e^{-x^2} (-2x) \cdot \sin \omega x \, dx$$

$$= \frac{1}{2} \sqrt{\frac{2}{\pi}} \left\{ \left[e^{-x^2} \sin \omega x \right]_0^\infty - \int_0^\infty e^{-x^2} \cos \omega x \times \omega dx \right\}$$

$$= -\frac{\omega}{2}\sqrt{\frac{2}{\pi}}\int_0^\infty e^{-x^2}\cos\omega x\, dx$$

$$= -\frac{\omega}{2}F(\omega)$$

$\therefore$ $$\frac{dF}{F} = -\frac{\omega}{2}d\omega$$

$\Rightarrow$ $$\log F = -\frac{w^2}{4} + \log a$$

$\therefore$ $$F = ae^{-\omega^2/4} = F(\omega)$$

In $$F_c(e^{-x^2}) = \sqrt{\frac{2}{\pi}}\int_0^\infty e^{-x^2}\cos\omega x\, dx = F(\omega),\ \text{let } x = 0;$$

We get $$F(0) = \sqrt{\frac{2}{\pi}}\int_0^\infty e^{-x^2}dx$$

Taking $$x^2 = t \Rightarrow x = \sqrt{t} \text{ so that } dx = \frac{dt}{2\sqrt{t}}$$

$$F(0) = \sqrt{\frac{2}{\pi}}\int_0^\infty e^{-t}t^{-1/2}\frac{dt}{2} = \frac{1}{2}\sqrt{\frac{2}{\pi}}\Gamma\left(\frac{1}{2}\right)$$

$$= \frac{1}{2}\sqrt{\frac{2}{\pi}}\sqrt{\pi} = \frac{1}{\sqrt{2}}$$

So, we obtain $$F(0) = a = \frac{1}{\sqrt{2}}$$

Hence, $$F_c(e^{-x^2}) = \frac{1}{\sqrt{2}}e^{-\omega^2/4}.$$

EXAMPLE 8.13

Find the Fourier sine transform of $\dfrac{e^{-ax}}{x}$.

Solution: We have $$F_s\left(\frac{e^{-ax}}{x}\right) = \sqrt{\frac{2}{\pi}}\int_0^\infty \frac{e^{-ax}}{x}\sin\omega x\, dx = F(\omega)$$

Differentiating both sides with respect to x, we get

$$\frac{dF}{dw} = \int_0^\infty \sqrt{\frac{2}{\pi}}\frac{e^{-ax}}{x} \times x \times \cos\omega x\, dx$$

$$= \sqrt{\frac{2}{\pi}} \int_0^{\infty} e^{-ax} \cos \omega x \, dx = \frac{a}{a^2+\omega^2} \sqrt{\frac{2}{\pi}}$$

$\therefore$ $$dF = \sqrt{\frac{2}{\pi}} \frac{a d\omega}{a^2+\omega^2}$$

Integrating, we obtain

$$F(\omega) = \sqrt{\frac{2}{\pi}} \tan^{-1}\left(\frac{\omega}{a}\right) + C$$

But, $$F(0) = 0$$

$\therefore$ $$C = 0$$

Hence, $$F(\omega) = \sqrt{\frac{2}{\pi}} \tan^{-1}\left(\frac{\omega}{a}\right).$$

EXAMPLE 8.14

Find the Fourier cosine transform of $f(x) = \dfrac{1}{1+x^2}$. Hence, derive Fourier sine transform of $\phi(x) = \dfrac{x}{1+x^2}$.

Solution: We have $$F_c(\omega) = \int_0^{\infty} \sqrt{\frac{2}{\pi}} \frac{\cos \omega x}{1+x^2} dx = F(\omega)$$

$\therefore$ $$\frac{dF}{d\omega} = \sqrt{\frac{2}{\pi}} \int_0^{\infty} -\frac{x \sin \omega x}{1+x^2} dx = \sqrt{\frac{2}{\pi}} \int_0^{\infty} -\frac{x^2 \sin \omega x \, dx}{x(1+x^2)}$$

$$= -\int_0^{\infty} \sqrt{\frac{2}{\pi}} \frac{(1+x^2)-1}{x(1+x^2)} \sin \omega x \, dx$$

$$= -\sqrt{\frac{2}{\pi}} \int_0^{\infty} \frac{\sin \omega x}{x} dx + \sqrt{\frac{2}{\pi}} \int_0^{\infty} \frac{\sin \omega x}{x(1+x^2)} dx$$

$$= \sqrt{\frac{2}{\pi}} \left(-\frac{\pi}{2} + \int_0^{\infty} \frac{\sin \omega x}{x(1+x^2) dx} \right)$$

$\therefore$ $$\frac{d^2F}{d\omega^2} = \sqrt{\frac{2}{\pi}} \int_0^{\infty} \frac{\cos \omega x}{1+x^2} dx = F(\omega)$$

which is $\frac{d^2F}{dw^2} - F = 0$ with general solution

$$F(\omega) = c_1 e^{\omega} + c_2 e^{-\omega}$$

Now, when $\omega = 0$

$$F(0) = \sqrt{\frac{2}{\pi}} \int_0^\infty \frac{dx}{1+x^2} = \sqrt{\frac{2}{\pi}} \left[\tan^{-1} x\right]_0^\infty = \sqrt{\frac{\pi}{2}}$$

Also,
$$\left.\frac{dF}{d\omega}\right|_{\omega=0} = -\sqrt{\frac{\pi}{2}}.$$

So, on using these, we have

$$C_1 + C_2 = \sqrt{\frac{\pi}{2}} \quad \text{and} \quad C_1 - C_2 = -\sqrt{\frac{\pi}{2}}$$

which gives
$$C_1 = 0 \quad \text{and} \quad C_2 = \sqrt{\frac{\pi}{2}}$$

Hence,
$$F(\omega) = \sqrt{\frac{\pi}{2}} e^{-\omega}$$

Now,
$$\frac{dF}{d\omega} = -\sqrt{\frac{\pi}{2}} e^{-\omega}$$

$\Rightarrow$
$$-\sqrt{\frac{2}{\pi}} \int_0^\infty \frac{x}{1+x^2} \sin \omega x \, dx = -\sqrt{\frac{\pi}{2}} e^{-\omega}$$

$\Rightarrow$
$$F_s\left(\frac{x}{1+x^2}\right) = \sqrt{\frac{\pi}{2}} e^{-\omega}.$$

EXAMPLE 8.15

Show that $F_s(xf(x)) = -\frac{d}{dw} F_c(\omega)$ and $F_c(xf(x) = \frac{d}{dw} F_s(\omega)$. Hence, find the Fourier sine and cosine transforms of xe^{-ax}.

Solution: We have
$$\frac{d}{d\omega} F_c(\omega) = \frac{d}{d\omega}\left(\int_0^\infty \sqrt{\frac{2}{\pi}} f(x) \cos \omega x \, dx\right)$$

$$= \sqrt{\frac{2}{\pi}} \int_0^\infty - f(x) \sin \omega x \, x dx$$

$$= -\sqrt{\frac{2}{\pi}}\int_0^\infty xf(x)\sin\omega x\,dx = -F_s(xf(x))$$

and $$\frac{d}{d\omega}F_s(\omega) = \frac{d}{d\omega}\left(\int_0^\infty \sqrt{\frac{2}{\pi}}\, f(x)\sin\omega x\,dx\right)$$

$$= \sqrt{\frac{2}{\pi}}\int_0^\infty f(x)\cos\omega x\, x dx$$

$$= \sqrt{\frac{2}{\pi}}\int_0^\infty xf(x)\cos\omega x\,dx = F_c(xf(x))$$

We have $$F_s(e^{-ax}) = \int_0^\infty \sqrt{\frac{2}{\pi}}\, e^{-ax}\sin\omega x\,dx$$

$$= \sqrt{\frac{2}{\pi}}\,\frac{\omega}{a^2+\omega^2}$$

and $$F_c(e^{-ax}) = \sqrt{\frac{2}{\pi}}\int_0^\infty e^{-ax}\cos\omega x\,dx = \sqrt{\frac{2}{\pi}}\,\frac{a}{a^2+\omega^2}$$

Now, $$F_c(e^{-ax}) = -\frac{d}{d\omega}F_s(e^{-ax})$$

$$= -\frac{d}{d\omega}\left(\sqrt{\frac{2}{\pi}}\,\frac{a}{a^2+\omega^2}\right) = \sqrt{\frac{2}{\pi}}\,\frac{2a\omega}{(a^2+\omega^2)^2}$$

and $$F_s(xe^{-ax}) = \frac{d}{d\omega} = F_c(e^{-ax})$$

$$= \frac{d}{d\omega}\left(\sqrt{\frac{2}{\pi}}\,\frac{\omega}{a^2+\omega^2}\right) = \sqrt{\frac{2}{\pi}}\,\frac{a^2-\omega^2}{(a^2+\omega^2)^2}.$$

EXAMPLE 8.16

Find the Fourier cosine transform of $xe^{-a^2x^2}$ and evaluate Fourier sine transform of $xe^{-a^2x^2}$.

Solution: We have $$F_c(e^{-a^2x^2}) = \sqrt{\frac{2}{\pi}}\int_0^\infty e^{-a^2x^2}\cos\omega x\,dx = F(\omega)$$

Differentiating both sides with respect to ω, we get

$$\frac{dF}{d\omega} = \sqrt{\frac{2}{\pi}}\,\frac{d}{d\omega}\left(\int_0^\infty e^{-a^2x^2}\cos\omega x\,dx\right)$$

$$= \sqrt{\frac{2}{\pi}} \int_0^\infty e^{-a^2x^2} \times -\sin \omega x \times x dx$$

$$= \sqrt{\frac{2}{\pi}} \frac{1}{a^2} \int_0^\infty e^{-a^2x^2} (-a^2x) \sin \omega x \, dx$$

$$= \frac{1}{2a^2} \sqrt{\frac{2}{\pi}} \left\{ \left[e^{-a^2x^2} \sin \omega x \right]_0^\infty - \int_0^\infty \cos \omega x \times \omega e^{-a^2x^2} dx \right\}$$

$$= \frac{1}{2a^2} \sqrt{\frac{2}{\pi}} \times -\omega \int_0^\infty e^{-a^2x^2} \cos \omega x \, dx$$

$$\frac{dF}{d\omega} = -\frac{\omega}{2a^2} \sqrt{\frac{2}{\pi}} F$$

$\Rightarrow$
$$\frac{dF}{F} = -\frac{1}{2a^2} \sqrt{\frac{2}{\pi}} \omega$$

Integrating, we obtain

$$\log F = -\sqrt{\frac{2}{\pi}} \frac{\omega^2}{4a^2} + \log C$$

$\Rightarrow$
$$F = Ce^{-\sqrt{\frac{2}{\pi}} \frac{\omega^2}{4a^2}}$$

When $\omega = 0$

$$F(0) = \sqrt{\frac{2}{\pi}} \int_0^\infty e^{-a^2x^2} dx$$

$$\left\{ \text{Taking } a^2x^2 = t \Rightarrow x = \frac{\sqrt{t}}{a} \text{ so that } dx = \frac{dt}{2a\sqrt{t}} \right.$$

$$= \sqrt{\frac{2}{\pi}} \int_0^\infty e^{-t} t^{-1/2} \frac{dt}{2a}$$

$$= \sqrt{\frac{2}{\pi}} \frac{1}{2a} \Gamma(1/2) = \sqrt{\frac{2}{\pi}} \frac{1}{2a} \sqrt{\pi} = \frac{1}{\sqrt{2}\,a}$$

Thus,
$$F(0) = \frac{1}{\sqrt{2}\,a} = C$$

Hence, $$F(\omega) = \frac{1}{\sqrt{2}\,a} e^{-\sqrt{\frac{2}{\pi}}\frac{\omega^2}{4a^2}}$$

We know that $$F_s(xf(x)) = -\frac{d}{d\omega}(F_c(f(x))$$

$\therefore$ $$F_s(xe^{-a^2x^2}) = -\frac{d}{d\omega}F_c(e^{-a^2x^2})$$

$$= -\frac{d}{d\omega}\left(\frac{1}{\sqrt{2}\,a} e^{-\sqrt{\frac{2}{\pi}}\frac{\omega^2}{4a}}\right)$$

$$= -\frac{1}{\sqrt{2}\,a} e^{-\sqrt{\frac{2}{\pi}}\frac{\omega^2}{4a}} \times -\sqrt{\frac{2}{\pi}}\frac{\omega}{2a}$$

$$= \frac{\omega}{2\sqrt{\pi}\,a^2} e^{-\sqrt{\frac{2}{\pi}}\frac{\omega^2}{4a}}.$$

EXAMPLE 8.17

Find the Fourier sine transform of $f(x) = \dfrac{1}{x(x^2+a^2)}$ and deduce cosine transform of $\dfrac{1}{x^2+a^2}$.

Solution: We have $$F_s[f(x)] = \sqrt{\frac{2}{\pi}}\int_0^\infty f(x)\sin\omega x\,dx$$

$$F(\omega) = \sqrt{\frac{2}{\pi}}\int_0^\infty \frac{1}{x(x^2+a^2)}\sin\omega x\,dx$$

Differentiating both sides with respect to ω, we get

$$\frac{dF}{d\omega} = \sqrt{\frac{2}{\pi}}\int_0^\infty \frac{\cos\omega x}{x^2+a^2}dx$$

We know that $$\int_0^\infty 2ye^{-(a^2+x^2)y^2}dy = \frac{1}{a^2+x^2}$$

Therefore, $$\sqrt{\frac{2}{\pi}}\int_0^\infty \frac{\cos\omega x}{a^2+x^2}dx = \sqrt{\frac{2}{\pi}}\int_0^\infty \cos\omega x \times \left(\int_0^\infty 2ye^{-(a^2+x^2)y^2}dy\right)dx$$

$$= \sqrt{\frac{2}{\pi}}\int_0^\infty 2ye^{-a^2y^2}\left\{\int_0^\infty e^{-x^2y^2}\cos\omega x\,dx\right\}dy$$

Now let $$I = \int_0^{\infty} e^{-x^2y^2} \cos \omega x \, dx$$

Differentiating both sides with respect to ω, we get

$$\frac{dI}{d\omega} = \int_0^{\infty} e^{-x^2y^2} (-x \sin \omega x) dx$$

$$= \frac{1}{2y^2} \int_0^{\infty} e^{-y^2x^2} (-2y^2x) \sin \omega x \, dx$$

$$= \frac{1}{2y^2} \left\{ \left[e^{-y^2x^2} \sin \omega x \right]_0^{\infty} - \int_0^{\infty} e^{-y^2x^2} (\cos \omega x) \omega dx \right\}$$

$$= -\frac{\omega}{2y^2} I$$

$$\Rightarrow \qquad \frac{dI}{I} = -\frac{\omega d\omega}{2y^2}$$

Integrating, we obtain

$$\log I = -\frac{\omega^2}{4y^2} + \log C_1$$

$$\Rightarrow \qquad I = C_1 e^{-\omega^2/4y^2}$$

when $$\omega = 0,\ I = \int_0^{\infty} e^{-x^2y^2} dx = \frac{\sqrt{\pi}}{2y}$$

$$\Rightarrow \qquad I(0) = c_1 = \sqrt{\pi}/2y$$

Thus, $$I = \frac{\sqrt{\pi}}{2y} e^{-\omega^2/4y^2}$$

Therefore, $$\sqrt{\frac{2}{\pi}} \int_0^{\infty} \frac{\cos \omega x}{x^2 + a^2} dx = \sqrt{\frac{2}{\pi}} \int_0^{\infty} 2y e^{-a^2y^2} \times \frac{\sqrt{\pi}}{2y} e^{-\omega^2/4y^2} dy$$

$$= \sqrt{2} \int_0^{\infty} e^{-a^2\left(y^2 + \frac{\omega^2}{4a^2y^2}\right)} dy$$

$$= \frac{\sqrt{\pi}}{\sqrt{2}\, a} e^{-\omega a}$$

$$\therefore \qquad \frac{d}{d\omega}F(\omega) = \frac{\sqrt{\pi}}{\sqrt{2}\,a}e^{-\omega a}$$

$$\Rightarrow \qquad dF = \frac{\sqrt{\pi}}{\sqrt{2}\,a}e^{-\omega a}d\omega$$

$$\therefore \qquad F(\omega) = \frac{\sqrt{\pi}}{\sqrt{2}\,a}\frac{e^{-\omega a}}{-a} + C_2$$

But when $\omega = 0$, we have

$$F_s(f(x)) = 0 \Rightarrow C_1 = \frac{\sqrt{\pi}}{\sqrt{2}\,a^2}$$

Thus, $$F_s(f(x)) = -\frac{\sqrt{\pi}}{\sqrt{2}\,a^2}e^{-\omega a} + \frac{\sqrt{\pi}}{\sqrt{2}\,a^2} = \frac{\sqrt{\pi}}{\sqrt{2}\,a^2}(1 - e^{-\omega a})$$

If $$f(x) = \frac{1}{x^2 + a^2} \text{ then}$$

$$F_c(f(x)) = \sqrt{\frac{2}{\pi}}\int_0^\infty \frac{\cos \omega x}{x^2 + a^2}dx = \sqrt{\frac{\pi}{2}}\,e^{-\omega a}.$$

EXAMPLE 8.18

Find the Fourier sine and cosine transform of x^{m-1}.

Solution: We have $$F_s(\omega) = \sqrt{\frac{2}{\pi}}\int_0^\infty x^{m-1}\sin \omega x\, dx$$

and $$F_c(\omega) = \sqrt{\frac{2}{\pi}}\int_0^\infty x^{m-1}\cos \omega x\, dx$$

Let $x = i\omega y$ in $\Gamma(m) = \int_0^\infty e^{-x}\, x^{m-1}dx$, we have

$$\Gamma(m) = \int_0^\infty e^{-i\omega y}(i\omega y)^{m-1}(i\omega dy)$$

$$= \omega^m \int_0^\infty e^{-i\omega y}\, y^{m-1}(i)^m\, dy$$

$$\Rightarrow \qquad \frac{\Gamma(m)}{\omega^m} = \int_0^\infty e^{j\omega y}\, y^{m-1}(e^{i\pi/2})^m\, dy$$

$$\Rightarrow \qquad e^{-\frac{im\pi}{2}} \frac{\Gamma(m)}{\omega^m} = \int_0^\infty e^{-i\omega x}\, x^{m-1} dx$$

$$\Rightarrow \qquad \frac{\Gamma(m)}{\omega^m}\left\{\cos\frac{m\pi}{2} - i\sin\frac{m\pi}{2}\right\} = \int_0^\infty (\cos\omega x - i\sin\omega x) x^{m-1} dx$$

Equating real and imaginary parts on both sides, we get

$$F_s(\omega) = \sqrt{\frac{2}{\pi}} \int_0^\infty x^{m-1} \sin\omega x\, dx = \sqrt{\frac{2}{\pi}}\, \frac{\Gamma(m)}{\omega^m} \sin\frac{m\pi}{2}$$

and

$$F_c(\omega) = \sqrt{\frac{2}{\pi}} \int_0^\infty x^{m-1} \sin\omega x\, dx = \sqrt{\frac{2}{\pi}}\, \frac{\Gamma(m)}{\omega^m} \cos\frac{m\pi}{2}.$$

EXAMPLE 8.19

Use the sine inversion formula to obtain $f(x)$ if $F_s(\omega) = \dfrac{\omega}{1+\omega^2}$.

Solution: We have

$$f(x) = \sqrt{\frac{2}{\pi}} \int_0^\infty F_s(\omega) \sin\omega x\, d\omega$$

$$= \sqrt{\frac{2}{\pi}} \int_0^\infty \frac{\omega}{1+\omega^2} \sin\omega x\, d\omega$$

$$= \sqrt{\frac{2}{\pi}} \int_0^\infty \frac{\omega^2+1-1}{\omega(1+\omega^2)} \sin\omega x\, d\omega$$

$$= \sqrt{\frac{2}{\pi}} \int_0^\infty \frac{\sin\omega x}{\omega} d\omega - \sqrt{\frac{2}{\pi}} \int_0^\infty \frac{\sin\omega x}{\omega(1+\omega^2)} d\omega$$

$$= \sqrt{\frac{2}{\pi}} \cdot \frac{\pi}{2} - \sqrt{\frac{2}{\pi}} \int_0^\infty \frac{\sin\omega x}{\omega(1+\omega^2)} d\omega \qquad (1)$$

$$\left\{\because \int_0^\infty \frac{\sin\omega x}{\omega} d\omega = \frac{\pi}{2}\right\}$$

$$\therefore \qquad \frac{df}{dx} = -\sqrt{\frac{2}{\pi}} \int_0^\infty \frac{\cos\omega x}{1+\omega^2} d\omega$$

and

$$\frac{d^2 f}{dx^2} = \sqrt{\frac{2}{\pi}} \int_0^\infty \frac{\omega \sin\omega x}{1+\omega^2} dw = f(x)$$

$$\Rightarrow \qquad \frac{d^2 f}{dx^2} - f = 0$$

which is a linear differential equation of order 2 with general solution given by

$$f(x) = C_1 e^{+x} + C_2 e^{-x}$$

Now when $$x = 0,\ f(0) = \sqrt{\frac{\pi}{2}} \quad \text{(from Eq. (1))}$$

and $$\frac{df}{dx} = -\sqrt{\frac{2}{\pi}} \int_0^\infty \frac{d\omega}{1+\omega^2} = -\sqrt{\frac{\pi}{2}}$$

$\therefore$ $$C_1 + C_2 = \sqrt{\frac{\pi}{2}} \quad \text{and} \quad C_1 - C_2 = -\sqrt{\frac{\pi}{2}}$$

$\Rightarrow$ $$C_1 = 0 \quad \text{and} \quad C_2 = \sqrt{\frac{\pi}{2}}$$

Hence, $$f(x) = \sqrt{\frac{\pi}{2}}\, e^{-x}.$$

Relation between Fourier and Laplace Transforms

If $$f(x) = \begin{cases} e^{-tx} g(x), & x > 0 \\ 0 \quad , & x < 0 \end{cases}$$

then $$F(f(x)) = L(g(x)) \times \frac{1}{\sqrt{2\pi}}$$

Proof: We have $$F(f(x)) = \frac{1}{\sqrt{2\pi}} \int_{-\infty}^{\infty} e^{i\omega x} f(x)dx$$

$$= \frac{1}{\sqrt{2\pi}} \int_0^\infty e^{i\omega x} \cdot e^{-tx} g(x)dx$$

$$= \frac{1}{\sqrt{2\pi}} \int_0^\infty e^{(j\omega - t)x} g(x)dx$$

$$= \frac{1}{\sqrt{2\pi}} \int_0^\infty e^{-\omega' x} g(x)dx$$

$$= \frac{1}{\sqrt{2\pi}} \int_0^\infty e^{-\omega t} g(x)\, dx \text{ where } \omega' = t - i\omega$$

$$= \frac{L(g(x))}{\sqrt{2\pi}}$$

Convolution: The convolution of two functions $f(x)$ and $g(x)$ over the interval $(-\infty, \infty)$ is defined as $(f * g)(x) = \frac{1}{\sqrt{2\pi}} \int_{-\infty}^{\infty} f(u)\, g(x-u)du = h(x)$.

Convolution Theorem for Fourier Transforms. The Fourier transform of the convolution of $f(x)$ and $g(x)$ is the product of their Fourier transform, that is

$$F((f * g)(x)) = F(f(x)) \cdot F(g(x))$$

Proof: We know that

$$(f * g)(x) = \frac{1}{\sqrt{2\pi}} \int_{-\infty}^{\infty} f(u) g(x-u) du$$

$$\therefore \quad F((f * g)(x)) = \frac{1}{\sqrt{2\pi}} \int_{-\infty}^{\infty} \left(\frac{1}{\sqrt{2\pi}} \int_{-\infty}^{\infty} f(u) g(x-u) du \right) e^{i\omega x} dx$$

$$= \frac{1}{2\pi} \int_{-\infty}^{\infty} f(u) \left(\int_{-\infty}^{\infty} g(x-u) e^{j\omega x} dx \right) du$$

(Changing the order of integration)

$$= \frac{1}{2\pi} \int_{-\infty}^{\infty} f(u) \left(\int_{-\infty}^{\infty} e^{j\omega(x-u)} g(x-u) d(x-u) \right) e^{i\omega u} dw$$

$$= \frac{1}{\sqrt{2\pi}} \int_{-\infty}^{\infty} f(u) e^{i\omega u} du \cdot \frac{1}{\sqrt{2\pi}} \int_{-\infty}^{\infty} e^{i\omega t} g(t) dt \text{ where } x - u = t$$

$$= F(f(u)) \cdot F(g(t))$$

$$= F_1(\omega) \cdot G_1(\omega)$$

Note. By inversion property, we have

$$F^{-1}(F_1(\omega) \cdot G_1(\omega)) = f * g = F^{-1}(F_1(\omega)) * F^{-1}(G_1(\omega)).$$

Parseval's Identity for Fourier Transforms

If the Fourier transform of $f(x)$ and $g(x)$ are $F(\omega)$ and $G(\omega)$ respectively, then

(a) $\int_{-\infty}^{\infty} F(\omega)\overline{G(\omega)}\, d\omega = \int_{-\infty}^{\infty} f(x)\overline{g(x)}\, dx$

(b) $\int_{-\infty}^{\infty} |F(\omega)|^2\, d\omega = \int_{-\infty}^{\infty} |f(x)|^2\, dx$

where bar implies the complex conjugate.

Proof: Now,

$$\int_{-\infty}^{\infty} f(x)\overline{g(x)}dx = \int_{-\infty}^{\infty} f(x)\left(\frac{1}{\sqrt{2\pi}}\int_{-\infty}^{\infty}\overline{G(\omega)e^{j\omega x}dx}\right)$$

(Using the inversion formula for Fourier transforms)

$$= \frac{1}{\sqrt{2\pi}}\int_{-\infty}^{\infty}\overline{G}(\omega)\left(\int_{-\infty}^{\infty} f(x)e^{i\omega x}dx\right)d\omega$$

$$= \int_{-\infty}^{\infty}\overline{G}(\omega)F(\omega)d\omega \text{ which is (a)}$$

Taking $g(x) = f(x)$ in (a), we get

$$\int_{-\infty}^{\infty} F(\omega)\overline{F(\omega)}d\omega = \int_{-\infty}^{\infty} f(x)\overline{f(x)}dx$$

$$\int_{-\infty}^{\infty} |F(\omega)|^2 d\omega = \int_{-\infty}^{\infty} |f(x)|^2 dx.$$

EXAMPLE 8.20

Using Parseval's identity, prove that

$$\int_0^{\infty}\frac{dx}{(a^2+x^2)(b^2+x^2)} = \frac{\pi}{2ab(a+b)}.$$

Solution: Let $f(x) = e^{-ax}$ and $g(x) = e^{-bx}$. Then

$$F_c(f(x)) = \frac{a}{a^2+\omega^2} \quad\text{and}\quad G_C(g(x)) = \frac{b}{b^2+\omega^2}$$

Using Parseval's identity, we obtain

$$\frac{2}{\pi}\int_0^{\infty} F_c(\omega)\overline{G_c(\omega)}\,d\omega = \int_0^{\infty} f(x)\,g(x)dx$$

$$\Rightarrow \qquad \frac{2}{\pi}\int_0^{\infty}\frac{a}{a^2+\omega^2}\cdot\frac{b}{b^2+\omega^2}d\omega = \int_0^{\infty} e^{-ax}e^{-bx}dx$$

$$= \int_0^{\infty} e^{-(a+b)x}dx = \left\{\frac{e^{-(a+b)x}}{-(a+b)}\right\}_0^{\infty} = \frac{1}{a+b}$$

Hence,

$$\int_0^{\infty}\frac{dx}{(a^2+x^2)(b^2+x^2)} = \frac{\pi}{2ab(a+b)}$$

EXAMPLE 8.21

Evaluate the following using Parseval's identity:

(a) $\int_0^\infty \frac{x^2}{(a^2+x^2)^2}dx \quad (a > 0)$

(b) $\int_0^\infty \frac{dx}{(a^2+x^2)^2} \quad (a > 0)$

Solution: Let $f(x) = e^{-ax}$ $(x > 0)$. Then

$F_c(f(x)) = \frac{a}{a^2+\omega^2}$ and $F_s(f(x)) = \frac{\omega}{b^2+\omega^2}$.

From, Parseval's identity, we obtain

$$\frac{2}{\pi}\int_0^\infty |F_s(\omega)|^2 \, d\omega = \int_0^\infty |f(x)|^2 \, dx$$

$$\Rightarrow \qquad \frac{2}{\pi}\int_0^\infty \frac{\omega^2}{(a^2+\omega^2)^2} d\omega = \int_0^\infty e^{-2ax} dx = \frac{1}{2a}$$

$$\therefore \qquad \int_0^\infty \frac{x^2}{(a^2+x^2)} dx = \frac{\pi}{4a}$$

Also, using Parseval's identity, we obtain

$$\frac{2}{\pi}\int_0^\infty |F_c(\omega)|^2 \, d\omega = \int_0^\infty |f(x)|^2 \, dx$$

$$\Rightarrow \qquad \frac{2}{\pi}\int_0^\infty \frac{a^2}{(a^2+\omega^2)} d\omega = \int_0^\infty e^{-2ax} dx = \frac{1}{2a}$$

$$\therefore \qquad \int_0^\infty \frac{d}{(a^2+\omega^2)} d\omega = \frac{\pi}{4a^3}.$$

EXAMPLE 8.22

Find the Fourier transform of $f(x)$ given by

$$f(x) = \begin{cases} a-|x|, & \text{for } |x| < a \\ 0, & \text{for } |x| > a \end{cases}$$

Hence, show that $\int_0^\infty \left(\frac{\sin x}{x}\right)^2 dx = \frac{\pi}{2}$ and $\int_0^\infty \left(\frac{\sin x}{x}\right)^4 dx = \frac{\pi}{2}$

Solution: Fourier transform of $f(x)$ is given by

$$F(f(x)) = \frac{1}{\sqrt{2\pi}} \int_{-\infty}^{\infty} f(x)e^{i\omega x}dx$$

$$= \frac{1}{\sqrt{2\pi}} \int_{-a}^{a} (a-|x|)e^{i\omega x}dx$$

$$= \frac{1}{\sqrt{2\pi}} \int_{-a}^{a} (a-|x|)\ (\cos\omega x + i\sin\omega x)dx$$

$$= \frac{2}{\sqrt{2\pi}} \int_{0}^{a} (a-x)\cos\omega x\ dx + 0$$

($\because$ $(a - |x|)$ cos ωx is an even function and $(a - |x|)$ sin ωx is an odd function).

$$= \frac{2}{\sqrt{2\pi}} \left[(a-x)\frac{\sin\omega x}{\omega} - \frac{\cos\omega x}{\omega^2}\right]_0^a$$

$$= \frac{2}{\sqrt{2\pi}} \frac{1-\cos\omega a}{\omega^2} = \frac{4\sin^2(\omega a/2)}{\sqrt{2\pi}\,\omega^2}$$

By inversion formula

$$f(x) = \frac{1}{\sqrt{2\pi}} \int_{-\infty}^{\infty} F(\omega)e^{i\omega x}d\omega = \frac{1}{\sqrt{2\pi}} \int_{-\infty}^{\infty} \frac{4\sin^2(\omega a/2)}{\sqrt{2\pi}\,\omega^2} e^{i\omega x}\, d\omega$$

Let $x = 0$ and $a = 2$, we get

$$f(0) = \frac{2}{\pi} \int_{-\infty}^{\infty} \frac{\sin^2\omega}{\omega^2} d\omega \qquad \{\because f(0) = a = 2\}$$

$$\int_{-\infty}^{\infty} \left(\frac{\sin x}{x}\right)^2 dx = 2\int_{0}^{\infty} \left(\frac{\sin x}{x}\right)^2 dx = \pi$$

$$\Rightarrow \qquad \int_{0}^{\infty} \left(\frac{\sin x}{x}\right)^2 dx = \frac{\pi}{2}$$

Using Parseval's identity, we get

$$\int_{-\infty}^{\infty} |F(\omega)|^2\, d\omega = \int_{-\infty}^{\infty} |f(x)|^2\, dx$$

$$\Rightarrow \quad \int_{-\infty}^{\infty} \frac{16}{2\pi}\left(\frac{\sin(\omega a/2)}{\omega}\right)^4 d\omega = \int_{-a}^{a}(a-|x|)^2 dx$$

$$= 2\int_{0}^{a}(a-x)^2 dx = 2\left[\frac{(a-x)^3}{-3}\right]_0^a = \frac{2}{3}a^3$$

Let $\omega a/2 = x$ so that $d\omega = \dfrac{2}{a}dx$, we get

$$\int_{-\infty}^{\infty} \frac{16}{2\pi}\left(\frac{\sin x}{2x/a}\right)^4 \frac{2}{a}dx = \frac{2}{3}a^3$$

$$\Rightarrow \quad \int_{0}^{\infty} \frac{16}{\pi}\left(\frac{\sin x}{x}\right)^4 \frac{a^4}{2^4}\cdot\frac{2}{a}dx = \frac{2}{3}a^3$$

$$\Rightarrow \quad \int_{0}^{\infty}\left(\frac{\sin x}{x}\right)^4 dx = \frac{\pi}{3}.$$

EXAMPLE 8.23

Find the Fourier transform of $f(x)$ defined by $f(x) = \begin{cases} 1, & |x| < a \\ 0, & |x| > a \end{cases}$.

and prove that $\displaystyle\int_{0}^{\infty}\frac{\sin^2 ax}{x^2}dx = \frac{\pi a}{2}$.

Solution: We have $\quad F(f(x)) = \dfrac{1}{\sqrt{2\pi}}\displaystyle\int_{-\infty}^{\infty} f(x)e^{i\omega x}dx$

$$= \frac{1}{\sqrt{2\pi}}\int_{-a}^{a} 1\cdot e^{j\omega x}dx = \frac{1}{\sqrt{2\pi}}\left(\frac{e^{i\omega a}-e^{-i\omega a}}{i\omega}\right)$$

$$= \sqrt{\frac{2}{\pi}}\frac{\sin a\omega}{\omega}$$

Using Parseval's identity, we obtain

$$\int_{-\infty}^{\infty}|f(x)|^2 dx = \int_{-\infty}^{\infty}|F(\omega)|^2 d\omega$$

$$\Rightarrow \qquad \int_{-a}^{a} (1)^2 dx = \int_{-\infty}^{\infty} \frac{2}{\pi\omega^2} \sin^2 a\omega \, d\omega$$

$$\Rightarrow \qquad \int_{0}^{a} dx = \frac{2}{\pi} \int_{0}^{\infty} \frac{\sin^2 a\omega}{\omega^2} d\omega$$

$$\Rightarrow \qquad \int_{0}^{\infty} \frac{\sin^2 ax}{x^2} dx = \frac{\pi a}{2}.$$

EXAMPLE 8.24

Show that the Fourier transform of

$$f(x) = \begin{cases} 1 - x^2, & \text{if } |x| < 1 \\ 0 \quad , & \text{if } |x| > 1 \end{cases} \text{ is } \frac{4}{\omega^3} \frac{(\sin\omega - \omega\cos\omega)}{\sqrt{2\pi}}.$$

Using Parseval's identity, prove that $\int_{0}^{\infty} \left(\frac{\sin x - x\cos x}{x^3} \right) dx \frac{\pi}{15}$

Solution: We have

$$F(f(x)) = \frac{1}{\sqrt{2\pi}} \int_{-\infty}^{\infty} f(x) e^{i\omega x} dx$$

$$= \frac{1}{\sqrt{2\pi}} \int_{-1}^{1} (1 - x^2) e^{j\omega x} dx$$

$$= \frac{1}{\sqrt{2\pi}} \left[(1 - x^2) \frac{e^{j\omega x}}{i\omega} + 2x \frac{e^{i\omega x}}{(i\omega)^2} - 2 \frac{e^{j\omega x}}{(i\omega)^3} \right]_{-1}^{1}$$

$$= \frac{1}{\sqrt{2\pi}} \left[-\frac{2}{\omega^2} (e^{i\omega} + e^{i\omega}) + \frac{2}{i\omega^3} (e^{i\omega} - e^{i\omega}) \right]$$

$$= \frac{1}{\sqrt{2\pi}} \left\{ -\frac{4\cos\omega}{\omega^2} + \frac{4\sin\omega}{\omega^3} \right\}$$

$$= \frac{4}{\omega^3} \frac{(\sin\omega - \omega\cos\omega)}{\sqrt{2\pi}}$$

Now, using Parseval's identity, we obtain

$$\int_{-\infty}^{\infty} |F(\omega)|^2 \, d\omega = \int_{-\infty}^{\infty} |f(x)|^2 \, dx$$

$$\int_{-\infty}^{\infty} \left(\frac{4(\sin\omega - \omega\cos\omega)}{\sqrt{2\pi}\;\omega^3} \right)^2 d\omega = \int_{-1}^{1} (1-x^2)^2 \, dx$$

$$\Rightarrow \qquad \int_{0}^{\infty} \frac{16}{2\pi} \left(\frac{\sin\omega - \omega\cos\omega}{\omega^3} \right)^2 d\omega = \int_{0}^{1} (1-x^2)^2 \, dx$$

$$= \int_{0}^{1} (1 + x^4 - 2x^2) dx$$

$$= \left(x + \frac{x^5}{5} - \frac{2x^3}{3} \right)_0^1 = 1 + \frac{1}{5} - \frac{2}{3} = \frac{8}{15}$$

$$\therefore \qquad \int_{0}^{\infty} \left(\frac{\sin x - x\cos x}{x^3} \right)^2 dx = \frac{\pi}{15}.$$

EXAMPLE 8.25

Using Parseval's identity, prove that

$$\int_{0}^{\infty} \frac{\sin ax}{x(a^2 + x^2)} dx = \frac{\pi}{2} \frac{1 - e^{-a^2}}{a^2}.$$

Solution: Let $\qquad f(x) = e^{-ax}$ and $g(x) = \begin{cases} 1, & 0 < x < a \\ 0, & x > a \end{cases}$

Then, $\qquad F_c(f(x)) = \dfrac{a}{a^2 + \omega^2}$ and $G_c(g(x)) = \dfrac{\sin a\omega}{\omega}$

Using Parseval's identity, we have

$$\frac{2}{\pi} \int_{-\infty}^{\infty} F_c(\omega)\, G_c(\omega) d\omega = \int_{-\infty}^{\infty} f(x)\, g(x) dx$$

$$\Rightarrow \qquad \frac{2}{\pi}\int_{-\infty}^{\infty}\frac{a}{a^2+\omega^2}\frac{\sin a\omega}{\omega}d\omega = \int_0^a e^{-ax}\cdot 1dx = \frac{1-e^{-ax}}{a}$$

Thus,
$$\int_0^{\infty}\frac{\sin ax}{x(a^2+x^2)}dx = \frac{\pi}{2a^2}(1-e^{-a^2}).$$

EXERCISE 8.1

Find the Fourier transform of $f(x)$ given by:

1. $f(x) = \begin{cases} 1 & |x| < a \\ 0 & |x| > a \end{cases}$

2. $f(x) = e^{-x^2/2}$.

3. $f(x) = \begin{cases} 0 & x < \alpha \\ x & \alpha \le x \le \beta \\ 0 & x > \beta \end{cases}$

4. $f(x) = \begin{cases} 1+\dfrac{x}{a} & \text{for } -a < x < 0 \\ 1-\dfrac{x}{a} & \text{for } 0 < x < a \\ 0 & \text{otherwise} \end{cases}$.

5. $f(x) = \begin{cases} \cos x & \text{for } x \le a \\ 0 & \text{for } x > a > 0 \end{cases}$

6. Find the Fourier sine and cosine transform of $f(x) = x$.

7. Find the Fourier sine and cosine transform of $f(x) = \begin{cases} 1 & 0 < x < 1 \\ 0 & x > 1 \end{cases}$

8. Find the Fourier cosine and sine transform of $e^{-ax}(a > 0)$ and by using the inversion formula, prove that $\int_0^{\infty}\frac{\cos ax}{a^2+x^2}dx = \frac{\pi}{2a}e^{-ax}$ and $\int_0^{\infty}\frac{x\sin ax}{a^2+x^2}dx = \frac{\pi}{2}e^{-ax}$

9. Find the Fourier sine transform of $\frac{x}{a^2+x^2}$ and Fourier cosine transform of $\frac{1}{a^2+x^2}$.

10. Find the Fourier sine and cosine transform of $f(x) = \begin{cases} k & \text{if } 0 < x < a \\ 0 & \text{if } x > a \end{cases}$.

11. Find the Fourier sine and cosine transform of $f(x)$ defined by $f(x) = \dfrac{e^{ax} + e^{-ax}}{e^{ax} - e^{-ax}}$.

12. Find the finite Fourier sine and cosine transform of $f(x) = 1$.

13. Find the finite Fourier sine and cosine transform of $f(x) = \sin ax$ in $(0, \pi)$.

14. Show that the Finite Fourier sine transform of $\dfrac{x}{\pi}$ is $\dfrac{(-1)^{\omega-1}}{\omega}$ $(0 < x < \pi)$.

15. Find $f(x)$ if $F_s(f(x)) = \dfrac{\omega}{1+\omega^2}$.

16. Find the inverse Fourier cosine transform $f(x)$ of $F_c(\omega) = \omega e^{-a\omega}$.

17. Find the inverse Fourier cosine transform of $F_c(\omega) = \begin{cases} \dfrac{1}{2a}\left(a - \dfrac{\omega}{2}\right) & \text{when } \omega < 2a \\ 0 & \text{when } \omega \geq 2a \end{cases}$.

18. Find the inverse Fourier cosine transform of $F_c(f(x)) = \dfrac{\sin a\omega}{\omega}$.

19. Find $f(x)$ is $F_s(f(x)) = e^{-a\omega}$.

20. Find the Fourier sine transform of $e^{-|x|}$ and evaluate $\displaystyle\int_0^\infty \frac{x \sin mx}{1+x^2} dx$.

21. Find $f(x)$ if $F_s(\omega) = e^{-a\omega/\omega}$.

22. Solve the integral equation $\displaystyle\int_0^\infty F(x) \cos(ax) dx = e^{-a}$.

23. Solve the integral equation $\displaystyle\int_0^\infty f(x) \cos(ax) dx = \begin{cases} 1-a & \text{if } 0 \leq a \leq \\ 0 & \text{if } a > 1 \end{cases}$. Show that $\displaystyle\int_0^\infty \left(\frac{\sin x}{x}\right)^2 dx = \frac{\pi}{2}$.

24. Use convolution theorem to find the inverse Fourier transform of $\dfrac{\omega}{(1+\omega^2)^2}$; given that $\dfrac{2}{1+\omega^2}$ is the Fourier transform of $e^{-|x|}$.

25. Evaluation (a) $\displaystyle\int_0^\infty \left(\frac{1-\cos x}{x}\right)^2 dx$, (b) $\displaystyle\int_0^\infty \frac{\sin^2 x}{x^2} dx$.

ANSWERS

Exercise 8.1

1. $F(\omega) = \dfrac{1}{\sqrt{2\pi}} \dfrac{2\sin a\omega}{\omega} \quad (\omega \neq 0).$

2. $F(\omega)= e^{-\omega 2/2}.$

3. $F(\omega) = \dfrac{1}{\sqrt{2\pi}} \left\{ \dfrac{1}{\omega}(\alpha e^{j\alpha\omega} - \beta e^{i\beta\omega}) + \dfrac{1}{\omega^2}(e^{i\omega\beta} - e^{j\omega\alpha}) \right\}.$

4. $F(\omega) = \dfrac{2\sqrt{2}}{\sqrt{\pi}} \dfrac{\sin^2\left(\dfrac{a\omega}{2}\right)}{a\omega^2}.$

5. $F(\omega) = \dfrac{1}{\sqrt{2\pi}} \dfrac{2}{1-w^2}$ (sin a cos $a\omega$ – ω cos a sin $a\omega$).

6. $F_c(\omega) = -1/\omega^2;\ F_s(\omega) = 0.$

7. $F_c(\omega) = \dfrac{\sin\omega}{\omega};\ F_s(\omega) = \dfrac{1-\cos\omega}{\omega}.$

8. $F_c(\omega) = \sqrt{\dfrac{2}{\pi}} \dfrac{a}{a^2+\omega^2},\ F_s(\omega) = \sqrt{\dfrac{2}{\pi}} \dfrac{\omega}{a^2+\omega^2}.$

9. $F_s(\omega) = \dfrac{\pi}{2} e^{-a\omega};\ F_c(\omega) = \dfrac{\pi}{2a} e^{-a\omega}.$

10. $F_c(\omega) = \sqrt{\dfrac{2}{\pi}} \dfrac{k\cdot\sin a\omega}{\omega};\ F_s(\omega) = \sqrt{\dfrac{2}{\pi}} \dfrac{k}{\omega}(1-\cos a\omega)$

11. $F_c(\omega) = \sqrt{\dfrac{2}{\pi}} \dfrac{\cos(a/2)(e^{\omega/2}+e^{-\omega/2})}{2\cos a + e^{\omega} + e^{-\omega}};\ F_s(\omega) = \dfrac{e^{\omega}-e^{-\omega}}{\sqrt{2\pi}\,(e^{\omega}+e^{-\omega}+2\cos a)}.$

12. $F_c(\omega) = \begin{cases} \dfrac{1}{\omega}(\sin\pi\omega) & (\omega\neq 0) \\ \pi & (\omega = 0) \end{cases}.$

13. $F_c(\omega) = \begin{cases} 0 & \text{if } a+\omega \text{ is even} \\ \dfrac{2a}{a^2-\omega^2} & \text{if } a+\omega \text{ is odd} \end{cases}.$

14. $F_s(\omega) = \begin{cases} 0 & \text{if } a\neq\omega \\ \dfrac{\pi}{2} & \text{if } a=\omega \end{cases}$ and a, ω are integers.

15. $f(x) = e^{-x}$.

16. $f(x) = \sqrt{\dfrac{2}{\pi}}\ \dfrac{n!\cos\left((n+1)\tan^{-1}\left(\dfrac{x}{a}\right)\right)}{(a^2+x^2)\dfrac{n+1}{2}}$.

17. $f(x) = \dfrac{\sin^2 ax}{a\sqrt{2\pi}\,x^2}$.

18. $f(x) = \begin{cases} 1 & \text{if } x < a \\ 0 & \text{if } x > a \end{cases}$.

19. $f(x) = \dfrac{\sqrt{2}\,\omega}{\sqrt{\pi}\,(a^2+\omega^2)}$.

20. $F_s(\omega) = \sqrt{\dfrac{2}{\pi}}\ \dfrac{\omega}{1+\omega^2}; \dfrac{\pi}{2}e^{-m}$.

21. $f(x) = \sqrt{\dfrac{2}{\pi}}\ \tan^{-1}\left(\dfrac{x}{a}\right)$.

22. $f(x) = \dfrac{2}{\pi(1+x^2)}$.

23. $f(x) = \dfrac{2(1-\cos x)}{\pi x^2}$.

24. $\dfrac{1}{4}\displaystyle\int_{-\infty}^{\infty} e^{-(|x+t|+|t|)}dt$.

25. (a) $\dfrac{\pi}{2}$ (b) $\dfrac{\pi}{2}$

{*Hint:* Use Parseval's identity for Fourier sine and cosine transforms of

$$f(x) = \begin{cases} 1 & 0 \le x \le 1 \\ 0 & x > 1 \end{cases}.$$

9

CHAPTER

Solution of Differential Equations using Fourier Transforms

In the earlier chapters, we have studied Laplace transform methods for the solution of differential equations and while discussing applications of partial differential equations, we have discussed the Fourier series solution of heat equation, wave equation and Laplace equation. In this chapter, we shall discuss the Fourier transform solution of some of these partial differential equations.

If the Fourier transform is applied with respect to one of the variables in the differential equation, then we obtain an ordinary differential equation in terms of the other variable. We solve this differential equation. The solution of the given boundary value problem is then obtained by taking the inverse Fourier transform.

The choice of sine or cosine Fourier transform is decided by the form of the boundary conditions and with the help of the following formulas

(a) Applying sine transform to $\frac{\partial^2 u}{\partial x^2}$, we get

$$\int_0^\infty \frac{\partial^2 u}{\partial x^2} \sin \omega x \, dx = \left[\frac{\partial u}{\partial x} \sin \omega x\right]_0^\infty - \int_0^\infty \omega \cos \omega x \frac{\partial u}{\partial x} dx$$

$$= -\omega \int_0^\infty \frac{\partial u}{\partial x} \cos \omega x \, dx \text{ if } \frac{\partial u}{\partial x} \to 0 \text{ as } x \to \infty$$

$$F_s(u_{xx}) = -\omega \left[[u \cos \omega x]_0^\infty + \omega \int_0^\infty u \sin \omega x \, dx \right]$$

$$= \sqrt{\frac{2}{\pi}}\, \omega u_{x=0} - \omega^2 F_s(u) \text{ as } u \to 0 \text{ as } x \to \infty$$

(b) Again, applying the cosine transform, we have

$$\int_0^\infty \cos \omega x \frac{\partial^2 u}{\partial x^2} dx = \left[\frac{\partial u}{\partial x} \cos \omega x\right]_0^\infty + \omega \int_0^\infty \sin \omega x \frac{\partial u}{\partial x} dx$$

$$= -\left(\frac{\partial u}{\partial x}\right)_{x=0} + \omega\left([u \sin wx]_0^\infty - \omega \int_0^\infty u \cos \omega x \, dx,\right) \text{ if } \frac{\partial u}{\partial x} \to 0 \text{ as } x \to \infty$$

$$\Rightarrow F_c(u_{xx}) = -\sqrt{\frac{2}{\pi}}\left(\frac{\partial u}{\partial x}\right)_{x=0} - \omega^2 F_c(u) \text{ if } u \to 0 \text{ as } x \to \infty.$$

In the problems in which the range of integration of one of the variables is finite, finite Fourier transform are applied.

(c) The finite Fourier sine and cosine transform of $\frac{\partial u}{\partial x}$ where, u is a function of x and t for $0 < x < l$, $t = 0$.

We have
$$F_s\left(\frac{\partial u}{\partial x}\right) = \int_0^l \frac{\partial u}{\partial x} \sin \frac{\omega \pi x}{l} dx$$

$$= \left[u \sin \frac{\omega \pi x}{l}\right]_0^l - \frac{\omega \pi}{l} \int_0^l u \cos \frac{\omega \pi x}{l} dx$$

$$\Rightarrow \quad F_s\left(\frac{\partial u}{\partial x}\right) = -\frac{\omega \pi}{l} F_c(u)$$

and
$$F_c\left(\frac{\partial u}{\partial x}\right) = \int_0^l \frac{\partial u}{\partial x} \cos \frac{\omega \pi x}{l} dx$$

$$= \left[u \cos \frac{\omega \pi x}{l}\right]_0^l + \frac{\omega \pi}{l} \int_0^l u \sin \frac{\omega \pi x}{l} dx$$

$$\Rightarrow \quad F_c\left(\frac{\partial u}{\partial x}\right) = \frac{\omega \pi}{l} F_s(u) - (u(0,t) - u(l,t) \cos \omega \pi)$$

(d) The finite Fourier sine and cosine transform of $\frac{\partial^2 u}{\partial x^2}$ where u is a function of x and t, for $0 < x < l$, $t > 0$.

$$F_s\left(\frac{\partial^2 u}{\partial x^2}\right) = \int_0^l \frac{\partial^2 u}{\partial x^2} \sin \frac{\omega \pi x}{l} dx$$

$$= \left\{\frac{\partial u}{\partial x} \sin \frac{\omega \pi x}{l}\right\}_0^l - \frac{\omega \pi}{l} \int_0^l \frac{\partial u}{\partial x} \cos \frac{\omega \pi x}{l} dx$$

$$= -\frac{\omega \pi}{l}\left[\left(u \cos \frac{\omega \pi x}{l}\right)_0^l + \frac{\omega \pi}{l} \int_0^l u \sin \frac{\omega \pi x}{l} dx\right]$$

$$= -\frac{\omega \pi}{l}\left[u(l,t) \cos \omega \pi - u(0,t) + \frac{\omega \pi}{l} F_s(u)\right]$$

$$\Rightarrow \quad F_s\left(\frac{\partial^2 u}{\partial x^2}\right) = -\frac{\omega^2 \pi^2}{l^2} F_s(u) + \frac{\omega \pi}{l}[u(0,t) - u(l,t) \cos \omega \pi]$$

and $$F_c\left(\frac{\partial^2 u}{\partial x^2}\right)=\int_0^l \frac{\partial^2 u}{\partial x^2}\cos\frac{\omega\pi x}{l}dx$$

$$=\left[\frac{\partial u}{\partial x}\cos\frac{\omega\pi x}{l}\right]_0^l+\frac{\omega\pi}{l}\int_0^l \frac{\partial u}{\partial x}\sin\frac{\omega\pi x}{l}dx$$

$$=-\left(\frac{\partial u}{\partial x}\right)_{x=0}+\left(\frac{\partial u}{\partial x}\right)_{x=l}\cos\omega\pi+\frac{\omega\pi}{l}\left[\left(u\sin\frac{\omega\pi x}{l}\right)_0^l-\frac{\omega\pi}{l}\int_0^l u\cos\frac{\omega\pi x}{l}dx\right]$$

$$\Rightarrow\quad F_c\left(\frac{\partial^2 u}{\partial x^2}\right)=-\frac{\omega^2\pi^2}{l^2}F_c(u)-\{u_x(0,t)-u_x(l,t)\cos\omega\pi\}.$$ ▲▲

EXAMPLE 9.1

Determine the distribution of temperature in the semi-infinite medium $x \geq 0$ when the end $x = 0$ is maintained at zero temperature and the initial distribution of temperature is $f(x)$.

Solution: Let $u(x, t)$ be the temperature at any point x and at any time t. We have to solve the following heat-flow equation

$$\frac{\partial u}{\partial t}=c^2\frac{\partial^2 u}{\partial x^2}\quad (x>0, t>0) \tag{1}$$

subject to the initial condition $u(x, 0) = f(x)$ and the boundary condition $u(0, t) = 0$. Taking Fourier sine transform of Eq. (1), we have

$$F_s\left(\frac{\partial u}{\partial t}\right)=c^2F_s\left(\frac{\partial^2 u}{\partial x^2}\right)$$

$$\Rightarrow\quad \frac{d}{dt}F_s(u)=c^2\left[\sqrt{\frac{2}{\pi}}\,\omega u(0,t)-\omega^2F_s(u)\right]$$

$$\Rightarrow\quad \frac{d}{dt}\overline{u}_s+c^2\omega^2\overline{u}_s=0$$

whose solution is

$$\overline{u_s}=c_1\,e^{-c^2\omega^2 t}$$

Also, $$F_s(u(x, 0)) = \overline{f_s}(\omega) \text{ at } t = 0$$

$$\therefore\quad \overline{u_s}=\overline{f_s}(\omega)e^{-c^2\omega^2 t}$$

Hence, taking its inverse Fourier sine transform, we obtain

$$u(x, t) = \sqrt{\frac{2}{\pi}}\int_0^\infty \overline{f_s}(\omega)e^{-c^2\omega^2 t}\sin x\omega\, d\omega.$$

EXAMPLE 9.2

Solve: $\frac{\partial u}{\partial t}=\frac{\partial^2 u}{\partial x^2}$ $(x>0, t>0)$ subject to the conditions $u(0, t) = 0$ and $u(x, 0) = \begin{cases} 1 & 0<x<1 \\ 0 & x\geq 1 \end{cases}$ and $u(x, t)$ is boundary.

Solution: On taking Fourier sine transform of the given partial differential equation, we obtain (as in Example 9.1)

$$\overline{u_s}=c_1e^{-\omega^2 t}$$

Now,
$$F_s(u(x,0))=\sqrt{\frac{2}{\pi}}\int_0^\infty u(x,0)\sin\omega x\,dx$$

$$=\sqrt{\frac{2}{\pi}}\int_0^1 1\cdot\sin\omega x\,dx$$

$$=\sqrt{\frac{2}{\pi}}\left[-\frac{\cos\omega x}{\omega}\right]_0^1$$

$$=\sqrt{\frac{2}{\pi}}\left(\frac{1-\cos\omega}{\omega}\right)$$

$$\therefore \quad \overline{u_s}=\sqrt{\frac{2}{\pi}}\left(\frac{1-\cos\omega}{\omega}\right)e^{-\omega^2 t}$$

Applying the inverse Fourier sine transform, we have

$$u(x,t)=\frac{2}{\pi}\int_0^\infty\left(\frac{1-\cos\omega}{\omega}\right)e^{-\omega^2 t}\sin\omega x\,d\omega$$

which is the required solution.

EXAMPLE 9.3

The temperature $u(x, t)$ in the semi-infinite rod $0\leq x<\infty$ is determined by the differential equation $\frac{\partial u}{\partial t}=k\frac{\partial^2 u}{\partial x^2}$. Subject to the condition $u(x, t) = 0$ and $u_x(0, t) = -\mu$ (a constant).

Making use of cosine transform, show that

$$u(x,t)=\frac{2\mu}{\pi}\int_0^\infty\frac{\cos\omega x}{\omega^2}(1-e^{-x\omega^2 t})d\omega$$

Solution: Taking the Fourier cosine transform, we have

$$F_c\left(\frac{\partial u}{\partial t}\right)=k\,F_c\left(\frac{\partial^2 u}{\partial x^2}\right)$$

$$\frac{d}{dt}\overline{u_c}=k\cdot\sqrt{\frac{2}{\pi}}\left(\frac{\partial u}{\partial x}\right)_{x=0}-\omega^2F_c(u_x)$$

$$\frac{d}{dt}\overline{u_c}=\sqrt{\frac{2}{\pi}}\,k\mu-k\omega^2\,\overline{u_c}$$

$$\Rightarrow \qquad \frac{d}{dt}\overline{u_c}+k\omega^2\overline{u_c}=\sqrt{\frac{2}{\pi}}\,k\mu.$$

which is a linear differential equation of first order with IF = $e^{\int k\omega^2 dt}=e^{k\omega^2 t}$

$\therefore$ Its solution is

$$e^{k\omega^2 t}\overline{u_c}=\sqrt{\frac{2}{\pi}}\,k\mu\int e^{k\omega^2 t}dt+C$$

$$\Rightarrow \qquad \overline{u_c}=\sqrt{\frac{2}{\pi}}\,k\mu\frac{1}{k\omega^2}+Ce^{-k\omega^2 t}$$

sine $u(x, 0) = 0$

$$\therefore \qquad C=-\sqrt{\frac{2}{\pi}}\frac{\mu}{\omega^2}$$

$$\therefore \qquad \overline{u_c}=\sqrt{\frac{2}{\pi}}\frac{\mu}{\omega^2}(1-e^{-k\omega^2 t})$$

Applying the inverse Fourier cosine transform, we have

$$u(x,t)=\frac{2\mu}{\pi}\int_0^\infty\frac{\cos\omega x}{w^2}(1-e^{-k\omega^2 t})d\omega$$

EXAMPLE 9.4

Solve $\frac{\partial u}{\partial t}=2\frac{\partial^2 u}{\partial x^2}$, if $u(x, 0) = 0$; $u(x, 0) = e^{-x}(x > 0)$ and $u(x, t)$ is bounded where, $x > 0, t > 0$.

Solution: As in the previous examples, taking Fourier sine transform, we obtain

$$\frac{d}{dt}\overline{u_s}=-2\omega^2\overline{u_s}$$

Integrating, we get

$$\overline{u_s} = ce^{-2\omega^2 t}$$

Since $\qquad u(x, 0) = e^{-x}(x > 0)$

$$\therefore \qquad \overline{u_s}(\omega,0) = \sqrt{\frac{2}{\pi}}\int_0^{\infty} e^{-x}\sin\omega x\,d\omega = \sqrt{\frac{2}{\pi}}\frac{\omega}{1+\omega^2}$$

$$\Rightarrow \qquad \overline{u_s} = \sqrt{\frac{2}{\pi}}\frac{\omega}{1+\omega^2}e^{-2\omega^2 t}$$

Now taking inverse Fourier sine transform, we get

$$u(x,t) = \frac{2}{\pi}\int_0^{\infty}\frac{\omega e^{-2\omega^2 t}}{1+\omega^2}\sin\omega x\,d\omega.$$

EXAMPLE 9.5

Using Fourier sine transforms, solve the partial differential equation $\frac{\partial u}{\partial t} = k\frac{\partial^2 u}{\partial x^2}$ for $x > 0$, $t > 0$ under the boundary conditions $u = u_0$ when $x = 0$, $t > 0$ and the initial-condition $u = 0$ when $t = 0$, $x > 0$.

Solution: Taking the Fourier sine transform, we have

$$\frac{d}{dt}\overline{u_s} + k\omega^2\overline{u_s} = k\omega\sqrt{\frac{2}{\pi}}u_0$$

which is a linear differential equation of first order with IF = $e^{+k\omega^2 t}$.

Hence, its solution is

$$\overline{u_s}\cdot e^{k\omega^2 t} = k\omega\sqrt{\frac{2}{\pi}}\,u_o\int e^{x\omega^2 t}dt + C$$

$$\Rightarrow \qquad \overline{u_s} = k\omega\sqrt{\frac{2}{\pi}}\,u_o\,\frac{1}{k\omega^2} + Ce^{-k\omega^2 t}$$

Since $\qquad u(x, 0) = 0 \Rightarrow \overline{u_s}(x,0) = 0$

$$\therefore \qquad C = -\sqrt{\frac{2}{\pi}}\frac{u_0}{\omega}$$

Thus, $\qquad \overline{u_s} = \sqrt{\frac{2}{\pi}}\frac{u_0}{\omega}(1 - e^{-k\omega^2 t})$

Now, applying the inverse Fourier transform, we have

$$u(x,t) = \frac{2}{\pi} u_0 \int_0^{\infty} \left(\frac{1 - e^{-k\omega^2 t}}{\omega} \right) \sin \omega x \, d\omega$$

$$= \frac{2u_0}{\pi} \left[\frac{\pi}{2} - \int_0^{\infty} \frac{e^{-x\omega^2 t}}{\omega} \sin \omega x \, d\omega \right]$$

$$\Rightarrow \quad u(x,t) = u_0 \left[1 - \frac{2}{\pi} \int_0^{\infty} \frac{e^{-k\omega^2 t}}{\omega} \sin \omega x \, d\omega \right].$$

EXAMPLE 9.6

If the flow of heat is linear, so that the variation of $u(x, t)$ temperature with x and y; may be neglected and if it is assumed that no heat is generated in the medium, then solve the differential equation $\frac{\partial u}{\partial t} = k \frac{\partial^2 u}{\partial x^2}$ (one-dimensional heat equation) where, $-\infty < x < \infty$ and $u(x, 0) = f(x)$, $f(x)$ being a given function of x.

Solution: Taking the Fourier transform, we obtain

$$\frac{d}{dt} \bar{u} = -k\omega^2 \bar{u}$$

whose solution is

$$u = Ce^{-k\omega^2 t}$$

Now, since

$$u(x, 0) = f(x)$$

$$\therefore \quad \bar{u}(x,0) = \sqrt{\frac{2}{\pi}} \int_0^{\infty} f(x) e^{i\omega x} x dx = \bar{f}(\omega)$$

$$\Rightarrow \quad c = \bar{f}(\omega)$$

$$\therefore \quad \bar{u} = \bar{f}(\omega) e^{-k\omega^2 t}$$

Taking the inverse Fourier transform, we have

$$u(x,t) = \sqrt{\frac{1}{2\pi}} \int_{-\infty}^{\infty} \bar{f}(\omega) e^{-k\omega^2 t} e^{j\omega x} d\omega.$$

EXAMPLE 9.7

Solve: $\frac{\partial^2 u}{\partial x^2} + \frac{\partial^2 u}{\partial y^2} = 0$ $(-\infty < x < \infty,\ y \geq 0)$ satisfying the conditions

(a) $u(x, t)$ and $u_x, u_y \to 0$ as $x \to \pm\infty$ and

(b) $u(x, 0) = f(x), \left(\dfrac{\partial u}{\partial y}\right)_{x=0} = 0.$

Solution: Taking Fourier transform, we obtain

$$\frac{d^2\bar{u}}{dy^2} + \omega^4 \bar{u} = 0$$

whose solution is

$$\bar{u} = A\cos(\omega^2 y) + B\sin(\omega^2 y)$$

since on

$$y = 0;\ u = f(x) \text{ and } \frac{\partial u}{\partial y} = 0$$

$\therefore$

$$\bar{u} = \frac{1}{\sqrt{2\pi}} \int_{-\infty}^{\infty} f(x) e^{j\omega x} dx = \bar{f}(\omega)$$

and

$$\frac{\partial \bar{u}}{\partial y} = 0$$

$\Rightarrow$

$$\bar{u} = \bar{f}(\omega)\cos(\omega^2 y)$$

By applying inverse Fourier transform, we obtain

$$u(x,t) = \frac{1}{\sqrt{2\pi}} \int_{-\infty}^{\infty} \bar{f}(\omega)\cos(\omega^2 y) e^{i\omega x} dx$$

EXAMPLE 9.8

Use a cosine transform to show that the steady temperature in the semi-infinite solid ($y > 0$), when the temperature on the surface ($y = 0$) and over the strip $|x| < a$ is kept at unity and at zero outside the strip, is

$$\frac{1}{\pi}\left[\tan^{-1}\left(\frac{a+x}{y}\right) + \tan^{-1}\left(\frac{a-x}{y}\right)\right]$$

$$\left(\text{The result } \int_0^{\infty} e^{-wx} x^{-1} \sin rx dx = \tan^{-1}\frac{r}{w} \ (r > 0, w > 0) \text{ may be assumed.}\right)$$

Solution: Here, the steady temperature $u(x, y)$ in the semi-infinite solid is governed by two-dimensional Laplace's equation

$$\frac{\partial^2 u}{\partial x^2} + \frac{\partial^2 u}{\partial y^2} = 0 \ \ (0 < y < \infty, -\infty < x < \infty)$$

subject to the condition

$u(x, 0) = 1$ when $|x| < a$ and $u(x, 0) = 0$ when $|x| > a$.

Taking Fourier cosine transform, we obtain

$$\frac{d^2}{dy^2} - w^2\overline{u_c} = 0 \text{ (if } u \to 0 \text{ as } x \to \infty\text{) whose solution is } \overline{u_c} = Ae^{wy} + Be^{-wy}$$

Now as $y \to \infty$ $\overline{u_c} \to 0 \Rightarrow A = 0$

$$\therefore \qquad \overline{u_c} = Be^{-wy}$$

Also, when $y = 0$

$$\overline{u_c} = \sqrt{\frac{2}{\pi}}\int_{-a}^{\infty} 1 \cdot \cos wx\, dx = \sqrt{\frac{2}{\pi}}\frac{\sin wa}{w}$$

$$\Rightarrow \qquad B = \sqrt{\frac{2}{\pi}}\frac{\sin wa}{w}$$

Hence,
$$\overline{u_c} = \sqrt{\frac{2}{\pi}}\frac{\sin wa}{w}e^{-wy}$$

Taking the inverse Fourier cosine transform, we have

$$u(x,y) = \frac{2}{\pi}\int_0^{\infty}\frac{\sin wa}{w}e^{-wy}\cos wx\, dw$$

$$= \frac{1}{\pi}\int_0^{\infty}\frac{e^{-wy}}{w}[\sin(a+x)w + \sin(a-x)w]dw$$

$$= \frac{1}{\pi}\int_0^{\infty}e^{-wy}w^{-1}\sin(a+x)wdy + \frac{1}{\pi}\int_0^{\infty}w^{-wy}w^{-1}\sin(a-x)wdy$$

$$= \frac{1}{\pi}\left[\tan^{-1}\left(\frac{a+x}{y}\right) + \tan^{-1}\left(\frac{a-x}{y}\right)\right]$$

using the result $\int_0^{\infty} e^{-wx}x^{-1}\sin rx\, dx = \tan^{-1}\frac{r}{w} \quad (r > 0, w > 0)$

EXAMPLE 9.9

Using finite Fourier transform, solve $\frac{\partial u}{\partial t} = \frac{\partial^2 u}{\partial x^2}$ given $u(0, t) = 0$, $u(4, t) = 0$ and $u(x, 0) = 2x$ where $0 < x < 4$, $t > 0$.

Solution: Since $u(xo, t) = 0$, we take finite Fourier sine transform, we obtain

$$\frac{d\overline{u_s}}{dt} = -\frac{w^2\pi^2}{4^2}\overline{u_s} + \frac{w\pi}{4}\left[u(0,t) - u(4,t)\cos w\pi\right]$$

$$\Rightarrow \quad \frac{d\overline{u_s}}{dt} = -\frac{w^2\pi^2}{16}\overline{u_s}$$

Integrating, we get

$$\overline{u_s} = Ce^{-w^2\pi^2 t/16}$$

Now,
$$\overline{u_s}(x,0) = \int_0^4 2x\sin\left(\frac{n\pi x}{4}\right)dx = -\frac{32}{n\pi}\cos n\pi$$

$$\Rightarrow \quad \overline{u_s} = -\frac{32}{n\pi}\cos \pi e^{-w^2\pi^2 t/16}$$

$$= -\frac{32}{n\pi}(-1)^n \cdot e^{-w^2\pi^2 t/16}$$

Now, taking inverse Fourier sine transform, we get

$$u(x,t) = \frac{2}{4}\sum_{n=1}^{\infty}\frac{32}{n\pi}(-1)^{n+1}e^{-w^2\pi^2 t/16}\sin\left(\frac{n\pi x}{4}\right)$$

$$= 16\sum_{n=1}^{\infty}\frac{(-1)^{n+1}}{n\pi}e^{-w^2\pi^2 t/16}\sin\left(\frac{n\pi x}{4}\right).$$

EXAMPLE 9.10

Use finite Fourier cosine transforms to solve $\frac{\partial u}{\partial t} = k\frac{\partial^2 u}{\partial x^2}$ with the boundary conditions $\frac{\partial u}{\partial t} = 0$ when $x = 0$ and $x = \pi$, $t > 0$ and the initial condition $u = f(x)$, when $t = 0$, $0 < x < \pi$.

Solution: Taking the finite Fourier cosine transform, we obtain

$$\frac{d\overline{u_c}}{dt} = k\left(-\frac{\omega^2\pi^2}{\pi^2}\overline{u_c} - \{u_x(0,t) - u_x(l,t)\cos\omega\pi\}\right)$$

$$\Rightarrow \quad \frac{d\overline{u_c}}{dt} = -k\omega^2\overline{u_c}$$

whose solution is $\overline{u_c} = Ae^{-k\omega^2 t}$

Now, $$\overline{u_c}(\omega,0)=\int_0^{\pi} f(x)\cos\left(\frac{n\pi x}{\pi}\right)dx = A$$

Hence, $$\overline{u_c}=e^{-k\omega^2 t}\int_0^{\pi} f(x)\cos(nx)dx$$

Taking the inverse finite Fourier cosine transform, we have

$$u(x,t)=\frac{1}{\pi}(\overline{u_c})_{t=0}+\frac{2}{\pi}\sum_{n=1}^{\infty}\overline{u_c}\cos\left(\frac{n\pi x}{\pi}\right)$$

$$=\left[\frac{1}{\pi}\int_0^{\pi} f(x)dx\right]+\left[\frac{2}{\pi}\sum_{n=1}^{\infty}\cos(nx)\int_0^{\pi} f(x)\cos(nx)dx\right]\times e^{-k\omega^2 t}$$

EXAMPLE 9.11

Use finite Fourier transform to solve $\frac{\partial u}{\partial t}=\frac{\partial^2 u}{\partial x^2}$; $0 < x < b$, $t > 0$ and $u_x(0, t) = 0$, $u_x(b, t) = 0$, $u(x, 0) = 2x$.

Solution: According to the boundary conditions, taking finite Fourier cosine transform, we obtain

$$\frac{d\overline{u_c}}{dt}=-\frac{\omega^2\pi^2}{b^2}\overline{u_c}-[u_x(0,t)-u_x(b,t)\cos\omega\pi]$$

$$\Rightarrow \qquad \frac{d\overline{u_c}}{dt}=-\frac{\omega^2\pi^2}{36}\overline{u_c}$$

whose solution is $$\overline{u_c}=Ae^{-\omega^2\pi^2 t/36}$$

Now at $$t = 0,\ u = 2x$$

$$\Rightarrow \qquad \overline{u_c}=\int_0^{6} 2x\cos\left(\frac{n\pi x}{6}\right)dx=\frac{72}{\omega^2\pi^2}(\cos n\pi-1)$$

$$\Rightarrow \qquad A=\frac{72}{\omega^2\pi^2}(\cos n\pi-1)$$

Hence, $$\overline{u_c}=\frac{72}{\omega^2\pi^2}(\cos\omega\pi-1)e^{-\omega^2\pi^2 t/36}$$

Taking the inverse finite Fourier cosine transform, we have

$$u(x,t)=\frac{1}{6}(\overline{u_c})_{t=0}+\frac{2}{6}\sum_{n=1}^{\infty}\overline{u_c}(n)\cos\left(\frac{n\pi x}{6}\right)$$

$$= \frac{1}{6}\int_0^6 2x dx + \frac{24}{\pi^2}\sum_{n=1}^{\infty}\left(\frac{\cos n\pi - 1}{\omega^2}\right)e^{-\omega^2 n^2 t/36} \times \cos\left(\frac{n\pi x}{6}\right)$$

$$\Rightarrow \qquad u(x,t) = 6 + \frac{24}{\pi^2}\sum_{n=1}^{\infty}\left(\frac{\cos n\pi - 1}{\omega^2}\right)e^{-\pi^2 n^2 t/36} \times \cos\left(\frac{n\pi x}{6}\right)$$

EXAMPLE 9.12

Find the steady temperature $u(x, y)$ in a long square bar of side π when one face is kept at constant and the other faces at zero temperature. Also, $u(x, y)$ is bounded.

Solution: Here, the steady temperature is governed by two-dimensional Laplace's equation

$$\frac{\partial^2 u}{\partial x^2} + \frac{\partial^2 u}{\partial y^2} = 0$$

with boundary conditions $u(0, y) = 0 = u(\pi, y)$ and $u(x, 0) = 0$ and $u(x, \pi) = u_0$ (constant).

Taking the finite Fourier sine transform, we obtain

$$F_s\left(\frac{\partial^2 u}{\partial x^2}\right) = -\frac{\omega^2\pi^2}{\pi^2}F_s(u) + \frac{\omega\pi}{\pi}\{u(0,t) - u(\pi,t)\cos\omega\pi\} = F_s\left(-\frac{d^2 u}{dy^2}\right)$$

$$\Rightarrow \qquad \frac{d^2\overline{u}_s}{dy^2} - \omega^2\overline{u}_s = 0$$

whose solution is $\qquad \overline{u}_s = A\cosh\omega t + B\sinh\omega y$

when $\qquad y = 0 \; u = 0 \Rightarrow \overline{u}_s = 0$

and when $\qquad y = \pi \; \overline{u}_s = \int_0^{\pi} u_0 \sin\omega x dx = u_0\left(\frac{1-\cos\omega\pi}{\omega}\right)$

$$\Rightarrow \qquad 0 = A \quad \text{and} \quad B = \frac{u_0}{\sinh\omega\pi}\left(\frac{1-\cos\omega\pi}{\omega}\right)$$

$$\therefore \qquad \overline{u}_s = \frac{u_0}{\sinh\omega\pi}\left(\frac{1-\cos\omega\pi}{\omega}\right)\sinh\omega y$$

Taking the inverse finite Fourier sine transform, we have

$$u(x,y) = \frac{2}{\pi}\sum_{\omega=1}^{\infty}\frac{u_0}{\sinh\omega\pi}\left(\frac{1-\cos\omega\pi}{\omega}\right)\sinh\omega y\sin\omega x$$

$$= \frac{4u_0}{\pi}\sum_{\omega=10}^{\infty}\frac{\sinh(2\omega+1)y\sin(2\omega+1)x}{(2\omega+1)\sinh(2\omega+1)\pi}$$

(Since other terms for which ω is even are zero)

EXAMPLE 9.13

An infinite string is initially at rest and that the initial displacement is $f(x)$ $(-\infty < x < \infty)$. Determine the displacement $u(x, t)$ of the string.

Solution: The equation for the vibrations of the string is

$$\frac{\partial^2 u}{\partial t^2} = c^2 \frac{\partial^2 u}{\partial x^2}$$

and the initial conditions are

$$u_t(x, 0) = 0 \quad \text{and} \quad u(x, 0) = f(x)$$

Taking Fourier transform, we obtain

$$\frac{d^2 \bar{u}}{dt^2} = c^2(-\omega^2 \bar{u})$$

whose solution is $\quad \bar{u} = A \cos c\omega t + B \sin c\omega t$

Also, $\quad \bar{u_t}\ (\omega, 0) = 0 \quad \text{and} \quad \bar{u} = F(u(x, 0)) = \bar{f}(\omega)$

$\therefore \quad A = 0 \quad \text{and} \quad B = \bar{f}(\omega)$

Thus, $\quad \bar{u} = \bar{f}(\omega) \cos c\omega t$

Now, taking inverse Fourier transform, we get

$$u(x,t) = \frac{1}{\sqrt{2\pi}} \int_{-\infty}^{\infty} \bar{f}(\omega) \cos c\omega t e^{i\omega x} d\omega$$

$$= \frac{1}{\sqrt{2\pi}} \int_{-\infty}^{\infty} \bar{f}(\omega) \left(\frac{e^{ic\omega t} - e^{-jc\omega t}}{2} \right) e^{i\omega x} d\omega$$

$$= \frac{1}{2\sqrt{2\pi}} \int_{-\infty}^{\infty} \left[\bar{f}(\omega) e^{-\omega(x-ct)} + \bar{f}(\omega) e^{j\omega(x+ct)} \right]$$

$$= \frac{1}{2} [f(x-ct) + f(x+ct)] \left\{ \because f(x) = \frac{1}{\sqrt{2\pi}} \int_{-\infty}^{\infty} \bar{f}(\omega) e^{i\omega x} d\omega \right\}$$

EXAMPLE 9.14

Solve the boundary value problem $\frac{\partial^2 u}{\partial t^2} = 9 \frac{\partial^2 u}{\partial x^2}$ subject to the conditions $u(0, t) = 0$, $u(2, t) = 0$, $u(x, 0) = x(2 - x)/2$ and $u_t(x, 0) = 0$ where $0 < x < 2$, $t > 0$.

Solution: Taking the finite Fourier sine transform, we obtain

$$\frac{d^2 \bar{u_s}}{dt^2} = 9 \cdot \frac{\omega^2 \pi^2}{2^2} \bar{u_s} + \frac{\omega\pi}{2} [u(0, t) - u(2, t) \cos \omega\pi]$$

$$\Rightarrow \qquad \frac{d^2\overline{u_s}}{dt^2} + \frac{9\omega^2\pi^2}{4}\overline{u_s} = 0$$

whose solution is $\overline{u_s} = A\cos\frac{3\omega\pi}{2}t + B\sin\frac{3\omega\pi}{2}t$

Now at $t = 0,\ u(x, 0) = x(2 - x)^{2/2}$ and $\left(\frac{\partial u}{\partial t}\right)_{t=0} = 0$

$$\therefore \qquad \overline{u_s}(\omega,0) = \int_0^2 \frac{x(2-x)}{2}\ \sin\frac{\omega\pi x}{2}dx = \frac{4}{5\omega^2\pi^2}(1-\cos\omega\pi)$$

and $\left(\frac{d\overline{u_s}}{dt}\right)_{t=0} = 0$

Thus, $A = \frac{4}{5\omega^2\pi^2}(1 - \cos\omega\pi)$ and $B = 0$

$$\Rightarrow \qquad \overline{u_s} = \frac{4}{5\omega^2\pi^2}(1-\cos\omega\pi)\ \cos\frac{3\omega\pi}{2}t$$

Taking inverse finite Fourier sine transform, we have

$$u(x,t) = \frac{2}{2}\sum_{\omega=1}^{\infty} \frac{4}{5\omega^2\pi^2}(1-\cos\omega\pi)\ \cos\frac{3\omega\pi}{2}t\ \sin\frac{\omega\pi x}{2}$$

$$= \frac{8}{5\pi^2}\sum_{\omega=1}^{\infty} \frac{1}{(2\omega-1)^2}\ \cos\frac{3(2\omega-1)\pi}{2}t\ \sin\frac{3(2\omega-1)\pi x}{2}$$

(Since other terms for which ω is even are zero)

EXERCISE 9.1

1. An infinitely long string having one end at $x = 0$, is initially at rest along the x-axis. The end $x = 0$, is given a transverse displacement $f(t)$ $(t > 0)$. Find the displacement of any point of this string at any time.

2. Solve the equation $\frac{\partial u}{\partial t} = \frac{\partial^2 u}{\partial x^2}$ $(x > 0,\ t > 0)$ subject to the condition

   (a) $u(0, t) =$ and

   (b) $u(x, 0) = \begin{cases} 1 & 0 \le x < 1 \\ 0 & x > 1 \end{cases}$ and

   (c) $u(x, t)$ is bounded.

3. In the initial temperature of an infinite bar is given by $u(x, 0) = \begin{cases} u_0 & \text{for } |x| < a \\ 0 & \text{for } |x| > a \end{cases}$ determine the temperature at any point x and at any instant t.
4. A tightly stretched flexible string has its ends fixed at $x = 0$ and $x = 1$. At time $t = 0$, the string is given a shape defined by $f(x) = \mu(l - x)$, where μ is a constant and then released. Find the displacement of any point x of the string at any time $t > 0$.
5. Solve $\dfrac{\partial u}{\partial t} = \dfrac{\partial^2 u}{\partial x^2}$ if $u_x(0, t) = 0$, $u(x, 0) = \begin{cases} x & 0 \le x \le 1 \\ 0 & x \ge 1 \end{cases}$ and $u(x, t)$ is boundary when $x > 0$, $t > 0$.
6. Solve $\dfrac{\partial u}{\partial t} = \dfrac{\partial^2 u}{\partial x^2}$; $0 < x < b$, $t > 0$ subject to the conditions $u(0, t) = 0$, $u(b, t) = 0$, $u(x, 0) = \begin{cases} 1 & 0 < x < 3 \\ 0 & 3 < x < 6 \end{cases}$ and interpret physically.
7. Solve the boundary value problem $\dfrac{\partial u}{\partial t} = \dfrac{\partial^2 u}{\partial x^2}$; $u(0, t) = 1$, $u(\pi, t) = 3$, $u(x, 0) = 1$ where, $0 < x < \pi$, $t > 0$ given the physical interpretation of the problem.
8. Using finite Fourier transform, find the solution of the wave equation - $\partial^2 u/\partial t^2 = 4\partial^2 u/\partial x^2$ subject to the conditions $u(0, t) = 0$, $u(\pi, t) = 0$, $u(x, 0) = (0.1) \sin x + (0.01) \sin 4x$ and $u_t(x, 0) = 0$ for $0 < x < \pi$, $t > 0$.
9. The temperature $u(x, t)$ in a semi-infinite bar $0 \le x < \infty$ is determined by the partial differential equation $\dfrac{\partial u}{\partial t} = c^2 \dfrac{\partial^2 u}{\partial x^2}$; subject to the conditions. (a) $u(x, 0) = 0$ and (b) $u_x(0, t) = k$ (a constant) show that $u(x, t) = \dfrac{2k}{\pi} \displaystyle\int_0^\infty (e^{-c^2\omega^2 t} - 1) \dfrac{\cos \omega x}{\omega^2} dx$.
10. Solve the differential equation $\partial u/\partial t = 3\partial^2 u/\partial x^2$ where, $u(\pi/2, t) = 0$, $(\partial u/\partial x)_{x=0} = 0$ and $u(x, 0) = 30 \cos 5x$.
11. Using suitable Fourier transform, solve the differential equation $\partial^2 u/\partial x^2 = \partial u/\partial t$ $(0 \le x \le \pi, t \ge 0)$ where, $u(0, t) = 0 = u(\pi, t)$ and $u(x, 0) = u_0$ (constant).
12. Solve the differential equation $\dfrac{\partial u}{\partial t} = \dfrac{\partial^2 u}{\partial x^2}$ $(-\infty < x < \infty, t > 0)$; find the temperature at any point of the bar at any point x.
13. A solid is initially at constant temperature u_0 while the ends $x = 0$ and $x = a$ are maintained at temperature zero. Determine the temperature at any point of the solid at any later time $t > 0$.
14. An infinite string is initially at rest along the x-axis. Its one end which is at $x = 0$, is given a synthetic transverse displacement $u_0 \sin \omega t (t > 0)$. Show

that the displacement of any point of the string any time is given by $u(x, t) = \begin{cases} u_0 \sin \omega(t - x/c) & (t < x) \\ 0 & (t > x/c) \end{cases}$ where c is the wave velocity.

15. An infinite string has an initial transverse displacement $u(x, 0) = f(x)$ $(-\infty < x < \infty)$ and is initially a rest. Show that $u(x, t) = \frac{1}{2}$ $[f(x + ct) + f(x - ct)]$.

ANSWERS

Exercise 9.1

1. $u(x, t) = \dfrac{1}{2\pi} \displaystyle\int_{x-j\omega}^{x+i\omega} e^{(1-\omega/c)x} \overline{f}(\omega) d\omega = f(t - x/c)$.

2. $u(x, t) = \displaystyle\int_0^\infty \left(\frac{1-\cos\omega}{\omega}\right) e^{-\omega^2 t} d\omega$.

3. $u(x, t) = \dfrac{u_0}{\pi} \displaystyle\int_0^\infty e^{-c^2\omega^2 t} \left(\frac{\sin(a+x)\omega + \sin(a-x)\omega}{\omega}\right) d\omega$.

4. $u(x, t) = \dfrac{8\mu l^2}{\pi^3} \displaystyle\sum_{n=1}^\infty \left(\frac{1}{(2n-1)^3} \sin\frac{(2n-1)\pi x}{l} \cos\frac{(2n-1)\pi ct}{l}\right)$.

5. $u(x, t) = \dfrac{2}{\pi} \displaystyle\int_0^\infty \left(\frac{\sin\omega}{\omega} + \frac{\cos\omega - 1}{\omega^2}\right) e^{-\omega^2 t} \times \cos\omega x d\omega$.

6. $u(x, t) = \dfrac{2}{6} \displaystyle\sum_{\omega=1}^\infty \frac{1}{\omega}(1 - \cos(\omega\pi/2)) e^{-\omega^2\pi^2 t/36} \times \sin\left(\frac{\omega\pi x}{6}\right)$.

7. $u(x, t) = \dfrac{4}{\pi} \displaystyle\sum_{\omega=1}^\infty \frac{\cos\omega\pi}{\omega} e^{-\omega^2 t} \sin\omega\pi + i + \frac{2x}{\pi}$.

8. $u(x, t) = (0.1)\cos 2t \sin x + (0.01)\cos 8t \sin 4x$.

10. $u(x, t) = 30e^{-75t} \cos 5x$.

11. $u(x, t) = \displaystyle\sum_{n=1}^\infty \frac{u_0}{n\pi} (1 - \cos n\pi) e^{-n^2 t} \sin xu_0$.

12. $u(x, t) = \dfrac{2}{\pi} \displaystyle\int_0^\infty \left(\frac{\omega\sin\omega + \cos\omega - 1}{\omega^2}\right) e^{-\omega^2 t} \cos\omega x\, dx$.

13. $u(x, t) = \dfrac{2}{a} \displaystyle\sum_{\omega=1}^\infty \frac{au_0}{\omega\pi}(1 - \cos\omega\pi) e^{-c^2\omega^2\pi^2/a^2} \times \sin\left(\frac{\omega\pi x}{a}\right)$.

▲▲

Bibliography

1. Kreyszig E, *Advanced Engineering Mathematics*, 10th edition, John Wiley and Sons, India, 2011.
2. Malik SC and Savita Arora, *Mathematical Analysis*, 5th edition, New Age International Publishers, India, 2018.
3. Rudin W, *Real and Complex Analysis*, 3rd edition, McGraw Hill Education, 2006 (Indian edition), 2009.
4. Jain RK, Iynengar SRK, *Advanced Engineering Mathematics*, 4th edition, Narosa Publishing House, New Delhi, 2013.
5. Coddington, EA, *An Introduction to Ordinary Differential Equations*, McGraw Hill Education, New Delhi, 2002.
6. Spigel, MR, *Laplace Transform Schaum Series*, McGraw Hill Education, New Delhi, India, 2008.
7. Spigel MR, *Advanced Calculus, Schaum Series*, McGraw Hill Education, New Delhi, India, 2008.
8. Ross, SL, *Introduction to Ordinary Differential Equations*, 3rd edition, John Wiley and Sons, India, 2004.
9. Ramana, BV, *Higher Engineering Mathematics*, McGraw Hill Education, Chennai, India, 2016.
10. Ahsan, Zafar, *Differential Equations and Their Applications*, 3rd edition, PHI Learning, Delhi, 2016.
11. Raisinghania, MD, *Advanced Differential Equations*, 18th edition, S. Chand and Company, New Delhi, 2014.
12. Jain, SK, *Integral Transforms*, Sarop and Sons, New Delhi, 2001.
13. Jain, SK, *Fourier Series and Fourier Transforms*, Sarop and Sons, New Delhi, 2001.
14. Sneddon IN, *Fourier Transforms*, McGraw Hill Education, New Delhi, India, 2008.
15. Srivastava, AN, Mohammad Ahmad, *Integral Transforms and Fourier Series*, Narosa Publishing House, New Delhi, 2012.
16. Dass, HK, *Advanced Engineering Mathematics*, 21st edition, S. Chand and Company, New Delhi, 2016.

17. Simmons, GF, Krantz, SG, *Differential Equations*, Tata McGraw Hill Education, New Delhi, India, 2003.
18. Sneddon IN, *Elements of Partial Differential Equation*, McGraw Hill, International edition, 1988.
19. Rai Singhania, MD, *Ordinary and Partial Differential Equations*, S. Chand and Company, New Delhi, 2014.
20. Piaggio, HTH, *An Elementary Treatise on Differential Equations, and Their Applications*, CBS Publishers and Distributors, Delhi, 1985.
21. Murray, DA, *Introductory Course on Differential Equations*, Orient Longman (India), 2003.
22. Ayres, Frank, *Theory and Problems of Differential Equations*, McGraw Hill Education, New Delhi, India, 2008.
23. Bronson, Richard, *Theory and Problems of Differential Equations*, McGraw Hill Education, New Delhi, India, 2015.
24. Cronin, Jane, Differential Equations: Introduction and Qualitative Theory, 2nd edition, Marcel Dekkar Inc. (India), 1994.
25. Rao, SB, Anuradha HR, *Differential Equations with Applications and Programs*, University Press (India), 1996.
26. Somasundram, D, *Ordinary Differential Equations, A First Course*, Narosa Publishing House, New Delhi, 2015.
27. Veeranjan, T, *Transforms and Partial Differential Equations*, 3rd edition, McGraw Hill Education, New Delhi, India, 2016.
28. Bronson, R, Gejji, V., *Differential Equations, Schaum's Outline Series*, 3rd edition, McGraw Hill Education, New Delhi, India, 2010.
29. Deo, SG, Lakshmikantham, V, and Raghvendra, V, "*Text Book of Ordinary Differential Equations*", 2nd edition., Tata McGraw Hill Education, New Delhi, India, 2010.
30. Veeranjan, T., *Linear Algebra and Partial Differential Equations*, McGraw Hill Education, New Delhi, India, 2018.

Index